T0176681

Fundamental Statistical Inference

Fundamental Statistical Inference

A Computational Approach

Marc S. Paolella

Department of Banking and Finance
University of Zurich
Switzerland

Registered Offices
John Wiley & Sons, Inc., 111 River Street, Hoboken, NJ 07030, USA
John Wiley & Sons Ltd, The Atrium, Southern Gate, Chichester, West Sussex, PO19 8SQ, UK

Editorial Office
9600 Garsington Road, Oxford, OX4 2DQ, UK

For details of our global editorial offices, customer services, and more information about Wiley products visit us at www.wiley.com.

Wiley also publishes its books in a variety of electronic formats and by print-on-demand. Some content that appears in standard print versions of this book may not be available in other formats.

Library of Congress Cataloging-in-Publication Data applied for

Hardback ISBN: 9781119417866

Cover design by Wiley
Cover images: Courtesy of Marc S. Paolella

Set in 10/12pt TimesLTStd by SPi Global, Chennai, India

Printed in Singapore by C.O.S. Printers Pte Ltd

10 9 8 7 6 5 4 3 2 1

Contents

Preface

Young people today love luxury. They have bad manners, despise authority, have no respect for older people, and chatter when they should be working.

(Socrates, 470–399 BC)

This book on statistical inference can be viewed as a continuation of the author's previous two books on probability theory (Paolella, 2006, 2007), hereafter referred to as Books I and II. Of those two, Book I (or any book at a comparable level) is more relevant, in establishing the basics of random variables and distributions as required to understand statistical methodology. Occasional use of material from Book II is made, though most of that required material is reviewed in the appendix herein in order to keep this volume as self-contained as possible. References to those books will be abbreviated as I and II, respectively. For example, Figure 5.1 in (Chapter 5 of) Paolella (2006) is referred to as Figure I.5.1; and similarly for equation references, where (I.5.22) and (II.4.3) refer to equations (5.22) and (4.3) in Paolella (2006) and Paolella (2007) respectively (and both are the Cauchy–Schwarz inequality).

Further prerequisites are the same as those for Book I, namely a solid command of basic undergraduate calculus and matrix algebra, and occasionally very rudimentary concepts from complex analysis, as required for working with characteristic functions. As with Books I and II, a solutions manual to the exercises is available.

Certainly, no measure theory is required, nor any previous exposure to statistical inference, though it would be useful to have had an introductory course in statistics or data analysis. The book is aimed at beginning master's students in statistics, though it is written to be fully accessible to master's students in the social sciences. In particular, I have in mind students in economics and finance, as I provide introductory coverage of some nonstandard topics, notably Chapter 9 on heavy-tailed distributions and tail estimation, and detailed coverage of the mixed normal distribution in Chapter 5.

Naturally, the book can be also used for undergraduates in a mathematics program. For the intended audience of master's students in statistics or the social sciences, the instructor is welcome to skip material that uses concepts from convergence and limit theorems if the target audience is not ready for such mathematics. This is one of the points of this book: such material is included so that, for example, accessible, detailed proofs of the Glivenko–Cantelli theorem and the limiting distribution of the maximum likelihood estimator can be demonstrated at a reasonably rigorous level. The vast majority of the book only requires simple algebra and basic calculus.

In this book, I stick to the independent, identically distributed (i.i.d.) setting, using it as a platform for introducing the major concepts arising in statistics without the additional overhead and complexities associated with, say, (generalized) linear models, survival analysis, copula methods, and time series. This also allows for more in-depth coverage of important topics such as bootstrap techniques, nonparametric inference via the empirical c.d.f., numerical optimization, discrete mixture models, bias-adjusted estimators, tail estimation (as a nice segue into the study of extreme value theory), and the method of indirect inference. A future project, referred to as Book IV, builds on the framework in the present volume and is dedicated to the linear model (regression and ANOVA) and, primarily, time series analysis (univariate ARMAX models), GARCH, and multivariate distributions for modeling and predicting financial asset returns.

Before discussing the contents of this volume, it is important to mention that, similar to Books I and II, the overriding goals are:

(i) to emphasize the practical side of matters by addressing computation issues;

(ii) to motivate students to actively engage in the material by replicating and extending reported results, and to read the literature on topics of their interest;

(iii) to go beyond the standard topics and examples traditionally taught at this level, albeit still within the i.i.d. framework; and

(iv) to set the stage for students intending to pursue further courses in statistical/econometric inference (and quantitative risk management), as well as those embarking on careers as modern data analysts and applied quantitative researchers.

Regarding point (i), I explain to students that computer programming skills are necessary, but far from sufficient, to be successful in applied research. In an occasional lecture dedicated to programming issues, I emphasize (not sarcastically – I do not test computer skills) that it is fully optional, and those students who are truly mathematically talented can skip it, explaining that they will always have programmers in their team (in industry) or PhD students and co-authors (in academics) as resources to do the computer grunt work implementing their theoretical constructs. Oddly, nobody leaves the room.

With respect to point (ii), the reader will notice that some chapters have few (or no) exercises (some have many). This is because I believe the nature of the material presented is such that it offers the student a judicious platform for self experimentation, particularly with respect to numerical implementation. Some of the material could have been packaged as exercises (and much is), though I prefer to illustrate important concepts, distributions, and methods in a detailed way, along with code and graphics, instead of banishing it to the exercises (or, far worse, littering the exercises with trite, useless algebraic manipulations devoid of genuine application) and instead encourage the student to replicate, complement, and extend the presented material. The reader will no doubt tire at my occasional explicit suggestions for this ("The reader is encouraged ... "). One of my role model authors is Hamilton (1994), whose book has *no* exercises, is twice the size of this book, and has been praised as an outstanding presentation of time series. Hamilton clearly intended to *teach* the material in a straightforward, clear way, with highly detailed and accessible derivations. I aspire to a similar approach, as well as adding numeric illustrations and Matlab code.[1]

Regarding point (iii), besides the obvious benefit of giving students a more modern viewpoint on methods and applications in statistics, having a large variety of such is useful for students (and instructors) looking for interesting, relevant topics for master's theses. An example of a nonstandard topic of interest is in Chapter 5, giving a detailed discussion on the problems associated with, and solutions to, estimating the (univariate) discrete mixed normal, via a variety of non-m.l.e. methods (empirical m.g.f., c.f., quantile-based methods, etc.), and the use of the EM algorithm with shrinkage, with its immediate extension to the multivariate case. For the latter, I refer to recent work of mine using the minimum covariance determinant (MCD) for parameter estimation, this also serving as an example of (i) what can be done when, here, the multivariate normal mixture is surely misspecified, and (ii) use of a most likely inconsistent estimator (which outperforms the m.l.e. in terms of density forecasting and portfolio allocation for financial returns data).

Particularly with the less common topics developed in Part III of this book, the result is, like Books I and II, a substantially larger project than some similarly positioned books. It is thus essential to understand that *not everything in the text is supposed to be (or could be) covered in the classroom*, at least not in one semester. In my opinion, students (even in mathematics departments, but particularly those in the social sciences) benefit from having clearly laid out explanations, detailed proofs, illustrative examples, a variety of approaches, introductions to modern techniques, and discussions of important, possibly controversial topics (e.g., the irrelevance of consistent estimators in light of the notion that, in realistic settings, the model is wrong anyway, and changing through time or space; and the arguable superfluousness, if not danger, of the typical hypothesis testing framework), as well as topics that could initially be skipped in a first course, but returned to later or assigned as outside reading, depending on the interests and abilities of the students.

I wish to emphasize that this book is for *teaching*, as (obviously) opposed to being a research monograph, or (less obviously) a dry regurgitation of traditional concepts and examples. An anonymous reviewer of Book I, when I initially submitted it to the publisher Wiley, remarked "it's too much material: It seems the author has written a brain dump." While I like to think I have much more in my head than what was written in that book, he (his gender was indeed disclosed to me) apparently believes that students (let alone instructors) are incapable of assessing what material is core, and what can be deemed "extra," or suitable for reading after the main concepts are mastered. It is trivial to just skip some material, whereas not having it at all results in an admittedly shorter book (who cares, besides arguably the publisher?) that accomplishes far less, and might even give the student a false sense of understanding and competence (which will be painfully revealed in a quant job interview). Fortunately, not everyone agrees with him: Besides heart-warming student feedback over the years on Book I (from master's students) and Book II (from doctoral

---

[1] While I am at it, Severini (2005) is another book I consider exemplary for teaching at the graduate level, as it is highly detailed and accessible, covers a range of important topics, and is at the same mathematical level as, and has some overlap with, my Book II. Though beware of the typos (which go far beyond his current errata sheet)!

students), I cherish the detailed, insightful, and highly positive reviews of Books I and II, by Harvill (2008, 2009). (I still need to send her flowers.)

The choice of precisely what material to cover (and what not to) is crucial. My decision is to blend "old" and "new," helping to emphasize that the subject has important roots going back over a century, and continues to develop unabated. (The reader will quickly see my adoration of Karl Pearson and Ronald Fisher, the founders of our subject; both fascinating, albeit complicated personalities, polymaths, and, at times, adversaries.)

Chapter 1 starts modestly with basic concepts of point estimation, and includes my diatribe on the unnecessary obsession with consistent estimators *in some contexts*. The same chapter then progresses to a very basic development of the single and double bootstrap for computing confidence intervals. If one were to imagine that the field of statistics somehow did not exist, I argue that a student versed in basic probability theory and with access to, and skills with, modern computing power would immediately discover on his/her own the (percentile, single, parametric) bootstrap as a natural way of determining a confidence interval. As such, it is presented before the usual asymptotic Wald intervals and analytic methods. The latter *are* important, as conceptual entities, and work well when applicable, but their relevance to the tasks and goals faced by the new generation of students dealing with modern, sophisticated models and/or big data applications is difficult to motivate.

Chapter 2 spends more time than usual on the empirical c.d.f., and shows, among other things, two simple, instructive proofs of the Glivenko–Cantelli theorem, as opposed to not mentioning it at all, or, perhaps worse, the dreaded "it can be shown . . . ." Besides being a fundamental result of enormous importance, this serves as a primer for students interested in point processes. The chapter also introduces the major concepts associated with hypothesis testing and *p*-values, within the context of distribution testing. I argue in the chapter that this is a very good platform for use of hypothesis testing, and then provide yet another diatribe about why I shy away from presenting the standard material on the subject when applied to parameters of a model.

The rest of Part I consists of three related chapters on parameter estimation. The five chapters of Part I are what I consider to be the core of fundamental statistical inference, and are best read in the order presented, though Chapter 4 can be studied independently of other chapters and possibly assigned as outside reading.

The cornerstone Chapter 3 introduces likelihood, and contains many standard examples, but also some nonstandard material, such as the MCD method to emphasize the relevance of robust statistics and the pernicious issue of masking. Chapter 4 is about numerical optimization, motivating the development of multivariate Hessian-based techniques via repeated application of simple, univariate methods that every student understands, such as bisection. This chapter also includes discussions, with Matlab code, for genetic algorithms and why they are of such importance in many applications.

Chapter 5 is rather unique, using the mixed normal distribution (itself of great relevance, notably now in machine learning) as a platform for showing numerous other methods of point estimation that can outperform the m.l.e. in smaller samples, serve as starting values for computing the m.l.e., or be used when the likelihood is not accessible. Chapter 5 also introduces the use of shrinkage as a penalty factor in the likelihood, and the EM algorithm in the context of the discrete mixed normal distribution.

The chapters of Part II are written to be more or less orthogonal. The instructor (or student working independently) can choose among them, based on his/her interests. The lengthy Chapter 6, on Q-Q plots and distribution testing, builds on the material in

Chapter 2. It emphasizes the distinction between one-at-a-time and simultaneous intervals, and presents various tests for composite normality, including a test of mine, conveniently abbreviated MSP: it is not the most powerful test against all alternatives (no such test yet exists), but its *development* illustrates numerous important concepts – and that is the point of the book.

Chapters 7 and 8 (and Section 3.2 on the univariate and multivariate Cramér–Rao lower bound) are the most "classic," on well-worn results for point and interval estimation, respectively, though Chapter 7 contains some more modern techniques for bias reduction and new classes of estimators. As most of this is standard textbook material at this level, the goal was to develop it in the clearest way possible, with accessible, detailed (sometimes multiple) proofs, and a large variety of examples and end-of-chapter algebraic exercises. There are now several excellent advanced books on mathematical statistics: Schervish (1995), Lehmann and Casella (1998), Shao (2003), and Robert (2007) come to mind, and it is pointless to compete with them, nor is it the goal of this book to do so.

The two chapters of Part III are more associated with financial econometrics and quantitative risk management, though I believe the material should be of interest to a general statistics audience. Chapter 9 covers much ground. It introduces the basics of tail estimation, with a simple derivation of the Hill estimator, discussion of its problems (along with customary Hill horror plots), and enough of a literature review for the interested student to pursue. Also in this chapter (and in Section A.16), the (univariate, asymmetric) stable Paretian distribution receives much attention: I dispel myths about its inapplicability or difficulty in estimation, and discuss several methods for the latter, as well as including recent work on *testing* the stability assumption.

The relatively short Chapter 10 introduces the concept and methodology of indirect inference, a topic rarely presented at this level but of fundamental importance in a variety of challenging contexts. One of the examples used for its demonstration involves the randomized response technique for dealing with awkward questions in surveys (this being notably a topic squarely within statistics, as opposed to econometrics). This elegant solution for obtaining point estimators appears to be new.

The appendix is primarily a review of important and useful facts from probability theory, condensed from Books I and II (where more detail can obviously be found), with its equations being referenced throughout, thus helping to keep this book as self-contained as possible. It also includes a large section of exercises, many of which are not in Books I or II and some of which are challenging, enabling the student to refresh, extend, and self-assess his/her abilities, and/or enabling the instructor to give an initial exam to determine if the student has the requisite knowledge. All the solutions are provided at the end of the appendix.

This appendix also includes some new material not found in Books I and II, such as (i) more results, with proofs, on convergence in distribution (as required for proving the asymptotic properties of the m.l.e.); (ii) a detailed section on expected shortfall (ES), including Stein's lemma, as required for illustrating the shrinkage estimator in Section 5.4; (iii) additional Matlab programs (not in Book II) for the p.d.f., c.d.f., quantiles and ES of the asymmetric stable; and (iv) among the exercises, some potentially useful ones, such as saddlepoint approximations and characteristic function inversion for computing the distribution and ES of a convolution of independent skew-normal random variables.

Numerous topics of relevance were omitted (and some notes deleted – which would delight my "brain dump" accuser), such as ancillarity, hierarchical models, rank and permutation tests, and, most notably, Bayesian methodology. For the latter, there are now many good textbooks on the topic, in both pure statistics and also econometrics, and the last thing I want is that the reader ignore the Bayesian approach. I think a solid grounding in basic principles, likelihood-based inference, and a strong command of computing serve as an excellent background for pursuing dedicated works on Bayesian methodology. Section 5.1.6 does introduce the idea of quasi-Bayesian estimation and its connection to shrinkage estimation, and illustrates (without needing to break the proverbial full Bayesian egg)[2] the effectiveness and importance of these methods.

With respect to computing, I chose (no doubt to the annoyance of some) Matlab as the vehicle for prototyping, though I strongly encourage readers versed in R to continue using R, or Python, or even to learn the relatively new and highly promising language Julia. Unlike with the Matlab codes in Book I, I do not (so far) provide R translations, though every attempt was made to use the most basic coding and data structures possible, so that translations should be straightforward, and also occasionally separating the very-specific-to-Matlab commands, such as for graphics.

No single book will ever cover every topic or aspect the author would like. As a complement to this book, I recommend students concurrently read some sections of Pawitan (2001) (with an updated and paperback version now available), Davison (2003), and Casella and Berger (2002), three books that I hold as exemplary; they cover additional topics I have omitted, and, in the case of the former two, contain far more examples with real data.

I recall a review of a book in financial econometrics (which I had best not name). Paraphrasing, the reviewer stated that academic books tend to have one of two purposes: (i) to teach the material; or (ii) to impress the reader and, particularly, colleagues with the authors' knowledge. The reviewer then went on to say how the book accomplished neither. My hope is that the reader and instructor understand my goal to be the former, with little regard for the latter: As emphasized above, the book contains much material, computer codes, and touches upon some recent developments. When proofs are shown, they are simple and detailed. I wrote the book for motivated students who want straightforward explanations, clear demonstrations, and discussions of more modern topics, particularly in a non-Gaussian setting. My guiding principle was to write the book that I would have killed for as a graduate student.

Some acknowledgments are in order. I owe an enormous amount of gratitude to the excellent scientists and instructors I worked with during and after my graduate studies. Alphabetically, these include professors Peter Brockwell, Ronald Butler, Richard Davis, Hariharan (Hari) Iyer, Stefan Mittnik, and Svetlozar (Zari) Rachev. All of these individuals also have textbooks that I highly recommend, and some of which will be mentioned in the preface to book IV. As the years go by, the proverbial circle starts to close, and I have my own doctoral students, all of whom have contributed in various ways to my book projects. Notable mention goes to Simon Broda, Pawel Polak (both of whom are now professors themselves) and (current PhD students) Marco Gambacciani and Patrick Walker, who, along with professors Kai Carstensen, Walter Farkas, Markus Haas, Alexander McNeil, Nuttanan (Nate) Wichitaksorn, and Michael Wolf, have read parts of this manuscript (and

[2] This refers to the oft-quoted statement in Savage (1961, p. 578) that Fisher's fiducial inferential method is "a bold attempt to make the Bayesian omelet without breaking the Bayesian eggs".

book IV) and helped tease out mistakes and improve the presentation. Finally, I am indebted to my copy editor Richard Leigh from Wiley, who read every line of the book, checked every graphic and bibliography reference, and made uncountable corrections and suggestions to the scientific English presentation, as well as (embarrassingly) caught a few math mistakes. I have obviously suggested to the editor to have him work on my book IV (and double his salary).

My gratitude to these individuals cannot be overstated.

Part I

Essential Concepts in Statistics

1

Introducing Point and Interval Estimation

> The discussions of theoretical statistics may be regarded as alternating between problems of estimation and problems of distribution. In the first place a method of calculating one of the population parameters is devised from common-sense considerations: we next require to know its probable error, and therefore an approximate solution of the distribution, in samples, of the statistics calculated.
>
> *(R. A. Fisher, 1922, reproduced in Kotz and Johnson, 1992)*

This chapter and the next two introduce the primary tools and concepts underlying most all problems in statistical inference. We restrict ourselves herein to the independent, identically distributed (i.i.d.) framework, in order to emphasize the fundamental concepts without the need for addressing the additional issues and complexities associated with the workhorse models of statistics, such as linear models, analysis of variance, design of experiments, and time series. The overriding goal is to extract relevant information from the available sample in order to learn about the underlying population from which it was drawn.

We begin with the basic definitions associated with point estimation, and introduce the maximum likelihood estimator (m.l.e.). We will have more to say about point estimation and m.l.e.s in Chapters 3 and 5. The remainder of the chapter is dedicated to individual parameter confidence intervals (c.i.s), restricting attention to the intuitive use of computer-intensive methods for their construction, as they are generally applicable and, for more complex problems, often the only available choice. In particular, a natural progression is made from simulation to the parametric bootstrap, to the nonparametric bootstrap, to the double nonparametric bootstrap, and finally to the double bootstrap with analytic inner loop, the latter using techniques from Chapter 8.

Fundamental Statistical Inference: A Computational Approach, First Edition. Marc S. Paolella.
© 2018 John Wiley & Sons Ltd. Published 2018 by John Wiley & Sons Ltd.

1.1 POINT ESTIMATION

To introduce the notion of parameter estimation from a sample of data, we make use of two simple models, the Bernoulli and geometric.

1.1.1 Bernoulli Model

Consider an idealized experiment that consists of randomly drawing a marble from an urn containing R red and W white marbles; its color is noted and it is then placed back into the urn. This is repeated n times, whereby n is a known, finite constant, *but R and W are unknown*. This corresponds to a sequence of Bernoulli trials with unknown probability $p = R/(R + W)$ or $p = W/(R + W)$, depending on what one wants to consider a "success." Assuming the former, let X_i, $i = 1, \ldots, n$, denote the outcomes of the experiment, with $X_i \overset{\text{i.i.d.}}{\sim} \text{Bern}(p)$, each with support $S = \{0, 1\}$. The ultimate goal is to determine the value of p. If n is finite (as reality often dictates), this will be an impossible task. Instead, we content ourselves with attempting to infer as much information as possible about the value of p.

As a starting point, in line with Fisher's "common-sense considerations" in the opening quote, it seems reasonable to examine the proportion of successes. That is, we would compute s/n, where s is the observed number of successes. The value s/n is referred to as a **point estimate** of p and denoted $\hat{p}$, pronounced "p hat." Sometimes it is advantageous to write $\hat{p}_n$, where the subscript indicates the sample size. From the way in which the experiment is defined, it should be clear that s is a realization from the binomial random variable (r.v.) $S = \sum_{i=1}^n X_i \sim \text{Bin}(n, p)$. To emphasize this, we also write $\hat{p} = S/n$ and call this a **point estimator** of p, the distinction being that a point estimator is a random variable, while a point estimate is a realization of this random variable resulting from the outcome of a particular experiment.

Note that the same notation of adding a "hat" to the parameter of interest is used to denote both estimate and estimator, as this is the common standard. However, the distinction between estimate and estimator is crucial when attempting to assess the properties of $\hat{p}$ (e.g., is it correct on average?) and compare its performance to other possible estimators (e.g., is one point estimator more likely to be correct than the other?). For instance, $\mathbb{E}[s/n] = s/n$, that is, s/n is a post-experiment constant, while $\mathbb{E}[S/n] = (np)/n = p$ can be computed before or after the experiment takes place. In this case, estimator S/n is said to be **(mean) unbiased**.

More formally, let $\hat{\theta}$ be a point estimator of the finite, fixed, unknown parameter $\theta \in \Theta \subset \mathbb{R}$ such that $\mathbb{E}[\hat{\theta}]$ exists. Then:

> The point estimator $\hat{\theta}$ is **(mean) unbiased** (with respect to the set Θ) if its expected value is θ (for all $\theta \in \Theta$); otherwise it is **(mean) biased** with bias $(\hat{\theta}) = \mathbb{E}[\hat{\theta}] - \theta$.

Generally speaking, mean unbiasedness is a desirable property because it implies that we are "correct on average," where the "average" refers to the hypothetical idea of repeating the experiment infinitely often – something that of course does not actually happen in reality. An impressive theoretical framework in mathematical statistics was developed, starting in the 1950s, for the derivation and study of unbiased estimators with minimum variance; see Chapter 7, especially Section 7.2. It is often the case, however, that estimators can be found that are biased, but, by virtue of having a lower variance, wind up having a lower mean squared error, as seen from (1.2) directly below. This concept is well known, and reflected, for example, in Shao and Tu (1995, p. 67), stating "We need to balance the advantage of unbiasedness against the drawbacks of a large mean squared error." Another type of unbiasedness involves using the median instead of the mean. See Section 7.4.2 for details on median-unbiased estimators.

For the binomial example, the variance of estimator $\hat{p}$ is

$$\mathbb{V}(\hat{p}) = \mathbb{V}\left(\frac{\sum_{i=1}^{n} X_i}{n}\right) = n\frac{p(1-p)}{n^2} = \frac{p(1-p)}{n}, \tag{1.1}$$

and it clearly goes to zero as the sample size increases. This is also desirable, because, as more samples are collected, the amount of information from which p is to be inferred is growing. This concept is referred to as consistency; recalling the definition of convergence in probability from (A.254) and the weak law of large numbers (A.255), the following definition should seem natural:

> An estimator $\hat{\theta}_n$ based on a sample of n observations is **weakly consistent** (with respect to Θ) if, as $n \to \infty$, $\Pr(|\hat{\theta}_n - \theta| > \epsilon) \to 0$ for any $\epsilon > 0$ (and all $\theta \in \Theta$).

Observe that an estimator can be (mean) unbiased but not consistent: if $X_i \overset{\text{i.i.d.}}{\sim} N(\mu, 1)$, $i = 1, \ldots, n$, then the estimator $\hat{\mu} = X_1$ is unbiased, but it does not converge to μ as the sample size increases.

Another popular measure of the quality of an estimator is its expected squared deviation from the true value, called *mean squared error*, or *m.s.e.*:

> The **mean squared error** of the estimator $\hat{\theta}$ is defined as $\mathbb{E}[(\hat{\theta} - \theta)^2]$.

An important decomposition of the m.s.e. is as follows. With $\Xi = \mathbb{E}[\hat{\theta}]$,

$$\begin{aligned} \text{m.s.e.}\,(\hat{\theta}) = \mathbb{E}[(\hat{\theta} - \theta)^2] &= \mathbb{E}[(\hat{\theta} - \Xi + \Xi - \theta)^2] \\ &= \mathbb{E}[(\hat{\theta} - \Xi)^2] + \mathbb{E}[(\Xi - \theta)^2] + \text{cross-term, which is zero} \\ &= \mathbb{E}[(\hat{\theta} - \Xi)^2] + (\Xi - \theta)^2 = \mathbb{V}(\hat{\theta}) + [\text{bias}\,(\hat{\theta})]^2. \end{aligned} \tag{1.2}$$

The reader should quickly verify that the cross-term is indeed zero. Note that, for an unbiased estimator, its m.s.e. and variance are equal. As the estimator $\hat{\theta}$ is a function of the data, it is itself a random variable. With $f = f_{\hat{\theta}}$ the p.d.f. of $\hat{\theta}$, we can write $\Pr(|\hat{\theta} - \theta| > \epsilon)$ for any $\epsilon > 0$ as

$$\int_{|t-\theta|>\epsilon} f(t)\,dt \le \int_{|t-\theta|>\epsilon} \frac{(t-\theta)^2}{\epsilon^2} f(t)\,dt \le \int_{-\infty}^{\infty} \frac{(t-\theta)^2}{\epsilon^2} f(t)\,dt = \frac{\mathbb{E}[(\hat{\theta}-\theta)^2]}{\epsilon^2},$$

so that $\hat{\theta}$ is weakly consistent if m.s.e. $(\hat{\theta}) \to 0$.

The estimator $\hat{p} = S/n$ for the Bernoulli model is rather intuitive and virtually presents itself as being a good estimator of p. It turns out that this $\hat{p}$ coincides with the estimator we obtain when applying a very general and powerful method of obtaining an estimator for an unknown parameter of a statistical model. We briefly introduce this method now, and will have more to say about it in Section 3.1.

The **likelihood function** $\mathcal{L}(\theta; \mathbf{x})$ is the joint density of a sample $\mathbf{X} = (X_1, \ldots, X_n)$ as a function of the (for now, scalar) parameter θ, for fixed sample values $\mathbf{X} = \mathbf{x}$. That is, $\mathcal{L}(\theta; \mathbf{x}) = f_{\mathbf{X}}(\mathbf{x}; \theta)$, where $f_{\mathbf{X}}$ is the p.m.f. or p.d.f. of $\mathbf{X}$. Let $\ell(\theta; \mathbf{x}) = \log \mathcal{L}(\theta; \mathbf{x})$,[1] and write just $\ell(\theta)$ when the data are clear from the context. Denote the first and second derivatives of $\ell(\theta)$ with respect to θ by $\dot{\ell}(\theta)$ and $\ddot{\ell}(\theta)$, respectively. The **maximum likelihood estimate**, abbreviated m.l.e. and denoted by $\hat{\theta}$ (or, to distinguish it from other estimates, $\hat{\theta}_{\mathrm{ML}}$), is that value of θ that maximizes the likelihood function for a given data set $\mathbf{x}$. The **maximum likelihood estimator** (as opposed to *estimate*) is the function of the X_i, also denoted $\hat{\theta}_{\mathrm{ML}}$, that yields the m.l.e. for an observed data set $\mathbf{x}$.

In many cases of interest (including the Bernoulli and geometric examples in this chapter), the m.l.e. satisfies $\dot{\ell}(\hat{\theta}) = 0$ and $\ddot{\ell}(\hat{\theta}) < 0$. For example, with $X_i \overset{\text{i.i.d.}}{\sim} \text{Bern}(\theta)$, $i = 1, \ldots, n$, the likelihood is

$$\mathcal{L}(\theta; \mathbf{x}) = \prod_{i=1}^{n} \theta^{x_i}(1 - \theta)^{1 - x_i} \mathbb{1}_{\{0,1\}}(x_i) = \theta^s (1 - \theta)^{n-s} \mathbb{1}_{\{0,1,\ldots,n\}}(s),$$

where $s = \sum_{i=1}^{n} x_i$. Then

$$\dot{\ell}(\theta) = \frac{s}{\theta} - \frac{n-s}{1-\theta} \quad \text{and} \quad \ddot{\ell}(\theta) = -\frac{s}{\theta^2} - \frac{n-s}{(1-\theta)^2},$$

from which it follows (by setting $\dot{\ell}(\hat{\theta}) = 0$ and confirming $\ddot{\ell}(\hat{\theta}) < 0$) that $\hat{\theta}_{\mathrm{ML}} = S/n$ is the m.l.e. It is easy to see that $\hat{\theta}_{\mathrm{ML}}$ is unbiased.

1.1.2 Geometric Model

As in the binomial case, independent draws with replacement are conducted from an urn with R red and W white marbles. However, now the number of trials is not fixed in advance; sampling continues until r red marbles have been drawn. What can be said about $p = R/(R + W)$? Let the r.v. X be the number of necessary trials. From the sampling structure, X follows a negative binomial distribution, $X \sim \text{NBin}(r, p)$, with p.m.f.

$$f_X(x; r, p) = \binom{x-1}{r-1} p^r (1-p)^{x-r} \mathbb{1}_{\{r, r+1, \ldots\}}(x). \tag{1.3}$$

Recall that X can be expressed as the sum of r i.i.d. geometric r.v.s, say $X = \sum_{i=1}^{r} G_i$, where $G_i \overset{\text{i.i.d.}}{\sim} \text{Geo}(p)$, each with support $\{1, 2, \ldots\}$.

This decomposition is important because it allows us to imagine that sampling occurs not necessarily consecutively in time until r successes occur, but rather as r independent (and possibly concurrent) geometric trials using urns with the same red to white ratio, that is, the same p. For example, interest might center on how long it takes a woman to become pregnant using a particular method of assistance (e.g., temperature measurements or hormone treatment). This is worth making an example, as we will refer to it more than once.

---

[1] Throughout this book, log refers to base e unless otherwise specified.

Example 1.1 (Geometric) *Let $G_i \overset{\text{i.i.d.}}{\sim} \text{Geo}(\theta)$, $i = 1, \dots, n$, with typical p.m.f.*

$$f_G(x; \theta) = \theta(1 - \theta)^{x-1} \mathbb{1}_{\{1,2,\dots\}}(x), \quad \theta \in \Theta = (0, 1).$$

Then $\ell(\theta; \mathbf{x})$, the log-likelihood of the sample $\mathbf{x} = (x_1, \dots, x_n)$, and its first derivative, $\dot{\ell}(\theta; \mathbf{x})$, are, with $s = \sum_{i=1}^{n} x_i$,

$$\ell(\theta; \mathbf{x}) = n \log(\theta) + \log(1 - \theta) \sum_{i=1}^{n} (x_i - 1), \quad \dot{\ell}(\theta; \mathbf{x}) = \frac{n}{\theta} - \frac{s - n}{1 - \theta}.$$

Solving the equation $\dot{\ell}(\theta; \mathbf{x}) = 0$ and confirming $\ddot{\ell}(\hat{\theta}) < 0$ gives $\hat{\theta}_{ML} = n/S = 1/\bar{G}$. We will see below and in Section 7.3 that the m.l.e. is not unbiased.[2] ∎

Imagine a study in which each of r couples (independently of each other) attempts to conceive each month until they succeed. In the $r = 1$ case, $X = G_1 \sim \text{Geo}(p)$ and, recalling that $\mathbb{E}[G_1] = 1/p$, an intuitive point estimator of p is $1/G_1$. Interest centers on developing a point estimator for the $r > 1$ case. Of course, in this simple structure, one would just compute the m.l.e. However, we use this easy case to illustrate how one might proceed when simple answers are not immediately available, and some thinking and creativity are required.

Based on the result for $r = 1$, one idea for the $r > 1$ case would be to use the average of the $1/G_i$ values, $r^{-1} \sum_{i=1}^{r} G_i^{-1}$, which we denote by $\hat{p}_1$. Another candidate is $\hat{p}_2 = 1/\bar{G} = r/\sum_{i=1}^{r} G_i = r/X$. This happens to be the m.l.e. from Example 1.2. Note that both of these estimators reduce to $1/G_1$ when $r = 1$. We also consider the nonobvious point estimator $\hat{p}_3 = (r-1)/(X-1)$. It will be derived in Section 7.3, and is only useful for $r > 1$.

Instead of algebraically determining the mean and variance of the $\hat{p}_i$, $i = 1, 2, 3$, we will begin our practice of letting the computer do the work. The program in Listing 1.1 computes the three point estimators for a simulated set of G_i; it repeats this $\text{sim} = 10,000$ times, and the resulting sample mean and variance of these simulated estimates approximate the true mean and variance.

To illustrate, Figure 1.1 shows the histograms of the simulated point estimators for the case with $p = 0.3$ and $r = 5$. From these, the large upward bias of $\hat{p}_1$ is particularly clear.

```
1  function [p1vec, p2vec, p3vec]=geometricparameterestimate(p,r,sim)
2  p1vec = zeros(sim,1); p2vec = p1vec; p3vec = p1vec;
3  for s=1:sim
4    gvec=geornd(p,[r 1]) +1;
5    p1 = mean(1./gvec); p2 = 1/mean(gvec); p3 = (r-1) / (sum(gvec)-1);
6    p1vec(s) = p1; p2vec(s) = p2; p3vec(s) = p3;
7  end
8  bias1 = mean(p1vec)-p, bias2 = mean(p2vec)-p, bias3 = mean(p3vec)-p
9  var1 = var(p1vec), var2 = var(p2vec), var3 = var(p3vec)
10 mse1 = var1+bias1^2, mse2 = var2+bias2^2, mse3 = var3+bias3^2
```

Program Listing 1.1: Simulates three point estimators for p in the i.i.d. geometric model. Calling the function with $p = 0.3$ and $r = 5$ corresponds to the true probability of success being 0.3 and using five couples in the experiment.

[2] We use the symbol ∎ to denote the end of proofs of theorems, as well as examples and remarks, acknowledging that it is traditionally only used for the former, as popularized by Paul Halmos.

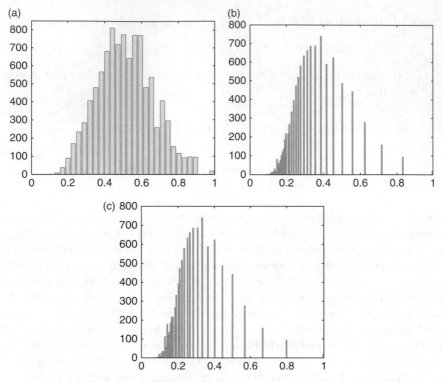

Figure 1.1 *Distribution of point estimators $\hat{p}_1$ (a), $\hat{p}_2$ (b), and $\hat{p}_3$ (c) using output from the program in Listing 1.1 with $p = 0.3$ and $r = 5$, based on simulation with 10,000 replications.*

The discrete nature of $\hat{p}_2$ and $\hat{p}_3$ arises because these two estimators first compute the sum of the observations and then take reciprocals, so that computation of their p.m.f.s is easy. As an example, $\hat{p}_3 = 0.4 \Leftrightarrow X = 11$, which, from (1.3), has probability 0.06, so that approximately 600 of the simulated values depicted in the histogram of $\hat{p}_3$ should be 0.4; there are 624 in the histogram. Similarly, $\hat{p}_3 = 0.8 \Leftrightarrow X = 6$, with probability 0.008505, and 94 in the histogram.

As p increases towards one, the number of points in the supports of $\hat{p}_2$ and $\hat{p}_3$ decreases. This is illustrated in Figure 1.2, showing histograms of $\hat{p}_2$ for $r = 10$ and four different values of p. The code used to make the plots is given in Listing 1.2. Observe how we avoid use of the FOR loop (as was used in Listing 1.1) for generating the 1 million replications, thus providing a significant speed increase. (The use of the `eval` command with concatenated text strings is also demonstrated.)

For the simulation of $\hat{p}_1$, $\hat{p}_2$, and $\hat{p}_3$ from Listing 1.1, with $p = 0.3$ and $r = 5$, the results are shown in the first numeric row of Table 1.1. We see that $\hat{p}_1$ has almost five times the bias of $\hat{p}_2$, while $\hat{p}_2$ has over 100 times the bias of $\hat{p}_3$. The variance of $\hat{p}_1$ is slightly larger than those of $\hat{p}_2$ and $\hat{p}_3$, which are nearly the same. By combining these according to (1.2), it is clear that the m.s.e. will be smallest for $\hat{p}_3$, as also shown in the table. The next row shows the results using a larger sample of 15 couples. While the bias of $\hat{p}_1$ stays the same, those of $\hat{p}_2$ and $\hat{p}_3$ decrease. For all point estimators, the variance decreases.

It turns out that, as the number of couples, r, tends towards infinity, the variance of all the estimators goes to zero, while the bias of $\hat{p}_1$ stays at 0.22 and that of $\hat{p}_2$ goes to zero.

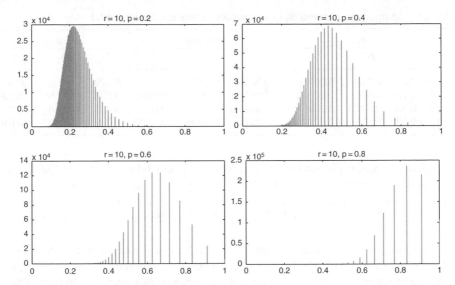

Figure 1.2 *Histogram of point estimator $\hat{p}_2$ for $r = 10$ and four values of p, based on simulation with 1 million replications.*

```
1  B=1e6; r=10;
2  for p=0.2:0.2:0.8
3    phatvec = 1./mean(geornd(p,[r B])+1); % the MLE
4    [histcount, histgrd]=hist(phatvec,1000);
5    figure, h1=bar(histgrd,histcount); set(gca,'fontsize',16), xlim([0 1])
6    title(['r=',int2str(r),'  p=' num2str(p)])
7    set(h1,'facecolor',[0.94 0.94 0.94],'edgecolor',[0.9 0.7 1])
8    eval(['print -depsc phatforgeogetsmorediscretep',int2str(10*p)])
9  end
```

Program Listing 1.2: Generates the graphs in Figure 1.2.

Hence, we say that $\hat{p}_2$ is **asymptotically unbiased**. We will see in Section 7.3 that $\hat{p}_3$ is unbiased – not just asymptotically, but for all $0 < p \le 1$ and any $r > 1$. This implies that the value 0.0004 in the $\hat{p}_3$ bias column of the table just reflects **sampling error** resulting from using only 10,000 replications in the simulation. In comparison, then, point estimator $\hat{p}_3$ seems to be preferred with respect to all three criteria.

The lower portion of Table 1.1 shows similar results using $p = 0.7$. Again, $\hat{p}_1$ is highly biased, while, comparatively speaking, the bias of $\hat{p}_2$ is much smaller and diminishes with growing sample size r. The bias of $\hat{p}_3$ appears very small and, as already mentioned, is theoretically zero. The interesting thing about this choice of p is that the variance of $\hat{p}_2$ is smaller than that of $\hat{p}_3$. In fact, this reduction in variance causes the m.s.e. of $\hat{p}_2$ to be smaller than that of $\hat{p}_3$ even though the bias of $\hat{p}_3$ is essentially zero. This demonstrates two important points:

(i) An unbiased point estimator need not have the smallest m.s.e.

(ii) The relative properties of point estimators may change with the unknown parameter of interest.

TABLE 1.1 Comparison of three point estimators for the geometric model

p	r	bias			variance			m.s.e.		
		$\hat{p}_1$	$\hat{p}_2$	$\hat{p}_3$	$\hat{p}_1$	$\hat{p}_2$	$\hat{p}_3$	$\hat{p}_1$	$\hat{p}_2$	$\hat{p}_3$
0.3	5	0.22	0.045	0.00040	0.023	0.018	0.017	0.070	0.020	0.017
0.3	15	0.22	0.015	0.00041	0.0077	0.0048	0.0045	0.055	0.0050	0.0045
0.7	5	0.13	0.040	0.00057	0.014	0.027	0.033	0.031	0.028	0.033
0.7	15	0.13	0.013	−0.00078	0.0046	0.0096	0.0102	0.022	0.0098	0.010

Having demonstrated these two facts using just the values in Table 1.1, it would be desirable to graphically depict the m.s.e. of estimators $\hat{p}_2$ and $\hat{p}_3$ as a function of p, for several sample sizes. This is shown in Figure 1.3, from which we see that m.s.e. $(\hat{p}_2) <$ m.s.e. $(\hat{p}_3)$ for (roughly) $p > 0.5$, but as the sample size increases, the difference in m.s.e. of the two estimators becomes negligible.

Facts (i) and (ii) mentioned above complicate the comparison of estimators. Some structure can be put on the problem if we restrict attention to unbiased estimators. Then, minimizing the m.s.e. is the same as minimizing the variance; this gives rise to the following concepts:

> An unbiased estimator, say $\hat{\theta}_{\text{eff}}$, is **efficient** (with respect to Θ) if it has the smallest possible variance of all unbiased estimators (for all $\theta \in \Theta$).
>
> The **efficiency** of an unbiased estimator $\hat{\theta}$ is $\text{Eff}(\hat{\theta}, \theta) = \mathbb{V}(\hat{\theta}_{\text{eff}})/\mathbb{V}(\hat{\theta})$.

We will see later (Chapter 7) that the estimator $\hat{p}_3$ used above is efficient.

In many realistic problems, there may be no unbiased estimators, or no efficient one; and if there is, like $\hat{p}_3$ above, it might not have the smallest m.s.e. over all or parts of Θ. This somewhat diminishes the value of the efficiency concept defined above. All is not lost, however. In many cases of interest, the m.l.e. has the property that, asymptotically, it is (unbiased and) efficient. As such, it serves as a natural benchmark with which to compare competing estimators. We expect that, with increasing sample size, the m.l.e. will eventually be as good as, or better than, all other estimators, with respect to m.s.e., for all $\theta \in \Theta$. This certainly does not imply that the m.l.e. is the best estimator in finite samples, as we see in Figure 1.3 comparing the m.l.e. $\hat{p}_2$ to the efficient estimator $\hat{p}_3$. (Other cases in which the

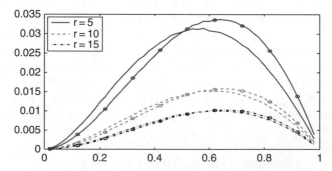

Figure 1.3 The m.s.e. of estimators $\hat{p}_2$ (lines) and $\hat{p}_3$ (lines with circles) for parameter p in the geometric model, as a function of p, for three sample sizes, obtained by simulation with 100,000 replications.

m.l.e. is not the best estimator with respect to m.s.e. in finite samples are demonstrated in Example 9.3 and Section 7.4.3.)

Before leaving this section, it is worth commenting on other facts observable from Figure 1.3. Note that, for any $r > 1$, the m.s.e.s for both estimators $\hat{p}_2$ and $\hat{p}_3$ approach zero as $p \to 0$ and $p \to 1$. This is because, for the former case, as $p \to 0$, $\mathbb{E}[X] = r/p \to \infty$, and $\hat{p} \to 0$. For the latter case, as $p \to 1$, $\Pr(X = r) = p^r \to 1$, so that $\hat{p} \to 1$. Also, the m.s.e. increases monotonically with p as p moves from 0^+ to (roughly) $p = 0.6$, and decreases monotonically back towards zero as $p \to 1$.

1.1.3 Some Remarks on Bias and Consistency

> The most effective way to discourage an applied statistician from using a model or method is to say that it gives asymptotically inconsistent parameter estimates. This is completely irrelevant for a fixed small sample; the interval of plausible values, not a point estimate, is essential. … If the sample were larger, in a properly planned study, the model could be different, so the question of a parameter estimate in some fixed model converging asymptotically to a "true" value does not arise.
>
> (J. K. Lindsey, 1999, p. 20)

Sections 1.1.1 and 1.1.2 introduced the fundamental concepts of point estimation, (mean) unbiasedness, consistency, m.s.e., likelihood, and efficiency. We demonstrated that a biased estimator might be preferred to an unbiased one with respect to m.s.e., such as the m.l.e., which, under certain conditions fulfilled in the vast majority of situations, is asymptotically unbiased and consistent (see Section 3.1.4 for the formalities of this).

While (mean) unbiasedness is an appealing property, in many modern applications, particularly in the context of big data and models with a relatively large number of parameters, unbiasedness is not only no longer a consideration, but biased estimates *are actually preferred*, via use of shrinkage estimation; see Chapter 5. Moreover, starting in the late twentieth century and continuing unabated, the Bayesian approach to inference has gained in attention and usage because of advances in computing power and recognition of some of its inferential benefits. In a sense, unbiasedness is the antithesis, or dual, of the Bayesian approach; see Noorbaloochi and Meeden (1983). So-called empirical Bayes methods form a link between pure Bayesian methods and shrinkage estimation, and yield a formidable approach to inference; see the references in Section 5.4.

As such, most researchers are now comfortable working with biased estimators, but will often still insist on such estimators being consistent. As consistency is an asymptotic notion, but reality deals with finite samples, one might also question its value, as suggested in the above quote from Lindsey (1999, p. 20). As a simple example of interest (particularly for anyone with a pension fund), consider a case from financial portfolio optimization. The basic framework of Markowitz (1952) (which led to him receiving the 1990 Nobel Memorial Prize in Economic Sciences) is still used today in industry, though (as was well known to Markowitz) the method is problematic because it requires estimating the mean vector and covariance matrix of past asset returns. This has been researched in a substantial body of literature, resulting in the established finding that shrinking the optimized portfolio weights towards the equally weighted vector (referred to as "$1/N$," where N is the number of assets under consideration) not only improves matters substantially (in terms of a risk–reward tradeoff), but *just taking the weights to be the shrinkage target* $1/N$ often results in better performance.[3] Alternatively, one can apply the Markowitz optimization framework, but in conjunction with shrinkage applied to the mean vector and/or the covariance matrix.[4]

The humbling result that one is better off forgoing basic statistical modeling and just putting equal amounts of money in each available asset (roughly equivalent to just buying an exchange traded fund) arises because of (i) the high relevance and applicability of shrinkage estimation in this setting; and (ii) the gross misspecification of the model underlying the multivariate distribution of asset returns, and how it evolves over time. More statistically sophisticated models for asset returns *do exist*, such that portfolio optimization *does* result in substantially better performance than use of $1/N$ (let alone the naive Markowitz framework), though unsurprisingly, these are complicated for people not well versed in statistical theory, and require more mathematical and statistical prowess than usually obtained from a course in introductory statistical methods for aspiring investors.[5] Book IV will discuss some such models.

Clearly, $1/N$ is not a "consistent" estimator of the optimal portfolio (as defined by specifying some desired level of annual return, and then minimizing some well-defined risk measure, such as portfolio variance, in the Markowitz setting). More importantly, this example highlights the fact that the model used (an i.i.d. Gaussian or, more generally, an elliptic distribution, with constant unknown mean vector and covariance matrix throughout time) for the returns is so completely misspecified, that the notion of consistency becomes vacuous in this setting.

Two further, somewhat less trivial cases in which inconsistent estimators are favored (and also in the context of modeling financial asset returns), are given in Krause and Paolella (2014) and Gambacciani and Paolella (2017).

1.2 INTERVAL ESTIMATION VIA SIMULATION

To introduce the concepts associated with interval estimation and how simulating from the true distribution can be used to compute confidence intervals, we use the Bernoulli model from Section 1.1.1. For a fixed sample size n, we observe realizations of $X_1, \dots, X_n$, where

---

[3] See, for example, DeMiguel et al. (2009a,b 2013) and the references therein.

[4] See, for example, Jorion (1986), Jagannathan and Ma (2003), Ledoit and Wolf (2003, 2004), Schäfer and Strimmer (2005), Kan and Zhou (2007), Fan et al. (2008); Bickel and Levina (2008), and the references therein.

[5] This result is also anathema to supposedly professional investment consultants and mutual fund managers, with their techniques for "stock picking" and "investment strategies." This was perhaps most forcefully and amusingly addressed by Warren Buffett (who apparently profits enormously from market inefficiency). "The Berkshire chairman has long argued that most investors are better off sticking their money in a low-fee S&P 500 index fund instead of trying to beat the market by employing professional stockpickers" (Holm, 2016). To quote Buffett: "Supposedly sophisticated people, generally richer people, hire consultants, and no consultant in the world is going to tell you 'just buy an S&P index fund and sit for the next 50 years.' You don't get to be a consultant that way. And you certainly don't get an annual fee that way. So the consultant has every motivation in the world to tell you, 'this year I think we should concentrate more on international stocks,' or 'this manager is particularly good on the short side,' and so they come in and they talk for hours, and you pay them a large fee, and they always suggest something other than just sitting on your rear end and participating in the American business without cost. And then, after they get their fees, they in turn recommend to you other people who charge fees, which ... cumulatively eat up capital like crazy" (Holm, 2016). See also Sorkin (2017) on (i) Buffett's views; (ii) why many high wealth individuals continue to seek highly paid consultants; and (iii) with respect to the concept of market efficiency, what would happen if most wealth were channeled into exchange traded funds (i.e., the market portfolio).

$X_i \overset{\text{i.i.d.}}{\sim}$ Bern (p), and compute the mean of the X_i, $\hat{p} = S/n$, as our estimator of the fixed but unknown p. Depending on n and p, it could be that $\hat{p} = p$, though if, for example, n is odd and $p = 0.5$, then $\hat{p} \neq p$. Even if n is arbitrarily large but finite, if p is an irrational number in $(0, 1)$, then with probability one (w.p.1), $\hat{p} \neq p$.

The point is that, for almost all values of n and p, the probability that $\hat{p} = p$ will be low or zero. As such, it would seem wise to provide a set of values such that, with a high probability, the true p is among them. For a univariate parameter such as p, the most common set is an interval, referred to as a **confidence interval**, or **c.i.**

Notice that a c.i. pertains to a parameter, such as p, and *not* to an estimate or estimator, $\hat{p}$. We might speak of a c.i. *associated with* $\hat{p}$, in which case it is understood that the c.i. refers to parameter p. It does not make sense to speak of a c.i. for $\hat{p}$.

To get an idea of the uncertainty associated with $\hat{p}$ for a fixed n and p, we can use simulation. This is easily done in Matlab, using its built-in routine `binornd` for simulating binomial realizations.

```
p=0.3; n=40; sim=1e4; phat=binornd(n,p,[sim,1])/n; hist(phat)
```

The following code is a little fancier. It makes use of Matlab's `tabulate` function, discussed in Section 2.1.4.

```
p=0.3; n=40; sim=1e4; phat=binornd(n,p,[sim,1])/n;
nbins=length(tabulate(phat)); [histcount, histgrd] =hist(phat,nbins);
h1=bar(histgrd,histcount); xlim([0 0.8])
set(h1,'facecolor',[0.64 0 0.24],'edgecolor',[0 0 0],'linewidth',2)
set(gca,'fontsize',16), title([ 'Using n=',int2str(n)])
```

Doing this with $p = 0.3$ and for sample sizes $n = 20$ and $n = 40$ yields the histograms shown in Figure 1.4. We see that, while the mode of $\hat{p}$ is at 0.3 in both cases, there is quite some variation around this value and, particularly for $n = 20$, a small but nonnegligible chance (the exact probability of which you can easily calculate) that $\hat{p}$ is zero or higher than 0.6.

We first state some useful definitions, and then, based on our ability to easily simulate values of $\hat{p}$, determine how to form a c.i. for p.

Consider a distribution (or statistical model) with unknown but fixed k-dimensional parameter $\theta \in \Theta \subseteq \mathbb{R}^k$. A **confidence set** $M(\mathbf{X}) \subset \Theta$ for θ with **confidence level** $1 - \alpha$ is any set such that

$$\Pr(\theta \in M(\mathbf{X})) \geq 1 - \alpha, \quad \forall \theta \in \Theta, \quad 0 < \alpha < 1, \tag{1.4}$$

where $M(\mathbf{X})$ depends on the r.v. $\mathbf{X}$, a realization of which will be observed, but does not depend on the unknown parameter θ. Typical values of α are 0.01, 0.05 and 0.10. The quantity $\Pr(\theta \in M(\mathbf{X}))$ is called the **coverage probability** and can depend on θ; its greatest lower bound

$$\inf_{\theta \in \Theta} \Pr(\theta \in M(\mathbf{X}))$$

is referred to as the **confidence coefficient** of $M(\mathbf{X})$.

It is imperative to keep in mind how (1.4) is to be understood: As θ is fixed and $M(\mathbf{X})$ is random, we say that, before the sample is collected, the set $M(\mathbf{X})$ will contain (or capture)

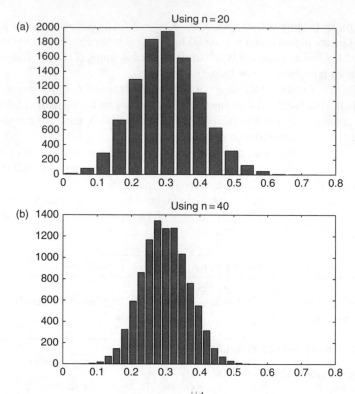

Figure 1.4 Simulations of $\hat{p} = S/n$ for $S = \sum_{i=1}^{n} X_i$, $X_i \overset{\text{i.i.d.}}{\sim} \text{Bern}(p)$, for $p = 0.3$ and $n = 20$ (a) and $n = 40$ (b), based on 10,000 replications.

the true θ with probability at least $1 - \alpha$. Once sample **x** is observed and $M(\mathbf{x})$ is computed, it either contains θ or not. For small values of α, with 0.10, 0.05 and 0.01 being typical in practice, we might be quite confident that $\theta \in M(\mathbf{x})$, but it no longer makes sense to speak of the probability of it being so.

Let the dimension of θ be $k = 1$, as in the Bernoulli case. In most (but not all) situations with $k = 1$, $M(\mathbf{X})$ will be an interval: denoting the left and right endpoints as $\underline{\theta} = \underline{\theta}(\mathbf{X})$ and $\bar{\theta} = \bar{\theta}(\mathbf{X})$, respectively, $M(\mathbf{X}) = (\underline{\theta}, \bar{\theta})$ is referred to a **confidence interval**, or **c.i.**, for θ with **confidence level** $1 - \alpha$ or, more commonly, a $100(1 - \alpha)\%$ c.i. for θ. It also makes sense to refer to a c.i. as an **interval estimator** of θ, which draws attention to its purpose in comparison to that of a point estimator.

To compute (say) a 95% c.i. for p in the i.i.d. Bernoulli model, a starting point would be to consider the 2.5% and 97.5% quantiles of the simulated $\hat{p}$-Values shown in Figure 1.4. These give us an interval in which $\hat{p}$ falls with 95% probability when the true p is 0.3. This is related, but not equivalent, to what we want: an interval $M(\mathbf{X}) = (\underline{p}, \bar{p})$ such that $\Pr(p \in M) = 0.95$ for all $p \in (0, 1)$. The first problem we have is that the aforementioned quantiles cannot be computed from data because we do not know the true value of p.

To address this problem, let us consider doing the following: From our data set of n i.i.d. Bern(p) realizations, we compute $\hat{p}_{\text{data}} = s/n$, where s is the sum of the observations, and, using it as a best guess for the unknown parameter p, simulate realizations of $\hat{p}_i = S_i/n$, $i = 1, \ldots, B$, each based on n i.i.d. Bern$(\hat{p}_{\text{data}})$ realizations, where B is a large number, say

```
1  n=20; p=0.3; B=1e4; alpha=0.05; sim=1e5; bool=zeros(sim,1);
2  for i=1:sim
3    phat0=binornd(n,p,[1,1])/n; % the estimate of p from Bin(n,p) data
4    phatvec=binornd(n,phat0,[B,1])/n; % B samples of S/n, S~Bin(n,phat0)
5    ci=quantile(phatvec,[alpha/2 1-alpha/2]); low=ci(1); high=ci(2);
6    bool(i) = (p>low) & (p<high); % is the true p in the interval?
7  end
8  actualcoverage=mean(bool)
```

Program Listing 1.3: Determines the actual coverage of the nominal 95% parametric single bootstrap c.i. for the Bernoulli model parameter p.

10,000. The 2.5th and 97.5th sample percentiles of these $\hat{p}_i$ are then computed. (This is referred to as a *percentile (single parametric) bootstrap c.i. method*, as will be explained soon below.)

While this is indeed some kind of c.i. for p, we have our second problem: It is likely, if not nearly certain, that this interval does not have the correct (in this case, 95%) coverage probability, because $\hat{p} \neq p$. To determine the **actual coverage probability** corresponding to the **nominal confidence level** of (in this case) $\alpha = 0.05$, we can "simulate the simulation," that is, repeat the aforementioned simulation of the B values of $\hat{p}_i$ for a given value of $\hat{p}_{\text{data}}$, for many draws of $\hat{p}_{\text{data}}$, all based on the same underlying value of p. The code in Listing 1.3 illustrates how to do this for $n = 20$, $p = 0.3$, and using $\texttt{sim} = 100,000$ replications.

The output is 0.844. Again, this is the **actual coverage probability** corresponding to the **nominal coverage probability** (confidence level) of 0.95. We can envision a function $s : (0, 1) \rightarrow (0, 1)$ mapping the nominal to actual coverage, one point of which is $0.844 = s(0.95)$. By repeating this exercise over a grid of nominal coverage probabilities, we obtain an approximation of function s (via, say, linear interpolation), and compute $\alpha_{\text{act}} = 1 - s(1 - \alpha_{\text{nom}})$. The code required for computing and plotting several values of s is shown in Listing 1.4. Based on this, we can approximate the nominal coverage probability that yields an actual one of 0.95, that is, we want $s^{-1}(0.95)$.

This mapping s, computed for $p = 0.3$ and $n = 20$, but also for $n = 40$, 80, and 1000, is shown in Figure 1.5(a). As n increases, the actual level approaches the nominal level.

```
1  n=20; p=0.3; B=1e4; nominal=0.90:0.002:0.998; sim=1e5;
2  nomlen=length(nominal); bool=zeros(sim,nomlen);
3  for i=1:sim
4    phat=binornd(n,p,[1,1])/n; art=binornd(n,phat,[B,1])/n;
5    for j=1:nomlen
6      alpha=1-nominal(j); ci=quantile(art,[alpha/2 1-alpha/2]);
7      bool(i,j)=(p>ci(1)) & (p<ci(2));
8    end
9  end
10 actual20=mean(bool); plot(nominal,actual20,'r-','linewidth',2)
11 set(gca,'fontsize',16), grid, xlabel('Nominal'), ylabel('Actual')
12 title(['Actual Coverage Probability for p=',num2str(p)])
```

Program Listing 1.4: Computes and plots the mapping s of the actual coverage probability in the Bernoulli model, as a function of the nominal coverage probability, based on single parametric bootstrap c.i.s.

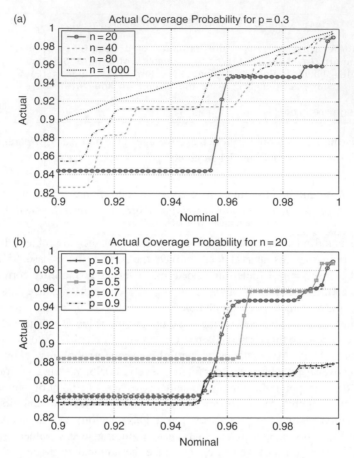

Figure 1.5 *Mapping s between nominal and actual coverage probabilities for c.i.s of the success parameter in the i.i.d. Bernoulli model, based on the (single) parametric bootstrap, each computed via simulation with 100,000 replications.*

From the graph for $p = 0.3$ and $n = 20$, we see that a nominal coverage level of $\alpha_{\text{nom}} \approx 0.015$ corresponds to an actual coverage level of $\alpha_{\text{act}} = 0.05$.

Note that, because of the discreteness of $\hat{p}$, the graphs for the smaller sample sizes are constant over certain ranges of the nominal values, and so s is not a bijection. For a specific value of α_{act}, say 0.05, we would choose α_{nom} to be the leftmost point along the graph such that the actual coverage probability is at least 0.95, to ensure that the resulting c.i. is the shortest possible one that still maintains the desired coverage. While this is graphically simple to see, such results are often expressed algebraically; for a given value of α_{act}, some thought shows that we should take

$$\alpha_{\text{nom}} = 1 - \inf(p \in (0, 1) \mid s(p) \geq 1 - \alpha_{\text{act}}) = \sup(p \in (0, 1) \mid s(1 - p) \geq 1 - \alpha_{\text{act}}).$$

From Figure 1.5(a), we see that the correct nominal value depends on the sample size n. Ideally, for a fixed sample size, the correct value of α_{nom} would not depend on the true p, so that the set of values $\alpha_{\text{nom}}(n)$ could be computed for various sample sizes "once and for all" for a given α_{act}. Figure 1.5(b) shows that this is unfortunately not the case; for a fixed value of n (here 20), we see that they depend strongly on the true p (though it appears that the behavior for p and $1 - p$ is the same). Thus, for a given data set of length n and observed $\hat{p}$, the simulation exercise above would need to be conducted to get the correct nominal coverage level, and then the c.i. for p could be delivered.

The method just described is an example of a very general and powerful technique for constructing c.i.s, called the **parametric bootstrap**. It serves as an introduction to the more popular, but somewhat less obvious, *nonparametric* bootstrap, which is discussed in the next section.

Before proceeding, we make two important remarks.

Remarks

(a) For this simple model, there exists an analytic method for constructing a c.i. for p (see Section 1.3.4 for demonstration and comparison, and Section 8.4.2 for development of the theory), thus obviating the need for simulation. The analytic method also delivers c.i.s that are shorter, on average, than the simulation method just described. For example, with $n = 20$, $p = 0.3$, and using $\alpha_{act} = 0.05$ (i.e., for 95% c.i.s with correct coverage probability), the simulation method yields an average c.i. length of 0.46, while the method developed in Section 8.4.2 yields an average c.i. length of 0.38. Thus, the above simulation method should not actually be used in practice for c.i.s for p in the Bernoulli model, and just serves as an introduction to the method in a simple setting. The real value of the parametric and nonparametric bootstrap methods arises in more complicated model settings for which analytical results are not available.

(b) The reader who has taken an introductory undergraduate course in statistics surely recalls the usual, simple, asymptotically valid c.i. formula for p given by

$$\hat{p} \pm z\hat{V}^{1/2}, \quad \hat{V} = \frac{\hat{p}(1 - \hat{p})}{n},$$

recalling (1.1), where z is the upper $1 - \alpha/2$ quantile of the standard normal distribution, such as 1.96 when $\alpha = 0.05$. This is referred to as a Wald interval, more details of which are given later; see (3.45). What might come as a shock and surprise is that this ubiquitous result turns out to behave rather poorly and erratically, even for reasonably large sample sizes, with the actual coverage potentially changing rather substantially when, for example, n is increased or decreased by one. A detailed discussion, and several alternative (non-bootstrap-based) intervals are provided in Brown et al. (2001). As such, the subsequent development of a bootstrap-based confidence interval for p, while designed for teaching the underlying concepts of bootstrap methodology using a simple example, could also be used in practice.

There are other approaches for addressing the problems that arise in the actual coverage of confidence intervals associated with discrete models. In particular, we recommend the use of the so-called **mid-p-values**; see Agresti (1992), Hwang and Yang (2001), and Agresti and Gottard (2005), as well as Butler (2007, Sec. 6.1.4). ∎

1.3 INTERVAL ESTIMATION VIA THE BOOTSTRAP

In the previous section, we used simulated Bernoulli r.v.s with success probability taken to be the point estimate of p from the observed data for constructing a confidence interval. A related way is to simulate not from this distribution, but rather, somewhat perversely, from the actual observed data. This is called the **nonparametric bootstrap**. The idea is to *treat the n observed data points as the entire underlying population*, and draw many n-length samples, or **resample** from it. There is a fundamental advantage to using this instead of the parametric one, as we will discuss below. We denote the number of resamples by B.

1.3.1 Computation and Comparison with Parametric Bootstrap

It is important to emphasize that each of the B samples for the nonparametric bootstrap is drawn *with replacement* from the observed data set. The reason is that our observed data set is being treated as the true underlying population, and thus an i.i.d. sample from it entails drawing n values from this population such that each draw is independent of the others and each of the n values in the population has an equal chance of being drawn. (If we draw n observations without replacement, we obtain exactly the original data set, just in permuted order.)

The total number of different unordered samples that can be drawn, with replacement, from a set of n observations, is $\binom{2n-1}{n}$ (see Section I.2.1 for derivation). This number quickly becomes astronomically large as n grows, already being over 90,000 for $n = 10$. If one were actually to use a complete enumeration of all these data sets, weighting them appropriately from their multinomial probabilities $n!/(n_1!n_2!\cdots n_n!\, n^n)$, where n_i denotes the number of times the ith observation occurs in the resample (see Section I.5.3.1), then it would yield what is called the **exact bootstrap distribution**. For n larger than about 20, such an enumeration is neither feasible in practice nor necessary. (See Diaconis and Holmes, 1994, and the references therein for details on the method of enumeration.) By randomly choosing a large number B of resamples, the exact bootstrap distribution can be adequately approximated.

The key to drawing with replacement is to generate n discrete r.v.s that have equal probability $1/n$ on the integers $1, 2, \ldots, n$. This is conveniently built into Matlab as function `unidrnd` (or `randi`). These then serve as indices into the array of original data. For example, the following code takes an i.i.d. Bernoulli data set and generates a single bootstrap sample `bsamp`:

```
1   n=100; p=0.3; data=binornd(1,p,[n,1]);
2   ind=unidrnd(n,[n 1]); bsamp=data(ind);
```

To help clarify the technique further, imagine you obtain a data set of $n = 100$ i.i.d. Bern(p) observations, with, say, 32 ones and 68 zeros, so that $\hat{p} = 0.32$. The parametric bootstrap can then be used to approximate the sampling distribution of $\hat{p}$ by computing B samples of $\hat{p}$, each based on n observations from a Bern(0.32) distribution. Likewise, the nonparametric bootstrap can be used to approximate the distribution of $\hat{p}$ by computing B samples of $\hat{p}$, each based on a resample of the actual data set. It should be clear from the simple structure of the model that, *in this case, the parametric and nonparametric distributions are theoretically identical*. (Of course, for finite B, they will not be numerically equal.)

```
1   n=100; p=0.3; data=binornd(1,p,[n 1]); % true data are Bern(p)
2   phat=mean(data) % point estimator of p, for use with parametric boot:
3   B=1e5; phatpara=sum(binornd(1,phat,[n B]))/n;
4   figure, hist(phatpara,5000), xlim([0.1 0.5])
5   phatnonpara=zeros(B,1);
6   for i=1:B % compute the nonparametric bootstrap distribution
7     ind=unidrnd(n,[n 1]); bsamp=data(ind); phatnonpara(i)=mean(bsamp);
8   end
9   figure, hist(phatnonpara,5000), xlim([0.1 0.5])
```

Program Listing 1.5: Compares the parametric and nonparametric bootstraps.

As further practice in programming the bootstrap, and also serving to confirm the equality of the parametric and nonparametric bootstraps in the Bernoulli model case, the code in Listing 1.5 should be studied and run, and the resulting histograms compared.

This implies, for example, that c.i.s based on the parametric and nonparametric bootstraps will have identical nominal coverage properties, and thus there is no need to perform the simulations for generating the plots in Figure 1.5 with the nonparametric bootstrap in order to compare their performance.

The equality of the parametric and nonparametric bootstrap distributions in this example is special for the Bernoulli (and, more generally, for the multinomial) distribution. The result does not hold for distributions with infinite countable support (geometric, Poisson, etc.) or uncountable support. Using the geometric as an example, one might imagine that the parametric bootstrap should be superior to the nonparametric bootstrap, because the point estimator of the geometric success probability parameter p contains all the information in the actual data set (we will qualify this notion in Section 7.3 with the concept of **sufficiency**) and thus samples drawn from a Geo($\hat{p}$) distribution will be of more value than resampled ones associated with the nonparametric bootstrap, which have their support limited to what happened to have been observed in the actual data set.

This conjecture is indeed true, and is demonstrated below in Section 1.4; though for large sample sizes (where "large" will depend on the model), the difference will be negligible. More importantly, however, the above reasoning is only valid *if you are sure about the parametric model that generated the data*. Rarely in real applications can one make such strong assumptions, and this is the reason why the nonparametric bootstrap is more often used; this point is illustrated in Section 2.2.

Remarks

(a) Two of the necessary conditions such that the nonparametric bootstrap leads to asymptotically correct inferential procedures are stated in Section 2.1.

(b) The term *bootstrap* was coined by one of the pioneers of the method, Bradley Efron, in the late 1970s. Via its analogy to a literary reference in which the main character, after falling into a lake, pulls himself out by his own shoelaces (bootstraps), the name indicates the self-referencing nature of the method. There are several textbooks that detail the theory, importance, and wide applicability of bootstrap, resampling, and so-called subsampling methods (as well as situations in which they do not work, and some possible solutions); an excellent starting point is Efron and Tibshirani

(1993), while a more advanced but still accessible and highly regarded treatment is given in Davison and Hinkley (1997). For emphasis on sub-sampling, see Politis et al. (1999).

(c) The method we show for computing bootstrap c.i.s is just one of several, and is referred to as the **percentile bootstrap method**. It is among the most intuitive methods, though not necessarily the most accurate or fastest (in terms of number of resamples required). In addition to the aforementioned references, see Efron (2003), Davison et al. (2003) and Efron and Hastie (2016, Ch. 11) for details on the other methods of bootstrap c.i. construction.   ■

1.3.2 Application to Bernoulli Model and Modification

Listing 1.6 shows how to simulate the actual coverage properties of the method, using a grid of values of parameter p. Observe that, by changing one line in the code, we can switch between the parametric and nonparametric bootstrap; though in this case, as discussed above, they are equivalent (and the faster method should then be chosen).

For each of four sample sizes n, we use it with 10,000 replications (passed as $\mathtt{sim}$) and $B = 10,000$ (called $\mathtt{B1}$ in the program) bootstrap resamples, to determine the actual coverage of the nominal 90% c.i.s, as a function of p. The results are shown in Figure 1.6(a). The actual coverage approaches 90% as the sample size increases, and is far below it for small sample sizes and extreme (close to zero or one) values of p. This latter artifact is easily explained: For small n and extreme, say small, p, there is a substantial chance that the data

```
1   function actual90 = bernoulliCIsingleboot(n,sim)
2   pvec=0.05:0.05:0.95; plen=length(pvec); actual90=zeros(plen,1);
3   bool0=zeros(sim,1); alpha=0.10; B1=1e4; bootphat=zeros(B1,1);
4   for ploop=1:plen
5     p=pvec(ploop), n
6     for i=1:sim
7       rand('twister',i) % data sets change 'smoothly' w.r.t. p
8       data=binornd(1,p,[n,1]); phat=mean(data);
9       for b1=1:B1
10         ind=unidrnd(n,[n,1]); bootsamp=data(ind); % nonparametric bootstrap
11         % bootsamp = binornd(1,phat,[n,1]);        %    parametric bootstrap
12         bootphat(b1)=mean(bootsamp);
13       end
14       ci=quantile(bootphat,[alpha/2 1-alpha/2]);
15       bool0(i) = (p>ci(1)) & (p<ci(2));
16     end
17     actual90(ploop)=mean(bool0);
18   end
```

Program Listing 1.6: Implementation of (single) parametric and nonparametric bootstrap for computing a c.i. in the Bernoulli model. As shown here, line 10 generates a resample from the data, and is thus using the nonparametric bootstrap. Line 11 (commented out) can be invoked instead to use the parametric bootstrap, in which case the code accomplishes the same thing as done in Listing 1.4 (with $\mathtt{B1}$ here set to 10,000, and using $\mathtt{nominal}$ equal to 0.90).

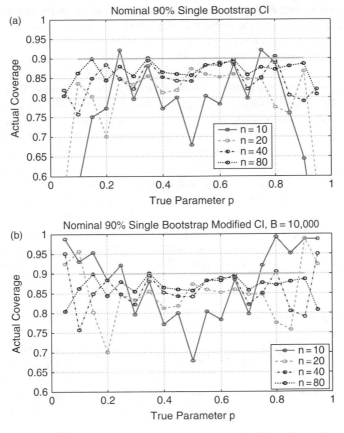

Figure 1.6 *(a) Actual coverage, based on simulation with 10,000 replications, of nominal 90% c.i.s using the (single) nonparametric bootstrap (with* `B1` *= 10,000). Graph is truncated at 0.6, with the actual coverage for n = 10 and p = 0.05 and p = 0.95 being about 0.4. (b) Same but using the modified c.i. in (1.5) and (1.6).*

set will consist of all zeros. If this happens, then clearly, both the parametric and nonparametric method will deliver a degenerate c.i. of the single point zero. This is unsatisfactory, given that, for small n and p, getting all zeros is not improbable.

A simple and appealing solution is to take a c.i. with lower bound zero and upper bound given by the smallest value of p such that the probability of getting n out of n zeros is less than or equal to the chosen value α_{act}. That is, with $S_n = \sum_{i=1}^{n} X_i \sim \text{Bin}(n, p)$, we suggest taking

$$\underline{p} = 0, \quad \bar{p} = \inf(p \in (0, 1) \mid \Pr(S_n = 0) \leq \alpha). \tag{1.5}$$

As this probability is a continuous function of p for a given n, it can be computed in Matlab as a solution of one equation in one unknown. Impressively, this can be accomplished with just one line of code (see the Matlab help file):

```
1   fzero(@(p) binopdf(0,n,p)−alpha,[1e−6 1−1e−6])
```

provided `alpha` is defined. A similar procedure yields the c.i. in the event that all the n observations are ones as

$$p = \sup(p \in (0, 1) \mid \Pr(S_n = n) \leq \alpha), \quad \bar{p} = 1. \tag{1.6}$$

When the data set does not consist of all zeros or ones, we proceed as before, with either the nonparametric or parametric bootstrap, and refer to the resulting c.i. as the **modified** c.i. for the Bernoulli parameter p.

Listing 1.7 gives the code for its implementation, and Figure 1.6(b) shows the actual coverage results with its use, again based on $\text{sim} = 10{,}000$ replications and $B = 10{,}000$ bootstrap samples. We see that the problem in the extreme cases has indeed been solved, with the resulting intervals for small n and extreme p being a bit too conservative (with higher actual than nominal coverage). For example, with $n = 10$, if all the observations in the data sample are zero, then we obtain, for $\alpha = 0.10$, $\bar{p} = 0.206$. Thus, the 90% c.i. will contain the true value of p whenever p is less than this value, for example, for $p = 0.05, 0.10, 0.15$ and 0.20, which are precisely those values in the graph that previously (Figure 1.6a) had lower actual coverage and now (Figure 1.6b) have higher actual coverage.

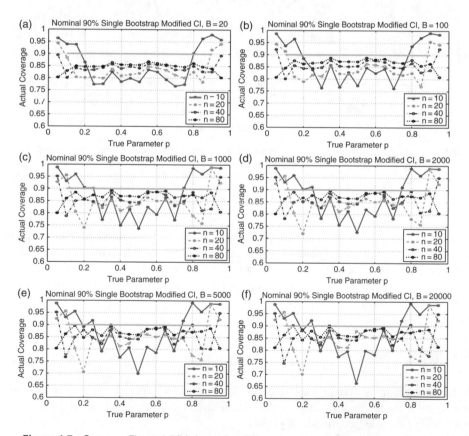

Figure 1.7 *Same as Figure 1.6(b), but using different numbers of bootstrap replications.*

Up to this point, we have used $B = 10,000$ resamples, this being a rather arbitrary choice that we hope is high enough such that the results are close to what would be obtained from the (unattainable) exact bootstrap distribution. A simple and intuitive way to heuristically determine if the choice of B is adequate is to use different choices of B, increasing it until the results no longer substantially change, where "substantial" is a relative term reflecting the desired precision. It is imperative here that each time the bootstrap is used (with the same or different B), the seed value is changed, so that a different sequence of draws with replacement is conducted. If this is not done, then, when changing B from, say, 900 to 1000, 90% of the draws will be the same in each set, so that the resulting object of interest (c.i. coverage probability, a standard error, a histogram of the approximate sampling distribution, etc.) will indeed look similar across both bootstrap runs, but not imply that $B = 1000$ is adequate.

As an illustration, Figure 1.7(a–f) is the same as in Figure 1.6(b), but based on different numbers of bootstrap replications (but still with $\mathtt{sim} = 10,000$). Figure 1.7(a) shows the case with only $B = 20$, and we see, perhaps surprisingly, that the actual coverage compared to the $B = 10,000$ case does not suffer much (and in a couple of cases, it is actually better). Figure 1.7(f) uses $B = 20,000$. The only discernable difference with the $B = 10,000$ case appears to be for $n = 10$, $p = 0.5$, and it is small; otherwise, they are identical, indicating that $B = 10,000$ is enough. In fact, if we can ignore the $n = 10$, $p = 0.5$ case, $B = 2000$ appears to be adequate.

```
1   function actual90 = bernoulliCIsingleboot(n,sim)
2   pvec=0.05:0.05:0.95; plen=length(pvec); actual90=zeros(plen,1);
3   bool0=zeros(sim,1); alpha=0.10; B1=1e4; bootphat=zeros(B1,1);
4   for ploop=1:plen
5     p=pvec(ploop), n
6     for i=1:sim
7       rand('twister',i) % data sets change 'smoothly' w.r.t. p
8       data=binornd(1,p,[n,1]); phat=mean(data);
9       if phat==1
10        ci(1)=fzero(@(p) binopdf(n,n,p)-alpha,[1e-6 1-1e-6]); ci(2)=1;
11      elseif phat==0
12        ci(1)=0; ci(2)=fzero(@(p) binopdf(0,n,p)-alpha,[1e-6 1-1e-6]);
13      else
14        for b1=1:B1
15          ind=unidrnd(n,[n,1]); bootsamp=data(ind); % nonpara boot
16          % bootsamp = binornd(1,phat,[n,1]);        %    para boot
17          bootphat(b1)=mean(bootsamp);
18        end
19        ci=quantile(bootphat,[alpha/2 1-alpha/2]);
20      end
21      bool0(i) = (p>ci(1)) & (p<ci(2));
22    end
23    actual90(ploop)=mean(bool0);
24  end
```

Program Listing 1.7: Same as the program in Listing 1.6, but treats the cases for which the sample data are all zeros, or all ones, in the modified fashion from (1.5) and (1.6).

1.3.3 Double Bootstrap

> The widespread availability of fast cheap computers has made [the bootstrap] a practical alternative to analytical calculation in many problems, because computer time is increasingly plentiful relative to the number of hours in a researcher's day.
>
> (Davison and Hinkley, 1997, p. 59)

Recall from Figure 1.5 that the actual coverage of the bootstrap (both parametric and non-parametric; they are identical in this case) can deviate substantially from the nominal coverage probability, becoming more acute as the sample size n decreases, and depending on the true parameter p. If we knew the true p, then we could just use simulation, as done to obtain those plots, to determine the mapping from nominal to actual coverage, and deliver a more accurate c.i. Of course, if we knew the true p, we would not have to bother with this exercise at all! The interesting question is how we can optimally choose the nominal coverage of the bootstrap c.i. *given only our data set*, and not the true p.

The answer is analogous to what we did above when we wished to assess the actual coverage: Just as we "simulated the simulation" there, we will *apply the bootstrap to the bootstrap* here. That is, onto each resampled bootstrap data set (referred to as an iteration of the *outer bootstrap loop*), we conduct the bootstrap procedure (called the *inner bootstrap loop*), for a range of nominal coverage probabilities, and keep track of the actual coverage. Then, for the actual data set, we use the nominal coverage that gives rise to the desired actual coverage. This is referred to as the **nested bootstrap** or **double bootstrap**. We use the convention that the outer bootstrap uses B_1 resamples, and each inner bootstrap uses B_2 resamples. Thus, $B_1 B_2$ resamples are needed in total. Pseudo-code Listing 1.1 gives the algorithm for the double bootstrap for parameter θ.

---

(1) From the data set under study, $\mathbf{y}^{obs}$, compute the estimate of parameter θ, say $\hat{\theta}^{obs}$ with a chosen method of estimation that is consistent, e.g., maximum likelihood, and where obs stands for "observed."

(2) FOR $b_1 = 1, \ldots, B_1$ DO

   a. Generate a resample of $\mathbf{y}^{obs}$, say $\mathbf{y}^{(b_1)}$ (or, for the parametric bootstrap, simulate data set $\mathbf{y}^{(b_1)}$ according to the presumed model and with parameter $\hat{\theta}^{obs}$).

   b. Compute $\hat{\theta}^{(b_1)}$ for data set $\mathbf{y}^{(b_1)}$ (using the chosen method of estimation).

   c. FOR $b_2 = 1, \ldots, B_2$ DO

     (i) Generate a resample of $\mathbf{y}^{(b_1)}$, say $\mathbf{y}^{(b_1,b_2)}$ (for the parametric bootstrap, simulate data set $\mathbf{y}^{(b_1,b_2)}$ according to the presumed model and with parameter $\hat{\theta}^{(b_1)}$).

     (ii) Compute $\hat{\theta}^{(b_1,b_2)}$ for data set $\mathbf{y}^{(b_1,b_2)}$ (using the same chosen method of estimation).

   d. FOR j over each nominal coverage probability in a grid of values, say $0.799, 0.801, \ldots, 0.999$, DO

(i) Compute the c.i. $\text{ci}^{(b_1, j)}$ as the corresponding lower and upper quantiles from the $(\hat{\theta}^{(b_1,1)}, \dots, \hat{\theta}^{(b_1,B_2)})$ values.

(ii) Record a one in the b_1th row and jth column of matrix `bool` if $\text{ci}^{(b_1 j)}$ contains $\hat{\theta}^{\text{obs}}$.

(3) Compute the average of each column of matrix `bool` to give a vector of actual coverage probabilities corresponding to the vector of nominal coverage probabilities $0.799, 0.801, \dots, 0.999$.

(4) Use the previous two vectors and linear interpolation to get the nominal level of coverage, say $1 - \alpha^*$, corresponding to an actual coverage probability of 90%.

(5) Deliver the c.i. for the actual data set $\mathbf{y}^{\text{obs}}$ as the $\alpha^*/2$ and $1 - \alpha^*/2$ quantiles of the outer bootstrap parameter values $(\hat{\theta}^{(1)}, \dots, \hat{\theta}^{(B_1)})$

Pseudo-code Listing 1.1: Algorithm for the double bootstrap. See Listing 1.8 for the associated Matlab program.

Listing 1.8 gives a program that implements the pseudo-code, and also simulates the double bootstrap to determine the actual coverage for a nominal coverage of 90%. It contains four nested FOR loops: The first is over a grid of nine values $(0.1, 0.2, \dots, 0.9)$ of the parameter p, and the second conducts, for each given value of p, a simulation of `sim` data sets, and keeps track of whether or not its c.i. covers the true value of p. The inner two FOR loops conduct the double nonparametric bootstrap for each given data set.

The execution time for computing a double bootstrap c.i. for a given data set will clearly be far longer than that of a single bootstrap. For the Bernoulli model, we require about 50 seconds (roughly irrespective of n for $20 \leq n \leq 80$) on (one core of) a 3 GHz PC when using (only) $B_1 = B_2 = 1000$. Based on these values for B_1 and B_2, the simulation study using nine values of p and `sim` = 1000 simulations for each value of p requires over 5 days of computing, for each sample size n.

The results are shown in Figure 1.8(a). The actual coverage for $n = 80$ and $p \in [0.1, 0.9]$ (and, in general, for larger n, and p close to 0.5) is quite close to the nominal value of 0.90 and better than the single bootstrap. However, the actual coverage breaks down as the sample size decreases and p moves away from 0.5, worse in fact than occurs for the single bootstrap. The reason, and the solution, are the same as discussed above.

Implementing (1.5) and (1.6) for each of the `sim` data sets, and also for the draws in the outer bootstrap (see Problem 1.2), yields the modified double bootstrap c.i.s for p. The results are shown in Figure 1.8(b). The modification via (1.5) and (1.6) clearly has helped, though compared to the coverage of the modified c.i.s using just the single bootstrap (Figure 1.6(b)), the improvement is not spectacular. This is presumably due in part to having used only $B_1 = B_2 = 1000$ bootstrap resamples (and `sim` = 1000 replications), whereas we used $B = 10,000$ (and `sim` = 10,000) for the single bootstrap. To be sure, we would have to rerun the calculations using larger values; though with $B_1 = B_2 = 10,000$, the calculation will take about 100 times longer, or 83 minutes per c.i. Doing this for nine values of p and, say, 10,000 replications would take over 14 years using a single core processor – for each sample size n. The next section presents a way around this problem.

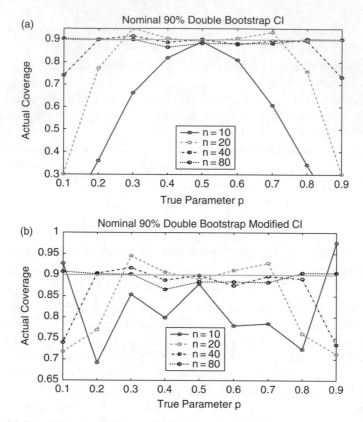

Figure 1.8 *(a) Actual coverage of nominal 90% c.i.s using the double bootstrap (truncated at 0.3), based on 1000 replications. (b) Same but using the modified c.i. in (1.5) and (1.6) applied to each simulated data set and to each bootstrap sample in the outer bootstrap loop.*

1.3.4 Double Bootstrap with Analytic Inner Loop

The previous calculation of computation time shows that simulating the performance of the double bootstrap c.i. over a grid of parameter values p, for several sample sizes, using relatively large values of B_1 and B_2, becomes burdensome, if not infeasible.

As already mentioned at the end of Section 1.2, there exists an analytic method for constructing a c.i. for p. Because of the discreteness of the data, it also does not have exactly correct actual coverage. However, given its speed of calculation, we could use it within a double bootstrap calculation, replacing the inner bootstrap loop with the analytic method. This will yield substantial time savings, because we avoid the B_2 resampling operations for each outer loop iteration, and also avoid the B_2 computations of the parameter estimator. Of course, in our case studied here, the estimator is just the sample mean of n Bernoulli realizations, and so is essentially instantaneously calculated. In general however, we might be using an m.l.e., obtained using multivariate numerical optimization methods, and thus taking vastly longer to compute.

The idea of replacing the inner bootstrap loop with an analytic result or (often a saddlepoint) approximation is very common; see, for example, the discussion in Davison

```
1    function actual90 = bernoulliCIdoubleboot(n,sim)
2    truenominal=0.90; % desired coverage probability;
3    %      change or pass as a parameter to the function
4    pvec=0.1:0.1:0.9; plen=length(pvec); actual90=zeros(plen,1);
5    B1=1e3; b1phat=zeros(B1,1); B2=1e3; b2phat=zeros(B2,1);
6    nominal=0.799:0.002:0.999; nomlen=length(nominal); bool0=zeros(sim,1);
7    bool1=zeros(B1,nomlen);
8    for ploop=1:plen    , p=pvec(ploop);
9      for i=1:sim        , i, p, n
10       rand('twister',i), data=binornd(1,p,[n,1]); % 'actual' data
11       %%%%%%%%%%%%%%%%%%%%%%%%%%%%%%%%%%%%%%%%%%%%%%%%%%%%%%%%%%%%%
12       phatdata=mean(data);
13       for b1=1:B1    %%%%% outer bootstrap loop
14         ind=unidrnd(n,[n,1]); b1samp=data(ind); b1phat(b1)=mean(b1samp);
15         for b2=1:B2 %%%%% inner bootstrap loop
16           ind=unidrnd(n,[n,1]); b2samp=b1samp(ind); b2phat(b2)=mean(b2samp);
17         end
18         for j=1:nomlen
19           alpha=1-nominal(j); ci=quantile(b2phat,[alpha/2 1-alpha/2]);
20           bool1(b1,j) = (phatdata>ci(1)) & (phatdata<ci(2));
21         end
22       end
23       bootactual=mean(bool1)+cumsum((1:nomlen)/1e10);
24       boot90=interp1(bootactual,nominal,truenominal);
25       alphanom=1-boot90; ci=quantile(b1phat,[alphanom/2,1-alphanom/2]);
26       % 'actual' CI
27       %%%%%%%%%%%%%%%%%%%%%%%%%%%%%%%%%%%%%%%%%%%%%%%%%%%%%%%%%%%%%
28       bool0(i) = (p>ci(1)) & (p<ci(2));
29     end
30     actual90(ploop)=mean(bool0); eval(['save c:\bernoulliCIdoublebootn',
31     int2str(n)])
32   end
```

Program Listing 1.8: Simulates, over a grid of values of p, the double nonparametric boot-strap for computing a 90% c.i. for parameter p in the Bernoulli model, to determine the actual coverage corresponding to a nominal coverage of 90%. Change variable `truenom-inal` to choose a different nominal coverage value. The algorithm for the double bootstrap is just the code between the two long comment lines; the rest is for the simulation. In line 21, observe that we add `cumsum((1:nomlen)/1e10)` to the empirical coverage values to force them to be increasing, so that the `interp1` command in the next line does not fail.

and Hinkley (1997) and the references therein, and Butler and Paolella (2002) for an application to computing c.i.s for certain quantities of interest associated with random effects models.

In what follows, we wish to treat the analytic method as a "black box," waiting until Section 8.4.2 to discuss why it works. The Matlab function implementing the method is called `binomCI.m`, and its contents are given in Listing 8.3. It inputs the sample size n, the number of successes s, and the value of α corresponding to the desired confidence level; it outputs the lower and upper limits of the c.i. The computation entails root searching over a function that involves the incomplete beta function (A.13), and so the method is not instantaneously calculated. In fact, it turns out to be considerably slower than using the inner bootstrap loop with $B_2 = 1000$. We will show a way of circumventing this issue below; but

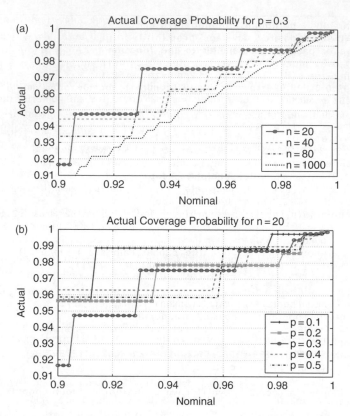

Figure 1.9 *Similar to Figure 1.5 (mapping between nominal and actual coverage probabilities for c.i.s of the success parameter in the i.i.d. Bernoulli model) except that, instead of using the single bootstrap for the c.i.s, this uses the analytic method. In figure b) actual coverage for a given p is identical to that for $1 - p$.*

first, we begin by computing the actual coverage of the analytic method itself, paralleling what we did in Figure 1.5 for the single bootstrap c.i.

We could use computer code as in Listing 1.4 and simulate `binomCI` as we did with the bootstrap, but this a waste of time and also less accurate than computing the *exact* nominal coverage values, which we can easily do in this case because, unlike the bootstrap, the analytic method is not stochastic. In other words, for a given n, s and α, the c.i. is always the same. Thus, for a given n and α, we can simply compute the actual coverage by using the law of total probability (A.30): with C the event that the c.i. covers the true value of p, $\Pr(C) = \sum_{s=0}^{n} \Pr(C \mid S = s) \Pr(S = s)$, where $S \sim \text{Bin}(n, p)$.

The code for this is shown in Listing 1.9, with output (now computed in a matter of seconds) shown in Figure 1.9. We see that, analogous to the bootstrap c.i., as the sample size increases, the nominal and actual coverage values converge, and are otherwise step functions. For a given sample size n and nominal coverage level $1 - \alpha$, the actual coverage for a specific p is identical to that for $1 - p$, as the reader can easily verify.

We now use the analytic method in place of the inner bootstrap loop of the double bootstrap. The procedure is the same as that outlined in Pseudo-code Listing 1.1, except for two changes. First, we delete step 2(c) and replace step 2(d) with the following:

```
1  function [nominal actual] = binomCIcheck (p,n)
2  nominal=0.9:0.002:0.998; nomlen=length(nominal); actual=zeros(nomlen,1);
3  for S=0:n, S
4    for j=1:nomlen
5      alpha=1-nominal(j); [lb,ub] = binomCI(n,S,alpha);
6      if (p>lb) && (p<ub), actual(j)=actual(j)+binopdf(S,n,p); end
7    end
8  end
```

Program Listing 1.9: Computes the mapping between the nominal and actual coverage probabilities for analytic c.i.s of the success parameter in the i.i.d. Bernoulli model. The function `binomCI` is given in Listing 8.3.

FOR j over each nominal coverage probability in a grid of values DO

(i) Compute the c.i. $\text{ci}^{(b_1,j)}$ from the analytic method using data $\mathbf{y}^{(b_1)}$.

(ii) Record a one in the b_1th row and jth column of matrix `bool` if $\text{ci}^{(b_1,j)}$ contains $\hat{\theta}^{\text{obs}}$.

Second, we replace step 5 with the following:

5. Deliver the c.i. for the actual data set $\mathbf{y}^{\text{obs}}$ computed using the analytic method (i.e., passing to `binomCI` the number of successes in the actual data set), and based on a confidence level of $1 - \alpha^*$.

Doing this, we discover that, instead of taking about 50 seconds to produce a c.i., as was the case with the regular double bootstrap (with $B_1 = B_2 = 1000$), we need about three times as long. This is because, as mentioned above, the analytic method is not of closed form. Thus, what was supposed to bestow upon us a massive time saving turns out to take longer!

This issue is easily resolved, however. Recall the crucial point that, for a given n, s and α, the c.i. is always the same. So, an idea that presents itself is to form a table in computer memory with $n - 1$ rows corresponding to the values that s can assume (recall that, with the modified c.i.s, we do not have to worry about the $s = 0$ and $s = n$ cases) and number of columns corresponding to the number of nominal α-values we entertain. Then, if a new s, α combination arises, we compute it with the `binomCI` function and add it to the table, and otherwise just read it from the table. This is a typical example of making a tradeoff between computer computation and memory, and we wish to emphasize this point:

> In many realistic statistical applications, a substantial amount of computation is required to obtain reliable inference. While computers are becoming ever faster, the amount of available memory is also increasing, and one can use a judicious tradeoff between the two in order to save time.

We will see another example of this idea in Section 9.3.3. In the case here, note that the required amount of memory, relative to what a modern desktop PC has, is trivial.

The implementation is done by modifying the code in Listing 1.8 as follows. First, the loop over b2 is of course removed, and the loop after that, `for j=1:nomlen`, is replaced with the following code:

```
1  for j=1:nomlen
2    alpha=1-nominal(j); [lb,ub]=localbinomCI(n,b1sum,alpha,j,nomlen);
3    bool1(b1,j) = (phatdata>lb) & (phatdata<ub);
4  end
```

The function `localbinomCI` is given in Listing 1.10, and should be appended to the bottom of the `bernoulliCIdoubleboot` program. Our only problem is how to instruct Matlab to keep the tables `tablo` and `tabhi` in memory, given that they are part of a function and thus temporal. The answer is to use global variables – these being typical in many programming languages. They are not reset at every function call, but rather stay in memory, and are available to any function, no matter what its scope. While that would work, newer versions of Matlab offer an even better solution, via use of so called *persistent* variables: these are not global, as they can only be "seen" by the function in which they are defined, but still do not get reset or lost when the evoked function is finished. That is precisely what we want.

We can do a similar tabulation for the calls to the routines that compute the c.i. using the modification via (1.5) and (1.6) in the outer bootstrap loop. This will be especially relevant when n is small, in which case the probability is relatively high that a resample can have all zeros or all ones. The reader is encouraged to implement this; the required code is, of course, very similar to that given in Listing 1.10.

Using this method, a c.i. with inner loop computed analytically, and using $B_1 = 1000$ outer loops, takes only 4 seconds. With this massive speed increase, we have the luxury of using $\text{sim} = 10,000$ replications over our grid of parameter values of p, and yielding the plot in Figure 1.10(a). We see that the actual coverage is now very close to the nominal value for all cases except for some values of p when $n = 10$, though even in this case the performance is admirable compared to the analogous results shown in Figure 1.8.

To assess the influence of the value of B_1 in this case, the method was run again, but having used 10 times the number of outer bootstrap iterations (i.e., $B_1 = 10,000$), though only $\text{sim} = 2500$ replications. Over 2 weeks of computing later, we have the results in Figure 1.10(b). The differences are extremely small, indicating that $B_1 = 1000$ is (possibly more than) enough.

```
1   function [lb,ub]=localbinomCI(n,s,alpha,j,maxj)
2
3   persistent tablo tabhi
4   if isempty(tablo), tablo=zeros(n-1,maxj)-1; end
5   if isempty(tabhi), tabhi=zeros(n-1,maxj)-1; end
6
7   if tablo(s,j)<0
8     [lb,ub]=binomCI(n,s,alpha); tablo(s,j)=lb; tabhi(s,j)=ub;
9   else
10    lb=tablo(s,j); ub=tabhi(s,j);
11  end
```

Program Listing 1.10: Used to store the c.i.s in memory once they are computed.

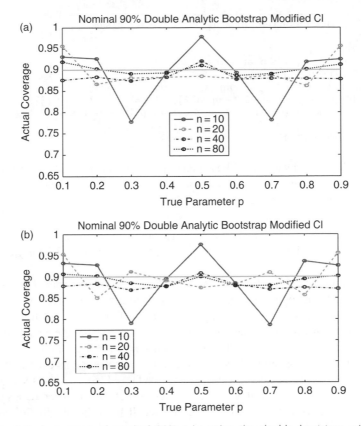

Figure 1.10 *Actual coverage of nominal 90% c.i.s using the double bootstrap with inner loop replaced by the analytic c.i., and having used the modification (1.5) and (1.6) in the outer bootstrap loop. (a) uses $B_1 = 1000$; (b) uses $B_1 = 10,000$.*

Remark. Observe that the analytic method for computing the c.i. of the Bernoulli parameter p is specifically designed for this sampling scheme, thus rendering it an *analytic parametric* c.i. As such, the parametric bootstrap can be viewed as an approximation to it; though for this particular model, the parametric and nonparametric bootstrap methods happen to be equivalent, as discussed above.

Problem 1.3 provides an example in which we have a method for computing an *analytic nonparametric* c.i. (for an order statistic of i.i.d. continuous data) and demonstrates that its performance is virtually identical to that of the nonparametric, but not the parametric, bootstrap c.i.   ∎

1.4 BOOTSTRAP CONFIDENCE INTERVALS IN THE GEOMETRIC MODEL

In this setting, we observe $G_1, \ldots, G_r$, where $G_i \overset{\text{i.i.d.}}{\sim} \text{Geo}(p)$, each with support $\{1, 2, \ldots\}$, and desire a c.i. for p.

Having already laid the groundwork for constructing c.i.s in the Bernoulli case, we just need to decide on a point estimator for p in the geometric model and, by changing a couple

```
1    r=10; p=0.3; B=1e4; alpha=0.05; sim=1e5; bool=zeros(sim,1);
2    for i=1:sim
3      if mod(i,1e3)==0, i, end
4      data=geornd(p,[r 1])+1; phat0=1/mean(data);
5      phatvec = 1./mean(geornd(phat0,[r B])+1);
6      % could instead use: r ./ ( nbinrnd(r,phat0,[B,1])+r )
7      ci=quantile(phatvec,[alpha/2 1-alpha/2]); low=ci(1); high=ci(2);
8      bool(i) = (p>low) & (p<high); % is the true p in the interval?
9    end
10   actualcoverage=mean(bool)
```

Program Listing 1.11: Determines the actual coverage of the nominal 95% parametric single bootstrap c.i. for the geometric model parameter p, using the m.l.e. as the point estimator of p. Compare with the code in Listing 1.3.

lines of code in Listing 1.3 above, we will have a program to compute a parametric bootstrap c.i. for p. We use the m.l.e. $\hat{p}_2 = 1/\bar{G} = r/\sum_{i=1}^{r} G_i$ from Section 1.1.2. This yields the code in Listing 1.11.

This can be augmented similarly to Listing 1.4 to use a grid of nominal coverage values and thus produce the mapping between the nominal and actual coverage levels. Figure 1.11 shows the results for four different sample sizes r and four choices of true parameter p. We see that, for each value of p, the nominal coverage approaches the actual coverage as r increases, similarly to the finding with the Bernoulli model as the sample size n increases. Observe also that, for a given r, the nominal and actual coverage levels are far closer for smaller values of p. This makes sense because of two facts: First, as indicated in Figure 1.3, m.s.c.$(\hat{p}) \to 0$ as $p \to 0$, so that the point estimates of p of the simulated data sets are very accurate. (This is also true for $p \to 1$.) Second, recall from the distribution of $\hat{p}$ shown in Figure 1.2 that, for a given r, as p increases, the number of points in the support of $\hat{p}$ decreases. This has the effect of causing the quantile function of $\hat{p}$ to become a step function with very few increments, rendering the c.i.s rather coarse.

As discussed in Section 1.3.1 above, we expect the parametric bootstrap to perform better than the nonparametric for the geometric model. The implementation of the nonparametric bootstrap in this context just requires modifying the code in Listing 1.11 to use resamples from `data`, as shown in Section 1.3.1. The results are shown in Figure 1.12, which parallels Figure 1.11. We indeed see that, for a given r and p, the actual coverage is considerably worse than that obtained with the parametric bootstrap.

We end this section with an example of constructing a c.i. not for the success probability p in Bernoulli trials, but rather the number of trials, n, conditional on knowing p.

Example 1.2 *According to the Wikipedia entry, Ibn Saud (1876–1953), the first monarch of Saudi Arabia, is estimated to have had 37 sons by 16 wives, though the actual number is not known, with other estimates stating the number of sons to be as high as 45, from 14 wives.[6] Let us say the number of sons is 40. If we assume that the gender of each child is an i.i.d. Bernoulli realization with probability $p = 0.5$ of getting a son, then an obvious and*

---

[6] The latter estimate was stated in the *Süddeutsche Zeitung*, March 23, 2002, page 3, in an interview with Prince Mamdouh, one of the younger of the 20 sons still alive at the time. Along with the exact number of sons being unknown, the number of daughters is even more of a mystery: With respect to this, the article states "von den Töchtern spricht hier keiner" ("no one here speaks of the daughters").

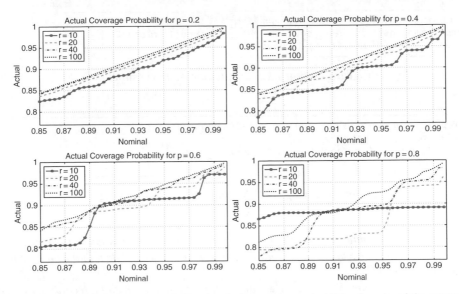

Figure 1.11 *Mapping between nominal and actual coverage probabilities for c.i.s of the success parameter p in the i.i.d. geometric model, using the parametric bootstrap. Based on 100,000 replications and $B = 10,000$.*

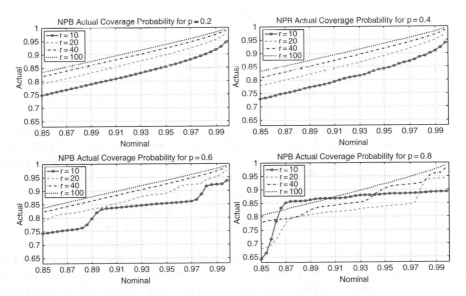

Figure 1.12 *Same as Figure 1.11, also with $B = 10,000$ and* sim $= 100,000$, *but with the nonparametric bootstrap (NPB).*

```
1   truep=0.5; boys=40; alpha=0.01; sim=1e5;
2   samp = sum( geornd(truep,[boys sim])+1 ) + round(1/truep)/2;
3   nstar = round(quantile(samp,1-alpha))
```

Program Listing 1.12: Calculation of n^* via simulation.

```
1   truep=0.5; boys=40; alpha=0.01; n=round(boys/truep); prob=0;
2   while prob < (1-alpha)
3     n=n+1; prob = 1-binocdf(boys-1,n,truep);
4   end, nstar = n
```

Program Listing 1.13: A second way of calculating n^*.

intuitive point estimate for the total number of children he fathered is 80. Less obvious is how to derive a value, say n^*, such that we can be, say, 99% confident that he had at most that number of children.

Here is one solution: We repeatedly simulate his reproductive history, recording for each replication the total number of engendered children, and report the 99% quantile of these values. The number of children begotten until a son arrives is a realization of a Geo(0.5) r.v., so one of his simulated histories consists of summing 40 Geo(0.5) r.v.s, and then adding on half of the number of children he would have produced, on average, between his 40th and 41st sons. Recall that the expected value of a Geo(p) r.v. is $1/p$, so we add on $1/(2p)$. This is easily implemented in Matlab, using the code in Listing 1.12.

For these values, we get $n^* = 104$. Repeating this numerous times yielded the same value, indicating that 100,000 replications is (more than) enough. If we are satisfied with only 90% confidence, we get $n^* = 93$. We can check our obtained value of $n^* = 104$ by verifying that $\Pr(S \geq 40) \geq 0.99$, where $S \sim \text{Bin}(104, 0.5)$. In Matlab, this is computed as `1-binocdf(boys-1, nstar, truep)`. In fact, this immediately gives us another way of computing n^* that does not require simulation: we find the smallest value of n such that $\Pr(S_n \geq 40) \geq 0.99$, where $S_n \sim \text{Bin}(n, 0.5)$. Code for this is given in Listing 1.13.

This results in $n^* = 103$ and $n^* = 92$ for $\alpha = 0.01$ and $\alpha = 0.10$, respectively, showing that the first algorithm is slightly too conservative.

Yet another method to compute n^* presents itself after a bit of contemplation: Given that we have the ability to compute a c.i. for p in the Bernoulli model for a fixed n, we can take n^* to be the largest value of n such that a $100(1 - 2\alpha)\%$ two-sided c.i. for p, given n, still contains 0.5. (We use 2α because we are only interested in a one-sided interval for n, but our program for c.i.s for p is two-sided – it inputs α and "shares" it in both tails by dividing by 2.) In particular, if we take $n = 80$ and have 40 successes, then the c.i. for p will obviously contain 0.5, but by increasing n (and always using 40 successes), eventually it would get so large that $p = 0.5$ becomes unlikely and will not be in the c.i. Starting from $n = 80$, we continue to increase n and take the last value such that its c.i. for p still contains 0.5. To increase speed, we use the analytic method for constructing a c.i. for p, implemented in function `binomCI` and discussed in Section 1.3.4. The code for computing n^* is given in Listing 1.14.

Running this results in $n^* = 104$ and $n^* = 93$ for $\alpha = 0.01$ and $\alpha = 0.10$, respectively, which are exactly the same values given by our first method.

```
1  truep=0.5; boys=40; alpha=0.01; n=round(boys/truep); upperbound=1;
2  while upperbound > truep
3     n=n+1; [lowerbound, upperbound] = binomCI(n,boys,2*alpha);
4  end
5  nstar = n-1 % we find the first value of n such that it is NOT true
6             % that 0.5 is in the c.i., so we need to subtract 1
```

Program Listing 1.14: Calculation of n^* using the analytic method for confidence intervals.

Likelihood-based inference on n for both known and unknown p is discussed in Aitkin and Stasinopoulos (1989). See Problem 1.1 for another "application." ∎

1.5 PROBLEMS

Opportunity is missed by most people because it is dressed in overalls and looks like work.

(Thomas Edison)

1.1 The yearly astrology meeting you are organizing takes place soon, but, as the stellar bodies would have it, you can only find the participation list for Sagittarius, which contains 39 members. You wish to be 99% sure of having enough seats (and other relevant paranormal paraphernalia) for you and all listed members. Being skilled in astrology, you find that, based on celestial divination and the Book of Revelation, you will require 666 seats. Confirm this without divination.

1.2 Write a program to compute the modified c.i.s with the double bootstrap and reproduce Figure 1.8(b).

1.3 Recall that the quantile ξ_p of the continuous r.v. X is the value such that $F_X(\xi_p) = p$ for given probability p, $0 < p < 1$. Let $Y_1 < Y_2 < \cdots < Y_n$ be the order statistics of an i.i.d. random sample of length n from a continuous distribution. Example II.6.7 showed that[7]

$$\Pr(Y_i \le \xi_p \le Y_j) = \sum_{k=i}^{j-1} \binom{n}{k} p^k (1-p)^{n-k} = F_B(j-1,n,p) - F_B(i-1,n,p), \quad (1.7)$$

where $B \sim \text{Bin}(n,p)$ and F_B is the c.d.f. of B. This can be used to obtain an analytic, nonparametric (observe that the distribution of X plays no role) c.i. for the quantile. If we attempt to get, say, a 95% c.i., then we first need to compute the inverse c.d.f. values $i = F_B^{-1}(0.025,n,p) + 1$ and $j = F_B^{-1}(1 - 0.025,n,p) + 1$, and then (because of the discreteness of the distribution), compute the true *nominal* coverage, say $1 - \alpha^*$, from (1.7). Write code to compute i, j, and $1 - \alpha^*$.

Next, and more substantially, let $X_i \overset{\text{i.i.d.}}{\sim} \text{Exp}(\lambda)$, $i = 1, \ldots, n$, each with density function $f_{X_i}(x; \lambda) = \lambda \exp(-\lambda x) \mathbb{I}_{(0,\infty)}(x)$. Take $p = 1/2$, so that we wish to construct a c.i. for the median. Recall from Example I.4.6 (or quickly check that), the median of an $\text{Exp}(\lambda)$ r.v. is $\log(2)/\lambda$. Write a program that determines, via simulation with sim replications, the *actual* coverage and average length of the c.i.s based on the order

---

[7] The equation stated in Example II.6.7 has a typo; it is correct here.

statistics. Likewise, the program should also compute the (single) nonparametric and parametric bootstrap c.i.s, using confidence level $1 - \alpha^*$, and determine their actual coverage and average length. The nonparametric bootstrap c.i. should, as usual, use resampling from the actual data, and also the nonparametric estimator of the median (that being just the sample median); whereas the parametric bootstrap should, as usual, draw samples from the exponential distribution using the m.l.e. for $\hat{\lambda}$ (this being $1/\bar{X}$; see Example 3.3), and use the m.l.e. as the parametric estimate of the median, $\log(2)/\hat{\lambda}$.

The results, computed over a grid of n-values, are shown in Figure 1.13. In Figure 1.13(a), the big dark circles show the true nominal coverage $1 - \alpha^*$. As expected, they approach the desired actual coverage level of 95% as n grows. For $n \geq 20$, the order statistics and nonparametric bootstrap c.i.s have virtually the same actual coverage and lengths. This was to be expected, as the former is just an analytic method for computing a nonparametric c.i. For $n \geq 100$, the actual coverages of the three c.i.s are virtually the same, yet the length of the parametric bootstrap c.i. is blatantly shorter. This is because it incorporates the knowledge that the underlying distribution is exponential.

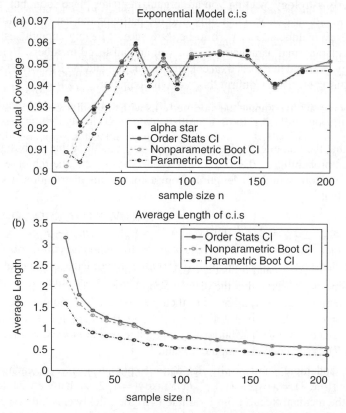

Figure 1.13   *(a) Actual coverage of the three types of c.i.s (lines), along with the true nominal coverage, $1 - \alpha^*$, from (1.7), as dark circles. (b) The average length of the c.i.s.*

2

Goodness of Fit and Hypothesis Testing

[N]o isolated experiment, however significant in itself, can suffice for the experimental demonstration of any natural phenomenon; for the "one chance in a million" will undoubtedly occur, with no less and no more than its appropriate frequency, however surprised we may be that it should occur to us. …

In relation to any experiment we may speak of this hypothesis as the "null hypothesis," and it should be noted that the null hypothesis is never proved or established, but is possibly disproved, in the course of experimentation. Every experiment may be said to exist only in order to give the facts a chance of disproving the null hypothesis.

(R. A. Fisher, The Design of Experiments, 1935)

Section 1.1 introduced the idea of estimating the unknown parameter in the Bernoulli and geometric i.i.d. cases. There are more realistic situations for which it is not clear what distributional or parametric form, if any, is appropriate for the data. There exist methods that, under some mild assumptions, are valid irrespective of the underlying population distribution. Such methods are referred to as nonparametric, which just means that they do not assume a particular parametric model.

These methods can also be used to assess the extent to which an estimated parametric model "fits" the data; this is referred to checking the **goodness of fit**.[1] We restrict ourselves to some basic concepts associated with ascertaining goodness of fit, and the central tool for doing so: the empirical c.d.f. We will detail how to construct and plot it with (pointwise) confidence intervals in both the discrete and continuous cases. Having seen the empirical

---

[1] It is not clear where this term originated, though it was brought to the forefront in the 1900 paper by Karl Pearson in which the χ^2 goodness-of-fit test was developed; see Plackett (1983) for a detailed account.

Fundamental Statistical Inference: A Computational Approach, First Edition. Marc S. Paolella.
© 2018 John Wiley & Sons Ltd. Published 2018 by John Wiley & Sons Ltd.

c.d.f., the Kolmogorov–Smirnov distance presents itself as a natural measure of goodness of fit. Its use motivates the need for, and major concepts associated with, significance and hypothesis testing.

2.1   EMPIRICAL CUMULATIVE DISTRIBUTION FUNCTION

Based on an i.i.d. sample $X_1, \dots, X_n$ from a distribution (discrete or continuous) with c.d.f. $F = F_X$, an approximation to $\Pr(X \le t)$ that suggests itself is the average of the occurrences of events $\{X_i \le t\}$. As a function of $t \in \mathbb{R}$, this expression is referred to as the **empirical c.d.f.**, or **e.c.d.f.**, and defined as

$$\widehat{F}(t) = \widehat{F}_X(t) = \widehat{\Pr}(X \le t) = n^{-1} \sum_{i=1}^{n} \mathbb{I}_{(-\infty,t]}(X_i) = n^{-1} \sum_{i=1}^{n} \mathbb{I}\{X_i \le t\}, \qquad (2.1)$$

where $\mathbb{I}$ denotes the indicator function (used in two interchangeable ways) and we suppress the subscript X on $\widehat{F}$ and F when it is clear from the context. We will also occasionally write $\widehat{F}_{\text{emp}}$ when it is useful to distinguish the e.c.d.f. from other such estimators. Observe that, if the X_i are i.i.d., then $\mathbb{I}_{(-\infty,t]}(X_i) \overset{\text{i.i.d.}}{\sim} \text{Bern}(p_t)$, where $p_t = F(t)$. It follows that $n\widehat{F}(t) = \sum_{i=1}^{n} \mathbb{I}_{(-\infty,t]}(X_i)$ is binomially distributed with parameters n and $F(t)$, so that $\mathbb{E}[\widehat{F}(t)] = F(t)$ and $\mathbb{V}(\widehat{F}(t)) = n^{-1}F(t)(1 - F(t))$. Problem 2.3 shows that $\text{Cov}(\widehat{F}(s), \widehat{F}(t)) = n^{-1}F(s)(1 - F(t))$ for $s \le t$.

The definition of the e.c.d.f. could be used for computing it for any given value of t, though it is more efficient to realize that, for a continuous distribution, if Y_i is the ith order statistic of the data, $Y_1 < Y_2 < \cdots < Y_n$, then $\widehat{F}(Y_i)$ is i/n. (That the order statistics Y_i are not equal, w.p. 1, is shown in (A.240).) To help account for the discreteness of the estimator in the continuous distribution case, Blom (1958) suggested using

$$\widehat{F}(Y_i) = \frac{i - \gamma}{n - 2\gamma + 1}, \qquad \text{for some } \gamma \in (0, 1), \qquad (2.2)$$

ideally where γ is dependent on n and i, but as a compromise, either $\gamma = 0.5$ or $\gamma = 3/8$, that is, $(i - 0.5)/n$ or $(i - 3/8)/(n + 1/4)$. The e.c.d.f. for all t is then formed as a step function. For discrete distributions, the e.c.d.f. will be formed from the tabulated frequencies of the elements from the support of the distribution that are actually observed.

The rest of this section is organized as follows. Section 2.1.1 presents the so-called Glivenko–Cantelli theorem and some related remarks, while Section 2.1.2 proves it – that can be initially skipped by readers more interested in applications. Sections 2.1.3 and 2.1.4 discuss using and plotting the e.c.d.f. in the continuous and discrete cases, respectively.

2.1.1   The Glivenko–Cantelli Theorem

As $n\widehat{F}(t)$ is the sum of n i.i.d. r.v.s in L_4 (existing fourth moments; see (A.220)), the strong law of large numbers (A.272) implies that, for each t, $\widehat{F}(t)$ converges almost surely to $F(t)$, that is, $\widehat{F}(t)$ converges almost surely to $F(t)$ *pointwise*. We can express this in terms of convergence in probability from (A.254) as

$$\lim_{n \to \infty} \Pr(|\widehat{F}_n(t) - F(t)| > \epsilon) = 0, \qquad (2.3)$$

for any $t \in \mathbb{R}$ and all $\epsilon > 0$, where we endow $\widehat{F}(t)$ with subscript n to denote the sample size it is based upon. The stronger result that $\widehat{F}(t)$ also converges almost surely to $F(t)$ *uniformly*, that is,

$$\sup_{t} |\widehat{F}_n(t) - F(t)| \overset{a.s.}{\to} 0, \quad \text{as } n \to \infty, \tag{2.4}$$

was shown in 1933 by both the Russian Valery Ivanovich Glivenko and the Italian Francesco Paolo Cantelli, and is now referred to as the *Glivenko–Cantelli theorem*, with basic proofs in the continuous case provided in Section 2.1.2.

Remarks

(a) We stated above that the X_i are i.i.d., from which we obtained the distribution of $\widehat{F}(t)$ and also the Glivenko–Cantelli theorem. While there are certainly many applications that involve, by design or by reasonable assumption, i.i.d. data, many interesting situations are such that the data will not be i.i.d. Prominent examples include data Y_i, $i = 1, \ldots, n$, whose mean is given by a function of a set of regressors $X_{1,i}, \ldots, X_{k,i}$, such as in a linear regression model (in which case the Y_i are not identically distributed and possibly not independent); and time series data Y_t, $t = 1, \ldots, T$, (in which case observations are neither identically distributed nor independent). In general, interest in these cases usually centers on conditional distributions: in the regression example, conditioning is on the regressors, while for (univariate) time series, conditioning is on the past observations $Y_1, Y_2, \ldots, Y_{T-1}$ and possibly other random variables as well.

Nevertheless, the *unconditional* distribution might still be important. As a case in point, daily financial asset returns are blatantly not independent (their volatility is highly persistent; see Example 3.9), but an important application includes predicting some type of risk measure (such as the variance) relatively far in the future. This is best computed by examining the unconditional properties of the time series process and, in essence, treating the data *as if they were* i.i.d. The use of the bootstrap in such cases is still valid under certain assumptions.[2]

(b) Section 1.3 discussed and applied the nonparametric bootstrap, and we continue to use it below in the context of the e.c.d.f., as well as elsewhere in the text. We state here only two of the necessary conditions that are required for its application, both of which should be intuitive. In order for the nonparametric bootstrap procedure to be applicable (meaning, among other things, that as the sample size tends to infinity, the resulting c.i.s have correct actual coverage), we require (among other technical conditions) that (i) the e.c.d.f. converges uniformly to the true c.d.f., which is the statement of the Glivenko–Cantelli theorem; and (ii) that the estimator being used (in this case, the order statistics, but it could also be the mean,

---

[2] See Shao and Tu (1995) and Mammen and Nandi (2004) and the references therein for details. Conditions include requiring that the time series process be strictly stationary, though convergence of the e.c.d.f. to the unconditional distribution will be slower, so that, for a particular desired accuracy, more observations would be required than under the i.i.d. case. This idea is analogous to the situation with extreme value theory and determination of the so-called extremal index; see, for example, McNeil et al. (2005, Sec. 7.1.3). For the theory and application of resampling methods to time series data, see Politis et al. (1999).

standard deviation, a single order statistic such as the median, the m.l.e. or some function of the m.l.e., or some other estimator of the distribution parameters, etc.) is consistent (see Section 1.1.1). Further information can be found in the books mentioned in Section 1.3 and the references therein.

(c) More precision can be given to the Glivenko–Cantelli theorem via the so-called DKW inequality, from Dvoretzky et al. (1956) and Massart (1990). This states that, for

$$D_n = \sqrt{n} \sup_t |\widehat{F}(t) - F(t)| \tag{2.5}$$

and for all $\lambda > 0$,

$$\Pr(D_n > \lambda) \leq 2 \exp(-2\lambda^2). \tag{2.6}$$

This could be used, for example, to find the smallest sample size n such that we can be, say, 95% confident that the absolute discrepancy between $\widehat{F}(t)$ and $F(t)$ is less than some $\epsilon > 0$, for all t. In particular, with $\lambda = \sqrt{n}\epsilon$, (2.6) states that $\Pr(\sup_t |\widehat{F}(t) - F(t)| > \epsilon) \leq 2 \exp(-2n\epsilon^2)$, so solving $0.05 = 2 \exp(-2n\epsilon^2)$ yields $n = \log(2/0.05)/(2\epsilon^2)$. For $\epsilon = 1/10$, this implies $n \geq 185$; for $\epsilon = 1/100$, we would require $n \geq 18,445$. ∎

Analogous to the contrast between the central limit theorem (A.160) and the strong law of large numbers (A.272), and the usefulness of the former for building confidence intervals, it is of value to find a function of sample size n such that, when $\sup_t |\widehat{F}(t) - F(t)|$ is multiplied by this, the resulting random variable converges (in distribution) to one that is not degenerate. In the continuous density case, this factor is precisely $\sqrt{n}$; and D_n, as given in (2.5), converges (rather quickly in fact) to what is now referred to as the *Kolmogorov distribution*, with p.d.f. and c.d.f. given by

$$f_K(x) = -8 \sum_{k=1}^{\infty} (-1)^k k^2 x \exp\{-2k^2 x^2\} \mathbb{1}_{(0,\infty)}(x) \tag{2.7}$$

and

$$F_K(x) = \Pr(K \leq x) = \left[1 + 2 \sum_{k=1}^{\infty} (-1)^k \exp\{-2k^2 x^2\} \right] \mathbb{1}_{(0,\infty)}(x), \tag{2.8}$$

respectively.[3]

To illustrate, Figure 2.1(a) shows the Kolmogorov distribution (2.7), and the actual (kernel) density of D_5 and D_{500}, as given in (2.22) below, obtained from simulation, based on standard normal data. Figure 2.1(b) is similar, but based on standard Cauchy data. It is apparent that the asymptotic distribution is correct, and also that it is irrelevant if the underlying distribution is thin-tailed, like the normal, or extremely fat-tailed, like the Cauchy – it just needs to be continuous. The reader is encouraged to replicate these results.

---

[3] The proof of (2.8) is straightforward, but would require an excursion into the basics of Brownian motion (in particular, the so-called Brownian bridge); see, for example, Wiersema (2008) for an introductory account. It is directly related to the fact that the Kolmogorov distribution is the distribution of the largest absolute value of a Brownian bridge; see, for example, Perman and Wellner (2014, p. 3112) and the references therein for an advanced presentation.

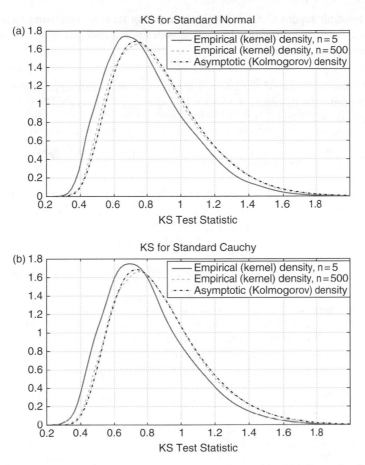

Figure 2.1 *The true distribution, obtained via simulation with 10,000 replications, of the Kolmogorov–Smirnov goodness-of-fit test statistic, and its asymptotic distribution (2.7) for the standard (location-zero, scale-one) normal (a) and Cauchy (b) distributions.*

Unfortunately, even though the asymptotic distribution appears to be a good approximation for relatively small sample sizes, matters change when we wish to test the adequacy of a distribution when its parameters are unknown. This is referred to as a *composite distribution hypothesis test*, and is discussed in detail below in Section 2.4. For the case at hand, we assumed knowledge of the location and scale terms of the normal and Cauchy distributions, which is not realistic in practice.

2.1.2 Proofs of the Glivenko–Cantelli Theorem

The Glivenko–Cantelli result is old, and serves as a starting point for studies in empirical processes. As such, many (excellent) rigorous probability theory books either provide a proof that is not much more than a brief comment, appealing to previously obtained, more general results throughout the text (e.g., Gut, 2005, p. 306) or assign it as an exercise after stating more general theory (e.g., Ash and Doléans-Dade, 2000, p. 331). As such, we provide two simple proofs, the first just "starting from scratch" and using basic principles from Section A.15; and the second which is well known and uses a procedure that is of value in more general settings.

Proof I. An instructive proof of (2.4) when F is continuous can be set up as follows. We begin by assuming the X_i are i.i.d. Unif(0, 1), $i = 1, \dots, n$. From (A.176), the p.d.f. of the ith order statistic $X_{i:n}, f_{X_{i:n}}(y)$, is

$$\frac{n!}{(i-1)!(n-i)!}F(y)^{i-1}[1-F(y)]^{n-i}f(y) = i\binom{n}{i}y^{i-1}(1-y)^{n-i}\mathbb{1}_{(0,1)}(y), \qquad (2.9)$$

and, from (A.178), $X_{i:n} \sim \text{Beta}(i, n-i+1)$, so that the mth moment is (see page I.246)

$$\mathbb{E}[X_{i:n}^m] = \frac{\Gamma(n+1)\Gamma(m+i)}{\Gamma(i)\Gamma(m+n+1)}, \qquad (2.10)$$

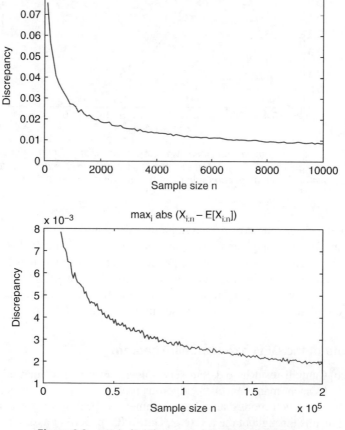

Figure 2.2 D_n^{**} *in (2.15) versus n based on simulation.*

and, in particular,[4]

$$\mathbb{E}[X_{i:n}] = i\frac{\Gamma(n+1)}{\Gamma(n+2)} = \frac{i}{n+1}, \quad \mathbb{V}(X_{i:n}) = \frac{i(n-i+1)}{(n+1)^2(n+2)}. \tag{2.11}$$

The latter attains its maximum at $i = (n+1)/2$, so that

$$\mathbb{V}(X_{i:n}) \le \frac{1}{4(n+2)}. \tag{2.12}$$

As $0 < X_i < 1$ and $\widehat{F}(X_{i:n})$ is a step function,

$$D_n = \max_{0<p<1}|\widehat{F}(p) - F(p)| = \max_{0<p<1}|\widehat{F}(p) - p| = \max_{1\le i\le n}|\widehat{F}(X_{i:n}) - F(X_{i:n})|, \tag{2.13}$$

and from the triangle inequality (A.222),

$$D_n = \max_{1\le i\le n}|\widehat{F}(X_{i:n}) - F(X_{i:n}) + \overbrace{F(\mathbb{E}[X_{i:n}]) - F(\mathbb{E}[X_{i:n}])}^{\text{zero}}| \le D_n^* + D_n^{**}, \tag{2.14}$$

where

$$D_n^* := \max_{1\le i\le n}|\widehat{F}(X_{i:n}) - F(\mathbb{E}[X_{i:n}])|$$

and, as $F(x) = x$ for $0 \le x \le 1$,

$$D_n^{**} := \max_{1\le i\le n}|F(\mathbb{E}[X_{i:n}]) - F(X_{i:n})| = \max_{1\le i\le n}|\mathbb{E}[X_{i:n}] - X_{i:n}|. \tag{2.15}$$

For D_n^*, using $\widehat{F}(X_{i:n}) = i/n$ and (2.11),

$$D_n^* := \max_{1\le i\le n}|\widehat{F}(X_{i:n}) - F(\mathbb{E}[X_{i:n}])| = \max_{1\le i\le n}\left|\frac{i}{n} - \frac{i}{n+1}\right| = \max_{1\le i\le n}\frac{i}{(n+1)n} = \frac{1}{n+1},$$

so that $D_n^* \to 0$ as $n \to \infty$.

For D_n^{**}, we first see what simulation suggests. Figure 2.2 plots D_n^{**} as a function of n, where, for each n, the mean of 200 replications of D_n^{**} is used, to help smooth the graphic. It appears to decrease monotonically with n.

Observe that, from countable subadditivity (A.239), Chebyshev's inequality (A.231), and (2.12),

$$\Pr(\max_i|X_{i:n} - \mathbb{E}[X_{i:n}]| \ge \epsilon) = \Pr\left(\bigcup_{i=1}^n |X_{i:n} - \mathbb{E}[X_{i:n}]| \ge \epsilon\right)$$

$$\le \sum_{i=1}^n \Pr(|X_{i:n} - \mathbb{E}[X_{i:n}]| \ge \epsilon) \le n\frac{\mathbb{V}(X_{i:n})}{\epsilon^2} \le \frac{1}{4\epsilon^2}, \tag{2.16}$$

---

[4] As an aside, the Beta(p,q) density is unimodal for $p, q > 1$, with mode $(p-1)/(p+q-2)$ (see, for example, Johnson et al., 1995, Ch. 21) so that Mode$(X_{i:n}) = (i-1)/(n-1)$ for $1 < i < n$. This is close to, but not the same as, $\mathbb{E}[X_{i:n}]$.

so that this form of inequality is not good enough to ensure that (2.14) converges to zero as $n \to \infty$. Instead, we require Markov's inequality (A.227) with $r = 4$, that is, for r.v. Z and $\epsilon > 0$,

$$\Pr(|Z| \geq \epsilon) \leq \frac{\mathbb{E}[|Z|^4]}{\epsilon^4}. \tag{2.17}$$

Using (2.10), relation $\mu_4 = \mu_4' - 4\mu_3'\mu + 6\mu_2'\mu^2 - 3\mu^4$ from (I.4.49), and a symbolic computing package, we can write $\mu_4 = \mathbb{E}[(X_{i:n} - \mathbb{E}[X_{i:n}])^4]$ as

$$\begin{aligned}
\mu_4 &= \left(\frac{i}{n+1} \frac{i+1}{n+2} \frac{i+2}{n+3} \frac{i+3}{n+4} \right) - 4 \left(\frac{i}{n+1} \frac{i+1}{n+2} \frac{i+2}{n+3} \right) \left(\frac{i}{n+1} \right) \\
&\quad + 6 \left(\frac{i}{n+1} \frac{i+1}{n+2} \right) \left(\frac{i}{n+1} \right)^2 - 3 \left(\frac{i}{n+1} \right)^4 \\
&= 6 \frac{i^3(i+1)}{(n+1)^3(n+2)} - 3 \frac{i^4}{(n+1)^4} - 4 \frac{i^2(i+1)(i+2)}{(n+1)^2(n+2)(n+3)} \\
&\quad + \frac{i(i+1)(i+2)(i+3)}{(n+1)(n+2)(n+3)(n+4)} \\
&< 3 \frac{i(i+2)}{n^4} \leq 3 \frac{(n+2)}{n^3} = \frac{3}{n^2} + O(n^{-3}),
\end{aligned}$$

where the penultimate inequality comes from replacing all denominator terms of the form $(n + h)$ with n and simplifying, and the last equality follows because $i \leq n$. Now, as n grows, (2.16) reads

$$\sum_{i=1}^n \Pr(|X_{i:n} - \mathbb{E}[X_{i:n}]| \geq \epsilon) \leq \frac{3}{n\epsilon^4},$$

and $D_n \overset{p}{\to} 0$.

Up to this point, we considered the specific case with X_i i.i.d. Unif$(0, 1)$, with its convenient c.d.f. expression. However, all the work is essentially done. Observe that, for any distribution with continuous c.d.f. F such that $F^{-1}(p)$ exists for all $0 < p < 1$, we have, from the **probability integral transform** from Section I.7.4.1 (sometimes referred to as the **inverse-transform method**; see, for example, Rubinstein and Kroese, 2017, p. 55) and $p = F(t) \Leftrightarrow t = F^{-1}(p)$,

$$D_n = \max_{t \in \mathbb{R}} |\widehat{F}(t) - F(t)| = \max_{0 < p < 1} |\widehat{F}(F^{-1}(p)) - F(F^{-1}(p))| = \max_{0 < p < 1} |\widehat{F}(F^{-1}(p)) - p|$$

and

$$\widehat{F}(F^{-1}(p)) = n^{-1} \sum_{i=1}^n \mathbb{I}\{X_i \leq F^{-1}(p)\} = n^{-1} \sum_{i=1}^n \mathbb{I}\{F(X_i) \leq p\}.$$

As $F(X_i)$, $i = 1, \dots, n$, is an i.i.d. sample from a Unif$(0, 1)$ distribution, D_n has the same distribution as $\max_{0<p<1} |\widehat{F}(p) - p|$, so that the previous analysis applies to any continuous distribution such that $F^{-1}(p)$ exists for all $0 < p < 1$. ∎

Proof II. This is the more common approach for the proof; see, for example, Sen and Singer (1993, p. 185), Zaman (1996, p. 221), the outline in Kallenberg (2002, p. 75), Venkatesh (2013, Ch. XVI.11), and Spokoiny and Dickhaus (2015, p. 13). As in the previous proof, to simplify things while still showing the main concept, we assume that F is continuous.

The key first step is similar to (2.13) above, namely, reducing the supremum over a continuous interval to a maximum over a finite set. Fix an $\epsilon \in \mathbb{R}_{>0}$. Next, similar to the proof of result (A.285), there exists an $m = m(\epsilon) \in \mathbb{N}$ for grid $x_0, x_1, \ldots, x_m$ such that $-\infty = x_0 < x_1 < x_2 < \cdots < x_m = \infty$ and $F(x_j) - F(x_{j-1}) \le \epsilon/2$, the latter being possible because F is continuous. For some $x \in \mathbb{R}$, there is an index $j \in \{1, \ldots, m\}$ such that $x_{j-1} \le x \le x_j$, so that, as F in general is nondecreasing (and for F continuous, is strictly increasing),

$$\widehat{F}(x) - F(x) \le \widehat{F}(x_j) - F(x_{j-1}) = [\widehat{F}(x_j) - F(x_j)] + [F(x_j) - F(x_{j-1})]$$

$$\le |\widehat{F}(x_j) - F(x_j)| + \epsilon/2 \le \max_{j \in \{1,\ldots,m\}} |\widehat{F}(x_j) - F(x_j)| + \epsilon/2.$$

As this holds for any x,

$$\sup_{x \in \mathbb{R}} |\widehat{F}(x) - F(x)| \le \max_{j \in \{0,\ldots,m\}} |\widehat{F}(x_j) - F(x_j)| + \epsilon/2. \tag{2.18}$$

Next, for fixed $j \in \{0, 1, \ldots, m\}$, (2.3) is equivalent to saying that, for any $\delta > 0$, there exists an $n_j \in \mathbb{R}$ such that, for all $n \ge n_j$, $\Pr(|\widehat{F}_n(x_j) - F(x_j)| \ge \epsilon/2) \le \delta$. We need to work with event $A_n = \max_{j \in \{0,\ldots,m\}} |\widehat{F}_n(x_j) - F(x_j)| \ge \epsilon/2$, but from (A.239),

$$\Pr(A_n) \le (1+m)\delta, \quad \forall n \ge \max_{j \subset \{0,\ldots,m\}} n_j, \tag{2.19}$$

that is, as $\delta > 0$ is arbitrary, $\lim_{n \to \infty} \Pr(A_n) \to 0$. Thus, from (2.18),

$$\Pr(\sup_{x \in \mathbb{R}} |\widehat{F}(x) - F(x)| \ge \epsilon) \le \Pr\left(\left[\max_{j \in \{0,\ldots,m\}} |\widehat{F}(x_j) - F(x_j)| + \epsilon/2 \right] \ge \epsilon \right)$$

$$= \Pr(A_n) \to 0,$$

as $n \to \infty$. $\blacksquare$

2.1.3 Example with Continuous Data and Approximate Confidence Intervals

To illustrate the computation of the e.c.d.f. for a continuous distribution, let $X_i \overset{\text{i.i.d.}}{\sim} \text{Exp}(1)$, $i = 1, \ldots, n$, with $F_{X_i}(x) = (1 - e^{-x}) \mathbb{I}_{(0,\infty)}(x)$. In Matlab, to plot the e.c.d.f. of a simulated sample of size n overlaid with the true c.d.f., we could use the code given in Listing 2.1. Observe that Matlab has a command, `stairs`, that is nicely suited for plotting a discrete c.d.f. The result of running this is shown in Figure 2.3(a). We see that the e.c.d.f. indeed hovers closely around the true c.d.f., shown as the dotted line.

```
1  n=50; rand('twister',3) % fix the seed value so we can replicate this
2  v=1:n; r=sort(exprnd(1,[n,1])); empcdf=(v-0.5)'/n;
3  truecdf=1-exp(-r); stairs(r,empcdf,'r-','linewidth',2), hold on
4  plot(r,truecdf,'b:','linewidth',3), hold off, set(gca,'fontsize',16)
```

Program Listing 2.1: Plots the e.c.d.f. and the true c.d.f. for a random i.i.d. sample of n observations from an Exp(1) distribution.

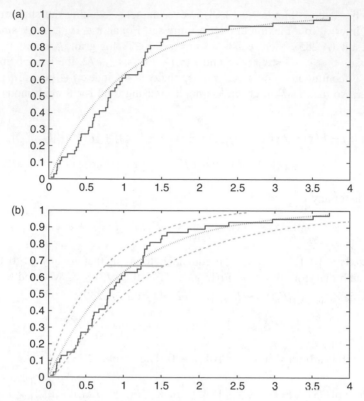

Figure 2.3   *(a) The e.c.d.f. (solid) based on 50 observations and true c.d.f. (dotted) of an Exp(1). (b) Same, but adds horizontal 95% error bounds obtained by simulation of order statistics using the true Exp(1) model. The horizonal line at the 34th order statistic just serves as a reminder that the bounds are to be understood horizontally.*

To make statistically precise what is meant by "close," bounds should be added to the graph that reflect the uncertainty of the estimator, otherwise the e.c.d.f. and the true c.d.f. shown in Figure 2.3(a) are not overly meaningful. Via simulation, we could obtain the 2.5% and 97.5% quantiles of each of the $n = 50$ order statistics from an i.i.d. Exp(1) sample. This parallels our starting point in Section 1.2 of using the quantiles from simulated realizations of $\hat{p}$ of the Bernoulli model with known p, as was shown in Figure 1.4. The code in Listing 2.2 carries this out (after having run the code in Listing 2.1) and plots the resulting bands as dashed lines, as shown in Figure 2.3(b).

```
1  sim=5e3; emat=zeros(sim,n);
2  for i=1:sim,   r=sort(exprnd(1,[1,n])); emat(i,:)=r; end
3  qlo=quantile(emat,0.025); qhi=quantile(emat,0.975);
4  hold on, plot(qlo,empcdf,'g—',qhi,empcdf,'g—','linewidth',2)
5  order=34; yy=(order-0.5)/n;
6  line([qlo(order) qhi(order)],[yy yy],'color','g','linewidth',3)
7  hold off, xlim([0,4])
```

Program Listing 2.2: Generates the bounds shown in Figure 2.3. First run Listing 2.1.

The ith of the $n = 50$ intervals pertains to the ith order statistic Y_i, for which the e.c.d.f. is $(i - 0.5)/n$, plotted on the y-axis. Thus, the overlaid intervals are to be interpreted horizontally; for example, for our data set and $i = 8$, the e.c.d.f. is $7.5/50 = 0.15$ at $Y_8 = 0.248$, with lower, or left, bound 0.074 (obtained as the 2.5% quantile of 5000 simulated Y_8-values) and upper, or right, bound 0.31 (as the 97.5% quantile), so that, for this data set, Y_8 exceeds its right bound. The e.c.d.f. also exceeds the right bound for $i = 14$ (e.c.d.f. value $13.5/50 = 0.27$). The horizonal line at the (arbitrarily chosen) 34th order statistic just serves to remind us of the horizontal nature of the bounds.

As in Section 1.2, the bounds shown in Figure 2.3(b) are not c.i.s because they depend on the true unknown parameter. They are not obtainable because the simulation used repeated sampling of the true model (in this case, Exp(1)), which obviously will be unknown in real settings. If we are willing to make the assumption that the data are i.i.d. exponential, but with unknown parameter θ, then we can proceed with the simulation but using the $\text{Exp}(\hat{\theta})$ model. This is the parametric bootstrap. However, this defeats the whole purpose of the e.c.d.f., which is to let the data "speak for themselves" and not (at least initially) impose a parametric structure. This can be achieved with the nonparametric bootstrap. That is, we replace the unknown population distribution with its estimate, the e.c.d.f. So, instead of drawing samples from the true, but unknown, Exp(1) distribution, *we treat the $n = 50$ observed data points as the entire underlying population*, and draw samples from it.

For the example at hand, the code in Listing 2.3 is nearly identical to that given in Listing 2.2 above, but instead of simulation with `sim` = 5000 replications from a Exp(1) distribution, the nonparametric bootstrap with B = 5000 resamples is applied. The output is shown in Figure 2.4(a). Observe that, for $i = 14$ (e.c.d.f. value $13.5/50 = 0.27$), the true c.d.f. falls outside the c.i., precisely as in Figure 2.3.

We now consider a second way of forming c.i.s for the e.c.d.f. that does not involve simulation. Recall that $\mathbb{V}(\hat{F}(t)) = n^{-1}F(t)(1 - F(t))$, which is unknown, but can be approximated by $\hat{\mathbb{V}}(\hat{F}(t)) = n^{-1}\hat{F}(t)(1 - \hat{F}(t))$. As $\hat{F}(t)$ is a sum of i.i.d. r.v.s and is unbiased (all we require here is that it is asymptotically unbiased), the central limit theorem implies that the distribution of $\hat{F}(t) - F(t)$, when suitably scaled, will resemble that of a standard normal as n grows. Thus, an approximate (and possibly not particularly accurate) 95% c.i. for $F(t)$ is given by $\hat{F}(t) \pm 1.96\widehat{\text{std}}(t)$, where $\widehat{\text{std}}(t) = [\hat{\mathbb{V}}(\hat{F}(t))]^{1/2}$ and 1.96 is, to three significant digits, the 97.5% quantile of the standard normal distribution.

The Matlab commands to add the asymptotic bounds to the plot are given in Listing 2.4, and the resulting graph is shown in Figure 2.4(b). The two plots are nearly identical, which is comforting, as it enhances our faith in the (possibly mysterious seeming)

```
1  B=5000;  bmat=zeros(B,n);
2  for i=1:B,   ind=unidrnd(n,[n,1]);  bsamp=r(ind);  bmat(i,:)=sort(bsamp);  end
3  qlo=quantile(bmat,0.025);  qhi=quantile(bmat,0.975);  hold on
4  stairs(qlo,empcdf,'g-','linewidth',2),  stairs(qhi,empcdf,'g-','linewidth',2)
5  order=34;  yy=(order-0.5)/n;
6  line([qlo(order) qhi(order)],[yy yy],'color','g','linewidth',3),  hold off,
7  xlim([0,4])
```

Program Listing 2.3: Generates the nonparametric bootstrap c.i.s shown in Figure 2.4(a).

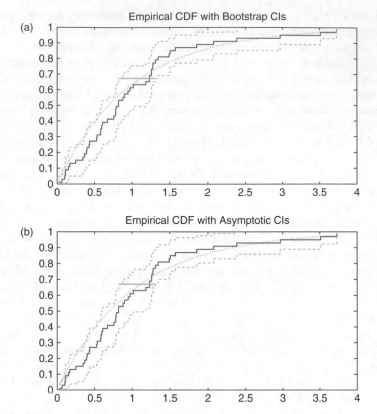

Figure 2.4 *The same e.c.d.f. (solid) and true Exp(1) c.d.f. (dotted) as in Figure 2.3 but with 95% nonparametric bootstrap c.i.s (a) and 95% asymptotic c.i.s (b).*

```
1  approxvar=empcdf.*(1-empcdf) / n;
2  low=empcdf-1.96*sqrt(approxvar); high=empcdf+1.96*sqrt(approxvar);
3  hold on, stairs(r,low,'g-','linewidth',2), stairs(r,high,'g-','linewidth',2)
4  order=34; yy=(order-0.5)/n;
5  line([qlo(order) qhi(order)],[yy yy],'color','g','linewidth',3), hold off,
6  xlim([0,4])
```

Program Listing 2.4: Generates the asymptotic c.i.s shown in Figure 2.4(b).

bootstrap procedure and the asymptotic method (which might perform poorly in small samples).

Remarks

(a) Matlab's statistics toolbox has function `ecdf` for computing and plotting the e.c.d.f. and its associated asymptotic-based error bounds.

(b) It is worth emphasizing the approximate nature of both the bootstrap and asymptotic normality-based c.i.s. For the latter: first, the formula for the variance, $\widehat{\mathbb{V}}(\widehat{F}(t))$, uses the estimated values of $\widehat{F}$, not their true ones; and second, the normality assumption is only valid asymptotically. For the bootstrap method, we know that, if certain conditions are fulfilled, then also asymptotically, it will deliver c.i.s with the correct nominal coverage; but as we have seen already in several examples, in small samples, the actual coverage probability of the resulting c.i.s can deviate substantially from the nominal coverage probability.

(c) To reiterate the point made in Section 1.2, keep in mind that it is not correct to say (irrespective of the approximate nature of the interval) that, with probability 0.95, $F(t)$ is contained within the interval: either the true but unknown value of $F(t)$ is in the interval, or it is not. What we can say is that, if we were to repeat this experiment many times, then about 95% of the constructed intervals for a given value of t would contain the true $F(t)$.

(d) There is another important point to keep in mind when working with c.i.s in this example: We have plotted a set of 50 c.i.s, each with (approximate) coverage probability 0.95, so that, if they were independent (they are not; we are working with order statistics) and the true underlying distribution is F, then we would *expect* 5%, or about two or three out of 50, not to contain the true $F(t)$. In the plot shown, precisely three values of the true c.d.f. $F(t)$ (around the value $t = 0.35$) lie outside the confidence bands. Such c.i.s are referred to as **pointwise confidence intervals**, or **one-at-a-time** or **individual c.i.s**, as opposed to **simultaneous c.i.s**, for which, if we had them, it would be the case that we could be 95% sure that *all* the intervals cover the true values. This is an important concept in statistics that appears in numerous contexts; see Chapter 6 for further discussion and a showcase application. ∎

2.1.4 Example with Discrete Data and Approximate Confidence Intervals

When the data are drawn from a discrete distribution, we need to tabulate the number of different realized values. For illustration we use the geometric distribution. Let $X_i \overset{\text{i.i.d.}}{\sim} \text{Geo}(p)$, with $f_X(x) = p(1-p)^{x-1}\mathbb{I}_{\{1,2,\dots\}}(x)$ and, as is easily verified,

$$F_X(x) = (1 - (1-p)^{\lfloor x \rfloor})\mathbb{I}_{[1,\infty)}(x), \qquad (2.20)$$

where $\lfloor x \rfloor$ is the floor function (e.g., $\lfloor 3.2 \rfloor = \lfloor 3.8 \rfloor = 3$). The code in Listing 2.5 computes the e.c.d.f. of a random sample of $n = 50$ observations from a Geo(0.4) distribution, using the handy built-in Matlab function `tabulate`.

Besides generating the graph, we allow the output from the `tabulate` function to be printed. It consists of a matrix with three columns. The first contains the sorted observed

```
1  rand('twister',4) % fix the seed value so we can replicate this
2  n=50; p=0.4; r=sort(geornd(p,[n,1])+1); t=tabulate(r)
3  empcdf=cumsum(t(:,3)/100); x=t(:,1); true=1-(1-p).^x;
4  stairs(x,empcdf,'r-','linewidth',3), hold on, stairs(x,true,'b:','linewidth',3)
5  set(gca,'fontsize',16), axis([1 10.2 0.26 1.04]), hold off
```

Program Listing 2.5: Code used for Figure 2.5(a).

values from the lowest to the highest, in increments of one. If the observed support contains only integers (the case for most discrete distributions we work with), then Matlab conveniently tabulates all integers between and including the observed lowest and highest. The second and third columns contain, respectively, their *actual frequency* and their *percentage frequency* (second column times 100, divided by n). For the generated data set used in the following illustration, the `tabulate` output (transposed to save space) is as follows:

1	2	3	4	5	6	7	8	9	10
21	15	3	6	1	1	1	0	0	2
42	30	6	12	2	2	2	0	0	4

Observe how values of 8 or 9 never occurred in this sample, but they are included in the output; this feature is crucial for our purpose. The e.c.d.f. is then plotted with the first column (here depicted as the first row) on the x-axis and the cumulative sum of the corresponding *relative frequencies* (obtained by applying the `cumsum` function to the third

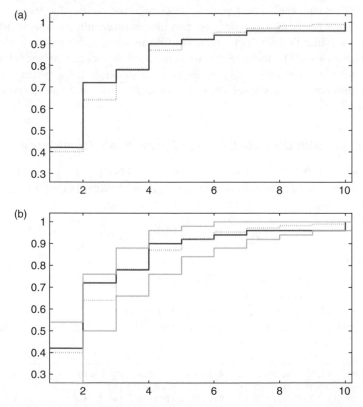

Figure 2.5 *(a) The e.c.d.f. (solid) based on 50 observations and true c.d.f. (dashed) of a Geo(0.4) distribution. (b) Same, but adds vertical 95% error bounds (these are not c.i.s) obtained by simulation of order statistics using the true Geo(0.4) model.*

column divided by 100), on the *y*-axis, as a step function. The code in Listing 2.5 generates the e.c.d.f. plot and overlays it with the true c.d.f.; see Figure 2.5(a).

We now wish to construct error bounds (not c.i.s) analogous to those shown in Figure 2.3. Figure 2.5(b) shows the 95% error bounds based on having simulated numerous *n*-length random samples from the true Geo(0.4) distribution, obtained as follows. First note that, with discrete distributions such as the geometric, we can anticipate the values a random sample will take (in this case, $1, 2, \ldots$, with an upper limit of, say, 20, when using $p = 0.4$). For each simulated Geo(0.4) sample, we store its e.c.d.f., which amounts to recording the relative frequencies of $x = 1$, $x = 2$, etc., up to $x = 20$. The 2.5% and 97.5% sample quantiles based on these values are then computed and plotted. This gives error bounds that are to be interpreted vertically (unlike in the continuous case).

The code to construct these bounds is given in Listing 2.6. It is just an implementation of the simple method just described, but importantly shows how to deal with the fact that we do not know in advance the maximal sample value over all the simulations (recall that the support of the geometric is not bounded). We could of course just set all values above, say, 20 to 20, but this is neither appealing nor necessary. Instead, a guess is made as to its expected size and then the size of the matrix for storage is updated dynamically as needed (quite a nontrivial luxury that is available, and we take for granted, in high-level programming languages such as Matlab and R). Crucially, this matrix is not initialized with zeros, as in almost all of our applications, but ones, so that the e.c.d.f. values beyond the maximum of any particular sample take on their correct default values.

The nonparametric bootstrap c.i.s are formed in a similar way, in which resamples of the actual data are used instead of draws from a Geo(0.4) distribution. Also, the maximum possible value in any resample is of course just the maximum of the original data set (which was 10 in our case), so the aforementioned issue regarding the unknown size of the storage matrix is no longer relevant. The code is given in Listing 2.7, and the output is shown in Figure 2.6(a). Similarly, Figure 2.6(b) shows the use of the asymptotic c.i.s. As with the exponential model case, there is virtually no difference between the bootstrap and asymptotic intervals.

```
1   maxr=16; % just a guess, it gets updated dynamically
2   sim=10000; ecdfmat=ones(sim,maxr);
3   for i=1:sim
4     r=sort(geornd(p,[n,1])+1);
5     if max(r)>maxr
6       ecdfmat=[ecdfmat ones(sim,max(r)-maxr)]; maxr=max(r)
7     end
8     t=tabulate(r); empcdf=cumsum(t(:,3)/100); ecdfmat(i,1:max(r))=empcdf';
9   end
10  x=1:maxr; qlo=quantile(ecdfmat,0.025); qhi=quantile(ecdfmat,0.975);
11  stairs(x,qlo,'g-','linewidth',2), stairs(x,qhi,'g-','linewidth',2)
12  hold off, set(gca,'fontsize',16), axis([1 10.2 0.26 1.04])
```

Program Listing 2.6: Code used to augment Figure 2.5(b) with error bounds, after having run the code in Listing 2.5.

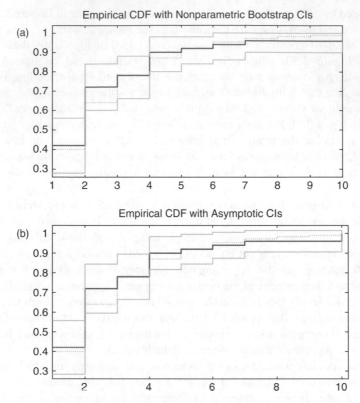

Figure 2.6 *The same e.c.d.f. (solid) and true $Geo(0.4)$ c.d.f. (dotted) as in Figure 2.5 but with 95% nonparametric bootstrap c.i.s (a) and 95% asymptotic c.i.s (b).*

```
1  B=10000;  maxr=max(r);  becdfmat=ones(B,maxr);
2  for i=1:B
3     ind=unidrnd(n,[n,1]);  bsamp=sort(r(ind));  t=tabulate(bsamp);
4     bcdf=cumsum(t(:,3)/100);  becdfmat(i,1:max(bsamp))=bcdf';
5  end
6  x=1:maxr;  qlo=quantile(becdfmat,0.025);  qhi=quantile(becdfmat,0.975);
7  hold on,  stairs(x,qlo,'g-','linewidth',2),  stairs(x,qhi,'g-','linewidth',2)
8  hold off,  set(gca,'fontsize',16),  axis([1 10 0.26 1.04])
```

Program Listing 2.7: Generates the bootstrap c.i.s shown in Figure 2.6(a).

2.2 COMPARING PARAMETRIC AND NONPARAMETRIC METHODS

Let me go a little more into detail about my agreements and disagreements with Fisher's views. The discovery of the property of sufficiency was a brilliant one, and yet it seemed to me that it could be dangerous in inexperienced hands. One or more than one statistic will contain all the information to be extracted from the data *if and only if* the probability density law in the population has the assumed mathematical form. But vital information to help check this assumption lies in the whole pattern of the observations within the sample. The use of such a phrase as "all the relevant information is contained in these statistics" may therefore mislead the non-mathematical research worker.

(Egon Sharpe Pearson, 1974, p. 7)

We saw that the empirical c.d.f. is unbiased, easy to compute, and obviates the need to specify a parametric model. As such, one might ask what the benefit could be of fitting a parametric model compared to use of the e.c.d.f. for estimating $\Pr(X \leq t)$. The answer depends on the extent to which the parametric form of the data is known.

Consider again the pregnancy example discussed in Section 1.1.2 and assume that inference on fertility is based on a study conducted with, say, 40 couples. If a woman is interested in knowing the probability of her becoming pregnant within the next four months, there are two possible estimators: The first is the nonparametric one based on the empirical c.d.f., which simply computes the fraction of the 40 couples who conceived within (less than or equal to) four months. The second is the parametric method, which involves (i) estimating the probability p associated with the geometric model (using, say, the point estimator $\hat{p}_2$) and (ii) computing the c.d.f. of a geometric r.v. with probability $\hat{p}_2$ for four months, denoted by $\widehat{F}_{\text{fit}}(4)$ or $\widehat{F}_X(4; \hat{p}_2)$. This is referred to as the **(parametrically) fitted c.d.f.**

If the model is correct, that is, the time to pregnancy for each couple in the study is indeed geometric with a constant probability p, then, generally speaking, the parametric method will lead to more accurate answers. *This is because information is added to the problem via specification of the parametric model.* In the nonparametric method, this information is not used; the data points are reduced to Bernoulli random variables (in this case, either ≤ 4 or > 4).

The price to pay for the gain in efficiency with the parametric model is this: If the assumed parametric model is wrong, then inference based on it could be (potentially highly) misleading.

In the previous example on number of months until pregnancy, let us compare the m.s.e. for computing $\Pr(X \leq 4)$ and $\Pr(X \leq 8)$ of the nonparametric approach with that of the two parametric approaches, the first based on the m.l.e. $\hat{p}_2$, and the second based on the efficient estimator $\hat{p}_3$. Figure 2.7 shows the results, based on simulation, for two sample sizes, $r = 10$ and $r = 40$ (see Problem 2.1). We see that the parametric estimators are everywhere better than the nonparametric, significantly so for smaller values of p. Comparing the two parametric estimators, we see that the one based on $\hat{p}_2$ is (at least with respect to m.s.e.) superior to that based on $\hat{p}_3$ for most of the parameter space, albeit not by that much, for the small sample size $r = 10$, while they are virtually equal for $r = 40$.

The following example illustrates what happens when the parametric model is misspecified, and also shows how the empirical c.d.f. can be used as a method for checking the suitability of the parametric fit.

Example 2.1 *Imagine that two independent pregnancy studies, A and B, both with 100 couples, will be conducted, each using a different method of assistance in conceiving. A histogram of the resulting 100 measurements (number of months until success) $G_1, \dots, G_{100}$, for the first study, A, is shown in the upper left panel of Figure 2.8, while the upper right panel shows the empirical c.d.f. (solid line) overlaid with the parametric c.d.f. of a geometric r.v. (dashed line) using the estimated value of p, which, based on the m.l.e. $\hat{p}_2$, was 0.34. It certainly appears from the c.d.f. plot that the empirical and fitted c.d.f.s are very close*

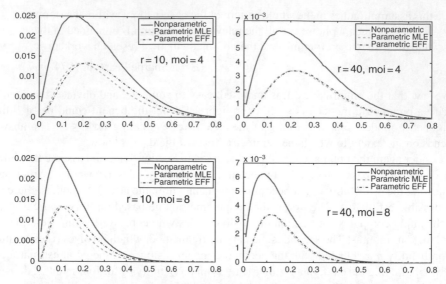

Figure 2.7   *The m.s.e. comparison of the three estimators nonparametric (solid), parametric using the m.l.e. $\hat{p}_2$ (dashed), and parametric using the efficient estimator $\hat{p}_3$ (dash-dotted), as a function of p, using sample size r, for estimating the probability of getting pregnant within (up to, and including) 4 (top) and 8 (bottom) months, where in the graphics* moi *stands for "month of interest." Based on simulation with 100,000 replications.*

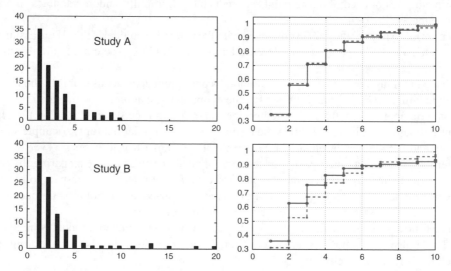

Figure 2.8   *Comparison of behavior of correctly specified (top) and misspecified (bottom) fitted c.d.f.s.*

and would deliver practically the same information. The data for study A were, in fact, 100 simulated realizations of a geometric r.v. with success probability 0.30. The c.d.f. plot was produced by the Matlab code in Listing 2.8.

The code in Listing 2.9 computes a bootstrap c.i. for p, with $B = 100,000$ bootstrap samples, for both the parametric and nonparametric 95% c.i.s, yielding $(0.296, 0.407)$ and

$(0.299, 0.403)$, *respectively. They are extremely close, and just barely include the true value of $p = 0.3$.*

Similarly, we can compute a c.i. for $\tau = \Pr(X \le 4)$ with the parametric bootstrap: Recalling the geometric c.d.f. (2.20), the parametric bootstrap uses the parametric estimator $1 - (1 - \hat{p}_2)^4$ (which is the m.l.e. of τ, because of the so-called invariance *property of the m.l.e.; see Section 3.1.1). The true value is $1 - (1 - p)^4 = 0.760$, and the parametric point estimate is 0.816. The resulting 95% parametric bootstrap c.i. for τ is $(0.754, 0.876)$, which just covers the true value.*

Likewise, we can use the nonparametric bootstrap, and use the nonparametric estimator of τ, $\sum_{i=1}^{r} \mathbb{I}_{(-\infty,4]}(G_i)$. The resulting 95% c.i. is $(0.730, 0.880)$, with point estimate 0.81. The nonparametric c.i. is somewhat wider, on the left side, than the parametric. The reader is encouraged to write the code to compute both of these intervals, as well as "crossing," that is, using the parametric bootstrap with the nonparametric point estimator of τ (yielding $(0.740, 0.890)$), and vice versa (yielding $(0.758, 0.873)$).

While both 95% bootstrap c.i.s covered the true value, the intervals themselves can tell us nothing about the actual coverage probability corresponding to the nominal value of 0.95. For that, we need simulation. The top left panel of Figure 2.9 shows the results, based on sim $= 10,000$ *replications and $B = 1000$ bootstrap samples, and having used several values of sample size r instead of just $r = 100$. For both the parametric and nonparametric bootstraps, the actual coverage is quite close to the nominal value of 95%, and improves (though not monotonically) as r increases. The bottom left panel is similar, but used a nominal coverage of 90%.*

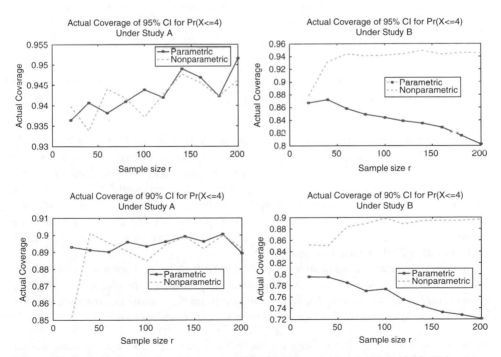

Figure 2.9 **Top**: *Actual coverage of 95% parametric (solid) and nonparametric (dashed) bootstrap c.i.s for $\tau = \Pr(X \le 4)$ as a function of sample size r under study A (left) and study B (right).* **Bottom**: *Same, but for 90% c.i.s.*

```
1  rand('twister',6) % set seed value so we can replicate results
2  p=0.3; r=100; y=geornd(p,[r,1])+1;
3  t=tabulate(y); ecdf=cumsum(t(:,3)/100); grd=t(:,1);
4  stairs(grd,ecdf,'b-o','linewidth',2), set(gca,'fontsize',16)
5  p2=1/mean(y), gcdf=1-(1-p2).^(grd);
6  hold on, stairs(grd,gcdf,'r-x','linewidth',2), hold off
7  grid, axis([0 10 0.3 1.05])
```

Program Listing 2.8: Code for the simulated data of study A.

```
1  rand('twister',6) % set seed value so we can replicate results
2  p=0.3; r=100; y=geornd(p,[r,1])+1; phat0=1/mean(y)
3  B=1e5; alpha=0.05;
4  phatvec = 1./mean(geornd(phat0,[r B])+1); % MLE
5  ci_para=quantile(phatvec,[alpha/2 1-alpha/2])    % parametric
6  for b1=1:B
7    ind=unidrnd(r,[r,1]); b1samp=y(ind); phatvec(b1)=1/mean(b1samp);
8  end
9  ci_nonpara=quantile(phatvec,[alpha/2 1-alpha/2]) % nonparametric
```

Program Listing 2.9: Code for computing parametric and nonparametric single bootstrap c.i.s for parameter p associated with the simulated data of study A.

```
1  rand('twister',6) % set to same seed value as before
2  r=100; p1=0.2; p2=0.5; y=zeros(r,1);
3  for i=1:r
4    if rand<0.3333, y(i)=geornd(p1,[1,1])+1;
5    else           y(i)=geornd(p2,[1,1])+1; end
6  end
```

Program Listing 2.10: Code for the simulated data of study B.

Let us now return to Figure 2.8, and consider the lower panels. They are similar to the upper ones, but correspond to study B, for which $\hat{p}_2 = 0.31$. Although the two parameter estimates are quite close, the histogram of the data for study B differs remarkably from that of study A. Moreover, there is a large discrepancy between the empirical and fitted c.d.f. The reason for this is that the geometric model for the data is wrong. *For study B, each of the 100 data points was generated as follows: with probability 1/3, it is a realization of a geometric r.v. with $p = 0.2$; with probability 2/3 it is geometric with $p = 0.5$. The Matlab code in Listing 2.10 was used to generate the data.*

*The resulting data are from what is called a **mixture model**, and is not terribly unrealistic. Imagine (somewhat simplistically of course) that couples having trouble conceiving have a probability of 0.2 of success. The treatment used in study B boosts the probability to 0.5, but it only works in two-thirds of women. This gives rise to the mixture.[5]*

---

[5] In fact, a mixture of geometric distributions is generally considered to be a good description of actual time until conception; see, for example, Ecochard and Clayton (2000) and the references therein. Another interesting and relevant issue in this modeling context is the estimation of p (assuming the simple geometric model) when the data are censored, that is, if the study of conception times stops at, say, 12 months, and some of the women are still not

Computing the parametric and nonparametric 95% c.i.s for parameter p for the data in study B yields $(0.269, 0.372)$ *and* $(0.255, 0.395)$*, respectively. They differ considerably, unlike those for study A, because the data are not i.i.d. geometric. The nonparametric interval is much wider, reflecting far more uncertainly about (the misspecified model parameter) p.*

The parametric and nonparametric c.i.s for $\tau = \Pr(X \le 4)$ *using the study B data are* $(0.714, 0.844)$ *and* $(0.750, 0.900)$ *respectively, with the true value being* $\frac{1}{3}(1 - (1 - 0.2)^4) + \frac{2}{3}(1 - (1 - 0.5)^4) = 0.822$*. The right panels of Figure 2.9 are similar to the left ones, but having used the mixture model from study B to generate the data. Now we see that, while the actual coverage of the nonparametric bootstrap c.i. converges to the nominal value as r increases, that of the parametric bootstrap worsens as r increases. As mentioned above, inference based on a misspecified parametric model can be highly misleading.*

The reader is encouraged to replicate the results in Figure 2.9.

2.3 KOLMOGOROV–SMIRNOV DISTANCE AND HYPOTHESIS TESTING

Recall that the e.c.d.f. for (i.i.d. realizations from) a continuous distribution, evaluated at the order statistics $Y_1 < Y_2 < \cdots < Y_n$, is, from (2.2), given by $\widehat{F}_{\mathrm{emp}}(Y_i) = (i - 0.5)/n$ or $(i - 3/8)/(n + 1/4)$, and is thus the step function formed from the points $t_i := (i - 3/8)/(n + 1/4)$, for $i = 1, \ldots, n$. Its counterpart is the fitted c.d.f. $\widehat{F}_{\mathrm{fit}}(y_i)$. The discrete case is similar, but requires tabulation. We saw from Figure 2.8 that, when the assumed parametric model is wrong, there will be a "large" discrepancy between the empirical and fitted c.d.f.s. As such, it seems natural to consider the maximal distance between the two as a goodness-of-fit measure. Such a measure has been extensively studied in the literature, and is referred to as the **Kolmogorov–Smirnov distance**, introduced by Kolmogorov in 1933.

This leads to the concept of significance and hypothesis testing, and the associated jargon, such as significance level, critical values, the null hypothesis, p-values, statistical significance, and power.

2.3.1 The Kolmogorov–Smirnov and Anderson–Darling Statistics

The Kolmogorov–Smirnov distance is defined as $D = \sup_{x \in \mathbb{R}} |\widehat{F}_{\mathrm{emp}}(x) - F_X(x)|$, where F_X is the hypothesized true distribution function of X. For a sample of (i.i.d.) univariate data from an underlying continuous distribution and with observed order statistics $y_1 < y_2 < \cdots < y_n$, it is calculated as

$$\mathrm{KD} = \max_i |\widehat{F}_{\mathrm{emp}}(y_i) - \widehat{F}_{\mathrm{fit}}(y_i)|, \tag{2.21}$$

(or sometimes 100 times this), where $t_i = \widehat{F}_{\mathrm{emp}}(y_i)$ is the e.c.d.f. (2.1), with $t_i = (i - 0.5)/n$ or $(i - 3/8)/(n + 1/4)$, and $\widehat{F}_{\mathrm{fit}}(y_i)$ is the parametrically fitted c.d.f. evaluated at y_i.

Another definition that helps account for the discreteness at the endpoints of the empirical c.d.f. takes $D^+ = \sup_x \{\widehat{F}_{\mathrm{emp}}(x) - F_X(x)\}$, $D^- = \sup_x \{F_X(x) - \widehat{F}_{\mathrm{emp}}(x^-)\}$, and

---

pregnant. These observations still contain information about p and cannot be discarded. The m.l.e. of p and other results are derived in Gan and Bain (1998). With regard to unfinished waiting times, see also Example 3.3.

$D = \max(D^+, D^-)$, where x^- denotes the left-hand-side limit of x. (Recall that F_X is right continuous, with $\lim_{x \downarrow x_0} F_X(x) = F_X(x_0)$, while $\lim_{x \uparrow x_0} F_X(x) = F_X(x_0^-)$.) For a given data set, this is calculated as

$$KS = \max(D^+, D^-), \tag{2.22}$$

where

$$D^+ = \max_i \left(\frac{i}{n} - \widehat{F}_{\text{fit}}(y_i) \right), \quad D^- = \max_i \left(\widehat{F}_{\text{fit}}(y_i) - \frac{i-1}{n} \right).$$

If, in Matlab, para denotes $\widehat{F}_{\text{fit}}(y_i)$, then the following code computes the KS statistic:

```
1  i=(1:n)'; ecdf1 = i/n; ecdf2 = (i-1)/n;
2  dd1=max(ecdf1-para); dd2=max(para-ecdf2); KS=max([dd1,dd2]);
```

Observe that we differentiate between (2.21) and (2.22) by calling the first KD and the second KS. In practice, there will not be much difference (in terms of the power of the test; see below) between the two, though simulations show that (2.22) is slightly superior. The short monograph by Durbin (1973) is dedicated to the topic of tests based on the e.c.d.f. and contains a wealth of information.

Related to KD is the **Anderson–Darling statistic**, based on Anderson and Darling (1952, 1954). It is a weighted form of the KD statistic, given by

$$AD = \max_i \frac{|\widehat{F}_{\text{emp}}(y_i) - \widehat{F}_{\text{fit}}(y_i)|}{\sqrt{\widehat{F}_{\text{fit}}(y_i)[1 - \widehat{F}_{\text{fit}}(y_i)]}}. \tag{2.23}$$

Observe that it divides by a term proportional to (an estimate of) the standard deviation of $\widehat{F}_{\text{emp}}(y_i)$, and so arguably weights discrepancies more appropriately across the support of the distribution. As emphasized by Anderson and Darling (1954, p. 767), as $\widehat{F}$ approaches zero or one, the reciprocal of the weight function becomes large, and thus this choice of weighting function weights the tails more heavily, increasing the sensitivity of the test in this area.

Remark. The AD test statistic is usually expressed somewhat differently than as stated in (2.23). In particular: (i) The $\widehat{F}_{\text{fit}}$ term in the numerator and denominator of (2.23) is replaced by the c.d.f. associated with a so-called simple distributional null hypothesis (whereby its parameters are known; the distinction between a *simple* and *composite* distributional hypothesis is discussed below in Section 2.4); (ii) the canonical case of the uniform distribution is assumed, this being essentially without loss of generality, because of the **probability integral transform**; (iii) the numerator is squared; and (iv) instead of use of the max as in (2.23), it is expressed in integral form, and is thus a weighted average of the c.d.f. discrepancies. That is, the AD statistic is formally given by

$$AD_n = n \int_0^1 \frac{[\widehat{F}_{\text{emp}}(x) - x]^2}{x(1-x)} \, dx = -n - \frac{1}{n} \sum_{k=1}^n (2k-1) \ln(X_{(k)}(1 - X_{(n+1-k)})), \tag{2.24}$$

where n denotes the sample size, and $X_{(1)} < X_{(2)} < \dots < X_{(n)}$ denote the order statistics. Considerable effort has gone into computing the c.d.f. of (2.24); see, for example, Giles (2001), Marsaglia and Marsaglia (2004), Chen and Giles (2008), Murakami (2009), Grace and Wood (2012), and the references therein. ∎

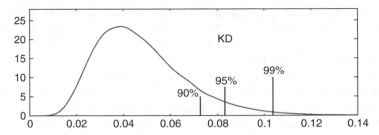

Figure 2.10 *Distribution of KD statistic for the i.i.d.* Geo(0.31) *model with* $r = 100$ *observations, with marked cutoff values 0.07273, 0.08304 and 0.10378.*

2.3.2 Significance and Hypothesis Testing

Whatever you do will be insignificant, but it is very important that you do it.

(Mahatma Gandhi)

The distribution of the Kolmogorov–Smirnov distance (2.22) for a particular model and sample size is most easily determined via simulation. To illustrate, Figure 2.10 shows the kernel density estimate of KD based on 100,000 realizations of the i.i.d. geometric model with $r = 100$ and $p = 0.31$ (which corresponds to the estimate of the misspecified model for study B in Example 2.1 above) and having used $\hat{p}_2$ to estimate p for use in $\widehat{F}_{\text{fit}}$ in (2.21). Marked on the plot are the 90%, 95%, and 99% quantiles; these are referred to as **critical values** or **cutoff values**. If, for example, $r = 100$ observed data points are drawn from an i.i.d. Geo(0.31) distribution, then there is a 1% chance that the corresponding KD value would exceed the 99% quantile.

> The random variable KD is a function of the observed data, and serves as a **test statistic** for testing the **null hypothesis**, denoted by H_0, that the $r = 100$ data points are a random sample from a Geo(0.31) distribution. The test with **significance level** (sometimes referred to as the *size*) α involves **rejecting** H_0 if KD equals or exceeds the $100(1 - \alpha)$% critical value of the KD statistic under H_0. Typical values of α are $0.01, 0.05$, and 0.10. If it does not exceed this value, then we **do not reject** H_0 at the 100α% level.

For the data set in Figure 2.8 associated with study B, the KD value is 0.103, which clearly exceeds the 90% and 95% quantiles, and is just barely lower than the 99% quantile. Based on the 100,000 simulated values, 1.07% exceeded the observed test statistic value of 0.103. This quantity, 0.0107, is an example of what is called the **significance probability** or, more commonly, the **p-value** (of the test).

> A **p-value** associated with a particular statistical test and data set is the probability of observing a value of the test statistic that is equal to, or more extreme than, the one actually observed, under the assumption that H_0 is true. If the significance level of a test (presumably agreed upon before the experiment) is α, and the p-value turns out to be less than α, then we say that the test is **statistically significant** at the 100α% level.

The origin of the p-value apparently goes back at least to Laplace, but is often credited to Karl Pearson, who used it in his 1900 paper with his χ^2 test. However, significance testing and the use of the p-value were popularized by Ronald Fisher. The name "p-value" is attributed to William Edwards Deming, from 1943.[6]

It is crucial to emphasize that a p-value is *not* the probability that H_0 is false. Denote the set of random variables that will comprise the data as **X**, and the observed values as **x**; and denote the test statistic as $T = T(\mathbf{X})$. Assuming T is such that large values are inconsistent with the null hypothesis, then $p = \Pr(T(\mathbf{X}) \geq T(\mathbf{x}) \mid H_0)$. It is *not* the posterior probability of the null, given the data, that is, $\Pr(H_0 \mid \mathbf{X} = \mathbf{x})$, which is arguably what is really desired. A p-value is typically used as a form of evidence regarding the extent to which the data do not support H_0, with smaller values indicating more evidence against it.

There is a large variety of misunderstandings of what a p-value represents, the most common one being, as mentioned above, that a p-value is (mistakenly) the probability that H_0 is false. Goodman (2008) provides a detailed discussion and 12 such misconceptions. The problem is exacerbated by the fact that, in numerous applied research articles, one or more such mistakes occur; see, for example, Krämer and Gigerenzer (2005), McShane and Gal (2017), and the numerous references therein, as well as Section 2.8 below. A possible reason for this is that the definition does not lend itself to probabilistic statements of what is actually of interest, nor does it lead to a capacity to make a decision (despite the erroneous "rule" that, if the p-value is less than 0.05, there is "significance"). As such, statisticians find themselves attempting to provide an interpretation in layman's terms, only to ultimately cause misunderstandings. This idea was embodied nicely by Berry (2017, p. 896):

> We have saddled ourselves with perversions of logic – p-values – and so we deserve our collective fate. I forgive non-statisticians who cannot provide a correct interpretation of $p < 0.05$. p-values are fundamentally un-understandable. I cannot forgive statisticians who give understand-able – and therefore wrong – definitions of p-values to their non-statistician colleagues. But I have some sympathy for their tack. If they provide a correct definition then they will end up having to disagree with an unending sequence of "in other words." And the colleague will come away confused and thinking that statistics is nutty.
>
> (Donald Berry, 2017)

More will be said about the use and misuse of p-values below in Section 2.8.

A p-value can also be seen as the smallest significance level α such that H_0 can be rejected. In the above example, based on the KD test, we could reject the Geo(0.31) null hypothesis at the $\alpha = 0.0107$ level, but no smaller, for example, not at the $\alpha = 0.01$ level. It is important to emphasize that, if one chooses to use this testing paradigm (as will be dis-couraged below in the typical context of testing parameter significance), then a test result that does not result in rejection of the null (this depending on the – often arbitrary – choice of α) can be interpreted as a form of evidence that **the data are not inconsistent with the null hypothesis**. *It does not lend support to the validity of H_0*; see, for example, Fisher (1929, p. 192).

---

[6] See David (1995), Stigler (1999), and the references therein for further details on this, original references, and the origins of other naming conventions in probability and statistics, as well as the Wikipedia entry "History of statistics," and the references and linked web pages therein.

The p-value, being a function of the data, is itself a random variable. Letting $p = P = P(\mathbf{X})$ to emphasize this, the **probability integral transform** shows that, under the null H_0, $P \sim \text{Unif}(0, 1)$. In particular, assuming the test statistic $T = T(\mathbf{X})$ is such that large values are inconsistent with the null, then, for $0 < p < 1$,

$$F_P(p \mid H_0) = \Pr(1 - F_T(T) \leq p \mid H_0) = \Pr(F_T(T) \geq 1 - p \mid H_0)$$

$$= \Pr(F_T^{-1}(F_T(T)) \geq F_T^{-1}(1 - p) \mid H_0) = \Pr(T \geq F_T^{-1}(1 - p) \mid H_0)$$

$$= 1 - F_T(F_T^{-1}(1 - p) \mid H_0) = 1 - (1 - p) = p, \tag{2.25}$$

showing $P \sim \text{Unif}(0, 1)$ under the null.

If the data-generating process is such that the null does not hold, but rather some specific alternative hypothesis, denoted H_A, and the distribution of T under the alternative is given by, say, G_T, then the same calculation up to the second line of (2.25) gives

$$F_P(p \mid H_A) = \Pr(T \geq F_T^{-1}(1 - p) \mid H_A), \tag{2.26}$$

but, because of H_A, this is now

$$F_P(p \mid H_A) = 1 - G_T(F_T^{-1}(1 - p)). \tag{2.27}$$

For the test statistic T to be meaningful (or have *power*; see below), it must be the case that (as we reject for large values of T) $1 - G_T(t) > 1 - F_T(t)$ for "almost all" t (with equality holding only on a set of measure zero), that is, F_T *(first-order) stochastically dominates G_T*, recalling the notation in Example I.4.21.

Use of the p-value in applied statistics is ubiquitous, and has the benefit of delivering a measure on the same scale, $(0, 1)$, with small values lending evidence against the null. It reflects how far the test statistic is in the tail of its distribution under the null. However, its use for model selection is controversial; more will be said about this in Sections 2.8 and 3.3. One point of contention is that, as a tail probability measure, it is (as stated in the definition) the probability of observing a value of the test statistic that is equal to, *or more extreme than*, the one actually observed; though one could argue that values of T *that were not observed* are not relevant for the analysis.

One analytic attempt at resolving this is to change from F_T to f_T, that is, from a tail area to the density at the observed value of the test statistic. To do this, first observe that, for Y a continuous r.v. with c.d.f. F_Y, as F_Y is strictly increasing,[7]

$$\frac{\partial F_Y^{-1}(t)}{\partial t} = \frac{1}{f_Y(F_Y^{-1}(t))}. \tag{2.28}$$

Thus, differentiating (2.27), using the chain rule and (2.28),

$$f_P(p \mid H_A) = \frac{\partial}{\partial p} F_P(p \mid H_A) = \frac{g_T(t)}{f_T(t)}, \quad t = F_T^{-1}(1 - p). \tag{2.29}$$

---

[7] Recall from calculus (see, for example, (I.A.32)) that, if f is a strictly increasing continuous function on a closed interval I, then the image of f is also a closed interval, and the inverse function g is defined as the function such that $g \circ f(x) = x$ and $f \circ g(y) = y$. It is also continuous and strictly increasing. If f is also differentiable in the interior of I with $f'(x) > 0$, then $g'(y) = 1/f'(g(y))$.

This is a ratio of the alternative density to the null one, evaluated at the observed test statistic, and motivates the study of the likelihood ratio and information criteria based on the likelihood, as discussed in Sections 3.3.2 and 3.3.3, respectively.

Continuing with Example 2.1, now that we have the three cutoff values, we might wish to know with what probability the KD statistic would exceed these values if the underlying process is precisely that of study B. In the jargon of hypothesis testing, this would be an example of an **alternative hypothesis**.

> For a given test statistic, a specific significance level α, and a specific alternative hypothesis, the probability of rejecting the null when the alternative hypothesis is true is referred to as the **power (of the test with significance level α)**.

The power is often denoted by β. It is a function of α, the test statistic T, and the distribution of the random variables associated with the data, $\mathbf{X}$, under the alternative hypothesis. Observe that

$$\beta = \Pr(P \leq \alpha \mid H_A) = F_P(\alpha \mid H_A) = \Pr(T \geq F_T^{-1}(1 - \alpha) \mid H_A),$$

from (2.26).

The power in the KD example is also trivially computed via simulation, and the results are given in the column of Table 2.1 marked KD, for $r = 100$, as a function of α, the significance level of the test (see Problem 2.2). Naturally, the power of the test decreases as the significance level of the test decreases. Still based on these cutoff values, we can determine the power if, instead of the $1/3$ and $2/3$ mixture of geometric r.v.s with probabilities $p_1 = 0.2$ and $p_2 = 0.5$, we use $p_1 = 0.25$ and $p_2 = 0.35$, which are both closer to the value of 0.31. For this different alternative hypothesis, the power corresponding to the three cutoff values is 0.122, 0.063, and 0.014. These values are barely larger than those we would get if $p_1 = p_2 = 0.31$, in which case the power values would be exactly 0.10, 0.05 and 0.01. In this case, we say that, for this sample size, **the test has low power against this alternative**.

If we return to the case with $p_1 = 0.2$ and $p_2 = 0.5$, but wish to consider the power for a different sample size r, then we have to recompute the cutoff values. This is easy once the short program to compute the cutoff and power values via simulation has been constructed. Doing so for several values of r yields the values in Table 2.1. As expected, the power increases with increasing sample size.

We can similarly obtain cutoff values and power for the Anderson–Darling statistic (2.23); the power results are shown in Table 2.1. From those, we see that the AD test has higher power for smaller samples and for the lower cutoff values. Of course, these results just apply to the model under consideration. We will see other cases below in which the

TABLE 2.1   Power of the KD and AD tests for the mixture of geometric distributions example

$\alpha \backslash r$	KD				AD			
	25	50	100	200	25	50	100	200
0.10	0.271	0.398	0.610	0.845	0.390	0.533	0.695	0.857
0.05	0.186	0.292	0.493	0.764	0.272	0.380	0.524	0.683
0.01	0.077	0.136	0.275	0.560	0.107	0.150	0.215	0.288

AD test performs much better (has higher power) and also much worse, than the KD test, showing that neither test dominates the other in all situations.

2.3.3 Small-Sample Correction

This section is more for aficionados and can be skipped upon a first reading of this chapter.

There is a caveat with the above procedure that is similar to the nominal and actual coverage probability of c.i.s discussed above. For the results of study A (which were based on 100 simulated geometric r.v.s, each with success probability 0.30), we obtained a point estimate for p of 0.34. If we applied the above simulation procedure to get the distribution of the KD statistic and the various cutoff values, we would have used $p = 0.34$ in the simulation, but the true value of p from which the data arose, and which we do not know, is different, 0.30. The method does not take into account the variability of the estimate, which could be substantial for smaller sample sizes.

To account for this, we could, as before, "simulate the simulation," generating data sets with the estimated parameter from the real data, and for each, compute its KD value (say, KD_0), simulate to get its distribution, and record if KD_0 exceeds the, say, 90% critical value. These Bernoulli r.v.s are then averaged, and the **actual acceptance probability** corresponding to the **nominal acceptance probability** for 90% is determined. This is best outlined step by step, as given in Pseudo-code Listing 2.1, for fixed numbers s_1 and s_2 representing the number of simulations to conduct.

---

(1) From the data set under study, compute the estimate of parameter θ, say $\hat{\theta}^{\text{obs}}$, with a chosen method of estimation (e.g., maximum likelihood), and where obs stands for "observed."

(2) FOR $i = 1, \dots, s_1$ DO

     a. Simulate a data set, say $\mathbf{y}^{(i)}$, according to the presumed parametric model and using parameter $\hat{\theta}^{\text{obs}}$.

     b. Estimate $\theta^{(i)}$ for data set $\mathbf{y}^{(i)}$ (using the same method of estimation) and calculate the KD statistic, say $KD^{(i)}$, based on the fitted c.d.f. using $\theta^{(i)}$ and the e.c.d.f. using $\mathbf{y}^{(i)}$.

     c. FOR $j = 1, \dots, s_2$ DO

         (i) Simulate a data set, say $\mathbf{y}^{(i,j)}$, according to the presumed model and using parameter $\hat{\theta}^{(i)}$.

         (ii) Estimate $\theta^{(i,j)}$ for data set $\mathbf{y}^{(i,j)}$ (using the same method of estimation) and calculate the KD statistic, say $KD^{(i,j)}$, based on the fitted c.d.f. using $\theta^{(i,j)}$ and the e.c.d.f. using $\mathbf{y}^{(i,j)}$.

     d. Calculate the 90% quantile of $(KD^{(i,1)}, \dots, KD^{(i,s_2)})$, say $q_{0.90}^{(i)}$.

     e. Record $B^{(i)}$ as one if $KD^{(i)} \le q_{0.90}^{(i)}$, and zero otherwise.

(3) Compute $\hat{\pi}_{0.90}$ as the mean of the $B^{(i)}$.

Pseudo-code Listing 2.1: Algorithm for the small-sample correction to take into account the variability of the estimate by using nested simulation to get the actual acceptance probabilities. The associated Matlab program is given in Listing 2.11.

Like the double bootstrap from Section 1.3.3, this is a nested simulation method, and will clearly take significantly longer to run. Here, $\pi_{0.90}$ denotes the true actual acceptance probability corresponding to the nominal acceptance probability level of 90%. This is done for several nominal probabilities between, say, 0.80 and 0.99, and then we would plot the nominal values versus the actual values; see below.

Ideally, we would use the true value of θ instead of its estimate, $\hat{\theta}^{\text{obs}}$, but obviously it is not available. If we *had* the true θ, then the above procedure would obtain (with s_1 and s_2 set high enough) $\pi_{0.90}$, the true actual acceptance probability for a 90% nominal level when we (i) observe a data set generated from the prospective model and true θ, (ii) estimate $\hat{\theta}$ and compute its KD statistic, and (iii) simulate to get the 90% cutoff based on the model with $\hat{\theta}$. As we do not know θ, we have to use its estimator, $\hat{\theta}^{\text{obs}}$, and $\hat{\pi}_{0.90}$ is only an estimate of $\pi_{0.90}$. Notice how, by drawing many data sets using $\hat{\theta}^{\text{obs}}$, we correctly account for the variability in the estimate; that is, the $\theta^{(i)}$ are, by design, being drawn from the actual distribution of $\hat{\theta}$ for the fixed sample size and true value $\hat{\theta}^{\text{obs}}$.

The code in Listing 2.11 implements the strategy for the geometric pregnancy example, but using $p = 0.31$ and $r = 25$ (instead of $r = 100$, because with the smaller sample size we would expect the difference between nominal and actual to be more pronounced). Observe how the code handles a vector of nominal probabilities, $0.80, 0.81, \ldots, 0.99$; we do not have to repeat the (already time-consuming) procedure for each nominal level. Figure 2.11(a) shows the output; we expect the plot of the nominal versus actual values to lie roughly on a $45°$ line, and in this case, they do. Thus, for this type of experiment and sample size, adjustment is not necessary.

```
1   r=25; p=0.31; sim1=2000; sim2=20000; nominalproblevels = 0.80:0.01:0.99;
2   pmat=zeros(sim1,length(nominalproblevels)); KD2=zeros(sim2,1);
3   for i=1:sim1
4     G1=geornd(p,[r 1])+1; i
5     % compute p-hat and KD of G1
6       t=tabulate(G1); x=t(:,1); ecdf=cumsum(t(:,3)/100);
7       phateff=(r-1)/(sum(G1)-1);
8       para=1-(1-phateff).^x; dd=abs(ecdf-para); KD1=max(dd);
9     % Now, with what probability does KD of G1 actually exceed its
10    % 90% quantile, empirically determined using its ESTIMATE
11    % of the model parameter theta?
12    for j=1:sim2
13      G2=geornd(phateff,[r 1])+1;
14      % compute p-hat and KD(j) of G2
15        t=tabulate(G2); x=t(:,1); ecdf=cumsum(t(:,3)/100);
16        phateff2=(r-1)/(sum(G2)-1);
17        para=1-(1-phateff2).^x; dd=abs(ecdf-para); KD2(j)=max(dd);
18    end
19    qvec=quantile(KD2,nominalproblevels); pmat(i,:) = (KD1 <= qvec);
20  end
21  actualprob = mean(pmat) % take averages, and answer the above question
22  plot(nominalproblevels,actualprob,'r-o','linewidth',2)
23  axis([0.79 0.996 0.79 0.996]), set(gca,'Fontsize',16), grid
24  line([0.8 1], [0.8 1],'linewidth',2,'color','g','linestyle','-')
25  title(['for r=',int2str(r),' and p=',num2str(p)])
```

Program Listing 2.11: Implements the nested simulation method to get the actual acceptance probabilities for the geometric pregnancy example.

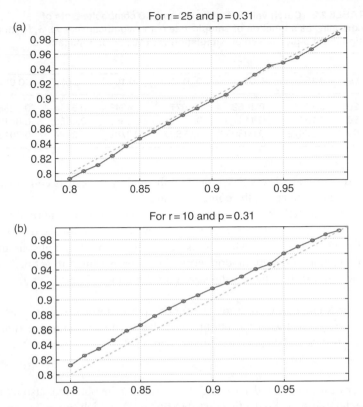

Figure 2.11 The actual acceptance probability A_p (x-axis) versus the nominal probabilities p (solid line), for the KD statistic and the geometric pregnancy example, for $p = 0.31$ and $r = 25$ (a) and $r = 10$ (b). The dashed 45° line indicates the case when nominal and actual are equal.

However, differences eventually arise if we continue to decrease the sample size. Figure 2.11(b) shows the case with $r = 10$, for which we see that, for the lower nominal levels, there is growing discrepancy. If, for example, we want to use a 90% cutoff probability for the KD statistic, then we should use a nominal value of about 0.883 instead of 0.90. That is, we would perform just the original, single-level simulation, based on $\hat{\theta}$ from the actual data, to get the distribution of the KD statistic, and compare the observed KD value to the 88.3% quantile of the simulated KD distribution. If it exceeds this value, we would say that this event should only occur with 10% probability if the postulated model is true, and we can reject the null hypothesis at the 10% significance level.

2.4 TESTING NORMALITY WITH KD AND AD

Here we wish to consider the null hypothesis that the data consist of an i.i.d. sample from *some* normal distribution – the actual parameter values μ and σ^2 are not part of the null hypothesis. In the context of goodness-of-fit testing, when the null hypothesis is a fully specified distribution, for example, normal with mean 1 and variance 2, we say that the null hypothesis is **simple**. When the null consists only of a family of distributions indexed by

TABLE 2.2   Cutoff values for the KD and AD composite tests of normality, as a function of sample size n and significance level α, to four significant digits, based on simulation with 10 million replications

	KD			AD		
$n\backslash \alpha$	0.10	0.05	0.01	0.10	0.05	0.01
20	0.1516	0.1669	0.1977	0.4648	0.5714	0.9288
50	0.1043	0.1144	0.1348	0.3562	0.4568	0.8419
100	0.07673	0.08395	0.09863	0.2789	0.3624	0.7013

an unspecified set of parameters θ, we say that the null hypothesis is **composite**. In most applications, interest centers on the composite case.

Consider using the KD and AD statistics for a composite test of normality, where the empirical c.d.f. is $\widehat{F}_{emp}(Y_i) = (i - 3/8)/(n + 1/4)$ and the fitted normal c.d.f. uses the m.l.e. $\hat{\mu} = \bar{X}$ and $\hat{\sigma}^2 = (n - 1)/n \times S^2$, where S^2 is the usual, unbiased, sample variance. A simulation with 10 million n-length random normal samples was used to get the cutoff values of the KD and AD statistics; these values are given in Table 2.2.

Remark.   The Kolmogorov–Smirnov test for composite normality, using the KS statistic (2.22), with correct cutoff values, is built into Matlab's statistics toolbox as function `lil-lietest`. It appears that there is no implementation of the test based on the AD statistic (or W^2 and U^2; see below).   ∎

Once the critical values are determined, the power can be obtained, also by simulation. For example, let us use the Student's t distribution (A.70), which serves as a representative model for data with tails heavier than the normal. An n-length random sample of Student's t data with v degrees of freedom is simulated, and its KD and AD statistics are computed using a fitted normal c.d.f. (remember: the composite null hypothesis is that the data are *normally* distributed). This is repeated 1 million times, and the fraction of statistics that exceed their respective cutoff values gives the power. This is done for $v = 1, 2, \ldots, 20$, for the three sample sizes $n = 20$, 50, 100, and for $\alpha = 0.05$. The resulting **power curves** are shown in the top left and middle left panels of Figure 2.12. We see that, for both sample sizes, the AD test has considerably higher power against the Student's t than the KD. This should not be surprising, recalling that, by construction, the AD statistic emphasizes deviations in the tails of the distribution.

The Student's t distribution is fat-tailed, but symmetric. To inspect the power of the KD and AD tests against distributions that are asymmetric but not fat-tailed, a similar exercise was done using the skew normal (A.115) as the alternative. Realizations from this distribution can be generated from the short program in Listing 2.12, based on the method given in Azzalini (1985, p. 172).

```
1  function Z=skewnormrnd(n,lambda)
2  X1=randn(n,1); X2=randn(n,1); b=(X2>lambda*X1); X1(b)=-X1(b); Z=X1;
```

Program Listing 2.12: Generates an $n \times 1$ vector of location-zero, scale-one Azzalini skew normal realizations with asymmetry parameter `lambda`.

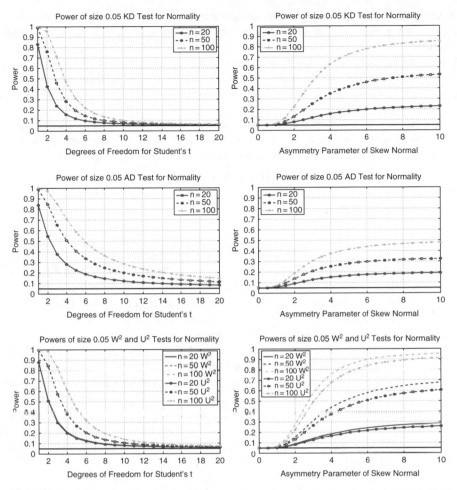

Figure 2.12 *(Top) Power of the KD test for normality, using significance level* $\alpha = 0.05$, *for three different sample sizes, and the Student's* t *alternative (left) and skew normal alternative (right), based on 1 million replications. (Middle) Same, but for the AD test. (Bottom) Same, but power of the* W^2 *(lines without circles) and* U^2 *(lines with circles) test for normality. The* W^2 *and* U^2 *power curves for the Student's* t *alternative are graphically indistinguishable.*

The top right and middle right panels of Figure 2.12 show the power curves as a function of the asymmetry parameter λ (for just positive values; the power curve is symmetric about zero). We see that, for this alternative, the KD test has considerably higher power than the AD. Thus, for testing the null hypothesis of composite normality, neither KD nor AD is more powerful for both fat-tailed and skewed alternatives. This happens often enough in applications to emphasize the point:

> In many statistical testing situations involving either a simple or composite null, and a composite alternative, there will not exist a test that is **uniformly most powerful** (UMP), meaning most powerful for all distributions within the set of alternative hypotheses.

Section 2.5 below will present two tests that both perform better (have higher power) than KD against the skew normal alternative, but do not perform as well as the AD test against the Student's t alternative. Chapter 6 will introduce further tests for composite normality. We will see that there is no single test that is most powerful against all alternatives, and that a test that performs best against a particular alternative can perform relatively poorly against a different alternative.

> A test of size α is **unbiased** (for sample sizes larger than N) if, for all values of parameter θ that index the distribution under the alternative hypothesis (and for all sample sizes $n > N$), the power is never less than α.
>
> A test of size α is **consistent** if, for all values of parameter θ that index the distribution under the alternative hypothesis, the power tends to 1 as the sample size tends to infinity.

Analogous to the idea that an unbiased point estimator may not have the smallest mean squared error, a test that is uniformly most powerful and unbiased (UMPU) need not have the highest power for all θ that index the alternative hypothesis. That is, an UMPU test need not be UMP. (For an example of this with a biased test that has higher power for a large and practical subset of the range of θ, and further discussion, see Suissa and Shuster, 1984.)

It appears from Figure 2.12 that the KD and AD tests for normality are unbiased for the two alternatives we considered, though we certainly cannot conclude that it is unbiased (for all alternatives). We will see in Chapter 6 that the AD test and the so-called Jarque–Bera test (Section 6.5.2) are both biased for certain alternatives.

Observe in Figure 2.12 how, for both tests, and all degrees of freedom v, the power of the tests increases when going from $n = 20$ to $n = 100$. Thus, we might postulate that the tests are consistent, though this does not constitute a proof. Recalling that, as v increases, the $t(v)$ distribution approaches the normal, it will be the case that, as v increases, so will the value, say $N(v, \epsilon)$, such that the power is greater than $1 - \epsilon$ for sample sizes larger than N, where $\epsilon > 0$ is fixed.

2.5 TESTING NORMALITY WITH W^2 AND U^2

There are several other goodness-of-fit measures related to the KD statistic; see, for example, Rahman and Chakrobartty (2004) for a comparison. We state two without derivation. These include the **Cramér–von Mises statistic**, commonly denoted W^2 and given by

$$W^2 = \sum_{i=1}^{n} \left(z_i - \frac{2i-1}{2n} \right)^2 + \frac{1}{12n}, \tag{2.30}$$

and **Watson's statistic**, given (the letter W having unfortunately already been taken) by

$$U^2 = W^2 - n\left(\bar{z} - \frac{1}{2} \right)^2, \tag{2.31}$$

where W^2 is as in (2.30), $z_i = F(y_i; \hat{\theta})$ refers to the parametrically fitted c.d.f., $y_1, \ldots, y_n$ are the order statistics, and $\bar{z}$ is the mean of the z_i. The measure (2.30) has its origin in

```
1   function [W2 U2]=W2U2(X)
2   n=length(X); y=reshape(X,n,1); i=(1:n)';
3   if ~issorted(y), y=sort(y); end % save time!
4   muhat=mean(y); sighat=std(y); para=normcdf(y,muhat,sighat);
5   W2=sum( (para − (2*i−1)/(2*n)).^2) + 1/(12*n); zbar=mean(para);
6   U2 = W2 − n*(zbar−0.5)^2;
```

Program Listing 2.13: The W^2 and U^2 goodness-of-fit measures for testing normality.

work by Cramér from 1928. For the original references of Kolmogorov, Smirnov, Cramér, and von Mises, as well as an exposition of goodness-of-fit tests giving (half a century ago) "fairly complete coverage of the history, development, present status, and outstanding current problems related to these topics," see Darling (1957). Discussion of Watson's statistic, and the original references, can be found in Durbin (1973). Listing 2.13 gives the Matlab implementation of both of these statistics. As the data set is standardized by the sample location and scale terms, the distribution of the test statistic for i.i.d. normal data, for a given sample size, is location- and scale-invariant.

By using cutoff values obtained via simulation, tests based on W^2 and U^2 with correct size can be conducted. Then, in the usual way, the power against particular alternatives can be computed. The bottom panels of Figure 2.12 shows the results. For the Student's t alternative, the powers of both tests are (to about 4 digits) identical, suggesting that the true, theoretical power curves are identical. For the skew normal alternative, W^2 is uniformly (over all sample sizes and λ-values) superior to U^2 (in disagreement with Durbin's expectation; see the remarks below).

Remarks

(a) With respect to the power of Watson's U^2 statistic compared with that of W^2, Durbin (1973, p. 36) wrote that "Although Watson introduced this statistic specifically for tests on the circle it can … be used for tests on the line and indeed can be expected to be more powerful than W^2."

(b) The W^2 and U^2 tests will also be used in Problem 6.3. in the context of the Laplace distribution; in that case, Durbin's expectation holds true.

(c) The asymptotic distribution of the U^2 statistic when used for testing in a multinomial distribution setting has been obtained by Freedman (1981). It is a weighted sum of independent χ^2 r.v.s, where the weights are eigenvalues of an easily computed matrix given in Freedman (1981). The exact p.d.f. and c.d.f. calculations of such a weighted sum, as well as saddlepoint approximations, are detailed in Section II.10.1.4. ∎

2.6 TESTING THE STABLE PARETIAN DISTRIBUTIONAL ASSUMPTION: FIRST ATTEMPT

Recall the discussion on the stable Paretian distribution in Section A.16. We wish to use the tools so far developed to construct tests to assess whether an i.i.d. data set comes from a composite stable Paretian distribution. This is valuable because the stable Paretian is a fundamental distribution in probability theory that contains the normal as a special case.

As many types of data exhibit heavy tails and asymmetry, it becomes a natural candidate for modeling such data.

The KD and AD tests from Section 2.3, as well as the Cramér–von Mises statistic (2.30) and Watson's statistic (2.31), can be used for testing the stable distribution. All require the values of the stable Paretian c.d.f. $z_i = F_S(y_i; \widehat{\theta})$ at the observed order statistics $y_1 < y_2 < \cdots < y_n$, and, with a composite test, the estimated parameter vector $\widehat{\theta}$ needs to be used. Recall that, when we used these statistics for testing composite normality, the appropriate cutoff values had to take into account that its two parameters are estimated; although, as these two parameters are just location and scale, the distribution of the test statistic (and, thus, the appropriate cutoffs) are invariant to their values. This is because we can easily transform the data to the standard (location-zero, scale-one) form of the location–scale normal distribution family, which is unique, that is, has no parameters.

In light of the previous statement, the following idea suggests itself for dealing with *all* of the unknown parameters associated with the stable Paretian distribution. Based on their estimates, we can apply the probability integral transform and compute $z_i = F_S(y_i; \widehat{\theta})$, $i = 1, \dots, n$, which should resemble a set of i.i.d. Unif$(0, 1)$ r.v.s if $Y_1, \dots, Y_n$ are i.i.d. stable, and $\widehat{\theta}$ is adequately close to θ. We can then compute the inverse c.d.f. of the normal distribution at the z_i to get an i.i.d. normal sample, and apply any of our composite normality tests. This procedure will certainly be flawed for at least one reason: The composite normality tests were calibrated under the assumption that the two parameters of the normal distribution were estimated; but the stable distribution has four unknown parameters. The predictable consequence of this is that the actual size will be lower than the nominal (say, 0.05) because four parameters were optimized to fit the data, but the test only accounts for two parameters having been fitted, and so the model "fits better than it should" with respect to the null distribution of the test statistic, resulting in fewer tail events than should occur. This important point will be elucidated in more detail in Section 6.2.2.

Simulations were conducted to investigate this procedure. The actual size of the tests was indeed, on average, well below 0.05. Thus, it seems we could just apply our usual trick to solve this problem – mapping the relation between the nominal and actual test sizes, as developed in Section 1.2. Unfortunately, and detrimentally, the actual size turns out to be quite dependent on the true tail index α.[8] It appears that this method is not promising, and we turn to another approach.

We require a strategy that correctly, or at least adequately, accounts for the parameter uncertainty of the stable tail index α (and, if we entertain the asymmetric case, also parameter β). To simplify matters at first, we begin by restricting ourselves to the case with $\beta = 0$, although in what follows, β will still be jointly estimated as an unknown parameter. We can then more easily assess the size and power properties just as a function of tail index α.

To get the critical values for a particular tail index α, say 1.5, and sample size n, we draw an n-length random sample of $S_{1.5,0}(0, 1)$ values, fit the four distributional parameters (using maximum likelihood) to get $\widehat{\theta}$, and compute the c.d.f. values z_i. (One could genuinely assume symmetry, and restrict β to zero, but we do not pursue this.) Then the four test statistics, KD, AD, W^2, and U^2, are computed. This is repeated a large number of times

---

[8] For example, as the tail index α moves from 1.1 to 1.9, the actual sizes of the KD, W^2, and U^2 tests decrease monotonically from about 0.1 to 0.01, while those of the AD and (as introduced in Section 6.4.3) MSP tests took on bathtub-like shapes, being relatively flat (at 0.045 for AD, 0.025 for MSP) for $1.2 < \alpha < 1.7$, and then increasing well past 0.05 as α decreases towards 1 or increases towards 2.

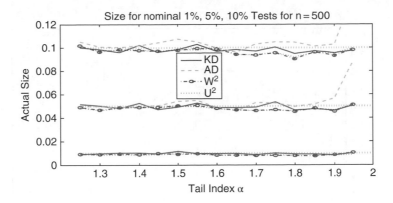

Figure 2.13 *Actual size of the four tests, for nominal size 0.05, based on 10,000 replications.*

(we used 100,000) and the 0.01, 0.05 and 0.10 quantiles of the test statistics are computed. This procedure is conducted for a grid of α-values, say $1.00, 1.01, 1.02, \ldots, 1.99, 2.00$, yielding, for each of the three cutoff probabilities, a vector of cutoff values as a function of α. (Note that it would be much faster to just use the true parameter vector θ instead of $\hat{\theta}$, but this would not take the estimation uncertainty into account and would not be useful for a composite test.)

Then, to calculate the test statistic for a particular data set of length n, estimate the four stable parameters and, based on $\hat{\alpha}$, use linear interpolation on the aforementioned vectors (as implemented in, say, Matlab's function `interp1`) to get the appropriate 0.01, 0.05 and 0.10 cutoff values. Finally, compare these to the observed test statistics to determine the test. Repeating this many times with symmetric stable draws provides a check on the actual size. While it should be close to the nominal size, it will not be exact because (besides having used a finite number of replications in all the simulations) α is estimated. In particular, even under the (correct) assumption that the m.l.e. of α is asymptotically unbiased, the fact that the cutoff value is determined from the estimate of α does not imply that the cutoff value is correct on average, even though $\hat{\alpha}$ itself might be correct on average. The discrepancy between the nominal and actual sizes will depend on the shape of the distribution of the test statistic near the cutoff values, for a given α, and how this changes with respect to α.

We attempt this for a sample size of $n = 500$. Cutoff values were computed as discussed above, and then the size check was performed, with 10,000 replications, based on the grid of α-values $1.25, 1.3, \ldots, 1.95$. Figure 2.13 shows the results. We immediately see that the AD test has very poor size properties as the tail index α increases towards 2, essentially disqualifying it from further study. In contrast, the actual size of the other tests is very accurate. Thus, the KD, W^2, and U^2 tests have approximately correct size under a large and relevant portion of the (but not the entire) parameter space of the tail index α. Further simulations show that, as α decreases towards 1, the actual size increases.

We wish to assess the power of this test when the true data are generated from a Student's $t(v)$ distribution, this also having heavy tails. For a particular v, we simulate numerous $t(v)$ data sets of length $n = 500$, fit the stable Paretian model to each of them, and protocol the resulting values of $\hat{\alpha}$. These need to fall in the range for which the size of the test is correct, as indicated in Figure 2.13. The results of this are shown in Figure 2.14(a), from which we see that the range $1.5 < v < 6$ is satisfactory. Obviously, this will not be so useful in

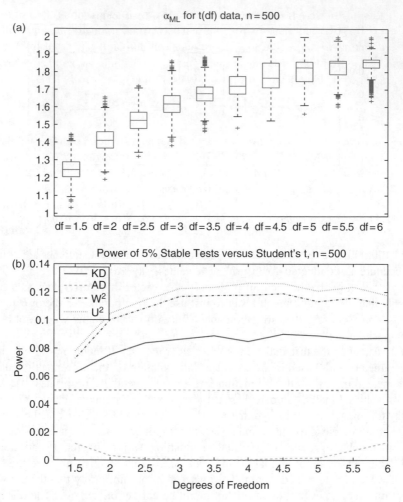

Figure 2.14 (a) Boxplots of $\hat{\alpha}_{ML}$ resulting when estimating all four parameters of the stable model, but with the data generated as Student's t with various degrees of freedom. (b) Power of the proposed set of tests against a Student's t alternative, for various degrees of freedom, and based on 10,000 replications.

practice, when we do not know the true distribution of the data. However, if we indeed find that one or more of the four tests have substantial power in our controlled experiment here, then we can justify expending energy in trying to improve the tests to adequately address the problem with the actual test size.

Figure 2.14(b) shows the power of the four tests against the Student's t alternative as a function of the degrees-of-freedom parameter v. The power of the AD test should be ignored anyway, based on its faulty size; W^2 and U^2 are very close in power, with the latter slightly and consistently larger, while the powrt of KD is, in comparison, not competitive. As the Student's t and symmetric stable coincide at the Cauchy case of $v = \alpha = 1$, the power should approach the size of the test (here, 0.05), as $v \to 1$. This indeed appears to be the case. The stable and Student's t also coincide as $v \to \infty$, so that as v increases, the power has to decrease down to the size. This will not be apparent from the power plot because the largest value of v used was 6, in order to maintain correct size.

Finally, and most importantly, we see that, for a test with significance level 0.05 and $n = 500$, we can expect a power of only around 0.12 against a Student's t alternative. We can temporarily console ourselves with the disappointment of this humbling outcome by recalling the fact that the symmetric stable and Student's t distributions are, except far into the tails, quite similar, and so should indeed be hard to distinguish in small samples. We will return to this problem in Section 9.5, armed with more sophisticated machinery needed to deal with this interesting and practical testing issue.

For a Laplace alternative with $n = 500$, simulations reveal (results not shown) that the estimated tail index lies between 1.4 and 1.85, so our tests are applicable. In this case, the power of the U^2 test was the highest, 0.89. While this is quite impressive, keep in mind that the Laplace has exponential tails, and thus differs remarkably from the stable Paretian, with its thick power tails. The reader is invited to confirm these results.

2.7 TWO-SAMPLE KOLMOGOROV TEST

Above, we used the Kolmogorov–Smirnov test to assess if an observed sample of i.i.d. data is consistent with a specific parametric distribution. We now consider the case in which interest centers on comparing if two independent samples (with possibly different sample sizes) are from the same underlying continuous distribution. In this case, the particular parametric form is not specified, and the null hypothesis is

H_0 : The two samples come from the same continuous distribution.

Let $\widehat{F}_A$ denote the e.c.d.f. of the first sample, of length n, and $\widehat{F}_B$ denote the e.c.d.f. of the second sample, of length m, and consider the least upper bound of the absolute distance between $\widehat{F}_A(x)$ and $\widehat{F}_B(x)$ over $x \in \mathbb{R}$,

$$D_{n,m} = \sup_{x \in \mathbb{R}} |\widehat{F}_A(x) - \widehat{F}_B(x)|. \tag{2.32}$$

Under the null hypothesis, use of the Glivenko–Cantelli theorem and the triangle inequality shows that, as $n \to \infty$ and $m \to \infty$, $D_{n,m}$ converges almost surely and uniformly to zero. Similarly to the one-sample case (2.5), there would ideally be a function of n and m such that, when $D_{n,m}$ is multiplied by this, the resulting product converges to a nondegenerate distribution. Indeed, Kolmogorov has shown that

$$K_{n,m} = \sqrt{\frac{nm}{n + m}} D_{n,m} \tag{2.33}$$

converges as $n, m \to \infty$ to the Kolmogorov distribution (2.7). Observe that, if $n = m$, then the sample size factor in (2.33) reduces to $\sqrt{n/2}$, which is precisely what one would expect it to have to be in light of (2.5) and the fact that we have not n, but $2n$ observations.

For the one-sample KS test, computation of the test statistic (2.22) is very simple, as seen from the Matlab code given just below it. For the two-sample case, notice that the e.c.d.f. of each data set must be computed at the union of points formed from both data sets. Program Listing 2.14 shows two ways of computing (2.33).

```
1   function KS=kstest2samp(x1,x2)
2   n1=length(x1); n2=length(x2);
3   if 1==2 % use definition of ecdf and a grid
4     x=sort([x1(:); x2(:)]); lo=x(1)-0.1; hi=x(end)+0.1;
5     inc=0.001; % better, but takes far longer: inc = min(diff(x));
6     tvec=lo:inc:hi; tlen=length(tvec); F1=zeros(tlen,1); F2=F1;
7     for g=1:tlen,  t=tvec(g); F1(g)=sum(x1<=t); F2(g)=sum(x2<=t); end
8     F1=F1/n1; F2=F2/n2; KS = sqrt(n1*n2/(n1+n2)) * max(abs(F1-F2));
9   else % code from Matlab's kstest2.m function
10    binEdges   =  [-inf; sort([x1;x2]); inf];
11    binCounts1 =  histc (x1, binEdges, 1); binCounts2 =  histc (x2, binEdges, 1);
12    sumCounts1 =  cumsum(binCounts1)./sum(binCounts1);
13    sumCounts2 =  cumsum(binCounts2)./sum(binCounts2);
14    sampleCDF1 =  sumCounts1(1:end-1); sampleCDF2 =  sumCounts2(1:end-1);
15    deltaCDF   =  abs(sampleCDF1 - sampleCDF2);
16    KS  =  sqrt(n1*n2/(n1+n2)) * max(deltaCDF);
17  end
```

Program Listing 2.14: Two ways of computing (2.33). The first is based on the definition of the e.c.d.f. and uses a tight grid of values over the range of the data. This can fail if the number of grid points is too small, unless one takes the grid increment to be the data-driven value as given in the commented-out code in line 5, but this tends to be far smaller than necessary, and computation time increases dramatically. The second method, taken from Matlab's function `kstest2`, is more elegant, far faster, and, most importantly, is always correct.

To illustrate, Figure 2.15 is similar to Figure 2.1, but using the two-sample test (2.33). We again see that the finite-sample distribution converges to the Kolmogorov distribution, and use of normal or Cauchy data makes no difference to the small-sample distribution of the test statistic.

2.8 MORE ON (MORON?) HYPOTHESIS TESTING

Hypothesis testing should immediately and forever be tossed onto the scrap heap of intellectual history and certainly never taught to the vulnerable.

(William Briggs, 2016, p. xiii)

Everyone is entitled to his own opinion, but not to his own facts.

(Daniel Patrick "Pat" Moynihan)

Significance and hypothesis testing form a cornerstone of statistical inference, and are almost always taught in an introductory statistics course, expanded upon in more advanced courses, and used extensively in applied research. A growing consensus is emerging that this indoctrination may not be wise, for reasons we now discuss. After a brief introduction to the issue, and mention of some preferred alternatives in Section 2.8.1, we discuss some problems associated with hypothesis testing and p-values in Sections 2.8.2 and 2.8.3, respectively. The latter section also elaborates upon Fisher's original intended use of p-values, with the goal of providing a partial, and careful, defense of their use, when appropriate.

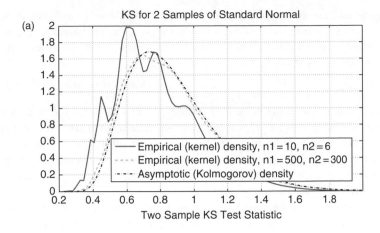

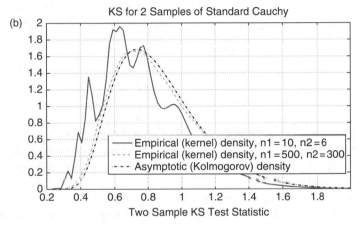

Figure 2.15 *The true distribution, obtained via simulation with 10,000 replications, of the Kolmogorov–Smirnov two-sample goodness-of-fit test statistic, and its asymptotic distribution (2.7) for the normal (a) and Cauchy (b) distributions.*

2.8.1 Explanation

One area where null hypotheses have quite high prior probabilities is model checking, including both goodness of fit and diagnostic checking. Here specific alternative hypotheses may not be well-formulated, and significance test *p*-values provide one convenient way to put useful measures on a standard scale. …

In general, for problems where the usual null hypothesis defines a specific value for a parameter, surely it would be more informative to give a confidence range for that parameter.

(David V. Hinkley, 1987, p. 128)

In a large majority of problems … hypothesis testing is inappropriate: Set up the confidence interval and be done with it!

(George Casella and Roger L. Berger, 1987, p. 134)

> [M]y own classroom presentations have largely abandoned [Neyman–Pearson]
> ideas when teaching regression, analysis of variance, statistical methods, or
> almost any applied course. I now teach Fisherian testing and confidence inter-
> vals based on Fisherian tests.
>
> (Ronald Christensen, 2005, p. 126)

Section 2.3.2 introduced the basic terminology associated with hypothesis testing. One differentiates between the use of *p*-values for evidence against the null, and the use of the Neyman–Pearson (NP) framework with its explicit alternative hypothesis. The former is referred to as "significance testing," and the latter as "hypothesis testing," terms established (to help differentiate between the two competing – and, formally, incompatible – paradigms) by Cox (1977).

During the development of statistical inference in the twentieth century, models tended to be rather small, both in terms of sophistication and number of parameters, and computation was prohibitive, so that the ideas of testing parameter significance made sense: they were feasible. Nowadays, with modern computing power, far more elaborate models can be estimated, and the actual issue of interest, say forecasting ability, can be directly assessed. For example, in a model with 1000 parameters (as is well possible in a large vector autoregression), it makes no sense to look at individual *t* statistics. Rather, modern methods combining shrinkage and selection are used. Such methods include the so-called *elastic net*, which is related to the least absolute shrinkage and selection operator, or lasso; see Bühlmann and van de Geer (2011), Hastie et al. (2015), and the references therein. Cross-validation for assessing in-sample predictive ability is also a very useful and related technique. Such ideas (and that of the bootstrap) most likely occurred to earlier statistical scientists such as Fisher, but they were computationally infeasible, and remained nothing but imaginative, unrealizable ideas.

The previous quotes, by such prominent names in statistics no less, serve to support the author's disinclination to discuss the NP hypothesis testing framework for testing a null value of a parameter of a model, say $\theta = 0$, versus an alternative, such as $\theta \neq 0$. Just the fact that, in a continuous setting (such as the popular test for the difference of the means of two Gaussian populations) the null hypothesis has measure zero should be cause for alarm. Detailed discussions supporting this position can be found in Johnson (1999), Hubbard and Bayarri (2003), Christensen (2005), Nguyen (2016), and the numerous references therein, as well as the forceful book-length treatments of Ziliak and McCloskey (2008), Reinhart (2015), and Briggs (2016).

Instead of statistical significance testing of parameters, we advocate reporting of confidence intervals for model parameters, effect sizes when appropriate (see Section 2.8.3 below), and, for determination of a suitable model or models, use of likelihood ratios (see Section 3.3.2) and, particularly, the more modern techniques mentioned above and use of information criteria (see Section 3.3.3). Every one of these "solutions" is potentially also problematic. As simple examples, a confidence interval can easily be interpreted to yield a dichotomous decision, while effect sizes are unobservable and can exaggerate the evidence for or against a particular claim. A paradigm for inference devoid of *p*-values and the NP hypothesis testing framework is developed in Briggs (2016, Sec. 9.8).

Alternatively (or, as some scientists do, in conjunction), the full Bayesian approach is adopted, with the use of posterior probability distributions, credible regions and intervals, Bayes factors, generalized Bayes rules, and (Bayesian) model averaging. See, for example, in order of mathematical complexity, Bolstad and Curran (2017), Gelman et al. (2013), and Robert (2007).

In contrast, and as alluded to by Hinkley above, the Fisherian significance testing framework (use of p-values) can have value when confidence intervals are not applicable, such as assessing distributional assumptions, as was done in this chapter, and in other aspects of data analysis, such as testing for independence of observations or residuals, or assessing the differences between two or more models in terms of out-of-sample density forecasting. Nevertheless, such use should still be accompanied by other measures linked to the purpose of the analysis. The use of p-values along the lines of how they were used by Fisher for accrual of scientific knowledge and planning of future studies can still have value, though some caveats are in order; see Section 2.8.3 below.

The formalities of, and common examples associated with, the NP hypothesis testing framework can be found in numerous texts; well-written classic introductory master's level presentations can be found in Mood et al. (1974), Hogg et al. (2014), and Rohatgi and Saleh (2015). Young and Smith (2005) and Davison (2003) additionally discuss use of higher-order asymptotics and the bootstrap. More mathematically advanced accounts can be found in Ferguson (1967), Schervish (1995), Shao (2003) and (perhaps ironically) Casella and Berger (2002). It is noteworthy that hypothesis tests can often be used to construct confidence sets for parameters; see the previous references, as well as Bhattacharya et al. (2016, Sec. 5.7). In most situations, however, application of the bootstrap will be easier and result in intervals or sets with acceptable nominal coverage. Via use of the nonparametric bootstrap, the nominal coverage might well be better than those of analytically derived c.i.s under a potentially incorrect parametric assumption.

2.8.2 Misuse of Hypothesis Testing

The Neyman–Pearson theories of testing hypotheses and of confidence interval estimation are sound theories of probable inference. … My *Logic of Statistical Inference* took vigorous issue with Neyman. This essay is a retraction.

(Ian Hacking, 1980, p. 141)

There is increasing concern that most current published research findings are false. … Moreover, for many current scientific fields, claimed research findings may often be simply accurate measures of the prevailing bias.

(John P. A. Ioannidis, 2005)

[W]e believe that the $p < 0.05$ bar is too easy to pass and sometimes serves as an excuse for lower quality research. We hope and anticipate that banning the NHSTP [null hypothesis significance testing procedure] will have the effect of increasing the quality of submitted manuscripts by liberating authors from the stultified structure of NHSTP thinking thereby eliminating an important obstacle to creative thinking.

(David Trafimow and Michael Marks, 2015, pp. 1–2)

There is not a single experiment reported in my 23-volume set of the standard edition of Freud nor is there a t test. But I would take Freud's clinical observations over most people's t tests any time.

I suggest to you that Sir Ronald has befuddled us, mesmerized us, and led us down the primrose path. I believe that the almost universal reliance on merely refuting the null hypothesis as the standard method for corroborating substantive theories in the soft sciences is a terrible mistake, is basically unsound, poor scientific strategy, and one of the worst things that ever happened in the history of psychology.

(Paul E. Meehl, 1978, p. 817)

One might think that formal hypothesis testing is, in a sense, the pinnacle of statistical theory for use in the progression of science. According to the famous philosopher of science Karl Popper,[9] a scientific hypothesis is valid only if it is falsifiable, that is, only if observations or experiments could demonstrate that it is false in the event that it is indeed false. If a hypothesis is falsifiable, then it is discarded when new evidence refutes it. A newer theory is formulated, and it in turn is subjected to testing, ad infinitum. (Contrast this with the quote from Fisher that opens this chapter: "it should be noted that the null hypothesis is never proved or established, but is possibly disproved, in the course of experimentation.")

This pivotal understanding of the progression of science certainly applies in full generality to all scientific inquiries, from grandiose theories in physics, to more mundane topics, though its use in complicated issues in the social, medical, and economic sciences has been argued by numerous scientists to be unnecessary, inappropriate, inapplicable, even misguided and harmful. The disturbing findings of Ioannidis (2005) mentioned in the above quote, and which were featured in The Economist (2013), corroborate this claim. See also Reinhart (2015) in this regard.

As concrete examples, imagine a hypothesis test applied to determine if the effectiveness of an old statin drug is the same as that of a new one; or if median incomes (or job satisfaction, or quality of life, etc.) of people from similar age cohorts and geographic regions differ based on having attended quality preschool (or amount of parental attention in the first 1000 days, or quality of secondary education, etc.). In both cases, of course they "differ," and what is important is how much, and the implications of the results.

The complexity of the aforementioned examples should preclude use of such one-dimensional considerations. In the former medical example, what about side effects, long-term effects, half-lives of the drugs, interactions with other substances, the effect size of their use in conjunction with dietary changes, and the costs and availability of the old and new drugs? For the second example, issues in the social sciences are yet more complex, and use of something so trite and banal as a hypothesis test for obvious questions that skirt the issues of genuine concern, no less in the guise of "objective, good science," should be frowned upon. This issue is alluded to in McShane and Gal (2016, p. 1715), where further discussion of the problems inherent in the null hypothesis significance testing paradigm can be found.

To add to the already misplaced analysis in such situations, hypothesis testing in the majority of "scientific inquiries" uses a parametric structure (such as normality), so that tests for the differences in treatment means, or means obtained by observational studies, are valid (formally at least) only under the assumption that the distributional assumptions are correct. (The same critique applies to confidence intervals formed under explicit parametric assumptions, but note that the nonparametric bootstrap can be deployed to help account for this.)

---

[9] A more accurate understanding of the relationship between this somewhat superficially stated Popperian criterion and its implementation in science can be found in Mulkay and Gilbert (1981), as a starting point.

While the use of hypothesis testing and indiscriminate reporting of p-values as evidence are deeply entrenched in science, there is a long list of critiques and naysayers since the 1960s, and possibly some signs of change in the twenty-first century; see Wasserstein and Lazar (2016) and the references therein. This is notably the case in psychology, with its large variety of compounding issues rendering application of "traditional" statistical inference difficult. In the words of the acclaimed professor of psychology Paul Everett Meehl (1920–2003): "In 10 minutes of superficial thought I easily came up with 20 features that make human psychology hard to scientize" (Meehl, 1978, p. 807). In fact, at least one journal in psychology has gone so far as to ban the reporting of p-values, as seen in the above quote; see also Krantz (1999) in this regard.

The first quote given above, by the acclaimed philosopher of science Ian Hacking, serves to provide some balance, and also to emphasize the difficult nature of statistical inference and the ongoing debate about how best to conduct it. He refers to his earlier work (Hacking, 1965) in which he previously supported what is referred to as the *law of likelihood*; see Section 3.3.2. One might be curious if, given the developments since then, he will retract his retraction.

2.8.3 Use and Misuse of p-Values

Far better an approximate answer to the *right* question, which is often vague, than an *exact* answer to the wrong question, which can always be made precise.

(John Tukey, 1962, p. 13)

The frequency theory of probability and statistics is the most misleading and irrelevant idea that has ever clouded our subject and ought to be forgotten.

(Dennis V. Lindley, 1968, p. 321)

[Rejecting the null hypothesis] simply indicates that the research design had adequate power to detect a true state of affairs, which may or may not be a large effect or even a useful effect.

(Roger E. Kirk, 1996, p. 747)

Objections are often raised to the Bayesian approach because of its dependence on the prior. It is not so often recognized that the p-value can equally be criticized because of its dependence on the sample space. … My personal view is that p-values should be relegated to the scrap heap and not considered by those who wish to think and act coherently.

(Dennis V. Lindley, 1999, p. 75)

Die, p-value, die die die.

(William Briggs, 2016, p. 178)

An apparent benefit of the p-value is that, no matter what the test statistic and its distribution under the null, it delivers a number between zero and one, with small values having the interpretation of evidence against the null, so that it appears to be an objective measure and universally applicable. This is, unfortunately, not the case, for which there are several powerful reasons.

Of the many critiques against the use of p-values or the discrepant NP paradigm, the "easy" ones include (i) the arbitrary selection of the significance level α, and (ii) the notion that the null hypothesis in many cases is surely false and somewhat of a straw man serving as an excuse for poor research; see the simple examples in Section 2.8.2 above, as well as Goodman and Royall (1988), Graybill and Iyer (1994), Anderson et al. (2000), and the references therein. A third critique is the problem associated with use of p-values for model selection amid a very large number of parameters. This issue involves controlling the so-called *family-wise error* or *false discovery rate*, and methods exist for this; see Bühlmann and van de Geer (2011, Ch. 11) and Efron (2013), though such methods are not germane to basic statistical science courses.

Now consider a "feature" of p-values that is not so readily apparent, involving the sample space issue mentioned by Lindley (1999) in the above quote. We use the well-known example based on Bernoulli trials; see, for example, Lindley (1993). Imagine an experiment consisting of a sequence of i.i.d. Bernoulli trials X_i with success probability p. You are told the experiment resulted in the sequence 111110, where 0 (1) denotes "failure" ("success"), and the null hypothesis is $p = 1/2$. If one assumes that $n = 6$ was fixed in advance, then the model is binomial, with sample space $S = \{0, 1, \ldots, 6\}$. The p-value associated with $S = \sum_{i=1}^{6} X_i \sim \text{Bin}(n, p)$ for $n = 6, p = 1/2$, and $s = 5$ is

$$\Pr(S \geq s \mid H_0) = \Pr(S = 5 \mid H_0) + \Pr(S = 6 \mid H_0) = \binom{6}{5}(1/2)^6 + \binom{6}{6}(1/2)^6$$

$$= 7/2^6 = 0.11$$

(or, in Matlab, `1-binocdf(4,6,0.5)`). If instead one assumes trials are repeated until failure, then the model is geometric, with sample space (for total number of successes) $S = \{0, 1, \ldots\}$, and, under the same null hypothesis of $p = 1/2$, the p-value associated with $T \sim \text{Geo}(p)$ is, with $f_T(t; p) = p(1 - p)^t \mathbb{I}_S(t)$,

$$\Pr(T \geq 5 \mid H_0) = 1 - \Pr(T \leq 4 \mid H_0) = (1 - 1/2)^5 = 0.031.$$

It is alarming that the exact same data, in the same order, give rise to two different p-values, and, in this devised example, tests at significance levels $\alpha = 0.10$ and $\alpha = 0.05$ would lead to different conclusions.

This canonical example illustrates the divergence between likelihood-based and frequentist-based inference. Inference based on the former embraces the so-called **likelihood principle**, which states that all the evidence in a sample related to model parameters is contained in the likelihood; an implication of this is the **law of likelihood**, as briefly discussed in Remark (d) in Section 3.3.2.

Another point of contention with the use of p-values is that they say nothing about **effect sizes**. Imagine a study comparing a traditional and a new treatment for diabetes that reports a p-value of 0.03 under the null that their efficacy is the same. Ready to switch?

If the test statistic compares the means within the two treatment groups of the survival times of a similarly aged cohort of people (a crude measure missing many important factors of interest, but still of some value), then knowing the effect size is not only more important than knowing the p-value, it is all that counts. If the null group lived an average of 23 additional years, while the alternative group lived an average of 24 additional years, is this practical difference relevant, particularly in light of all the other missing factors of interest (such as that the alternative involves a meticulous dietary plan that dieticians and doctors know most people will never accurately follow)? If one argues that $p = 0.03$ is not small enough, pick 0.001 – the same argument holds. Of course the null hypothesis is not true, and a large enough sample size will reveal this. It does not change the necessity of viewing the effect sizes and confidence intervals and incorporating medical and other relevant knowledge for a decision to be made.

As another example, imagine comparing a traditional method of assessing the 10-day value at risk of a major financial institution to a newer method. Again, say we have a p-value of 0.03. Should the entire risk management division revamp its entire infrastructure and adopt the new method? The effect size is what is needed to make an intelligent decision, factoring in the various aspects associated with adopting the new method, as well as the cost of doing so (and the cost of not doing so). The p-value offers nothing in the way of decision-making. Strong arguments for reporting effect sizes (instead of p-values), and numerous examples of studies from different disciplines using effect sizes, can be found in, for example, Kirk (1996), Coe (2002), Gelman and Stern (2006), Sullivan and Feinn (2012), and Barry et al. (2016). Robinson and Wainer (2002) also advocate reporting effect sizes, but temper a potentially dogmatic insistence on their use with examples for which they may not be necessary or applicable.

One might counter that a p-value may not provide a basis for action, but it does indicate that the data are not consistent with the null. Wrong again. Arguably the strongest critique against the use of p-values is that *they exaggerate the evidence against the null hypothesis*. This was most forcefully demonstrated by Berger and Sellke (1987), who showed, among other things, that in certain common cases, a p-value of 0.05 provides essentially no evidence against the null, and 0.01 very little. These findings hinge on the aforementioned fact that the p-value is not what is actually desired, this being namely $\Pr(H_0 \mid \mathbf{X} = \mathbf{x})$. See also Hubbard and Lindsay (2008) and the references therein for discussion of this point, as well as several other issues against the use of p-values as measures of evidence.

We now turn to how p-values can be used – and a partial defense of Fisher. As emphasized in Tukey (1969) and Robinson and Wainer (2002), Fisher advocated **replication**. His idea of strong evidence in favor of rejecting the null came not from a single experiment (irrespective of whether $p < 0.05$ or even $p < 0.001$), but rather from obtaining low p-values (he advocated below 0.05) on *repeated trials of an experiment*, ideally from independent researchers, each challenging the null hypothesis with a similar treatment. He recognized that scientific investigations of genuine interest typically start with a small-scale study whose limited sample size will not give rise to high power, but rather *suggest avenues for future research*.

Another way of seeing the importance of repeated trials is to realize that one set of data delivers one p-value. The p-value itself is a statistic, whose null distribution is Unif(0, 1). So, essentially, you have one data point and wish to judge if its distribution is uniform or not. As values of p are equally likely to be anywhere on $(0, 1)$, one observation does not give much guidance. With, say, 10 independent p-values available, a much clearer picture emerges about the veracity of the null hypothesis.

Fisher was aware that the cost to society of missing a useful result is large. If an idea (a new agricultural or medical treatment, or sociological explanation, etc.) is actually false, then repeated studies (wishing to confirm it and ultimately capitalize on it) would confirm its nonviability, whereas if a genuine improvement was at hand, and the threshold for the p-value was set too low (say, 0.001), then the initial signal of its viability would be suppressed and it might remain undiscovered. Hence the decision to advocate use of 0.05 as the heuristic threshold. If the null is rejected, then others will replicate the experiment, further small p-values will arise most of the time, and science progresses. The key is replication: "A scientific fact should be regarded as experimentally established only if a properly designed experiment *rarely fails* to give this level of significance" (Fisher, 1926, p. 504, emphasis in original).

The use of a fixed α of 0.05 as the standard level in one-shot reported studies is patently absurd, and just gives the "researchers" the comfort of appeal to authority and (misguided) tradition, without actually contributing anything useful to science. In particular, the chosen level should reflect the relevance of the research. If, for example, an initial experiment for a novel, simple cure for cancer resulted in a p-value of 0.06, should the results just be discarded? The issue is not whether to declare "significance" or not, but rather to use the results as an indication that "something might be there," and engage in further trials that are perhaps better funded, and thus possibly better designed, and with larger sample sizes, etc. (Indeed, if the p-value is between 0.05 and 0.2, Fisher advocated repeating the experiment, possibly with a better design.)

Observe how Fisher's understanding of the advancement of science differs from the modern usage: a single experiment is run (or a data set is analyzed that cannot be replicated, as often with economic data), and if the p-value is less than 0.05, "significance" is declared (and the paper is submitted for publication). This false approach to scientific progress is what gives rise to the alarming results of Ioannidis (2005) quoted above. This author (along with others, such as Hubbard and Lindsay, 2008, p. 69) believes that Fisher would react similarly (perhaps more pungently) to the misuse of p-values and the erosion of his ideas to such low-quality, "stultified" (recall the above quote by Trafimow and Marks, 2015) misrepresentations of scientific progress, and continue to advocate the concept of replication, as opposed to one off trials for which commonly the null hypothesis is anyway just a straw man.

The now famous quote by Tukey (1962) above, written well before the academics in the social, medical, economic, and other sciences made reporting of p-values and outcomes of hypothesis tests mandatory requirements for publishing "scientifically rigorous" studies, is still fully applicable and distills these notions perfectly. The entire article is still well worth reading.

2.9 PROBLEMS

2.1 Write programs to produce Figures 1.3 and 2.7 for studying the point estimators of p for the geometric model.

2.2 Write a program to generate Figure 2.10 and the entries in Table 2.1.

2.3 Let $\widehat{F}(\cdot)$ be the empirical c.d.f. of an i.i.d. sample of size n.
   (a) For $s < t$, show that $Cov(\widehat{F}(s), \widehat{F}(t)) = n^{-1}F(s)(1 - F(t))$, from which it follows that

$$\text{Corr}(\widehat{F}(s), \widehat{F}(t)) = \sqrt{\frac{F(s)(1 - F(t))}{F(t)(1 - F(s))}}$$

and $\mathbb{V}(\widehat{F}(t) - \widehat{F}(s)) = n^{-1}(F(t) - F(s))(1 - F(t) + F(s))$.

(b) Observe that the standardized values $(\widehat{F}(t) - \mathbb{E}[\widehat{F}(t)]) / \sqrt{\mathbb{V}(\widehat{F}(t))}$ will not be i.i.d. because $\widehat{F}$ is computed based upon order statistics, which are correlated. One might expect that, if we pre-multiply $(\widehat{F}(t) - \mathbb{E}[\widehat{F}(t)])$ by $\mathbf{\Sigma}^{-1/2}$, where $\mathbf{\Sigma}$ is the covariance matrix of the $\widehat{F}(t)$, the resulting vector would be uncorrelated. Construct a program to do this, using simulated data from a standard normal distribution. What do you find?

3

Likelihood

Likelihood is the single most important concept of statistics. Fisher (1912) was the first to delineate the concept specifically and unequivocally.

(Ole E. Barndorff-Nielsen, 1991, p. 232)

3.1 INTRODUCTION

The importance of likelihood can hardly be overstated as one of the primary – if not the core – concepts underlying statistical inference. It is also the major common ground in the frequentist and Bayesian schools of thought, as well as (obviously) the likelihood-based paradigm. Regarding these three schools of statistical thought, see Royall (1997) and Efron and Hastie (2016) for very useful and readable accounts of their histories, their differences, and what questions each one answers.

Recall from Section 1.1.1 that the likelihood function $\mathcal{L}(\theta; \mathbf{Y})$ is the joint density of a sample $\mathbf{Y} = (Y_1, \dots, Y_n)$, viewed as a function of the k-dimensional parameter $\theta \in \Theta \subset \mathbb{R}^k$, and the m.l.e. is $\widehat{\theta}_{\mathrm{ML}} = \arg\max \mathcal{L}(\theta; \mathbf{Y})$ over $\theta \in \Theta$. Essentially, $\widehat{\theta}_{\mathrm{ML}}$ can be loosely interpreted as the value of the parameter such that the probability of the observed data (for discrete r.v.s) is the largest. (The modification required for the continuous case is discussed below around (3.1).) In light of such an intuitive and appealing property, the central role of inference based on the likelihood, and the m.l.e. as a point estimator of the unknown parameter, in statistical inference should come as no surprise.

Fundamental Statistical Inference: A Computational Approach, First Edition. Marc S. Paolella.
© 2018 John Wiley & Sons Ltd. Published 2018 by John Wiley & Sons Ltd.

Unlike our rule of using a capital letter for an r.v., but in line with standard convention, we use the same notation $\mathcal{L}$ for the likelihood function $\mathcal{L}(\theta; \mathbf{y})$ based on a given set of data $y_1, \ldots, y_n$, and for the random variable $\mathcal{L}(\theta; \mathbf{Y})$. Similarly, $\ell(\theta; \mathbf{y})$ and $\ell(\theta; \mathbf{Y})$ denote the (natural) log of the likelihood function.

In most problems we encounter, the likelihood is **regular**, meaning that it satisfies a set of technical conditions called the **regularity conditions**. Among other things, these conditions imply that $\mathcal{L}(\theta; \mathbf{Y})$ is second-order differentiable on an open set in $\Theta \subset \mathbb{R}^d$ containing the m.l.e. $\hat{\theta}_{\mathrm{ML}}$, so that solving $\partial \ell(\theta)/\partial \theta_i = 0$, $i = 1, \ldots, d$, yields an extremum; see, for example, Lehmann and Casella (1998) for more detail. In univariate problems $(d = 1)$ with a regular likelihood, a standard calculus exercise often reveals the maximum of the (log of the) likelihood function. An example of a nonregular model is one for which the support of the r.v. depends on the parameter θ. In this case calculus cannot be used, and the m.l.e. will often involve the minimum or maximum of the sample. As an example, let X_i, $i = 1, \ldots, n$, be an i.i.d. sample from the $\mathrm{Unif}(0, \theta)$ distribution, with p.d.f. $f_X(x; \theta) = \theta^{-1} \mathbb{I}_{(0,\theta)}(x)$ for $\theta \in \mathbb{R}_{>0}$ with order statistics $Y_1 < Y_2 < \cdots < Y_n$. The likelihood is $\mathcal{L}(\theta) = \theta^{-n} \mathbb{I}_{[Y_n, \infty)}(\theta)$ (which the reader is encouraged to quickly plot), and $\hat{\theta}_{\mathrm{ML}} = Y_n$. See also Problem 7.11.

In many problems of interest, the likelihood is regular, but the solution to the equations $\partial \ell(\theta)/\partial \theta_i = 0$ will be algebraically intractable, so that one has to resort to numerical techniques to determine the maximum instead of obtaining a closed-form expression for the m.l.e. Methods for this purpose will be discussed, and code given, in Chapter 4. In this chapter, we will mostly restrict ourselves to cases for which closed-form solutions are available, in order to more easily illustrate the theory.

It is important to emphasize that the likelihood function is not a density or mass function, in that it does not integrate to 1. (Recall that it is a function of θ.) It is also a relative measure, in the sense that its value at, say, an argument $\theta_1 \in \Theta$ only has meaning in relation to its value at another argument, say $\theta_2 \in \Theta$, via the ratio $\mathcal{L}(\theta_1; \mathbf{x})/\mathcal{L}(\theta_2; \mathbf{x})$. This follows from its definition as the p.d.f. or p.m.f. of $\mathbf{X}$, viewed as a function of θ. In particular, if $Y = g(X)$ is a bijective function defined on the support of X, then the likelihood based on the sample $x_1, \ldots, x_n$ should contain exactly the same information as that based on the sample $y_1, \ldots, y_n$; but as

$$f_Y(y; \theta) = f_X(x; \theta) \left| \frac{\mathrm{d}x}{\mathrm{d}y} \right| \quad \text{and} \quad \mathcal{L}(\theta; \mathbf{y}) = \mathcal{L}(\theta; \mathbf{x}) \left| \frac{\mathrm{d}x}{\mathrm{d}y} \right|,$$

in order for $\mathcal{L}(\theta; \mathbf{x})$ and $\mathcal{L}(\theta; \mathbf{y})$ to be equivalent in terms of information, it must be the case that only the ratio of likelihoods is of value, that is,

$$\frac{\mathcal{L}(\theta_1; \mathbf{x})}{\mathcal{L}(\theta_2; \mathbf{x})} = \frac{\mathcal{L}(\theta_1; \mathbf{y})}{\mathcal{L}(\theta_2; \mathbf{y})}.$$

This means, for example, that terms in the likelihood that do not depend on θ can be ignored, as they would cancel out in the ratio. Furthermore, it is common to work with the standardized likelihood $\mathcal{L}(\theta; \mathbf{x})/\mathcal{L}(\hat{\theta}_{\mathrm{ML}}; \mathbf{x})$, whose maximum is 1.

There is an important, but usually ignored, issue with the likelihood if the Y_i are continuous r.v.s, namely that $\Pr(\mathbf{Y} = \mathbf{y}) = 0$. This is alleviated by recognizing that all collected data will be observable up to a finite precision, say the gth digit after the decimal point. For example, a reported value of $y = 0.123$ is, when stored in a computer using the usual double-precision eight-byte format, 0.12300000000000, with precision between 14 and 15 decimal digits, and thus its accuracy is far overstated. The value of 0.123 would be reported for any observation that lies in the interval $[0.1225, 0.1235)$. This motivates the following remarks. Let $\epsilon = 10^{-g}$. Then, precisely as in Fisher (1922), the contribution to the likelihood from observation y is

$$\int_{y-\epsilon/2}^{y+\epsilon/2} f_Y(y)\, dy = F_Y(y + \epsilon/2) - F_Y(y - \epsilon/2) \approx \epsilon f_Y(y), \tag{3.1}$$

where the latter approximation follows from a special case of the mean value theorem for integrals (see, for example, p. I.369). In particular, if f is continuous on $[a, b]$, then $\exists c \in [a, b]$ such that $\int_a^b f = (b - a)f(c)$.[1] So, in place of $f_Y(y)$, we use $\{F_Y(y + \epsilon/2) - F_Y(y - \epsilon/2)\}/\epsilon$. Note that, in the maximization of the likelihood, dividing the latter expression by ϵ is theoretically irrelevant, as it is just a constant, but in practice it is valuable when g is large, in which case (3.1) can be very small. We refer to this as the g-**rounded likelihood**.

If the y_i are measured to reasonable accuracy, then the approximation in (3.1), and the fact that ϵ is a constant that does not depend on θ, justify using $f_Y(y; \theta)$ for $\mathcal{L}(\theta; y)$ in the continuous setting. *Unless otherwise specified, we will always use $\mathcal{L}(\theta; y) = f_Y(y; \theta)$.* An example in Section 5.1 shows that use of (3.1) can have some unexpected features.

3.1.1 Scalar Parameter Case

In some respects, the log-likelihood $\ell(\theta; \mathbf{X}) = \log \mathcal{L}(\theta; \mathbf{X})$ is of more theoretical value than the likelihood itself, as will be seen below. As in Section 1.1.1, let $\dot{\ell}(\theta; \mathbf{X})$ denote the first derivative of $\ell(\theta; \mathbf{X})$ with respect to θ. This is referred to as the **score function** or score equation. If the likelihood is regular, then $\dot{\ell}(\hat{\theta}_{\mathrm{ML}}; \mathbf{X}) = 0$. Similarly, for regular models, we denote the second derivative of $\ell(\theta; \mathbf{X})$ by $\ddot{\ell}(\theta; \mathbf{X})$, and let $I(\theta; \mathbf{X}) = -\ddot{\ell}(\theta; \mathbf{X})$. As the likelihood at the m.l.e. is a maximum, $I(\hat{\theta}_{\mathrm{ML}}) > 0$; this is referred to as the **observed (Fisher) information** or the **curvature** of the likelihood at the m.l.e. It measures the precision of the m.l.e., with relatively high curvature implying a relatively peaked likelihood and, thus, greater certainty regarding the true parameter value. It plays a fundamental role in likelihood-based statistical inference, as we shall see below.

Recall that, for function $f : D \to \mathbb{R}$, with $D \subset \mathbb{R}$ an open interval and such that $f^{(n)}(x)$ exists for all $x \in D$, if $c \in D$, then Taylor's formula for f at c with (the Lagrange form of the) remainder is, for $n = 0, 1, \ldots,$

$$f(x) = \sum_{k=0}^{n} \frac{f^{(k)}(c)}{k!}(x - c)^k + \frac{f^{(n+1)}(\zeta)}{(n+1)!}(x - c)^{n+1}, \quad \zeta \in (x, c), \tag{3.2}$$

as in (I.A.132). Letting $\ell(\theta) = \ell(\theta; \mathbf{X})$ for notational convenience, applying (3.2) to $\ell(\theta)$ about the point $\hat{\theta}_{\mathrm{ML}}$, and recalling that $\dot{\ell}(\hat{\theta}_{\mathrm{ML}}) = 0$ and $I(\theta) = -\ddot{\ell}(\theta)$, we have

$$\ell(\theta) = \ell(\hat{\theta}_{\mathrm{ML}}) - \frac{1}{2}I(\hat{\theta}_{\mathrm{ML}})(\theta - \hat{\theta}_{\mathrm{ML}})^2 + \frac{1}{6}\ell^{(3)}(\zeta)(\theta - \hat{\theta}_{\mathrm{ML}})^3, \quad \zeta \in (\theta, \hat{\theta}_{\mathrm{ML}}). \tag{3.3}$$

Thus, when θ is close to $\hat{\theta}_{\mathrm{ML}}$,

$$\log \frac{\mathcal{L}(\theta; \mathbf{x})}{\mathcal{L}(\hat{\theta}_{\mathrm{ML}}; \mathbf{x})} \approx -\frac{1}{2}I(\hat{\theta}_{\mathrm{ML}})(\theta - \hat{\theta}_{\mathrm{ML}})^2. \tag{3.4}$$

---

[1] The approximation in (3.1) is graphically obvious: draw a thin rectangle horizontally between $y - \epsilon/2$ and $y + \epsilon/2$, and vertically from zero to $f_Y(y)$.

Example 3.1 (Poisson)   *Let $X_i \overset{\text{i.i.d.}}{\sim} \text{Poi}(\theta)$, $i = 1, \ldots, n$, $\mathbf{x} = (x_1, \ldots, x_n)$, and $s = \sum_{i=1}^{n} x_i$. Then*

$$\mathcal{L}(\theta; \mathbf{x}) = \prod_{i=1}^{n} f_{\text{Poi}}(x_i; \theta) = \prod_{i=1}^{n} \frac{e^{-\theta} \theta^{x_i}}{x_i!} = \frac{e^{-n\theta} \theta^{s}}{\prod_{i=1}^{n} x_i!},$$

and $\ell(\theta) = -n\theta + s \log \theta - \log(x_1! \cdots x_n!)$. It is easy to check that $\dot{\ell}(\theta) = -n + s/\theta$ and $\ddot{\ell}(\theta) = -s\theta^{-2}$, yielding $\hat{\theta}_{\text{ML}} = s/n = \bar{x}_n$. In terms of the r.v.s X_i, $\hat{\theta}_{\text{ML}} = S/n = \bar{X}_n$.

The standardized log-likelihoods $\ell(\theta; \mathbf{x}) - \ell(\hat{\theta}_{\text{ML}}; \mathbf{x})$, for true $\theta = 3$ and random samples of sizes $n = 3$ and $n = 100$ are shown in Figure 3.1, overlaid with the quadratic approximation (3.4). It clearly improves as n increases.

A simple calculation verifies that $\hat{\theta}_{\text{ML}}$ is unbiased. The curvature is s/θ^2; it increases with s, the sum of the observations, which grows as more observations are obtained, and so is intuitively plausible. Of course, the sum would also automatically be large for large values of θ (recall that the mean of the Poisson is θ), implying that m.l.e. accuracy would increase with θ. However, the reverse is true, because the curvature divides s not by θ, but by θ^2. This is, however, quite plausible: the variance of a $\text{Poi}(\theta)$ random variable is θ, so

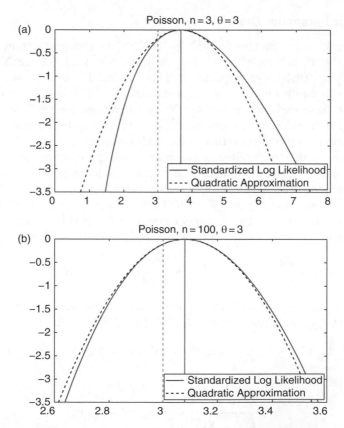

Figure 3.1   *Standardized log-likelihoods (solid) and quadratic approximation (3.4) (dashed) for the Poisson with $n = 3$ (a) and $n = 100$ (b), with solid and dashed vertical lines showing the m.l.e. and true parameter, respectively.*

that, for a given sample size, the larger θ is, the larger is the spread of obtained sample values, which tends to decrease the accuracy of the estimate.

Example 3.2 (Normal) *Let $X_i \overset{\text{i.i.d.}}{\sim} N(\mu, \sigma^2)$, $i = 1, \ldots, n$, with μ unknown but σ known. Then $\mathcal{L}(\mu; \mathbf{x}) = \prod_{i=1}^{n} f_N(x_i; \mu, \sigma^2)$ and*

$$\ell(\mu; \mathbf{x}) = -\frac{n}{2}\log(2\pi) - \frac{n}{2}\log(\sigma^2) - \frac{1}{2\sigma^2}\sum_{i=1}^{n}(x_i - \mu)^2, \tag{3.5}$$

with the first two terms being irrelevant as they are not functions of μ. Setting $\dot{\ell}(\mu; \mathbf{x}) = \sigma^{-2}\sum_{i=1}^{n}(x_i - \mu)$ to zero yields $\hat{\mu}_{\text{ML}} = \bar{x}$. It is easy to verify that, as a random variable, $\hat{\mu}_{\text{ML}} = \bar{X}$ is unbiased. Also, the observed Fisher information is easily seen to be $I(\hat{\mu}_{\text{ML}}; \mathbf{x}) = n/\sigma^2$. Observe that $\mathbb{V}(\bar{X}) = \sigma^2/n = 1/I(\hat{\mu}_{\text{ML}}; \mathbf{x})$ and, after a little simplification, that

$$\ell(\mu; \mathbf{x}) - \ell(\hat{\mu}_{\text{ML}}; \mathbf{x}) = -\frac{1}{2\sigma^2}\sum_{i=1}^{n}(x_i - \mu)^2 - \left(-\frac{1}{2\sigma^2}\sum_{i=1}^{n}(x_i - \bar{x})^2\right)$$

$$= -\frac{1}{2}\frac{n}{\sigma^2}(\mu - \bar{x})^2 = -\frac{1}{2}I(\hat{\mu}_{\text{ML}})(\mu - \hat{\mu}_{\text{ML}})^2,$$

showing that the quadratic approximation (3.4) is exact.

Recall that, in terms of $\mathbf{X}$, $\mathcal{L}(\theta; \mathbf{X})$ is a random variable, so that expectations of (functions of) it can be computed. As an important example, if we assume that the support of the X_i does not depend on parameter θ, we have

$$\mathbb{E}[\dot{\ell}(\theta; \mathbf{X})] = \int_{\mathbb{R}^n}\left(\frac{\mathrm{d}}{\mathrm{d}\theta}\log f_{\mathbf{X}}(\mathbf{x}; \theta)\right)f_{\mathbf{X}}(\mathbf{x}; \theta)\mathrm{d}\mathbf{x}$$

$$= \int_{\mathbb{R}^n}\left(\frac{1}{f_{\mathbf{X}}(\mathbf{x}; \theta)}\frac{\mathrm{d}}{\mathrm{d}\theta}f_{\mathbf{X}}(\mathbf{x}; \theta)\right)f_{\mathbf{X}}(\mathbf{x}; \theta)\mathrm{d}\mathbf{x}$$

$$= \frac{\mathrm{d}}{\mathrm{d}\theta}\int_{\mathbb{R}^n}f_{\mathbf{X}}(\mathbf{x}; \theta)\mathrm{d}\mathbf{x} = 0, \tag{3.6}$$

as $\mathrm{d}1/\mathrm{d}\theta = 0$, and where, under the stated assumption, we can justify interchanging the derivative and integral. (See Pawitan, 2001, p. 216, and the reference therein, for further required assumptions to allow this.) That is, under the stated condition on the support, $\mathbb{E}[\dot{\ell}(\theta; \mathbf{X})] = 0$, and the m.l.e. is such that $\dot{\ell}(\hat{\theta}_{\text{ML}}) = 0$.

The m.l.e. possesses a very important property referred to as **invariance**: the m.l.e. of a one-to-one function of parameter θ, say $g(\theta)$, is given by $g(\hat{\theta}_{\text{ML}})$; see, for example, Zehna (1966) or Casella and Berger (2002, p. 320). Thus, in Example 3.2, the m.l.e. of μ^2 is just $\bar{X}^2$. The result also holds for functions that are not one-to-one; see Pal and Berry (1992).

Example 3.3 *Recall that the exponential distribution can be useful in modeling the time until a living creature or electrical component "fails" or, more optimistically, until a specific task is reached, such as gaining employment. Let $X_i \overset{\text{i.i.d.}}{\sim} \text{Exp}(\lambda)$, $i = 1, \ldots, n$, with $f_{X_i}(x; \lambda) = \lambda\exp(-\lambda x)\mathbb{I}_{(0,\infty)}(x)$, be a sample of such times, but assume that only the first k failures, $k \leq n$, are observed. (Do you really want to wait until all the light bulbs burn* *out?) That is, only the first k order statistics, $\mathbf{Y} = (Y_1, \ldots, Y_k)$, are available. From (A.181), the likelihood of $\mathbf{Y}$ is*

$$\mathcal{L}(\lambda; \mathbf{Y}) = \frac{n!}{(n-k)!}[1 - F_X(y_k; \lambda)]^{n-k} \prod_{i=1}^{k} f_X(y_i; \lambda) \qquad (3.7)$$

$$= \frac{n!}{(n-k)!} \exp(-\lambda(n-k)y_k) \, \lambda^k \exp\left(-\lambda \sum_{i=1}^{k} y_i\right), \qquad (3.8)$$

which can be written as

$$\mathcal{L}(\lambda; \mathbf{Y}) \propto \lambda^k \exp(-\lambda T), \quad T := \sum_{i=1}^{k} y_i + (n-k)y_k.$$

Setting $\dot{\ell}(\lambda)$ to zero and solving yields $\hat{\lambda}_{\mathrm{ML}} = k/T$. If $k = n$, then $\hat{\lambda}_{\mathrm{ML}}$ simplifies to $1/\bar{X}_n$.
  If the exponential distribution is instead parameterized as a usual scale model, with p.d.f. $\theta^{-1} \exp(-x/\theta)$, then the invariance property of the m.l.e. implies that $\hat{\theta}_{\mathrm{ML}} = 1/\hat{\lambda}_{\mathrm{ML}} = T/k$. Using (A.185),

$$\mathbb{E}[T; \lambda] = \sum_{i=1}^{k} \mathbb{E}[Y_i] + (n-k)\mathbb{E}[Y_k] = \frac{1}{\lambda}\left(\sum_{h=1}^{k}\sum_{j=0}^{h-1}\frac{1}{n-j} + (n-k)\sum_{j=0}^{k-1}\frac{1}{n-j}\right).$$

The big term in parentheses is, interestingly enough, just k (Problem 3.8) so that $\mathbb{E}[T/k] = 1/\lambda$. That is, $1/\hat{\lambda}_{\mathrm{ML}}$ is unbiased for $1/\lambda$. As such, $\hat{\lambda}_{\mathrm{ML}}$ is not unbiased for λ; from Jensen's inequality, $\mathbb{E}[\hat{\lambda}_{\mathrm{ML}}] > \lambda$. Simulation can be used to determine the bias; see Problem 3.8.

Remark.   We have now seen that, for Poisson, normal, and (recalling Section 1.1.1) Bernoulli i.i.d. samples, the unknown parameter is the expected value (or, in the Bernoulli case, a scale factor of it) and the m.l.e. turned out to be the sample mean. This was also the case for i.i.d. samples from the exponential distribution with parameterization $\theta^{-1} \exp(-x/\theta)$. This is not always the case. For an i.i.d. sample from Laplace data with unknown location parameter, that is, $f_X(x) = \exp(-|x - \mu|)/2$, $\mathbb{E}[X] = \mu$ but $\hat{\mu}_{\mathrm{ML}}$ is the median of the sample. Similarly, for the location parameter of Student's t data with $v > 1$ degrees of freedom, or for the location parameter of stable Paretian data with tail index $1 < \alpha < 2$, $\mathbb{E}[X] = \mu$ but the m.l.e. is not the sample mean (and needs to be found numerically).
  Hoyt (1969) gave the following two examples. Let the r.v. X have p.m.f.

$$f_X(x; \theta) = \frac{1}{3}(5 - 2\theta)^{(3-x)(2-x)/2}(\theta - 1)^{(x-1)(4-x)/2}\mathbb{1}_{\{1,2,3\}}(x), \quad 1 \leq \theta \leq 2.5.$$

It is straightforward to confirm that $\mathbb{E}[X] = \theta$ but that, for an i.i.d. sample of size n, the m.l.e. of θ is

$$\hat{\theta} = \frac{15}{4}\left(\frac{\sum_{i=1}^{n} x_i}{n}\right) - \frac{3}{4}\left(\frac{\sum_{i=1}^{n} x_i^2}{n}\right) - 2.$$

Similarly, let the r.v. X have p.d.f.

$$f_X(x; \theta) = \frac{\theta}{1-\theta} x^{(2\theta-1)/(1-\theta)}\mathbb{1}_{(0,1]}(x), \quad \frac{1}{2} < \theta < 1.$$

Then $\mathbb{E}[X] = \theta$ but the m.l.e. of θ (for an i.i.d. sample of size n) is

$$\hat{\theta} = \frac{1}{1 - \sum \log x_i / n}.$$

See also Philippou and Dahiya (1970). ∎

The **(Fisher) information** of an i.i.d. sample $\mathbf{X} = (X_1, \dots, X_n)$ is given by $J(\theta) :=$ $-\mathbb{E}[\ddot{\ell}(\theta; \mathbf{X})]$. It is the expected value of the observed (Fisher) information. For the Poisson example, this yields n/θ, showing clearly the tradeoff in this measure of "information" as a function of sample size and θ. Provided that the support of X does not depend on θ,

$$\mathbb{E}[\dot{\ell}^2(\theta; \mathbf{X})] = -\mathbb{E}[\ddot{\ell}(\theta; \mathbf{X})] = J(\theta), \tag{3.9}$$

seen as follows. Differentiate (3.6) to get, using the abbreviations $\dot{\ell} = \dot{\ell}(\theta; \mathbf{x})$, $\ddot{\ell} = \ddot{\ell}(\theta; \mathbf{x})$, $f = f_{\mathbf{X}}(\mathbf{x}; \theta)$, and $\dot{f} = df_{\mathbf{X}}(\mathbf{x}; \theta)/d\theta$,

$$0 = \frac{d}{d\theta} \int_{\mathbb{R}^n} \dot{\ell} f d\mathbf{x} = \int_{\mathbb{R}^n} \frac{d(\dot{\ell} f)}{d\theta} d\mathbf{x} = \int_{\mathbb{R}^n} \left(\dot{\ell} \frac{df}{d\theta} + f \frac{d\dot{\ell}}{d\theta} \right) d\mathbf{x} = \int_{\mathbb{R}^n} (\dot{\ell} \dot{f} + \ddot{\ell} f) d\mathbf{x}. \tag{3.10}$$

As

$$\dot{\ell} \dot{f} = \left(\frac{d}{d\theta} \log f \right) \frac{df}{d\theta} = \frac{1}{f} \frac{df}{d\theta} \frac{df}{d\theta} = \left(\frac{1}{f} \frac{df}{d\theta} \right)^2 f = \left(\frac{d}{d\theta} \log f \right)^2 f = \dot{\ell}^2 f, \tag{3.11}$$

(3.10) and (3.11) imply

$$\int_{\mathbb{R}^n} \dot{\ell}^2 f d\mathbf{x} + \int_{\mathbb{R}^n} \ddot{\ell} f d\mathbf{x} = 0,$$

from which (3.9) follows. As differentiating tends to eliminate terms, while squaring tends to increase and complicate them, the latter form in (3.9) is most often easier to resolve.

Recalling from (3.6) that $\mathbb{E}[\dot{\ell}(\theta; \mathbf{X})] = 0$, (3.9) implies

$$\mathbb{V}(\dot{\ell}(\theta; \mathbf{X})) = J(\theta). \tag{3.12}$$

Example 3.4 *Let $X_i \overset{\text{i.i.d.}}{\sim} \text{Bern}(\theta)$, $i = 1, \dots, n$, for $\theta \in (0, 1)$ a fixed but unknown parameter value. The support of the density of $\mathbf{X}$ does not depend on θ. The likelihood based on $\mathbf{X} = \mathbf{x}$ is then*

$$\mathcal{L}(\theta; \mathbf{x}) = \prod_{i=1}^{n} \theta^{x_i} (1 - \theta)^{1-x_i} \mathbb{I}_{\{0,1\}}(x_i) = \theta^s (1 - \theta)^{n-s} \mathbb{I}_{\{0,1,\dots,n\}}(s),$$

where $s = \sum_{i=1}^{n} x_i$ and $S = \sum_{i=1}^{n} X_i \sim \text{Bin}(n, \theta)$, with $\mathbb{E}[S] = n\theta$. Then

$$\dot{\ell}(\theta) = \frac{s}{\theta} - \frac{n-s}{1-\theta}, \quad \text{and} \quad \ddot{\ell}(\theta) = -\frac{s}{\theta^2} - \frac{n-s}{(1-\theta)^2},$$

yielding $\hat{\theta}_{\text{ML}} = S/n$. As

$$\mathbb{E}[\ddot{\ell}(\theta; \mathbf{X})] = -\frac{n\theta}{\theta^2} - \frac{n - n\theta}{(1-\theta)^2} = -\frac{n}{\theta(1-\theta)},$$

the information is $J(\theta) = -\mathbb{E}[\ddot{\ell}(\theta; \mathbf{X})] = n/[\theta(1-\theta)]$.

Using the first term in (3.9) instead, $\mathbb{E}[S^2] = \mathbb{V}(S) + (\mathbb{E}[S])^2 = n\theta(1-\theta) + n^2\theta^2$, *so that*

$$\ddot{\ell}^2(\theta; \mathbf{x}) = \frac{s^2 - 2s\theta n + \theta^2 n^2}{\theta^2(1-\theta)^2}, \quad \mathbb{E}[\dot{\ell}^2(\theta; \mathbf{X})] = \frac{n}{\theta(1-\theta)},$$

confirming that $\mathbb{E}[\dot{\ell}^2(\theta; \mathbf{X})] = -\mathbb{E}[\ddot{\ell}(\theta; \mathbf{X})]$ *in this case.*

Let $X_1, \dots, X_n$ be i.i.d. with common p.m.f. or p.d.f. $f_X(\cdot; \theta)$. Then the information $J(\theta)$ can also be calculated as follows. Letting $\vartheta_i(\theta) = \ell(\theta; X_i)$, we can write $\ell(\theta; \mathbf{X}) = \sum_{i=1}^n \vartheta_i(\theta)$. Then, from the i.i.d. nature of the X_i,

$$\mathbb{E}[\ddot{\ell}(\theta; \mathbf{X})] = \sum_{i=1}^n \mathbb{E}[\ddot{\vartheta}_i(\theta)] = n\mathbb{E}[\ddot{\vartheta}_1(\theta)]. \tag{3.13}$$

By denoting $\vartheta_1(\theta)$ in (3.13) as just ϑ, we can express the result compactly as

$$J(\theta) = -\mathbb{E}[\ddot{\ell}(\theta; \mathbf{X})] = -n\mathbb{E}[\ddot{\vartheta}]. \tag{3.14}$$

A similar result holds using the score function; as verified in Problem 3.4,

$$\mathbb{E}[\dot{\ell}^2(\theta; \mathbf{X})] = n\mathbb{E}[\dot{\vartheta}^2]. \tag{3.15}$$

3.1.2 Vector Parameter Case

Let the (fixed but unknown) set of parameters be given by the k-length column vector $\theta = (\theta_1, \dots, \theta_k)'$ and such that θ is **identifiable**, the formal definition of which is given and discussed in Section 5.1.1.

The m.l.e. of θ is the vector of θ_i-values that jointly maximize the likelihood for a sample $\mathbf{X}$. For regular likelihood functions, the standard conditions from multivariate calculus for the maximum of a multivariate differentiable function hold. The score $\dot{\ell}(\theta; \mathbf{X})$ is a k-length column vector with ith element given by $\dot{\ell}_i(\theta; \mathbf{X}) = \partial \ell(\theta; \mathbf{X})/\partial \theta_i$; it satisfies

$$\mathbb{E}[\dot{\ell}(\theta; \mathbf{X})] = \mathbf{0}_k, \tag{3.16}$$

generalizing (3.6). The information is a matrix, $\mathbf{J}(\theta)$, referred to appropriately as the **information matrix**, defined analogously as in the $k = 1$ case as $-\mathbb{E}[\ddot{\ell}(\theta; \mathbf{X})]$, but where (dropping argument $\mathbf{X}$ for notational convenience) $\ddot{\ell}(\theta)$ is the $k \times k$ symmetric matrix with (i, j)th element given by $\ddot{\ell}_{ij}(\theta) = \partial^2 \ell(\theta)/\partial \theta_i \partial \theta_j$. The matrix $\ddot{\ell}(\theta)$ is also known as the **Hessian**. Note that (further abbreviating $\ddot{\ell}_{ij}(\theta; \mathbf{X})$ by $\ddot{\ell}_{ij}$, etc.)

$$\mathbb{E}[\ddot{\ell}_{ij}] = \mathbb{E}\left[\frac{\partial \dot{\ell}_j}{\partial \theta_i}\right] = \mathbb{E}\left[\frac{\partial}{\partial \theta_i}\left(\frac{1}{f}\frac{\partial f}{\partial \theta_j}\right)\right] = \mathbb{E}\left[\frac{1}{f}\frac{\partial^2 f}{\partial \theta_i \partial \theta_j} - \frac{1}{f^2}\frac{\partial f}{\partial \theta_i}\frac{\partial f}{\partial \theta_j}\right]$$

$$= \mathbb{E}\left[\frac{1}{f}\frac{\partial^2 f}{\partial \theta_i \partial \theta_j} - \dot{\ell}_i \dot{\ell}_j\right].$$

Thus, if

$$0 = \mathbb{E}\left[\frac{1}{f}\frac{\partial^2 f}{\partial \theta_i \partial \theta_j}\right] = \int_{\mathbb{R}^n} \frac{\partial^2 f}{\partial \theta_i \partial \theta_j} d\mathbf{x} \tag{3.17}$$

for all i and j, then $\mathbb{E}[\dot{\ell}_i\dot{\ell}_j] = -\mathbb{E}[\ddot{\ell}_{ij}]$ or, in vector terms,

$$\mathbb{E}[\dot{\ell}\dot{\ell}'] = -\mathbb{E}[\ddot{\ell}]. \tag{3.18}$$

Relation (3.17) is true if we can interchange the integral and derivative; that is, if, letting $g(\mathbf{x}; \boldsymbol{\theta}) = \int_{\mathbb{R}^n} f(\mathbf{x}; \boldsymbol{\theta}) d\mathbf{x}$,

$$\int_{\mathbb{R}^n} \frac{\partial^2 f}{\partial\theta_i\partial\theta_j} d\mathbf{x} = \frac{\partial^2 g(\mathbf{x}; \boldsymbol{\theta})}{\partial\theta_i\partial\theta_j} = 0,$$

because $g(\mathbf{x}; \boldsymbol{\theta}) = 1$. From (3.16) and (3.18),

$$\mathbb{V}(\dot{\ell}(\boldsymbol{\theta})) = \mathbf{J}(\boldsymbol{\theta}), \tag{3.19}$$

generalizing (3.12). The multivariate extension of (3.14) holds: with $\boldsymbol{\vartheta} = \boldsymbol{\vartheta}(\boldsymbol{\theta})$,

$$-\mathbb{E}[\ddot{\ell}] = -n\mathbb{E}[\ddot{\boldsymbol{\vartheta}}]. \tag{3.20}$$

Example 3.5 (Normal, cont.) *As in Example 3.2, let $X_i \overset{\text{i.i.d.}}{\sim} N(\mu, \sigma^2)$, $i = 1, \ldots, n$, but with both μ and σ^2 unknown, so that $\boldsymbol{\theta} = (\theta_1, \theta_2)' = (\mu, \sigma^2)'$. Problem 3.1 shows that $\hat{\mu}_{\text{ML}} = \bar{X}$, $\hat{\sigma}^2_{\text{ML}} = n^{-1} \sum_{i=1}^n (X_i - \bar{X})^2$, and*

$$\mathbf{J}^{-1}(\boldsymbol{\theta}) = \frac{1}{n} \begin{bmatrix} \sigma^2 & 0 \\ 0 & 2\sigma^4 \end{bmatrix}, \tag{3.21}$$

which does not depend on μ.

Example 3.6 *Let $X_i \overset{\text{i.i.d.}}{\sim} \text{Gam}(\alpha, \beta)$, $i = 1, \ldots, n$, with $\boldsymbol{\theta} = (\alpha, \beta)'$. The log-likelihood is given by*

$$\ell(\boldsymbol{\theta}; \mathbf{x}) = n\alpha \log \beta - n \log \Gamma(\alpha) + (\alpha - 1) \sum_{i=1}^n \log x_i - \beta \sum_{i=1}^n x_i + \log \prod_{i=1}^n \mathbb{I}_{(0,\infty)}(x_i),$$

with first derivatives

$$\dot{\ell}_1(\boldsymbol{\theta}; \mathbf{x}) = n \log \beta - n \psi(\alpha) + \sum_{i=1}^n \log x_i \quad \text{and} \quad \dot{\ell}_2(\boldsymbol{\theta}; \mathbf{x}) = \frac{n\alpha}{\beta} - \sum_{i=1}^n x_i, \tag{3.22}$$

where $\psi(\alpha) = (\text{d}/\text{d}\alpha) \log \Gamma(\alpha)$ is the digamma function (A.18). From $\dot{\ell}_2$, $\hat{\beta}_{\text{ML}} = \alpha/\bar{X}$ and, substituting this into $\dot{\ell}_1$, $\hat{\alpha}_{\text{ML}}$ is given implicitly as the solution to

$$\log(\alpha/\bar{X}) - \psi(\alpha) + \bar{y} = 0, \quad y_i = \log x_i, \tag{3.23}$$

which needs to be solved numerically. With $\mathbb{E}[X] = \alpha/\beta$, we immediately have $\mathbb{E}[\dot{\ell}_2(\boldsymbol{\theta}; \mathbf{X})] = 0$. As $\mathbb{E}[\log X] = \psi(\alpha) - \log \beta$ (see Problem II.1.3), $\mathbb{E}[\dot{\ell}_1(\boldsymbol{\theta}; \mathbf{X})] = 0$. To compute the information matrix, with

$$\dot{\boldsymbol{\vartheta}} = (\dot{\vartheta}_1, \dot{\vartheta}_2) = (\log \beta - \psi(\alpha) + \log x, \ \alpha/\beta - x)',$$

and $\ddot{\vartheta}_{ij}$ denoting the (ij)th element of $\mathbb{E}[\ddot{\boldsymbol{\vartheta}}]$, we have

$$\ddot{\vartheta}_{11} = -\psi'(\alpha) = \psi^2(\alpha) - \frac{\Gamma''(\alpha)}{\Gamma(\alpha)}, \quad \ddot{\vartheta}_{22} = -\alpha\beta^{-2}, \quad \ddot{\vartheta}_{12} = \ddot{\vartheta}_{21} = \beta^{-1},$$

so that

$$\mathbf{J}(\theta) = n \begin{bmatrix} \psi'(\alpha) & -\beta^{-1} \\ -\beta^{-1} & \alpha\beta^{-2} \end{bmatrix},$$

without having to take expectations.

Example 3.7 *Let*

$$\begin{bmatrix} X \\ Y \end{bmatrix} \sim N\left(\begin{bmatrix} \mu_1 \\ \mu_2 \end{bmatrix}, \begin{bmatrix} \sigma_1^2 & \rho\sigma_1\sigma_2 \\ \rho\sigma_1\sigma_2 & \sigma_2^2 \end{bmatrix} \right),$$

in short, $(X, Y)' \sim N(\boldsymbol{\mu}, \boldsymbol{\Sigma})$, with density

$$f_{Y_1,Y_2}(x, y) = K \exp\left\{ -\frac{\tilde{x}^2 - 2\rho\tilde{x}\tilde{y} + \tilde{y}^2}{2(1-\rho^2)} \right\},$$

where

$$K = \frac{1}{2\pi\sigma_1\sigma_2(1-\rho^2)^{1/2}}, \quad \tilde{x} = \frac{x - \mu_1}{\sigma_1}, \quad \tilde{y} = \frac{y - \mu_2}{\sigma_2}.$$

Let $(X_i, Y_i)' \overset{\text{i.i.d.}}{\sim} N(\boldsymbol{\mu}, \boldsymbol{\Sigma})$, $i = 1, \ldots, n$, with all five parameters unknown. It is straightforward to show that the solution to the system of equations $\dot{\ell} = \mathbf{0}$ yields

$$\hat{\mu}_1 = \bar{X}, \quad \hat{\mu}_2 = \bar{Y}, \quad \hat{\sigma}_1^2 = n^{-1}\sum_{i=1}^{n}(X_i - \bar{X})^2, \quad \hat{\sigma}_2^2 = n^{-1}\sum_{i=1}^{n}(Y_i - \bar{Y})^2, \tag{3.24}$$

and the m.l.e. for $\rho = \text{Corr}(X, Y)$ given by

$$\hat{\rho} = \frac{n^{-1}\sum_{i=1}^{n}(X_i - \bar{X})(Y_i - \bar{Y})}{\hat{\sigma}_1\hat{\sigma}_2}; \tag{3.25}$$

see, for example, Stuart et al. (1999, pp. 76–77) for a direct derivation, or Example 3.8 below for the general multivariate case. From (3.24), only $X_1, \ldots, X_n$ are required for estimating μ_1 and σ_1^2, and $Y_1, \ldots, Y_n$ are not used, even though they are correlated with the X_i. This – perhaps initially surprising – result is indeed intuitive, recalling that the linear correlation between two r.v.s is not dependent on their location or scale, which are what the μ_i and σ_i represent in this case. As a result, the marginal densities of $\hat{\mu}_i$ and $\hat{\sigma}_i^2$ are the same as those in the univariate case and the independence between them also still holds (see Section II.3.7). However, $\hat{\mu}_1$ and $\hat{\mu}_2$ are not independent, nor are $\hat{\sigma}_1^2$, $\hat{\sigma}_2^2$ and $\hat{\rho}$ from one another.

Obtaining the density of $R = \hat{\rho}$ is far more difficult. Fisher succeeded in deriving an expression for it in 1915, though a different expression, due to Hotelling (1953) and detailed in Stuart and Ord (1994, pp. 559–565), lends itself better to computation; it is

$$f_R(r; \rho) = \mathbb{I}_{[-1,1]}(r) \times \frac{n-2}{n-1} \frac{(1-\rho^2)^{(n-1)/2}}{\sqrt{2}\, B(1/2, n - 1/2)}$$

$$\times (1 - r^2)^{(n-4)/2}(1 - \rho r)^{(3/2)-n} \, {}_2F_1\left(\frac{1}{2}, \frac{1}{2}, n - \frac{1}{2}, \frac{1 + \rho r}{2} \right), \tag{3.26}$$

where the ${}_2F_1$ function is given in (A.17).

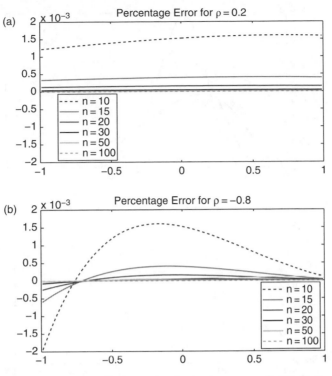

Figure 3.2 *Percentage error of (3.26) when using the Laplace approximation to the $_2F_1$ function for $\rho = 0.2$ (a) and $\rho = -0.8$ (b) for sample sizes $n = 10, 15, 20, 30, 50,$ and 100 (lines from top to bottom). The x-axis indicates the value of r in (3.26)*

Section II.5.3 gives a Laplace approximation from Butler and Wood (2002) to $_2F_1$. It is far faster to calculate, and accurate enough for most purposes in statistical inference. As an illustration, Figure 3.2 shows the incurred percentage error of this approximation, for two values of ρ and several sample sizes.

Figure 3.3(a) illustrates the density for $\rho = -0.4$ and four sample sizes, using the program in Listing 3.1. The c.d.f. can be computed by numerically integrating (3.26). For $\rho = 0$, (3.26) reduces to

$$f_R(r;0) = \frac{1}{B(1/2, (n-2)/2)}(1 - r^2)^{(n-4)/2}\mathbb{I}_{[-1,1]}(r), \tag{3.27}$$

(see Problem 3.7) which is clearly symmetric about zero.

The mean of R can be expressed as (Stuart and Ord, 1994, pp. 565–567)

$$\mathbb{E}[R;\rho] = \frac{\rho(1-\rho^2)^{(n-1)/2}\Gamma^2(n/2)}{\Gamma((n-1)/2)\Gamma((n+1)/2)} \; _2F_1\left(\frac{n}{2},\frac{n}{2},\frac{n+1}{2},\rho^2\right) \tag{3.28}$$

$$= \rho\left(1 - \frac{1-\rho^2}{2n} + O(n^{-2})\right), \tag{3.29}$$

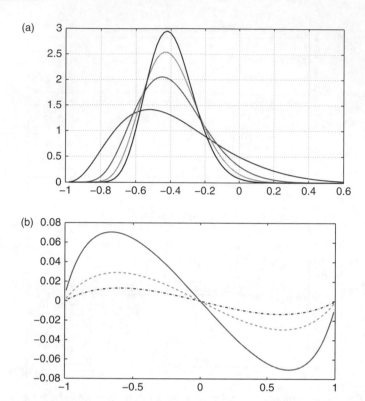

Figure 3.3 (a) Density of $\hat{\rho}$ for $\rho = -0.4$ and $n = 10, 20, 30$, and 40. (b) Bias of $R = \hat{\rho}$ as given in (3.25) and (3.28), for $n = 4$ (solid), $n = 8$ (dashed), and $n = 16$ (dash-dotted) as a function of ρ. There is no graphical difference when using the Laplace approximation for the $_2F_1$ function instead of its exact values.

with (3.28) implemented in Listing 3.2. This shows that $\mathbb{E}[R] = 0$ for $\rho = 0$; and from (3.29), for all $\rho \neq 0$, R is biased towards zero. This is most easily seen graphically; Figure 3.3(b) plots $\mathbb{E}[R; \rho]$ as a function of ρ for three sample sizes.

Example 3.8 *Now consider the generalization of the bivariate normal distribution in Example 3.7 to the d-variate case, with distribution given by*

$$f_N(\mathbf{y}; \boldsymbol{\mu}, \boldsymbol{\Sigma}) = \frac{1}{|\boldsymbol{\Sigma}|^{1/2}(2\pi)^{d/2}} \exp\left\{-\frac{1}{2}[(\mathbf{y} - \boldsymbol{\mu})'\boldsymbol{\Sigma}^{-1}(\mathbf{y} - \boldsymbol{\mu})]\right\}, \tag{3.30}$$

for $\boldsymbol{\mu} \in \mathbb{R}^d$ and $\boldsymbol{\Sigma}$ a symmetric, positive definite $d \times d$ matrix. Let $\mathbf{Y}_i \overset{i.i.d.}{\sim} N(\boldsymbol{\mu}, \boldsymbol{\Sigma})$, $i = 1, \ldots, n$, for $n > d$. Generalizing (3.24) and (3.25), the m.l.e.s of $\boldsymbol{\mu}$ and $\boldsymbol{\Sigma}$ are given by

$$\hat{\boldsymbol{\mu}} = \bar{\mathbf{Y}} = \frac{1}{n}\sum_{i=1}^{n} \mathbf{Y}_i, \quad \hat{\boldsymbol{\Sigma}} = \mathbf{S} = \frac{1}{n}\sum_{i=1}^{n}(\mathbf{Y}_i - \bar{\mathbf{Y}})(\mathbf{Y}_i - \bar{\mathbf{Y}})', \tag{3.31}$$

```
1   function [f,F]=corrcoefpdf(rvec,rho,n,approx)
2   if nargin<4, approx=1; end
3   p=rho; F=zeros(length(rvec),1); f=F; A= (n-2) / (n-1) / sqrt(2) /
4     beta(0.5,n-0.5);
5   B = (1-p^2).^((n-1)/2) * (1-rvec.^2).^((n-4)/2).* (1-rho*rvec).^(3/2-n);
6   if approx==1 % the Laplace approximation from page II.197
7     f = A * B.* f21(0.5,0.5,n-0.5,(1+rho*rvec)/2);
8   else
9     if 1==2 & exist('hypergeom','file')
10      % symbolic toolbox in Matlab release 2008b. Very slow compared to
11      % numeric integration for 2F1
12      f = A * B.* hypergeom([0.5,0.5],n-0.5,(1+rho*rvec)/2);
13    else % do the integration; see below for f_21
14      for i=1:length(rvec), r=rvec(i);
15        f(i) = A * B(i) * f_21(0.5,0.5,n-0.5,(1+rho*r)/2);
16      end, f=f';
17    end
18  end
19  if nargout>1
20    if ~all(rvec==sort(rvec)), F=[]; 'Requires that rvec is sorted', return, end
21    F(1) = quadgk(@(rvec)corrcoefpdf(rvec,rho,n,approx),-1,rvec(1));
22    for i=2:length(rvec),  r=rvec(i);
23      F(i) = F(i-1) + quadgk(@(rvec)corrcoefpdf(rvec,rho,n,approx),
24        rvec(i-1),rvec(i));
25    end
26  end
27
28  function f=f_21(a,b,c,z)
29  s=quadgk(@(y) f_21_int(y,a,b,c,z),0,1); f = s * gamma(c) / gamma(a) /
30    gamma(c-a);
31
32  function f=f_21_int(y,a,b,c,z), f=y.^(a-1).* (1-y).^(c-a-1)./ (1-z*y).^b;
```

Program Listing 3.1: Computes the p.d.f. (3.26) and, optionally, the c.d.f. by numeric integration of the p.d.f. Use of the Laplace approximation for the $_2F_1$ function is about 42 times faster than using numeric integration to compute it; and use of the built-in Matlab routine (at least in version 7.8) to compute it is about 150 times *slower* than using numeric integration.

```
1   function themean=corrcoefmean(rho,n,approx)
2   if nargin<3, approx=1; end
3   rl=length(rho); themean=zeros(rl,1);
4   for i=1:rl
5     p=rho(i);
6     K = p*(1-p^2)^(0.5*(n-1)) * (gamma(n/2))^2 / gamma((n-1)/2) / gamma((n+1)/2);
7     if approx==1 % the Laplace approximation from page II.197
8       themean(i) = K * f21(n/2,n/2,(n+1)/2,p^2);
9     else
10      if 1==2 & exist('hypergeom','file'), themean(i) = K * hypergeom([n/2,n/2],
11        (n+1)/2,p^2);
12      else themean(i) = K * f_21(n/2,n/2,(n+1)/2,p^2);
13      end
14    end
15  end
```

Program Listing 3.2: Computes (3.28).

as now shown. From (3.30), with $\mathbf{Y} = (\mathbf{Y}_1, \ldots, \mathbf{Y}_n)$, *the log-likelihood is*

$$\ell(\boldsymbol{\mu}, \boldsymbol{\Sigma}; \mathbf{Y}) = -\frac{nd}{2} \log(2\pi) - \frac{n}{2} \log|\boldsymbol{\Sigma}| - \frac{M}{2}, \quad M = \sum_{i=1}^{n} (\mathbf{Y}_i - \boldsymbol{\mu})' \boldsymbol{\Sigma}^{-1} (\mathbf{Y}_i - \boldsymbol{\mu}).$$

Using basic properties of the matrix trace function, scalar $M = \mathrm{tr}(M)$ *is easily seen to be expressible as*

$$M = \mathrm{tr}\left[\boldsymbol{\Sigma}^{-1} \sum_{i=1}^{n} (\mathbf{Y}_i - \boldsymbol{\mu})(\mathbf{Y}_i - \boldsymbol{\mu})' \right]$$

$$= \mathrm{tr}\left[\boldsymbol{\Sigma}^{-1} \sum_{i=1}^{n} (\mathbf{Y}_i - \bar{\mathbf{Y}} + \bar{\mathbf{Y}} - \boldsymbol{\mu})(\mathbf{Y}_i - \bar{\mathbf{Y}} + \bar{\mathbf{Y}} - \boldsymbol{\mu})' \right],$$

or

$$M = \mathrm{tr}\left[\boldsymbol{\Sigma}^{-1} \left\{ \sum_{i=1}^{n} (\mathbf{Y}_i - \bar{\mathbf{Y}})(\mathbf{Y}_i - \bar{\mathbf{Y}})' + n(\bar{\mathbf{Y}} - \boldsymbol{\mu})(\bar{\mathbf{Y}} - \boldsymbol{\mu})' \right\} \right]$$

$$= n\,\mathrm{tr}[\boldsymbol{\Sigma}^{-1} \{ \mathbf{S} + (\bar{\mathbf{Y}} - \boldsymbol{\mu})(\bar{\mathbf{Y}} - \boldsymbol{\mu})' \}]$$

$$= n\,\mathrm{tr}[\boldsymbol{\Sigma}^{-1} \mathbf{S}] + n(\bar{\mathbf{Y}} - \boldsymbol{\mu})' \boldsymbol{\Sigma}^{-1} (\bar{\mathbf{Y}} - \boldsymbol{\mu}),$$

so that $\ell(\boldsymbol{\mu}, \boldsymbol{\Sigma}; \mathbf{Y})$ *is*

$$-\frac{nd}{2} \log(2\pi) - \frac{n}{2} \log|\boldsymbol{\Sigma}| - \frac{n}{2}\mathrm{tr}[\boldsymbol{\Sigma}^{-1} \mathbf{S}] - \frac{n}{2}(\bar{\mathbf{Y}} - \boldsymbol{\mu})' \boldsymbol{\Sigma}^{-1} (\bar{\mathbf{Y}} - \boldsymbol{\mu}). \qquad (3.32)$$

While matrix differentiation can be used to show (3.31), as detailed, for example, in Schott (2005, pp. 373–374) and Abadir and Magnus (2005, pp. 387–389), the method in Watson (1964) (cf. Flury, 1997, pp. 240–242) is interesting and only requires a simple application of basic univariate calculus. Maximizing (3.32) is equivalent to determining

$$\{\hat{\boldsymbol{\mu}}_{\mathrm{ML}}, \hat{\boldsymbol{\Sigma}}_{\mathrm{ML}}\} = \arg\min \ell^*(\boldsymbol{\mu}, \boldsymbol{\Sigma}; \mathbf{Y}),$$

where $\ell^*(\boldsymbol{\mu}, \boldsymbol{\Sigma}; \mathbf{Y}) = \log|\boldsymbol{\Sigma}| + \mathrm{tr}[\boldsymbol{\Sigma}^{-1} \mathbf{S}] + (\bar{\mathbf{Y}} - \boldsymbol{\mu})' \boldsymbol{\Sigma}^{-1} (\bar{\mathbf{Y}} - \boldsymbol{\mu})$. *Provided* $\boldsymbol{\Sigma}$ *(and, thus,* $\boldsymbol{\Sigma}^{-1}$*) is positive definite,* $(\bar{\mathbf{Y}} - \boldsymbol{\mu})' \boldsymbol{\Sigma}^{-1} (\bar{\mathbf{Y}} - \boldsymbol{\mu}) \geq 0$ *and is precisely zero for* $\boldsymbol{\mu} = \bar{\mathbf{Y}}$*, so that, as long as* $\hat{\boldsymbol{\Sigma}}_{\mathrm{ML}}$ *is positive definite,* $\hat{\boldsymbol{\mu}}_{\mathrm{ML}} = \bar{\mathbf{Y}}$. *To minimize* $\ell^*(\bar{\mathbf{Y}}, \boldsymbol{\Sigma}; \mathbf{Y}) = \log|\boldsymbol{\Sigma}| + \mathrm{tr}[\boldsymbol{\Sigma}^{-1} \mathbf{S}]$ *with respect to* $\boldsymbol{\Sigma}$*, subtract* $\log|\mathbf{S}|$ *from both sides, so that we wish to minimize*

$$\ell^*(\bar{\mathbf{Y}}, \boldsymbol{\Sigma}; \mathbf{Y}) - \log|\mathbf{S}| = \log|\boldsymbol{\Sigma}| + \mathrm{tr}[\boldsymbol{\Sigma}^{-1} \mathbf{S}] - \log|\mathbf{S}| = \mathrm{tr}[\boldsymbol{\Sigma}^{-1} \mathbf{S}] - \log\frac{|\mathbf{S}|}{|\boldsymbol{\Sigma}|},$$

or, using basic properties of determinants, we seek, with $\mathbf{A} = \mathbf{S}^{1/2} \boldsymbol{\Sigma}^{-1} \mathbf{S}^{1/2}$,

$$\hat{\boldsymbol{\Sigma}}_{\mathrm{ML}} = \arg\min\{\mathrm{tr}[\boldsymbol{\Sigma}^{-1} \mathbf{S}] - \log|\boldsymbol{\Sigma}^{-1} \mathbf{S}|\} = \arg\min\{\mathrm{tr}\,\mathbf{A} - \log|\mathbf{A}|\}.$$

As $\mathbf{A}$ *is symmetric and positive definite,* $\mathrm{tr}(\mathbf{A}) = \sum_{i=1}^{d} \lambda_i$ *and* $|\mathbf{A}| = \prod_{i=1}^{d} \lambda_i$*, where* $\{\lambda_i\} = \mathrm{eig}(\mathbf{A})$*, with* $\lambda_i > 0$*,* $i = 1, \ldots, d$*, so that we seek* $\min \sum_{i=1}^{d} (\lambda_i - \log \lambda_i)$*. However, it is easily confirmed that* $h(x) = x - \log x$ *is minimized at* $x = 1$*, and as the objective function is just a sum of d such functions, we require* $\lambda_i = 1$*,* $i = 1, \ldots, d$*, that is,* $\mathbf{A} = \mathbf{I}_d$*, or* $\hat{\boldsymbol{\Sigma}}_{\mathrm{ML}} = \mathbf{S}$*. It can be shown that, provided* $n > d$*,* $\mathbf{S}$ *is positive definite with probability one, confirming (3.31).*

Example 3.9 (Example 3.8, cont.)   *We fit the multivariate normal distribution to the 1945 daily returns on each of the $d = 30$ stocks composing the Dow Jones Industrial Average (DJIA) stock market index from June 2001 to March 2009. One way of assessing the goodness of fit is to plot the **Mahalanobis distance** based on estimates of μ and Σ, where the Mahalanobis distance between $\mathbf{y}$ and μ with covariance matrix Σ is*

$$m(\mathbf{y}; \mu, \Sigma) = \sqrt{(\mathbf{y} - \mu)'\Sigma^{-1}(\mathbf{y} - \mu)} = \|(\mathbf{y} - \mu)\|_{\Sigma}. \tag{3.33}$$

Under the null hypothesis of normality, and based on the true parameters, $m^2(\mathbf{Y}; \mu, \Sigma) \sim \chi_d^2$, from (A.201). Figure 3.4(a) plots $m(\mathbf{y}_i; \widehat{\mu}, \widehat{\Sigma})$, $i = 1, \ldots, 1945$, along with the 97.5% cutoff value. As these are one-at-a-time tests, we expect $1945/40 \approx 49$ values to exceed the cutoff. It is apparent that the returns do not follow a multivariate normal distribution.

The next section illustrates a flaw with this procedure – the extent of the nonnormality is far greater than it appears.

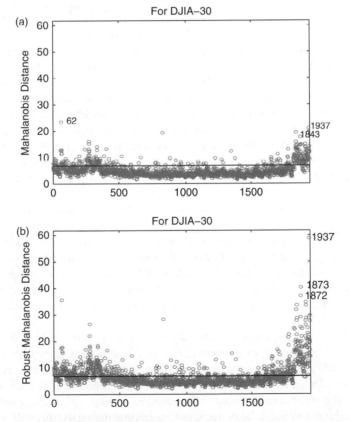

Figure 3.4   *(a) The traditional Mahalanobis distances (3.33) based on the m.l.e.s $\widehat{\mu}$ and $\widehat{\Sigma}$ for the 1945 observations of the returns on the components of the DJIA 30 index. Fifteen percent of the observations lie above the cutoff line. (b) Similar, but having used the robust Mahalanobis distance (3.34) based on the mean vector and covariance matrix from the m.c.d. method, resulting in 33% of observations above the cutoff line.*

3.1.3 Robustness and the MCD Estimator

> Econometric textbooks reveal a pronounced lack of concern for the foundations
> of probability in regard to economic phenomena, while focusing on myopic
> accounts of estimation and inference in some well-specified abstract models.
>
> (Omar Hamouda and Robin Rowley, 1996, p. 133)

> One symbol of robustness is that parameter estimates are not being unduly
> influenced by extreme observations. Some applied statisticians are haunted by
> the possibility of outliers. They believe that they must modify or eliminate
> them for inferences to be valid, instead of modifying the model (if the obser-
> vations are not in error) or accepting them as rare occurrences. Outliers are
> often the most informative part of a data set, whether by telling us that the
> data were poorly collected or that our models are inadequate. Inference proce-
> dures should not automatically mask or adjust for such possibilities but should
> highlight them.
>
> (J. K. Lindsey, 1999, p. 23)

The use of the Mahalanobis distance (3.33) is sensitive to the presence of **outliers** in
the data, or observations that do not appear to belong to the dominant group or groups of
the observed data. This is particularly the case when their number is relatively large and/or
when they are of extreme magnitudes, because they have a strong and deleterious effect on
the estimates of the mean and covariance matrix, with the pernicious effect of allowing them
to **mask** themselves. There exist several methods of **robust estimation** that can mitigate
this, and thus yield a potentially far better assessment of the actual extent of nonnormality
and more accurately identify the observations that are driving it. The idea of the **minimum
covariance determinant** (m.c.d.) method is to deliver estimates of μ and Σ for a set of data
that are (purported to be) i.i.d. from a multivariate normal distribution such that they are
resistant to outliers.

Such methods, in all contexts in statistics, including the i.i.d. setting, but also multiple and
multivariate regression, principal component analysis, discriminant analysis, classification,
and (univariate and multivariate) time series, among others, are referred to as robust techniques (methods, procedures, estimators) and the field itself is called (statistical) robustness.
Books on the methodology include Rousseeuw and Leroy (1987), Maronna et al. (2006),
Huber and Ronchetti (2009), and Aggarwal (2013), while shorter overviews can be found in
Hubert et al. (2008) and Rousseeuw and Hubert (2011). Earlier works on testing for outliers
include Ferguson (1961).

The importance of robust methods cannot be overstated, as real data will often contain
outliers. The quote above from Lindsey (1999, p. 23) alludes to their relevance in practice, and also indicates correctly that some outliers might be just recording errors. Robust
techniques can be used to help find such errors or, if the recorded data are not erroneous,
to help locate the observations that can be deemed as outliers. Of course, plotting the data
can help reveal obvious potential outliers, but for multivariate data, this will not always be
easy to do. Inspection of a plot of statistics that reduces multivariate data to a univariate
dimension, such as the Mahalanobis distance, may be highly unreliable in this regard, due
to the masking effect.

For a given d-dimensional data set with n observations, the idea underlying the m.c.d.
method is to determine the group of h observations, $d + 1 \leq h \leq n$, such that this subset has

the smallest determinant of its sample covariance matrix. The exact algorithm for the m.c.d. estimator was introduced by Rousseeuw (1984), and involves inspecting the determinant of all the $\binom{n}{h}$ subsets of size h, $h = d + 1, \ldots, n$. The m.c.d. estimator for the location vector is then the usual plug-in estimator (sample mean), but applied to the h-subset. That of the covariance matrix is similar, but scaled by a consistency factor. The properties of the exact m.c.d. estimator were subsequently studied by Lopuhaa and Rousseeuw (1991) and Butler et al. (1993).

The dimension of the subsets, h, determines the tolerance for outlier contamination, given as $1 - \alpha := 1 - h/n$. In robust statistics, the **breakdown point** corresponds to the maximum percentage of outliers an estimator can tolerate before failing or breaking down. The m.c.d. estimator achieves the so-called highest breakdown point, 50%, when $h = (n + d + 1)/2$, or, in terms of proportion, when $\alpha \approx 0.5$ for large n/d. However, high outlier resistance comes at the cost of lower efficiency. This tradeoff is dealt with via the tuning parameter $\alpha \in (0.5, 1)$, the most common choice of which is $\alpha = 0.75$, resulting in an acceptable balance between efficiency and robustness to outliers. As $\alpha \to 0.5$, the estimator attains the maximum breakdown point but loses efficiency, while $\alpha = 1$ results in the classical estimator, with $h = n$. For reliable estimation, a typical suggestion is to ensure that $n > 5d$.

Application of the exact m.c.d. method becomes computationally prohibitive as n increases. To address this, an algorithm resulting in a substantial decrease in computation time, termed FASTMCD, was developed by Rousseeuw and Van Driessen (1999), and further improved in terms of speed by Hubert et al. (2012) with the so-called DetMCD algorithm. The two algorithms differ in how the initial estimates are computed, with the former using a sampling procedure of $d + 1$ of the h-subsets, while the latter uses a well-defined set of robust estimators. The benefit of the former is that it inherits the high-breakdown property of the m.c.d. estimator and is affine equivariant, but its estimators are stochastic in the sense that they rely on random sampling of subsets. The benefit of the latter is that the resulting m.c.d. mean and covariance estimators are always the same for a given data sample. However, unlike the former, it is only imperfectly affine equivariant. Matlab code has been kindly made available by the research group ROBUST@Leuven. Verboven and Hubert (2005) give more detail on the method, while Fauconnier and Haesbroeck (2009) provide a discussion on choosing the m.c.d. tuning parameters.

Similar to the Mahalanobis distance (3.33), outliers are identified in the m.c.d. method via the so-called **robust Mahalanobis distance** (r.M.d.), given by

$$\text{RMD}(\mathbf{y}; \alpha) = \sqrt{(\mathbf{y} - \hat{\boldsymbol{\mu}}_r)' \hat{\mathbf{S}}_r^{-1} (\mathbf{y} - \hat{\boldsymbol{\mu}}_r)}, \tag{3.34}$$

where $\mathbf{y}$ is a d-dimensional column vector, and $\hat{\boldsymbol{\mu}}_r$ and $\hat{\mathbf{S}}_r$ are robust estimators of the location vector and scatter matrix, respectively. Observe that the r.M.d. depends on the choice of α – it controls the amount of outlier contamination. These r.M.d. values are compared to a pre-specified cutoff level, given by the square root of the q-quantile of a $\chi^2(d)$ distribution. Its exact finite-sample distribution is unknown. Hardin and Rocke (2005) introduce an approximation of the exact distribution, while a further extension for an outlier identification methodology has been proposed by Cerioli et al. (2009) and Cerioli (2010).

Figure 3.4(b) shows the r.M.d. (3.34), indicating that the extent of nonnormality of the DJIA stock returns data is much stronger than the traditional Mahalanobis distance indicates. In particular, 33% of the observations lie above the cutoff line, and the plot also shows that the violations occur predominantly at the end of the time period. This period is characterized by massive price drops and high volatility, as occurred during the banking and liquidity crisis starting around mid-2007.[2] This robust estimator is clearly a superior tool in situations such as this, in which a substantial number of observations are present that deviate from the overall typical behavior, and cause genuine outliers to be masked.

3.1.4 Asymptotic Properties of the Maximum Likelihood Estimator

> If you need to use asymptotic arguments, do not forget to let the number of observations tend to infinity.
>
> (Lucien Le Cam, 1990, p. 165)[3]

In order to emphasize the underlying concepts associated with the asymptotic properties of the m.l.e., we provide detail for only the scalar parameter case, with the extension to the vector parameter case being conceptually similar. References are given below for the reader interested in seeing the proofs for the latter case.

Under the appropriate regularity conditions, the estimator $\hat{\theta}_{\mathrm{ML}}$ is strongly consistent. Furthermore, the asymptotic distribution of the m.l.e. of univariate θ with true value θ_0 is given by

$$(J(\theta_0))^{1/2}(\hat{\theta}_{\mathrm{ML}} - \theta_0) \xrightarrow{d} N(0, 1) \quad \text{or} \quad \hat{\theta}_{\mathrm{ML}} \overset{\text{asy}}{\sim} N(\theta_0, \ 1/J(\theta_0)), \tag{3.35}$$

where the latter term is informal notation. Both expressions convey that $\hat{\theta}_{\mathrm{ML}}$ is asymptotically unbiased and normally distributed with variance given by the reciprocal of the information. From (3.14) and (3.20) we see that, in the i.i.d. setting, the information is linear in n, the number of observations. Thus, (3.35) and Chebyshev's inequality (A.231) imply $\hat{\theta}_{\mathrm{ML}}$ is weakly consistent, recalling the definition of weak consistency given in Section 1.1.1. We now demonstrate this more rigorously, first showing that $\hat{\theta}_{\mathrm{ML}}$ is strongly consistent, and then proving (the first – formally correct – expression in) (3.35).

The starting point for demonstrating consistency is setting up the expectation of the log of the likelihood ratio. Let X be a continuous random variable with p.d.f. $f(x; \theta)$ and support S. (The proof is analogous in the discrete case.) The **Kullback–Leibler information** (number or criterion) with respect to X is defined as

$$K_X(\theta_0, \theta_1) = \mathbb{E}_{\theta_0}\left[\ln \frac{f(X; \theta_0)}{f(X; \theta_1)}\right], \tag{3.36}$$

where $\mathbb{E}_{\theta_0}[g(X)]$ denotes the expectation of the function $g(X)$ when $\theta = \theta_0$. Observe that, from Jensen's inequality (A.49) and as $\ln(x)$ is concave,

$$-K_X(\theta_0, \theta_1) = \mathbb{E}_{\theta_0}\left[\ln \frac{f(X; \theta_1)}{f(X; \theta_0)}\right] \leq \ln \mathbb{E}_{\theta_0}\left[\frac{f(X; \theta_1)}{f(X; \theta_0)}\right],$$

---

[2] The starting date for the sub-prime crisis is often taken to be early August, 2007; see Covitz et al. (2013) and the references therein.

[3] Lucien Le Cam (1924–2000) is recognized as one of the most important twentieth-century mathematical statisticians, most notably for his work on the so-called local asymptotic normality (LAN) condition. See the obituary by Rudolf Beran and Grace Yang for further details; https://www.stat.berkeley.edu/~rice/LeCam/obituary.html.

with strict inequality holding unless $f(X; \theta_1)/f(X; \theta_0) = 1$, which will not be the case if $\theta_0 \neq \theta_1$ and the family f is identified; recall the definition in Section 5.1.1. As

$$\mathbb{E}_{\theta_0} \left[\frac{f(X; \theta_1)}{f(X; \theta_0)} \right] = \int_S \frac{f(x; \theta_1)}{f(x; \theta_0)} f(x; \theta_0) \, \mathrm{d}x = 1,$$

it follows that $K_X(\theta_0, \theta_1) \geq 0$. Fix $\delta > 0$ and let

$$\mu_1 = \mathbb{E}_{\theta_0} \left[\ln \frac{f(X; \theta_0 - \delta)}{f(X; \theta_0)} \right] \leq 0, \quad \mu_2 = \mathbb{E}_{\theta_0} \left[\ln \frac{f(X; \theta_0 + \delta)}{f(X; \theta_0)} \right] \leq 0.$$

Denote the log-likelihood $\ell(\theta; \mathbf{X})$ based on the n i.i.d. observations $\mathbf{X} = (X_1, \ldots, X_n)$ by $\ell_n(\theta)$. The strong law of large numbers (A.272) implies $n^{-1}[\ell_n(\theta_0 - \delta) - \ell_n(\theta_0)] \xrightarrow{a.s.} \mu_1$ and $n^{-1}[\ell_n(\theta_0 + \delta) - \ell_n(\theta_0)] \xrightarrow{a.s.} \mu_2$, so that, w.p. 1, as $n \to \infty$, $\ell_n(\theta_0 - \delta) < \ell_n(\theta_0)$ and $\ell_n(\theta_0 + \delta) < \ell_n(\theta_0)$. Thus, as $n \to \infty$, there exists an estimator $\hat{\theta}_n$ that maximizes the log-likelihood on $(\theta_0 - \delta, \theta_0 + \delta)$ for any $\delta > 0$. As θ_0 is the true parameter and $\delta > 0$ is arbitrary, this implies that, as $n \to \infty$, the likelihood function will be larger at θ_0 than at any other value (again assuming f is identified), and thus the m.l.e. is strongly consistent.

We now prove (3.35). Assume as above that θ_0 is the true parameter, and that $\ell(\theta; \mathbf{X})$ is twice differentiable on a neighborhood of θ_0. Suppressing the dependence of the log-likelihood on $\mathbf{X}$ for notational convenience, use of the zero-order Taylor series (3.2) applied to $\dot{\ell}(\theta)$ about $\hat{\theta}_{\mathrm{ML}}$ gives $\dot{\ell}(\theta) = \dot{\ell}(\hat{\theta}_{\mathrm{ML}}) + \ddot{\ell}(\theta^*)(\theta - \hat{\theta}_{\mathrm{ML}})$ for $\theta^* \in (\theta, \hat{\theta}_{\mathrm{ML}})$. Using $\theta = \theta_0$ and $\dot{\ell}(\hat{\theta}_{\mathrm{ML}}) = 0$ yields

$$\hat{\theta}_{\mathrm{ML}} - \theta_0 = -\frac{\dot{\ell}(\theta_0)}{\ddot{\ell}(\theta^*)}, \quad \text{for } \theta^* \in (\theta_0, \hat{\theta}_{\mathrm{ML}}). \tag{3.37}$$

Recalling the Fisher information $J(\theta) = -\mathbb{E}[\ddot{\ell}(\theta; \mathbf{X})]$, we have

$$(J(\theta_0))^{1/2}(\hat{\theta}_{\mathrm{ML}} - \theta_0) = \frac{\dot{\ell}(\theta_0)}{(J(\theta_0))^{1/2}} \times \frac{\ddot{\ell}(\theta_0)}{\ddot{\ell}(\theta^*)} \times \left(-\frac{\ddot{\ell}(\theta_0)}{J(\theta_0)} \right)^{-1}. \tag{3.38}$$

As the X_i are i.i.d., the central limit theorem (A.160) in conjunction with (3.6) and (3.9) shows that the first term converges in distribution to a N(0, 1) random variable. We need to show that the other two converge in probability to 1, so that application of Slutsky's theorem (A.281) yields the desired result (3.35). This is the case for the last term in (3.38) from the i.i.d. nature of the X_i and the weak law of large numbers (A.255).

The middle term appears intuitively to converge to 1, as $\theta^* \in (\theta_0, \hat{\theta}_{\mathrm{ML}})$ and $\hat{\theta}_{\mathrm{ML}}$ is consistent. More rigorously, for sample size n, the reciprocal of the middle term minus one can be expressed as

$$\frac{\ddot{\ell}(\theta^*)}{\ddot{\ell}(\theta_0)} - 1 = \frac{\ddot{\ell}(\theta^*) - \ddot{\ell}(\theta_0)}{n} \times \left(\frac{\ddot{\ell}(\theta_0)}{n} \right)^{-1}$$

$$= \frac{\ddot{\ell}(\theta^*) - \ddot{\ell}(\theta_0)}{n} \times \left(\frac{\ddot{\ell}(\theta_0)}{J(\theta_0)} \right)^{-1} \times \left(\frac{J(\theta_0)}{n} \right)^{-1}, \tag{3.39}$$

with the middle term in (3.39) converging in probability to 1 as in (3.38), and the latter term, recalling (3.14), converging in probability to $-1/\mathbb{E}[\ddot{\vartheta}_1(\theta)]$. It remains to show that the first term in (3.39) converges to zero. To this end, first recall the mean value theorem. Let f be a differentiable function on (a, b). Then $\exists \xi \in (a, b)$ such that $f(b) - f(a) = f'(\xi)(b - a)$. Then, (i) if there is a function $g(x)$ such that $\mathbb{E}_{\theta_0}[|g(X)|] < \infty$, (ii) assuming the third derivative $\dddot{\ell}(\theta; X)$ exists, and (iii) assuming there is an open neighborhood G of θ_0 such that, for all $\theta \in G$, $\dddot{\ell}(\theta; x) \leq g(x)$ for all $x \in S_X$, then the mean value theorem implies

$$\frac{\ddot{\ell}(\theta^*) - \ddot{\ell}(\theta_0)}{n} \leq |\theta^* - \theta_0| \times \frac{\sum_{i=1}^n g(X_i)}{n}. \tag{3.40}$$

The weak law of large numbers ensures that the last term in (3.40) converges to a constant, while term $|\theta^* - \theta_0|$ converges to zero because of the consistency of the m.l.e. and since $\theta^* \in (\theta_0, \hat{\theta}_{ML})$ implies $|\theta^* - \theta_0| < |\hat{\theta}_{ML} - \theta_0|$. We are finally done. It should now be clear to the reader why most introductory accounts, including this one, do not formally list all the required regularity conditions!

Remarks

(a) More rigorous proofs and the precise regularity conditions, also for the vector parameter case and functions of $\hat{\theta}_{ML}$ (both stated below), are given, for example, in Ferguson (1996), Lehmann and Casella (1998), and Bhattacharya et al. (2016). See also Pawitan (2001) for a book-length treatment of the mechanics and applications of likelihood, at the same level as here, and Severini (2000) for a much more advanced treatise.

(b) In practice, as θ_0 is not known, $J(\theta_0)$ in (3.35) is taken to be the expected information $J(\hat{\theta}_{ML})$ or the observed information $I(\hat{\theta}_{ML}; \mathbf{X}) = -\ddot{\ell}(\hat{\theta}_{ML}; \mathbf{X})$ for computing the standard error and building approximate confidence intervals; see (3.45) below. Use of the observed information will clearly be much easier in many problems, and tends to result in more accurate inference, with the canonical reference being Efron and Hinkley (1978).

(c) The opening quote by Le Cam (1990) serves as a reminder that the asymptotic distribution (3.35) is just that, and it says nothing about the quality of the approximation in finite samples. It is a remarkable result that can be useful for inference for "adequate" sample sizes, though the bootstrap is more reliable for constructing confidence intervals, which is why it was presented so early, in Chapter 1.

(d) We have assumed throughout that the distributional family of the X_i is correctly specified – this being formal language for "it is known," and the notion of knowing the true data-generating process up to parameter uncertainty is, in most contexts of genuine interest (particularly econometrics and other social sciences), an absurdity.

To help counter this, one can investigate the asymptotic behavior of the m.l.e. when the model is misspecified. Suppose the data are i.i.d. with distribution $g \neq f$, and we wrongly assume f. Then $\ell(\theta; \mathbf{X})$ is referred to as the **quasi-log-likelihood**; the Kullback–Leibler information criterion (3.36) can be expressed in obvious notation as $K(g, f_\theta)$; and $\hat{\theta}_{ML}$ is the value of θ that minimizes $K(g, f_\theta)$.

Informally, from (3.37) and using the k-variate case (as the result is nearly always stated in the vector parameter case in the literature) with $\mathbf{H} := \ddot{\ell}(\hat{\boldsymbol{\theta}}_{ML})$

as the Hessian, or negative of the observed information matrix, at the m.l.e., and $\mathbf{O} := \dot{\ell}(\widehat{\theta}_{ML})\dot{\ell}(\widehat{\theta}_{ML})'$ the outer product of the gradient at the m.l.e., we have

$$\mathbb{V}(\widehat{\theta}_{ML}) \approx \mathbf{V} := (-\mathbf{H}^{-1})\mathbf{O}(-\mathbf{H}^{-1}), \tag{3.41}$$

where $\approx$ informally denotes notation that asymptotically, under the true data-generating process, the ratio of the corresponding elements of the left- and right-hand sides of (3.41) tends to 1 as $n \to \infty$, recalling (3.18) and (3.19). The matrix $\mathbf{V}$ is referred to as the **(Huber) sandwich estimator**, with the square roots of its diagonal elements serving as robust standard errors against misspecifications of f, or sometimes "Huber–White standard errors," in honor of Huber (1967) and White (1982), where formal proofs can be found that this variance estimator makes sense when the assumed f is not the true data-generating process. (See also Chow, 1984, regarding a critique of White's paper in the non-i.i.d. case.)

The interpretation of $\widehat{\theta}_{ML}$ is less clear, being the parameter under f that minimizes $K(g, f_\theta)$. As Freedman (2006, p. 299) states regarding the use of (3.41), "the sandwich may help on the variance side, but the parameters being estimated by the m.l.e. are likely to be meaningless." A review and further original references of the sandwich estimator can be found in Hardin (2003).

Importantly, the difference between $\mathbf{H}^{-1}$ and $\mathbf{O}$ can be used to detect model misspecification; see McCabe and Leybourne (2000), Golden et al. (2016), and the references therein. ∎

Example 3.10   *Let $X_i \overset{\text{i.i.d.}}{\sim} \text{Cau}(\mu, 1)$, where μ is a location parameter. With*

$$f_X(x) = \frac{1}{\pi} \cdot \frac{1}{1 + (x - \mu)^2}, \quad \vartheta(\mu) = -\log \pi - \log(1 + (x - \mu)^2),$$

it is easy to check that

$$\dot{\vartheta}(\mu) = \frac{2(x - \mu)}{1 + (x - \mu)^2}, \quad \ddot{\vartheta}(\mu) = 2\frac{(x - \mu)^2 - 1}{(1 + (x - \mu)^2)^2},$$

from which we see that, unless n is extremely small, the m.l.e. of μ needs to be obtained by numerically maximizing the likelihood; see Chapter 4. Problem 3.6 shows that

$$J(\mu) = -\mathbb{E}[\ddot{\ell}(\mu; \mathbf{X})] = -n\mathbb{E}[\ddot{\vartheta}] = \frac{n}{2}, \tag{3.42}$$

which implies (using our informal notation for asymptotic convergence with n on the right-hand side)

$$\widehat{\mu}_{ML} \overset{\text{asy}}{\sim} N\left(\mu, \frac{2}{n}\right),$$

assuming, correctly, that the regularity conditions are satisfied.

Asymptotic consistency and normality hold also for the more general case of the (not necessarily one-to-one) differentiable function $\tau(\widehat{\theta}_{ML})$, referred to as the **delta method**, with

$$\tau(\widehat{\theta}_{ML}) \overset{\text{asy}}{\sim} N(\tau(\theta), \dot{\tau}^2/J). \tag{3.43}$$

See the references in Remark (a) above for proof.

Example 3.11   *Let $X_i \overset{\text{i.i.d.}}{\sim} \text{Exp}(\lambda)$, $i = 1, \ldots, n$. Example 3.3 showed that the m.l.e. of λ is $\bar{X}^{-1}$, while straightforward calculation shows $J^{-1} = \lambda^2/n$, yielding (in our informal notation) $\bar{X}^{-1} \overset{\text{asy}}{\sim} N(\lambda, \lambda^2/n)$.*

From the invariance property, the m.l.e. of $\tau(\lambda) = \exp(-K\lambda)$ is just $\exp(-K/\bar{X})$, with $\dot{\tau}^2/J = \lambda^2 K^2 e^{-2K\lambda}/n$, so that $\exp\{-K/\bar{X}\} \overset{\text{asy}}{\sim} N(e^{-K/\lambda}, \lambda^2 K^2 e^{-2K\lambda}/n)$.

Result (3.35) can be generalized to the vector parameter case. The asymptotic distribution of the m.l.e. of $\boldsymbol{\theta}$ is given by

$$\hat{\boldsymbol{\theta}}_{\text{ML}} \overset{\text{asy}}{\sim} N_k(\boldsymbol{\theta}, \mathbf{J}^{-1}). \tag{3.44}$$

Similarly to the univariate case, as $\boldsymbol{\theta}$ is not known, $\mathbf{J}(\boldsymbol{\theta})$ is taken to be $\mathbf{J}(\hat{\boldsymbol{\theta}}_{\text{ML}})$.

Example 3.12   *For the two-parameter gamma model in Example 3.6 with $\boldsymbol{\theta} = (\alpha, \beta)'$, it follows from (3.44) that*

$$\begin{bmatrix} \hat{\alpha}_{\text{ML}} \\ \hat{\beta}_{\text{ML}} \end{bmatrix} \overset{\text{asy}}{\sim} N_2 \left(\begin{bmatrix} \alpha \\ \beta \end{bmatrix}, \frac{1}{n(\alpha \psi'(\alpha) - 1)} \begin{bmatrix} \alpha & \beta \\ \beta & \beta^2 \psi'(\alpha) \end{bmatrix} \right)$$

(again using the informal notation with n appearing in the asymptotic density).

Example 3.13 (Bivariate normal)   *From (3.44),*

$$\hat{\boldsymbol{\theta}}_{\text{ML}} = (\hat{\mu}_1, \hat{\mu}_2, \hat{\sigma}_1^2, \hat{\sigma}_2^2, \hat{\rho})' \overset{\text{asy}}{\sim} N(\boldsymbol{\theta}, \mathbf{J}^{-1}),$$

with standard calculations showing that

$$\mathbf{J}^{-1} = \begin{bmatrix} \mathbf{U}^{-1} & \mathbf{0} \\ \mathbf{0} & \mathbf{V}^{-1} \end{bmatrix}, \quad \mathbf{U}^{-1} = \frac{1}{n} \begin{bmatrix} \sigma_1^2 & \rho \sigma_1 \sigma_2 \\ \cdot & \sigma_2^2 \end{bmatrix},$$

and

$$\mathbf{V}^{-1} = \frac{1}{n} \begin{bmatrix} 2\sigma_1^4 & 2\rho^2 \sigma_1^2 \sigma_2^2 & \rho(1 - \rho^2)\sigma_1^2 \\ \cdot & 2\sigma_2^4 & \rho(1 - \rho^2)\sigma_2^2 \\ \cdot & \cdot & (1 - \rho^2)^2 \end{bmatrix}.$$

See, for example, Stuart et al. (1999, pp. 78–79) for details.

Notice that (i) when $\rho = 0$, the result simplifies to the case of two independent univariate normal samples via (3.21), and (ii) as $|\rho| \to 1$, $\mathbb{V}(\hat{\rho}) \to 0$. More detail on estimation of ρ will be given in Section 7.4.3.

Denoting the ith diagonal element of $\mathbf{J}^{-1}$ by j_{ii}, $i = 1, \ldots, k$, where k is the number of model parameters, an approximate, asymptotically valid, 95% **Wald**[4] confidence interval for the ith element of $\boldsymbol{\theta}$ is

$$\hat{\theta}_i \pm 1.96 j_{ii}^{1/2}, \tag{3.45}$$

---

[4] After Abraham Wald (1902–1950). Use of the Wald interval for parameter p of the binomial model goes back to Laplace. See the Wikipedia entry on him for further information, including the posthumous arguments between Ronald Fisher and Jerzy Neyman over his work.

where $\hat{\theta}_i$ is the ith element of $\hat{\theta}_{ML}$, $i = 1, \ldots, k$. The value 1.96 is just the usual three-digit approximation of the 0.025 quantile of the standard normal distribution, and can be replaced by the appropriate quantile corresponding to the desired nominal coverage probability. These c.i.s can be computed for any set of the k parameters, though it is crucial to emphasize that they are one-at-a-time or individual c.i.s, and do not represent a joint or simultaneous confidence region.

Most generally, for the function $\tau(\theta) = (\tau_1(\theta), \ldots, \tau_m(\theta))'$ from $\mathbb{R}^k$ to $\mathbb{R}^m$,

$$\tau(\hat{\theta}_{ML}) \stackrel{asy}{\sim} N(\tau(\theta), \; \dot{\tau}J^{-1}\dot{\tau}'), \tag{3.46}$$

where $\dot{\tau} = \dot{\tau}(\theta)$ denotes the matrix with (i, j)th element $\partial \tau_i(\theta)/\partial \theta_j$.

Example 3.14 (Normal, cont.) *It follows from Example 3.5 and (3.44) that*

$$\hat{\theta}_{ML} = \begin{bmatrix} \hat{\mu}_{ML} \\ \hat{\sigma}^2_{ML} \end{bmatrix} \stackrel{asy}{\sim} N_2 \left(\begin{bmatrix} \mu \\ \sigma^2 \end{bmatrix}, \frac{1}{n} \begin{bmatrix} \sigma^2 & 0 \\ 0 & 2\sigma^4 \end{bmatrix} \right).$$

*Let τ be the **coefficient of variation**, defined by $\tau(\theta) = \sigma/\mu$ for $\mu \neq 0$, and discussed, for example, in Snedecor and Cochran (1967, pp. 62–64). Then $k = 2$, $m = 1$, and, keeping in mind that the asymptotic distribution is in terms of σ^2 (and not σ),*

$$\dot{\tau} = \left[-\frac{\sigma}{\mu^2}, \; \frac{1}{2\sigma\mu} \right], \quad \dot{\tau}J^{-1}\dot{\tau}' = \frac{2\sigma^4 + \sigma^2\mu^2}{2\mu^4 n} =: V,$$

and $\tau(\hat{\theta}_{ML}) \stackrel{asy}{\sim} N(\sigma/\mu, V)$ from (3.46). Figure 3.5 compares the actual density of the m.l.e. of the coefficient of variation (obtained via simulation and kernel density estimation) and the asymptotic distribution based on the true parameters for two sample sizes. The reader is encouraged to replicate these graphs. In practice, estimates of σ and μ would have to be used in the expression for V.

3.2 CRAMÉR–RAO LOWER BOUND

This remarkably simple result, which we abbreviate by CRlb, gives a lower bound on the variance of an unbiased estimator of $\tau(\theta)$ and was obtained independently by several authors in the 1940s. While most of the English literature uses the name Cramér–Rao lower bound or Cramér–Rao inequality, some books, such as Gourieroux and Monfort (1995), refer to it (arguably correctly) as the Fréchet–Darmois–Cramér–Rao inequality.[5]

---

[5] Reference to the Cramér–Rao inequality appears in Neyman and Scott (1948), recognizing the influential book by Cramér in 1946, and article by C. R. Rao in 1945, in which it appeared. Savage in 1954 drew attention to the work of Fréchet in 1943 and of Darmois in 1945, where the result also appeared, and proposed calling it just the **information inequality**. He was partially successful: a more general form of the inequality has since been developed and is indeed referred to as such; see, for example, Lehmann and Casella (1998, Sec. 2.6). See David (1995), Stigler (1999), and the references therein for further details on this and the origins of other naming conventions in probability and statistics, as well as the Wikipedia entry "History of statistics", and the references and linked web pages therein.

According to the interview by Bera (2003, p. 344), C. R. Rao worked on the problem one night after having been asked a question the previous day by a student on the Calcutta University master's program, V. M. Dandekar (who later became one of India's leading economists). Due to the suspension of certain publications in India during the Second World War, the result was first published in 1945.

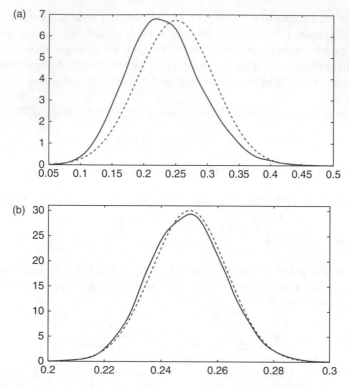

Figure 3.5 *Kernel density estimate using 10,000 replications of the coefficient of variation based on $\mu = 100$ and $\sigma = 25$ (solid) and the asymptotic normal distribution (dashed), for $n = 10$ (a) and $n = 200$ (b).*

3.2.1 Univariate Case

Cramér–Rao lower bound (single-parameter case) Let $U = U(\mathbf{X})$ be an unbiased estimator for θ. Then $\mathbb{V}(U) \geq 1/J(\theta)$. More generally, if $U(\mathbf{X})$ is unbiased for $\tau(\theta)$, then

$$\mathbb{V}(U) \geq \frac{\dot{\tau}^2(\theta)}{J(\theta)} = \frac{(d\tau / d\theta)^2}{J(\theta)}. \tag{3.47}$$

Recalling (3.43), it is imperative to note that, asymptotically, the m.l.e. reaches the CRlb, again underscoring its prominence as an estimator.

The conditions for (3.47) to hold include that (i) θ lies in an open subset Θ_* of $\Theta \subset \mathbb{R}$, (ii) the sample observations are i.i.d. with common p.d.f. $f_X(\cdot; \theta)$ and support S_X, (iii) S_X does not depend on θ, and (iv) $\log f_X(x; \theta)$ possesses second derivatives for all $\theta \in \Theta_*$ and almost all $x \in S_X$. Additionally, τ is a differentiable function of θ for all $\theta \in \Theta_*$. Condition

(iii) is usually replaced with

$$\frac{\partial}{\partial \theta} \int_{\mathbb{R}^n} h(\mathbf{x}) f_{\mathbf{X}}(\mathbf{x}; \theta) d\mathbf{x} = \int_{\mathbb{R}^n} h(\mathbf{x}) \frac{\partial}{\partial \theta} f_{\mathbf{X}}(\mathbf{x}; \theta) d\mathbf{x}, \qquad (3.48)$$

for any function $h : S_X^n \to \mathbb{R}$ such that $\mathbb{E}[|h(\mathbf{X})|] < \infty$.[6]

Proof. Let $U(\mathbf{X})$ be unbiased for $\tau(\theta)$. As $\mathbb{E}[\dot{\ell}(\theta; \mathbf{X})] = 0$ from (3.6) and condition (3.48),

$$\text{Cov}(U, \dot{\ell}) = \mathbb{E}[U\dot{\ell}]$$

$$= \int_{\mathbb{R}^n} U \frac{\dot{\ell}}{\mathcal{L}} f_{\mathbf{X}}(\mathbf{x}; \theta) d\mathbf{x} = \int_{\mathbb{R}^n} U\dot{\mathcal{L}} \, d\mathbf{x} = \frac{d}{d\theta} \int_{\mathbb{R}^n} U\mathcal{L} \, d\mathbf{x} = \frac{d}{d\theta} \mathbb{E}[U] = \dot{\tau}(\theta).$$

Then, as $\mathbb{V}(\dot{\ell}(\theta; \mathbf{X})) = J(\theta)$ from (3.12), the Cauchy–Schwarz inequality (A.54) implies

$$\mathbb{V}(U) \geq \frac{[\text{Cov}(U, \dot{\ell})]^2}{\mathbb{V}(\dot{\ell})} = \frac{\dot{\tau}^2(\theta)}{J(\theta)}.$$

The discrete case and its relevant assumptions follow analogously.   ∎

If an unbiased estimator U has a variance that coincides with the CRlb, then it must be what is called the **uniformly minimum variance unbiased estimator**, (u.m.v.u.e.; see Section 7.2), where "uniform" means for all $\theta \in \Theta_*$. In general, an u.m.v.u.e. may not reach the bound. If the variance of an unbiased estimator U is precisely the CRlb, then U is also referred to as the **minimum variance bound estimator** (m.v.b.e.).

Example 3.15 (Poisson, cont.)   *From Example 3.1, $J(\theta) = n/\theta$ and the CRlb is $1/J = \theta/n$. This coincides with the variance of the unbiased estimator $\bar{X}_n$, so that $\bar{X}_n$ is the u.m.v.u.e.*

Example 3.16 (Bernoulli, cont.)   *Example 3.4 derived the information $J(\theta)$, so that the CRlb of θ is $1/J = \theta(1 - \theta)/n$. This coincides with the variance of $\bar{X}$ (which is unbiased), showing that $\bar{X}$ is the u.m.v.u.e. of θ.*

For the CRlb of $\tau(\theta) = \theta(1 - \theta)$, we compute $\dot{\tau}(\theta) = 1 - 2\theta$, yielding a lower bound of $\dot{\tau}^2/J = (1 - 2\theta)^2 \theta(1 - \theta)/n$. Problem 7.13. derives the u.m.v.u.e. for $\tau(\theta)$ and compares its variance to its CRlb.

The CRlb is attained if $U(\mathbf{X})$ and $\dot{\ell}(\theta; \mathbf{X})$ are linearly related as, say,

$$\dot{\ell}(\theta; \mathbf{X}) - \mathbb{E}[\dot{\ell}(\theta; \mathbf{X})] = k(\theta) \cdot (U(\mathbf{X}) - \mathbb{E}[U(\mathbf{X})]),$$

---

[6] Technically speaking, this restriction can be relaxed because the proof will require only two cases of h in (3.48), namely taking h to be $U(\mathbf{X})$, where $\mathbb{E}[U(\mathbf{X})] = \tau(\theta)$, and $h = 1$, which is necessary to show that $\mathbb{E}[\dot{\ell}(\theta; \mathbf{X})] = 0$. No examples will be presented here that require this added flexibility.

for some k, which may be a function of θ (and any other constant parameters, such as n), but not of $\mathbf{X}$. As $\mathbb{E}[U(\mathbf{X})] = \tau(\theta)$ and $\mathbb{E}[\dot{\ell}(\theta; \mathbf{X})] = 0$, this can be written simply, with $U = U(\mathbf{X})$ and $\dot{\ell} = \dot{\ell}(\theta; \mathbf{X})$, as

$$\dot{\ell} = k(\theta)(U - \tau(\theta)). \tag{3.49}$$

For the Poisson model with $s = \sum_{i=1}^{n} x_i$, $\dot{\ell} = s/\theta - n$, so that, taking $U(\mathbf{x}) = s/n$ and $\tau(\theta) = \theta$,

$$\frac{s}{\theta} - n = k(\theta)\left(\frac{s}{n} - \theta\right) \quad \text{for } k(\theta) = \frac{n}{\theta}. \tag{3.50}$$

Recall from (A.35) that a family of distributions indexed by parameter vector $\theta = (\theta_1, \ldots, \theta_k)'$ belongs to the **exponential family** if it can be algebraically expressed as

$$f(x; \theta) = a(\theta)b(x) \exp\left\{\sum_{i=1}^{k} c_i(\theta)d_i(x)\right\}, \tag{3.51}$$

where $a(\theta) \geq 0$ and $c_i(\theta)$ are real-valued functions of θ but not x; and $b(x) \geq 0$ and $d_i(x)$ are real-valued functions of x but not θ. The Poisson is easily seen to be in the exponential family, for which the general result is true: relation (3.49) holds if and only if $f_{\mathbf{X}}$ belongs to the one-parameter exponential family.

Proof. ($\Leftarrow$) Let $f_{\mathbf{X}}(\mathbf{x}; \theta) = a(\theta)b(\mathbf{x}) \exp\{c(\theta)d(\mathbf{x})\}$ so that

$$\dot{\ell}(\theta) = \frac{d}{d\theta} \log f_{\mathbf{X}}(\mathbf{x}; \theta) = \frac{d}{d\theta}(\log a(\theta) + \log b(\mathbf{x}) + c(\theta)d(\mathbf{x})) = \frac{a'(\theta)}{a(\theta)} + c'(\theta)d(\mathbf{x})$$

$$= c'(\theta)\left[d(\mathbf{x}) - \left(-\frac{a'(\theta)/a(\theta)}{c'(\theta)}\right)\right] = c'(\theta)[U - \tau(\theta)]. \tag{3.52}$$

($\Rightarrow$) Integrate both sides of (3.49) with respect to θ (which requires adding a constant of integration, say $h(\mathbf{x})$, to one side) to get

$$\int_{\Theta} \frac{d}{d\theta} \log f_{\mathbf{X}}(\mathbf{x}; \theta)\, d\theta = U(\mathbf{x})\int_{\Theta} k(\theta)\, d\theta - \int_{\Theta} k(\theta)\tau(\theta)\, d\theta + h(\mathbf{x}),$$

or

$$f_{\mathbf{X}}(\mathbf{x}; \theta) = \exp\left\{U(\mathbf{x})\int_{\Theta} k(\theta)\, d\theta - \int_{\Theta} k(\theta)\tau(\theta)\, d\theta + h(\mathbf{x})\right\},$$

which is a one-parameter exponential family with $a(\theta) = \exp\left\{-\int_{\Theta} k(\theta)\tau(\theta)\, d\theta\right\}$, $b(\mathbf{x}) = \exp\{h(\mathbf{x})\}$, $c(\theta) = \int_{\Theta} k(\theta)\, d\theta$ and $d(\mathbf{x}) = U(\mathbf{x})$. ∎

It is easy to see that, if (3.49) holds for some $\tau(\theta)$, then it will also hold for linear functions of $\tau(\theta)$. Otherwise, there is no other $\tau(\theta)$ for which it holds.

Example 3.17 (Scale gamma) *Let $X_i \overset{\text{i.i.d.}}{\sim} \text{Gam}(\alpha, \beta)$ with α known and β an (inverse) scale parameter, so that $f_{\mathbf{X}}$ belongs to the exponential family with $a(\beta) = \beta^{\alpha n}$, $b(\mathbf{x}) = b(\mathbf{x}; \alpha, n) = \exp\left\{(\alpha - 1)\sum \log x_i\right\}/\Gamma(\alpha)^n$, $c(\beta) = -\beta$ and $d(\mathbf{x}) = \sum_{i=1}^{n} x_i$.*

From (3.52), take $U = S = \sum_{i=1}^{n} X_i$ (with expected value $n\mathbb{E}[X] = n\alpha/\beta$) and

$$\tau(\beta) = -\frac{a'(\beta) \, / \, a(\beta)}{c'(\beta)} = n\frac{\alpha}{\beta}$$

to get the function of β and its unbiased estimator that reaches the CRlb. That is, $U = S$ is the u.m.v.u.e. of $\mathbb{E}[S] = n\alpha/\beta$ or, as a linear function, $S/(n\alpha) = \bar{X}/\alpha$ is the u.m.v.u.e. and m.v.b.e. of $1/\beta$. This implies that an m.v.b.e. will not exist for other (nonlinear) functions, such as β itself. An u.m.v.u.e. for β does exist (and is indeed quite close to $\alpha/\bar{X}$); see Problem 7.12, which derives it and compares its variance to the CRlb of $\tau(\beta) = \beta$.

Example 3.18 (Normal, cont.)   *The calculation of J is given in Example 3.5; for μ, $J = n/\sigma^2$, but $\mathbb{E}[\bar{X}] = \mu$ and $\mathrm{Var}(\bar{X}) = \sigma^2/n = 1/J$, so that $\bar{X}$ is the u.m.v.u.e. and m.v.b.e. Similarly, for σ^2, $J = n/(2\sigma^4)$, but from (A.210), $\mathbb{V}(S_n^2) = (2\sigma^4)/(n-1)$, which is quite close to $1/J$, but does not reach it. That does not necessarily mean that there is no unbiased estimator of σ^2 that can reach it. However, from Example 3.5,*

$$\dot{\ell}_{\sigma^2} = -\frac{n}{2\sigma^2} + \frac{\sum_{i=1}^{n}(x_i - \mu)^2}{2\sigma^4} = \frac{n}{2\sigma^4}\left(\frac{\sum_{i=1}^{n}(x_i - \mu)^2}{n} - \sigma^2\right),$$

and (3.49) then shows that the m.v.b.e. of σ^2 is $\sum_{i=1}^{n}(x_i - \mu)^2/n$. This requires μ to be known; if not, the bound cannot be reached for finite n. As σ is a nonlinear function of σ^2, even if μ is known, no estimator of σ exists that reaches the CRlb of $\tau(\sigma^2) = \sigma$. The u.m.v.u.e. of σ is given via use of (A.207).

3.2.2 Multivariate Case

Cramér–Rao lower bound (multi-parameter case) For parameter vector $\theta = (\theta_1, \ldots, \theta_k)'$ and with $\tau(\theta) = (\tau_1(\theta), \ldots, \tau_m(\theta))$, let $\mathbf{U}(\mathbf{X}) = (U_1(\mathbf{X}), \ldots, U_m(\mathbf{X}))'$ be unbiased for $\tau(\theta)$. With $\mathbf{J}(\theta) = -\mathbb{E}[\ddot{\ell}(\theta; \mathbf{X})] = \mathbb{E}[\dot{\ell}\dot{\ell}']$ the $k \times k$ information matrix discussed in Section 3.1.2, and letting $\mathbf{D}$ be the $m \times k$ Jacobian matrix with (i, j)th element $\partial\tau_i(\theta)/\partial\theta_j$

$$\mathbb{V}(\mathbf{U}) \geq \mathbf{D}\mathbf{J}^{-1}\mathbf{D}', \tag{3.53}$$

which means that $\mathbb{V}(\mathbf{U}) - \mathbf{D}\mathbf{J}^{-1}\mathbf{D}'$ is positive semi-definite. From (3.46), the CRlb in (3.53) is the asymptotic variance–covariance matrix of the m.l.e.,

Similarly to the scalar parameter case, one of the required regularity conditions is that (3.48) holds for each θ_i:

$$\frac{\partial}{\partial\theta_i}\int_{\mathbb{R}^n} h(\mathbf{x})f_{\mathbf{X}}(\mathbf{x}; \theta)d\mathbf{x} = \int_{\mathbb{R}^n} h(\mathbf{x})\frac{\partial}{\partial\theta_i}f_{\mathbf{X}}(\mathbf{x}; \theta)d\mathbf{x}, \quad i = 1, \ldots, k. \tag{3.54}$$

Before showing the proof of (3.53), let us recall from matrix algebra that, for an $n \times n$ matrix $\mathbf{A}$, the **leading principle minor of order** k is the determinant of the $k \times k$ matrix consisting of the first k rows and first k columns of $\mathbf{A}$; and if $\mathbf{A}$ is positive semi-definite, then all leading principle minors of $\mathbf{A}$ are positive semi-definite. This is easy to see. Let $\mathbf{A}$ be positive semi-definite so that, by definition, $\mathbf{x}'\mathbf{A}\mathbf{x} \geq 0$ for all $\mathbf{x} \in \mathbb{R}^n$. As a special case of $\mathbf{x}$, let the last i elements be zero, so $\mathbf{x} = (\mathbf{x}_1', \mathbf{0}')'$. Then, using a corresponding partition of $\mathbf{A}$,

$$\mathbf{x}'\mathbf{A}\mathbf{x} = [\mathbf{x}_1' \ \ \mathbf{0}'] \begin{bmatrix} \mathbf{A}_{11} & \mathbf{A}_{12} \\ \mathbf{A}_{12}' & \mathbf{A}_{22} \end{bmatrix} \begin{bmatrix} \mathbf{x}_1 \\ \mathbf{0} \end{bmatrix} = \mathbf{x}_1'\mathbf{A}_{11}\mathbf{x}_1,$$

showing that the $(n-i) \times (n-i)$ matrix $\mathbf{A}_{11}$ is also positive semi-definite. Doing this for $i = 1, \dots, n-1$ shows the result.

Proof. (Dhrymes, 1982, pp. 125–126)   As $\mathbf{U}$ is unbiased for $\tau(\theta)$,

$$\tau_i(\theta) = \int_{\mathbb{R}^n} U_i(\mathbf{X}) f_{\mathbf{X}}(\mathbf{x}; \theta) d\mathbf{x},$$

so that, using (3.54), each element in $\mathbf{D}$ can be written as

$$\frac{\partial \tau_i(\theta)}{\partial \theta_j} = \int_{\mathbb{R}^n} U_i(\mathbf{x}) \frac{\partial}{\partial \theta_j} f_{\mathbf{X}}(\mathbf{x}; \theta) d\mathbf{x} = \int_{\mathbb{R}^n} U_i(\mathbf{x}) \dot{\ell}_j f_{\mathbf{X}}(\mathbf{x}; \theta) d\mathbf{x} = \mathbb{E}[U_i \ \dot{\ell}_j]. \qquad (3.55)$$

Then, with $\mathbf{Z} = (\mathbf{U}', \dot{\ell}')'$, (3.55) and (3.19) imply that

$$\mathbb{V}(\mathbf{Z}) = \begin{bmatrix} \mathbb{V}(\mathbf{U}) & \mathbf{D} \\ \mathbf{D}' & \mathbf{J} \end{bmatrix},$$

which, being a variance–covariance matrix, is positive semi-definite. Defining

$$\mathbf{C} = \begin{bmatrix} \mathbf{I}_m & -\mathbf{D}\mathbf{J}^{-1} \\ \mathbf{0}_{k \times m} & \mathbf{I}_k \end{bmatrix},$$

the matrix $\mathbf{C}\mathbb{V}(\mathbf{Z})\mathbf{C}'$ is clearly also positive semi-definite, and as

$$\mathbf{C}\mathbb{V}(\mathbf{Z})\mathbf{C}' = \begin{bmatrix} \mathbb{V}(\mathbf{U}) - \mathbf{D}\mathbf{J}^{-1}\mathbf{D}' & \mathbf{0}_{m \times k} \\ \mathbf{0}_{k \times m} & \mathbf{J} \end{bmatrix}, \qquad (3.56)$$

$\mathbb{V}(\mathbf{U}) - \mathbf{D}\mathbf{J}^{-1}\mathbf{D}'$ is also positive semi-definite.   ∎

The bound is reached if $\mathbb{V}(\mathbf{U}) = \mathbf{D}\mathbf{J}^{-1}\mathbf{D}'$, which is equivalent to $\mathbb{V}(\mathbf{Y}) = \mathbf{0}$, where $\mathbf{Y}$ is the $m \times 1$ random variable with covariance given in the upper left corner of (3.56),

$$\mathbf{Y} = [\ \mathbf{I}_m \ \ -\mathbf{D}\mathbf{J}^{-1} \ ] \begin{bmatrix} \mathbf{U} \\ \dot{\ell} \end{bmatrix} = \mathbf{U} - \mathbf{D}\mathbf{J}^{-1}\dot{\ell}.$$

As $\mathbb{V}(\mathbf{Y}) = \mathbf{0}$ is equivalent to $\Pr(\mathbf{Y} = \mathbb{E}[\mathbf{Y}]) = 1$, this gives

$$\mathbf{U} - \mathbf{D}\mathbf{J}^{-1}\dot{\ell} = \mathbb{E}[\mathbf{U}] - \mathbf{D}\mathbf{J}^{-1}\mathbb{E}[\dot{\ell}] = \tau(\theta)$$

from the assumption on $\mathbf{U}$ and (3.16), or

$$\mathbf{U} = \tau(\theta) + \mathbf{D}\mathbf{J}^{-1}\dot{\ell}. \qquad (3.57)$$

As a simple univariate example of (3.57), take the Poisson model with $\dot{\ell} = s/\theta - n$, $J = n/\theta$ and $\tau(\theta) = \theta$ so that

$$\tau(\theta) + \frac{d\tau}{d\theta}J^{-1}\dot{\ell} = \theta + \frac{\theta}{n}\left(\frac{s}{\theta} - n\right) = \frac{s}{n}.$$

Thus, $U(\mathbf{x}) = s/n$ is the m.v.b.e. for $\tau(\theta) = \theta$. This was also demonstrated in (3.50).

Example 3.19 *For the $N(\mu, \sigma^2)$ model, the results from Example 3.5 are*

$$\dot{\ell}(\theta) = \left[\sigma^{-2} \sum_{i=1}^{n}(x_i - \mu), -n(2\sigma^2)^{-1} + (2\sigma^4)^{-1} \sum_{i=1}^{n}(x_i - \mu)^2\right]'$$

and

$$\mathbf{J}(\theta) = \begin{bmatrix} n/\sigma^2 & 0 \\ 0 & n/(2\sigma^4) \end{bmatrix}.$$

For $\tau(\theta) = \theta$, we have $\mathbf{D} = \mathbf{I}_2$, and the right-hand side of (3.57) is

$$\begin{bmatrix} \mu \\ \sigma^2 \end{bmatrix} + \begin{bmatrix} \sigma^2/n & 0 \\ 0 & 2\sigma^4/n \end{bmatrix} \begin{bmatrix} \sigma^{-2} \sum_{i=1}^{n}(x_i - \mu) \\ -n(2\sigma^2)^{-1} + (2\sigma^4)^{-1} \sum_{i=1}^{n}(x_i - \mu)^2 \end{bmatrix}$$

$$= \begin{bmatrix} \mu \\ \sigma^2 \end{bmatrix} + \begin{bmatrix} n^{-1} \sum_{i=1}^{n}(x_i - \mu) \\ -\sigma^2 + n^{-1} \sum_{i=1}^{n}(x_i - \mu)^2 \end{bmatrix}$$

$$= \begin{bmatrix} \bar{X} \\ n^{-1} \sum_{i=1}^{n}(x_i - \mu)^2 \end{bmatrix}.$$

As this is not a statistic (it is a function of μ), we see that, for this choice of τ, the CRlb cannot be reached. This agrees with the results of Example 3.18.

Example 3.20 *For the $\mathrm{Gam}(\alpha, \beta)$ model, the results from Example 3.6 are*

$$\dot{\ell}_1(\theta) = n \log \beta - n \, \psi(\alpha) + \sum_{i=1}^{n} \log x_i, \quad \dot{\ell}_2(\theta) = \frac{n\alpha}{\beta} - \sum_{i=1}^{n} x_i$$

and

$$\mathbf{J}(\theta) = n \begin{bmatrix} \psi'(\alpha) & -\beta^{-1} \\ -\beta^{-1} & \alpha\beta^{-2} \end{bmatrix}.$$

For $\tau(\theta) = \alpha/\beta = \mathbb{E}[X]$,

$$\mathbf{D} = \begin{bmatrix} \dfrac{d\tau}{d\alpha} & \dfrac{d\tau}{d\beta} \end{bmatrix} = \begin{bmatrix} \beta^{-1} & -\alpha\beta^{-2} \end{bmatrix},$$

so that the right-hand side of (3.57) is

$$\frac{\alpha}{\beta} + \frac{1}{n} \begin{bmatrix} \beta^{-1} & -\alpha\beta^{-2} \end{bmatrix} \begin{bmatrix} \psi'(\alpha) & -\beta^{-1} \\ -\beta^{-1} & \alpha\beta^{-2} \end{bmatrix}^{-1} \begin{bmatrix} n \log \beta - n \, \psi(\alpha) + \sum_{i=1}^{n} \log x_i \\ n\alpha/\beta - \sum_{i=1}^{n} x_i \end{bmatrix}.$$

The reader can confirm that this simplifies to just $n^{-1} \sum_{i=1}^{n} x_i$, showing that $\bar{X}$ is the m.v.b.e. for α/β, with variance computed from the CRlb as

$$\mathbf{D}\mathbf{J}^{-1}\mathbf{D} = \frac{1}{n} \begin{bmatrix} \beta^{-1} & -\alpha\beta^{-2} \end{bmatrix} \begin{bmatrix} \psi'(\alpha) & -\beta^{-1} \\ -\beta^{-1} & \alpha\beta^{-2} \end{bmatrix}^{-1} \begin{bmatrix} 1/\beta \\ -\alpha/\beta^2 \end{bmatrix} = \frac{\alpha}{n\beta^2},$$

or directly as $\mathbb{V}(\bar{X}) = \mathbb{V}(X)/n = \alpha/(n\beta^2)$.

The earlier result on attaining the CRlb and membership in the one-parameter exponential family can be extended to the vector parameter case; see, for example, Zacks (1971, pp. 194–201) and Čencov (1982, pp. 219–225) for details.

3.3 MODEL SELECTION

> The purpose of the analysis of empirical data is not to find the "true model" – not at all. Instead, we wish to find a best approximating model, based on the data, and then develop statistical inferences from this model. ... Data analysis involves the question, "What level of model complexity will the data support?" and both under- and over-fitting are to be avoided.
>
> (Kenneth P. Burnham and David R. Anderson, 2002, p. 143)

> A traditional approach to statistical inference is to identify the true or best model first with little or no consideration of the specific goal of inference in the model identification stage.
>
> (Yuhong Yang, 2005, p. 937)

One of the most important parts of conducting a statistical analysis is the choice of what model to use, if any, to approximate the (usually unknown and complicated) stochastic process that generated the data, referred to as the **data-generating process**. An example of a model is an n-length sequence of i.i.d. Bernoulli trials, with unknown value of the success probability p, which is to be estimated from the data. More complicated models deviate from the i.i.d. assumption, such as (generalized) linear and time series models.

While it might, at first blush, appear that the extent of a model's misspecification is an analytic concept that has nothing whatsoever to do with the amount of data available for estimation, the demands of reality when working with a nontrivial data-generating process suggest that the two are indeed intimately linked. The above quote from Burnham and Anderson (2002) expresses this well: essentially, the amount of available data decisively dictates the possible complexity of the model.

The quote from Yang (2005) helps emphasize one of the primary messages of this book: The point of model selection should be intimately tied to the application. In particular, if prediction of a time series, or prediction of weather using spatial models, is desired, then the model used should be chosen and calibrated (estimated) to (statistically correctly) maximize this performance. See also the discussion in Section 2.8.

We begin with a simple illustration of the concept of model misspecification, and then discuss the likelihood ratio statistic and the information criteria. The interested reader is encouraged to explore the modern literature in model selection; starting points include Burnham and Anderson (2002), Hastie et al. (2009), Murphy (2012), and Efron (2013, 2014).

3.3.1 Model Misspecification

A parametric statistical model is said to be **misspecified** if (up to parameter uncertainty) it does not match that of the true data-generating process underlying the data of interest. We mentioned in Remark (d) in Section 3.1.4 that, in many contexts and applications, this is more the rule than the exception.

One common form of misspecification occurs when a variable or parameter is left out of the model, in which case it is **under-specified**. Examples include omission of relevant regressors in a linear model and omission of lagged variables with nonzero coefficients in a time series model. In the i.i.d. case, an example would be fitting a random sample of Student's t data using a normal distribution: the degrees-of-freedom parameter v is missing from the model and mistakenly taken to be infinity. If, in this latter example, interest centers on estimating the expected value, then use of the m.l.e. under normality, $\bar{X}$, will be a very poor choice if v is small (see Example 4.3). One of the most insidious problems with model under-specification is that, even asymptotically, that is, with unbounded sample sizes, the results can be flawed. A simple example is using $\bar{X}$ to estimate the location parameter for a random sample of i.i.d. Student's t data with true degrees of freedom equal to 1 (i.e., Cauchy).

Model **over-specification** occurs when parameters are included that need not be there. In regression analysis, for example, the addition of superfluous regressors to an otherwise correctly specified model unnecessarily decreases the available degrees of freedom and results in larger confidence intervals for all the parameters. The estimates of the coefficients in this regression case, however, remain unbiased.

Example 3.21 *We investigate the performance of certain estimated quantities when using correct and misspecified models. We begin by using 500 sets of $n = 100$ i.i.d. Student's $t(5)$ observations (with location zero and scale one) and, for each data set, fitting (via maximum likelihood) four different i.i.d. models with location μ and scale σ. The first two models are the normal and Student's t, with the latter assuming the degrees-of-freedom parameter $v \in \mathbb{R}_{>0}$ is unknown and jointly estimated with the location and scale parameters. The third distribution is the normal–Laplace convolution: we say that $Z \sim \text{NormLap}(c, 0, 1)$ if $Z = cX_0 + (1 - c)Y_0$ with $X_0 \sim N(0, 1)$ independent of $Y_0 \sim \text{Lap}(0, 1)$ and $0 \le c \le 1$. The density of Z is given in (A.154). The fourth distribution is the symmetric stable Paretian, $Z = \mu + \sigma Y$, for $Y \sim S_\alpha(0, 1)$, with $\varphi_Y(t; \alpha) = \exp\{-|t|^\alpha\}$, $0 < \alpha \le 2$; see Section A.16.*

We use the abbreviations N, T, L and S for the four models, in obvious notation. Note that the normal model has just two parameters, μ and σ, for which the m.l.e. takes on a closed form, while the other three models each have an additional parameter besides the location and scale that dictates the shape of the density (in particular, the thickness of the tails), so that the likelihood needs to be numerically maximized.

Several quantities could be used to measure the effect of misspecification. One is the m.s.e. of the location parameter. As each of the four densities considered is symmetric about zero, the location parameter is also the expected value of a random variable following one of these distributions (assuming for the Student's t model that $v > 1$, and for the stable Paretian that $\alpha > 1$). Using the 500 sets of simulated Student's $t(5)$ data, we obtain the estimates of the m.s.e. given by (in obvious notation) $\text{mse}_N(\mu) = 0.0190$, $\text{mse}_T(\mu) = 0.0145$, $\text{mse}_L(\mu) = 0.0150$, and $\text{mse}_S(\mu) = 0.0147$. As expected, T performs best, being the true model. Also, while N performs relatively poorly, there is little difference when using any of the three fat-tailed assumptions. Another candidate for measuring the location parameter is a nonparametric estimator, abbreviated NP, such as the sample mean or median. We use instead the trimmed mean with trimming parameter calculated from the results of Example 4.3 and based on the m.l.e. of the degrees-of-freedom parameter in the T model. This gives $\text{mse}_{NP}(\mu) = 0.0147$, which is also quite good compared to use of the true model.

Quantile estimates are also of interest, for which we consider the 1%, 5%, and 10% values, that is, $q = 0.01, 0.05, 0.10$. The true quantile is $F_T^{-1}(q; v)$, where F_T^{-1} denotes the inverse c.d.f. of a Student's $t(v)$ random variable. For a given model M, $M \in \{N, T, L, S\}$, the estimated quantiles are given by $\hat{\sigma} F_M^{-1}(q; \hat{\theta}) + \hat{\mu}$, where θ refers the shape parameter in the T, L, and S models, for example, α for the stable Paretian. For the nonparametric estimator NP, the qth quantile is estimated from the sample quantile, easily obtained using

TABLE 3.1   Mean squared error values for five models. The true model is Student's *t*, column T, with five degrees of freedom

q	N	T	L	S	NP
0.01	0.295	0.304	**0.194**	0.710	1.372
0.05	0.103	**0.048**	0.053	0.049	0.124
0.10	0.095	**0.032**	0.032	0.033	0.060

the `prctile(x,100*q)` *command in Matlab. The resulting m.s.e. values are shown in Table 3.1, the smallest of which are indicated in bold.*

For $q = 0.01$, the m.s.e. for the T *model is about the same as that of* N, *while it is higher than that of* L, *which might be surprising. This can be attributed to the fact that estimation of the degrees-of-freedom parameter is, relative to the location and scale parameters, less accurate, in the sense of having a much higher variance, because it is a measure of the heaviness of the tails (see Chapter 9 for further discussion).*

The S *model performs poorly, as would be expected for small q, given its extremely fat tails (and also relative inaccuracy in estimating α, again being a tail index, like the degrees-of-freedom parameter in the Student's t model). Also,* NP *performs relatively poorly, which is understandable at the 1% level using only 100 observations per sample. For both $q = 0.05$ and $q = 0.10$,* T *exhibits the smallest m.s.e.,* N *performs relatively poorly,* S *becomes "competitive,"* L *remains good, and* NP *improves considerably.*

This same exercise was then repeated but using samples drawn from L *with $c = 0.25$, and* S *with $α = 1.7$, with results shown in Tables 3.2 and 3.3, respectively. The last row in each, labelled μ, gives 1000 times the m.s.e. for the location parameter.*

Unsurprisingly, we see from Table 3.2 that model L *performs best when it also coincides with the data-generating process. That* L *also outperforms* S *at the 1% quantile when* S *is the true model, in Table 3.3, may not have been expected, and is, again, due to the relatively higher estimation uncertainty of the tail thickness parameter α.*

TABLE 3.2   Mean squared error values for five models. True model is NormLap, column L, with $c = 0.025$

q	N	T	L	S	NP
0.01	0.272	0.239	**0.117**	1.015	0.429
0.05	0.046	0.039	**0.037**	0.043	0.103
0.10	0.057	0.024	**0.023**	0.027	0.054
μ	12.6	9.17	**8.87**	9.92	9.29

TABLE 3.3   Mean squared error values for five models. True model is symmetric stable, column S, with $α = 1.7$

q	N	T	L	S	NP
0.01	11.95	1.74	**0.986**	1.984	59.2
0.05	7.33	0.14	0.170	**0.101**	0.212
0.10	4.76	0.057	0.080	**0.053**	0.082
μ	71.4	22.9	24.5	**22.6**	23.5

While conclusions of a very general nature should not be drawn, this result appears to suggest that, for fat-tailed i.i.d. data with unknown data-generating process and sample size near 100, the normal–Laplace convolution model appears to yield reliable estimates of the 1% quantile.

3.3.2 The Likelihood Ratio Statistic

Recall the discussion in Section 3.1 about how the likelihood is only a relative measure: it has meaning only as a ratio, say $\mathcal{L}(\theta_1; \mathbf{x})/\mathcal{L}(\theta_2; \mathbf{x})$, for distinct values θ_1 and θ_2. This motivates the study of the ratio at two parameter values of interest, with one of them being a null hypothesis, and examining the resulting p-value of the test. Another motivation for this structure came from (2.29), based on the study of the behavior of the p-value under the alternative hypothesis. Bearing in mind the comments about hypothesis testing in Section 2.8, its use in this way for model selection should be cause for some scepticism. However, see Remark (d) below.

For a parametric model with $\theta = (\theta_1, \dots, \theta_k)' \in \Theta$, where Θ is an open subset of $\mathbb{R}^k$, consider the null hypothesis $H_0 : \theta \in \Theta^0$, where

$$\Theta^0 = \{\theta \in \Theta : \theta_1 = \theta_1^0, \dots, \theta_m = \theta_m^0\}, \quad 1 \le m \le k,$$

for $\theta_1^0, \dots, \theta_m^0$ fixed constants, versus the alternative that θ is unrestricted. Interest centers on assessing whether the **reduced**, or **restricted**, model is adequate in some sense, and traditionally serves as the null hypothesis. Observe how the restricted model is nested within the unrestricted one – this can be relaxed; see Remark (c) below.

The **likelihood ratio statistic** and its asymptotic distribution are given by

$$\Lambda = \Lambda(\mathbf{X}; \Theta, \Theta^0) = \frac{\sup_{\theta \in \Theta} \mathcal{L}(\theta; \mathbf{X})}{\sup_{\theta \in \Theta^0} \mathcal{L}(\theta; \mathbf{X})}, \quad 2 \log \Lambda \xrightarrow{d} \chi_m^2, \tag{3.58}$$

as attributed to Wilks (1938), building on work by Neyman and Pearson (1928). Note that some presentations use the reciprocal of the ratio in (3.58). We prove (3.58) in the $k = m = 1$ case. This requires the same regularity conditions as were invoked to prove the asymptotic normality of $\hat{\theta}_{\mathrm{ML}}$. Again from (3.2),

$$\ell(\theta^0) = \ell(\hat{\theta}_{\mathrm{ML}}) + \dot{\ell}(\hat{\theta}_{\mathrm{ML}})(\theta^0 - \hat{\theta}_{\mathrm{ML}}) + \frac{1}{2}\ddot{\ell}(\theta^*)(\theta^0 - \hat{\theta}_{\mathrm{ML}})^2, \quad \theta^* \in (\theta^0, \hat{\theta}_{\mathrm{ML}}),$$

with $\dot{\ell}(\hat{\theta}_{\mathrm{ML}}) = 0$, so that $2 \log \Lambda = 2(\ell(\hat{\theta}_{\mathrm{ML}}) - \ell(\theta^0)) = -\ddot{\ell}(\theta^*)(\hat{\theta}_{\mathrm{ML}} - \theta^0)^2$ or

$$2 \log \Lambda = J(\theta^0)(\hat{\theta}_{\mathrm{ML}} - \theta^0)^2 \times \frac{\ddot{\ell}(\theta^*)}{\ddot{\ell}(\theta^0)} \times \frac{\ddot{\ell}(\theta^0)}{-J(\theta^0)}.$$

Then, as shown in Section 3.1.4, the latter two terms converge in probability to 1, and the former term is, from the continuous mapping theorem (A.280), asymptotically the square of a standard normal, so that Slutsky's theorem (A.281) gives the desired result.

Remarks

(a) Proofs for the general vector parameter case, can be found in, for example, Ferguson (1996, Ch. 22), Severini (2000, Sec. 4.3), and Bhattacharya et al. (2016, Sec. 8.3). An adjustment to the ratio such that the asymptotic distribution is more accurate in small samples is known as the **Bartlett correction**, going back to work by Maurice S. Bartlett in 1937. It has been subsequently shown to be related to the saddlepoint approximation; see Reid (1988) and Butler (2007, Sec. 7.1.2) for discussion and original references. One can also apply the bootstrap to numerically obtain the required correction; see, for example, Davison and Hinkley (1997, p. 149).

(b) The classic framework requires Θ to be an open subset of $\mathbb{R}^k$, as stated above. Inference in the case such that Θ is not necessarily an open set and a parameter in θ^0 is on the boundary has been addressed by several authors; see, for example, Andrews (2001), Molenberghs and Verbeke (2007), and Cavaliere et al. (2017).

(c) It is possible to form a meaningful likelihood ratio such that the two models are nonnested, as studied in Vuong (1989) and the references therein. As a showcase example, we will use this structure in Section 9.5 to test the stable distribution hypothesis (see also Section 2.6).

(d) Related to the previous remark, the likelihood ratio can be used more generally than the nested model framework in (3.58), and also in a conceptually different inferential paradigm. The **law of likelihood** states:

> If one hypothesis, H_1, implies that a random variable X takes the value x with probability $f_1(x)$, while another hypothesis, H_2, implies that the probability is $f_2(x)$, then the observation $X = x$ is evidence supporting H_1 over H_2 if $f_1(x) > f_2(x)$, and the likelihood ratio, $f_1(x)/f_2(x)$, measures the strength of that evidence.[7]

See, for example, Hacking (1965) and Royall (1997). In this setup, neither hypothesis is the null and it need not be the case that one model is the true one, nor that one is nested in the other. The ratio prescribes how to interpret evidence for one hypothesis in relation to another.

The fundamental distinction of this use of a likelihood ratio is that it is based on the observations themselves, as opposed to (3.58), which considers the distribution of the ratio under a declared null and generates a decision based on how extreme the observed statistic is under this distribution, that is, the p-value. In particular, the latter incorporates probabilities of X that were not observed. This is the key distinction, and is used to clarify the nature of *evidence*, via the law of likelihood, and *uncertainty*, as measured by probabilities. Royall (1997) provides numerical values such that likelihood ratios above it or below its inverse constitute "strong" and "very strong" evidence in favor of one H over the other; see also Royall (2000) and the references therein.

As emphasized in Royall (1997), this likelihood-based framework helps answer the questions "what do the data tell me?," and "how should I interpret this set of observations as evidence?", as opposed to "what should I now

---

[7] Recall that, for discrete random variable X, the mass function $f_X(x)$ gives the probability that $X = x$, while in the continuous case, this interpretation holds in the sense that $f_X(x)\Delta x$ gives the probability that X is in some small Δ interval of x.

believe, given the data?" or "what action should be taken, given the data?" These questions are the domain of Bayesian statistical inference and decision theory.

A critique of this paradigm is that a distribution can always be found that gives rise to a higher likelihood compared to another. For example, adding regressors to a linear model, or increasing the number of components of a discrete mixture of normals, will induce this. Such considerations lead to the idea of somehow "penalizing" the likelihood for the number of parameters. One way of doing this is considered next. ∎

3.3.3 Use of Information Criteria

Let k denote the number of model parameters, T the sample size, and, as usual, $\ell(\widehat{\theta}_{ML})$ the log-likelihood evaluated at the m.l.e. The **Akaike information criterion** (or AIC) is given by

$$\mathrm{AIC} = -2\ell(\widehat{\theta}_{ML}) + 2k, \tag{3.59}$$

and embodies a tradeoff between a model that maximizes the likelihood (which can always be increased by just adding parameters, such as more shape parameters in a distribution, more regressors in a linear model, or more lags in an autoregressive time series model), and the number of parameters. It is remarkable that such a simple formula drops out of the derivation. It involves use of the Kullback–Leibler information criterion (3.36) and asymptotic arguments; see the references below.

It is important to note that the choice of models need not be nested, as opposed to the classic likelihood ratio test (3.58). Operationally, from a selection of (nested or nonnested) models, the one with the smallest AIC is chosen. Observe how there is no null hypothesis, and also that it says nothing about the quality of the chosen model. If all the models entertained have the same number of parameters, then use of the AIC is equivalent to choosing the model with the highest obtained likelihood. A bias-corrected version of AIC, denoted AICc or AICC, is given by

$$\mathrm{AICc} = \mathrm{AIC} + \frac{2k(k+1)}{T-k-1}, \tag{3.60}$$

and is favored in a Gaussian context for linear models and time series model selection, particularly for relatively small T. Observe that, for fixed k, they are asymptotically equivalent.
   (Schwarz's) Bayesian information criterion (BIC), given by

$$\mathrm{BIC} = -2\ell(\widehat{\theta}_{ML}) + k\log(T), \tag{3.61}$$

looks superficially similar in spirit to AIC, but with a different (harsher) penalty, though formally its derivation does not make use of concepts from information theory but rather (obviously) Bayesian arguments. Details on the justification, derivation, and asymptotic properties of these and other criteria, as well as original references, can be found in Konishi and Kitagawa (2008), Brockwell and Davis (1991, Sec. 9.3), McQuarrie and Tsai (1998), and Burnham and Anderson (2002).
   Finally, an alternative method, called the **minimum description length** (MDL), has gained substantially in prominence, and is related to the field of stochastic complexity,

though it has some nontrivial similarities to information criteria, Bayesian statistical inference, and the BIC. Similar to the AIC, it recognizes that the "true model" is possibly, and most likely, not among the set of models considered, and embraces the notion that the complexity of the chosen model is a function of the amount of data available. A good starting point for understanding the MDL is Grünwald (2007).

3.4 PROBLEMS

> Every day you face battles – that is the reality for all creatures in their struggle to survive. But the greatest battle of all is with yourself – your weaknesses, your emotions, your lack of resolution in seeing things through to the end. You must declare unceasing war on yourself.
>
> (Robert Greene, 2006, p. xx)

3.1 Recall Example 3.5. Derive $\dot{\ell}_1$ and $\dot{\ell}_2$, set to zero and confirm that $\hat{\mu}_{ML} = \bar{X}$ and $\hat{\sigma}^2_{ML} = n^{-1} \sum_{i=1}^{n} (X_i - \bar{X})^2$. How do we know this is a *maximum* of the likelihood? Using $\dot{\ell}_1$ and $\dot{\ell}_2$, compute $\dot{\vartheta}$ and verify that $\mathbb{E}[\dot{\vartheta}] = (0, 0)'$. Compute $\dot{\vartheta}\,\dot{\vartheta}'$ and its expected value. Confirm that $\mathbb{E}[\ddot{\vartheta}\,\dot{\vartheta}'] = -\mathbb{E}[\ddot{\vartheta}]$. Finally, verify (3.21).

3.2 Consider the i.i.d. Cauchy model with location parameter μ.

(a) For $n = 2$ observations, it can be shown (quite easily using a symbolic mathematical software package such as Maple) that there are three possible estimators for the m.l.e. of μ:

$$\mu_1 = \bar{x}, \qquad \mu_{2,3} = \bar{x} \pm \frac{1}{2}\sqrt{(x_1 - x_2)^2 - 4}.$$

(i) Does the fact that $\bar{X}$ satisfies the likelihood equation make any sense?

(ii) When are the latter two valid?

(b) Evaluating the second derivative of ℓ yields, with $a = (x_1 - x_2)^2$,

$$\left.\frac{d^2\ell}{d\mu^2}\right|_{\substack{n=2\\ \mu=\mu_1}} = 16\frac{a - 4}{(a + 4)^2},$$

$$\left.\frac{d^2\ell}{d\mu^2}\right|_{\substack{n=2\\ \mu=\mu_2}} = \frac{-32(a - 4)}{(a + (x_1 - x_2)\sqrt{a - 4})^2(x_1 - x_2 - \sqrt{a - 4})^2},$$

$$\left.\frac{d^2\ell}{d\mu^2}\right|_{\substack{n=2\\ \mu=\mu_3}} = \frac{-32(a - 4)}{(a + (x_2 - x_1)\sqrt{a - 4})^2(x_1 - x_2 + \sqrt{a - 4})^2}.$$

Discuss which estimator you would use.

(c) For $\mu = 0$, choose a constant value for x_1 and plot the three estimators for several values of x_2 between -16 and 16. Interpret the behavior of the resulting graph.

(d) Derive $f_{X_{2:3}}(x)$, the density of the median for $n = 3$ observations and graphically compare the tail behavior of $f_{X_{2:3}}$ to that of a Cauchy with scale parameter $1/2$ and that of a Laplace with scale parameter 2. Can the mean and variance of $f_{X_{2:3}}(x)$ be easily calculated?

3.3 Let $X_i \stackrel{\text{i.i.d.}}{\sim} \text{Exp}(a, b)$, $i = 1, \ldots, n$, with p.d.f. $f_X(x) = b \exp(-b(x - a))\mathbb{I}_{(a,\infty)}(x)$ for $b > 0$.

(a) Compute $\mathbb{E}[X]$.

(b) Compute the c.d.f. of X.

(c) Derive $\hat{a}_{\text{ML}}$ and $\hat{b}_{\text{ML}}$.

(d) Compute $\mathbb{E}[\hat{a}_{\text{ML}}]$.

(e) For a random variable Y, a zero-order approximation to $\mathbb{E}[1/Y]$ is given by $1/\mathbb{E}[Y]$, using (A.159). Compute a zero-order approximation to $\mathbb{E}[\hat{b}_{\text{ML}}]$.

3.4 Prove (3.15).

3.5 Calculate the asymptotic distribution of $\tau(\hat{\theta}_{\text{ML}}) = (\mu, \sigma)$ using the setup in Example 3.14.

3.6 Show (3.42).

3.7 Recall Example 3.7, function $_2F_1$ from (A.15), and its integral expression (A.17). Simplify the constant of integration in (3.26) for $\rho = 0$ to obtain (3.27). Hint: Use the fact that

$$_2F_1\left(\frac{1}{2}, \frac{1}{2}, n - \frac{1}{2}, \frac{1}{2}\right) = \frac{2^{\frac{3}{2}-n}\sqrt{\pi}\Gamma\left(n - \frac{1}{2}\right)}{\Gamma(n/2)\Gamma(n/2)}, \tag{3.62}$$

which is a special case of the result given in Abramowitz and Stegun (1972, Eq. 15.1.26), and use Legendre's duplication formula (A.8).

As an aside, using the integral formula (A.17), it is easy to show that (3.62) also implies

$$\int_0^1 y^{-1/2}(1 - y)^{n-2}\left(1 - \frac{y}{2}\right)^{-1/2} dy = \frac{2^{\frac{3}{2}-n}\pi}{(n - 1)\, B(n/2, n/2)},$$

which is (apparently) valid for all real $n > 1$.

3.8 Recall Example 3.3.

(a) Show that

$$k = \sum_{h=1}^k \sum_{j=0}^{h-1} \frac{1}{n - j} + (n - k)\sum_{j=0}^{k-1}\frac{1}{n - j}.$$

(b) While it is somewhat challenging to algebraically compute $\mathbb{E}[k/T]$, it is numerically quite easy to determine the bias for given values of n and k: just simulate the experiment a large number of times, compute $\hat{\lambda}_{\text{ML}}$ for each, and take the average. Do so for $\lambda = 0.25$, $n = 20$ and $k = 2, 4, \ldots, 20$.

(c) To compute $\mathbb{V}(T)$, write $T = Y_1 + Y_2 + \cdots + Y_{k-1} + (n - k + 1)Y_k =: \mathbf{a}'\mathbf{Y}$, where $\mathbf{a}$ is so defined and use (A.187) to write

$$\mathbb{V}(T; \lambda) = \sum_{i=1}^k \sum_{j=1}^k a_i a_j \text{Cov}(Y_i, Y_j) = \frac{1}{\lambda^2}\sum_{i=1}^k \sum_{j=1}^k a_i a_j \sum_{p=0}^{\min(i,j)-1} \frac{1}{(n - p)^2},$$

which is easily computable. Then $\mathbb{V}(1/\hat{\lambda}_{\mathrm{ML}}) = \mathbb{V}(T/k) = \mathbb{V}(T)/k^2$. Plot this as a function of k. Overlay onto the graph the variance values as determined by simulation. They should agree with the theoretical ones.

(d) Now consider the mean and variance of $\hat{\lambda}_{\mathrm{ML}}$. Determine the exact values by simulation for $\lambda = 0.25$, $n = 20$ and $k = 2, 4, \ldots, 20$. Compare these values to those obtained from the Taylor series approximation (A.159) in conjunction with the theoretical mean and variance for $1/\hat{\lambda}_{\mathrm{ML}}$.

4

Numerical Optimization

This chapter presents methods for numerically determining the m.l.e. The existence of a closed-form solution to the m.l.e. is far more the exception than the rule: it is often defined only implicitly, and numerical methods will be necessary to locate it. This can be quite challenging in general, particularly if a large number of parameters are involved.

When the usual regularity conditions are satisfied, we can obtain the m.l.e. either by finding the roots of the score function, or by directly maximizing the likelihood function using numerical optimization algorithms. Both of these methods will be discussed, and their computer implementation will be given. There is one interesting exception, such that, even with a large number of parameters and no closed-form solution, the m.l.e. can be obtained very quickly. This is the case when the so-called expectation-maximization (EM) algorithm is applicable; this is discussed in Chapter 5.

4.1 ROOT FINDING

Root finding refers to locating, algebraically or numerically, the values of a parameter for which the function of interest takes the value zero. In general, when numerical methods are employed, it is important to keep in mind that the desired accuracy of the outcome, typically specified by way of one or more **tolerance parameters**, will be limited by the accuracy with which the calculations can be performed. Thus, when we speak of "numerically solving" the equation $\dot{\ell}(\theta; \mathbf{x}) = \mathbf{0}$, where, for our purposes, $\dot{\ell} : \mathbb{R}^k \rightarrow \mathbb{R}^k$ will be a vector of derivatives of the log-likelihood function ℓ, what is essentially meant is finding a value $\widehat{\theta}$ such that, for a specified value $\epsilon > 0$, $\| \dot{\ell}(\widehat{\theta}; \mathbf{x}) \| < \epsilon$, where $\| ! \cdot \|$ is some norm, or distance from zero, typically either the Euclidean norm or maximum absolute value (supremum norm).

Fundamental Statistical Inference: A Computational Approach, First Edition. Marc S. Paolella.
© 2018 John Wiley & Sons Ltd. Published 2018 by John Wiley & Sons Ltd.

In order to accomplish this numerically, the evaluation of $\dot{\ell}$ must also satisfy a certain degree of accuracy depending on ϵ. In practice, a high degree of accuracy will be difficult to reach if evaluation of $\dot{\ell}$ is based on numerical differentiation of ℓ. Thus, for root finding to be successful, an analytic expression for $\dot{\ell}$ (or a way of evaluating it to machine precision) is highly desirable. When this is the case, and $\dot{\ell}$ is continuous, and the solution does not lie on a boundary of the parameter space Θ, then root finding is often numerically quite stable and leads successfully to the desired solution.

4.1.1 One Parameter

When $k = 1$, the simplest method, in terms of both theoretical underpinnings and required programming effort, is bisection. However, other methods exist that have faster convergence properties, that is, they can locate the solution with a smaller number of function evaluations. Such routines are commonly implemented in high-level programming languages. In Matlab for instance, the canned routine `fzero` can be used. For $k > 1$, function `fsolve` is required.

Example 4.1   *Let $X \sim \text{Cau}(0, c)$, a scaled Cauchy with p.d.f.*

$$f_X(x; c) = \frac{1}{\pi c} \cdot \frac{1}{1 + x^2/c^2}, \quad c > 0.$$

For n i.i.d. observations, $\ell(c; \mathbf{x}) = -n \log(\pi) - \sum_{i=1}^{n} \log(c + x_i^2/c)$ and we wish to calculate $\hat{c}_{\text{ML}}$ such that $\dot{\ell} = 0$, that is, we need to solve

$$\sum_{i=1}^{n} \frac{1 - x_i^2/c^2}{1 + x_i^2/c^2} = 0. \tag{4.1}$$

Except for extremely small n, this must be solved numerically. The program in Listing 4.1 shows how to simulate an i.i.d. Cauchy sample and calculate $\hat{c}_{\text{ML}}$. The built-in Matlab function `fzero` is used to compute the zero of (4.1). It requires a lower and upper bound on $\hat{c}_{\text{ML}}$, obtained as two values such that the left-hand side of (4.1) differs in sign. For what follows, we used values 0.1 and 20; in general, a smarter method of obtaining bounds on $\hat{c}$ would be required.

With true value $c = 3$, 10,000 data sets were simulated for each of the three different sample sizes $n = 10$, 50, and 100, and the corresponding $\hat{c}_{\text{ML}}$ were computed. The means of $\hat{c}_{\text{ML}}$ were 3.35, 3.06 and 3.03, respectively, showing that, while the m.l.e. is biased, the bias diminishes as n grows.

Figure 4.1 shows the kernel density estimates of $\hat{c}_{\text{ML}}$ for each sample size. We see that, for small n, the density of $\hat{c}_{\text{ML}}$ is quite skewed, while, for $n = 100$, $\hat{c}_{\text{ML}}$ appears nearly normally distributed. These findings agree with the general asymptotic results in Section 3.1.4. Note, however, that knowledge of the asymptotic distribution of $\hat{c}_{\text{ML}}$ does not give any indication as to what sample size is required such that the unbiasedness and normal approximation can be relied upon. ∎

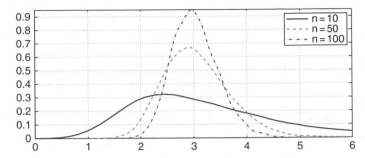

Figure 4.1 *Kernel density estimates of the m.l.e. of scale parameter c based on n i.i.d. Cauchy observations, n = 10, 50, and 100. The larger n is, the more mass is centered around the true value of c = 3.*

```
1  function chat=cauchylik(c,n)
2  c=3; x=c * randn(n,1) ./ randn(n,1);
3  opt = optimset('disp','none','TolX',1e-6);
4  chat = fzero(@(c) ff(c,x), [0.1 20],opt);
5  function f=ff(c,x), num = 1-x.^2/c^2; den = 1+x.^2/c^2; f=sum(num./den);
```

Program Listing 4.1: Simulates i.i.d. Cauchy samples and estimates $\hat{c}$.

An analysis similar to the previous example will be conducted for the location parameter μ of an i.i.d. $\mathrm{Cau}(\mu, 1)$ sample or, without loss of generality, for $X_i \sim \mathrm{Cau}(\mu, c)$ with $c > 0$ known, as $X_i/c \sim \mathrm{Cau}(\mu/c, 1)$. Some simple algebra shows that the m.l.e. is obtained by solving

$$\sum_{i=1}^{n} \frac{x_i - \mu}{1 + (x_i - \mu)^2} = 0. \tag{4.2}$$

There exist other candidate estimators for μ besides the m.l.e. In particular, given the extremely heavy tails of the Cauchy, the median should be a good estimator of μ, while the mean $\bar{X}$ is useless. Both the mean and median are special cases of the **trimmed mean**, $\bar{X}_{(\alpha)}$, $0 \leq \alpha < 100$, which is computed as the mean of the X_i after dropping the smallest and largest $\alpha/2\%$ values of the sample. For $\alpha = 0$, nothing is dropped, so that $\bar{X}_{(0)} = \bar{X}$, while for $\alpha = 100^-$, that is, close enough to 100, $\bar{X}_{(100^-)}$ reduces to the median. Values of α between these extremes provide varying degrees of "outlier removal."

Example 4.2 *We presume that the quality of the estimators is invariant to the true value of μ, for which we use $\mu = 0$. For sample size n = 20 and based on 5000 simulations, the top left corner of Figure 4.2 shows boxplots of $\hat{\mu}$ based on the m.l.e., $\bar{X}_{(100)}$ (median), $\bar{X}_{(10)}$ (denoted Trim 10), $\bar{X}_{(20)}$ (denoted Trim 20) and $\bar{X}_{(0)}$, denoted Average. As expected, the latter performs terribly; in fact, numerous values were observed outside of $(-25, 25)$, but the boxplot was truncated to allow for better comparison. Both trimmed means perform much better than $\bar{X}$, but the m.l.e. and median perform best.*

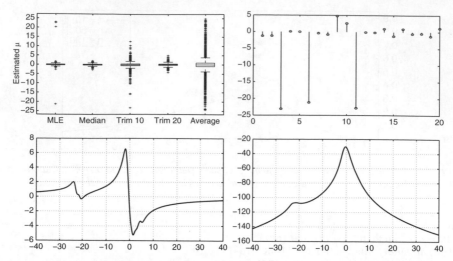

Figure 4.2 *Estimation results for the location parameter of a Cauchy model and illustration of a like-lihood with multiple roots.*

Note, however, the four values of $\hat{\mu}_{ML}$ that lie very far from zero. These are the result not of numerical error, but rather of the existence of multiple roots of $\dot{\ell}$ (i.e., multiple maxima of ℓ and picking the wrong one). For example, the top right panel of Figure 4.2 shows the sample that gave rise to $\hat{\mu}_{ML} = -21$, while the bottom left panel plots the left-hand side of equation (4.2) for values of μ between -40 and 40 and the bottom right panel plots the log-likelihood

$$\ell(\mu; \mathbf{x}) \propto - \sum_{i=1}^{n} \log(1 + (x_i - \mu)^2).$$

While the global maximum is clearly near zero, we see, firstly, that multiple maxima are possible and, secondly, that a numeric search procedure just might find the wrong one. Figure 4.3(a) shows the same boxplots of the m.l.e. and the median after removing the cases with a false global maximum. We see that the m.l.e. has a smaller variance. Figure 4.3(b) is similar, but uses a sample size of $n = 200$ instead. This time, no aberrant values of the m.l.e. were observed, and the variance of the m.l.e. is again smaller than that of the median.   ∎

Regarding the multiple zeros in the previous example, note that they could all have been avoided by restricting the search of $\hat{\mu}_{ML}$ to a smaller range around zero instead of $(-30, 30)$. However, in practice, the very fact that μ is unknown makes such an arbitrary specification difficult. Fortunately, the researcher will often have some idea of where the parameter should lie. This is one way in which such "prior information" can (and should) be used. This is particularly important when there are many parameters involved, so that a simple plot of the log-likelihood is not feasible.

The next example is similar, but uses Student's t data with varying degrees of freedom, v. For $v = 1$, we just saw that the median is practically as good as the m.l.e., while, for very large v, the m.l.e. should approach $\bar{X}$. Thus, we might expect that, for any given $1 \le v < \infty$,

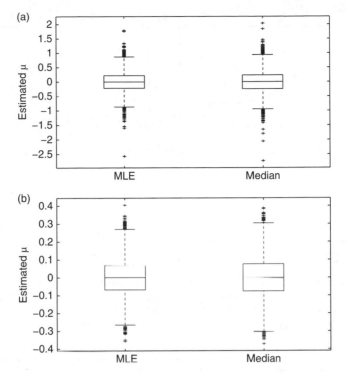

Figure 4.3 *Comparison of the m.l.e. of μ and the median of Cauchy samples with n = 20 (a) and n = 200 (b).*

there will exist an optimal value $\alpha(v)$ such that the trimmed mean $\bar{X}_{(\alpha(v))}$ has the smallest m.s.e. An algebraic determination of $\alpha(v)$ does not seem possible, but use of simulation is straightforward.

Example 4.3   *Let $X_i \overset{\text{i.i.d.}}{\sim} t(v, \mu, \sigma)$, that is, Student's $t(v)$ with location μ and scale σ. If v is known and $\sigma = 1$, then*

$$f_{\mathbf{X}}(\mathbf{x}; \mu) \propto \prod_{i=1}^{n} (1 + (x - \mu)^2/v)^{-\frac{v+1}{2}}, \quad \log f \propto \sum_{i=1}^{n} \log(1 + (x_i - \mu)^2/v),$$

and, writing $\dot{\ell}_\mu$ for $\dot{\ell}(\mu; \mathbf{x}, v)$,

$$\dot{\ell}_\mu = \frac{d \log f}{d\mu} \propto \sum_{i=1}^{n} \frac{x_i - \mu}{1 + (x_i - \mu)^2/v}.$$

For $n > 2$, the value $\hat{\mu}_{\mathrm{ML}}$ for which $\dot{\ell}_\mu = 0$ needs to be obtained numerically. In addition (and numerically easier), the trimmed mean $\bar{X}_{(\alpha)}$ can be computed for various α. For a given value of v and sample size $n = 1000$, a set of $s = 50{,}000$ simulated samples was generated, using (without loss of generality) $\mu = 0$.

For each of the s samples, and each integer value of α ranging from 0 to 99, $\bar{X}_{(\alpha),i}$ was computed, $i = 1, \ldots, s$. The m.s.e. as a function of α was then approximated as $s^{-1} \sum_{i=1}^{s} \bar{X}_{(\alpha),i}^2$.

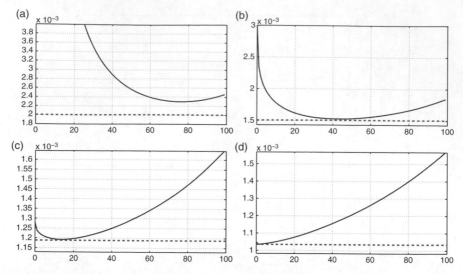

Figure 4.4 *The m.s.e. of $\bar{X}_{(\alpha)}$ versus α as an estimator of the location parameter μ of Student's t data with known scale 1 and degrees of freedom 1 (a), 3 (b), 10 (c), and 50 (d), based on a sample size of $n = 1000$ observations. The vertical axis was truncated to improve appearance. The dashed line in each plot is the m.s.e. of $\hat{\mu}_{ML}$.*

```
1   function alphahat = tlocsim(df,n,sim)
2   trimchoice = 0:99; ML1=zeros(sim,1);
3   trim=zeros(sim,length(trimchoice));
4   opt = optimset('disp','none','TolX',1e-6);
5   for i=1:sim % use location-zero, scale-one Stud t
6     x=trnd (df,n,1);  ML1(i) = fzero(@(mu) ff(mu,df,x), [-10,10], opt);
7     trim(i,:) = mytrim(x,trimchoice);
8   end
9   mu=0; ML1mse = mean((ML1 - mu).^2); trimmse = mean((trim - mu).^2);
10  plot(trimchoice, trimmse,'b-','linewidth',2), set(gca,'fontsize',16), grid
11  tmax = max(trimmse); tmin - 0.95*min(ML1mse,min(trimmse));
12  axis([0 100 tmin tmax])
13  line([0 100],[ML1mse ML1mse],'linestyle','-','color','r','linewidth',3);
14  mm = min(trimmse); alphahat = trimchoice(trimmse==mm);
15
16  function z=ff(mu,df,x),  y = x-mu; z = sum(y ./ (1+y.^2/df));
```

Program Listing 4.2: Generate the graphs in Figure 4.4. The function `mytrim` is given in Listing 4.3.

To illustrate, for $v = 1, 3, 10$, and 50, Figure 4.4 plots the m.s.e. as a function of α, overlaid with a horizontal line indicating the m.s.e. of $\hat{\mu}_{ML}$ (based on the same s samples). Indeed, as v increases from 1 to 50, the optimal value of α, say $\hat{\alpha}(v)$, decreases from 76 down to 2.

There are two other points to be observed from the figure: with increasing v, the m.s.e. of $\hat{\mu}_{ML}$ decreases, and the m.s.e. of $\bar{X}_{\hat{\alpha}(v)}$ approaches that of $\hat{\mu}_{ML}$. The code used to produce each graph is given in Listing 4.2.

This procedure was then conducted for the 99 values $v = 1, 1.5, 2, \dots, 50$, and the optimal value of α, $\hat{\alpha}(v)$, was determined. The results are shown in the top panels of Figure 4.5,

```
1   function m=mytrim(x,pvec)
2   x=sort(x); n=length(x); pl=length(pvec); m=zeros(1,pl);
3   for i=1:pl
4       p=pvec(i)/2; lo=round(n*p/100); lo=max(1,lo) + 1;
5       hi=round(n*(100-p)/100); m(i) = mean(x(lo:hi));
6   end
```

Program Listing 4.3: Calculates the trimmed means of vector x at the percent values given in pvec. When pvec is a vector, this is much faster than the built-in Matlab function trimmean.

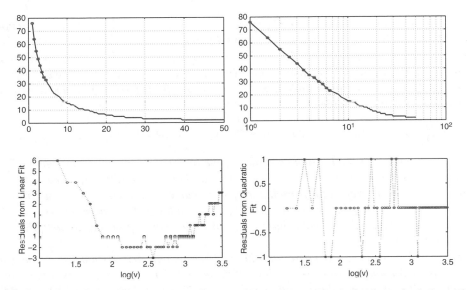

Figure 4.5 *The top left panel plots $\hat{\alpha}(v)$ versus v for $n = 1000$, each obtained via simulation using 25,000 replications. The top right is the same, but using a log scale. The bottom panels show the least squares residuals for the linear (left) and quadratic (right) fits for $3 \leq v \leq 33$.*

which plots v versus $\hat{\alpha}(v)$ and shows that the $\hat{\alpha}$ behave as expected. The top right is a log plot, which reveals an almost linear structure of $\hat{\alpha}$ for low values of v. This can be used to construct a simple approximation to the relationship between v and $\hat{\alpha}(v)$ that will be useful for obtaining an accurate interpolated value of α when $v \notin \{1, 1.5, 2, \dots, 50\}$. In particular, the first five observations ($v = 1, 1.5, \dots, 3$), are virtually perfectly modeled as

$$\hat{\alpha}(v) \approx \breve{\alpha}(v) := \text{round}(75.8264 - 29.2699 \log(v)), \quad 1 \leq v \leq 3, \qquad (4.3)$$

where the coefficients are obtained via least squares and we round off because α is typically an integer. This resulted in a regression R^2 of 0.9993, with the five residuals given by $0, 0, -1, 0, 0$. This means, for example, that $\breve{\alpha}(2) = 55$, while the simulation results imply $\hat{\alpha}(2) = 56$. Recalling the plots in Figure 4.4, this difference is not important.[1]

---

[1] Moreover, given the highly appealing linear form of $\hat{\alpha}(v)$, $1 \leq v \leq 3$, and the fact that the simulations are also subject to error (having used only 25,000 replications), the value 55 might, in fact, be closer to the theoretical optimal value.

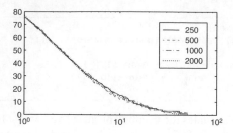

Figure 4.6   *Same as the top right panel in Figure 4.5 but for three additional sample sizes.*

The remaining observations could be modeled by fitting a low-order polynomial function of t. However, from the top left plot in Figure 4.5, we see that $\hat{\alpha}(v) = 3$ for $33 \leq v < 50$. Thus, it is much simpler just to take $\breve{\alpha}(v) = 3$ for v in this range. For $3 < v \leq 33$, a linear fit did not fully capture the behavior of the $\hat{\alpha}(v)$; the residuals are shown in the bottom left panel of Figure 4.5 and suggest additional use of the quadratic term $(\log v)^2$. Doing so (via least squares) resulted in the fit

$$\hat{\alpha}(v) \approx \breve{\alpha}(v) = \mathrm{round}\,(81.6637 - 40.5658\ \log v + 5.1540\ (\log v)^2), \quad 3 < v \leq 33,$$

with an R^2 of 0.9983, and residuals shown in the bottom right panel of the figure. They appear random (i.e., no obvious pattern exists that could be further exploited) and are bounded by 1, so that very little degradation of performance will result if $\breve{\alpha}(v)$ is used instead of $\hat{\alpha}(v)$. Finally, for $v \geq 50$, we could just set $\breve{\alpha}(v) = 2$, although for v large enough, it might be the case that the true optimal value $\hat{\alpha}$ drops to 1 and, as v approaches ∞, to 0. Further simulation would reveal this.

It must be emphasized that the results obtained pertain only to $n = 1000$, but should also be adequate for sample sizes close to 1000. This severely limits the usefulness of the study, which needs to be repeated for multiple values of n. This was done for $n = 250$, $n = 500$, and $n = 2000$, with results graphically shown in the log plot of Figure 4.6. It appears in this case that the optimal values of α do not depend on sample size (at least for n between 250 and 2000). If they were dependent on n, all would not be lost; one could also construct an approximate function of v and n. Fortunately, in this case, the simple regressions used above suffice.   ∎

Remark.   This method of estimation for μ was conditional on knowing v and σ, and as such, does not appear overly practical. However, it can be used in conjunction with joint parameter estimation, as discussed in Section 4.1.2. Furthermore, it can be used as the basis for accelerating the estimation of nontrivial models, and particularly in situations where a large number of them need to be estimated; see, for example, Krause and Paolella (2014).

Finally, the same concept could be used for estimation of the location term in other heavy-tailed distributions, such as the stable Paretian. The reader is encouraged to repeat the above analysis for the symmetric stable distribution, where the tail index, unfortunately typically called α, and not to be confused with the trimming factor, replaces the use of the degrees-of-freedom parameter in Student's t. This idea was used to advantage in Paolella (2016b).   ∎

4.1.2 Several Parameters

The residuals of many common models in regression and time series are often assumed to be normally distributed, appealing to the central limit theorem (A.160), whereby the error term is envisioned to be the sum of a large number of small factors not accounted for by the model. However, in some applications, particularly with financial returns data, the residuals tend to exhibit leptokurtic and asymmetric behavior. Assuming for now only the former, the t distribution offers a reasonable way of modeling such data and includes the normal as a (limiting) special case. However, the number of degrees of freedom v has to be estimated from the data, and is then allowed to be any positive real number.

The location-μ, scale-σ Student's $t(v)$ density is given by

$$f(x; v, \mu, \sigma) = \frac{1}{\sigma} \frac{\Gamma\left(\frac{v+1}{2}\right) v^{\frac{v}{2}}}{\sqrt{\pi}\, \Gamma\left(\frac{v}{2}\right)} \left(v + \left(\frac{x-\mu}{\sigma}\right)^2\right)^{-\frac{v+1}{2}},$$

so that, for $X_i \overset{\text{i.i.d.}}{\sim} t(v, \mu, \sigma)$ and $\theta = (v, \mu, \sigma)'$, the log-likelihood for $X_1, \ldots, X_n$ is

$$\ell(\theta; \mathbf{x}) = -n \log \sigma + n \log \Gamma\left(\frac{v+1}{2}\right) + \frac{nv}{2} \log v$$

$$- \frac{n}{2} \log \pi - n \log \Gamma\left(\frac{v}{2}\right) - \frac{v+1}{2} \sum_{i=1}^{n} \log y_i,$$

where $y_i = v + z_i^2$ and $z_i = (x_i - \mu)/\sigma$. The score function for v is

$$\dot{\ell}_v(\theta; \mathbf{x}) = \frac{n}{2} \psi\left(\frac{v+1}{2}\right) + \frac{n}{2}(1 + \log v)$$

$$- \frac{n}{2} \psi\left(\frac{v}{2}\right) - \frac{v+1}{2} \sum_{i=1}^{n} y_i^{-1} - \frac{1}{2} \sum_{i=1}^{n} \log y_i,$$

where $\psi(\cdot)$ is the digamma function. We would like to compute $\widehat{\theta}_{\mathrm{ML}} = (\hat{v}_{\mathrm{ML}}, \hat{\mu}_{\mathrm{ML}}, \hat{\sigma}_{\mathrm{ML}})'$. If values $\hat{\mu}_{\mathrm{ML}}$ and $\hat{\sigma}_{\mathrm{ML}}$ are available, then solving $\dot{\ell}_v(v, \hat{\mu}_{\mathrm{ML}}, \hat{\sigma}_{\mathrm{ML}}; \mathbf{x}) = 0$ and $\ddot{\ell}_v(v, \hat{\mu}_{\mathrm{ML}}, \hat{\sigma}_{\mathrm{ML}}; \mathbf{x}) > 0$ will yield $\hat{v}_{\mathrm{ML}}$. Similarly, given $\hat{v}_{\mathrm{ML}}$ and $\hat{\mu}_{\mathrm{ML}}$, solving $\dot{\ell}_\sigma(\hat{v}_{\mathrm{ML}}, \hat{\mu}_{\mathrm{ML}}, \sigma; \mathbf{x}) = 0$ and $\ddot{\ell}_\sigma(\hat{v}_{\mathrm{ML}}, \hat{\mu}_{\mathrm{ML}}, \sigma; \mathbf{x}) > 0$ will yield $\hat{\sigma}_{\mathrm{ML}}$, where

$$\dot{\ell}_\sigma(\theta; \mathbf{x}) = -\frac{n}{\sigma} + \frac{v+1}{\sigma} \sum_{i=1}^{n} \frac{z_i^2}{y_i} \propto n - (v+1) \sum_{i=1}^{n} \frac{z_i^2}{y_i}.$$

Lastly, given $\hat{v}_{\mathrm{ML}}$ and $\hat{\sigma}_{\mathrm{ML}}$, solving $\dot{\ell}_\mu(\hat{v}_{\mathrm{ML}}, \mu, \hat{\sigma}_{\mathrm{ML}}; \mathbf{x}) = 0$ will yield $\hat{\mu}_{\mathrm{ML}}$, where

$$\dot{\ell}_\mu(v, \mu, \sigma; \mathbf{x}) = \frac{v+1}{\sigma} \sum_{i=1}^{n} \frac{z_i}{y_i} \propto \sum_{i=1}^{n} \frac{z_i}{y_i}.$$

The problem is that, in general, all three m.l.e. values will be unknown. The three equations $\dot{\ell}_v = 0$, $\dot{\ell}_\mu = 0$, and $\dot{\ell}_\sigma = 0$ need to be solved simultaneously in order to obtain $\widehat{\theta}_{\mathrm{ML}}$.

Numeric methods do exist to solve a system of $k > 1$ nonlinear equations, but they are considerably more complicated than for the $k = 1$ case. In fact, the methods required are quite similar to those employed for unconstrained maximization of multivariate functions, so that we could just as well compute $\hat{\theta}_{ML} = \arg\max \ell(\theta; \mathbf{x})$ directly; this is discussed further in Section 4.3 below. Moreover, for $k = 1$, one can easily plot the function whose zero is to be found and, while it is not the most efficient method, we could find the solution by a simple bisection approach. This does not carry over for $k > 1$.

A way does exist, however, that can still use the three equations for each $\dot{\ell}$ separately and obtain a value $\hat{\theta}$ satisfying $\dot{\ell} = \mathbf{0}$. This is by *iterating* on the three equations until convergence, and proceeds as follows.

(1) Using v_0, μ_0, and c_0 as initial guesses for v, μ, and c, respectively, the first step solves $\dot{\ell}_v(v, \mu_0, \sigma_0; \mathbf{x}) = 0$ to get v_1, which improves upon v_0 and can be viewed as $\hat{v}_{ML}$ given $\mu = \mu_0$ and $c = c_0$. Observe that this step is also equivalent to solving $\dot{\ell}_v(v, 0, 1; \mathbf{z}) = 0$, where $z_i = (x_i - \mu_0)/\sigma_0$, so that simple bisection could be used.

(2) The next step involves solving $\dot{\ell}_\mu(v_1, \mu, \sigma_0; \mathbf{x}) = 0$; notice how the updated value v_1 is used instead of v_0. A simple transformation shows that $W = X/\sigma \sim t(v, \mu/\sigma, 1)$, so that we could also solve $\dot{\ell}_\mu(v_1, \mu^*, 1; \mathbf{w}) = 0$ to get, say, μ_1^*, and set $\mu_1 = \mu_1^* \sigma_0$.

(3) The last step solves $\dot{\ell}_\sigma(v_1, \mu_1, \sigma; \mathbf{x}) = 0$ to get σ_1; again notice how the updated values of v and μ from the previous two steps are used. This step is also the same as solving $\dot{\ell}_\sigma(v_1, 0, \sigma; \mathbf{x} - \mu_1) = 0$.

(4) The process then begins "from the top" again, that is, we would solve $\dot{\ell}_v(v, \mu_1, \sigma_1; \mathbf{x}) = 0$ to get v_2, etc.

Let $\theta_i = (v_i, \mu_i, \sigma_i)'$. In order for this scheme to work, it must be the case that the sequence $\theta_1, \theta_2, \dots$, converges; and that it converges to $\hat{\theta}_{ML}$. Assuming this to be the case, the process is stopped when $\| \theta_{i-1} - \theta_i \| < \epsilon$ for some small value ϵ, say 10^{-5}, and where $\| \cdot \|$ is some distance measure, usually Euclidean.

This iterative method also has another advantage over attempting to solve the system of three equations simultaneously or to maximize $\ell(\theta; \mathbf{x})$, in that the individual parameter constraints are easy to impose; for example, with the t distribution, that $v > 0$ and $\sigma > 0$. Without additional programming effort, unconstrained maximization routines could (and sometimes do) try invalid parameter values, for example, a negative scale parameter, which can directly lead to disaster. Methods of preventing this from occurring do exist, a simple one of which will be discussed in Section 4.3.2.

A final advantage of the iterative method arises in the Student's t case being considered: The second step of solving $\dot{\ell}_\mu(v_i, \mu, \sigma_{i-1}; \mathbf{x}) = 0$ to obtain μ_i can be approximated by setting $\mu_i = \bar{X}_{\hat{\alpha}(v_i)}$, the trimmed mean with the m.s.e.-optimal value of α computed by the expressions for $\breve{\alpha}$ in Example 4.3. We have seen that, for $v \geq 3$, $\bar{X}_{\hat{\alpha}(v)}$ is virtually as good as the m.l.e. and, being a closed-form expression for μ_i, will save computing time.

The one issue that remains is the choice of starting values v_0, μ_0, and c_0. In many situations, one can find relatively simple functions of the data that can serve as reasonable estimates for the unknown parameters. This is true for Student's t data. For μ_0, we simply take the trimmed mean with $\alpha = 50$, which, recalling Figure 4.4, seems to be an acceptable compromise for most values of v. Deriving simple and effective estimators for v and σ,

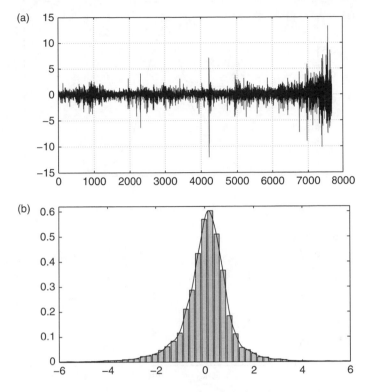

Figure 4.7   *Daily returns for the NASDAQ index.*

however, is more difficult. We use the results of Singh (1988), who proposed taking

$$v_0 = 2\frac{2a-3}{a-3} \quad \text{and} \quad \sigma_0^2 = \frac{3\sum_{i=1}^{n} r_i^2}{(2a-3)(n+1)}, \qquad (4.4)$$

where $a = n\left(\sum_{i=1}^{n} r_i^4\right)/\left(\sum_{i=1}^{n} r_i^2\right)^2$ and $r_i = x_i - \mu_0$. As fourth powers of the r_i are involved, this procedure can only work for $v > 4$ and, for values of v not much greater than 4, will most likely perform poorly.

The reader is invited to implement the estimation procedure in Problem 3.1, while Problem 3.2 conducts a simulation to investigate and compare the properties of the estimators.

Example 4.4   *Consider the daily returns (taken to be the continuously compounded percentage returns, $r_t = 100(\log P_t - \log P_{t-1})$, where P_t denotes the index level at time t), on the NASDAQ stock index covering the period from its inception in February 1971 to June 2001, yielding 7681 observations. A plot of the returns is given in Figure 4.7, along with a scaled histogram overlaid with a kernel density plot.*

We apply the previous procedure to fit the three parameters of the location–scale Student's t. The initial estimates were $\hat{v}_0 = 4.4$, $\hat{\mu}_0 = 0.099$ and $\hat{\sigma}_0 = 0.12$. The latter value needed to be quadrupled, that is, set to 0.48, for $\dot{\ell}_v$ to have a sign change between $v = 1$ and $v = 60$. The iterative scheme to obtain the m.l.e. converged after 14 iterations using a

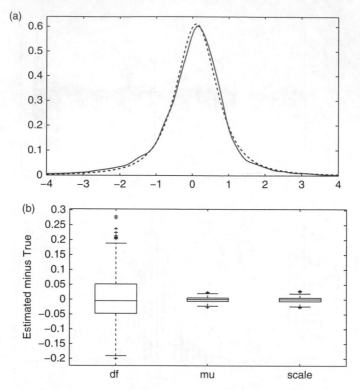

Figure 4.8 *(a) Kernel density (solid) and fitted Student's t density (dashed) of the NASDAQ returns. (b) Simulation results of the m.l.e. for the Student's t model, based on $n = 7681$ observations and true parameter values taken to be the m.l.e. of the t model for the NASDAQ returns. The boxplots show their differences.*

tolerance of 10^{-5}, with values $\hat{\upsilon}_{\mathrm{ML}} = 2.38$, $\hat{\mu}_{\mathrm{ML}} = 0.105$ and $\hat{\sigma}_{\mathrm{ML}} = 0.588$. Observe how $\hat{\mu}_{\mathrm{ML}}$ is quite close to $\hat{\mu}_0$, while the other starting values based on (4.4) were not very good, even though we have 7681 observations. The occurrence of poor starting values for υ and σ is not too surprising, given that $\hat{\upsilon}_{\mathrm{ML}}$ is much less than 4, thus providing evidence that $\upsilon < 4$, so that the values given by (4.4) will not be reliable.

Figure 4.8(a) plots the kernel density of the NASDAQ returns, overlaid with the fitted t density. (It is truncated to $[-4, 4]$, though the lowest and highest returns are -12.04 and 13.25, respectively.) Except for the fact that the t cannot capture the mild asymmetry of the returns, the fit appears very good. However, recalling Figure 4.7(a), the returns deviate blatantly from being i.i.d., and instead appear to have a time-varying scale term. One can say that the Student's t model is a reasonable approximation for the unconditional distribution of the data. ■

Example 4.5 (Example 4.4, cont.)   *In light of the findings from Problem 3.2 regarding the relative imprecision of $\hat{\upsilon}$, we simulated 500 i.i.d. samples, each of length $n = 7681$ (the same as for the NASDAQ return series) and with parameter vector equal to the m.l.e. of that series, and estimated the t model. All 500 estimations converged successfully. Boxplots* *of the deviations of the $\hat{\theta}_{ML}$ from their true values are shown in Figure 4.8(b). Observe that $\hat{\upsilon}_{ML}$ is far more imprecise than the location and scale parameters. The reason for this is discussed below in Example 4.6.*

Notice also that the density of $\hat{\upsilon}_{ML}$ is slightly skewed to the right. The 2.5% and 97.5% sample quantiles from the 500 values of $\hat{\upsilon}_{ML}$ were 2.23 and 2.54, respectively. These serve as a 95% confidence interval for the degrees-of-freedom parameter in the model for the NASDAQ returns, formed from a (single, parametric) bootstrap. ∎

4.2 APPROXIMATING THE DISTRIBUTION OF THE MAXIMUM LIKELIHOOD ESTIMATOR

For situations in which the m.l.e. of a parameter is available in closed form, its distribution can be (theoretically, at least) obtained via transformation. For example, with i.i.d. Bernoulli data, the m.l.e. of p is S/n, which is a scaled binomial random variable. However, as the previous examples illustrated, closed-form expressions for the m.l.e. often do not exist, which will often imply that the exact distribution of the m.l.e. will be impossible to determine. Moreover, for more complicated models often encountered in practice, simple simulation-based studies, such as those undertaken in the previous examples, will not be feasible. In such cases, the asymptotic properties of the m.l.e. (see Section 3.1.4) can be used to approximate the behavior of the m.l.e. in finite samples. This requires knowledge of the information matrix $\mathbf{J}$. It is, moreover, usually the case that $\mathbf{J}^{-1}$ is too difficult to analytically derive, so that the inverse of a numerical approximation of $\mathbf{J}$ is used.

One possibility is the following: With $\theta = (\theta_1, \dots \theta_k)'$ and $\kappa_i = (0, \dots, h_i, \dots, 0)'$, that is, the zero vector with ith element h_i, the (i, j)th element of the Hessian matrix $\mathbf{J}$ can be approximated by $H_{i,j}$, where

$$H_{j,i} = H_{i,j} = \frac{\ell(\theta_{(i,j)}) - \ell(\theta_{(i)}) - \ell(\theta_{(j)}) + \ell(\theta)}{h_i h_j}, \tag{4.5}$$

$\theta_{(i)} = \theta + \kappa_i$, $\theta_{(i,j)} = \theta + \kappa_i + \kappa_j$, and $h_i = \varepsilon \theta_i$ is a small perturbation relative in size to θ_i, with $\varepsilon = 0.01$ being a reasonable choice. A program to compute this is given in Listing 4.4.

Example 4.6 (Example 4.4, cont.)   *The approximation in (4.5) was applied to the Student's t model for the NASDAQ data. The negative of the inverse of $\mathbf{H}$ is*

$$10^{-3} \cdot \begin{pmatrix} 6.5336 & -0.0679 & 0.4956 \\ -0.0679 & 0.0722 & -0.0078 \\ 0.4956 & -0.0078 & 0.0905 \end{pmatrix}. \tag{4.6}$$

From this, several quantities of interest can be computed. For example, the (approximate) correlation between $\hat{\upsilon}_{ML}$ and $\hat{\mu}_{ML}$ is −0.10, while that between $\hat{\upsilon}_{ML}$ and $\hat{\sigma}_{ML}$ is 0.64. This substantial amount of positive correlation is reasonable because an increase in the scaling of the t density can be partially offset by an increase in the degrees of freedom. In general, an absolute correlation value between two parameters that is higher than, say, 0.9, indicates that one of the two parameters is close to being redundant.

Approximate standard deviations associated with the parameter estimates are given by the square roots of the diagonal elements. In this case, to two significant digits, these are

```
1   function H = hessian(f,theta,varargin)
2   x=theta; % easier to work with letter x
3   k=length(x); tol1=1e-2; tol2=1e-3; h = tol1*max(abs(x),tol2);
4   fx = feval(f,x,varargin{:}); g = zeros(k,1);
5   for i=1:k % get gradient type elements
6       perturb = zeros(k,1); perturb(i)=h(i); g(i) = feval(f,x+perturb,varargin{:});
7   end
8   for i=1:k, for j=i:k
9       perturb = zeros(k,1); perturb(i)=h(i); perturb(j)=perturb(j)+h(j);
10      fxij = feval(f,x+perturb,varargin{:});
11      H(i,j) = ( fxij - g(i) - g(j) + fx ) / (h(i) * h(j)); H(j,i) = H(i,j);
12  end, end
```

Program Listing 4.4: Numerically approximates the Hessian matrix of the function f at the parameter vector theta. Additional required parameters of the log-likelihood function can be supplied as varargin

0.081, 0.0085, and 0.0095. This means, for example, that an approximate 95% confidence interval for the degrees of freedom v is $\hat{v}_{ML} \pm 1.96 \cdot 0.081 = (2.22, 2.54)$. This is virtually the same result obtained in Example 4.5, but having used far less calculation, and arises because of the large sample size. Observe that the heavy-tailed nature of Student's t has nothing per se to do with the quality of asymptotic results on the m.l.e. The data could be Cauchy, without existence of a mean or variance, but the asymptotic distribution of the m.l.e. of the location and scale parameters of the Cauchy are well behaved; in particular, they are unbiased, normally distributed, and reach the CRlb discussed in Section 3.2.

This should not be confused with the relatively large uncertainty (determined via confidence intervals) of v compared to that of μ and σ. This is an artifact of the model, in that v is also the tail index, and governs the maximally existing moment. By its nature, being a tail index, its estimate is determined from tail observations and, by definition, there are not many observations in the tail. See Chapter 9 for details on estimating the maximally existing moment and tail estimation. ∎

4.3 GENERAL NUMERICAL LIKELIHOOD MAXIMIZATION

Similar to root finding, all numeric methods designed to maximize a (multivariate) function require that the function (in our context, the log-likelihood ℓ) can be evaluated very accurately relative to the desired accuracy of the m.l.e. Some methods require expressions for (or approximations to) the gradient and Hessian, others need only the gradient, while yet others need neither. Recall from Section 4.1.2 that, at each step of the method of iterating on the score functions, the maximum of the likelihood with respect to a single component in the parameter vector is calculated. Multivariate methods attempt to move "diagonally" in an optimal fashion and should thus be faster. There are two general rules of thumb: (1) the more information used, that is, gradient and Hessian, the faster the convergence; (2) the more ℓ deviates from "perfect smoothness" and/or is plagued with multiple local maxima, the worse gradient- and Hessian-based routines tend to perform.

4.3.1 Newton–Raphson and Quasi-Newton Methods

For the scalar parameter case, using the Taylor series (3.2) applied to the score function about θ_0 and omitting the remainder term, we obtain the approximation

$$\dot{\ell}(\theta;\mathbf{x}) \approx \dot{\ell}(\theta_0;\mathbf{x}) + \ddot{\ell}(\theta_0;\mathbf{x})(\theta - \theta_0),$$

which grows in accuracy as $\theta \to \theta_0$. For θ close to the m.l.e., $\dot{\ell}(\theta;\mathbf{x}) \approx 0$, so that, rearranging, $\theta \approx \theta_0 - \dot{\ell}(\theta_0;\mathbf{x})/\ddot{\ell}(\theta_0;\mathbf{x})$. This suggests the iterative scheme

$$\hat{\theta}_{(i)} = \hat{\theta}_{(i-1)} - \frac{\dot{\ell}(\hat{\theta}_{(i-1)};\mathbf{x})}{\ddot{\ell}(\hat{\theta}_{(i-1)};\mathbf{x})},$$

which is the famous **Newton–Raphson** algorithm. The so-called **method of scoring**, as proposed by Fisher, is the same, but using the expected information instead of the observed information.

The multivariate case with k parameters is similar, resulting in

$$\hat{\boldsymbol{\theta}}_{(i)} = \boldsymbol{\theta}_{(i-1)} - \dot{\boldsymbol{\ell}}(\hat{\boldsymbol{\theta}}_{(i-1)};\mathbf{x})'\ddot{\boldsymbol{\ell}}(\hat{\boldsymbol{\theta}}_{(i-1)};\mathbf{x})^{-1}.$$

A closed-form expression for the inverse of the Hessian $\ddot{\boldsymbol{\ell}}(\hat{\boldsymbol{\theta}}_{(i-1)};\mathbf{x})$ will often not exist, and numerical evaluation will be too costly. For these cases, there exist so-called *quasi-Newton* algorithms that replace the computation of the inverse of the Hessian with an iterative-based approximation that is faster to evaluate, and tends to the true inverse Hessian as $\hat{\boldsymbol{\theta}}_{(i)} \to \hat{\boldsymbol{\theta}}_{\mathrm{ML}}$.

In the following, to be consistent with existing literature on numerical optimization, rather than maximizing the log-likelihood, we minimize the negative of the log-likelihood, and its first derivative, the score function, is then negated. In particular, let

$$\mathbf{g}(\hat{\boldsymbol{\theta}}_{(i)}) = -\dot{\boldsymbol{\ell}}(\hat{\boldsymbol{\theta}}_{(i)};\mathbf{x}), \tag{4.7}$$

where we suppress the dependence of $\mathbf{g}$ on $\mathbf{x}$ to simplify the notation, and the letter g is used to signify the gradient. Let $\widehat{\mathbf{H}^{-1}}_{(i)}$ denote the approximation to the inverse of the Hessian of the negative log-likelihood at the ith iteration. The iterative formula that is common to all quasi-Newton algorithms is

$$\hat{\boldsymbol{\theta}}_{(i)} = \boldsymbol{\theta}_{(i-1)} - \lambda_{(i)}\mathbf{g}(\hat{\boldsymbol{\theta}}_{(i-1)})'\widehat{\mathbf{H}^{-1}}_{(i-1)}, \tag{4.8}$$

where $-\mathbf{g}(\hat{\boldsymbol{\theta}}_{(i-1)})'\widehat{\mathbf{H}^{-1}}_{(i-1)}$ is the **search direction** at the ith iteration and $\lambda_{(i)}$ is the corresponding **step size**. It satisfies

$$\lambda_{(i)} = \underset{\lambda}{\operatorname{argmin}} \left[-\ell\left(\hat{\boldsymbol{\theta}}_{(i)}(\lambda)\right) \right] = \underset{\lambda}{\operatorname{argmin}} \left[-\ell\left(\boldsymbol{\theta}_{(i-1)} - \lambda\mathbf{g}(\hat{\boldsymbol{\theta}}_{(i-1)})'\widehat{\mathbf{H}^{-1}}_{(i-1)}\right) \right]. \tag{4.9}$$

The $\lambda_{(i)}$ are found via a separate univariate line search algorithm on each iteration. Depending on the choice of $\widehat{\mathbf{H}^{-1}}$, several algorithms can be constructed. For example, taking $\widehat{\mathbf{H}^{-1}}_{(i)}$ to be the matrix of all zeros except with a 1 in the rth diagonal position, iteratively for $r = 1, \ldots, k$, results the aforementioned method of iterating on the score functions individually, whereas with

$$\widehat{\mathbf{H}^{-1}}_{(i-1)} = \widehat{\mathbf{H}^{-1}} = \mathbf{I},$$

the so-called **method of steepest descent** is obtained. The name stems from the fact that, in each iteration, the search direction is the negative of the gradient, that is, the direction of the steepest descent of the objective function.

More sophisticated methods build an approximation to $\mathbf{H}^{-1}$ from the history of the gradients encountered in earlier iterations. For example, the so-called BFGS (after the authors Broyden, Fletcher, Goldfarb, and Shanno) algorithm uses

$$\widehat{\mathbf{H}_{(i)}^{-1}} = \widehat{\mathbf{H}_{(i-1)}^{-1}} + \left(\frac{1 + \mathbf{q}_{(i)}' \widehat{\mathbf{H}_{(i-1)}^{-1}} \mathbf{q}_{(i)}}{\mathbf{q}_{(i)}' \mathbf{p}_{(i)}} \right) \frac{\mathbf{p}_{(i)} \mathbf{p}_{(i)}'}{\mathbf{p}_{(i)}' \mathbf{q}_{(i)}} - \frac{\mathbf{p}_{(i)} \mathbf{q}_{(i)}' \widehat{\mathbf{H}_{(i-1)}^{-1}} + \widehat{\mathbf{H}_{(i-1)}^{-1}} \mathbf{q}_{(i)} \mathbf{p}_{(i)}'}{\mathbf{q}_{(i)}' \mathbf{p}_{(i)}}, \quad i \geq 2,$$

where $\mathbf{q}_{(i)} = \mathbf{g}(\theta_{(i)}; \mathbf{x}) - \mathbf{g}(\theta_{(i-1)}; \mathbf{x})$, $\mathbf{p}_{(i)} = \hat{\theta}_{(i)} - \hat{\theta}_{(i-1)}$, and $\widehat{\mathbf{H}_{(1)}^{-1}} = \mathbf{I}$.

Figure 4.9 illustrates the resulting paths drawn out by the different iterative methods for a particular two-parameter model and data set. It can be expected that, the better the approximation to the Hessian, the fewer iterations will be required for convergence, and this is indeed the case.

One of the major advantages of this method is that a numerically calculated gradient can be used if an analytic formula for the gradient is not available, or even if its algebraic determination is difficult or its numerical evaluation is costly. The jth element of the gradient

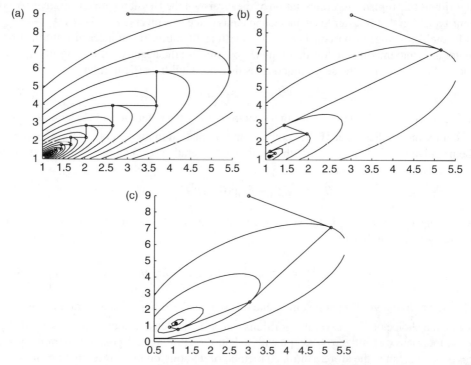

Figure 4.9   *Convergence of the method of iterating on the score functions (a), method of steepest descent (b), and the BFGS algorithm (c), for the log-likelihood of a Gam(1, 1) sample of size 100. The number of iterations required to arrive at the m.l.e. of $\hat{\theta}_{\mathrm{ML}} = (1.0797, 1.1595)'$ with the same accuracy was 56, 16, and 11, respectively.*

g can simply be approximated by

$$g_j \approx -\frac{\ell(\boldsymbol{\theta} + \boldsymbol{\kappa}_j) - \ell(\boldsymbol{\theta})}{h_j}, \qquad (4.10)$$

$j = 1, \ldots, k$, where $\boldsymbol{\kappa}_j = (0, \ldots, h_j, \ldots, 0)'$, $h_j = \varepsilon \theta_j$ is a small perturbation relative in size to θ_j, the jth element of $\boldsymbol{\theta}$, and the minus sign in (4.10) is because of our sign convention in (4.7). Because of the approximate nature of **g**, the "path" from the starting value of $\boldsymbol{\theta}$ to the m.l.e. taken by the minimization algorithm will most likely be longer than that using the exact gradient, so that more log-likelihood function evaluations and, hence, more time, will be required for estimation. To offset this, more accurate methods could be used to calculate the numerical derivative, for example, the two-sided expression

$$g_j \approx -\frac{\ell(\boldsymbol{\theta} + \boldsymbol{\kappa}_j) - \ell(\boldsymbol{\theta} - \boldsymbol{\kappa}_j)}{2h_j},$$

but this involves more function evaluations and, thus, correspondingly more time.[2] For each problem, there might indeed be an optimal tradeoff between these two factors that leads to minimal estimation time; however, it might be a function of k, the likelihood, and its computational complexity, and also of the desired accuracy of the m.l.e., the unknown value of $\hat{\theta}_{\text{ML}}$, and possibly even the quality of the starting value of $\boldsymbol{\theta}$, which is also unknown before estimation.

The BFGS algorithm is implemented in Matlab's function `fminunc`, which returns the parameter vector for which the target function assumes its *minimum*. For this reason, our target function is the *negative* of the log-likelihood function. The program in Listing 4.5 shows the simplest implementation for the Student's t model. (We will improve upon this shortly; for now, it serves to understand the basic structure of optimization in Matlab.) As nothing prevents `fminunc` from trying negative values for the degrees of freedom or scale

```
1   function MLE = tlikmax0 (x, initvec)
2   tol=1e-5;
3   opts=optimset('Disp','none','LargeScale','Off', ...
4          'TolFun',tol ,'TolX',tol ,'Maxiter',200);
5   MLE = fminunc(@(param) tloglik (param,x), initvec , opts);
6
7   function ll=tloglik (param,x)
8   v=param(1); mu=param(2); c=param(3);
9   if v<0.01, v=rand, end % An ad hoc way of preventing negative values
10  if c<0.01, c=rand, end %      which works, but is NOT recommended!
11  K=beta(v/2,0.5) * sqrt(v); z=(x−mu)/c;
12  ll = −log(c) −log(K) −((v+1)/2) * log (1 + (z.^2)/v); ll = −sum(ll);
```

Program Listing 4.5: Attempts to maximize the log-likelihood of the i.i.d. Student's t model using data vector x, starting values `initvec=[df location scale]`, a convergence tolerance of 0.00001 and allowing at most 200 function evaluations.

---

[2] Notice that, with (4.10), $\ell(\boldsymbol{\theta})$ can be used for each i, so that a total of $k + 1$ log-likelihood functions are required, while with the two-sided calculation, $2k$ evaluations are needed. More elaborate expressions (involving more function evaluations) exist for approximating the gradient with higher accuracy. Consult a numerical analysis book for details.

parameter, a primitive method is (temporarily) used to circumvent this: if a negative value is proposed, it is replaced by a standard uniform random number. This works better than using a constant close to zero, in which case the algorithm sometimes never leaves the disallowed parameter space.

Example 4.7 (NASDAQ, cont.)   *Use of the optimization method as given in Listing 4.5 for the m.l.e. of the i.i.d. location–scale Student's t model applied to the NASDAQ return series yields* $\hat{v}_{ML} = 2.38$, $\hat{\mu}_{ML} = 0.105$, *and* $\hat{\sigma}_{ML} = 0.588$, *with a log-likelihood at the m.l.e., denoted by* ℓ_{ML}, *of* $-10,286.14$. *These are the same values, to three significant digits, as found in Example 4.4.*   ∎

4.3.2 Imposing Parameter Restrictions

Consider maximum likelihood estimation of parameters θ_i, $i = 1, \ldots, k$, each of which is restricted to lie in the respective interval

$$-\infty < a_i < \theta_i \le b_i < \infty,$$

for known constants $a_i, b_i, i = 1, \ldots, k$. (Observe that $a_i < \theta_i$ is a strict inequality; see (4.11) below.) These are sometimes referred to as **box constraints**. Numerically, this can often lead to complications, most notably when the numerical algorithm, unaware of the constraints, decides to try a value of θ outside its possible range, such as a negative variance parameter. One way to impose the restriction is to maximize the likelihood function with respect to another set of parameters, say ϕ_i, that are related to the θ_i and such that $-\infty < \phi_i < \infty$ for each i. A useful transformation to achieve this is

$$\phi_i = +\sqrt{\frac{b_i - \theta_i}{\theta_i - a_i}}, \quad i = 1, \ldots, k. \tag{4.11}$$

Solving for θ_i in terms of ϕ_i yields

$$\theta_i = \theta_i(\phi_i) = \frac{b_i + \phi_i^2 a_i}{1 + \phi_i^2}. \tag{4.12}$$

Inspection shows that $\lim_{|\phi_i| \to \infty} \theta_i(\phi_i) = a_i$ and $\phi_i = 0 \Leftrightarrow \theta_i = b_i$. Notice that the upper bound b_i can be attained, but not the lower bound a_i. To see that the ranges of the θ_i are indeed restricted to the interval $a_i < \theta_i \le b_i$ provided $a_i < b_i$, let $\epsilon_i = b_i - a_i > 0$ and note that $\phi_i^2 \ge 0$,

$$\theta_i = \frac{b_i + \phi_i^2(b_i - \epsilon_i)}{1 + \phi_i^2} = b_i - \frac{\phi_i^2}{1 + \phi_i^2}\epsilon_i \le b_i,$$

and, similarly,

$$\theta_i = \frac{\epsilon_i + a_i + \phi_i^2 a_i}{1 + \phi_i^2} = a_i + \frac{1}{1 + \phi_i^2}\epsilon_i > a_i.$$

The log-likelihood can be numerically maximized with respect to the ϕ parameterization, so that $\hat{\phi}_{ML}$ and $\mathbf{J}^{-1} = \mathbf{J}^{-1}(\phi)$ can be obtained. The original point estimates θ_i are computed from (4.12), while their approximate variance–covariance matrix can be computed

```
1   function [param,stderr,iters ,loglik ,Varcov] = tlikmax(x,initvec)
2
3   %%%%%%%   df    mu    c
4   bound.lo=    [1    1    0.01];
5   bound.hi=    [100   1    100 ];
6   bound.which=[1     0    1   ];
7   % In this case, as bound.which for mu is zero, mu will not be
8   % restricted. As such, the values for .lo and .hi are irrelevant
9
10  maxiter=100; tol=1e-3; % change these as you see fit
11  opts=optimset('Display','notify-detailed','Maxiter',maxiter, ...
12               'TolFun',tol, 'TolX',tol, 'LargeScale','Off');
13   [pout,fval, exitflag ,theoutput ,grad ,hess]= ...
14   fminunc(@(param) tloglik (param,x,bound), einschrk (initvec ,bound), opts);
15  V=inv(hess); % Don't negate: we work with the neg of the loglik
16   [param,V]=einschrk(pout ,bound,V); % Transform back, apply delta method
17  param=param'; Varcov=V;
18  stderr=sqrt(diag(V));    % Approx std err of the params
19  loglik=-fval;            % The value of the loglik at its maximum.
20  iters=theoutput.iterations;   % Number of loglik function evals
21
22  function ll=tloglik (param,x,bound)
23  if nargin<3, bound=0; end
24  if isstruct(bound), paramvec=einschrk(real(param),bound,999);
25  else paramvec=param;
26  end
27
28  v=paramvec(1); mu=paramvec(2); c=paramvec(3);
29  K=beta(v/2,0.5)*sqrt(v); z=(x-mu)/c;
30  ll= -sum(-log(c)-log(K)-((v+1)/2)*log (1 + (z.^2)/v));
```

Program Listing 4.6: An improved version of the program in Listing 4.5 that restricts the parameters in a smarter fashion and delivers approximate standard errors of the estimated parameters. The variable bound is referred to as a "structural array."

via (3.46). In particular, $\dot{\tau}$ will be a diagonal matrix with ith element

$$\frac{\partial \theta_i(\phi_i)}{\partial \phi_i} = \frac{2\phi_i(a_i - b_i)}{(1 + \phi_i^2)^2},$$

so that $\hat{\theta}_{\mathrm{ML}} \overset{\mathrm{asy}}{\sim} N_k(\theta, \dot{\tau}\mathbf{J}^{-1}\dot{\tau}')$. Also,

$$\widehat{\mathbb{V}}(\hat{\theta}_i) = 4\phi_i^2(a_i - b_i)^2(1 + \phi_i^2)^{-4}\mathbf{J}_{i,i}^{-1},$$

where $\mathbf{J}_{i,i}^{-1}$ denotes the (i, i)th element of $\mathbf{J}^{-1}(\phi)$. Of course, $\mathbf{J}(\hat{\theta}_{\mathrm{ML}})$ can also be directly approximated via (4.5).

The final value of the Hessian matrix computed from fminunc can also be output from the function. To illustrate, the program in Listing 4.6 implements the above method of parameter restriction for the Student's t model. It works by calling an extra program that can convert back and forth between θ and ϕ. This is given in Listing 4.7.

Remark. For constrained optimization, one could instead use Matlab's more general routine fmincon, which allows not only for box constraints, but also for additional linear

```
1   function [pout,Vout]=einschrk(pin,bound,Vin)
2   lo=bound.lo;  hi=bound.hi;  welche=bound.which;
3   if nargin < 3
4     trans=sqrt((hi-pin) ./ (pin-lo));   pout=(1-welche).* pin + welche .* trans;
5     Vout=[];
6   else
7     trans=(hi+lo.*pin.^2) ./ (1+pin.^2); pout=(1-welche).* pin + welche .* trans;
8     % now adjust the standard errors
9     trans=2*pin.*(lo-hi) ./ (1+pin.^2).^2;
10    d=(1-welche) + welche .* trans; % either unity or delta method.
11    J=diag(d); Vout = J * Vin * J;
12  end
```

Program Listing 4.7: Function `einschrk` (from the German word *einschränken*, "to restrict") used to perform conversion between θ and ϕ. If `Vin` is specified, then `pout` is transformed back, and so refers to θ; otherwise `pout` is transformed and refers to ϕ. Use of foreign words as function and variable names is occasionally useful, as Matlab, being in English, will surely not have a similarly named function, thus preventing potential problems. In the function, the word `welche` is German for "which", this being a reserved word in Matlab.

and nonlinear constraints. Other unconstrained optimization techniques that are not reliant on the gradient and Hessian, such as the simplex method via Matlab's `fminsearch`, or other methods, as discussed below, can be used in place of `fminunc`, but still with our custom-implemented box constraints, whereas the Hessian-based optimization technique in `fmincon` cannot be changed. ∎

Example 4.8 (Example 4.7, cont.)   *Using the program in Listing 4.6 for estimation, the same m.l.e. as in Example 4.7 was obtained (up to the requested precision), as well as the approximate Hessian matrix. It is quite similar to that computed directly, yielding an approximate variance–covariance matrix of the m.l.e. of 10^{-3} times*

$$\begin{pmatrix} 6.3529 & -0.0627 & 0.5057 \\ -0.0627 & 0.0707 & -0.0063 \\ 0.5057 & -0.0063 & 0.0932 \end{pmatrix}.$$

This results in the same 95% Wald c.i. of $\hat{v}_{\mathrm{ML}}$ as that based on (4.6). ∎

Example 4.9 (Example 4.8 cont.)   *To accommodate the asymmetry and allow for more flexibility in the tail shape of the NASDAQ returns, we use the generalized asymmetric t, or GA t, distribution, as discussed in Section A.8. All five parameters (location, scale, and the three shape parameters d, v, θ) need to be estimated. We impose an upper limit on $\hat{v}$ of 100, as it would otherwise tend to infinity using truly Gaussian data. The program for estimating the GAt is easily constructed by (i) modifying the one for the Student's t distribution in Listing 4.6, (ii) using the code in Listing 4.8 for computing the p.d.f., and (iii) remembering to correctly incorporate the location and scale parameters, as:*

```
1   z=(x-mu)/c;  pdf = GAt(z,d,v,theta)/c;  llvec=log(pdf);  ll=-sum(llvec);
```

It is given in Listings 4.9 and 4.10.

```
1  function [pdf, cdf] = GAt(xvec,d,v,theta)
2  ll=length(xvec); xvec=reshape(xvec,ll,1); pdf = zeros(ll,1);
3  konst = 1 / ( beta(1/d,v) * v^(1/d) * (theta+1/theta) / d );
4  k = find(xvec<0);
5  if any(k), y=xvec(k); pdf(k) = ( 1 + (-y*theta).^d / v ).^(-(v+1/d)); end
6  k = find(xvec>=0);
7  if any(k), y=xvec(k); pdf(k) = ( 1 + (y/theta).^d / v ).^(-(v+1/d)); end
8  pdf = konst * pdf;
9  if nargout>1
10    cdf = zeros(length(xvec),1); k = find(xvec<0);
11    if any(k)
12      y=xvec(k); L = v./(v+(-y*theta).^d); cdf(k) = betainc(L,v,1/d)/(1+theta^2);
13    end
14    k = find(xvec==0); if any(k), y=xvec(k); cdf(k) = 1/(1+theta^2); end
15    k = find(xvec>0);
16    if any(k)
17      y=xvec(k); top=(y/theta).^d; U=top./(v+top);
18      cdf(k) = 1/(1+theta^2) + betainc(U,1/d,v)/(1+theta^(-2));
19    end
20  end
```

Program Listing 4.8: The GAt p.d.f. and c.d.f.

```
1  function [param,stderr,iters,loglik,Varcov]=GAtestimation(x,vlo,initvec,fixd)
2  if nargin<2, vlo =[]; end, if isempty(vlo), vlo=0.01; end
3  if nargin<3, initvec =[]; end, if nargin <4, fixd =[]; end
4  if isempty(initvec)
5    versuch=3; vhi=4; loglik=-Inf; vvec=linspace(vlo+0.02,vhi,versuch);
6    for i=1:versuch
7      vv=vvec(i);
8      if isempty(fixd), initvec = [2 vv 0.98 0 3];
9      else initvec = [vv 0.98 0 3];
10     end
11     [param0,stderr0,iters0,loglik0,Varcov0]=GAtestimation(x,vlo,initvec,fixd);
12     if loglik0 >loglik
13        loglik=loglik0; param=param0; stderr=stderr0;
14        iters=iters0; Varcov=Varcov0;
15     end
16   end
17   return
18 end
19 if isempty(fixd) % d     v    theta   mu     c
20        bound.lo=    [0.1 vlo    0.2   -1  1e-4];
21        bound.hi=    [ 30  100     3    2  1e+4];
22        bound.which=[   1    1     1    0     1];
23 else     %         v    theta  mu     c
24   bound.lo=    [vlo    0.2   -1  1e-4];
25   bound.hi=    [100     3    2  1e+4];
26   bound.which=[   1     1    0     1];
27 end
```

Program Listing 4.9: Computes the m.l.e. of the GAt distribution, with the ability to pass to the function the lower bound for $\hat{v}$, an initial parameter vector, and a fixed value of parameter d (so that it is not estimated), with the latter feature used in the program in Listing 10.5. The value `versuch` (German for "try", noting that "try" is a keyword in Matlab) indicates the length of a grid of v-values to be used to form initial parameter estimates (if no initial parameter is passed) so that possible local but non-global maxima can be avoided. Continued in Listing 4.10.

```
1  nobs=length(x);  maxiter=length(initvec)*100;
2  tol=1e-8;  MaxFunEvals=length(initvec)*400;
3  opts=optimset('Display','none','Maxiter',maxiter,'TolFun',tol,'TolX', ...
4              tol,'MaxFunEvals',MaxFunEvals,'LargeScale','Off');
5  if 1==1
6    [pout,fval,exitflag,theoutput,grad,hess]= ...
7      fminunc(@(param) GAtloglik(param,x,fixd,bound), ...
8          einschrk(initvec,bound),opts);
9  else
10   [pout,fval,exitflag,theoutput]= ...
11     fminsearch(@(param) GAtloglik(param,x,fixd,bound), ...
12         einschrk(initvec,bound),opts);
13   hess=eye(length(pout));
14 end
15 V=inv(hess)/nobs;  [param,V]=einschrk(pout,bound,V);
16 param=param';  Varcov=V;
17 stderr=sqrt(diag(V));  loglik=-fval*nobs;
18 iters=theoutput.iterations;
19
20 function ll=GAtloglik(param,x,fixd,bound)
21 if nargin<4, bound=0; end
22 if isstruct(bound),  paramvec=einschrk(real(param),bound,999);
23 else paramvec=param;
24 end
25 if isempty(fixd)
26   d=paramvec(1); v=paramvec(2); theta=paramvec(3);
27   mu=paramvec(4); c=paramvec(5);
28 else
29   d=fixd; v=paramvec(1); theta=paramvec(2);
30   mu=paramvec(3); c=paramvec(4);
31 end
32 z=(x-mu)/c;  pdf = GAt(z,d,v,theta) / c;
33 llvec=log(pdf);  ll=-mean(llvec);
34 if isinf(ll), ll=1e5; end
```

Program Listing 4.10: Continued from Listing 4.9.

Doing so yields an ℓ_{ML} value of $-10{,}248.52$, which is far higher than for the plain Student's t model (which gave an ℓ_{ML} of $-10{,}286.14$), and parameter estimates $\hat{d} = 1.828$, $\hat{v} = 1.409$, $\hat{\theta} = 0.881$, location $\hat{\mu} = 0.210$, and scale $\hat{c} = 0.813$. These imply a maximally existing moment of $\widehat{\mu d} = \hat{\mu}\hat{d} = 2.58$. This is larger than that implied by the Student's t model, and makes sense because the fatter left tail can now be accommodated by the asymmetry parameter.

The fitted density is shown in Figure 4.10(a); compare to that of Figure 4.8. More relevant for risk management purposes is the "view from the tails," which is shown in Figure 4.10(b), for the left tail, along with the Student's t fit. The latter underestimates the probability of large negative returns.   ∎

Example 4.10 (Example 4.9, cont.)   *Now consider fitting the asymmetric stable Paretian distribution as discussed in Section A.16, denoted by $S_{\alpha,\beta}(\mu, c)$, to the NASDAQ data, where α is the tail index with $0 < \alpha \leq 2$, β is the asymmetry parameter, $|\beta| \leq 1$, and μ and c are location and scale parameters, respectively.*

Once a method is available for computing the stable p.d.f., the program to maximize the likelihood is virtually identical to the one developed for the GAt distribution in Example 4.9,

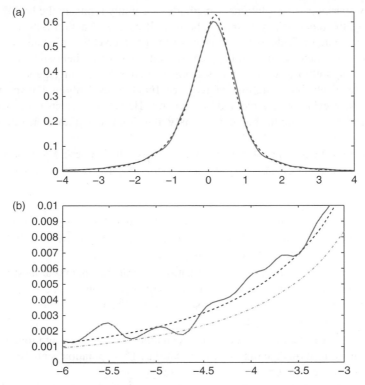

Figure 4.10 *(a) Kernel density (dashed) and fitted GAt density of the NASDAQ returns. (b) Same, but showing only the left tail and including the Student's t fit (dash-dotted).*

except that the stable p.d.f. is used, and the constraints on α, β, and c are incorporated. As for computing the stable p.d.f., see Section A.16.

We obtain $\hat{\alpha}_{ML} = 1.471$, $\hat{\beta}_{ML} = -0.258$, $\hat{\mu}_{ML} = -0.003$, and $\hat{c}_{ML} = 0.485$. This results in a log-likelihood of $-10,291.09$, which is lower than the GAt log-likelihood by about 43, and, interestingly, even lower than the Student's t fit (by about 5), despite the clear asymmetry in the data. This provides evidence against the stable model as the unconditional distribution for this data set, though recalling the blatant non-i.i.d. nature of the returns, as shown in Figure 4.7, any i.i.d. model is highly inappropriate for daily density forecasting. ∎

4.4 EVOLUTIONARY ALGORITHMS

The author is grateful to Jochen Krause for helping to develop this section and, particularly, for constructing the Matlab implementations.

Evolutionary algorithms epitomize general approaches for minimizing some function of m variables $f : \mathbb{R}^m \to \mathbb{R}$ and, as the name implies, usually follow some notion of biological evolution. Function values thus are *fitness values*, parameter vectors are called *individuals* and the complete set of individuals constitutes the *population*. Its designation as a general approach implies that it is suited for the optimization of possibly nonconvex, nonsmooth or even noncontinuous functions, precisely the situation in which gradient- and Hessian-based methods are prone to fail.

Generally speaking, all evolutionary methods essentially replicate the biological evolution targeting the function's minimum. The overall course of action is to make it easier for well-performing individuals to "survive" or to "produce children" than for poorly performing ones. The algorithms can be seen as well-constructed heuristics, as opposed to deterministic algorithms, and as a consequence, their rate of convergence is unknown. This and a relatively large number of required function evaluations are the two major disadvantages of evolutionary methods in general. However, even in cases when deterministic algorithms are applicable, evolutionary algorithms can outperform them in some situations.

The most popular class of evolutionary algorithms is called **genetic algorithms**. Methods from this class share an evolutionary framework that consists of the following successive steps:

- Selection: The selection operator chooses individuals for the subsequent step of evolution and can be interpreted as a weighting scheme. Any individual that gets a zero weight is removed from the evolution.

- Crossover: The crossover (or recombination) operator combines at least two individuals and generates or *spawns* a new one. This operation is crucial for the process of evolution and is target-oriented, so that future generations are, probabilistically speaking, closer to the target.

- Mutation: The mutation operator is a special kind of crossover operation, where single individuals are subject to random modifications. This operation is also elementary for the process of evolution.

- Survival: At the last stage of every evolutionary step, the newly spawned individuals are either accepted or rejected. If a new individual is accepted, then it replaces an old one, otherwise the old individual survives and the young one is discarded.

To form an algorithm, the above steps are repeated until some stopping condition is reached. The method of differential evolution described below is a typical example from this class.

The second most important class of evolutionary algorithms is referred to as **evolutionary strategies**. Algorithms from this class are characterized by a nonstandard adaption of the population towards the minimum of the function. As such, evolutionary strategy algorithms can implement any biologically motivated strategy. The so-called CMAES algorithm detailed below is an example from this class.

In the following, any parameter that is part of an optimization method, as opposed to belonging to the function to be minimized, is called a *meta-parameter*, sometimes referred to as a *tuning* parameter.

4.4.1 Differential Evolution

The Differential Evolution (DE) algorithm proposed in Storn and Price (1995) is a genetic algorithm for finding the minimizer of a function f. It consists of the three typical operations of selection, mutation, and crossover. In its original formulation, many meta-parameters have to be specified (or calibrated for the particular problem under consideration), before the algorithm can be applied. Among many different settings the most popular one is commonly referred to as "DE/best/1 with jitter." We focus on this particular variant of DE, because the

```
1    function [x,fx]=fminde(f,x0,opt,varargin)
2    maxtol=1e-6; maxtolfun=1e-6; maxiter=intmax; maxfevals=100000;
3    x0=x0(:)'; dim=length(x0); psize=max(4+3*fix(log(dim)),10);
4    maxrestarts=5; w_mut=0.8; w_xo=0.5;
5    two2np1=2:psize; n1=psize-1; stop=false;
6    iter=0; fevals=0; restarts=0;
7    [indiv,fitness,fn]=rndsimplex(psize,f,x0,varargin{:});
8    fevals=1+fn; [fitness,perm]=sort(fitness); indiv=indiv(perm,:);
9    while (true)
10     if (stop)
11       if (maxrestarts <= restarts), break; else restarts=restarts+1; end
12       [indiv,fitness,fn]=rndsimplex(psize,f,indiv(1,:),varargin{:});
13       fevals=fevals+fn; [fitness,perm]=sort(fitness); indiv=indiv(perm,:);
14     end
15     iter=iter+1; oldies=indiv;
16     for i=1:psize
17       a=randi(psize); while (a==i); a=randi(psize); end
18       b=randi(psize); while (b==i || b==a); b=randi(psize); end
19       c=randi(psize); while (c==i || c==a || c==b); c=randi(psize); end
20       jitter=1e-4*rand(1,dim);
21       newbie=oldies(1,:)+(w_mut+jitter).*(oldies(a,:)-oldies(b,:));
22
23       selection=(rand(1,dim) < w_xo); inverseSelection=(1-selection);
24       newbie=newbie.*selection+oldies(i,:).*inverseSelection;
25       newbieScore=feval(f,newbie,varargin{:});
26       if (newbieScore <= fitness(i))
27         indiv(i,:)=newbie; fitness(i)=newbieScore;
28       end
29     end
30     fevals=fevals+psize; [fitness,perm]=sort(fitness); indiv=indiv(perm,:);
31     rad_x=max(max(abs(repmat(indiv(1,:),n1,1) - indiv(two2np1,:))));
32     rad_fx=max(abs(fitness(1)-fitness(two2np1)));
33     if (rad_fx<=max(maxtolfun,10*eps(fitness(1)))); stop=true;
34     elseif (rad_x<=max(maxtol,10*eps(max(indiv(1,:))))); stop=true;
35     else stop=false;
36     end
37     stop=(stop || iter >= maxiter || fevals >= maxfevals);
38   end
39   x=indiv(1,:); fx=fitness(1);
```

Program Listing 4.11: An implementation of the DE algorithm.

operation of the algorithm remains almost the same for all settings. Listing 4.11 provides a basic implementation.

Initialization

In common with some of the DE literature, we use k to denote the iterations, or steps, of the algorithm, and m is the number of unknown function parameters to be determined.

At $k = 0$, a population $S^k = \{s_j^k \in \mathbb{R}^m, j = 1, \dots, p\}$ is generated to serve as starting values. Several suggestions for how to define a good initial population exist in the literature. The usual way, however, is to repeatedly create copies of the initial individual, while modifying a random number of its properties, until sufficiently many different individuals have been created. This spawning of the population is the first part of DE that is subject to the influence of meta-parameters. One way to initialize a random simplex is given in Listing 4.12.

```
1   function [x,fx,fevals] = rndsimplex(n,f,x0,varargin)
2   x0 = x0(:)'; dim = length(x0); x = zeros(n,dim); fx = zeros(n,1);
3   x(1,:) = x0; fx(1) = feval(f,x0,varargin{:});
4   for i=2:n
5     y = x0;
6     for k=1:randi(dim)
7       z = randn(1,1); while (z == 0), z = randn(1,1); end
8       j = randi(dim);
9       if (y(j) == 0), y(j) = z * 0.00025; else y(j) = z * 1.05 * y(j); end
10    end
11    x(i,:) = y; fx(i) = feval(f,y,varargin{:});
12  end
13  fevals = n;
```

Program Listing 4.12: Generation of a random simplex.

Selection, Mutation, and Crossover

In all subsequent iterations, $k \geq 1$, the following steps are applied to the ith individual $\mathbf{s}_i^k \in S^k$. First, DE randomly selects three different individuals, $\mathbf{a}_i^k, \mathbf{b}_i^k, \mathbf{c}_i^k \in S^k$, $\mathbf{s}_i^k \notin \{\mathbf{a}_i^k, \mathbf{b}_i^k, \mathbf{c}_i^k\}$, for the subsequent manipulations from the current population. In the chosen setting, one of the selected individuals is set to be the best individual from the population; without loss of generality, let $\mathbf{c}_k$ be the best-performing individual. Next, a "mutant" (new individual) is assembled from the selected three. In DE, the kth mutant, $\mathbf{d}_i^k \in \mathbb{R}^m$, is defined as

$$d_{i,j}^k = c_{i,j}^k + (\lambda + \tau \varepsilon_{i,j}^k) \cdot (a_{i,j}^k - b_{i,j}^k),$$

where $\lambda \in [0, 1]$ denotes the magnitude of the mutation, $\varepsilon_{i,j}^k \sim \text{Unif}(0, 1)$, and $\tau \in \mathbb{R}_{>0}$ is a small constant controlling the "jitter" of the mutation, this being a small, random perturbation. DE adds the weighted difference between two individuals to a third one. For the chosen setting, $\lambda = 0.8$ and $\tau = 10^{-4}$ are standard values. Different values may be picked; for example, the jitter might even be removed ($\tau = 0$) and more than only a single difference between individuals could be used.

Third, the mutant $\mathbf{d}_i^k$ is randomly combined with the individual $\mathbf{s}_i^k$ in a form of genetic crossover, and their "child," $\mathbf{e}_i^k \in \mathbb{R}^m$, is produced according to the rule

$$e_{i,j}^k = \begin{cases} d_{i,j}^k, & \text{if } v_{i,j}^k < \gamma, \\ s_{i,j}^k, & \text{otherwise,} \end{cases}$$

where $\gamma \in [0, 1]$ denotes the crossover probability and $v_{i,j}^k \sim \text{Unif}(0, 1)$.

Finally, only the better performing of the two individuals $\mathbf{s}_i^k$ and $\mathbf{e}_i^k$ survives, and the new population is

$$S^{k+1} = \begin{cases} S^k \setminus \{\mathbf{s}_i^k\} \cup \{\mathbf{e}_i^k\}, & \text{if } f(\mathbf{e}_i^k) > f(\mathbf{s}_i^k), \\ S^k, & \text{otherwise.} \end{cases}$$

In some DE variants this modification takes place immediately, while in others it is applied after the loop over the population has been completed (one evolutionary step).

Stopping Conditions

Besides the two trivial stopping criteria of the maximum permitted number of iterations and number of function evaluations, we state two criteria that account for specific properties of the population. Let $\mathbf{b}^k \in S^k$ denote the best individual (the one with the smallest function value). Then the algorithm stops if one of the following two conditions,

$$\max\{\|\mathbf{b}^k - \mathbf{s}^k\|_\infty : \mathbf{s}^k \in S^k\} \leq \delta_{\mathrm{I}},$$
$$\max\{\|f(\mathbf{b}^k) - f(\mathbf{s}^k)\|_\infty : \mathbf{s}^k \in S^k\} \leq \delta_{\mathrm{II}},$$

holds, where $\delta_{\mathrm{I}}, \delta_{\mathrm{II}} \in \mathbb{R}_{>0}$ denote the minimal tolerances commonly set to values less than or equal to 10^{-4}. In addition, a restart option is often used that restarts the algorithm a fixed number of times.

Example 4.11 *Consider the function*

$$f(x_1, x_2) = 3(1 - x_1)^2 \exp\{-x_1^2 - (1 + x_2)^2\}$$
$$-10\left(\frac{1}{5}x_1 - x_1^3 - x_2^5\right)\exp\{-x_1^2 - x_2^2\} - \frac{1}{3}\exp\{-(1 + x_1)^2 - x_2^2\}, \quad (4.13)$$

with three local minima and minimum $\mathbf{x}_{\min} = (0.228, -1.625)$. Figure 4.11 shows the evolution of the DE population (10 individuals) over time at selected states using the starting value $(1, 1)$.

The algorithm converged to the correct solution, having required 46 iterations and 471 function evaluations. If the initial population does not comprise individuals "near" the global minimum, it can also terminate at one of the wrong minima. ∎

4.4.2 Covariance Matrix Adaption Evolutionary Strategy

The so-called Covariance Matrix Adaption Evolutionary Strategy (CMAES) algorithm, introduced in Hansen and Ostermeier (1996), is an evolutionary strategy for function minimization. It combines statistical methodology with adaptive control and searches for the minimizer of the function f by repeatedly resampling the parameter space based on an adaptively updated multivariate normal distribution. It gives an estimator of the (global) minimum of f that minimizes the variance of its own estimate.

To motivate its optimality, recall the CRlb from Section 3.2. This states that the inverse Fisher information matrix (the inverse negative Hessian matrix) of the estimator's log-likelihood function is a lower bound on the variance of any unbiased estimator. Accordingly, if the variance of the estimate is minimized, then the Fisher information is maximized. Thus, given the regularity condition that the gradient (score) is zero, the positive (negative) Hessian implies that the function is minimized (maximized), albeit possibly locally, at an inferior local extremum. In summary, CMAES adaptively updates the two parameters of the normal distribution based on local samples from the parameter space such that the covariance shrinks towards zero when the mean is the minimum of f. Listing 4.13 gives an implementation of the basic CMAES algorithm.

```
1   function [x,fx]=fmincmaes(f,x0,opt,varargin)
2   maxtol=1e-6; maxtolfun=1e-6; maxiter=intmax; maxfevals=100000;
3   x0=x0(:)'; dim=length(x0); psize=max(4+3*ceil(log(dim)),10);
4   esize=floor(psize/2); w=zeros(esize,1);
5   for i=1:esize, w(i) = log(esize+1) − log(i+1); end;
6   w=w./sum(w); psi=1/sum(w.^2);
7   s=1; c_nu=4/(dim+4); c_q=psi; c_rho=(psi+2)/(dim+(psi+3));
8   c_c = 2/(dim+sqrt(2))^2/c_q + (1−(1/c_q))*min(1,(2*psi−1)/(psi+(dim+2)^2));
9   c_sig = 1 + 2 * max(0,sqrt((psi−1)/(dim+1))−1) + c_rho;
10  euclnorm=sqrt(dim)*(1−1/(4*dim)+1/(21*dim*dim)); % approx.
11  C=eye(dim); B=eye(dim); D=ones(dim,1);
12  path_rho=zeros(dim,1); path_nu=zeros(dim,1);
13  one2e=1:esize; two2np1=2:psize; n1=psize−1; stop=false; iter=0; fevals=0;
14  [indiv,fitness,fn]=rndsimplex(psize,f,x0,varargin{:});
15  fevals=1+fn; [fitness,perm]=sort(fitness); indiv=indiv(perm,:);
16  xmean=indiv(one2e,:)'*w;
17  while (true)
18    if (stop); break; else; iter=iter+1; end
19    [B,D]=eig(C); D=sqrt(abs(diag(D)));
20    fac=1;
21    if (min(D) < 1e−8); fac=1/max(D); end
22    if (max(D) > 1e+8); fac=1/max(D); end
23    if (fac ~= 1); s=s/fac; path_nu=path_nu*fac; D=D*fac; C=C*fac^2; end
24    for i=2:psize;
25      indiv(i,:)=xmean+s*B*(D.*randn(dim,1));
26      fitness(i)=feval(f,indiv(i,:),varargin{:});
27    end
28    fevals=fevals+psize; [fitness,perm]=sort(fitness); indiv=indiv(perm,:);
29
30    xmean_old=xmean; xmean=indiv(one2e,:)'*w;
31    BDz=sqrt(psi)*(xmean−xmean_old)/s; temp=B'*BDz./D;
32    for k=1:dim
33      path_rho(k)=(1−c_rho)*path_rho(k) + ...
34          sqrt(max1(c_rho))*B(k,:)*temp;
35    end
36    path_nu=(1−c_nu)*path_nu+sqrt(max1(c_nu))*BDz;
37    C=(1−c_c)*C+(c_c/c_q)*(path_nu*path_nu');
38    for i=1:esize
39      cq =(indiv(i,:)−xmean_old')/s; C=C+c_c*(1−1/c_q)*w(i).*cq'*cq;
40    end
41    s=s*exp(((sqrt(path_rho'*path_rho) / euclnorm)−1)*c_rho/c_sig);
42    if (max(D)/min(D) > 1e+8),
43      warning('found ill−conditioned covariance matrix'); break;
44    end
45    rad_x = max(max(abs(repmat(indiv(1,:),n1,1) − indiv(two2np1,:))));
46    rad_fx = max(abs(fitness(1)−fitness(two2np1)));
47    if (rad_fx<=max(maxtolfun,10*eps(fitness(1)))); stop=true;
48    elseif (rad_x<=max(maxtol,10*eps(max(indiv(1,:))))); stop=true;
49    else stop=false;
50    end;
51    stop=(stop || iter >= maxiter || fevals >= maxfevals);
52  end
53  x=indiv(1,:); fx=fitness(1);
54
55  function [y]=max1(x), y=x*(2−x);
```

Program Listing 4.13: An implementation of the CMAES algorithm.

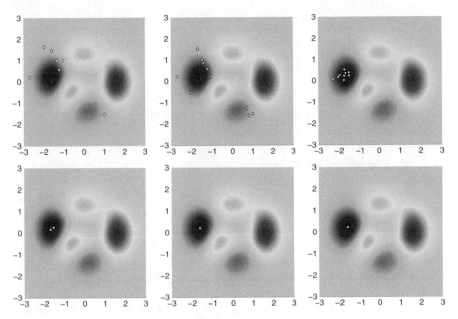

Figure 4.11 *Evolution of the DE population over time for selected iteration states, showing (from left to right, top to bottom) iterations 1, 10, 20, 30, 40, and 46.*

Initialization

At each iteration $k \geq 1$, the revelation of information and progression of the algorithm are fundamentally based on the weighted adaption of the underlying multivariate normal distribution, $N(\boldsymbol{\mu}_k, \boldsymbol{\Sigma}_k)$, where $\boldsymbol{\mu}_k \in \mathbb{R}^m$ denotes the mean, and the covariance matrix $\boldsymbol{\Sigma}_k$ is the product of a matrix $\mathbf{C}_k \in \mathbb{R}^{m \times m}$ and a scale term $\sigma_k \in \mathbb{R}_{>0}$. At $k = 0$, the start of the procedure, the population consists of p individuals in the parameter space,

$$S^k = \{\mathbf{s}_j^k \sim N(\boldsymbol{\mu}_k, \sigma_k^2 \mathbf{C}_k)\}, \quad j = 1, \ldots, p,$$

which is initially spawned using $\boldsymbol{\mu}_k = \mathbf{s}_1^0$ (initial starting value), $\mathbf{C}_k = \mathbf{I}_m$ and $\sigma_k = 1$. The initial population could be determined by other means, depending on the problem and prior knowledge of the optimum.

Evolution

At each iteration, first, a new population is created from the sampling distribution, $\mathbf{s}_i^k \sim N(\boldsymbol{\mu}_{k-1}, \sigma_{k-1}^2 \mathbf{C}_{k-1}), j = 1, \ldots, p$, according to mean $\boldsymbol{\mu}_{k-1}$, covariance $\mathbf{C}_{k-1}$ and scale parameter σ_{k-1}. Then the fitness of each individual, $f(\mathbf{s}_i^k)$, is evaluated and the population is split into two parts, where only the best fraction of size $q \leq p$ (e.g., $q = \lfloor p/2 \rfloor$) is selected for the subsequent adaption process. In addition, exponentially decaying weights, $\mathbf{w} \in \mathbb{R}_{>0}^q$, $\sum_{j=1}^q w_j = 1$, are assigned to the selected individuals according to the rule: the better the fitness, the larger the weight. By default, super-linearly decaying weights, that is,

$$w_j = \frac{\log(q + 1) - \log(j + 1)}{\displaystyle\sum_{j=1}^{q} \log(q + 1) - \log(j + 1)},$$

are used in CMAES. Besides this selection operation, other genetic operations, such as mutation and recombination, are not explicitly implemented at the level of individuals, but they implicitly take place in the following mean and covariance update scheme.

The mean is updated (this serving as recombination) and set to the weighted sample mean of the newly drawn population, $\mu_k = \sum_{j=1}^{q} w_j s_j^k$, where no advanced control mechanism is involved. The actual tracking and control of the mean, however, is then performed as part of the covariance update. As the covariance is closely related to the expected mean of the next generation, the control of the mean is realized by the adaptive scale parameter σ_k which additionally increases or decreases the whole covariance. For this purpose, an adaptive step size control is introduced based on the idea of line search algorithms. But instead of applying some derivative-based descent direction (as is done in, say, the BFGS algorithm, discussed above in Section 4.3.1), a so-called *evolution path* based on the sequence $\mu_0, \mu_1, \ldots$ is utilized. Starting with $\tau_0 = (0, \ldots, 0) \in \mathbb{R}^m$, the (evolutionary) step size control path $\tau_0, \tau_1, \ldots, \tau_k \in \mathbb{R}^m$, is defined by

$$\tau_k = \underbrace{(1 - c_\tau)\tau_{k-1}}_{\text{adaption}} + \underbrace{\sqrt{c_\tau(2 - c_\tau)}\psi}_{\text{normalization}} \cdot \underbrace{\sigma_{k-1}^{-1}}_{\text{rescaling}} \cdot \underbrace{\mathbf{C}_{k-1}^{-\frac{1}{2}}}_{\text{decorrelation}} \cdot \underbrace{(\mu_k - \mu_{k-1})}_{\text{update}},$$

where $c_\tau \in (0, 1)$ is an adaption coefficient (meta-parameter) and $\psi = (\sum_{j=1}^{q} w_j^2)^{-1}$. The update term is multivariate normal with zero mean and covariance $\sigma_{k-1}^2 \mathbf{C}_{k-1}$ (up to normalization), and $\tau_k \sim N(\mathbf{0}, \mathbf{I}_m)$, as $(1 - c_\tau)^2 + \sqrt{c_\tau(2 - c_\tau)}^2 = 1$ and $\sum_{j=1}^{q} w_j N_j(\mathbf{0}, \mathbf{\Sigma}) \sim \psi^{-\frac{1}{2}} N(\mathbf{0}, \mathbf{\Sigma})$.

The path expresses the cumulated length and direction of all past decorrelated and descaled evolutionary steps and is defined adaptively to prevent backward steps that may directly cancel out the latest path updates. The step size control is chosen as follows: If the path is short due to the cancelation of many updates, the step size is decreased. Otherwise, if the path is long, the step size is increased. In CMAES, this is realized by taking

$$\sigma_k = \sigma_{k-1} \exp\left\{ \frac{c_\tau}{c_\sigma} \left(\frac{||\tau_k||_2}{\mathbb{E}[||N(\mathbf{0}, \mathbf{I}_m)||_2]} - 1 \right) \right\},$$

where $c_\sigma \in \mathbb{R}_{>0}$ is a damping coefficient (meta-parameter) and the expected value of the Euclidean norm of a standard multivariate normal distributed random vector of size m is given by

$$\mathbb{E}[||N(\mathbf{0}, \mathbf{I}_m)||_2] = \sqrt{2}\Gamma\left(\frac{m + 1}{2}\right)\left[\Gamma\left(\frac{m}{2}\right)\right]^{-1}$$

(this being just $\mathbb{E}[X^{1/2}]$ with $X \sim \chi_m^2$, which is given in, for example, (I.7.42)). Observe that if the length of the path equals its expected value, then the step size is set to one, $\sigma_k = 1$.

The covariance adaption is then given by

$$\mathbf{C}_k = \underbrace{(1 - c_C)\mathbf{C}_{k-1}}_{\text{adaption}} + \underbrace{\frac{c_C}{c_q}(\boldsymbol{\eta}_k \boldsymbol{\eta}_k')}_{\text{rank-1 update}} + \underbrace{\left(c_C - \frac{c_C}{c_q}\right)\mathbf{C}_q}_{\text{rank-}q\text{update}},$$

where $c_C, c_q \in (0, 1)$ are adaption coefficients (meta-parameters) of the rank-1 and rank-q update, respectively. Here, the step size is incorporated via the control path of the rank-1 update $\boldsymbol{\eta}_k \in \mathbb{R}^m$, defined by

$$\boldsymbol{\eta}_k = \underbrace{(1 - c_\eta)\boldsymbol{\eta}_{k-1}}_{\text{adaption}} + \underbrace{\sqrt{c_\eta(2 - c_\eta)\psi}}_{\text{normalization}} \cdot \underbrace{\sigma_{k-1}^{-1}}_{\text{rescaling}} \cdot \underbrace{(\boldsymbol{\mu}_k - \boldsymbol{\mu}_{k-1})}_{\text{update}},$$

where $c_\eta \in (0, 1)$ is an adaption coefficient (meta-parameter) and $\boldsymbol{\eta}_k \sim N(\mathbf{0}, \mathbf{C}_k)$. Furthermore, $\mathbf{C}_q$ is a weighted version of the sample covariance based on the best fraction of the current population,

$$\mathbf{C}_q = \sum_{j=1}^{q} w_j(\mathbf{s}_j^k - \boldsymbol{\mu}_{k-1})(\mathbf{s}_j^k - \boldsymbol{\mu}_{k-1})',$$

and is called a rank-q update. Of course, different covariance estimators could be employed at this point, and the choice made here symbolizes just another meta-parameter. Having introduced many meta-parameters, the authors of CMAES propose the default values

$$c_q = \psi, \quad c_\eta = \frac{4}{m + 4}, \quad c_\sigma = 1 + 2\max\left(0, \sqrt{\frac{\psi - 1}{m + 1}} - 1\right) + c_\tau,$$

$$c_\tau = \frac{\psi + 2}{m + \psi + 3}, \quad c_C = \frac{2}{c_q(m + \sqrt{2})^2} + \left(1 - \frac{1}{c_q}\right)\min\left(1, \frac{2\psi - 1}{\psi + (m + 2)^2}\right).$$

Finally, the whole population is discarded and, starting with the spawn of the new generation, the procedure is repeated until one of the stopping conditions holds. In some variants of CMAES, the best individual is kept alive over the generations, because new populations may not include better individuals; recall that they are randomly drawn.

The sampling of new individuals is usually done via the spectral decomposition $\mathbf{C} = \mathbf{BDB}^{-1}$, where $\mathbf{B}$ is orthonormal and $\mathbf{D}$ is the diagonal matrix of eigenvalues. Thus, given $\mathbf{Z} \sim N(\mathbf{0}, \mathbf{I}_m)$, new individuals can be created (mutation) by drawing from $N(\boldsymbol{\mu}_k, \sigma_k \mathbf{B}_k \sqrt{\mathbf{D}_k})$, as

$$\mathbb{V}(\sigma \mathbf{B}\sqrt{\mathbf{D}}\mathbf{Z}) = \mathbb{E}[(\mathbf{B}\sigma\sqrt{\mathbf{D}}\mathbf{Z})(\mathbf{B}\sigma\sqrt{\mathbf{D}}\mathbf{Z})']$$

$$= \sigma^2 \mathbf{B}\sqrt{\mathbf{D}}\mathbb{E}[\mathbf{Z}\mathbf{Z}']\sqrt{\mathbf{D}}\mathbf{B}^{-1} = \sigma^2 \mathbf{BDB}^{-1} = \sigma^2 \mathbf{C}.$$

Stopping Conditions

In addition to the stopping conditions from the previous section on DE, the CMAES iterations are stopped if the variance of all dimensions falls below a certain threshold. This is the case when the max norm between two successive mean vectors $\| \boldsymbol{\mu}_{k-1} - \boldsymbol{\mu}_k \|_\infty$ is close to zero. In that case, the method has converged to some point that must be a minimizer of the objective function. However, as is common to any numerical algorithm, it can also break down due to numerical problems before it converges. In particular, the decomposition of the covariance matrix may numerically break down due to an ill-conditioned matrix. In that case, the algorithm cannot continue and must be restarted or its result is worthless. For that reason, such situations are traced by the matrix condition number and the method is stopped if the condition number becomes exceptionally large.

In practice, this can easily be implemented through the Euclidean matrix norm for which the condition number of C_k is given by d_{max}/d_{min}. Thus, the two additional stopping rules for CMAES are

$$\max(\mathbf{d}) \leq \delta_{\mathrm{I}} \quad \text{and} \quad d_{max}/d_{min} \geq \delta_{\mathrm{II}},$$

where $\mathbf{d} \in \mathbb{R}^m$ is the diagonal of matrix $\mathbf{D}$ from the eigenvalue decomposition of $\boldsymbol{\Sigma}_k$, d_{min} and d_{max} are smallest and largest eigenvalue, respectively, $\delta_{\mathrm{I}} \in \mathbb{R}_{>0}$ denotes the minimal tolerance for the variances (e.g., $\delta_{\mathrm{I}} = 10^{-4}$), and $\delta_{\mathrm{II}} \in \mathbb{R}_{>0}$ is the upper bound on the condition number of the covariance matrix (e.g., $\delta_{\mathrm{II}} = 10^5$).

Example 4.12 *Using the CMAES algorithm for function (4.13), Figure 4.12 shows the evolution of the population (10 individuals) over time at selected states using the initial value $(1, 1)$. As with DE, the global minimum is found correctly, but CMAES requires only 28 iterations and 291 function evaluations, as it recovers more information about the function f per iteration.* ∎

Building on the success of CMAES, new algorithms of a similar kind have been proposed and developed. The most competitive of these is the **Efficient Natural Evolution Strategy** (ENS) algorithm, an evolutionary algorithm that also employs the multivariate

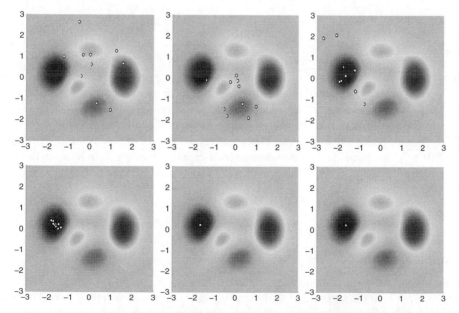

Figure 4.12 *Evolution of the CMAES population over time for selected iteration states, showing (from left to right, top to bottom) iterations 1, 5, 10, 15, 20, and 28.*

normal distribution and targets the maximization of the Fisher information; see Wierstra et al. (2008) and Sun et al. (2009a,b).

4.5 PROBLEMS

4.1 In Section 4.1.2 we discussed an iterative method for computing $\widehat{\theta}_{\mathrm{ML}} = (\hat{v}_{\mathrm{ML}}, \hat{\mu}_{\mathrm{ML}}, \hat{\sigma}_{\mathrm{ML}})'$, given $X_1, \ldots, X_n$ with $X_i \overset{\text{i.i.d.}}{\sim} t(v, \mu, \sigma)$. Construct a program to implement the method. Some experimenting with simulated data revealed that the initial estimates (4.4) can be rather poor and occasionally are such that $\dot{\ell}_v(v, \mu_0, \sigma_0; \mathbf{x})$ does not cross zero for $1 \leq v \leq 60$, confirmed if

$$\operatorname{sgn}(\dot{\ell}_v(1, \mu_0, \sigma_0; \mathbf{x})) = \operatorname{sgn}(\dot{\ell}_v(60, \mu_0, \sigma_0; \mathbf{x})).$$

This can often be remedied by altering the initial value σ_0. The program should first double σ_0, then quadruple it, etc., until $\dot{\ell}_v$ changes sign. If after doubling it, say, six times, $\dot{\ell}_v$ still does not cross zero, then halving, quartering, etc., should be tried.

4.2 Using the program developed in Problem 4.1, we wish to determine the small-sample properties of $\widehat{\theta}_{\mathrm{ML}} = (\hat{v}_{\mathrm{ML}}, \hat{\mu}_{\mathrm{ML}}, \hat{\sigma}_{\mathrm{ML}})'$ for a particular set of parameters and sample size n. For, say, $v = 8$, $\mu = 0$, $\sigma = 2$ and $n = 200$, draw 500 i.i.d. samples, compute the m.l.e., and examine the properties of the estimates.

When we did this, in 49 of the 500 cases, a zero of $\dot{\ell}_v$ (and, hence, the m.l.e.) could not be found. The reason is that, with such a relatively small sample size, these 49 data sets contained too little information about v. For given values of μ and σ, the likelihood function of one of these data sets never attains a maximum, but instead slowly continues to increase with v. If we did not know the true data-generating process, we would simply conclude that the data are normally distributed.[3]

For each of the three parameters, construct boxplots of the final parameter estimates corresponding to the successful cases (we had 451 such cases). Also compare the use of the trimmed mean $\bar{X}_{\hat{a}(v_i)}$ to that of the exact m.l.e. via the solution to $\dot{\ell}_\mu(v_i, \mu, \sigma_{i-1}; \mathbf{x}) = 0$.

4.3 Recall Example 3.6, in which $X_i \overset{\text{i.i.d.}}{\sim} \mathrm{Gam}(\alpha, \beta)$, $i = 1, \ldots, n$, with $\theta = (\alpha, \beta)'$. The m.l.e. for α was implicitly given in (3.23) as the solution to a nonlinear equation. By expressing the equation as $f(\alpha) = 0$ and obtaining values α_1 and α_2 such that $f(\alpha_1)$ and $f(\alpha_2)$ differ in sign, fzero can be used to locate $\hat{\alpha}_{\mathrm{ML}}$.

Compute the m.l.e. for each of 1000 simulated data sets for which $\alpha = \beta = 5$, using the three sample size values $n = 20$, 40, and 80. Construct a table with rows for each n, and the first two columns containing the means of the 1000 m.l.e. values for the two parameters.

Compute $\mathbf{J}^{-1}(\hat{\theta}_{\mathrm{ML}})$ for each replication and compare the median (of each component of) the resulting matrices with the true value of $\mathbf{J}^{-1}(\theta)$. Do the same for the

---

[3] For some of these 49 data sets, it might actually be the case that a zero of $\dot{\ell}_v$ does exist for some $\hat{v}$ greater than 60, the upper limit imposed in the simulation. However, because t distributions with more than 60 degrees of freedom are, for most practical purposes, very similar to Gaussian, inference based on the model will not appreciably change if $\hat{v}$ is just taken to be 60 (or replaced with the Gaussian assumption).

approximate Hessian (4.5). Which do you expect to be better? Finally, also examine the sample variance–covariance matrix of the 1000 values of $\hat{\theta}_{ML}$.

Show that the normality assumption itself suffers for $n \leq 40$ by plotting the kernel density of the 1000 estimates along with an inscribed normal density with parameters determined from the mean and variance of the 1000 estimates.

4.4   In the two-parameter gamma model in Problem 4.3, the burden of maximizing the likelihood for two parameters was reduced to solving a single nonlinear equation and the Hessian matrix was available analytically. For the two-parameter Weibull model, we again are able to obtain a single nonlinear equation for the likelihood, but the analytic Hessian appears intractable, so that numeric methods, such as (4.5), would need to be used.

Let $X_i \overset{\text{i.i.d.}}{\sim} \text{Weib}(b, 0, s)$, $i = 1, \ldots, n$, with density

$$f_{X_i}(x; b, s) = \frac{b}{s}\left(\frac{x}{s}\right)^{b-1} \exp\left\{-\left(\frac{x}{s}\right)^b\right\} \mathbb{1}_{(0,\infty)}(x), \quad b, s > 0.$$

Construct a program to compute the method of moments estimator (m.m.e.) from Section 5.2.1. This involves using function `fsolve` in Matlab for the system of two equations.

Compute the m.l.e. for b and s. In particular, substitute the expression for $\hat{s}_{ML}$ for known b into $0 = \partial \ell / \partial b$.

Conduct a simulation for $b = s = 1$, using $n = 20$ observations and 2000 replications and compare the m.m.e. and m.l.e. performance. Are the estimators approximately normally distributed?

4.5   Recalling the discussion on estimating parameters constrained to lie in an interval, assume that θ is restricted to be positive but has no upper limit. One way to impose this restriction is to estimate $\phi = \sqrt{\theta}$ resulting in $\hat{\phi}_{ML}$ and $J^{-1}(\phi)$. How is $\hat{\theta}_{ML}$ asymptotically distributed?

4.6   For small sample sizes, particular observations can have a large impact on the m.l.e. Write a program that does the following:

1. Generate T i.i.d. $t(v)$ realizations and sort them, giving $\mathbf{X} = (X_1, \ldots, X_T)$.
2. Let $\mathbf{X}_{(i)} = (X_1, X_2, \ldots, X_{i-1}, X_{i+1}, \ldots, X_T)$, that is, $\mathbf{X}$ with the ith observation removed. Calculate the m.l.e. for each subsample $\mathbf{X}_{(1)}, \ldots, \mathbf{X}_{(T)}$.
3. Make three graphs: (1) a histogram of $\mathbf{X}$; (2) a plot of $\hat{v}_{ML}$ based on sample $\mathbf{X}_{(i)}$ versus X_i; and (3) a plot of the log-likelihood (based on sample $\mathbf{X}_{(i)}$) scaled by $T - 1$ versus X_i. Add a line on the previous two plots indicating the value using the whole sample $\mathbf{X}$.

Do this for $T = 200$ and $T = 2000$, each with degrees of freedom $v = 3$, which is small enough to ensure that a few "outliers" occur in the sample, and interpret the results.

5

Methods of Point Estimation

Chapter 3 presented the basic theory of maximum likelihood, while Chapter 4 detailed the numerical means for computing the m.l.e. This chapter continues the theme of likelihood-based inference, emphasizing the numeric issues associated with its use for mixture distribution models, in particular, the mixed normal. This model class is important in its own right, and also serves as a showcase example of the EM algorithm for computing the m.l.e.

After this material is detailed in Section 5.1, Section 5.2 presents several non-likelihood-based methods for obtaining a point estimator of an unknown parameter vector. These are of value in many contexts when the likelihood is not (easily) accessible. Section 5.3 provides a performance comparison of the various methods in the (univariate) mixed normal context. Finally, Section 5.4 provides a short introduction to the theory of shrinkage estimation.

5.1 UNIVARIATE MIXED NORMAL DISTRIBUTION

5.1.1 Introduction

The univariate two-component mixed normal distribution is a good candidate for illustrating the various methods of estimation discussed in this chapter, owing to its enormous popularity in a wide variety of disciplines (Titterington et al., 1985; McLachlan and Peel, 2000, see, for example,) and also because of a particular numeric issue that arises when computing the m.l.e. It has a long and illustrious history in statistics, with its use

---

for modeling non-Gaussian phenomena dating back to the astronomical studies by the Canadian mathematician Simon Newcomb.[1]

The k-component mixed normal p.d.f. is given by

$$f_{\text{MixN}}(x; \boldsymbol{\mu}, \boldsymbol{\sigma}, \boldsymbol{\lambda}) = \sum_{i=1}^{k} \lambda_i f_{\text{N}}(x; \mu_i, \sigma_i^2), \quad \lambda_i \in (0, 1), \quad \sum_{i=1}^{k} \lambda_i = 1, \tag{5.1}$$

for parameters $\boldsymbol{\mu} = (\mu_1, \dots, \mu_k)$, $\boldsymbol{\sigma} = (\sigma_1, \dots, \sigma_k)$, and $\boldsymbol{\lambda} = (\lambda_1, \dots, \lambda_k)$. We will write $Y \sim \text{Mix}_k \, \text{N}_1(\boldsymbol{\mu}, \boldsymbol{\sigma}, \boldsymbol{\lambda})$ to help emphasize that there are k components for one-dimensional Y, or just $Y \sim \text{MixN}(\boldsymbol{\mu}, \boldsymbol{\sigma}, \boldsymbol{\lambda})$ when the context is clear. A necessary condition in order to estimate the parameters of the model is that they are **identifiable**. We first state a definition valid for parametric families that are *not* finite mixtures.

> A parametric class of distributions with support S, indexed by parameter $\boldsymbol{\theta} \in \boldsymbol{\Theta}$, is **identifiable** if different $\boldsymbol{\theta}$ determine distinct members in the class. That is, class $\{f(\cdot; \boldsymbol{\theta})\}$ is identifiable if $f(\cdot; \boldsymbol{\theta}_1) = f(\cdot; \boldsymbol{\theta}_2) \Leftrightarrow \boldsymbol{\theta}_1 = \boldsymbol{\theta}_2$ almost everywhere (i.e., the equality holds over the entire support of the distribution, with the possible exception of a set in S of measure zero in the continuous case). When the parametric density is clear from the context, we just say that the parameter vector $\boldsymbol{\theta}$ is identifiable.

As a quick illustration with a rather trivial example, let $Y_1, \dots, Y_n \overset{\text{i.i.d.}}{\sim} \text{N}(\mu_1 + \mu_2, 1)$, where μ_1 and μ_2 are both unknown. It should be apparent that they are not identifiable. (This example is a special case of a linear model with design matrix not of full rank; the linear combination $\mu_1 + \mu_2$ is said to be **estimable**.)

> If, for a given parametric distributional class $\{f(\cdot; \boldsymbol{\theta})\}$ with support S, $\boldsymbol{\theta}$ is *not* identifiable, then $\exists \boldsymbol{\theta}_1, \boldsymbol{\theta}_2 \in \boldsymbol{\Theta}$ and the set $\Upsilon \in S$ with nonzero measure such that $\boldsymbol{\theta}_1 \neq \boldsymbol{\theta}_2$ and $f(\mathbf{y}; \boldsymbol{\theta}_1) = f(\mathbf{y}; \boldsymbol{\theta}_2)$ for all $\mathbf{y} \in \Upsilon$, in which case we say that $\boldsymbol{\theta}_1$ and $\boldsymbol{\theta}_2$ are **observationally equivalent** (with respect to f and Υ).

In model (5.1) for $k = 2$, observe that $\boldsymbol{\theta} = (\mu_1, \mu_2, \sigma_1, \sigma_2, \lambda_1)'$ is observationally equivalent to $\boldsymbol{\theta}^{=} = (\mu_2, \mu_1, \sigma_2, \sigma_1, \lambda_2)'$, so for mixture distributions the definition of identifiability has to be modified to account for this. The idea is simple; we just need good notation and clear definitions. Let $\mathcal{F}$ denote a set of distributions (not necessarily from the same parametric family), and let Λ be the set of possible mixture weights, given by

$$\Lambda = \left\{ \lambda = (\lambda_1, \dots, \lambda_k) \middle| k \in \mathbb{N}; \, \lambda_i \in (0, 1), i = 1, \dots, k; \, \sum_{i=1}^{k} \lambda_i = 1 \right\}. \tag{5.2}$$

The set of finite mixtures of $\mathcal{F}$ contains the distributions that can be expressed as a sum, with weights from Λ, of distributions from $\mathcal{F}$,

$$\mathcal{G} = \left\{ G \middle| \exists \lambda = (\lambda_1, \dots, \lambda_k) \in \Lambda, \, \exists f_1, \dots, f_k \in \mathcal{F} \, : \, G = \sum_{i=1}^{k} \lambda_i f_i \right\}. \tag{5.3}$$

---

[1] The use of the mixed normal distribution has become very popular within machine learning – see, for example, Bishop (2006, Ch. 9) and Murphy (2012, Ch. 11) – as well as in financial econometrics, in conjunction with GARCH-type structures – see, for example, Venkataraman (1997), Haas et al. (2004a,b, 2009, 2013), Alexander and Lazar (2006), Wu and Lee (2007), Bauwens et al. (2007), Giannikis et al. (2008), Buckley et al. (2008), Haas and Paolella (2012), Broda et al. (2013), Wirjanto and Xu (2013), Paolella (2015a), and the references therein.

The mixture set $\mathcal{G}$ is identifiable if the following condition holds:

For $G, \tilde{G} \in \mathcal{G}$ with $G = \sum_{i=1}^{k} \lambda_i f_i$ and $\tilde{G} = \sum_{i=1}^{\tilde{k}} \tilde{\lambda}_i \tilde{f}_i$, we have $G = \tilde{G}$ if and only if $k = \tilde{k}$ and $(\lambda_1, f_1), \ldots, (\lambda_k, f_k)$ is a rearrangement, or permutation, of $(\tilde{\lambda}_1, \tilde{f}_1), \ldots, (\tilde{\lambda}_k, \tilde{f}_k)$.

Determining sufficient conditions such that a mixture set is identifiable is a topic of ongoing research, with the seminal paper being from Teicher (1963), who showed that the class of mixtures of normals is indeed identifiable, with extensions (e.g., to the multivariate normal case) by Yakowitz and Spragins (1968). See Atienza et al. (2006) for further references and results, as well as the discussions in the aforementioned textbooks. In addition to having required a modified definition of identifiability for finite mixture distributions, we also have to distinguish between identifiability of the class of distributions, and of the parameter vector θ. If the distributional class is identified, then the associated parameter vector can also be identified with the help of some type of restriction, for example, $\lambda_1 \leq \lambda_2 \leq \cdots \leq \lambda_k$.

Section II.7.3.1 illustrated how to compute the raw moments of a general discrete mixture; Example II.7.14 showed the first two moments of the mixed normal to be

$$\mathbb{E}[Y] = \sum_{i=1}^{k} \lambda_i \mu_i, \quad \mathbb{E}[Y^2] = \sum_{i=1}^{k} \lambda_i (\mu_i^2 + \sigma_i^2),$$

from which $\mathbb{V}(Y) = \mathbb{E}[Y^2] - (\mathbb{E}[Y])^2$ can be computed.

In the sections below, we will use as our primary or showcase example the two-component mixed normal with parameter vector $\theta = (\mu_1, \mu_2, \sigma_1, \sigma_2, \lambda_1)'$ consisting of

$$\mu_1 = 0.1, \quad \mu_2 = -0.6, \quad \sigma_1 = 1, \quad \sigma_2 = 3, \quad \lambda_1 = 0.7, \tag{5.4}$$

with $\lambda_2 = 1 - \lambda_1$. The p.d.f. is shown, along with those of its two components, in Figure 5.1(a). Figure 5.1(b) shows the histogram of a simulated data set from the model, with $n = 100$ observations, and having used seed value 50 (see below) so it can be replicated. This data set is interesting because it contains an **outlier**, or an observation that is (subjectively deemed to be) noticeably separated from the main cloud of points. Recall the discussion in Section 3.1.3.

To assess the quality of the m.l.e. and various other estimators we will use the log sum of squares as the summary measure, that is,

$$M(\hat{\theta}, \theta) := \log (\hat{\theta} - \theta)'(\hat{\theta} - \theta). \tag{5.5}$$

As model (5.4) is observationally equivalent to that with $\theta^= = (\mu_2, \mu_1, \sigma_2, \sigma_1, \lambda_2)'$, we have to take as our measure of discrepancy

$$M^*(\hat{\theta}, \theta) = \min\{M(\hat{\theta}, \theta), M(\hat{\theta}, \theta^=)\}. \tag{5.6}$$

(Notice that it is not adequate to first convert $\hat{\theta}$ to ensure that, say, $\hat{\lambda}_1 \leq \hat{\lambda}_2$, and then just use $M(\hat{\theta}, \theta)$: as an example, imagine if $\lambda_1 = 0.5$ and the two means are very far apart.)

A second model of interest takes

$$\mu_1 = \mu_2 = 0, \quad \sigma_1 = 1, \quad \sigma_2 = 4, \quad \lambda_1 = 0.95. \tag{5.7}$$

It represents a population that is, 19 times out of 20, from a standard normal distribution, and 5% of the time from a N(0, 16) distribution, and thus has a nonnegligible probability

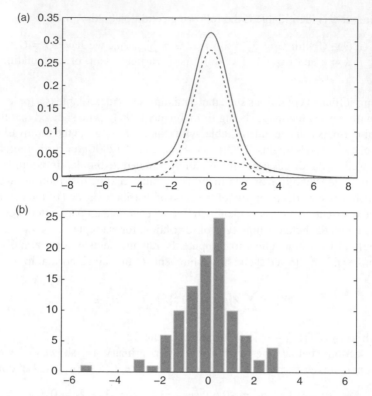

Figure 5.1 *(a) Mixed normal density with parameters (5.4) shown as the solid line. The two compo-nents (multiplied by their respective mixture weights) are shown as dashed lines. (b) Simulated data set from model (5.4) using $n = 100$ and seed value 50, illustrating the possibility of an outlier.*

of producing what appear to be outliers. This is sometimes referred to as a **contaminated normal** distribution.

5.1.2 Simulation of Univariate Mixtures

We also require a method to simulate from the mixed normal distribution. As in Section II.7.3.1, let the discrete r.v. C have p.m.f. $f_C(c) = \Pr(C = c) = \lambda_c$, for $c \in \{1, \dots, k\}$, and zero otherwise, where $\lambda_c \in (0, 1)$ are the weights in the mixed normal distribution, with $\sum_{c=1}^{k} \lambda_c = 1$. The realization of C thus indicates the component from which Y is to be drawn. Let $Y_c \sim \mathrm{N}(\mu_c, \sigma_c^2)$, $c = 1, \dots, k$, and $Y \sim \mathrm{MixN}(\boldsymbol{\mu}, \boldsymbol{\sigma}, \boldsymbol{\lambda})$. A mixed normal realiza-tion can then be generated as that of Y_c with probability $\lambda_c, c = 1, \dots, k$. This is true because, when sampled this way, and using the law of total probability (A.30),

$$\Pr(Y \in (y + \triangle y)) = \sum_{c=1}^{k} \Pr(Y \in (y + \triangle y) \mid C = c) \Pr(C = c)$$

$$= \sum_{c=1}^{k} \lambda_c \Pr(Y_c \in (y + \triangle y)) \approx \sum_{c=1}^{k} \lambda_c f_{Y_c}(y) \triangle y = f_Y(y) \triangle y.$$

To implement this, we first need the value of C. This is a draw from a multinomial distribution with vector of probabilities $\lambda = (\lambda_1, \ldots, \lambda_k)$. As was detailed on page I.187, this can be accomplished with the function

```
1   function C=randmultinomial(p), pp=cumsum(p); u=rand; C=1+sum(u>pp);
```

Assume we want an i.i.d. sample of size n from the MixN$(\boldsymbol{\mu}, \boldsymbol{\sigma}, \boldsymbol{\lambda})$ distribution. In "vectorized" computer languages such as Matlab and R, it turns out to be more efficient to generate the set of k vectors of $Y_c, c = 1, \ldots, k$. In particular, let $\mathbf{Y}$ be the $n \times k$ matrix with columns consisting of n-length vector realizations of $Y_1, Y_2, \ldots, Y_k$. Then, we multiply elementwise (Hadamard product) the matrix pick generated as

```
1   n=100; k=3; pick=zeros(n,k); lambda=[0.4,0.3,0.3];
2   for i=1:N, c=randmultinomial(lambda); pick(i,c)=1; end
```

with the $\mathbf{Y}$ matrix, generated as, say,

```
1   k=3; mu=[-2,1,1]; sigma=[1,1,3]; Y=[];
2   for c=1:k,  Y=[Y mu(c)+sigma(c)*randn(N,1)]; end
```

and compute the realized normal mixture vector as Y=pick.*Y; y=sum(Y,2);.[2] The function in Listing 5.1 implements this, and allows a seed value to be used so that the output can be reproduced.

5.1.3 Direct Likelihood Maximization

5.1.3.1 Three Problems

Maximization of the log-likelihood

$$\ell(\mathbf{y}; \boldsymbol{\theta}) = \sum_{i=1}^{n} \log f_{\text{MixN}}(y_i; \boldsymbol{\theta}) \tag{5.8}$$

is not as straightforward as for the simpler models presented in Chapter 3. In particular, there are three major problems associated with it. The first of these problems is unique to maximum likelihood; the other two are also potentially shared by the other estimation methodologies we discuss below in Section 5.2.

The first problem stems from the fact that, if y is any of the observed data points $y_1, \ldots, y_n$, then

$$\lim_{\substack{\mu_j \to y \\ \sigma_j \to 0}} \ell(\mathbf{y}; \boldsymbol{\theta}) = \infty, \quad j = 1, \ldots, k, \tag{5.9}$$

---

[2] The .* notation is for elementwise multiplication in Matlab. The 2 in the sum function indicates along which dimension to sum. Use of a 1, which is the default, sums along columns, and 2 is for rows. Less elegant, and slower, is to use y=sum(Y')'.

```
1   function y = mixnormsim(mu,sig,lam,n,seed)
2   if nargin<5, seed=0; end
3   k=length(mu); if k<=1, error('at least 2 components'), end
4   if abs(sum(lam)-1) > 1e-10, error('lambda does not sum to one'), end
5   if any(lam<=0), error('lambda out of range'), end
6   if any(lam>=1), error('lambda out of range'), end
7   if any(sig<=0), error('sigma out of range'), end
8   pick=zeros(n,k); if seed ~= 0, rand('twister',seed), end
9   for i=1:n, mult=randmultinomial(lam); pick(i,mult)=1; end
10  X=[];
11  for j=1:k
12    if seed ~= 0, Z=normlocal(n,j*seed+1); else Z=randn(n,1); end
13    X=[X mu(j)+sig(j)*Z];
14  end
15  X=pick.*X; y=sum(X,2);
16
17  function x=normlocal(n,seed), rand('twister',seed), u=rand(n,1); x=norminv(u);
```

Program Listing 5.1: Program to simulate n i.i.d. mixed normal realizations, where mu is the vector of means, sig is vector of scale terms (not the variances), and lam is the vector of weights. If seed is zero, then the faster method for generating normals is used, but the output is not reproducible; otherwise, the same output is returned when the same seed value is used. The term j*seed+1 ensures that the seed value for the multinomial draws, and those for the normal draws, never coincide, otherwise the results are flawed.

each point of which is referred to as a (log-likelihood) **singularity**. Thus, a global maximum of the likelihood cannot be found, and the desired solution is a local maximum (or one of the local maxima). Zaman (1996, p. 219) provides a detailed discussion of this issue.

Given that the statistical theory of maximum likelihood is based on having maximized the likelihood, which is not possible in this case, there appears to be both a numerical and a theoretical problem. The latter has been addressed by Kiefer (1978b), who demonstrated that a local maximum of $\ell(\mathbf{y}; \theta)$ is strongly consistent and asymptotically efficient. In what follows, we discuss the numerical issues relevant for avoiding the singularities and locating the appropriate local maximum. Note that the singularities arising from (5.9) are not measure-zero quantities in the parameter space, but rather small punctured neighborhoods in which the log-likelihood exceeds the maximum at the desired local maximum, which we call $\widehat{\theta}_{ML}$. Once the numeric algorithm finds its way into one of these n neighborhoods, it is like a black hole – there is no way out.

The second problem associated with estimating θ is actually more insidious than the singularities. As pointed out by Day (1969), any group of two or more data points that are sufficiently close together and sufficiently distinct from the larger bell-shaped cluster of observations can also give rise to a false, yet bounded, local maximum such that all the components are nondegenerate (have positive variances). Even a single outlying observation might "attract" the optimizer and lead to a degenerate component. This happens, for example, with the data set shown in Figure 5.1(b), in which case the most extreme negative observation forms its own group. Of course, knowing the true model in this case, we know that this point does not constitute an observation from a low-weighted component centered around −6; and in practice, such an observation would almost certainly be deemed either an error or just a rather improbable, but valid, outcome. Observe, however, that such a decision is purely subjective and ultimately draws on the researcher's prior knowledge of the distribution under study. Even more severe is the case in which two or more observations are clustered in such a fashion, so that more extensive prior input is required on behalf of the researcher to decide if the apparent cluster is spurious or genuine.

The last of the three pitfalls that can arise in computing the m.l.e. is that there is more than one *plausible* local maximum of the likelihood. If one manages to avoid the n singularities arising from (5.9), and any of the local maxima driven by spurious density components associated with a small cluster of observations, then there still might remain more than one local maximum of the likelihood that gives rise to a plausible solution. As an example, for the data set generated by

```
1  n=100; mu=[-0.6 0.1]; scale=[3 1]; lam=[0.3 0.7];
2  y = mixnormsim(mu,scale,lam,n,87); % 87 is the seed value
```

with histogram shown in Figure 5.2(a), two sets of m.l.e.s, say M_1 and M_2, were found.[3]

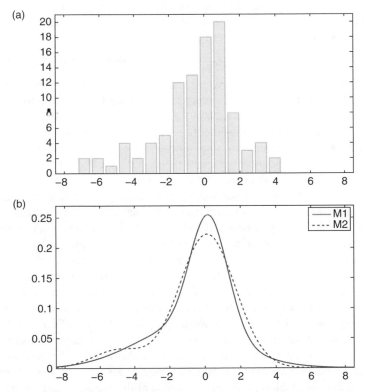

Figure 5.2   *(a) Realization of model (5.4) with $n = 100$. (b) Fitted models (5.10), both being from local maxima of the likelihood.*

[3] This example was found simply by simulating until a sizeable discrepancy was found between the m.l.e.s obtained from the direct and the EM approach, described below, *despite having used the same starting values.*

The parameter values are

$$\mu_1 = 0.223, \quad \mu_2 = -1.223, \quad \sigma_1 = 1.062, \quad \sigma_2 = 2.889, \quad \lambda_1 = 0.523 \quad (M_1),$$
$$\mu_1 = 0.124, \quad \mu_2 = -4.899, \quad \sigma_1 = 1.581, \quad \sigma_2 = 1.509, \quad \lambda_1 = 0.882 \quad (M_2).$$

(5.10)

Their p.d.f.s are shown in Figure 5.2(b); fortunately, they do not differ greatly, though the differences are readily apparent. As a further check that these are two genuine maxima, using M_1 as the starting value (or with a small perturbation) induces both the direct and EM algorithms (see below) to converge to M_1; and similarly for M_2. The log-likelihood associated with M_1 is -215.43; that for M_2 is -215.51. In this case, they happen to be nearly identical, though that need not be the case in general. One usually just chooses the parameter vector associated with the largest likelihood, though plausible solutions with very close values of the likelihood evaluated at the estimates should all be entertained, and possibly combined in an appropriate manner such as via techniques associated with **model averaging**; see the monograph by Claeskens and Hjort (2008) for a detailed presentation.

5.1.3.2 *Solution via Imposing Box Constraints*

> It is my impression that rather generally, not just in econometrics, it is con-
> sidered decent to use judgement in choosing a functional form but indecent to
> use judgement in choosing a coefficient. If judgement about important things
> is quite all right, why should it not be used for less important ones as well?
>
> (Tukey, 1978, p. 52; quoted in Poirier, 1995, p. 524)

The first two aforementioned problems associated with the m.l.e. for the normal mixture model – singularities and spurious clusters – can be mitigated by using **box constraints**, or fixed lower and upper bounds on each of the model parameters that are enforced during the numeric maximization of the log-likelihood; see Section 4.3.2 for details.

The neighborhoods containing the singularities in the likelihood are then easy to avoid by using a positive lower bound on each of the σ_i. Observe that in any case we need to employ a positive lower bound on each of the σ_i to prevent the optimization routine from attempting a zero or negative scale parameter. Likewise, λ_1 in the two-component case, and $\lambda_1, \ldots, \lambda_{k-1}$ in the k-component case, have to be restricted to lie between zero and one.

The more difficult question concerns the choice of bounds to prevent the optimizer from selecting a spurious cluster of points as one of the k density components. As mentioned above, this is more difficult, if not impossible to automate for all situations, and requires input from the researcher. With dedicated effort, some creative linguistic acrobatics can be used to avoid the words "prior information" or "subjective judgement," but ultimately the researcher will have to specify bounds that are informative in the sense that they rule out what he or she knows, or believes to be, bogus solutions.[4] See also Section 5.1.6 below.

Generally applicable and uninformative bounds for the μ_i are easy to construct. Irrespective of what the true μ_i are, the m.l.e. of the μ_i cannot take on values below the minimum of the data or above the maximum of the data, so it would seem wise to prevent the optimization algorithm from even attempting values outside of this range. Clearly, if the researcher

---

[4] This idea is anathema to some statisticians, as they argue that it violates an "objective" analysis of the data. Just the fact that a particular distribution, the mixed normal, has been chosen to model the data should be evidence enough that the notion of objectivity on behalf of the researcher can be safely banished, though that just scratches the surface of how much implicit "information" enters into all but the most trivial of empirical analyses.

somehow knows that, say, $2 < \mu_1 < 3$, then this constraint should be incorporated, though such precise information is often not available.

The bounds for the scale and λ terms require more subjective input. Use of the trivial constraints $\lambda_1 \in (0, 1)$ and $\sigma_i > 0$ is insufficient to prevent the optimization routine from settling in on a singularity of the likelihood function. We suggest formulating a set of sensible constraints that should apply to the kind of data the researcher expects to encounter in the study. For our model (5.4), which is designed to loosely mimic the behavior of financial asset returns (rather leptokurtic, mildly asymmetric), we expect the component associated with the lower λ weight to have a negative mean and higher variance relative to the other component, and a weight that is at least, say, 10%. For the σ_i, an upper bound given by the sample standard deviation, S_n, of the data would seem reasonable, though simulation using the model (5.4) confirms that every so often, the (correct) m.l.e. of one of the σ_i indeed exceeds this by a small amount. So, one might choose $2S_n$. A reasonable lower bound that should help prevent spurious components is, say, $S_n/10$.

Finally, for the component weights, if our knowledge and expectations of the model are along the lines of model (5.4), then one might consider, say, $p < \lambda_i < 1 - p$, where $p = n^{-1/2}$. This can best be interpreted with respect to a model with its two components being highly separated (the μ_i being far apart relative to the σ_i). Thus, for $n = 100$, $p = 1/10$, so that the smaller component can be established by not less than 10 observations. For $n = 1000$, a separate component would need at least 32 observations. These are clearly just rules of thumb that are loosely tailored to model (5.4) and will not be appropriate in all situations. In general, the bounds used for the data associated with an actual empirical study should reflect the prior knowledge of the researcher.

We will make use of the following four sets of box constraints. The data set is $\mathbf{y}$, of length n, and S_n is its sample standard deviation:

$$0 < \sigma_i < 20S_n, \qquad 0 < \lambda_i < 1, \qquad \text{(Constraint 0)} \qquad \text{(5.11a)}$$

$$S_n/100 < \sigma_i < 6S_n, \qquad p < \lambda_i < 1 - p, \quad p = n^{-1/2}/4, \qquad \text{(Constraint 1)} \qquad \text{(5.11b)}$$

$$S_n/10 < \sigma_i < 2S_n, \qquad p < \lambda_i < 1 - p, \quad p = n^{-1/2}, \qquad \text{(Constraint 2)} \qquad \text{(5.11c)}$$

$$S_n/3 < \sigma_i < 1.1S_n, \qquad p < \lambda_i < 1 - p, \quad p = 2n^{-1/2}, \qquad \text{(Constraint 3)} \qquad \text{(5.11d)}$$

and, for each, $\min(\mathbf{y}) < \mu_i < \max(\mathbf{y})$, $i = 1, 2$. Note that constraint 0 in (5.11a) is essentially noninformative and simply prevents $\boldsymbol{\theta}$ from assuming values that are not contained in the allowable parameter space $\boldsymbol{\Theta}$.

Optimization routines require a starting value of the parameter vector. In all the simulation studies throughout Section 5.1, we choose, rather arbitrarily,

$$\mu_1 = \mu_2 = \frac{\min(\mathbf{y}) + \max(\mathbf{y})}{2}, \quad \sigma_1 = S_n/2, \quad \sigma_2 = S_n, \quad \lambda_1 = 0.6. \qquad \text{(5.12)}$$

A simulation exercise with our showcase model (5.4), $n = 50$ and the (rather weak) constraint (5.11b) results in 25 out of 10,000 cases failing (and 6 out of 10,000 for $n = 100$). For those 25 cases, what happens is that the optimization algorithm is enticed towards the boundary of the allowable parameter space (dictated from the box constraints) by a locally high and ever-increasing likelihood value that – unbeknown to the greedy but ignorant optimization routine – leads to a singularity (and the algorithm fails). This can be prevented by checking in the objective function whether the negative log-likelihood is negative infinity

(-Inf in Matlab) and, if so, replacing it with a large but finite value to the optimizer. (We use the negative log-likelihood because Matlab's implementation is to minimize, not maximize.) In all 25 cases this check-and-replace worked, and the m.l.e. was a valid parameter vector (which agreed with the value obtained via the EM algorithm, when it worked; see below). In fact, we find that just using check-and-replace, even with the noninformative constraint (5.11a) is also enough to ensure convergence to some valid local likelihood maximum. It is important to keep in mind that if the box constraints are too strict and the genuine m.l.e. is not contained in the box, then the method will fail to deliver the genuine m.l.e. For example, model (5.7) should not be estimated using constraint (5.11c) or (5.11d).

We will now demonstrate the impact of imposing each of the four constraints (5.11) on the m.l.e. based on the data set shown in Figure 5.1(b) (and always using check-and-replace for the infinite likelihood, as discussed above). The plots in Figure 5.3 show the true density (thicker line) and 100 fitted densities (thin lines), each based on the m.l.e., which was constrained to lie in $C_i \subset \Theta$, where C_i is the space indicated by the ith box constraint from (5.11), $i = 0, 1, 2, 3$, and having used a different starting value, randomly chosen from C_i. From this exercise and the plots, we immediately see that the number of local maxima decreases as we increase the severity of the box constraints, as expected. Note, though, that even under the rather restrictive box constraint 3, there are still three plausible local maxima.

Of the 100 fitted densities, the one corresponding to the highest likelihood was preserved, and is shown in Figure 5.4(a), which overlays this density for each of the four sets of constraints. We see that, under constraints 0 and 1, the "best" solution is a single normal density for all observations except the single outlier at -5.7 (apparent in Figure 5.1), while under constraints 2 and 3, the more plausible mixture density is chosen.

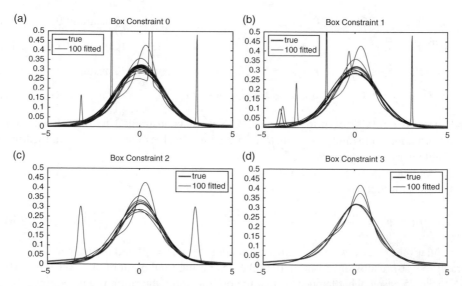

Figure 5.3   *Using showcase model (5.4) with $n = 100$ and the data set shown in Figure 5.1b, the plots show the true density (thick solid line; the same as in Figure 5.1a) and 100 fitted densities (thin lines), each having used a different (randomly chosen) starting value. The box constraint numbers $0, \ldots, 3$ are given in (5.11), and increasingly place more restrictions on the allowable parameter space.*

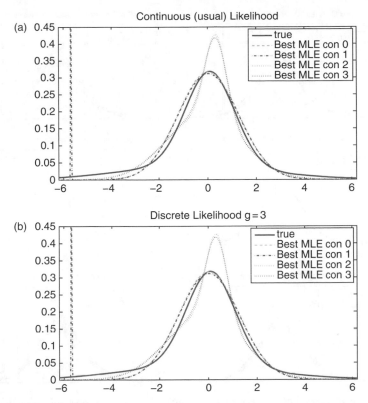

Figure 5.4 *(a) Of the 100 fitted densities shown in Figure 5.3, this shows the one corresponding to the highest likelihood, for each of the four constraints. (b) Same, but from the densities shown in Figure 5.5(a)–(d), which are based on the g-rounded likelihood for g = 3.*

5.1.3.3 Accounting for Finite Precision

Recall the discussion in Section 3.1 regarding the possible relevance of accounting for the finite precision of the observations. Doing so in the context of the mixed normal distribution will, in principle, resolve the singularity problem in (5.9) because each data point y_i no longer represents a measure-zero point mass, but rather an interval from, in the notation of (3.1), $y_i \pm \epsilon/2$.

We now perform the likelihood calculation accounting for finite data accuracy, using $g = 3$ digits after the decimal point. The top four panels of Figure 5.5(a–d) are the analog of those in Figure 5.3, except having used (3.1) with $g = 3$. Comparing Figures 5.3(a) and 5.5(a), we see that, with only the bare-bones box constraint 0, use of the g-rounded likelihood results in fewer local maxima of the likelihood, so its 0 intended improvement is indeed realized.

However, imposing the other box constraints leads to rather erratic behavior of the likelihood, which is now apparently plagued by numerous local maxima. One might expect this behavior to diminish as g is increased, though from the bottom four panels of Figure 5.5(e–h), which are the same as the top four (a–d) except having used $g = 6$, we see that the problem of local maxima is significantly exacerbated. It is odd that, with box constraint 3, which otherwise *helps* us to locate the genuinely desired m.l.e., the surface of

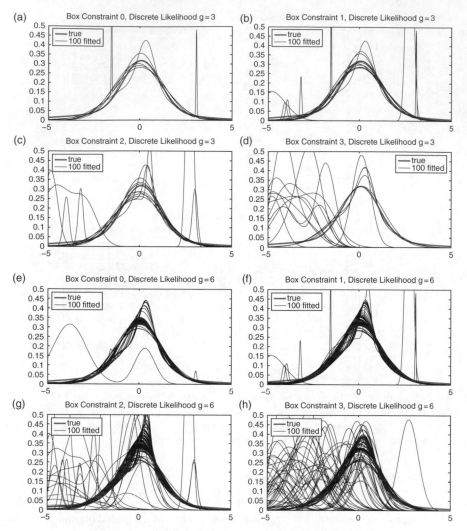

Figure 5.5 *Same as Figure 5.3 but using the g-rounded likelihood from (3.1) with g = 3 (a–d) and g = 6 (e–h).*

the likelihood apparently becomes more jagged and pocketed than the surface of the moon. The only way to find the genuine m.l.e. is by repeated calculation of the m.l.e. based on a large number and wide variety of starting values.

Figure 5.4(b) is the same as Figure 5.4(a), but based on the results for the g-rounded likelihood for $g = 3$. The densities corresponding to the highest likelihood are graphically indistinguishable from those based on the usual, p.d.f.-based likelihood calculation. Thus, they are indeed among the various fitted densities shown in Figure 5.5.

As an aside, if we simply round off the data to three places after the decimal (in Matlab, this is done with `y=round(y*1e3)/1e3;`) and perform the exercise that led to Figure 5.3, using the conventional, p.d.f.-based likelihood, then we obtain results virtually identical to Figure 5.3, confirming that it is the use of (3.1), and not the loss of significant digits *per se*, that leads to the disturbing results in Figure 5.5.

5.1.4 Use of the EM Algorithm

> Like many popular methods in statistics, however, [the EM algorithm] has also been misused or misperceived. For example, one [misconception] is to regard EM as an estimation procedure … and to study the properties of "EM estimators," forgetting that it is simply a computation method intended for computing maximum likelihood estimates.
>
> (Meng, 1997)

Instead of, or in addition to, using a generic optimizer to maximize the log-likelihood, one can use the **EM algorithm**, where EM stands for **expectation-maximization**. It is, like generic optimization routines, an iterative process, but uses a technique that is specifically catered to the model – and is thus not always applicable. It is straightforward to derive in the mixture-of-normals case, and is in fact one of the showcase examples of the EM algorithm. Its use in this context goes back to work by Newcomb in 1886; see McLachlan and Krishnan (2008, Sec. 1.8) for the history of the algorithm and original references. As in the above quote, it is crucial to emphasize that the EM algorithm is not an estimation methodology (like least squares, method of moments, etc.), but rather a numerical technique to obtain the m.l.e. To distinguish it from the use of generic optimization routines applied to the log-likelihood, we denote the latter as the "direct" method.

The basis for the EM algorithm is to compute the m.l.e. in the face of missing data. It does so by iterating on its two steps: the first, or E-step, computes the expectation of the **complete data** log-likelihood, which embodies both the observed data and the missing data, conditional on the observed data and the current parameter value $\hat{\theta}$; and the second, or M-step, is to maximize it, providing an updated value of $\hat{\theta}$. This is iterated until the change in the observed data log-likelihood and/or the changes in the sequence of the $\hat{\theta}$ are below some threshold.

The enormous popularity and applicability of the EM algorithm stems from the fact that, in many models, there are no actual missing data, but the structure of the model lends itself to use of a **latent** or unobservable variable that can be treated as missing. If those data are available, then the M-step is often numerically vastly simpler, and so an otherwise difficult maximization can be reduced to a sequence of iterations using an easier objective function to maximize. This is precisely the case with the mixed normal distribution with n observations: the latent variable is the set of n values that indicate from which component observation y_i came. If we know that, then we simply have k sets of independent normal r.v.s, and the estimation of the μ_i and σ_i^2 is trivial, as are the weights, λ_i.

Let $Y_i \overset{\text{i.i.d.}}{\sim} \text{Mix}_k \, \text{N}_1(\boldsymbol{\mu}, \boldsymbol{\sigma}, \boldsymbol{\lambda})$ with observed data $y_1, \dots, y_n$ and parameter vector $\boldsymbol{\theta} = (\boldsymbol{\mu}' \ \boldsymbol{\sigma}' \ \boldsymbol{\lambda}')'$, and denote the latent, or hidden, variable associated with the ith observation Y_i by $\mathbf{H}_i = (H_{i1}, \dots, H_{ik})$, where $H_{ij} = 1$ if Y_i came from the jth component, and zero otherwise. Thus, in the notation of Section I.5.3.1, $\mathbf{H}_i = (H_{i1}, \dots, H_{ik}) \sim \text{Multinom}(1, \lambda_1, \dots, \lambda_k)$ with p.d.f.

$$f_{\mathbf{H}_i}(\mathbf{h}_i; \boldsymbol{\theta}) = f_{\mathbf{H}_i}(\mathbf{h}_i; \boldsymbol{\lambda}) = \prod_{j=1}^{k} \lambda_j^{h_{ij}} \mathbb{I}_{\{0,1\}}(h_{ij}) \mathbb{I}\left(\sum\nolimits_{j=1}^{k} h_{ij} = 1\right).$$

Conditional on $\mathbf{H}_i$ being $\mathbf{h}_i$ with jth element unity, $f_{Y_i|\mathbf{H}_i}(y, \mathbf{h}_i; \theta) = f_N(y; \mu_j, \sigma_j^2)$, so that we can write $f_{Y_i|\mathbf{H}_i}(y, \mathbf{h}_i; \theta) = \prod_{j=1}^{k} [f_N(y; \mu_j, \sigma_j^2)]^{h_{ij}}$. Thus, the joint density of Y_i and $\mathbf{H}_i$ is

$$f_{Y_i|\mathbf{H}_i}(y, \mathbf{h}_i; \theta) f_{\mathbf{H}_i}(\mathbf{h}_i; \theta) = \prod_{j=1}^{k} [\lambda_j f_N(y; \mu_j, \sigma_j^2)]^{h_{ij}} \mathbb{I}_{\{0,1\}}(h_{ij}) \mathbb{I}\left(\sum_{j=1}^{k} h_{ij} = 1\right), \qquad (5.13)$$

and $f_{\mathbf{H}_i|Y_i}(\mathbf{h}_i, y; \theta)$ is

$$\frac{f_{Y_i,\mathbf{H}_i}(y, \mathbf{h}; \theta)}{f_{Y_i}(y; \theta)} = \frac{\prod_{j=1}^{k} [\lambda_j f_N(y; \mu_j, \sigma_j^2)]^{h_{ij}}}{\sum_{j=1}^{k} \lambda_j f_N(y; \mu_j, \sigma_j^2)} \mathbb{I}_{\{0,1\}}(h_{ij}) \mathbb{I}\left(\sum_{j=1}^{k} h_{ij} = 1\right),$$

or

$$\Pr(H_{ij} = 1 \mid Y_i = y) = \frac{\lambda_j f_N(y; \mu_j, \sigma_j^2)}{\sum_{j=1}^{k} \lambda_j f_N(y; \mu_j, \sigma_j^2)}, \quad j = 1, \dots, k. \qquad (5.14)$$

The complete data log-likelihood, denoted ℓ_c, is, from (5.13), with $\mathbf{Y} = (Y_1, \dots, Y_n)$ and $\mathbf{H} = (\mathbf{H}_1, \dots, \mathbf{H}_n)$,

$$\ell_c(\theta; \mathbf{Y}, \mathbf{H}) = \sum_{i=1}^{n} \sum_{j=1}^{k} H_{ij} \log \lambda_j + \sum_{i=1}^{n} \sum_{j=1}^{k} H_{ij} \log f_N(Y_i; \mu_j, \sigma_j^2). \qquad (5.15)$$

We require the m.l.e. of θ from the complete data likelihood. Note that the parameters λ and (μ, σ) are disjoint in (5.15), so that their m.l.e.s can be determined separately. For λ, given their linear constraint, we use a Lagrange multiplier ξ to maximize the function

$$g(\lambda, \xi; \mathbf{H}) = \sum_{i=1}^{n} \sum_{j=1}^{k} H_{ij} \log \lambda_j - \xi\left(\sum_{j=1}^{k} \lambda_j - 1\right).$$

Differentiating,

$$\frac{\partial g}{\partial \lambda_j} = \frac{1}{\lambda_j} \sum_{i=1}^{n} H_{ij} - \xi, \quad j = 1, \dots, k,$$

so that, as $\lambda_j \neq 0$, we set this equal to zero and solve, to get $\sum_{i=1}^{n} H_{ij} = \lambda_j \xi, j = 1, \dots, k$. Adding these, using the fact that $\sum_{j=1}^{k} \sum_{i=1}^{n} H_{ij} = n$ and $\sum_{j=1}^{k} \lambda_j = 1$, yields $n = \xi$. Substituting this into $\sum_{i=1}^{n} H_{ij} = \lambda_j \xi$ gives the appealing and, *ex post facto* perhaps obvious, result

$$\hat{\lambda}_j = \frac{1}{n} \sum_{i=1}^{n} H_{ij}, \quad j = 1, \dots, k, \qquad (5.16)$$

in agreement with that stated without derivation in McLachlan and Krishnan (2008, Eq. 1.31). Now for $(\mu', \sigma')'$, observe from the last term in (5.15) that, given the binary nature of the H_{ij} and that $\sum_{j=1}^{k} H_{ij} = 1$, we have k independent normal populations, each with $\sum_{i=1}^{n} H_{ij}$ observations. Thus, from Example 3.5, the m.l.e. is

$$\hat{\mu}_j = \frac{\sum_{i=1}^{n} H_{ij} Y_i}{\sum_{i=1}^{n} H_{ij}}, \quad \hat{\sigma}_j^2 = \frac{\sum_{i=1}^{n} H_{ij}(Y_i - \hat{\mu}_j)^2}{\sum_{i=1}^{n} H_{ij}}, \quad j = 1, \dots, k. \qquad (5.17)$$

The estimators (5.16) and (5.17) of course cannot be computed because we do not observe the $\mathbf{H}_i$. The E-step of the EM algorithm provides the way to use what we do observe, namely the Y_i, and the value of $\widehat{\boldsymbol{\theta}}$, say $\boldsymbol{\theta}^{(s)}$, in the sth step of the iterative scheme, to produce values of the $\mathbf{H}_i$. In general, the E-step requires

$$Q(\boldsymbol{\theta}; \boldsymbol{\theta}^{(s)}) := \mathbb{E}_{\boldsymbol{\theta}^{(s)}}[\ell_c(\boldsymbol{\theta}; \mathbf{Y}, \mathbf{H}) \mid \mathbf{Y} = \mathbf{y}], \tag{5.18}$$

the conditional expectation of the complete data log-likelihood with respect to the hidden r.v.s, given the observed data $\mathbf{y}$, and using as parameter $\boldsymbol{\theta}$ the current value $\boldsymbol{\theta}^{(s)}$. In our case, as ℓ_c in (5.15) is linear in the H_{ij}, we need only the expectation of the H_{ij}, conditional on $\mathbf{Y} = \mathbf{y}$, and using $\boldsymbol{\theta}^{(s)}$. This is given by (5.14).

Based on some initial value $\boldsymbol{\theta}^{(0)}$, the sth iteration of the EM algorithm for the mixed normal distribution consists of the following steps: Calculate H_{ij} from (5.14), $i = 1, \dots, n, j = 1, \dots, k$, then calculate (5.16) and (5.17) to get $\boldsymbol{\theta}^{(s+1)}$. The program in Listing 5.2 performs the computation in the $k = 2$ case, with the extension to any $k > 2$ being straightforward.

Observe how, by the nature of the algorithm, $\lambda_i \in [0, 1]$ and $\sigma_i \geq 0$ automatically. Thus, its use should be equivalent to the use of the direct method for obtaining the m.l.e. using the noninformative constraint (5.11a). However, the algorithm can, and on occasion does, allow a σ_i to be precisely zero, in which case the algorithm fails. This will happen if there is (at least) one outlier without any other observations near it, such as depicted in Figure 5.1(b). Our implementation in Listing 5.2 is such that, if the algorithm fails, then the most extreme observation is removed. This is repeated until it works. The EM algorithm can also be augmented to support parameter constraints, though we do not pursue this.[5]

One of the most important properties of the EM algorithm is that, at each iteration, the log-likelihood $\ell(\boldsymbol{\theta}; \mathbf{Y})$ cannot decrease. The proof is straightforward, using Jensen's inequality, and is shown in many textbooks; see, for example, McLachlan and Krishnan (2008, Sec. 3.2), Givens and Hoeting (2013, Sec. 4.2.1), and Hogg et al. (2014, Sec. 6.6), where also numerous basic examples applying the EM algorithm can be found.

We estimated via the EM algorithm the same 10,000 data sets as before, keeping track of which data sets caused the EM method to fail. For $n = 50$, there were 324 such cases. In each of those data sets, there was at least one outlier. By removing the extreme observation from the sample (as the implementation in Listing 5.2 does), the EM algorithm converges in 311 of these 324 cases, and in the remaining 13, two outliers had to be removed. (This problem does not arise with the direct method because of the box constraints.) Using $n = 100$, only 36 of the 10,000 data sets needed to have an outlier removed.

Recall Figure 5.3(a), which shows 100 fitted densities, based on 100 random starting values, using the direct approach for maximizing the likelihood and imposing constraint 0 from (5.11a). Figure 5.6(a) is the same, again via the direct approach, but having used 1000 such fitted densities (and enforcing the high tolerance of 10^{-8} for convergence). Comparing this to Figure 5.6(b), which shows the same result but having used the EM algorithm for estimation, we see that, while both the direct method and EM algorithm get stuck in one of the numerous local maxima as a result of using different starting values, the number of potential solutions found by the EM algorithm appears to be considerably smaller.

---

[5] See Hathaway (1986) and Ciuperca et al. (2003) for the case of univariate normal mixtures; the general discussion and references in McLachlan and Krishnan (2008, Sec. 3.5.4); and Tian et al. (2008) for a general treatment in the multivariate normal and Student's t case.

```
1   function [solvec,crit,iter] = mixnormEM (y,init,tol,maxit)
2   if nargin < 4, maxit=5e4; end, if nargin < 3, tol=1e-6; end
3   outlier=0;
4   while 1
5     if nargin < 2
6       m=min(y); M=max(y); sd=std(y); mid=(m+M)/2; init = [mid mid sd/2 sd 0.6]';
7     end
8     [solvec,crit,iter] = hartley(y,init,tol,maxit);
9     if all(~isnan(solvec)), break
10    else
11      y=sort(y); left=y(2)-y(1); right=y(end)-y(end-1);
12      if left>right, y=y(2:end); else y=y(1:end-1); end
13      'removing an outlier', outlier=outlier+1;
14    end
15  end
16
17  function [solvec,crit,iter] = hartley (y,init,tol,maxit)
18  old=init; new=zeros(5,1); iter = 0; crit=0;
19  while 1
20    iter=iter+1;
21    mu1=old(1); mu2=old(2); s1=old(3); s2=old(4); lam = old(5);
22    mixn =lam*normpdf(y,mu1,s1)+(1-lam)*normpdf(y,mu2,s2);
23    H1=lam*normpdf(y,mu1,s1) ./ mixn; H2=1-H1; N1=sum(H1); N2=sum(H2);
24    new(1) = sum (H1 .* y) / N1; new(2) = sum (H2 .* y) / N2;
25    new(3) = sqrt( sum (H1.*((y-new(1)).^2)) / N1 );
26    new(4) = sqrt( sum (H2.*((y-new(2)).^2)) / N2 );
27    new(5) = mean(H1);
28    crit = max (abs (old-new)); solvec=new; if any(isnan(solvec)), break, end
29    if (crit < tol) || (iter >= maxit), break, end
30    old=new;
31  end
```

Program Listing 5.2: The function `hartley` is the EM algorithm for computing the m.l.e. of the two-component mixed normal, as was given in Hartley (1978). Parameter `tol` is the desired tolerance for each parameter to assume convergence; `maxit` is the allowed number of iterations before giving up. In our implementation here, in the header program `mixnormEM`, if during the iterations one of the σ_i is very close or equal to zero, the most extreme observation is removed from the data set and the estimation is attempted again.

This otherwise appealing fact is of little value if the EM algorithm still can settle on local maxima that involve spurious components, which we see, from the eight or so large spikes in the graph, is possible. As such, the best strategy when faced with a real data set appears to be to use the direct method for the m.l.e., with as stringent constraints as deemed acceptable, and compute it using a variety of starting values, producing graphs such as those in Figure 5.3. Visual inspection of these, as well as their corresponding likelihood values, can be used to *subjectively* choose the appropriate value for $\hat{\theta}$.

There are two further disadvantages of the plain EM algorithm presented here. The first concerns its relative speed of convergence. It is well known (see, for example, Everitt, 1984) that the direct method of optimization tends to converge much faster to the optimum than the EM algorithm as the algorithms approach the solution. This statement is obviously not very precise, and could be made so in terms of estimation time in order to increase the number of significant digits in $\hat{\theta}$ from, say, two to six. In fact, with regard to the EM algorithm, Redner

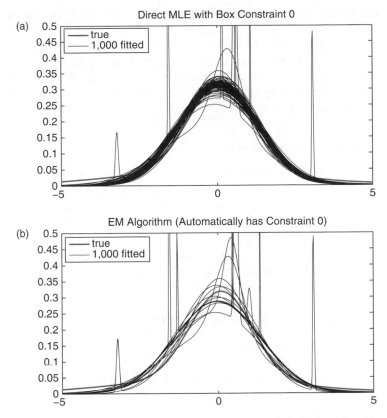

Figure 5.6 *(a) Same as Figure 5.3a but having used 1000 instead of 100 fitted densities. (b) Same, but having used the EM algorithm (which implicitly imposes the same constraints on the λ_i and σ_i as does our constraint 0) with 1000 fitted densities.*

and Walker (1984, p. 203) in their extensive paper go so far as to say that "unfortunately, its convergence can be maddeningly slow in simple problems which are often encountered in practice."

This problem can be ameliorated by using more sophisticated EM algorithms that combine features of the basic EM algorithm with quasi-Newton (Hessian-based) general optimization algorithms; see Givens and Hoeting (2013), McLachlan and Krishnan (2008), and the references therein for details. As those authors show, this is a rich and continually developing field of research that has provided nonobvious solutions to many interesting problems. Nevertheless, in the context of the univariate mixed normal distribution, as well as other applications in which the number of parameters to be estimated is not excessive, the use of black box optimization routines (with appropriate box constraints, and use of several random starts) can often provide a faster solution – in terms of speed of programming, and speed of estimation.

An advantage of the direct method of optimization (at least for the gradient- and Hessian-based algorithms) not shared by the plain EM algorithm is that they automatically return an approximation to the variance covariance matrix of the parameter vector, and thus approximate confidence intervals for its elements; see Section 5.1.7 below. The EM algorithm can be augmented to also yield approximate parameter standard errors; see, for example, the survey article of Ng et al. (2004), McLachlan and Krishnan (2008), and Givens and Hoeting (2013, Ch. 4). Alternatively, using either method, the bootstrap can be used to compute approximate c.i.s for the parameters, as was detailed in Section 1.3. We expect this to be more accurate, particularly in smaller samples, when the normal approximation of the sampling distribution of $\hat{\theta}$ might be questionable. This will be examined in Section 5.1.7 below.

5.1.5 Shrinkage-Type Estimation

Shrinkage estimation is a method such that the m.s.e. of an estimator, say $\hat{\theta}$, is reduced by using the weighted average $\hat{\theta}_{a,k} = (1-a)\hat{\theta} + ak$, where $0 \le a \le 1$ is the **shrinkage weight** and k is the nonstochastic **shrinkage target**. If $\hat{\theta}$ is unbiased, then clearly $\hat{\theta}_{a,k}$ will be biased (unless $\theta = k$), but, recalling the relationship (1.2) among the m.s.e., bias, and variance, the reduction in variance obtained by combining $\hat{\theta}$ with a constant (zero- variance) estimator can significantly outweigh the increase in squared bias, even if (and most notably when) $k \ne \theta$. A substantial formal presentation of the theory of shrinkage estimation, along with original references, can be found in Lehmann and Casella (1998). Section 5.4 below provides an introduction to the main concept.

In our context, we might use $(1-a)\hat{\boldsymbol{\theta}}_{\text{ML}} + a\boldsymbol{\theta}_0$, where $\boldsymbol{\theta}_0 = (\mu_{1,0}, \mu_{2,0}, \sigma_{1,0}, \sigma_{2,0}, \lambda_{1,0})$ is the shrinkage target or parameter vector to which we wish to shrink. In this section, we pursue a modification of the original formulation of shrinkage, taking

$$\hat{\boldsymbol{\theta}}_{\text{ML-Shr}} = \arg\min_{\boldsymbol{\theta}}\{-\ell(\mathbf{y};\boldsymbol{\theta}) + \tau P(\boldsymbol{\theta},p)/1000\}, \tag{5.19}$$

where $\ell(\mathbf{y};\boldsymbol{\theta})$ is the log-likelihood (5.8),

$$P(\boldsymbol{\theta},p) = \|\boldsymbol{\theta} - \boldsymbol{\theta}_0\|_p$$
$$= |\mu_1 - \mu_{1,0}|^p + \cdots + |\lambda_1 - \lambda_{1,0}|^p \tag{5.20}$$

is the penalty term with $p > 0$, and $\tau \ge 0$ dictates the strength of the shrinkage.[6] (The division of $P(\boldsymbol{\theta},p)$ in (5.19) by 1000 is arbitrary and allows τ to be in a more comfortable range for the examples shown in Figure 5.7.)

The benefit of this formulation, as compared to $(1-a)\hat{\boldsymbol{\theta}}_{\text{ML}} + a\boldsymbol{\theta}_0$, is that the latter requires the m.l.e., which we know can be numerically problematic to obtain without parameter constraints, whereas the shrinkage estimator (5.19) for $\tau > 0$ avoids the singularities (5.9).

In what follows, we will use the shrinkage target given by $\boldsymbol{\theta}_0 = (0, 0, 1, 1, \cdot)$, with the target for λ not specified. That is, each component is shrunk towards the standard normal distribution, and $P(\boldsymbol{\theta},p) = |\mu_1|^p + |\mu_2|^p + |\sigma_1 - 1|^p + |\sigma_2 - 1|^p$. The top two panels of Figure 5.7 show boxplots of the m.s.e. measure M^* from (5.6) for the m.l.e. and the shrinkage estimator (5.19) for a variety of values of τ, and the two values $p = 1$ and $p = 2$, using our showcase model (5.4). For both values of p, we see a small improvement in the

---

[6] The penalty term (5.20) with power $p = 2$ is the same structure as that used in **ridge regression**, a very important method of shrinkage in regression analysis (see, for example, Ravishanker and Dey, 2002, Sec. 8.3.3); while for $p = 1$, this resembles the **lasso** (least absolute shrinkage and selection operator) from Tibshirani (1996).

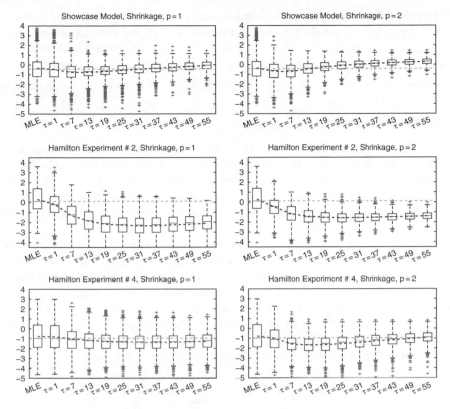

Figure 5.7 *Comparison of log total m.s.e. for θ_{ML} (leftmost boxplot in all six panels) and $\hat{\theta}_{shr}$ from (5.19), using shrinkage form (5.20), for $p = 1$ (left) and $p = 2$ (right), for a grid of values of τ, dictating the strength of the shrinkage. The top panels are for the showcase constellation (5.4) with $n = 100$ observations. The middle and bottom panels correspond to (5.21) and (5.22), respectively. The simulation is based on 1000 replications. The horizontal dashed lines show the median m.l.e. value of M^* from (5.6). The other dashed line traces the mean of M^*.*

m.s.e. as τ increases, reaching an optimum around $\tau = 7$, and then gradually becoming worse than the m.l.e. as τ continues to increase. The latter fact was to be expected, given that, except for σ_1, the true θ and θ_0 do not coincide.

The middle panels of Figure 5.7 are similar, but based on the model

$$\mu_1 = 0, \quad \mu_2 = 1, \quad \sigma_1 = 1, \quad \sigma_2 = 1.5, \quad \lambda_1 = 0.3, \quad n = 100, \qquad \text{(experiment 2)} \qquad (5.21)$$

which we refer to as "experiment 2" because it was the second model entertained by Hamilton (1991) in a simulation exercise, as discussed in the next section. Now, the improvement is quite substantial, with the choice of $p = 1$ performing better than $p = 2$. Finally, the bottom panels are based on Hamilton's "experiment 4," given by

$$\mu_1 = 0.5, \quad \mu_2 = 1.5, \quad \sigma_1 = 0.5, \quad \sigma_2 = 1.5, \quad \lambda_1 = 0.5, \quad n = 50. \qquad \text{(experiment 4)} \quad (5.22)$$

The improvement is clear, but not as impressive as with the previous case, and there is mild preference for use of $p = 2$ instead of $p = 1$.

Shrinkage estimation is useful in a variety of settings, particularly when faced with a large number of parameters, and such that these parameters are related in terms of what they are capturing; the n elements of a vector of means is the classic example, and the application to the $\binom{n}{2}$ covariance terms in an n-dimensional variance–covariance matrix is also very popular (see, for example, Ledoit and Wolf, 2003, 2004; and Schäfer and Strimmer, 2005).

5.1.6 Quasi-Bayesian Estimation

We entertain a rather clever idea from Hamilton (1991) for estimating the mixed normal distribution with shrinkage, applicable in both the univariate and multivariate cases. It is also applicable to other distributions; see Paolella (2015a). This method (i) maintains the benefit of using the likelihood, (ii) helps avoid the singularities in a different way than, and independently of, the use of box constraints, (iii) can be used with the direct method or EM algorithm, and (iv) allows the researcher to explicitly incorporate prior information about the parameters into the estimation in such a way that is much more straightforward to interpret (they are in terms of fictitious observations coming from the prior distribution). Observe that the method of shrinkage estimation in (5.19) above possesses only features (i) and (ii).

As its name implies, it is not a proper Bayesian method (with specification of a prior distribution and use of Bayesian inferential tools), though it is considerably easier to implement than a genuine Bayesian analysis of the normal mixture model (see Evans et al., 1992; Robert, 1996; Roeder and Wasserman, 1997; and Richardson and Green, 1997).[7]

For the general k-component mixture, Hamilton (1991) proposed maximizing

$$\tilde{\ell}(\boldsymbol{\theta}; \mathbf{y}) = \ell(\boldsymbol{\theta}; \mathbf{y}) - \sum_{i=1}^{k} \frac{a_i}{2} \log \sigma_i^2 - \sum_{i=1}^{k} \frac{b_i}{2\sigma_i^2} - \sum_{i=1}^{k} \frac{c_i(m_i - \mu_i)^2}{2\sigma_i^2}, \tag{5.23}$$

where $\ell(\boldsymbol{\theta}; \mathbf{y})$ is the usual log-likelihood for the observations, and the m_i and $a_i, b_i, c_i \geq 0$, $i = 1, \ldots, k$, are fixed values indicating the prior information, as now described. We refer to the resulting parameter estimator as the quasi-Bayesian estimator (q-B.e.), denoted $\hat{\boldsymbol{\theta}}_{\mathrm{qB}}$. The prior serves to mimic the situation in which we have also observed n_i independent observations, $y_{i,1}, \ldots, y_{i,n_i}$, *known to have been drawn from the ith component*, $i = 1, \ldots, k$.

Let $\bar{y}_i$ and s_i^2 be the sample mean and variance of these hypothetical observations. The choices of $\bar{y}_i$ and s_i^2 indicate our prior on the μ_i and σ_i^2, while the choice of n_i signals the strength of that prior. (A genuine Bayesian analysis would take the prior of $\boldsymbol{\theta}$ to be a proper probability distribution and not just point values, as done here. However, one might argue that, realistically, a researcher might actually feel more comfortable specifying his or her beliefs in terms of the $\bar{y}_i$, s_i^2, and n_i than specifying marginal distributions for each of the parameters, let alone a more complicated multivariate distribution for $\boldsymbol{\theta}$.)

If the $y_{i,j}$ were really observed (and knowing from which component they came), then we would add to the usual log-likelihood $\ell(\boldsymbol{\theta}; \mathbf{y})$ the quantity

$$-\frac{n_i}{2} \log(2\pi) - \frac{n_i}{2} \log \sigma_i^2 - \frac{1}{2\sigma_i^2} \sum_{j=1}^{n_i} (y_{i,j} - \mu_i)^2,$$

---

[7] This idea of solving a problem that arises with likelihood- or frequentist-based inference but without a formal Bayesian approach is reminiscent of Fisher's fiducial inferential approach, about which Savage (1961, p. 578) remarked, in what is now an often-used quote, that it was "a bold attempt to make the Bayesian omelet without breaking the Bayesian eggs."

$i = 1, \ldots, k$; or, omitting the constant and rewriting,

$$-\frac{n_i}{2} \log \sigma_i^2 - \frac{n_i s_i^2}{2\sigma_i^2} - \frac{n_i(\bar{y}_i - \mu_i)^2}{2\sigma_i^2}. \tag{5.24}$$

Comparing (5.23) with (5.24), we see that $a_i = c_i = n_i$ represents the weight of the ith prior, with $m_i = \bar{y}_i$ its mean, and b_i/a_i its variance. The values a_i and c_i need not be the same, with a relatively smaller c_i corresponding to less information on μ_i than on σ_i^2; nor do they need to be integer, and can also be less than 1, so that we have only the weight contributed by a fraction of a direct observation. If $a_i, b_i, c_i > 0$, then $\tilde{\ell}(\theta; \mathbf{y}) \to -\infty$ when $\sigma_i \to 0$, thus serving to penalize the likelihood and prevent the optimization routine from settling in on a singularity.

The q-B.e. can be computed by directly maximizing (5.23) or modifying the EM algorithm as follows. The calculation of the conditional expectation of the H_{ij} is the same as before, given in (5.14). The parameters corresponding to the maximum of Q in (5.18) are then updated: the $\hat{\lambda}_j$ are as before, from (5.16), while (5.17) is modified to

$$\hat{\mu}_j = \frac{c_j m_j + \sum_{i=1}^{n} H_{ij} Y_i}{c_j + \sum_{i=1}^{n} H_{ij}}, \quad \hat{\sigma}_j^2 = \frac{b_j + \sum_{i=1}^{n} H_{ij}(Y_i - \hat{\mu}_j)^2 + c_j(m_j - \hat{\mu}_j)^2}{a_j + \sum_{i=1}^{n} H_{ij}}, \tag{5.25}$$

$j = 1, \ldots, k$, as given in Hamilton (1991, p. 29). Observe how these reduce to just the prior if there are no observations.

In his comparisons, Hamilton used the EM algorithm to compute the m.l.e., and discarded samples for which it settled on a singularity. In our comparisons, we use the direct method with box constraint (5.11b) both for the m.l.e. and for optimizing (5.23), so that the methods can be fairly compared.

His suggested prior information is

$$a_i = 0.2, \ b_i = 0.2, \ c_i = 0.1, \quad m_1 = m_2 = 0, \tag{5.26}$$

which serves to shrink both means towards zero, and both scale terms towards unity, and does so with a weight of (only) 0.1 or 0.2 observations. A simulation exercise based on model (5.21), computing both $\widehat{\theta}_{\mathrm{ML}}$ and $\widehat{\theta}_{\mathrm{qB}}$, with 1000 replications, yields an average $(\widehat{\theta} - \theta)'(\widehat{\theta} - \theta)$ of 3.90 and 1.43 for the m.l.e. and q-B.e., respectively, thus demonstrating a sizeable improvement from the latter. (The values of $(\widehat{\theta} - \theta)'(\widehat{\theta} - \theta)$ were computed the same as in (5.6), just without the log.)

Now consider the same shrinkage target (sometimes called the **shrinkage prior**) of zero means and unit variances, but now as a function of hyperparameter ω, with

$$a_1 = a_2 = b_1 = b_2 = c_1 = c_2 = \omega, \quad m_1 = m_2 = 0, \tag{5.27}$$

so that ω indicates the strength of the prior. We would expect improvements in the m.s.e. of the parameters as ω increases from zero, though because some of the true parameters are not the same as the prior ($\mu_2 \neq 0, \sigma_2 \neq 1$), it must be the case that, for large enough ω, the q-B.e. will perform worse than the m.l.e. Similarly to the illustration in Section 5.1.5 above, we investigate this via simulation using a grid of ω-values, ranging from 1 to 28 – these being vastly larger than the weights of 0.1 and 0.2 as in (5.26) – and based on 1000 replications. The top panel of Figure 5.8 is analogous to the middle panels of Figure 5.7, juxtaposing

```
1   n=100; mu=[0 1]; scale=[1 1.5]; lam=[0.3 0.7]; % experiment 2
2   true1=[mu scale lam(1)]'; true2=[mu(2) mu(1) scale(2) scale(1) lam(2)]';
3   priorvec=1:3:28; pl=length(priorvec); sim=1e3; disc=zeros(sim,pl+1);
4   for i=1:sim, i
5     y = mixnormsim(mu,scale,lam,n,i); ML = MixN2estimation(y);
6     dd1 = sum((ML-true1).^2); dd2 = sum((ML-true2).^2); disc(i,1) = min(dd1,dd2);
7     for j=1:pl
8        TH = MixN2estimation(y,[],7,priorvec(j)); dd1 = sum((TH-true1).^2);
9        dd2 = sum((TH-true2).^2); disc(i,j+1) = min(dd1,dd2);
10    end
11  end
12  disclog=log(disc); figure, lab={'MLE'};
13  for i=1:pl, str=['ω=',num2str(priorvec(i))]; lab = cat(2,lab,str); end
14  boxplot(disclog,'labels',lab), set(gca,'fontsize',16)
15  % adjust the font size and shape in the boxplot labels
16  h = get(get(gca,'Children'),'Children'); ht = findobj(h,'type','text');
17  set(ht,'fontsize',18,'Rotation',30)
18  set(ht,'Interpreter','latex','horizontalalignment','right')
19
20  yy=median(disclog(:,1)); ax=axis; xx=[ax(1) ax(2)]; yy=[yy yy];
21  line(xx,yy,'linestyle','-','color','g','linewidth',2)
22  title('Hamilton Experiment # 2, Shrinkage Prior'), ylim([-4.5 4.5])
23  hold on, plot(1:pl+1,mean(disclog),'b--','linewidth',2), hold off
```

Program Listing 5.3: Code used to produce the top panel of Figure 5.8.

the resulting set of boxplots of the m.s.e. measure M^* from (5.6). Just as with the shrinkage estimator (5.19), the extent of the improvement is fascinating, if not alarming. The code used to generate the plot is given in Listing 5.3.

The bottom panel of Figure 5.8 is similar, but uses experiment 4 in (5.22). For this model, neither of the two means or two variances coincide with the parameters of the shrinkage prior, so initially, one might expect that, for *any* $\omega > 0$, the q-B.e. will be inferior to the m.l.e. As we see from the bottom panel, this is far from the case. However, the value of ω for which the m.l.e. overtakes the q-B.e. is indeed much less than it is for experiment 2.

Recall that the shrinkage estimator adds bias (unless the shrinkage target happens to coincide with the true parameter) proportional to the strength of the shrinkage, but reduces the m.s.e. We investigate this further in the context of the q.-B.e. Figure 5.9(a) shows the bias for each of the five model parameters (e.g., the first boxplot is $\hat{\mu}_{1,\text{ML}} - \mu_1$) when using the model from experiment 2, with $n = 100$, using the m.l.e. (left half) and the q-B.e. with shrinkage prior (5.27) for $\omega = 10$ (right half). We used the rule that, if $M(\hat{\theta}, \theta^=) < M(\hat{\theta}, \theta)$, then $\hat{\theta}$ is converted to $\hat{\theta}^=$. We see that the m.l.e. in this case is quite biased, while the q-B.e. with $\omega = 10$ is also biased, but with a substantially smaller variance, yielding its much lower m.s.e. Figure 5.9(b) is the same except for having used $n = 10,000$ observations. With this larger sample size, the m.l.e. is virtually unbiased, yet is still inferior to the q-B.e., with its lower variance, even though ω remains at 10, which is now dwarfed by the sample size of $n = 10,000$.

5.1.7 Confidence Intervals

This section can be skimmed: The results are neither very good nor conclusive.

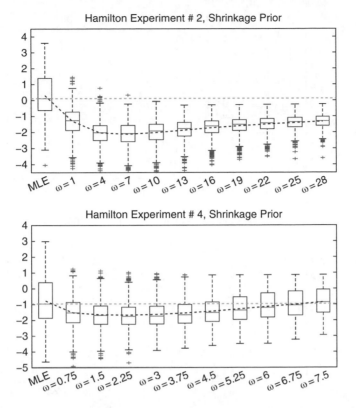

Figure 5.8 Comparison of total m.s.e. for $\hat{\theta}_{ML}$ (leftmost boxplot in both panels) and $\hat{\theta}_{qB}$ using, for the latter, prior (5.27) with varying strength ω. The simulation is based on 1000 replications. The horizontal dashed lines show the median m.l.e. value of M^* from (5.6). The other dashed line traces the mean of M^*.

So far, we have only addressed point estimators of θ. Interval estimation turns out to be more difficult, partially because of the need to address the observational equivalence (sometimes called the label switching) issue. We consider use of the asymptotic Wald interval and the bootstrap.

5.1.7.1 Use of Asymptotic Normality

Recall the Wald confidence interval (3.45), easily obtained as a by-product of direct maximum likelihood estimation. We begin with a new mixed normal parameter configuration, given by

$$\mu_1 = -2, \quad \mu_2 = 2, \quad \sigma_1 = 1, \quad \sigma_2 = 2, \quad \lambda_1 = 0.5. \quad \text{(well separated)} \quad (5.28)$$

It is valuable because the two density components are reasonably well separated and have equal weights. As such, we might expect to get accurate parameter estimates. Consider four estimators: the m.l.e.; the q-B.e. with Hamilton's suggested shrinkage prior (5.26), denoted qB(0.1) for simplicity; and the q-B.e.s with shrinkage prior (5.27) and strengths $w = 1$ and $w = 4$, denoted qB(1) and qB(4), respectively.

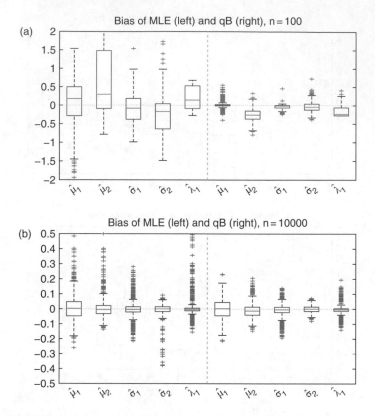

Figure 5.9   *Bias of the m.l.e. (left half of both panels) and q.B.e. (right half of both panels) based on $\omega = 10$, for two sample sizes $n = 100$ (a) and $n = 10,000$ (b), all based on 1000 replications.*

```
1  function V=MixNswitch(Theta)
2  V=zeros(size(Theta));
3  V(1)=Theta(2); V(2)=Theta(1); V(3)=Theta(4); V(4)=Theta(3); V(5)=1-Theta(5);
```

Program Listing 5.4: Switches between the two equivalent parameterizations of the $\text{Mix}_k N_1(\mu, \sigma, \lambda)$ parameter.

From a simulation exercise with $n = 100$ and based on 10,000 replications, the average of the total parameter squared error $(\hat{\theta} - \theta)'(\hat{\theta} - \theta)$ values, hereafter just m.s.e., was, for the four estimators just mentioned, 0.63, 0.55, 0.34, and 0.38, respectively, where, as always, these were computed the same as in (5.6), just without the log, that is, correctly accounting for the two possible parameter configurations. As expected, the m.s.e. values for the m.l.e. and qB(0.1) are close, with preference for the latter, while that for qB(4) is higher than that for qB(1), indicating that the optimal value of w is somewhere between 0.1 and 4 (assuming monotonic behavior of the m.s.e. with respect to ω). The Matlab code to perform the simulation (for one of the estimators) is given in Listing 5.5; it uses the short function in Listing 5.4.

```
1   n=100; mu=[-2 2]; scale=[1 2]; lam=[0.5 0.5]; % well separated
2   true1=[mu scale lam(1)]';  true2=MixNswitch(true1);  sim=1e4;
3   boolmat=zeros(sim,5);  disc=zeros(sim,4);
4   for i=1:sim,  if mod(i,1e3)==0, i,  end
5     y = mixnormsim(mu,scale,lam,n,i);
6     [Theta1,stderr1] = MixN2estimation(y,[],1); % the MLE
7     dd1 = sum((Theta1-true1).^2); dd2 = sum((Theta1-true2).^2);
8     disc(i,1) = min(dd1,dd2);
9     if dd2<dd1, Theta1=MixNswitch(Theta1); stderr1=MixNswitch(stderr1); end
10    for j=1:5
11      lo=Theta1(j)-1.96*stderr1(j); hi=Theta1(j)+1.96*stderr1(j);
12      boolmat(i,j) = (true1(j)>lo) && (true1(j)<hi);
13    end
14  end
15  coverage_asymp_mle=mean(boolmat), mean(disc)
```

Program Listing 5.5: Performs the simulation for computing the actual coverage of the 95% nominal c.i. via the asymptotic normal distribution for the $\text{Mix}_k N_1(\mu, \sigma, \lambda)$ models. The code for `MixNswitch` is given in Listing 5.4.

TABLE 5.1 Empirical coverage of one-at-a-time 95% c.i.s of mixed normal models (5.28) (left) and (5.4) (right) based on $n = 100$ observations

Estimator	Well-separated model					Showcase model				
	μ_1	μ_2	σ_1	σ_2	λ_1	μ_1	μ_2	σ_1	σ_2	λ_1
m.l.e.	0.90	0.85	0.89	0.86	0.87	0.92	0.90	0.85	0.84	0.84
qB(0.1)	0.92	0.87	0.92	0.87	0.90	0.93	0.91	0.86	0.86	0.86
qB(1)	0.95	0.95	0.90	0.92	0.95	0.94	0.94	0.00	0.00	0.00
qB(4)	0.77	0.99	0.87	0.98	0.98	0.96	0.92	0.93	0.77	0.94

The left half of Table 5.1 shows, for each of the four mixed normal models under study and for each of the $d = 5$ parameters, the fraction of the ten thousand 95% c.i.s that contained the true parameter. Those based on the m.l.e. are all under 0.95, showing that the c.i.s delivered with that method (and this model, sample size, etc.) are, on average, too short. As expected, the fractions for the c.i.s based on the m.l.e. and qB(0.1) are close, but those for the latter are, for all five parameters, better. In comparison, the c.i.s based on qB(1) are more accurate, while those for qB(4) are far less so, with some being far too short and others too long.

Based on what we have so far seen, we might (*very* cautiously) hypothesize that the optimal approximate standard errors are those based on the estimator within the class of shrinkage m.l.e.s that, on average, delivers the most accurate point estimates.

Let us do a second study, using our showcase model (5.4), that is, with $\mu_1 = 0.1$, $\mu_2 = -0.6$, $\sigma_1 = 1$, $\sigma_2 = 3$, and $\lambda_1 = 0.7$, again with $n = 100$. Now the m.s.e. values for the four estimators are 2.42, 1.86, 0.92, and 0.66. They are much higher than those for the well-separated normal mixture, showing that the parameters of the showcase model are (for the same sample size n) overall more difficult to estimate. Also, for this model, qB(4) performs the best, showing that, in comparison to the well-separated model, more shrinkage is required to obtain further accuracy. The right half of Table 5.1 shows the results for the

TABLE 5.2   Similar to Table 5.1 but for the experiment 2 and contaminated models

Estimator	Experiment 2 model					Contaminated model				
	μ_1	μ_2	σ_1	σ_2	λ_1	μ_1	μ_2	σ_1	σ_2	λ_1
m.l.e.	0.57	0.63	0.65	0.65	0.57	0.77	0.90	0.76	0.56	0.72
qB(0.1)	0.77	0.86	0.82	0.85	0.80	0.90	0.90	0.86	0.61	0.74
qB(1)	0.88	0.95	0.91	0.94	0.92	0.93	0.98	0.89	0.56	0.88
qB(4)	0.97	0.85	0.98	0.93	0.77	0.95	0.99	0.91	0.33	0.95

actual c.i. coverage at a 95% nominal level. As with the well-separated model, the actual coverages for all five parameters are better for qB(0.1) than for the m.l.e. All four models do a poor job with the c.i. for σ_2, which is the larger of the two σ_i and also the one associated with the lower component weight. If we could ignore the c.i. quality for σ_2, then qB(4) would be the best performer, though otherwise qB(1) leads to the overall most satisfactory coverage.

We repeated the same exercise for the experiment 2 and contaminated models, both using $n = 100$. For the former, the m.s.e. values of the four estimators were 3.87, 1.44, 0.39, and 0.17, respectively; for the latter, 14.33, 10.35, 5.36, and 4.32. Thus, in both cases, qB(4) is the best among the estimators entertained. Table 5.2 shows the corresponding actual c.i. coverage results.

For experiment 2, as with all models we have examined, the c.i.s from the m.l.e. are highly inaccurate, while those for qB(0.1) offer an (in this case, enormous) improvement over the m.l.e. Those for qB(1) are, similar to the other models, reasonably good and best overall, while those for qB(4) are, similar to the other models, more erratic. Virtually the same comments apply to the contaminated model. In this case, σ_2 is four times higher than σ_1 and is associated with a component with very low weight (0.05). Indeed, all estimators yield poor actual coverage for σ_2, the worst being from qB(4). However, qB(4) exhibits correct coverage for μ_1 and λ_1.

5.1.7.2 Use of the Bootstrap

In this setting, the percentile bootstrap can be used to construct approximate c.i.s for each of the model parameters (as well as more advanced bootstrap methods; see Remark (c) in Section 1.3.1). Unlike with use of the previous method for computing the c.i.s, which is essentially instantaneous and obtained as a by-product of the estimation via the direct method of maximizing the likelihood using the BFGS algorithm, we need to compute the estimator B times (for which we choose $B = 1000$). Thus, a simulation study like the one above will be quite time-consuming if the estimations are not fast.

The estimation times required for the m.l.e. and q-B.e., as well as each of the estimation methods discussed in Section 5.2 below, are reported in Table 5.3. Clearly, the method of moments estimator (m.m.e.), as detailed in Example 5.7 below, is the fastest, followed by direct m.l.e. and the q-B.e.

As in Section 1.3, for each of the B resamples from the actual data, $\mathbf{y}^{(b)}, b = 1, \dots, B$, we compute the point estimator $\hat{\theta}^{(b)}$ and store it. Of course, in reality, the true value of θ is not known, so, for each $\hat{\theta}^{(b)}$, we choose the model representation that is closer to the estimate from the actual data (say $\hat{\theta}_{ML}$), that is, if $M(\hat{\theta}^{(b)}, \hat{\theta}^=_{ML}) < M(\hat{\theta}^{(b)}, \hat{\theta}_{ML})$, then $\hat{\theta}^{(b)}$

TABLE 5.3   The time required to estimate 100 of the contaminated model data sets (5.7), each with $n = 100$, on a standard 3.2 GHz PC, and given in seconds unless otherwise specified. All methods using the generic optimizer are based on a convergence tolerance of 10^{-8}, while the EM algorithm used a convergence tolerance of 10^{-6}. The calculation of the direct m.l.e. is just denoted by m.l.e., whereas EM indicates the use of the EM algorithm, and q-B.e. denotes the quasi-Bayesian estimator with shrinkage prior and strength $w = 4$

Model	m.m.e.	m.l.e.	EM	q-B.e.	QLS	X_P^2	KD	m.g.f.	c.f.
Time	0.34	4.2	13.0	7.1	22 min	50	4.8	15	160

TABLE 5.4   Actual coverage of nominal one-at-a-time c.i.s based on the bootstrap, for four models and two estimation methods

Model	Based on m.m.e.					Based on m.l.e.				
	μ_1	μ_2	σ_1	σ_2	λ_1	μ_1	μ_2	σ_1	σ_2	λ_1
Well separated	0.88	0.82	0.96	0.55	0.80	0.96	0.96	0.96	0.92	0.97
Showcase	0.95	0.82	1.00	0.86	0.99	0.96	0.93	0.95	0.81	0.98
Experiment 2	0.97	0.64	0.94	0.46	0.82	0.98	0.95	0.97	0.81	0.99
Contaminated	0.87	0.62	0.91	0.42	0.91	0.89	0.81	0.86	0.30	0.81

TABLE 5.5   Similar to Table 5.4 but using the qB(1) and qB(4) estimation methods

Model	Based on qB(1)					Based on qB(4)				
	μ_1	μ_2	σ_1	σ_2	λ_1	μ_1	μ_2	σ_1	σ_2	λ_1
Well separated	0.88	0.96	0.96	0.92	0.96	0.97	0.96	0.94	0.90	0.98
Showcase	0.96	0.93	0.94	0.80	0.98	0.95	0.93	0.94	0.80	0.98
Experiment 2	0.95	0.92	0.94	0.86	0.98	0.95	0.93	0.94	0.87	0.98
Contaminated	0.95	0.85	0.86	0.30	0.74	0.95	0.87	0.85	0.30	0.75

is converted to $\widehat{\theta}^{(b),=}$ and stored in the bth row of a $B \times d$ matrix. When done, the sample 0.025 and 0.975 quantiles from each of the columns are computed, and we record whether that interval contains the true parameter value or not. This is repeated sim times. The code to do this with the m.m.e. is given in Listing 5.6. The code to perform the simulation with the bootstrap c.i.s using the m.l.e. or q-B.e. is similar but simpler: we do not need the while loop because those latter methods do not fail like the m.m.e. For the m.l.e. and q-B.e., we impose constraint 1 from (5.11b).

The results using the m.m.e. with sim $= 10{,}000$ and $B = 1000$, are shown in the left half of Table 5.4, for each of the four normal mixture parameter configurations under consideration (and always with $n = 100$). The right panel contains the results based on using the m.l.e. with sim $= 2000$, while those in Table 5.5 correspond to qB(1) and qB(4), respectively, also based on 2000 replications.

The coverage based on the m.m.e. is clearly unsatisfactory overall, though for some parameters in some of the models it works well. Overall, the coverage based on the m.l.e. and q-B.e. is somewhat better than those for the m.m.e., though not uniformly better. The coverage properties of qB(1) and qB(4) are very close, so there is little to choose

```
1   n=100; mu=[-2 2]; scale=[1 2]; lam=[0.5 0.5]; % well separated
2   true1=[mu scale lam(1)]'; true2=[mu(2) mu(1) scale(2) scale(1) lam(2)]';
3   sim=1e4; boolmat=zeros(sim,5); B=1000; thetamat=zeros(B,5);
4   for i=1:sim, if mod(i,100)==0, i, end
5      Theta1=[];
6      while isempty(Theta1)
7         y=mixnormsim(mu,scale,lam,n,0); Theta1=MixNmme(y)';
8      end
9      dd1 = sum((Theta1-true1).^2); dd2 = sum((Theta1-true2).^2);
10     if dd2<dd1, Theta1=MixNswitch(Theta1); end
11     Theta2=[Theta1(2) Theta1(1) Theta1(4) Theta1(3) 1-Theta1(5)]';
12     for j=1:B
13        Thetab=[];
14        while isempty(Thetab)
15           ind=unidrnd(n,[n 1]); yb=y(ind); Thetab = MixNmme(yb)';
16        end
17        dd1 = sum((Thetab-Theta1).^2); dd2 = sum((Thetab-Theta2).^2);
18        if dd2<dd1, Thetab=MixNswitch(Thetab); end
19        thetamat(j,:)=Thetab;
20     end
21     for j=1:5
22        lo=quantile(thetamat(:,j),0.025); hi=quantile(thetamat(:,j),0.975);
23        if (true1(j)>lo) && (true1(j)<hi), boolmat(i,j)=1; end
24     end
25  end
26  mean_boolmat_mme = mean(boolmat)
```

Program Listing 5.6: Performs the simulation for computing the actual coverage of the 95% nominal c.i. via the bootstrap for the mixed normal models. The code for MixNswitch is given in Listing 5.4 and that for mixnormsim is given in Listing 5.1. The function MixNmme is very easy to write based on the formulas in Section 5.2.1.

among them. It is not obvious how a meaningful summary measure should be devised, for example, whether the contaminated model should be discounted, or whether the intervals for σ_2 should be omitted in the comparison. Possibly, to get reasonably accurate c.i.s for all the parameters, the double bootstrap would be required. At least for the m.m.e., this would not be too time-consuming, and the reader is encouraged to investigate its accuracy.

5.2 ALTERNATIVE POINT ESTIMATION METHODOLOGIES

> Statistics is an applied science and deals with finite samples. Asymptotic theorems refer to limits which, by definition, are never attained. There is no mathematical demonstration that there is any method of estimation that assures that the estimate obtained is best in an operational sense in all circumstances, particularly not the method of maximum likelihood, which was abandoned by Gauss.
>
> (Joseph Berkson, 1980, p. 458)

While the m.l.e. and its variations form an important, if not the most important, class of estimators, we present a variety of alternative methods of point estimation that can be significantly better than the m.l.e. in terms of, say, m.s.e., for some situations, depending on the model, the true parameter, and the sample size. Also, because of potential numerical issues related with computing the m.l.e., such as those seen above in the mixed normal case, or when the likelihood itself is complicated to compute (as with, for example, the stable Paretian distribution), methods not sharing these negative features could be preferred in practice. Finally, alternative methods that are relatively fast to evaluate can be employed to obtain good starting values for use with the optimization methods used to compute the m.l.e.

The same numerical methods that we discussed in Chapter 4 for maximizing the likelihood can be used to minimize the objective function associated with an estimator of θ. In all cases with the mixed normal model and all estimators except the m.m.e., the (rather weak) constraint (5.11b) is imposed.

5.2.1 Method of Moments Estimator

The rth **raw sample moment** computed from sample $\mathbf{Y} = (Y_1, \dots, Y_n)$ is

$$\hat{\mu}_r'(\mathbf{Y}) := n^{-1} \sum_{i=1}^{n} Y_i^r, \qquad r = 1, 2, \dots, \tag{5.29}$$

abbreviated as $\hat{\mu}_r'$ when the sample from which it is computed is clear from the context.[8] When the Y_i are an i.i.d. sample with finite rth moment, it is an estimate of (what we will refer to as) the rth theoretical moment $\mu_r' = \mathbb{E}[Y^r]$. When $r = 1$, $\hat{\mu}_1'$ is referred to as the sample mean and commonly denoted by either $\bar{Y}$ or $\bar{Y}_n$, the latter explicitly including the sample size, and estimates the expectation of Y, which we denote by μ. It can be shown (see, for example, Stuart and Ord, 1994, p. 347) that, for $r \in \mathbb{N}$ and assuming the existence of the $2r$th moment,

$$\mathbb{E}[\hat{\mu}_r'] = \mu_r' \quad \text{and} \quad n\mathbb{V}(\hat{\mu}_r') = \mu_{2r}' - (\mu_r')^2. \tag{5.30}$$

The rth **central sample moment** is defined by

$$\hat{\mu}_r(\mathbf{Y}) = n^{-1} \sum_{i=1}^{n} (Y_i - \bar{Y})^r \tag{5.31}$$

and estimates $\mu_r = \mathbb{E}[(Y - \mu)^r]$. The sample variance is obtained by taking $r = 2$, although the unbiased version, usually denoted by S^2 or S_n^2, is more popular, and is obtained by dividing by $n - 1$ instead of by n, that is,

$$S_n^2 = \frac{n}{n-1}\hat{\mu}_2 = \frac{1}{n-1} \sum_{i=1}^{n} (Y_i - \bar{Y}_n)^2 = \frac{1}{n-1} \left(\sum_{i=1}^{n} Y_i^2 - n\bar{Y}_n^2 \right). \tag{5.32}$$

The variances of S_n^2 in the general and Gaussian cases are given in (A.209) and (A.210), respectively.

It is almost invariably the case that particular theoretical moments of a random variable are functions of the parameters of the underlying population distribution. Equating sample

---

[8] The notation using the prime symbol is quite standard and unfortunately coincides with the most popular notation for the first derivative, and vector and matrix transpose. Not only will the meaning of the prime symbol be clear from the context, but keep in mind that we use bold type for vectors and matrices.

and theoretical moments and solving for the unknown parameters is an intuitive and often simple way of obtaining point estimates. We refer to such an estimator as a **method of moments estimator** (m.m.e.), and denote the m.m.e. of the parameter θ by $\hat{\theta}_{\text{MM}}$. From (5.30), the variance of $\hat{\mu}'_r$ depends on the $2r$th moment and so, in most cases, will increase with r. Thus, it makes sense to use the lowest possible (integer) moments. For example, with one unknown parameter, we use $r = 1$; for two unknown parameters, we use $r = 1$ and $r = 2$, etc.

Example 5.1   Let $X_i \overset{\text{i.i.d.}}{\sim} \text{Bern}(p)$, $i = 1, \ldots, n$. Equating the sample mean $\bar{X}$ with the theoretical mean $p = \mathbb{E}[X]$ gives the m.m.e. $\hat{p}_{\text{MM}} = \bar{X}$. This agrees with the intuitive estimator discussed in Section 1.1.1 for the binomial model, and also coincides with the m.l.e., as derived in Example 3.4.

Example 5.2   Let $X_i \overset{\text{i.i.d.}}{\sim} \text{Geo}(p)$, $i = 1, \ldots, n$, where the X_i have support $\{1, 2, \ldots\}$. By equating the sample mean $\bar{X}$ with the theoretical mean $1/p$, the m.m.e. of p is $\hat{p}_{\text{MM}} = 1/\bar{X}$. This is also the m.l.e., derived in Example 1.1.

Example 5.3   Let $X_i \overset{\text{i.i.d.}}{\sim} \text{Exp}(\lambda)$, with $f_{X_i}(x; \lambda) = \lambda \exp(-\lambda x) \mathbb{I}_{(0,\infty)}(x)$, for $i = 1, \ldots, n$. As $\mathbb{E}[X] = 1/\lambda$, the m.m.e. of λ is $\hat{\lambda}_{\text{MM}} = 1/\bar{X}$. This coincides with the m.l.e. for λ, derived in Example 3.3.

Example 5.4   Let $X_i \overset{\text{i.i.d.}}{\sim} \text{Exp}(\mu, 1)$, $i = 1, \ldots, n$, that is, an n-length i.i.d. sample from an exponential distribution with scale 1 and location μ. As $f_{X_i}(x) = \exp\{-(x - \mu)\} \mathbb{I}_{(\mu,\infty)}(x)$, take $y = x - \mu$ so that

$$\mathbb{E}[X] = \int_{\mu}^{\infty} x e^{-x+\mu}\, dx = \int_{0}^{\infty} (y + \mu) e^{-y}\, dy = 1 + \mu,$$

and $\hat{\mu}_{\text{MM}} = \bar{X} - 1$. The m.l.e. in this case turns out to be $\min(X_i)$ and is superior to the m.m.e., with respect to m.s.e., for $n > 2$.

Example 5.5   Let $X_i \overset{\text{i.i.d.}}{\sim} \text{Lap}(\mu, 1)$, $i = 1, \ldots, n$, with $f_{X_i}(x) = \exp\{-|x - \mu|\}/2$. It is easy to see that $\mathbb{E}[X] = \mu$, so that $\hat{\mu}_{\text{MM}} = \bar{X}$. The m.l.e. is the sample median of the X_i, and has a lower m.s.e. for all $n > 2$. See also Problem 5.6, which constructs and compares the m.m.e. and m.l.e. for the location term.

The previous five examples all involved basic distributions, and their m.m.e.s all involved $\bar{X}$. The next example provides an exception and, even more interestingly, yields an estimator that outperforms the m.l.e. for small sample sizes.

Example 5.6   An easily computable expression for the rth absolute moment of the stable Paretian random variable $X \sim S_{\alpha,\beta}(0, 1)$ is given in (A.306), namely

$$\mathbb{E}[|X|^r] = \kappa^{-1} \Gamma\left(1 - \frac{r}{\alpha}\right)(1 + \tau^2)^{r/2\alpha} \cos\left(\frac{r}{\alpha} \arctan \tau\right), -1 < r < \alpha,$$

where $\tau = \beta \tan(\pi\alpha/2)$; and $\kappa = \pi/2$ for $r = 1$ and $\kappa = \Gamma(1 - r)\cos(\pi r/2)$ otherwise. Note that, for $\beta = 0$, this reduces to $\mathbb{E}[|X|^r] = \Gamma(1 - r/\alpha)/\kappa$.

```
1   function ahat=stablealphamme(data,m)
2   if nargin<2, m=1; end
3   s=mean(abs(data).^m); tol=1e-4; opt = optimset('disp','none','TolX',tol);
4   lo=fff(m+0.001,s,m); hi=fff(2,s,m);
5   if lo*hi>=0
6     ahat=NaN;
7   else
8     ahat = fzero(@(a) fff(a,s,m), [m+0.001, 2], opt);
9   end
10
11  function d=fff(a,s,m), d = s - stabmom(m,a,0);
```

Program Listing 5.7: Solves a moment equation for α in the symmetric stable model, assuming the data are $X_1, \ldots, X_n$, where X_i i.i.d $\sim S_{\alpha,\beta}(\mu, c)$, with $\mu = \beta = 0$, $c = 1$. The program first checks whether the objective function crosses zero; if not, an NaN (denoting not-a-number in Matlab) is returned. The function `stabmom` is given in Listing II.8.2.

In many applications, such as modeling the distribution of daily financial asset returns, the data are such that the empirical distribution (inspected from a histogram or kernel density) is quite leptokurtic, but reasonably close to being symmetric about zero. Thus, the location parameter μ and asymmetry parameter β will both be close to zero, and only the tail index α and the scale c need to be estimated. For the demonstration here, we further assume that the scale parameter is known and, without loss of generality, use $c = 1$. Then we can numerically solve the moment equation to get the m.m.e. of α using a value of r that we know to be less than α, recalling that moments of order α and higher do not exist for the stable distribution when $0 < \alpha < 2$.

Assuming $1 < \alpha \leq 2$ (realistic for most data), we use $r = 1$. The code for doing this is shown in Listing 5.7. The m.s.e. of the resulting estimator, denoted by $\hat{\alpha}_{MM}$, is determined via simulation over a grid of α-values, and the result is shown in Figure 9.10 (on page 344) where the m.s.e.s for several estimators are compared. We see that, for a sample size of $n = 50$, the m.m.e. has the lowest m.s.e. of all the estimators for the sizeable and important range $1.5 < \alpha < 1.85$. For $n = 500$, the m.m.e. is strongly dominated by two of the other estimators across all values of α. As might have been anticipated, by decreasing r, the m.s.e. decreases for values of α near 1, but increases for values closer to 2.

Estimating the tail index α and the scale parameter c simultaneously is more interesting, and is considered in Problem 5.1.

The next example also involves a less trivial distribution, the two-component mixed normal, as introduced above, and involves not one, but five unknown parameters.

Example 5.7   *The m.m.e. for a two-component mixed normal was first addressed by Karl Pearson in a paper published in 1894.[9] He showed that the estimates are given via the solution to a ninth-degree polynomial – the solution of which, in his day, was far from trivial. Cohen (1967), building on that and subsequent work, provided a considerable*

---

[9] The title of his paper, "Contributions to the mathematical theory of evolution", clearly reveals his intended area of application. This is perhaps not so surprising, given that Pearson was Sir Francis Galton's protégé, and Galton was Charles Darwin's half-cousin. Interestingly, mixture distributions have come to play a prominent role in biology, genetic studies, and medicine; see, for example, Schlattmann (2009).

simplification, allowing for straightforward computation of the estimator, as well as detailing the useful special cases when the variances are equal, when symmetry holds $(\lambda_1 = 1/2, \sigma_1 = \sigma_2)$, *and when the means are equal. For the two-component case, we can let* $\lambda = \lambda_1$. *Assume we have a sample* $y_1, \ldots, y_n$. *Define* $v_j = \sum_{i=1}^{n} (y_i - \bar{y})^j$, *for* $j = 2, \ldots, 5$, $k_4 = v_4 - 3v_2^2$, *and* $k_5 = v_5 - 10v_2v_3$.

We begin with the case in which the means are constrained to be equal, with $\mu_1 = \mu_2 =: \mu$. *There exists a potential solution when* $k_4 > 0$. *In particular, with*

$$k_6 = v_6 - 15v_4v_2 - 10v_3^2 + 30v_2^3 \quad \text{and} \quad t_1, t_2 = \frac{1}{2}\left(\frac{k_6}{5k_4} \mp \sqrt{\left(\frac{k_6}{5k_4}\right)^2 + \frac{4k_4}{3}} \right),$$

Cohen (1967) showed that, if $t_1 + v_2 > 0$, *then*

$$\hat{\sigma}_i^2 = t_i + v_2, \quad \hat{\lambda} = \frac{t_2}{t_2 - t_1}, \quad \hat{\mu} = \bar{y}.$$

To compare the performance of this m.m.e. with the m.l.e., we use the contaminated normal model (5.7). For $n = 100$, *we required 1493 draws to amass 1000 cases such that the restricted m.m.e. existed. The performances of the m.l.e. calculated via the direct method, the m.l.e. from the EM algorithm, the restricted m.m.e., and, for the 824 out of the 1000 cases in which it existed, the unrestricted m.m.e. (as discussed below) are compared by juxtaposing boxplots of* M^* *from (5.6) for each, as shown in Figure 5.10(a). We see that, when it exists, the restricted m.m.e. performs much better than the m.l.e. The unrestricted m.m.e. also outperforms the m.l.e., though obviously cannot do as well as the restricted m.m.e. Of course, if the researcher is indeed sure that the means* μ_1 *and* μ_2 *are (close enough to be deemed) equal, then the m.l.e. (calculated via the direct method or the EM algorithm) can be easily computed incorporating the constraint that* $\mu_1 = \mu_2$, *and will perform better than its unrestricted counterpart.*

The perhaps unexpectedly good performance of the m.m.e. (restricted and unrestricted) in this case is actually in line with the intuition of Johnson (1978), who argued that, as the m.l.e. is a function of the moments of the data when $\lambda = 0$, *one might expect the m.m.e. to perform particularly well when* λ *is close to zero (or one). The superiority of the restricted (but not the unrestricted) m.m.e. continues to hold even for a sample size of* $n = 1000$, *as shown in Figure 5.10(b), though the difference with the m.l.e. is less. For this sample size, the restricted m.m.e. existed in 100% of the cases (86% for the unrestricted).*

Regarding the two methods of calculating the m.l.e., note that the EM algorithm performs slightly better than the direct method of calculating the m.l.e. for the $n = 100$ *case. However, this is not due to a better performance of the EM algorithm per se, but rather because, in 27% of the cases, the EM algorithm tended to a singularity in the likelihood, resulting in one of the* σ_i *being too close to zero. We recovered from this problem by removing the most extreme observation from the data set. In 10 out of the 1000 cases, two such observations had to be removed. This singularity issue would surely have been severe in the direct m.l.e. method if it were not for the imposed box constraints. For* $n = 1000$, *the EM algorithm never failed.*

For the general case, let v *be the solution to* $0 = \sum_{i=0}^{9} a_i v^i$, *where* $a_0 = -24v_3^6$, $a_1 = -96v_3^4 k_4$, $a_2 = -63v_3^2 k_4^2 - 72v_3^3 k_5$, $a_3 = 288v_3^4 - 108v_3 k_4 k_5 + 27k_4^3$, $a_4 = 444k_4 v_3^2 - 18k_5^2$,

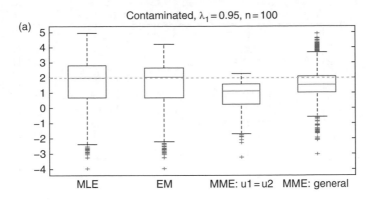

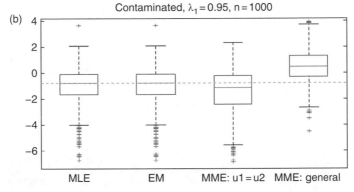

Figure 5.10 *(a) For the contaminated normal model (5.7) with $n = 100$, measure M^* from (5.6) for the m.l.e. computed via the direct method (denoted MLE), the m.l.e. computed via the EM algorithm (denoted EM), the m.m.e. restricted to have equal means, and the unrestricted m.m.e. for the 824 out of 1000 data sets for which the unrestricted (and restricted) m.m.e. existed. The horizontal dashed line shows the median m.l.e. value of M^*. (b) Same, but using sample size $n = 1000$.*

$a_5 = 90k_4^2 + 72k_5v_3$, $a_6 = 36v_3^2$, $a_7 = 84k_4$, $a_8 = 0$ *and* $a_9 = 24$. *Cohen (1967) showed that only real negative roots lead to solutions. With*

$$r = \frac{-8v_3v^3 + 3k_5v^2 + 6v_3k_4v + 2v_3^3}{2v^4 + 3k_4v^2 + 4v_3^2v},$$

$m_1, m_2 = (r \mp \sqrt{r^2 - 4v})/2$, *and* $\beta = (2r - v_3/v)/3$, *the m.m.e. is given by*

$$\hat{\mu}_i = m_i + \bar{y}, \quad \hat{\sigma}_i^2 = m_i\beta + v_2 - m_i^2, \quad \hat{\lambda} = \frac{m_2}{m_2 - m_1}.$$

When there is more than one negative real root, v, choosing the one for which the sixth central sample and theoretical moments are closest has been suggested. As the sixth (central) sample moment has a very high sampling variance, we instead take the solution with the largest likelihood value (a luxury that was not available to Cohen).

For showcase model (5.4) and $n = 100$, we required 1236 draws to amass 1000 cases with an m.m.e., so that it exists about 81% of the time for this model and sample size.

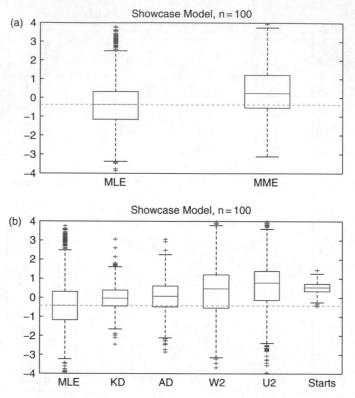

Figure 5.11 (a) For showcase model (5.4), measure M* from (5.6) for the m.l.e. and the m.m.e., using n = 100, and based on 1000 replications. (b) Same, but using the four goodness-of-fit measures in Section 5.2.2.

Figure 5.11(a) is similar to those in Figure 5.10. In this case, the m.l.e. is clearly superior.

The last special case is when $\sigma_1 = \sigma_2 = \sigma$ and $\mu_1 \neq \mu_2$, which is quite common in genetics (Roeder, 1994; Mendell et al., 1993) and other disciplines (Titterington et al., 1985, Ch. 2). For this, one takes v to be the negative real root of $2v^3 + k_4v + v_3^2 = 0$, which exists and is unique when $v_3 \neq 0$. Then, with $r = -v_3/v$, $\hat{\mu}_1$, $\hat{\mu}_2$, and $\hat{\lambda}$ are obtained as in the general case and $\hat{\sigma}^2 = v + v_2$. Redner and Walker (1984) discuss the m.m.e. when $k > 2$, while Lindsay and Basek (1993) proposed an easily computed, consistent and efficient m.m.e. for the multivariate mixed normal case with the same covariance matrix in each component.

5.2.2 Use of Goodness-of-Fit Measures

It would seem natural to consider an estimator that minimizes some function of the absolute discrepancy between the empirical and fitted c.d.f.s, for which the KD and AD statistics from Section 2.3, and the W^2 and U^2 statistics from Section 2.5, suggest themselves. This can be operationalized by using the same optimization routine we used for maximizing the likelihood to get the m.l.e., but instead we simply replace the objective function by one of the goodness-of-fit measures and minimize it.

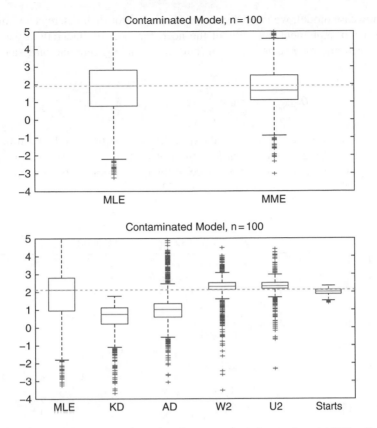

Figure 5.12 *Same as Figure 5.11, but using the contaminated normal model (5.7) with n = 100.*

To assess the performance, we use our showcase normal mixture model (5.4) in a simulation study, using $n = 100$ and based on 1000 replications. There were no numeric problems with any of the four goodness-of-fit estimators, and we see from Figure 5.11(b) that they all are inferior to the m.l.e., with the KD performing the best among the goodness-of-fit estimators. For comparison, the rightmost boxplot corresponds to the chosen starting values. Figure 5.12 is similar, but having used the contaminated model (5.7), for $n = 100$. Now the KD and AD estimators are highly superior to the m.l.e. (as are our arbitrary starting values), with preference for the KD.

Other related measures could also be entertained, such as minimizing not the maximum of $|\widehat{F}_{\text{emp}}(y) - F(y; \widehat{\theta})|$ as with the KD, but rather

$$\int_{-\infty}^{\infty} \{\widehat{F}_{\text{emp}}(y) - F(y; \widehat{\theta})\}^2 \, dy, \tag{5.33}$$

as was considered for the normal mixture model by Clarke and Heathcote (1994).

5.2.3 Quantile Least Squares

As a sort of "dual" to the use of the empirical and fitted c.d.f.s, one could use the empirical and fitted quantile function. In light of the poor performance of KD and AD in this regard

for the showcase model, we instead consider a measure of their discrepancy more in line
with (5.33). To implement this without the need for numeric integration, we choose an
equally spaced grid of q values between (but not including) zero and one, say $g_1, \ldots, g_q$,
and take

$$\hat{\theta}_{\text{QLS},q} = \arg\min_\theta \sum_{i=1}^{q} \{F_Y^{-1}(g_i; \theta) - \hat{F}_{\text{emp}}^{-1}(g_i)\}^2, \tag{5.34}$$

where $Y \sim \text{MixN}(\mu, \sigma, \lambda)$. We call this the **quantile least squares** (q.l.s.) estimator (not to
be confused with the method of *quantile regression*). Figure 5.13 is similar to Figure 5.11,
illustrating in the top panel the performance of the q.l.s. estimator for several values of q.
There is little change in performance after $q = 52$.

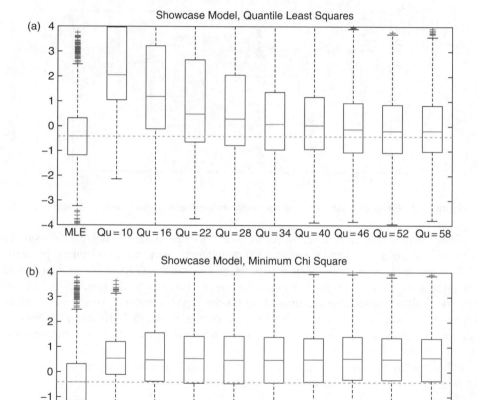

Figure 5.13 *(a) Same as Figure 5.11 but using the m.l.e. and q.l.s. estimators for several values of q,
applied to the showcase model (5.4) for $n = 100$. (b) Same, but for the X_p^2 estimator for several values
of m.*

The benefit of q.l.s. compared to maximum likelihood is that, while the likelihood needs to be computed for each candidate of θ in the numeric optimization, we only need to compute the empirical quantiles of the data set *once*. Of course, during the optimization with q.l.s., the quantiles of the distribution corresponding to each candidate θ need to be computed, which might turn out to take longer than evaluation of the likelihood. In principle, however, the desired set of quantiles could be pre-computed for a large array of θ-values and stored, so that the optimization with a particular data set amounts to finding that θ from the table that yields the smallest performance measure, say $(\hat{\theta} - \theta)'(\hat{\theta} - \theta)$. Obviously, if the parameter space is an uncountably infinite subset of $\mathbb{R}^k$, then not all values of θ can be tabulated, but nor is it necessary for the method to be useful: the desired coarseness of the grid of tabulated θ-values will be a function of several factors, including (i) the necessary accuracy of the estimator (which itself will depend on the application), (ii) available computer memory, (iii) the speed of table lookup and the desired speed of the estimation process.

This idea of tabulating the quantiles will be of potential use for (i) models whose likelihood is potentially plagued with several local optima; (ii) models such that the calculation of the likelihood is slow because it is numerically costly to evaluate and/or that use a large set of data, so that repeated calculation of the likelihood is slow; or (iii) models for which numeric likelihood optimization is potentially numerically problematic (e.g., mixture distributions, with their singularities). Moreover, there are applications in which the estimation procedure is not done just once, but many times, so that estimation speed becomes crucial. Examples include any problem in which the bootstrap or cross-validation is used, or in time series models, in which a moving window of observations is used in backtest forecasting exercises.

Section 9.3 illustrates a useful case in point for estimating the shape parameters of the noncentral Student's t distribution. In that case, evaluation of the likelihood is very slow, but use of the q.l.s. with pre-tabulated quantiles performs admirably, and is nearly instantaneous.

5.2.4 Pearson Minimum Chi-Square

Pearson's X_P^2 test is formally introduced in Section 6.5.4 in the context of assessing the goodness of fit of a model for categorical data. It works by summing $(O_i - E_i)^2/E_i$, where O_i and E_i are the observed and expected number of observations in the ith category, respectively. We can use the same principle in reverse, and take as a point estimator that value which minimizes X_P^2. That is, we take

$$\hat{\theta}_{X_P^2,m}(\mathbf{y}) = \arg \min_\theta \sum_{i=1}^m \frac{\{O_i(\mathbf{y}) - E_i(\theta)\}^2}{E_i(\theta)}, \tag{5.35}$$

where m is the number of categories, $\mathbf{y}$ is the observed data set, and E_i is written as a function of θ to indicate its dependence on the parameter vector. We call this the X_P^2 estimator, where the P reminds us that it is the original form proposed by Pearson. Since then, several other forms of the estimator have been proposed, such as using O_i instead of E_i in the denominator; see, for example, Berkson (1980) for a list of these. More interestingly, Berkson (1980), and the comments that accompany his paper, provide some very educational, lively, and heated discussion of the merits and pitfalls of the X_P^2 estimator versus the m.l.e.

As a basic demonstration, recall the Bernoulli model we studied in Section 1.1.1, in which we observe s out of n successes. Thus, with $O_0 = O_0(s) = n - s$ and $E_0 = E_0(p) = n(1 - p)$ denoting the observed and expected number of failures, respectively, and similarly for O_1 and E_1, the Pearson minimum χ^2 estimate is given by

$$\hat{p}_{X_P^2} = \arg\min_p \left(\frac{(O_0 - E_0)^2}{E_0} + \frac{(O_1 - E_1)^2}{E_1} \right)$$

$$= \arg\min_p \left(\frac{(n - s - n(1 - p))^2}{n(1 - p)} + \frac{(s - np)^2}{np} \right) = \arg\min_p \frac{(s - np)^2}{n(1 - p)p}.$$

The usual calculus exercise then reveals that $\hat{p}_{X_P^2} = s/n$, which agrees with the m.m.e. and m.l.e.

For our showcase mixed normal model, or for an application to any continuous model, the implementation of the X_P^2 method requires the data to be grouped or binned. We restrict ourselves to equal-size bins, such that the lower endpoint of the leftmost bin is $\min(\mathbf{y}) - 0.1$, and the upper endpoint of the rightmost bin is $\max(\mathbf{y}) + 0.1$, where $\mathbf{y}$ is the observed data sample. Use of the X_P^2 estimator for the $\text{Mix}_2 N_1$ is mentioned in Cohen (1967, p. 21), while Fryer and Robertson (1972) have investigated it in more theoretical detail, showing that it is generally more accurate than the m.m.e., and comparable to estimation via maximum likelihood with grouped data.

During the optimization, it is possible that values of θ far from the optimal solution will be tried and such that one or more of the E_i will be zero or very small. We replace those E_i less than $\epsilon = 0.0001$ with ϵ. This was found to be effective both in preventing failure of the routine (via division by zero), and in serving as enough of a penalty to encourage the optimizer to "look elsewhere." Note that this estimator also shares the feature of q.l.s. in which the E_i could be stored for a large array of θ vectors.

Figure 5.13(b) shows our usual boxplots comparing the performance of the m.l.e. and the X_P^2 estimators for several values of m. We see that its performance is not particularly good compared to the m.l.e. or q.l.s. estimator. A possible way of improvement is to use shrinkage estimation from Section 5.1.5. In particular, note that, unlike the specific method of the q-B.e., with shrinkage prior (5.26), we can use the shrinkage term (5.20) with any estimator. In this case, we use a generalization of (5.35),

$$\hat{\theta}_{X_P^2,m,\tau}(\mathbf{y}) = \arg\min_\theta \left\{ \sum_{i=1}^m \frac{\{O_i(\mathbf{y}) - E_i(\theta)\}^2}{E_i(\theta)} + \tau P(\theta, p) \right\}, \tag{5.36}$$

with $P(\theta, p)$ given in (5.20) and $\tau \geq 0$ dictating the strength of the shrinkage of both means towards zero and both scale terms towards one.

We use (5.36) with $m = 16$ bins and power $p = 2$, for a range of τ-values, on our showcase model (5.4) and contaminated model (5.7) with $\lambda_1 = 0.90$. The results are in Figure 5.14, and clearly show that the method not only improves dramatically on the usual $\tau = 0$ case, but also demonstrably beats the m.l.e. It appears that the choice of $\tau = 0.01$ is nearly optimal in both models.

A similar procedure can also be used with the q.l.s. estimator. We did not pursue it because its calculation takes considerably longer than that for the X_P^2 estimator.

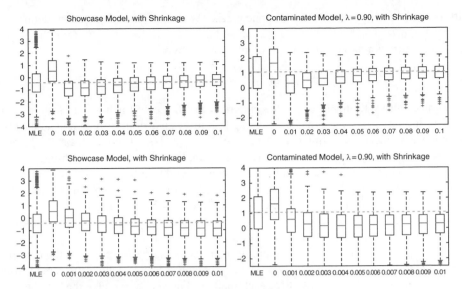

Figure 5.14 *Left: Similar to Figure 5.13a, using the X_p^2 estimator applied to our showcase model (5.4),* sim *= 1000 replications and sample size n = 100, for fixed number of bins m = 16, and penalized according to (5.36) for p = 2 and two sets of τ-values (top and bottom).* **Right**: *Same, but for the contaminated normal model (5.7), with λ_1 = 0.90,* sim *= 1000, n = 100.*

5.2.5 Empirical Moment Generating Function Estimator

Recall from Section A.7 that the m.g.f. of r.v. X is given by $\mathbb{E}[e^{tX}]$, and is finite for all t in its convergence strip. If the convergence strip contains an open interval around zero, then all the positive moments of $|X|$ exist. With this in mind, it might seem valuable to use it for estimation, because it embodies *all* positive moments, as opposed to just choosing one or more particular moments for estimation, as is used with the m.m.e. This can be operationalized by choosing a set of values of t in the convergence strip, say $t_1, \ldots, t_m$, $m \geq d$, and searching for the value of $\boldsymbol{\theta} \in \mathbb{R}^d$ that minimizes the sum of the m squared or absolute distances between the empirical m.g.f. $n^{-1} \sum_{i=1}^{n} \exp(t_j x_i)$ and its theoretical counterpart $\mathbb{E}[\exp\{t_j X\}]$, $j = 1, \ldots, m$. The performance of the method will depend on the number and choice of the $t_1, \ldots, t_m$, the optimal values of which might depend on the true but unknown model parameters.

A similar approach can be taken by using the characteristic function (c.f.) instead of the m.g.f. While the two methods are clearly related, we consider use of the c.f. in the next section, because the fact that $\mathbb{E}[e^{itX}]$ exists for all t allows for a different method of estimation.

It appears that Press (1972) and Paulson et al. (1975) were the first to use the empirical c.f. for estimation (in the context of the stable distribution), while in a 1976 conference presentation Kumar, Nicklin, and Paulson used the empirical c.f. to estimate a mixture of normals (see Kumar et al., 1979, p. 52). Apparently unaware of the aforementioned work, Quandt and Ramsey (1978) proposed and studied the use of the empirical m.g.f. for estimating the parameters of the mixed normal distribution. The m.g.f. of the two-component mixed normal is

$$\mathbb{M}_X(t; \boldsymbol{\theta}) = \mathbb{E}[e^{tX}] = \lambda_1 \exp\left(\mu_1 t + \frac{\sigma_1^2 t^2}{2}\right) + \lambda_2 \exp\left(\mu_2 t + \frac{\sigma_2^2 t^2}{2}\right),$$

with obvious extension to the k-component case. With x_i the ith observation, define $\epsilon_j = n^{-1} \sum_{i=1}^{n} \exp(t_j x_i) - \mathbb{M}_X(t_j; \boldsymbol{\theta}), j = 1, \dots, m$, and let $\epsilon = (\epsilon_1, \epsilon_2, \dots, \epsilon_m)'$. Then the empirical m.g.f. estimator based on m m.g.f. points is given by

$$\hat{\boldsymbol{\theta}}_{\text{MGF},m} = \arg \min_{\boldsymbol{\theta}} \epsilon' \epsilon. \tag{5.37}$$

As pointed out by Kiefer (1978a), the empirical m.g.f. method is a generalization of the ordinary m.m.e. in the sense that, via the m.g.f., it makes use of information in all the positive integer moments, recalling that $\mathbb{M}_X(t) = \sum_{k=0}^{\infty} t^k \mathbb{E}[X^k]/k!$. The weights $t^k/k!$ decrease rapidly, which is appropriate given the large sampling variance of the higher integer moments, whereas the m.m.e. simply places equal weight on the first five moments. The choice of the t_j can thus be interpreted as choosing how one wishes to weight the moments. Quandt and Ramsey (1978) recommend taking the t_j to be $(-0.2, -0.1, 0.1, 0.2, 0.3)$, noting that values too large (say $|t_j| > 0.75$) cause numeric overflow, while values too small ($|t_j| < 0.04$) result in ϵ_j being uninformative.

Via simulation, using our usual setup, we can assess the performance of the empirical m.g.f. method as a function of m. As with all methods in this section, the same starting values, given in (5.12), were used for each method. For $m = 5$, we take the t_j to be those suggested by Quandt and Ramsey (1978) given above, otherwise, we use an equally spaced grid of points between -0.3 and 0.3 (which will not include the value zero for even values of m; in Matlab, use `linspace(-0.3,0.3,m)` to construct this). Figure 5.15(a) shows the results. The method performs very well, and does best for $m = 5$. The bottom panel (b) is similar, but having used the model associated with experiment 2 from (5.21). In this case, $\hat{\boldsymbol{\theta}}_{\text{MGF},5}$ is blatantly superior to $\hat{\boldsymbol{\theta}}_{\text{ML}}$. It is worth repeating that the same starting values (5.12) were used for both the m.l.e. and the empirical m.g.f. estimator.

It turns out that the performance of the empirical m.g.f. is far more dependent on the convergence criteria chosen for the optimization routine, compared to the other estimation methodologies. In particular, as there are five parameters to be estimated, if we use $m = 5$, then minimizing $\epsilon' \epsilon$ can be thought of as solving five nonlinear equations in five unknowns, so that convergence occurs when $\epsilon' \epsilon$ is arbitrarily close to zero (as opposed to just being minimized, which is the case when $m > 5$). The results we show were obtained by requesting the optimization routine to declare convergence when the changes in the objective function, $\epsilon' \epsilon$, were less than 10^{-8}. (This was done for the m.l.e. as well, though the results are indistinguishable from those with use of 10^{-4}.) This is quite an extreme tolerance that may not work for some problems in practice because the objective function (usually the log-likelihood) cannot be evaluated to such high precision for certain reasons. For this simple model, the log-likelihood is easily evaluated to machine precision, as is the objective function associated with the empirical m.g.f., and so in this case such an extreme tolerance is feasible.

When using a tolerance of "only" 10^{-4} (a value that is otherwise considered fully adequate for many problems, and precisely the value used by Quandt and Ramsey, 1978, in their simulation study, as stated in their footnote 6), we find that the quality of the method suffers, particularly for $m = 5$. Worse yet, when using a tolerance of 10^{-4}, $\hat{\boldsymbol{\theta}}_{\text{MGF},5}$ is highly influenced by the starting value of $\boldsymbol{\theta}$. Thus, if we use the m.l.e. as the starting value, $\hat{\boldsymbol{\theta}}_{\text{MGF},5}$ is virtually the same as the m.l.e., and, if we use the *true value of* $\boldsymbol{\theta}$, then the m.s.e. boxplots indicate that $\hat{\boldsymbol{\theta}}_{\text{MGF},5}$ is clearly superior to the m.l.e.!

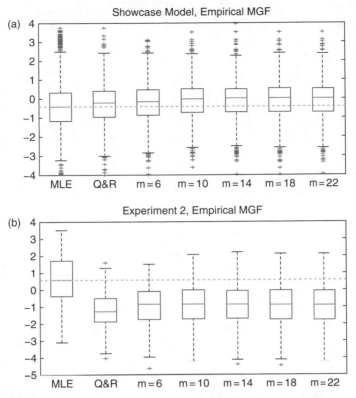

Figure 5.15 *(a) Same as Figures 5.11 and 5.13, but using the m.l.e. and the empirical m.g.f. estimator (5.37) for several values of m. Q&R denotes the use of m = 5, with the t_j being those suggested by Quandt and Ramsey (1978). (b) Same, but using the model for experiment 2 in (5.21).*

This finding presumably explains, and significantly tempers, if not invalidates, the quite favorable results reported by Quandt and Ramsey (1978), who used the *true parameter values as starts*, and the 10^{-4} tolerance. To exacerbate the problem, they did not enforce *any* box constraints, and instead just discarded any data set for which the optimization routine attempted values of λ_1 outside of $(0, 1)$ or if $\sigma_i \leq 0$, $i = 1, 2$. Moreover, they also discarded the data set if the m.m.e. did not yield a solution. This massive **sample selection bias** results in only "very well-behaved" data sets that are possibly not representative of the true sampling distribution. To add to the list of problems (no doubt owing to the limited computational feasibility at the time), they did this until amassing (only) 50 data sets with both an m.m.e. and the empirical m.g.f. estimator.

Remark.   Realizing that the ϵ_j are correlated and have unequal variances, Schmidt (1982) proposed correcting the objective function to account for this, yielding what he deems the modified m.g.f. estimator. One takes $\widetilde{\theta}_{\text{MGF},m} = \arg\min_\theta \epsilon' \Omega^{-1} \epsilon$, where Ω was actually derived in, but not used by, Quandt and Ramsey (1978). Its (i,j)th element is given by $\mathbb{M}_X(t_i + t_j; \theta) - \mathbb{M}_X(t_i; \theta)\mathbb{M}_X(t_j; \theta)$. Disappointingly, this idea actually led to a slight decrease in performance.   ∎

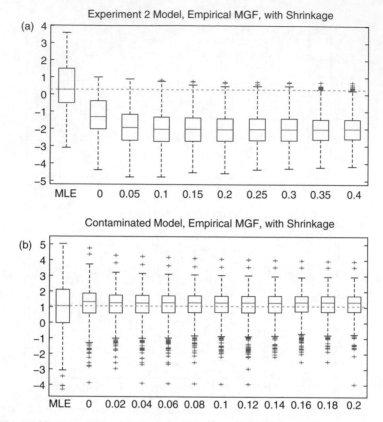

Figure 5.16   *(a) Similar to Figure 5.15(b), but using the empirical m.g.f. estimator, with $m = 5$, with shrinkage, for $p = 2$ and a set of shrinkage values τ, as in (5.36), and based on 1000 replications. The x-axis gives the value of τ times 10^6, that is, the values of τ are very close to zero. (b) Same, but for the contaminated normal model (5.7), with $\lambda_1 = 0.90$, $n = 100$.*

The objective function (5.37) can also be extended to support the shrinkage method as used in (5.36). The choice of weight τ in (5.36) needs to be far lower than what was used with the X_P^2 estimator. Figure 5.16 shows the results, based on $m = 5$ and, for the shrinkage, $p = 2$ and a set of τ-values given by the value indicated on the x-axis, multiplied by 10^{-6}. There is a substantial improvement with the experiment 2 model, for which the empirical m.g.f. estimator without shrinkage already significantly improves upon the m.l.e. For the contaminated model, there is little improvement, even though various values of τ, other than the ones shown, were also tried.

5.2.6 Empirical Characteristic Function Estimator

Even though both the c.f. and the m.g.f. carry, in principle, the same information about the distribution of the underlying random variable, Kumar et al. (1979) discuss why the use of the empirical c.f. should be superior to the m.g.f. One obvious reason is the fact that the c.f. is bounded (see Section II.1.2.3), so that *any* value of t_j could be used, as opposed to use of the m.g.f., whereby numerical instability can arise as $|t_j|$ increases. Once any value can be used, one can contemplate how to use *all* of them. Paulson et al. (1975) propose taking

$$\hat{\theta}_{\text{CF}} = \arg\min_{\theta} \int_{-\infty}^{\infty} |\hat{\varphi}_{\text{emp}}(u; \mathbf{x}) - \varphi_X(u; \theta)|^2 \exp(-u^2)\, du, \qquad (5.38)$$

where $|\hat{\varphi}_{\text{emp}}(u; \mathbf{x}) - \varphi_X(u; \theta)|$ is the modulus of the difference of the empirical and theoretical c.f.s. This is similar to (5.33), but uses the factor $\exp(-u^2)$ to dampen the contribution as $|u| \to \infty$. With respect to the mixed normal model, its performance is shown below in Section 5.3, in comparison to the other estimators entertained.

Remarks

(a) Section 9.4.5 illustrates the good performance of a c.f.-based estimator for the four parameters of the i.i.d. stable Paretian model, this being a perfect case in point, as the stable c.f. is easily and quickly computed, while its density (and thus the likelihood) is not. Another example is their use in asset pricing models in finance; see Singleton (2001) and the references therein. Yu (2004) provides an overview of empirical c.f. estimators and an illustration of various applications of the method in financial econometrics.

(b) Instead of squaring the modulus in (5.38), one could consider using different power values. Simulation with our showcase model indicated that, for powers between 0.6 and 2.0, there was virtually no difference in performance. Instead, one could try different damping functions, such as the one in (9.11), as used by Matsui and Takemura (2008) in the context of testing the stable Paretian distribution. The reader is invited to use it for estimation of the mixed normal parameters and compare the resulting m.s.e. values. ∎

5.3 COMPARISON OF METHODS

> The best advice to those about to embark on a very large simulation is often the same as Punch's famous advice to those about to marry: Don't!
>
> (Paul Bratley, Bennett L. Fox, and Linus E. Schrage, 1987, p. 34)

Having introduced several estimation strategies, it seems appropriate to stage a final showdown, or "horse race," between them. For our showcase model (5.4) based on sample size $n = 100$ and $\texttt{sim} = 1000$ replications, we use the m.l.e.; the q-B.e. with Hamilton's shrinkage prior (5.26) with prior strength $w = 4$, denoted qB-4 (this w being far higher than used by Hamilton, who took $w = 0.2$); the estimator based on KD (the best of the goodness-of-fit measures); the quantile estimator with $q = 52$; the X_P^2 estimator with $m = 16$; the shrinkage X_P^2 estimator (5.36) with $m = 16$ and $\tau = 0.01$, denoted X2P-S; the empirical m.g.f. estimator using $m = 5$ and the suggested values of the t_j from Quandt and Ramsey (1978); and the empirical c.f. estimator (5.38). We omit the m.m.e. as it often does not exist. The top left panel of Figure 5.17 shows the results.

The other panels are similar; the top right panel corresponds to the contaminated normal model (5.7) with $n = 100$, except having used $\lambda_1 = 0.90$ instead of 0.95; the bottom left

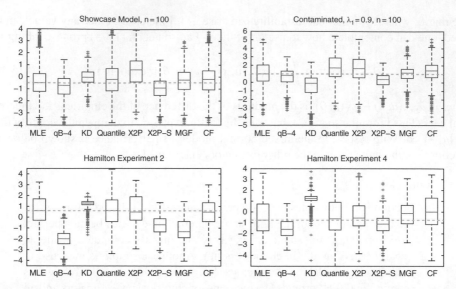

Figure 5.17   *Horse race between the various methods of estimation for the models considered throughout the chapter. All are based on* `sim` = *1000 replications and sample size* n = 100 – *except experiment 4, which uses* n = 50.

panel uses experiment 2, with $\mu_1 = 0$, $\mu_2 = 1$, $\sigma_1 = 1$, $\sigma_2 = 1.5$, $\lambda_1 = 0.3$, and $n = 100$; and the bottom right panel uses experiment 4, with $\mu_1 = 0.5$, $\mu_2 = 1.5$, $\sigma_1 = 0.5$, $\sigma_2 = 1.5$, $\lambda_1 = 0.5$, and, unlike the other three models, $n = 50$. As noted previously, this case is interesting because the shrinkage prior, with its zero means and unit variances, does not coincide with the true values.

The only estimators that perform better than the m.l.e. in all four cases examined are qB-4 and X2P-S. It is noteworthy that, for the contaminated model, KD is the winner, even beating the two shrinkage estimators.

5.4   A PRIMER ON SHRINKAGE ESTIMATION

> In closing, we would like to elaborate on the likely reason for the forecasting success of our approach, which relies heavily on a broad interpretation of the shrinkage principle. The essence of our approach is intentionally to impose substantial a priori structure, motivated by simplicity, parsimony, and theory, in an explicit attempt to avoid data mining and hence enhance out-of-sample forecasting ability. … Here we interpret the shrinkage principle as the insight that imposition of restrictions, which will of course degrade in-sample fit, may nevertheless be helpful for out-of-sample forecasting, even if the restrictions are false.
>
> (Francis X. Diebold and Canlin Li, 2006, p. 362)

This short section illustrates the – initially nonintuitive – result that shrinkage estimation can deliver a point estimate for a set of population means superior to the m.l.e. with respect to mean squared error, such that the populations are *independent*. Given independence, one would intuitively think that application of an estimator to each of the independent data sets (such as the m.l.e., this resulting in the joint m.l.e.) is both optimal and obvious, though this turns out not to be the case.

The explanation for this apparent oddity is that we are concerned with the m.s.e. of the *vector* of parameters. The intuition that using an independent sample, say **Y**, to improve upon an estimator for (say, the mean of) population **X** is useless *is correct*, but when the m.s.e. of the *vector* of parameters is of concern, their overall m.s.e. can be reduced. In particular, it is indeed *not* the case that the m.s.e. of each of the individual mean estimators is reduced, but rather that the reduction in m.s.e. of one or more of the individual estimators (which ones not being ascertainable) is greater in magnitude than the total increase in m.s.e. of the remaining ones, that is, the overall m.s.e. is reduced.

The origins of the result go back to work by Charles Stein in 1956, though, like other results in statistics and science in general, once the result is known, simpler demonstrations often become available. We now detail what is referred to as Stein's example, from Stein (1981) (which can also be found in the Wikipedia entry "Proof of Stein's example"), which illustrates the concept taking $\mathbf{X} = (X_1, \ldots, X_p)' \sim N(\theta, \mathbf{I}_p)$ for $p \geq 3$.

It is useful to express the problem in terms of the language associated with **decision theory**. The **loss function** $L(\hat{\theta}, \theta)$ is a measure of discrepancy between θ and $\hat{\theta}$, such as the squared error, $\|\hat{\theta} - \theta\|^2$, while the **risk function** averages this over all possible **X**, namely $R(\hat{\theta}, \theta) = \mathbb{E}[L(\hat{\theta}, \theta)]$. For estimators $\hat{\theta}$ and $\tilde{\theta}$, if $R(\hat{\theta}, \theta) \leq R(\tilde{\theta}, \theta)$ for all θ, with strict inequality for at least one θ, then $\hat{\theta}$ is said to **strictly dominate** $\tilde{\theta}$, and $\tilde{\theta}$ is deemed to be **inadmissible**. If estimator $\hat{\theta}$ is not strictly dominated by any other estimator of θ, then $\hat{\theta}$ is said to be **admissible**. We will see that, for $p \geq 3$, the m.l.e. of θ is inadmissible.

The simple demonstration hinges on Stein's lemma (A.120), namely that, for $Z \sim N(0, 1)$ and differentiable function $h : \mathbb{R} \to \mathbb{R}$ such that $\mathbb{E}[h'(Z)] < \infty$ and $|h(0)| < \infty$, $\mathbb{E}[Zh(Z)] = \mathbb{E}[h'(Z)]$. Slightly more generally, for differentiable function $h : \mathbb{R}^p \to \mathbb{R}$ such that $\mathbb{E}\left[\frac{\partial h(\mathbf{X})}{\partial x_i}\right] < \infty$, $i = 1, \ldots, p$, and $|h(0)| < \infty$, if $\mathbf{X} \sim N(\theta, \mathbf{I}_p)$, then

$$\mathbb{E}[(X_i - \theta_i)h(\mathbf{X})] = \mathbb{E}\left[\frac{\partial h(\mathbf{X})}{\partial X_i}\right]. \tag{5.39}$$

Let (the **decision rule**) $d_a(\mathbf{X}) = (1 - a/\|\mathbf{X}\|^2)\mathbf{X}$, $a \geq 0$. For $\|\mathbf{X}\|^2 > a > 0$, $d_a(\mathbf{X})$ shrinks **X** towards zero. Let the loss function be $L(\hat{\theta}, \theta) = \|\hat{\theta} - \theta\|^2$, so that the risk is the m.s.e. Clearly, $R(d_0(\mathbf{X}), \theta) = \sum_{i=1}^{p} \mathbb{E}[(X_i - \theta_i)^2] = p$, while, in general,

$$R(d_a(\mathbf{X}), \theta) = \mathbb{E}[\|\mathbf{X} - \theta\|^2] - 2a\mathbb{E}\left[\frac{\mathbf{X}'(\mathbf{X} - \theta)}{\|\mathbf{X}\|^2}\right] + a^2\mathbb{E}\left[\frac{1}{\|\mathbf{X}\|^2}\right]. \tag{5.40}$$

Invoking (5.39),

$$\mathbb{E}\left[\frac{\mathbf{X}'(\mathbf{X} - \theta)}{\|\mathbf{X}\|^2}\right] = \mathbb{E}\left[\sum_{i=1}^{p} \frac{X_i(X_i - \theta_i)}{\sum_j X_j^2}\right] = \sum_{i=1}^{p} \mathbb{E}\left[\frac{\partial h(\mathbf{X})}{\partial X_i}\right], \quad h(\mathbf{X}) = \frac{X_i}{\epsilon + \sum_j X_j^2}, \quad \epsilon = 0,$$

$$\tag{5.41}$$

that is,

$$\sum_{i=1}^{p} \mathbb{E}\left[\frac{\partial h(\mathbf{X})}{\partial X_i}\right] = \sum_{i=1}^{p} \mathbb{E}\left[\frac{\sum_j X_j^2 - 2X_i^2}{\left(\sum_j X_j^2\right)^2}\right] = \mathbb{E}\left[\frac{p-2}{\|\mathbf{X}\|^2}\right],$$

where for $p = 3$, the last equality is confirmed from direct calculation,

$$\mathbb{E}\left[\frac{-X_1^2 + X_2^2 + X_3^2}{\left(\sum_j X_j^2\right)^2}\right] + \mathbb{E}\left[\frac{X_1^2 - X_2^2 + X_3^2}{\left(\sum_j X_j^2\right)^2}\right] + \mathbb{E}\left[\frac{X_1^2 + X_2^2 - X_3^2}{\left(\sum_j X_j^2\right)^2}\right] = \mathbb{E}\left[\frac{1}{\sum_j X_j^2}\right],$$

and for $p = 4$, the reader should check that $\mathbb{E}\left[2/\sum_j X_j^2\right]$ indeed results. Thus, (5.40) is

$$R(d_a(\mathbf{X}), \boldsymbol{\theta}) = p - [2a(p-2) - a^2]\mathbb{E}\left[\frac{1}{\|\mathbf{X}\|^2}\right],$$

so that $R(d_a(\mathbf{X}), \boldsymbol{\theta}) < p \equiv R(d_0(\mathbf{X}), \boldsymbol{\theta})$ for $0 < a < 2(p-2)$, showing that $d_0(\mathbf{X})$ (the m.l.e.) is inadmissible for $p \geq 3$. The only caveat with this proof is that h as defined in (5.41) does not satisfy $|h(\mathbf{0})| < \infty$ nor is differentiable at $\mathbf{0}$, though modifying it such that $\epsilon > 0$ and then taking the limit as $\epsilon \to 0$ leads to the result.

We illustrated the concept using the oldest, simplest (though still highly relevant) example, but shrinkage estimation can be justified in numerous settings, as alluded to in the above quote by Diebold and Li (2006). A highly readable account and application of shrinkage estimation, aimed at a general scientific audience, is given by Efron and Morris (1977). More detailed theoretical developments, further relevant and original references, and the important connection between Stein's paradox (being a purely frequentist result that stunned the statistical community) and empirical Bayes methods are provided in Lehmann and Casella (1998), Robert (2007, Sec. 2.8.2, 10.5), and Efron (2013, Ch. 1).

5.5 PROBLEMS

> The world ain't all sunshine and rainbows. It's a very mean and nasty place, and I don't care how tough you are, it will beat you to your knees and keep you there permanently if you let it. You, me or nobody is gonna hit as hard as life. But it ain't about how hard you hit. It's about how hard you can get hit, and keep moving forward. How much you can take and keep moving forward. That's how winning is done.
>
> (Rocky Balboa, *Rocky VI*)

5.1 Building on Example 5.6, we wish to estimate both the tail index α and the scale parameter c. If $X \sim S_{\alpha,\beta}(0, 1)$, then $Y = cX \sim S_{\alpha,\beta}(0, c)$ and $\mathbb{E}[|Y|^r] = c^r\mathbb{E}[|X|^r]$. So, we need to choose two different moments, r_1 and r_2, and numerically solve the two equations for the two unknowns, α and c. Write a program to do this. Hint: In Matlab, you will need the `fsolve` command – the multivariate version of `fzero` – for solving systems of equations. It takes only a starting value; you cannot provide it with a range on the parameters as in `fzero`. Thus, in the objective function, you will need to constrain α to lie in $(\alpha_0, 2)$ for $0 < \max(r_1, r_2) < \alpha_0 < 2$ and c to be nonnegative. See the help on `fsolve` to see how the `exitflag` variable is assigned. This is needed to determine whether a solution was found.

Fix $r_1 = 1$, and use simulation to determine the optimal value of r_2. As an example, Figure 5.18 shows the m.s.e. for α and c over a grid of r_2-values, based on 1000 replications, sample size $n = 50$, and having used true values $\alpha = 1.5$ and $c = 1$ (and $\beta = 0$, $\mu = 0$). It appears that the choice of $r_2 = 0.3$ is optimal in this case.

As a comparison, from Figure 9.10(a) (on page 344), we see that, in the less realistic case when c is known, the m.s.e. of $\hat{\alpha}$ is about 0.044, this being substantially less than the minimal m.s.e. of 0.067 in Figure 5.18(a).

Show by constructing similar plots, say, with $\alpha = 1.6$ and $\alpha = 1.7$, that the optimal value of r_2 changes with α. What would you recommend using? What if, based on the estimate of α, you choose r_2? This could be iterated until convergence.

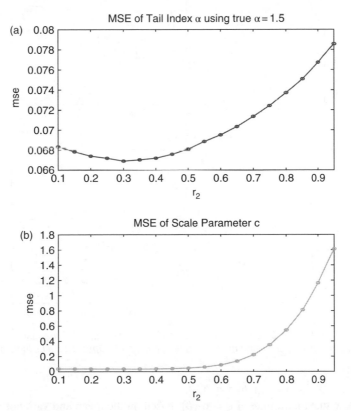

Figure 5.18 *Mean squared error, as a function of r_2, based on simulation with 1000 replications, of $\hat{\alpha}_{MM}$ (a) and $\hat{c}_{MM}$ (b) for the m.m.e. using two moment equations, with $r_1 = 1$. It is based on data $X_1, \ldots, X_n$, $n = 50$, where $X_i \overset{\text{i.i.d.}}{\sim} S_{\alpha,\beta}(\mu, c)$, with $\mu = \beta = 0$, and α and c are to be estimated. True values are $\alpha = 1.5$, $c = 1$.*

5.2 Recall from (A.304) for the stable distribution that, for $\alpha < 2$ and $X \sim S_{\alpha,\beta}(0,1)$,

$$\beta = \lim_{x \to \infty} \frac{\Pr(X > x) - \Pr(X < -x)}{\Pr(X > x) + \Pr(X < -x)}.$$

This can be used to construct a simple estimator of β, say Tailx, where x denotes the choice of x and $\Pr(X > x)$ is approximated by the e.c.d.f. Write a program that inputs α, β, a range of x-values, say $4, 5, 6, 7$, sample size n, and number of replications s, and then simulates the performance of the estimators, showing boxplots and the m.s.e. as output. Sample output is shown in Figure 5.19. It also shows the performance of the McCulloch estimator, denoted McC, which is a simple estimator based on sample quantiles, and is presented in Section 9.4.4.

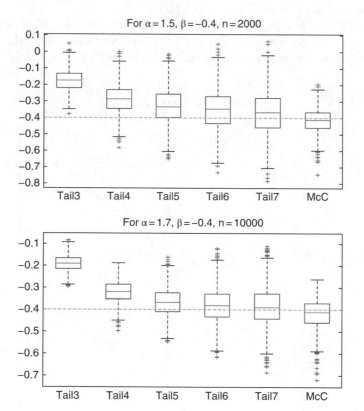

Figure 5.19 *Boxplot of 1000 values of $\hat{\beta}$ using the Tailx estimator (and the last boxplot being the McCulloch quantile estimator), using the true values of α, β and sample size n as indicated in the titles of the plots.*

5.3 Let $X_1, \ldots, X_n \overset{\text{i.i.d.}}{\sim} \text{Beta}(a, b)$.

(a) Show that the m.m.e. of $\theta = (a, b)'$ based on the mean and variance is given by

$$\hat{a} = m \frac{m(1 - m) - v}{v} \quad \text{and} \quad \hat{b} = \hat{a} \frac{1 - m}{m},$$

where $m = \bar{X}_n$ and $v = (n - 1)S_n^2/n$.

(b) Is $\hat{b}$ consistent, that is, does $\hat{b} \overset{p}{\to} b$ as $n \to \infty$?

5.4 Let $X_1, \ldots, X_n \overset{\text{i.i.d.}}{\sim} \text{Gam}(a, b)$. Derive an m.m.e. of $\theta = (a, b)'$ based on the mean and variance.

5.5 Let $X_i \overset{\text{i.i.d.}}{\sim} \text{Weib}(b, 0, s)$ for $i = 1, \ldots, n$ with density

$$f_{X_i}(x; b, s) = \frac{b}{s}\left(\frac{x}{s}\right)^{b-1} \exp\left\{-\left(\frac{x}{s}\right)^b\right\} \mathbb{1}_{(0,\infty)}(x).$$

Assume that the shape parameter b is a known constant. Derive method of moments and maximum likelihood estimators of S.

5.6 Let $X_i \overset{\text{i.i.d.}}{\sim} \text{Lap}(0, b)$, $i = 1, \ldots, n$, with $f_{X_i}(x; b) = (2b)^{-1} \exp(-|x|/b)$ for $b > 0$.
(a) Compute a method of moments estimator.
(b) Compute the m.l.e. of b.
(c) Write a program that simulates Laplace data and calculates the estimators $\hat{b}_{\text{ML}}$ and $\hat{b}_{\text{MM}}$ for a given value of b. For fixed b, by repeatedly simulating and estimating with different random draws, estimate $r = \text{m.s.e.}(\hat{b}_{\text{ML}})/\text{m.s.e.}(\hat{b}_{\text{MM}})$ for $n = 1, 2, \ldots, 20$ and plot r versus n. What do you notice about the graph for different values of b?

5.7 Recall Example 5.7, in which we showed that, for the contaminated normal model, the m.m.e. outperformed the m.l.e. One might then conjecture that the empirical m.g.f. method will also do well in this case and, unlike the m.m.e., should always exist. Investigate this as follows. Use $n = 100$ and the empirical m.g.f. estimator with $m = 5$, and the Quandt and Ramsey (1978) suggestion for the t_j, but change the distributional weights in the mixture from those in model (5.7) so that $\mu_1 = \mu_2 = 0, \sigma_1 = 1, \sigma_2 = 4$, and $\lambda_1 = 0.86, 0.88, \ldots, 0.96$.

Figure 5.20 shows the result – our usual boxplot comparisons (based on 1000 replications). Indeed, as λ_1 approaches 1, the empirical m.g.f. estimator outperforms the m.l.e., while both estimators improve, on average, as λ_1 moves towards 0.5.

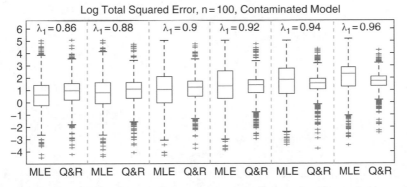

Figure 5.20 *Measure M^* from (5.6) for the m.l.e. and empirical m.g.f. estimator (with $m = 5$) using the contaminated normal model (5.7) but for different values of λ_1.*

Part II

Further Fundamental Concepts in Statistics

6

Q-Q Plots and Distribution Testing

This chapter, like Chapter 2, is concerned with distribution testing based on the e.c.d.f., but emphasizes plots for assessing goodness of fit, starting with P-P and Q-Q plots in Section 6.1. Historically, such plots were used as an informal graphical method for determining the appropriateness of a particular distribution for a set of data, either assumed i.i.d. or, being model residuals, approximately i.i.d. In Section 6.2 we illustrate how such informal methods are of little use without correct error bounds, and then detail the method for computing such bounds. It involves forming the mapping between pointwise (or one-at-a-time) and simultaneous significance levels. The distinction between the two is not only crucial in this context, but also highly relevant in many other statistical inference problems; see, in particular, the outstanding monograph of Efron (2013) and the references therein.

Armed with correct error bounds on the Q-Q plot, we can use it to deliver a size-correct test statistic for composite normality, as detailed in Section 6.3. These ideas are extended to other graphical methods, resulting in the MSP and Fowlkes-MP tests, in Section 6.4. Further normality tests are briefly illustrated in Section 6.5. Having presented several tests for normality, Section 6.6 presents a way of combining tests to yield new tests with correct size and potentially higher power, summarizes a comparison between the various tests, and introduces the notion of a power envelope for specific alternatives.

6.1 P-P PLOTS AND Q-Q PLOTS

Given an i.i.d. sample $X_1, \ldots, X_n$ from a continuous distribution with order statistics $Y_1 < Y_2 < \cdots < Y_n$, we might presume that the underlying distribution has c.d.f. F, but with an

Fundamental Statistical Inference: A Computational Approach, First Edition. Marc S. Paolella.
© 2018 John Wiley & Sons Ltd. Published 2018 by John Wiley & Sons Ltd.

unknown parameter vector θ that we can estimate as $\widehat{\theta}$. The KD statistic in (2.21) compared the empirical c.d.f., denoted by $\widehat{F}_{\text{emp}}$, with the fitted c.d.f., denoted by $\widehat{F}_{\text{fit}}$ or $F(\cdot;\widehat{\theta})$ or $\widehat{F}_{\text{fit}}(\cdot;\widehat{\theta})$, at the observed x_i (and took its maximum absolute difference).

Recall that the e.c.d.f. evaluated at the order statistics is given by $\widehat{F}_{\text{emp}}(Y_i) = (i - 0.5)/n$ or $(i - 3/8)/(n + 1/4)$, and is thus the step function formed from the points $t_i :=$ $(i - 3/8)/(n + 1/4)$, for $i = 1, \ldots, n$. It is reasonable to consider plotting $\widehat{F}_{\text{emp}}(Y_i)$ versus $\widehat{F}_{\text{fit}}(Y_i)$, $i = 1, \ldots, n$. When $\widehat{F}_{\text{emp}}$ is on the x-axis and $\widehat{F}_{\text{fit}}$ is on the y-axis, this is referred to as a *percentile–percentile plot, percent–percent plot,* or (arguably best) **probability–probability plot,** or **P-P plot** for short. It is so called because, for each point y_i, both the empirical (nonparametric) and fitted (parametric) c.d.f. are estimating the probability $F_X(y_i)$, where F_X is the true, underlying, unknown, c.d.f. If the fitted c.d.f. is the correct one, then we expect the plotted points to lie close to a $45°$ line in the unit box (a square with coordinates $(0,0)$, $(0,1)$, $(1,0)$, and $(1,1)$).

While such plots are indeed used, a more popular procedure is to invert the empirical and fitted c.d.f.s to get the corresponding quantile functions, and plot these. This is referred to as a **quantile–quantile plot** or **Q-Q plot**. In particular, instead of plotting $\widehat{F}_{\text{emp}}$ versus $\widehat{F}_{\text{fit}}$, we plot $F_{\text{fit}}^{-1}(\widehat{F}_{\text{emp}}(Y_i);\widehat{\theta}) = F_{\text{fit}}^{-1}(t_i;\widehat{\theta})$ on the x-axis and the sorted data $Y_i = \widehat{F}_{\text{fit}}^{-1}(\widehat{F}_{\text{fit}}(Y_i))$ on the y-axis. As with the P-P plot, if the fitted c.d.f. is the correct one, then the points will lie on a $45°$ line in a box with coordinates (y_1, y_1), (y_1, y_n), (y_n, y_1), and (y_n, y_n).

Remark.   The parameters of some distributions (e.g., normal, exponential, Laplace and Cauchy) are just location and scale parameters. From (A.130), quantiles preserve location–scale transformations. Thus, for a Q-Q plot, it is not necessary to estimate the location and scale parameters for use in $F^{-1}(\cdot;\widehat{\theta})$: The plot will still be linear if the underlying distribution is the same (up to location and scale) as the assumed one. This is useful if, as a very typical example, we wish to know whether the observed data come from a normal distribution; once we are assured it is, the estimation of its two parameters is routine.

This fact is not helpful if we wish to use Q-Q plots in contexts that involve distributions possessing additional shape parameters (e.g., beta, gamma, Weibull, Student's t, Pareto, stable Paretian). In such cases, the parameters have to be estimated. Because of the prevalence of distributions used in practice that have more than just location and scale parameters, in all our applications of Q-Q plots below, even in the examples using location–scale families such as the normal, we estimate all unknown parameters.   ∎

Given normal data, generated by, say, `data=10+2*randn(100,1)` in Matlab, a bare-bones piece of code to construct a Q-Q plot based on a normal distribution is as follows:

```
1   n=length(data); themean=mean(data); thestd=std(data,1); i=1:n;
2   t=(i-3/8)/(n+1/4); x=norminv(t,themean,thestd); plot(x,sort(data),'r+')
```

Matlab offers the command `qqplot` to construct a Q-Q plot, with default fitted c.d.f. referring to the standard normal distribution.

6.2 NULL BANDS

6.2.1 Definition and Motivation

As mentioned, when the parametric c.d.f. used for the Q-Q plot corresponds to the distribution from which the data were sampled, the points will fall around the 45° line, but of course not lie precisely on it. The question is then how much deviation from the line should be tolerated before one should doubt the distributional assumption. To illustrate, both panels of Figure 6.1 show Q-Q plots corresponding to the same data set of $n = 1000$ standard Cauchy observations, and having used the true parametric c.d.f., $F_{\mathrm{Cau}}(c; 0, 1) = 1/2 + \arctan(c)/\pi$.

The code used to generate the plots is given in Listing 6.1. The top Q-Q plot indeed looks nearly perfectly linear, as it theoretically should for a location–scale model, but it does not show all of the 1000 data points. The second plot is the same, differing only in that it shows more of the data points; it is anything but linear, and the understandable knee-jerk reaction of the unsuspecting data analyst would be to reject the claim that the data have arisen from the parametric c.d.f. used to make the plot. In this case, the deviation should come as little surprise given the extremely fat-tailed nature of the Cauchy distribution. Nevertheless, it should be clear that Q-Q plots are not meaningful without some indication of the range that the data can take on under the null hypothesis that the data were generated from the claimed distribution. In this context, we will refer to such a range as **null bands**.

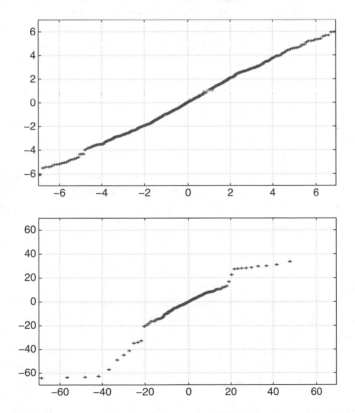

Figure 6.1 *Q-Q plots for the same Cauchy data set, just differing by the range on the x- and y-axes.*

```
1  rand('twister',6), n=1000; data=norminv(rand(n,1))./norminv(rand(n,1));
2  i=1:n;  t=(i-3/8)/(n+1/4);  x=tan(pi*(t-0.5));
3  plot(x,sort(data),'r+','linewidth',2), k=7; axis([-k k -k k])
4  set(gca,'fontsize',16), grid
```

Program Listing 6.1: Code for generating Figure 6.1. Recall that a Cauchy realization can be generated as the ratio of two independent standard normals. We use `norminv(rand(n,1))` instead of `randn(n,1)` because we want to be able to replicate this run, using the same seed value (here 6).

6.2.2   Pointwise Null Bands via Simulation

A natural starting point would be to construct pointwise c.i.s for each $F^{-1}(t_i; \theta), i = 1, \ldots, n$, similar to what we did for the e.c.d.f. in Section 1.2. This is easily accomplished with simulation, now described and illustrated for normally distributed data. Based on the true parameter (in this case, $\theta = (\mu, \sigma^2)'$ for the normal distribution), we generate a large number of normal random samples (say, $s = 20{,}000$) of length n with parameters μ and σ^2, sort each one, and store them (in an $s \times n$ matrix). Then, for $i = 1, \ldots, n$, the 0.05 and 0.95 sample quantiles are computed from the set of s simulated ith order statistics. Next, the usual Q-Q plot is made, along with the bands corresponding to the 0.05 and 0.95 quantiles obtained. These are called 90% **pointwise null bands**. For each $F^{-1}(t_i; \theta), i = 1, \ldots, n$, this gives a range such that it contains the ith sorted data value, on average, 90% of the time.

The construction of these bands is analogous to the intervals obtained from Figure 1.4 in that they make use of the true parameter value and are thus valuable as a theoretical starting point, but not directly applicable to the realistic situation in which the true parameter vector is unknown. We will now investigate the consequences of replacing θ with $\hat{\theta}$.

The program in Listing 6.2 implements this method, and was used with the code in Listing 6.3 to produce the two Q-Q plots shown in the top panels of Figure 6.2. Notice that both Q-Q plots refer to the same data set. The first plot uses the estimated values of μ and σ from the data; these are $\hat{\mu} = 10.484$, $\hat{\sigma} = 1.852$. The second plot uses the true values of $\mu = 10$ and $\sigma = 2$. In the first plot, only one point out of the 50 exceeds the 90% band, and none exceeds the 95% band. If the plotted points were independent (they are not; they are order statistics), then we would *expect* about 10% of the points, or 5 in this case, to exceed the 90% bounds, and 2 or 3 points to exceed the 95% bounds.

Of course, perhaps we "just got lucky" with this data set, but repeating the exercise shows that, more often than not, very few points exceed the bounds when we use the estimated parameters. Looking at the top right Q-Q plot in Figure 6.2, which uses the same data set but the true parameter values, we see that there are several points that exceed the bounds. Thus, it appears that, by fitting the parameters and drawing the pointwise null bands, we get a false sense of the goodness of fit. Indeed, this makes sense: *By fitting the parameters, we alter the shape of the parametric distribution we are entertaining in such a way that it best accommodates the observed data.* In practice, we naturally do not know the true parameters and will need to estimate them. So, we need a way of accounting for this statistical artifact.[1] This is done in Section 6.2.4 below in the more useful context of simultaneous bands.

---

[1] Recall that the word "artifact" has several meanings. The ones we have in mind are (i) "a spurious observation or result arising from preparatory or investigative procedures" and (ii) "any feature that is not naturally present but is a product of an extrinsic agent, method, or the like."

```
1   function normqqplotwithpointwise(data,trueparams,siglevel)
2   % pass trueparams as [] to estimate them, otherwise pass [truemu,truesig]
3   % pass siglevel as, say, 0.10, to use 90% pointwise null bands
4   %   or omit, and default of 90% and 95% are shown
5   n=length(data); data=sort(data);
6   if nargin>1, shouldestimate = isempty(trueparams); else shouldestimate = 1; end
7   if ~shouldestimate, themean=trueparams(1); thestd=trueparams(2);
8   else themean=mean(data); thestd=std(data,1); end % the MLE
9   i=1:n; t=(i-3/8)/(n+1/4); x=norminv(t,themean,thestd);
10  plot(x,data,'k+','linewidth',3), hold on
11  %%%%%%%%%%%%%%%%%%%%%%%%%%%%%%%%%%%%%%%%%%%%%%%%%%%%%%%
12  sim1=20000; ymat=zeros(sim1,n);
13  for i=1:sim1, ymat(i,:)=sort(themean+thestd*randn(1,n)); end
14  if nargin<3
15    q05=quantile(ymat,0.05); q95=quantile(ymat,0.95); % 90% indiv CI
16    q025=quantile(ymat,0.025); q975=quantile(ymat,0.975); % 95% indiv CI
17    plot(x,q05,'b-',x,q95,'b-',x,q025,'r--',x,q975,'r--','linewidth',2)
18  else
19    p=siglevel/2; qlo=quantile(ymat,p); qhi=quantile(ymat,1-p);
20    plot(x,qlo,'b-',x,qhi,'b-','linewidth',2)
21  end
22  %%%%%%%%%%%%%%%%%%%%%%%%%%%%%%%%%%%%%%%%%%%%%%%%%%%%%%%%%%%%%
23  hold off, set(gca,'fontsize',16)
24  rangex=max(x)-min(x); adj=0.03*rangex; xlim([min(x)-adj,max(x)+adj])
25  if ~shouldestimate, title('Using the true \mu and \sigma')
26  else title('Using the MLE for \mu and \sigma'), end
```

Program Listing 6.2: Generates a Q-Q plot, assuming normal data, with pointwise null bands. If thesecond argument is passed to the function, its two elements are used as the mean and standarddeviation of the parametric normal distribution; otherwise, they are fitted from the dataset passed to the function.

```
1   n=50; mu=10; sig=2;
2   rand('twister',6), data = mu+sig*norminv(rand(n,1));
3   figure(1), normqqplotwithpointwise(data)
4   figure(2), normqqplotwithpointwise(data,[mu,sig])
```

Program Listing 6.3: Code segment to produce the two Q-Q plots in the top panels of Figure 6.2.

6.2.3 Asymptotic Approximation of Pointwise Null Bands

A much faster method to produce these plots is to replace the simulation of the quantiles with the use of the asymptotic distribution of the order statistics (A.189). Approximation (A.189) is easily implemented by replacing the code in lines 12–21 of Listing 6.2 with the following (for a passed value of `siglevel`):

```
1   h=siglevel/2; nc=norminv(1-h);
2   sd = sqrt( t .* (1-t)/n ) ./ normpdf(x,themean,thestd);
3   plot(x,x-nc*sd,'b-',x,x+nc*sd,'b-','linewidth',2)
```

The two bottom panels of Figure 6.2 show the resulting Q-Q plots using (A.189) instead of simulation. The difference is barely noticeable; the only discernible difference is that

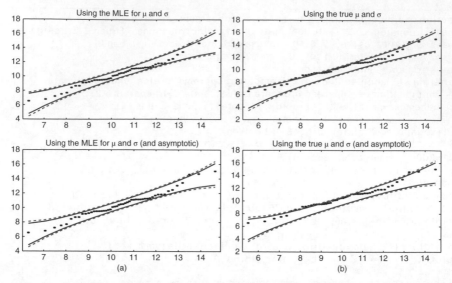

Figure 6.2   *Q-Q plot for a random $N(10,2)$ sample of size $n = 50$ with 10% and 5% pointwise null bands obtained via simulation (top panels), using the estimated parameters (left) and the true parameters (right) of the data. The bottom panels are similar, but based on the asymptotic distribution in (A.189).*

the simulated null bands in the left tail are slightly thinner than their asymptotically based counterparts. This agrees with the remark concerning the values of p for which approximation (A.189) will be accurate.

This convenient result unfortunately breaks down in applications with data generated from distributions with a power tail (e.g., Pareto, Student's t, stable Paretian). In this case, the correct null bands, obtained from simulation, will be highly asymmetric and extend deeply into the tails. The Cauchy example in Figure 6.1 serves as an extreme example: Using the same data set, the top panels of Figure 6.3 show the Q-Q plots with null bands

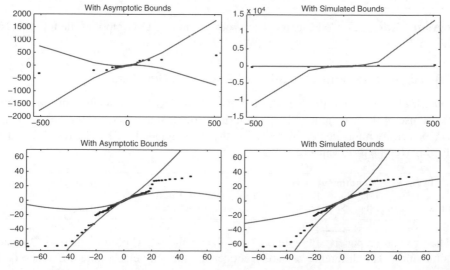

Figure 6.3   *Q-Q plots with pointwise null bands, using a size of 0.05, for the same Cauchy data as shown in Figure 6.1.*

formed from (A.189) (left) and simulation (right), both using a pointwise significance level of 0.05. The difference in the tails is striking, though it also exists in the more central part of the distribution, as shown in the bottom panels, which are the same plots, just magnified. Observe how the lower null band using (A.189) is not extreme enough, while its upper null band is too liberal.

6.2.4 Mapping Pointwise and Simultaneous Significance Levels

In addition to addressing the problem of having to use estimated parameter values, we also wish to design $100(1-\alpha)\%$ **simultaneous null bands**, where α is, say, 0.10, 0.05 or 0.01. This means that, when using, say, 90% simultaneous null bands corresponding to the true distribution, if we were to repeatedly conduct the experiment, we would find that, on average, 90% of the cases are such that *no* points fall outside the null bands.

We first address the issue of not knowing the true parameters. Recall in Section 1.2 how we determined the actual coverage probability of a confidence interval corresponding to a particular nominal coverage probability, and in Section 2.3.3 how we accounted for parameter uncertainty in the context of hypothesis testing with the KD statistic: by "simulating the simulation." Precisely the same logic applies here. We use a point estimator from the real data set to give us a best guess for the true parameter vector, and then use a nested simulation to correctly account for parameter estimation. The exact logistics of this will be shown below in Listing 6.4, in the context of constructing simultaneous null bands.

Now we turn to the construction of these simultaneous bands. It is not at all obvious how such a region can be optimally constructed, in the sense of having the smallest area while still possessing the correct desired simultaneous significance level. The problem can be made feasible and still maintain the latter constraint by restricting the search to pointwise null bands on the $F^{-1}(t_i; \theta)$, $i = 1, \ldots, n$, such that each has the same significance level. More formally, let $\mathbb{U} = (0, 1)$ denote the open unit interval. Then we wish to construct a mapping, say $s : \mathbb{U} \times \mathbb{N} \to \mathbb{U}$, such that, for a pointwise significance level $p \in \mathbb{U}$ and a sample size $n \in \mathbb{N}$, $s(p, n)$ is the simultaneous significance level. For fixed p and n, s can be determined by (you guessed it) simulation; and if this is done for a suitably chosen grid

```
1   function sslv = normqqplotsimultan(n,mu,sig,pslv)
2   if nargin<4, pslv=0.0025:0.0025:0.1; end
3   sim1=5000; sim2=20000;
4   psllen=length(pslv); viol=zeros(psllen,1); ymat=zeros(sim2,n);
5   for k=1:sim1, k
6     data = sort(mu+sig*randn(1,n));
7     themean=mean(data); thestd=std(data,1);
8     for i=1:sim2, ymat(i,:)=sort(themean+thestd*randn(1,n)); end
9     for j=1:psllen, psl=pslv(j);
10      qlo=quantile(ymat,psl/2); qhi=quantile(ymat,1-psl/2);
11      viol(j) = viol(j) + ( any(data<qlo) | any(data>qhi) );
12    end
13  end
14  sslv=viol/sim1; plot(pslv,sslv,'b-','linewidth',2)
```

Program Listing 6.4: Computes, via simulation, the mapping $s_{\mathrm{Norm}}(p, n)$ between pointwise and simultaneous significance levels for a random sample of size n from a $N(\mu, \sigma^2)$ population, taking into account that the true parameters are unknown and need to be estimated.

of p and n points, say G, then interpolation can be used to approximate s for any p and n contained in the range spanned by G. This can then be used to get what we are actually interested in, namely $p = s^{-1}(s_0, n)$. That is, given a desired *simultaneous* significance level s_0 and fixed sample size n, we want that value of p such that $s_0 = s(p, n)$.

For normally distributed data using a fixed sample size n, and a grid of values p, the method to determine the pointwise to simultaneous mapping $s_{Norm}(p, n)$ is implemented in the program in Listing 6.4. It simple and short, but quite slow, because it involves nested simulation with $sim1 \times sim2 = 10^8$ random n-length draws for the values of $sim1$ and $sim2$ used. (For $n = 500$, this corresponds to 5 days on a 3 GHz PC.) Note that, for each simulated data set in the outer loop, we have to estimate the parameters of the model. For the normal, this is not expensive because the estimators are closed-form; but for other models, such as the beta, gamma, Weibull, Student's t, and stable Paretian, numerical methods are required for obtaining the m.l.e., rendering this procedure even more time-consuming.

The procedure was run for $n = 10, 20, 50, 100$, and 500, each using a different vector of pointwise significance levels p (given by $pslv$ in the program) to best capture the range of interest for the simultaneous levels. The results are plotted in Figure 6.4(a) (we used $\mu = 0$ and $\sigma = 1$, but the results are invariant to these values). Then we can compute $s_{Norm}^{-1}(0.01, 100)$ to be 0.0080 (obtained from $interp1(sslv100, pslv100, 0.01)$), that is, to achieve a simultaneous significance level of $s_0 = 0.01$ for sample size $n = 100$, we would use a pointwise value of $p = 0.0080$. Similarly, for $n = 50, 20$, and 10, we would use $0.01375, 0.0330$, and 0.0750, respectively.

As is evident from envisioning a $45°$ line in the plot, we see that, perhaps somewhat unexpectedly, $s_{Norm}(p, n)$ can be either less than or greater than p, depending on both p and n. Our intuition would suggest that we would need to take p, the pointwise significance level, to be very close to zero (meaning that the null bands are very wide), in order to get the simultaneous significance level s to be a typical value, say 0.05. That is, we expect $s_{Norm}(p, n) > p$. This is indeed the case as n gets larger, but for small n it is just the opposite – we should use rather narrow pointwise intervals in order to get a simultaneous level of, say, 0.05. What appears to be a paradox (or a mistake) is easily resolved, recalling that the parameters are estimated. In particular, in small samples, they will be relatively inaccurate, reflecting the random characteristics of the small sample, and thus giving rise to a spuriously better-fitting Q-Q plot, as in the left panels of Figure 6.2.

Example 6.1   *A similar exercise was carried out, but using random samples of Weibull data, with typical p.d.f.*

$$f_{Weib}(x; \beta, 0, 1) = \beta x^{\beta-1} \exp\{-x^\beta\} \mathbb{1}_{(0,\infty)}(x).$$

The location parameter was fixed at zero, but a scale parameter, σ, was introduced, so that there are two unknown parameters to be estimated, β and σ. Simulations were done using $\sigma = 1$ and three different values of β: 0.5, 1, and 2. For each, two sample sizes, $n = 20$ and $n = 50$, were used. The results are shown in Figure 6.4(b). As with the normal case, the s-curves corresponding to the larger sample size $n = 50$ lie above those for $n = 20$. For $n = 20$, the value of β makes a small but noticeable difference in the function $s_{Weib}(p, \beta, n)$, whereas for $n = 50$, the difference is no longer discernible. This again reflects the fact that, for small sample sizes, the effect of having to estimate unknown parameters is more acute.

See Problem 6.2 for these calculations, and Problem 6.3(c) for application to the Laplace distribution.

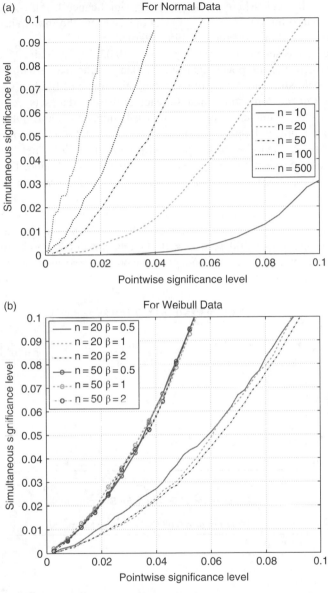

Figure 6.4 *The mapping between pointwise and simultaneous significance levels, for normal data (a) and Weibull data (b) using sample size n.*

6.3 Q-Q TEST

Having obtained the mapping from pointwise to simultaneous significance levels for a given sample size n and a particular parametric distribution, we can use it to test whether the data are in accordance with that distribution. In particular, we would reject the null hypothesis of, say, normality at significance level α if any points in the normal Q-Q plot exceed their pointwise null band, where p, the pointwise significance level, is chosen such that $\alpha = s(p, n)$. We will refer to this as the **Q-Q test** of size α. Furthermore, via simulation, we can obtain the power of this test for a specific alternative.

For example, and as seen in Figure 6.4, $p = s_{\text{Norm}}^{-1}(0.05, 50)$ is obtained via interpolation to be about 0.03816. For the power against a Student's t alternative with $n = 50$ and size $\alpha = 0.05$, we would simulate, say s_1 times, a random sample of Student's t data of length n, with v degrees of freedom, sort it, and, for each of these s_1 data sets, compute the m.l.e. corresponding to the parameters of the *normal* distribution, and then simulate, say s_2 times, a *normal* random sample of length n with that m.l.e., sort it, and store it. From those s_2 sorted series, we would compute the empirical $p/2$ and $1 - p/2$ quantiles. Then, we would record if any of the n sorted Student's t data points exceeds its pointwise bound. This is repeated s_1 times, and the mean of these Bernoulli r.v.s is the (approximate) power. Problem 6.5 invites the reader to implement this.

As a check on the size, doing this exercise with normal data instead of Student's t, we confirm that the test has the correct significance level; that is, for $\alpha = 0.05$, $n = 50$ and $p = 0.03816$, and using $s_1 = 2000$ and $s_2 = 10,000$, the power is 0.050, to three significant digits. Plotting the power as a function of v gives the power curve, as shown in Figure 6.5(a). Overlaid are also the power curves corresponding to $n = 20$ and $n = 100$. Figure 6.5(b) is

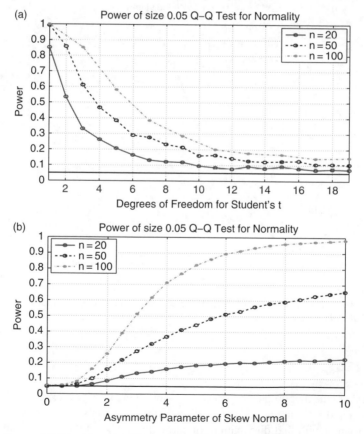

Figure 6.5 *Power of Q-Q test for normality, for three different sample sizes, and Student's t alternative (a) and skew normal alternative (b), based on simulation with 1000 replications.*

similar, but uses the skew normal distribution (A.115) as the alternative, indexed by its asymmetry parameter λ. For $\lambda = 0$, the power coincides with the size; otherwise, the power is greater. The power also increases with sample size, for a given $\lambda > 0$. It appears that the Q-Q test for normality against Student's t and skew normal alternatives is unbiased and consistent, recalling their definitions from Section 2.4. Comparison with the power of the KD and AD statistics (using size 0.05) shown in Figure 2.12 reveals that the Q-Q test is almost as powerful as the AD test for the Student's t, and more powerful than the KD test for the skew normal alternatives.

6.4 FURTHER P-P AND Q-Q TYPE PLOTS

> Although formal testing procedures allow an objective judgment of normality, … they do not generally signal the reason for rejecting a null hypothesis, nor do they have the ability to compensate for masking effects within the data which may cause acceptance of a null hypothesis.
>
> (Henry C. Thode, 2002, p. 15)

We discuss two less common variations of P-P and Q-Q plots that are arguably more useful as graphical devices for indicating potential deviation from normality. Moreover, we augment them in such a way as to yield tests that are vastly simpler to compute than the Q-Q test, have correct size, and turn out to have impressive power properties against relevant alternatives.

6.4.1 (Horizontal) Stabilized P-P Plots

Recall (again) that the e.c.d.f. is a step function with $\widehat{F}_{\text{emp}}(Y_i) = t_i$ at the order statistics $Y_1, \ldots, Y_n$, and $t_i = (i - 3/8)/(n + 1/4)$, $i = 1, \ldots, n$, and that the P-P plot has the e.c.d.f. $\widehat{F}_{\text{emp}}$ on the x-axis and $\widehat{F}_{\text{fit}}(y_i)$ on the y-axis.

Michael (1983) proposed a simple and effective transformation that stabilizes the variance (renders the variance of $\widehat{F}_{\text{fit}}$ nearly uniform over the support of X). He terms this the **stabilized probability plot**, which we abbreviate as S-P plot. It plots g_i (on the x-axis) versus h_i, where

$$g_i = \frac{2}{\pi} \arcsin(t_i^{1/2}) \quad \text{and} \quad h_i = \frac{2}{\pi} \arcsin(\widehat{F}_{\text{fit}}^{1/2}(y_i)). \tag{6.1}$$

To see why this works, let $U \sim \text{Unif}(0, 1)$ and $S = (2/\pi)\arcsin(U^{1/2})$. Then, recalling the trigonometric identity $\sin(x - y) + \sin(x + y) = 2 \sin x \cos y$ (see, for example, Example I.A.5), % the jth row. works without squeeze

$$f_S(s) = f_U(u) \left| \frac{du}{ds} \right| = \mathbb{I}_{(0,1)} \left(\sin^2 \left(\frac{\pi s}{2} \right) \right) \left| \frac{du}{ds} \right| = \left| \frac{du}{ds} \right|$$

$$= \pi \sin \left(\frac{\pi s}{2} \right) \cos \left(\frac{\pi s}{2} \right) = \frac{\pi}{2} \sin(\pi s) \mathbb{I}_{(0,1)}(s)$$

and

$$F_S(s) = \mathbb{I}_{(0,1)}(s) \int_0^s \frac{\pi}{2} \sin(\pi t) \, dt + \mathbb{I}_{[1,\infty)}(s) = \mathbb{I}_{(0,1)}(s) \frac{1}{2}[1 - \cos(\pi s)] + \mathbb{I}_{[1,\infty)}(s).$$

```
1   n=length(data); data=sort(data); themean=mean(data); thestd=std(data,1);
2   i=1:n; t=(i-3/8)/(n+1/4); g=arcsintransform(t);
3   u=normcdf(data,themean,thestd); h=arcsintransform(u)';
4   plot(g,h,'k+','linewidth',3), sim1=20000; ymat=zeros(sim1,n);
5   for i=1:sim1, ddd=sort( themean+thestd*randn(1,n) );
6     u = normcdf(ddd,themean,thestd); ymat(i,:)= arcsintransform(u);
7   end
8   p=0.05/2; qlo=quantile(ymat,p); qhi=quantile(ymat,1-p);
9   hold on, plot(g,qlo,'b-',g,qhi,'b-','linewidth',2), hold off
```

Program Listing 6.5: Produces the normal S-P plot. The short program `arcsintrans-form` is required to compute `(2/pi) * asin(sqrt(x))` for input vector x.

The range of s follows because, for $v = u^{1/2}$, $0 < v < 1 \Rightarrow 0 < \arcsin(v) < \pi/2$ so $0 < s < 1$ for $s = (2/\pi) \arcsin(u^{1/2})$. Integration by parts reveals that $\mathbb{E}[S] = 1/2$ and $\mathbb{V}(S) = 1/4 - 2/\pi^2$. Now consider its ith order statistic Y_i out of n. Using (A.176), the p.d.f. turns out to be

$$f_{Y_i}(y) = \frac{\pi\, n!}{2^n (n-i)!(i-1)!} [1 - \cos(\pi y)]^{i-1} [1 + \cos(\pi y)]^{n-i} \sin(\pi y) \mathbb{I}_{(0,1]}(y).$$

Algebraically expressing $\mathbb{E}[Y_i]$ and $\mathbb{V}(Y_i)$ appears difficult, so we use simulation to investigate $\mathbb{V}(Y_i)$ as n grows. Indeed, as claimed (but not proven) in Michael (1983), $n\mathbb{V}(Y_i)$ approaches $1/\pi^2$ *for all i*. The reader is encouraged to numerically verify this.

Pointwise null bands are formed via simulation in an analogous way as for Q-Q plots. Bare-bones code to generate the normal S-P plot for data set `data` is given in Listing 6.5.

As a striking example of the effect of the transformation, Figure 6.6(a) shows the Cauchy S-P plot applied to the same Cauchy data as was used in Figures 6.1 and 6.3. (In this case, we replace the call to `normcdf` above with `u=0.5+atan(data)/pi`.) Notice that, with 1000 data points, (i) the S-P plot has a large amount of wasted space; (ii) it is somewhat difficult to see the outliers; and (iii) it is very difficult to see any curvature in the null bands. As such, one is behooved to plot the points on a straight line. This is accomplished simply by plotting g_i versus $h_i - g_i$ (and also the null bands, minus g_i), resulting in Figure 6.6(b). We deem this the *horizontal S-P plot*. Now, among other things, we can see that the null bands are close to, but not of equal width, especially in the tails.

The top panels in Figure 6.7 show the normal S-P plot using the same normal data sample as was used in the Q-Q plots of Figure 6.2, with size 0.10 and 0.05 pointwise null bands obtained via simulation. Comparison with the top panels in Figure 6.2 shows that the informational content of the Q-Q and S-P plots is identical, in the sense that the location of each of the 50 points is the same with respect to the null bands – either inside or outside. This holds for the case when we estimate the parameters, and when we use the true parameters. Figure 6.8 is the same, but using the horizontal format.

6.4.2 Modified S-P Plots

For a given distribution (normal, Cauchy, etc.), a specific parameter vector θ, a sample size n, and a pointwise significance level p, the null bands of the Q-Q plot could be computed once via simulation and stored as two vectors in a lookup table, but doing this for a variety of sample sizes and significance levels would become unwieldy. With S-P plots, if we are willing to assume that the width of the band is constant over $(0, 1)$ (it is not, as shown in Figures 6.6 and 6.8; we deal with the consequences of this below), then *all we need*

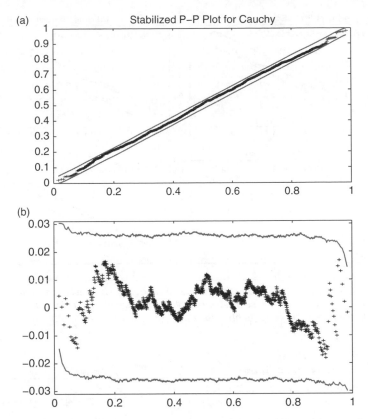

Figure 6.6 *(a) Cauchy S-P plot with null bands, obtained via simulation, using a pointwise signifi-cance level of 0.01. (b) Same, but using the horizontal format.*

to store is a single number. That is, for a given p, n, and θ, we would calculate the null bands as in Figure 6.7, and record only, say, the median of the n widths depicted in the plot; call this $w(p; n, \theta)$. To illustrate its use, the bottom panels in Figures 6.7 and 6.8 use these constant-width null bands.

Once $w(p; n, \theta)$ is obtained, we no longer have to simulate to get the null bands, but instead just generate them simply as

$$g_i \pm w(p; n, \theta)/2. \tag{6.2}$$

It should be clear that, for location–scale families, the values of the location and scale parameters do not change the widths, and so we can drop the dependence of w on θ. Figure 6.9(a) plots the width as a function of p for the normal distribution and three sample sizes, computed using 50,000 replications and using a tight grid of values of p from 0.002 to 0.15. We could store the w-values as a function of p and n in a lookup table, but there is an even better way. Some trial and error shows that each curve is virtually perfectly fitted (with a regression R^2 of over 0.9999) using the function of the pointwise significance level p given by

$$w(p; n) \approx b_1 + b_2 p + b_3 p^{1/2} + b_4 p^{1/3}, \tag{6.3}$$

where the coefficients depend on the sample size n and are given in Table 6.1.

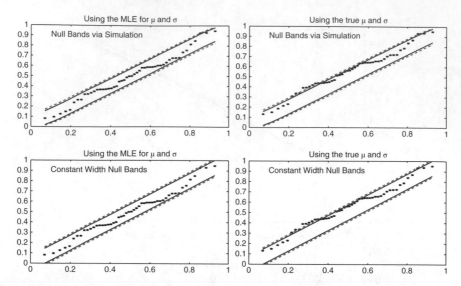

Figure 6.7   *(Top) Stabilized P-P plot using the same random $N(10,2)$ sample of size $n = 50$ as in Figure 6.2 with 10% and 5% pointwise null bands obtained via simulation, using the estimated parameters (left) and the true parameters (right) of the data. (Bottom) Same as top, but with constant-width null bands.*

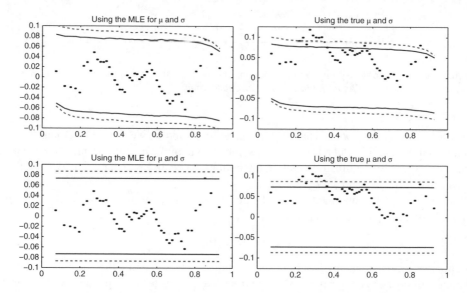

Figure 6.8   *Same as Figure 6.7, but plotted in horizontal format.*

Thus, for each sample size, only four numbers need to be stored to get the null bands corresponding to any pointwise significance level in the range from 0.002 to 0.15. We can do this for numerous sample sizes, and store the values. But again there is a better way: If each of the resulting b_i coefficients, $i = 1, 2, 3, 4$, as a function of n, is "smooth enough", then we can fit each one as a function of n. This turns out to be the case, and so they can be

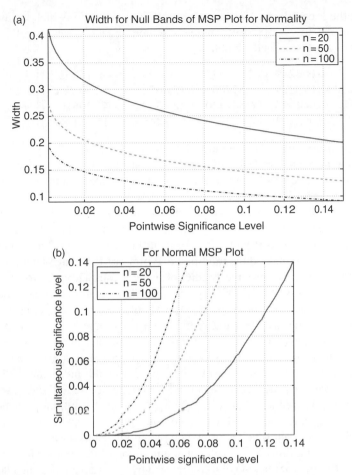

Figure 6.9 *(a) The solid, dashed, and dash-dotted lines are the widths for the pointwise null bands of the normal MSP plot, as a function of the pointwise significance level p, computed using simulation with 50,000 replications. The overlaid dotted curves are the same, but having used the instantaneously computed approximation from (6.4) and (6.3). There is no optical difference between the simulation and the approximation. (b) For the normal MSP plot, the mapping between pointwise and simultaneous significance levels using sample size n.*

TABLE 6.1 Coefficients in regression (6.3)

n	b_1	b_2	b_3	b_4
20	0.53830	−0.41414	1.25704	−1.43734
50	0.35515	−0.33988	0.95000	−1.02478
100	0.26000	−0.24000	0.70463	−0.76308

used to give the b_i coefficients for any n in the chosen range, for which we used a grid of values from $n = 10$ to $n = 500$. Even more conveniently, each b_i can be well modeled with the same set of regressors, and we obtain

$$
\begin{bmatrix} b_1 \\ b_2 \\ b_3 \\ b_4 \end{bmatrix} \approx
\begin{bmatrix}
-0.002249 & 2.732572 & -1.080870 & -0.750621 \\
-0.072228 & -0.577005 & -18.465466 & 52.692956 \\
0.044074 & 6.544420 & 8.157051 & -41.295102 \\
-0.006993 & -8.042349 & 1.344564 & 15.781063
\end{bmatrix}
\begin{bmatrix} 1 \\ n^{-1/2} \\ n^{-1} \\ n^{-3/2} \end{bmatrix}. \tag{6.4}
$$

These b_i are then used in (6.3) to get the width.

Observe that there are two levels of approximation, (6.3) and (6.4). To confirm that the method works, overlaid in Figure 6.9(a) are the approximate widths obtained from using (6.3) and (6.4). There is no optical difference. Use of this approximation allows us to instantly compute the S-P plot for normal data (with sample size between 10 and 500) with null bands corresponding to any pointwise significance level in $[0.002, 0.15]$.

We call the horizontal S-P plot with constant-width null bands computed using the outlined approximation method the **modified S-P plot** or **MSP plot**.[2] One apparent caveat of the method is that it assumes the validity of using a constant width for the null bands, which from Figure 6.8 is clearly not fully justified. This fact, however, becomes irrelevant if we wish to construct the mapping $s(p, n)$ to simultaneous coverage, because, through the simulation to get $s(p, n)$, the actual simultaneous coverage corresponding to a chosen value of p is elicited, even if this value of p would be slightly different if we were to use the correct pointwise null bands. Looked at in another way, once $s(p, n)$ is computed using the (approximation via (6.3) and (6.4) to the) constant-width null bands, for a given s_0, we recover that value p_0 such that, when used for the pointwise null bands with constant width, we get the desired simultaneous coverage. It is irrelevant that p_0 is not precisely the true pointwise significance level for each of the n points – remember, we just use *some* pointwise null bands to get what we want: correct simultaneous coverage.[3]

The benefit of using this method is that the inner loop of the nested simulation (see the program in Listing 6.4 for the Q-Q plot) is replaced by an instantaneous calculation, so that the determination of $s(p, n)$ for a vector of p and fixed n takes about a minute, from which we obtain the pointwise value p_0 from linear interpolation. The results for three sample sizes are shown in Figure 6.9(b). For example, for $n = 100$, we should use a pointwise significance level of $p_0 = 0.03575$ to get a simultaneous one of 0.05. Not shown is the graph for $n = 200$; we would use 0.02322.

6.4.3 MSP Test for Normality

Continuing the discussion in the preceding section, a test of size α for normality simply consists of rejecting the null hypothesis if any of the plotted points in the MSP plot lie outside the appropriate simultaneous null bands. We call this the **MSP test (for composite normality)**, as developed in Paolella (2015b).

---

[2] The function to generate the MSP plot (for assessing the normal distribution), such as those in Figures 6.7 and 6.8 (with traditional or horizontal format, and with constant-width null bands, or using simulated, nonconstant-width bands) is called `MSPnormplot` (code not listed here), and is available in the book's collection of programs.
[3] This, in turn, has its own caveat: while the simultaneous coverage will indeed be correct, the *power* of the resulting test will not be as high as having used the actual, nonconstant null bands. As an extreme case, imagine using the Q-Q test with constant null bands: the power would presumably be abysmal.

While the calculation of the s function is indeed fast, there is, yet again, a better way. For a fixed α, we compute the pointwise significance values p_0 corresponding to each sample size in a tight grid of n-values, using a high precision (we used 500,000 replications) and then fit the resulting values as a linear function of various powers of n. For example, with $\alpha = 0.05$, this yields (and requires all the significant digits shown)

$$p_0 = 0.01149009 - 0.00000496n - 0.12946828n^{-1/2} + 5.91761206n^{-2/2}$$

$$- 26.24384775n^{-3/2} + 51.04722197n^{-4/2} - 35.10748101n^{-5/2}, \quad 10 \le n \le 520.$$
$$(6.5)$$

This was also done for $\alpha = 0.01$ and $\alpha = 0.10$; the method is implemented in function `MSPnormtest` (not shown in the text). It inputs the data set and desired significance level α, and returns a zero if the null hypothesis of normality cannot be rejected, and a one otherwise. The computation takes a fraction of a second; in particular, about 1.5×10^{-4} seconds on a 3 GHz PC. For values of α different than 0.01, 0.05 and 0.10, or sample sizes outside the range $10 \le n \le 500$, simulation is used to get the correct value of p_0.

With such an enormous improvement in speed for calculating the null bands, we can also perform the power calculations similar to those shown in Figure 6.5 for the Q-Q test, but now in a matter of seconds. Figure 6.10 shows the results based on a significance

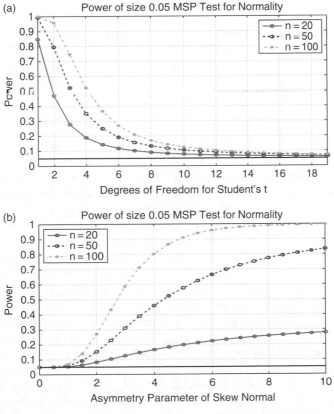

Figure 6.10 *Power of the MSP test for normality, for three different sample sizes, and Student's t alternative (a) and skew normal alternative (b), based on 1 million replications.*

level of 0.05, as in Figure 6.5. From the plot, we immediately confirm that the MSP test has the correct size, confirming the discussion above regarding use of the (approximate) constant-width null bands and as a check on all the approximations used to calculate p_0. Comparing Figures 6.5 and 6.10, we see that the Q-Q test has higher power against the Student's t, but the MSP test has higher power against the skew normal. In fact, the MSP test has even higher power than the JB test for normality (see Section 6.5.2 below) against this alternative for larger values of λ, as seen by comparing with Figure 6.18(b) below. For example, MSP has substantially higher power for $n = 50$ and $\lambda > 3$, and for $n = 100$ and $\lambda > 3.5$.

In addition to modeling p_0 as a function of n in order to obtain the correct widths to conduct a test at levels 0.10, 0.05, and 0.01, it is useful also to have (an approximation to) the p-value of the test. One way of accomplishing this would be to obtain p_0 not just for α-levels 0.10, 0.05, and 0.01, but for a large grid of significance levels between 0 and 1, and take the p-value to be the smallest significance level such that the null hypothesis of normality can be rejected. Unless the grid of α-values has a very high resolution (entailing quite some computation and storage), the resulting p-value will not be very accurate. Instead, we consider another way.

Denote the value of the MSP test by `hyp`. It is one (reject the null hypothesis) if any of the plotted points in the MSP plot lie outside the appropriate simultaneous null bands (6.2), and zero otherwise. This is accomplished in Matlab as

```
qlo=g-w/2; qhi=g+w/2; hyp=(any(h<qlo) | any(h>qhi));
```

This is equivalent to `hyp` $= $ `T>w/2`, where $T = T_{\mathrm{MSP}}$ is the test statistic defined by

$$T_{\mathrm{MSP}} = \max|\mathbf{h} - \mathbf{g}|, \tag{6.6}$$

and $\mathbf{h}$ and $\mathbf{g}$ are the vectors formed from values in (6.1).

For a given sample size n, we simulate a large number of test statistics, generated under the null. Then, for an actual data set $\mathbf{X} = (X_1, \dots, X_n)$, its p-value is the fraction of those simulated test statistics that exceed $T_{\mathrm{MSP}}(\mathbf{X})$. Doing this takes about 150 seconds with 1 million replications and, worse, will deliver a different p-value each time the method is used, with the *same* data set $\mathbf{X}$. Fortunately, there is a better way: Figure 6.11 shows the kernel density (solid line) of $n \times T_{\mathrm{MSP}}$, for two sample sizes, $n = 10$ and $n = 50$. Remarkably, and as an amusing coincidence, the distribution strongly resembles that of a location–scale skew normal. The dashed lines in the plots show the best fitted location–scale skew normal densities, with the match being striking. In particular, using the m.l.e. and the numerical methods discussed in Section 4.3, we obtain asymmetry parameter $\hat{\lambda} = 2.6031$, location parameter $\hat{\mu} = 0.6988$, and scale parameter $\hat{c} = 0.3783$ for $n = 10$; and $\hat{\lambda} = 2.7962$, $\hat{\mu} = 2.2282$, and $\hat{c} = 1.0378$ for $n = 50$. By using other sample sizes, up to $n = 500$, we confirm that the skew normal yields an extremely accurate approximation to the true distribution of T_{MSP} under the null for all sample sizes between 10 and (at least) 500.

Thus, for a given sample size, we only need to store the three parameters. Then, for an actual data set $\mathbf{X} = (X_1, \dots, X_n)$, the p-value is $1 - F_{\mathrm{SN}}(T_{\mathrm{MSP}}(\mathbf{X}); \hat{\lambda}_n, \hat{\mu}_n, \hat{c}_n)$. As with the modeling of p_0 as a function of n, we can compress this information further by conducting this simulation for a range of sample sizes, obtaining the skew normal m.l.e. for each, and then fitting a polynomial model in n to each of the three parameters $\hat{\lambda}_n$, $\hat{\mu}_n$, and $\hat{c}_n$. This

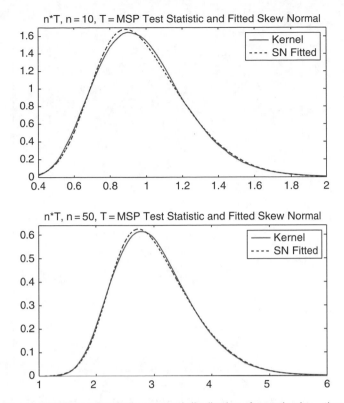

Figure 6.11 *Kernel density and fitted skew normal distribution of sample size n times the MSP test statistic (6.6), computed under the null, and based on 1 million replications.*

was successful, using the same regressors as were used for p_0, and function MSPnormtest incorporates this to also return the p-value.

To assess the quality of the approximation, we simulate 1 million p-values, under the null of normality, based on $n = 50$, using the following code:

```
1   n=50; s=1e6; pv=zeros(sim,1);
2   for i =1:s, x=randn(n,1); [g1,g2,g3,p]=MSPnormtest(x,0.05); pv(i)=p; end
```

The resulting histogram is shown in Figure 6.12(a). Having used so many replications, we are able to discern a pattern in the bars, and thus a deviation from the uniform distribution, though the approximation is still clearly accurate, and certainly adequate for our purposes. Figure 6.12(b) shows the resulting p-values when having used a Student's t distribution with 8 degrees of freedom instead of the normal (and again $n = 50$). Now the p-values pile up closer zero. The fraction of these less than 0.05, in this case 0.130, gives the power of the 5% test with this sample size and alternative.

Remark. Notice that, once we can approximate the distribution of (n times) the test statistic, its $1 - \alpha$ quantile, divided by n, is $w/2$, and we could do away with the approximation for p_0 in (6.5) for $\alpha = 0.10, 0.05$, and 0.01. However, while the skew normal approximation

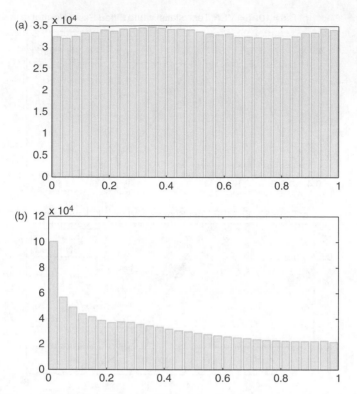

Figure 6.12 *One million p-values from the MSP test with $n = 50$, under the null (a) and for a Student's t with $v = 8$ degrees of freedom alternative (b).*

to T_{MSP} is very good, it is not exact, and so this method is not as accurate as using our initial method to get the correct p_0 and w for the three most common values of α. Thus, we use it only for values of α passed to the function that are *not* equal to 0.01, 0.05, or 0.10. (To see the difference it makes, just call the function with, say, $\alpha = 0.05$ and $\alpha = 0.05000001$.) ∎

6.4.4 Modified Percentile (Fowlkes-MP) Plots

Recall (for the last time) that the e.c.d.f. evaluated at the order statistics $Y_1 < Y_2 < \cdots < Y_n$ from a continuous distribution is given by $\widehat{F}_{\text{emp}}(Y_i) = (i - 0.5)/n$ or $(i - 3/8)/(n + 1/4)$ and is the step function formed from the points $t_i := (i - 3/8)/(n + 1/4)$, for $i = 1, \ldots, n$. Plotting $t_i = \widehat{F}_{\text{emp}}(y_i)$ on the x-axis and $\widehat{F}_{\text{fit}}(y_i)$ on the y-axis gives the P-P plot, while plotting $F_{\text{fit}}^{-1}(t_i; \widehat{\theta})$ on the x-axis and the y_i on the y-axis gives the Q-Q plot.

Fowlkes (1979) pointed out that the normal Q-Q plot is not particularly sensitive to a mixture of (two) normal distributions when the means of the components are not well separated. To address this, he suggested plotting the standardized order statistics $z_i = (y_i - \widehat{\mu})/\widehat{\sigma}$ versus $\Phi(z_i) - t_i$, $i = 1, \ldots, n$, where Φ is the standard normal c.d.f. Observe how the z_i are quantiles, while $\Phi(z_i)$ and t_i are probabilities, so in a sense this is a cross between a P-P and Q-Q plot. We call this (differing from Fowlkes, but in line with Roeder, 1994, p. 488) the (normal) **Fowlkes modified percentile plot**, or just the (normal) **Fowlkes-MP plot**.

As with all such goodness-of-fit plots, their utility is questionable without having sensible null bands. No such procedure was suggested by Fowlkes to remedy this.

Observe that, unlike the P-P, Q-Q, and MSP plots, the quantities on the x-axis of the Fowlkes-MP plot are functions of the Y_i (and not just n), so that the usual method we used for simulating to get the null bands would result in a band not for a $t_i = \hat{F}_{emp}(Y_i)$ (as in P-P) or $F_{fit}^{-1}(t_i; \hat{\theta})$ (as in Q-Q), but rather for an order statistic, z_i, whose x-coordinate changes with each sample. The resulting null bands are extremely wide and of little value. A feasible alternative is to compute lower and upper horizontal lines, as with the MSP plot, that have the desired simultaneous significance level.

This idea is incorporated in the `FMPnormplot` function (not shown, but in the collection of available programs). It approximates the widths of the null bands corresponding to simultaneous significance levels of 0.01, 0.05, and 0.10 as a function of the sample size, similar to what was done for the MSP test. In addition to optionally generating the Fowlkes-MP plot (such as the ones in Figure 6.13), it returns the result of the hypothesis test at the three aforementioned levels, which we refer to as the **Fowlkes-MP test** for normality.

To illustrate, the top left panel of Figure 6.13 shows the Fowlkes-MP plot, with null bands, for a normal random sample with $n = 100$. The top right shows the MSP plot for the same data. The random sample of normal data was found with trial and error (use `rand('twister',14); y = norminv(rand(100,1));` to duplicate it) such that the MSP normal test rejects at the 10% level, but not at the 5%. This can be seen from the MSP plot (top right), in which one data point exceeds the lower 10% line. In the corresponding Fowlkes-MP plot (top left) using the same data, there is one data point which is indeed very close to its lower 10% line.

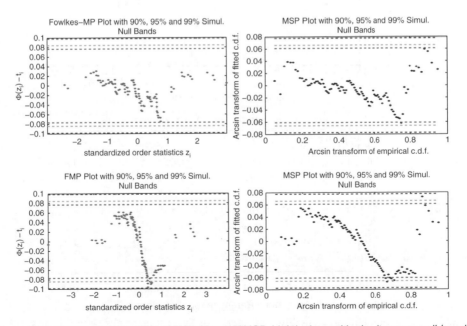

Figure 6.13 *Normal Fowlkes-MP (left) and normal MSP (right) plots, with simultaneous null bands, for normal data (top) and mixed normal data (bottom).*

The bottom two panels of Figure 6.13 are similar, but using 100 observations from a two-component MixN(μ, σ, λ) distribution (5.1), with $\mu_1 = -0.1$, $\mu_2 = 0$, $\sigma_1 = 3$, $\sigma_2 = 1$, $\lambda_1 = 0.4$, and $\lambda_2 = 0.6$ (these parameters being typical of daily financial returns data). We see that both plots are able to signal that the data are not normally distributed. The reader is encouraged to produce the programs necessary to replicate these results.

Now turning to power comparisons, our usual two plots are given in Figure 6.14. With respect to the Student's t alternative, the Fowlkes-MP test performs the same as the KD test, which, we recall, was not particularly good. Against the skew normal, the Fowlkes-MP test again performs the same as the KD test, which was dominated by the U^2, W^2, and MSP tests. Based on these alternatives, the Fowlkes-MP test is not impressive. The fact that its power is identical to that of KD might behoove us to consider whether they are theoretically identical. Indeed, given the fact that the Fowlkes-MP test is based on precisely the fitted and empirical c.d.f.s, this must be the case. In fact, the equality of their powers for the Student's t and skew normal cases provides confirmation that the Fowlkes-MP test was implemented correctly.

In light of Fowlkes's motivation for his goodness-of-fit plot, it remains to be seen how the normal Fowlkes-MP test (or, equivalently, the KD test) performs with mixed normal

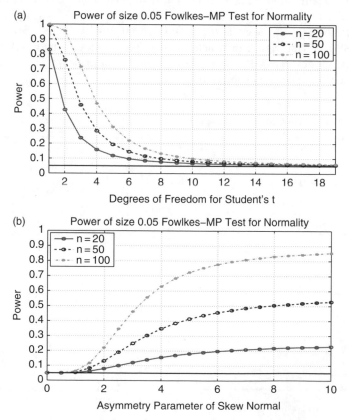

Figure 6.14 *Power of Fowlkes-MP test for normality, for three different sample sizes, and Student's t alternative (a) and skew normal alternative (b), based on 1 million replications.*

TABLE 6.2   Comparison of power for various normal tests of size 0.05, using the two-component mixed normal distribution as the alternative, obtained via simulation with 1 million replications for each model, and based on two sample sizes, $n = 100$ and $n = 200$. Model #0 is the normal distribution, used to serve as a check on the size. The entry with the highest power for each alternative model and sample size n appears in bold. Entries with power lower than the nominal/actual value of 0.05, indicating a biased test, are given in *italic*

Model	n	KD	AD	W^2	U^2	MSP	F-MP	JB	X_P^2
#0 (Normal)	100	0.050	0.050	0.050	0.050	0.050	0.050	0.050	0.050
#1 (Finance)	100	0.799	0.712	0.916	**0.924**	0.799	0.798	0.890	0.635
#2 (Equal means)	100	0.198	0.322	0.298	0.309	0.216	0.197	**0.417**	0.127
#3 (Equal vars)	100	0.303	*0.001*	0.400	**0.439**	0.250	0.302	*0.039*	0.191
#0 (Normal)	200	0.050	0.050	0.050	0.050	0.050	0.050	0.050	0.050
#1 (Finance)	200	0.983	0.881	0.997	**0.998**	0.973	0.983	0.994	0.940
#2 (Equal means)	200	0.358	0.430	0.523	0.545	0.328	0.358	**0.642**	0.225
#3 (Equal vars)	200	0.593	*0.000*	0.756	**0.789**	0.492	0.594	0.428	0.394

alternatives. We restrict attention to the two-component mixed normal, with three sets of parameters. The first, denoted #1, uses the parameters mentioned above, which are typical of financial returns data. Parameter set #2 takes $\mu_1 = \mu_2 = 0$, $\sigma_1 = 2$, $\sigma_2 = 1$, $\lambda_1 = \lambda_2 = 0.5$, and #3 takes $\mu_{1,2} = \pm1.25$, $\sigma_1 = \sigma_2 = 1$, $\lambda_1 = \lambda_2 = 0.5$. The results are shown in Table 6.2. We immediately confirm that the KD and Fowlkes-MP tests are identical. The U^2 test clearly dominates in models 1 and 3, while JB has the highest power for model 2, though U^2 also performs well in this case.

With respect to the tests that derive from graphical methods, MSP and Fowlkes-MP have virtually equal power for #1, are very close for #2, and are mildly different for #3, with Fowlkes-MP being better. Thus, *as a graphical tool*, Fowlkes's method, augmented with correct error bounds, does have value for detecting the presence of mixtures. In terms of power, however, it is equivalent to the KD, and is clearly dominated by U^2. The Pearson X_P^2 test also lends itself to graphical inspection, though its low power severely limits its value.

Finally, observe how, in model #3 (different means, equal variances), the AD (for both sample sizes) and JB (just for $n = 100$) tests have power lower than the size of the test; these are our first examples of biased tests. Their performance is justifiable: For the AD test, recall that it places more weight, relative to the KD statistic, onto the tails of the distribution, and therefore less in the center, where the deviation from normality for this model is most pronounced. Indeed, the opposite effect holds under the fatter-tailed alternative #2, in which case AD has higher power than the KD. Similarly, the JB statistic relies on skewness and kurtosis, but these are not features of this model.

6.5   FURTHER TESTS FOR COMPOSITE NORMALITY

After motivating in Section 6.5.1 the need for powerful, easily computed normality tests with delivered p-values, Section 6.5.2 outlines one of the oldest, simplest, and among the most powerful (but not *the* most powerful), tests of (composite) normality for heavy-tailed alternatives, namely what we will refer to as the Jarque–Bera test. Section 6.5.3 briefly discusses three more recent tests, showing their power performance in our usual set of graphics. Section 6.5.4 presents Pearson's X_P^2 test, even though it was not designed with normality testing in mind (and despite having among the worst power properties among the tests for normality). It is an important conceptual idea and arises in numerous contexts, particularly with categorical data, and thus deserves at least a basic elucidation.

6.5.1 Motivation

Before commencing, it is worth mentioning our motivation for such an in-depth study of normality testing, in light of modern, big-data applications and the growing recognition, and attention paid to, non-Gaussian processes; and the desire for normality tests that are, along with p-values, calculated nearly instantaneously, in light of ever-faster computing hardware. These two factors, particularly in combination, would seem to support the notion that such pursuits will gradually belong to the dustbin of statistical history. This is anything but the case, and we give one representative example, in a highly non-Gaussian setting, for which availability of such techniques is highly beneficial.

Consider a multivariate model for hundreds or even thousands of financial assets (as is common for large financial institutions), measured at daily or higher frequency. A piece of one such series is shown in Figure 4.7. The time-varying scale term is often modeled as a generalized autoregressive conditional heteroskedasticity (GARCH) type of model, with the resulting filtered innovations being approximately i.i.d., but exhibiting strong leptokurtic and significant asymmetric behavior; Book IV will give a detailed presentation of GARCH and the use of non-Gaussian innovation sequences. These are typically modeled with an asymmetric Student's t (see Kuester et al., 2006; Krause and Paolella, 2014; Paolella and Polak, 2015a, and the references therein), or, more generally, a special or limiting case of the generalized hyperbolic, as discussed in Chapter II.9. Recall the generalized hyperbolic distribution from Section II.9.5, and how it was expressed as a continuous normal mixture. The multivariate generalized hyperbolic (MGHyp) distribution is similar, taking $(\mathbf{Y} \mid G = g) \sim N(\boldsymbol{\mu}, g\boldsymbol{\Sigma})$ with G (univariate) generalized inverse Gaussian (GIG). As a special case, the multivariate Laplace distribution takes $G \sim \text{Gam}(b, 1)$. Endowing Y, G, $\boldsymbol{\mu}$, and $\boldsymbol{\Sigma}$ with time subscripts (and the diagonal elements of $\boldsymbol{\Sigma}_t$ with a GARCH structure) gives rise to the so-called COMFORT time series model proposed in Paolella and Polak (2015b,c, 2018b). The crucial point is that, conditional on G_t (and these are obtained numerically via filtering with an EM algorithm), $\mathbf{Y}_t \mid (G_t = g_t)$ is multivariate Gaussian.

A different model for large sets of asset returns data, not using GARCH-type filters, but which instead is i.i.d., is the multivariate discrete mixed normal, as developed in Paolella (2015a) and Gambacciani and Paolella (2017), both of which advocate using $k = 2$ (normal) mixture components. In the former, the components can be approximately separated based on the output of the EM algorithm, while in the latter, the MCD method, as discussed in Section 3.1.3, is used for explicit separation and subsequent estimation.

In both the COMFORT and discrete mixed normal settings, interest centers on the adequacy of a multivariate normal approximation. Testing multivariate normality, particularly in very high dimensions, is difficult, and one can compromise by applying univariate tests to all the univariate margins. The resulting set of p-values can be inspected as, say, a boxplot. This can also be done through time in moving windows exercises, and a plot of a set of empirical quantiles of the p-values versus time can be delivered, thus showing the evolution of the quality of the distributional assumption.

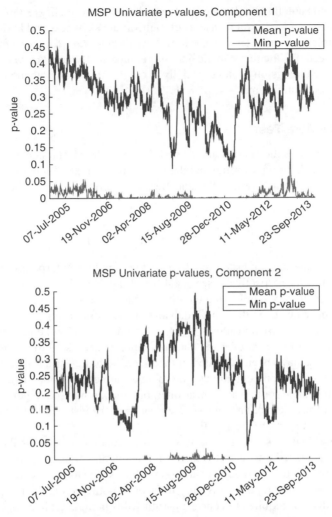

Figure 6.15 *The average and smallest p-values of the MSP univariate test of normality from Section 6.4.3, for the $d = 30$ stocks comprising the DJIA, in each of the two separated mixed normal components, and based on moving windows of sample size $v = 250$.*

This was done in Gambacciani and Paolella (2017), with the results replicated in Figure 6.15. The plots show, for each of the two components, the average and minimum p-values, based on the (log percentage) returns of the $d = 30$ components of the Dow Jones Industrial Average index (DJIA), through moving windows of length $v = 250$. Recalling that, under the null hypothesis of normality, the p-values should follow a uniform distribution on $(0, 1)$, we see that the average p-value over the windows is lower than 0.5 in both components, more so for the second one, and particularly so during the global financial crisis period.[4] These plots over time correspond to the results (shown in

---

[4] Thus, it appears that the two-component multivariate mixed normal distribution is not a highly accurate approximation to the actual distribution of returns. This is true, but that was known before having done the analysis – of

Gambacciani and Paolella, 2017) based on the multivariate normality test of Mardia (1971, 1974), which is based (only) on measures of multivariate skewness and kurtosis.

Thus, (multiple) *normality* tests can be useful in a (highly) *non-Gaussian* framework, and computation speed (for the test statistic, but, more importantly, for its *p*-value) is essential, given a very large dimension *d*, and especially in conjunction with potentially thousands of moving windows through time.

6.5.2 Jarque–Bera Test

This section introduces the popular and powerful Jarque–Bera test for composite normality, named after the authors who formulated the test in the context of econometric regression models, Jarque and Bera (1980, 1987) and Bera and Jarque (1981).[5] The test statistic is given by

$$ JB = \frac{n}{6}\left(skew^2 + \frac{(kurt - 3)^2}{4} \right), \tag{6.7} $$

where *n* is the sample size, and skew and kurt are the sample counterparts of the theoretical skewness and kurtosis, respectively; see (I.4.42) and (I.4.45). The test statistic is obviously trivial to program; the code in Listing 6.6 gives a function for this (we will need this code in an application below, even though Matlab already has it built in.)

As $n \to \infty$, the test statistic follows a $\chi^2(2)$ distribution under the null hypothesis, but deviates from this in finite samples, so that simulation is necessary to get the exact cutoff values. While these have been tabulated in the older literature for a set of sample sizes, a modern computing environment allows a far more accurate tabulation of such values, over finer grids, so that table lookup and linear interpolation can be used to get highly accurate cutoff values and also deliver an approximate *p*-value of the test. Matlab has done this, and it is implemented in its `jbtest` function.

Unfortunately, Matlab's `jbtest` only returns *p*-values that are less than or equal to 0.5. While this is indeed adequate for the traditional application of applying the test to a particular data sct and then using the resulting *p*-value as a measure of evidence against the null (values above 0.5 being unequivocally in favor of not rejecting the null), we detail an application below that requires the correct *p*-value (this being a realization of a Unif(0, 1)

```
1  function jb = jbteststat(x)
2  n=length(x); z = (x−mean(x))/std(x,1);
3  skew = sum(z.^3)/n; kurt = sum(z.^4)/n − 3; jb = n*(skew^2/6 + kurt^2/24);
```

Program Listing 6.6: Computes the JB test statistic (6.7).

course financial asset returns are not mixed normal! The real question is how adequate the approximation is. The use of the mixed normal allows for asymmetry and leptokurtosis, and, also crucially, for very easy calculation of downside risk measures and, thus, portfolio optimization. The out-of-sample portfolio results in Gambacciani and Paolella (2017) show that the model demonstrably beats (in terms of total cumulative returns, and risk-adjusted measures) the equally weighted portfolio, and the Markowitz allocation based on both an i.i.d. assumption and having used the so-called DCC-GARCH model of Engle (2002, 2009). To accommodate the non-Gaussianity but preserve the ease of computation of downside risk measures, one idea, as developed in Paolella (2015a), is to use a discrete mixture of Laplace distributions, for which an EM algorithm for estimation is also available.

[5] The idea of using the sample skewness and kurtosis for testing normality goes back at least to D'Agostino and Pearson (1973) and Bowman and Shenton (1975), and it is thus also referred to as the D'Agostino–Pearson or Bowman–Shenton test. It almost surely has its origins in the work of Karl Pearson; see the discussion in Section A.3. Takemura et al. (2006) provide theoretical insight into why this test performs well.

random variable under the null). To obtain this, we can proceed as follows. For a particular sample size, say $n = 50$, simulate a large number, b, of JB test statistics under the null, $J_1, \ldots, J_b$. We take b to be 10 million. Then, for any given data set of length $n = 50$ and JB test statistic T, the approximate JB p-value is given by the fraction of the J_i that exceed T. This is accomplished with the following code:

```
1  sim=1e7; JBstat=zeros(sim,1); n=50;
2  for i=1:sim, x=randn(n,1); JBstat(i)=jbteststat(x); end
3  JBstat=sort(JBstat); save JB1e7
4  % Then, for a particular data set, x, of length 50
5  jb = jbteststat(x); pvalue=mean(JBstat>jb);
```

While this works, computing the latter fraction is relatively slow, and so a simulation study that uses this method takes comparatively long. Instead, we can attempt to fit a very flexible parametric density to the distribution of the J_i, similar to having fitted the skew normal to the MSP test statistic in Section 6.4.3. We first consider use of the GAt distribution, with p.d.f. (A.125), along with location and scale parameters.

Figure 6.16(a) shows the kernel density estimate of *the log of* the 10 million J_i values, and a fitted (by maximum likelihood; see Example 4.9) GAt density, with estimated parameters

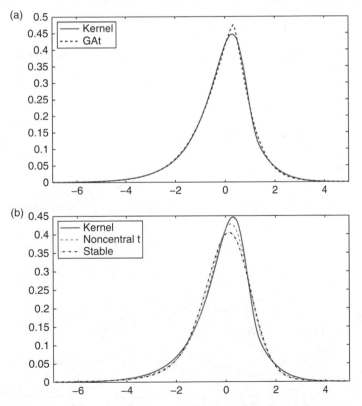

Figure 6.16 *Kernel density estimate (solid) of the log of the JB test statistic, under the null of normality and using a sample size of $n = 50$, based on 10 million replications (and having used Matlab's* `ksden- sity` *function with 300 equally spaced points). (a) fitted GAt density (dashed). (b) fitted noncentral t (dashed) and asymmetric stable (dash-dotted).*

$\hat{d} = 1.3727$, $\hat{v} = 9.2617$, $\hat{\theta} = 0.7633$, $\hat{\mu} = 0.3432$, and $\hat{c} = 1.0971$. The fit appears excellent. To compare, Figure 6.16(b) shows, along with the kernel density, a fitted asymmetric stable density and a fitted noncentral Student's t. The GAt is clearly superior. Moreover, unlike these and other competing fat-tailed, asymmetric distributions (see Chapters II.7–II.9 for a large variety of candidates), the GAt has a closed-form expression for its c.d.f. that is also computed in the function in Listing 4.8, and is thus far faster to evaluate. In particular, the p-value corresponding to a JB test statistic T of any data set of length $n = 50$ can be (virtually instantly) approximated as $1 - F_{\text{GA}t}(\log T; \hat{d}, \hat{v}, \hat{\theta}, \hat{\mu}, \hat{c})$.

To assess the quality of the approximation, Figure 6.17(a) shows a histogram of 1 million p-values from the JB test under the null, based on $n = 50$ and the GAt approximation. Similar to the histogram of the MSP p-values in Figure 6.12, having used this many replications, its deviation from uniformity is apparent. In this case however, it is clearly not as accurate as the approximation for the MSP p-values. Fortunately, attaining more accuracy at little or no cost is easy: we can fit a *mixture* of two GAt distributions (this having 11 parameters), precisely as was used in Section 9.5.1. Its p.d.f. and c.d.f. are just weighted sums of GAt p.d.f.s and c.d.f.s, respectively, so that evaluation of the c.d.f. is no more involved than that of the GAt. The code in Listing 6.7 shows the final parameter estimates (observe that the first component is essentially Gaussian-tailed, with $\hat{v}_1 \approx 460$, but its other shape parameters are not indicative of Gaussianity); running it and plotting the histogram yields

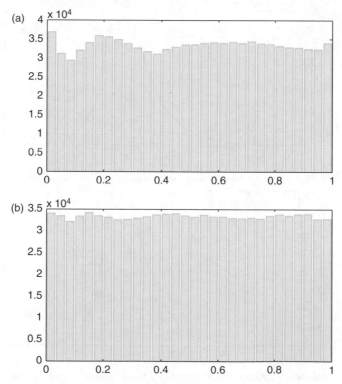

Figure 6.17 *Simulated p-values of the JB test statistic, based on 1 million replications, using the GAt approximation (a) and the two-component GAt mixture (b).*

```
 1  n=50; sim=1e6; pv=zeros(sim,1);
 2  d1= 1.79881; v1= 459.33799; theta1= 1.25036; mu1= 0.21159; c1= 1.20031;
 3  d2= 1.69948; v2=   6.87790; theta2= 0.50097; mu2= 0.49357; c2= 0.99361;
 4  lam1= 0.36429;
 5  for i=1:sim
 6    x=randn(n,1); jb = jbteststat(x); ljb=log(jb);
 7    z1=(ljb-mu1)/c1; [garb, cdf1]= GAt(z1,d1,v1,theta1);
 8    z2=(ljb-mu2)/c2; [garb, cdf2]= GAt(z2,d2,v2,theta2);
 9    cdf=lam1*cdf1+(1-lam1)*cdf2; pv(i)=1-cdf;
10  end
```

Program Listing 6.7: Final parameter estimates of GA*t* mixture for Jarque–Bera test.

Figure 6.17(b), demonstrating that the fit of the two-component GA*t* mixture is much better than that of the single GA*t*.

This exercise could be conducted over a range of sample size values *n*, and a mapping formed from *n* to the 11 estimated parameters. The reader is encouraged to explore this.

We turn now to the power of the JB test with size 0.05. (Note that, in our simulation to compute the power, we do not require the method for computing approximate *p*-values; we only need to compare the JB test statistic to the appropriate cutoff value, which is already conveniently and accurately provided by Matlab's function.) Our customary power plots are given in Figure 6.18. Among all the normality tests so far presented, the JB test has the highest power against the Student's *t* alternative, and fares well against the skew normal, though it does not dominate the Q-Q, W^2, and MSP tests. See Section 6.6.2 below for an ordering of the tests in terms of power.

0.5.3 Three Powerful (and More Recent) Normality Tests

6.5.3.1 Ghosh MGF Graphical Test

Another graphical method for i.i.d. normality, based on the (third derivative of the log of the empirical) m.g.f., was proposed and studied by Ghosh (1996).[6] It also yields a test statistic, and, as it is based on the m.g.f., is consistent. The test is asymptotically size-correct, and simulation shows the actual sizes of the 5% test are 0.030, 0.050, and 0.060, for sample sizes $n = 20$, 50, and 100, respectively. The associated power curves are given in Figure 6.19. Considering the $n = 50$ case, as it has nearly correct size, we see that the power is a little lower compared to the JB test for the Student's *t* alternative, while for the skew normal, the power is very close to that of JB. Compared to the MSP test, Ghosh has considerably greater power against the Student's *t*, and higher power for the skew normal for very small values of λ, though as λ increases, MSP dominates.

6.5.3.2 Stengos–Wu Information-Theoretic Distribution Test

Stengos and Wu (2010) provide two easily computable test statistics for normality, denoted KL1 and KL2, based on concepts from maximum entropy and Kullback–Leibler information. The tests are asymptotically size-correct, and, from the bottom panel of the power graphs in Figure 6.20(b), for $\lambda = 0$, we can see that the actual size for KL1 is

---

[6] The author wishes to thank Sucharita (Rita) Ghosh for providing valuable feedback on an initial draft of Paolella (2015b), in which the MSP test is developed.

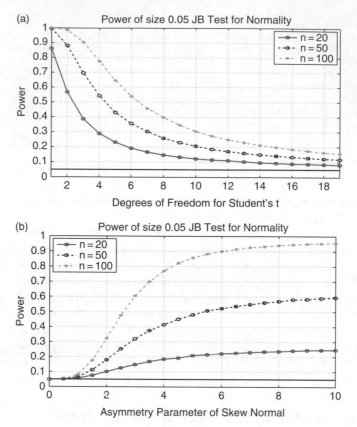

Figure 6.18 *Power of JB test for normality, for three different sample sizes, and Student's t alternative (a) and skew normal alternative (b), based on 100,000 replications.*

very close to the nominal. The KL2 test had slightly lower power against Student's t, and virtually the same power against the skew normal. As such, we omit the graphs for KL2. With respect to power against Student's t, KL1 and Ghosh perform very similarly, with neither completely dominating the other. For skew normal, while not fully dominating, KL1 overall performs better. Compared to the MSP test for the skew normal, we find that, for small λ, KL1 has slightly higher power, but as λ grows, MSP dominates. Related tests (not studied here) include those in Noughabi and Arghami (2013), who develop distribution tests based on the minimum Kullback–Leibler distance, applied to the beta, Laplace, Weibull, and uniform distributions.

6.5.3.3 Torabi–Montazeri–Grané ECDF-Based Test

Building on ideas in Noughabi and Arghami (2013), Torabi et al. (2016) propose, in our notation, the following test statistic for composite normality:

$$T_{\text{TMG}} = D(\widehat{F}_{\text{fit}}, \widehat{F}_{\text{emp}}) = \int_{-\infty}^{\infty} h\left(\frac{1 + \widehat{F}_{\text{fit}}(x; \hat{\mu}, \hat{\sigma})}{1 + \widehat{F}_{\text{emp}}(x)}\right) \, d\widehat{F}_{\text{emp}}(x) = \sum_{i=1}^{n} h\left(\frac{1 + \widehat{F}_{\text{fit}}(y_i; \hat{\mu}, \hat{\sigma})}{1 + i/n}\right),$$

$$(6.8)$$

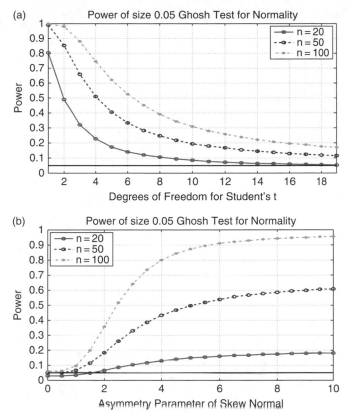

Figure 6.19 *Power of the Ghosh (1996) test for normality, for three different sample sizes, and Student's t alternative (a) and skew normal alternative (b), based on 100,000 replications.*

where $h : (0, \infty) \rightarrow [0, \infty)$ is a continuous convex function, decreasing on $(0, 1)$, increasing on $(1, \infty)$, and such that $h(1) = 0$. Measure (6.8), when used with p.d.f.s, is referred to as a ϕ-disparity measure; see Pardo (2006). Observe that, under the null hypothesis, T_{TMG} will be close to zero and cannot be negative, so we reject for large values of T_{TMG}. Also, for each i, the ratio of (one plus) the c.d.f.s lies in $[1/2, 2]$, so that h of that ratio lies in $[0, \max\{h(1/2), h(2)\}]$. While several candidate functions h exist, with $h(x) = x \ln x - x + 1$ being associated with the Kullback–Leibler divergence measure, Torabi et al. (2016) suggest use of $h(x) = (x - 1)^2/(x + 1)^2$, as it gives rise to higher power for their test.

Cutoff values for the usual test significance levels can be determined via simulation. Torabi et al. (2016) provide a table of some, based on 100,000 replications, but report them to only one significant digit. As such, we calculate the ones we require (for $\alpha = 0.05$ and sample sizes $n = 20$, 50, and 100), based on 10 million replications, assuring three significant digits. For example, with $n = 20$, the 95% quantile of (6.8) (recall that we reject for large values of the test statistic) under the null is 0.000784, in contrast to the value of 0.0007 as reported by Torabi et al. (2016). For $n = 50$ and $n = 100$, the 95% quantiles are 0.000306 and 0.000152, respectively.

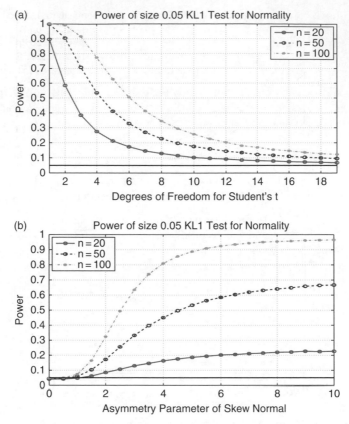

Figure 6.20 *Power of KL1 test for normality, for three different sample sizes, and Student's t alternative (a) and skew normal alternative (b), based on 100,000 replications.*

Figure 6.21 shows the results of our usual power experiment for statistic T_{TMG} in (6.8), hereafter just TMG. We concentrate on the power performance for the skew normal alternative, as Torabi et al. (2016) remark that its power against heavy-tailed alternatives is less than that of JB. Inspection shows that, while not uniform, TMG has higher power than the Ghosh and KL1 tests shown in Figures 6.20 and 6.19 respectively, particularly for the smaller sample size $n = 20$ and as λ grows. In their power simulation studies, Torabi et al. (2016) compare 40 tests of normality (though not including MSP, or those of Ghosh, 1996, and Stengos and Wu, 2010) for a variety of alternatives and demonstrate that their test is the most powerful against asymmetric alternatives. Our results lend confirmation to their conclusion.

The reader is encouraged to repeat the method used for the MSP or JB test to determine approximate, quickly computed p-values for the TMG test. This could then be used to produce a combined test with possibly higher power, as described in Section 6.6.1.2 below.

6.5.4 Testing Goodness of Fit via Binning: Pearson's X_P^2 Test

For categorical data, an apparently natural measure to assess the appropriateness of a postulated distribution would be to sum up (some function of) the absolute deviations between

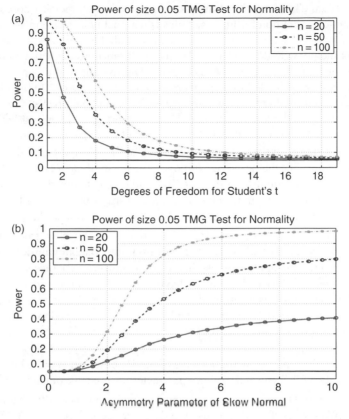

Figure 6.21 *Power of the Torabi et al. (2016) (TMG) test for normality, for three different sample sizes, and Student's t alternative (a) and skew normal alternative (b), based on 100,000 replications.*

what is observed and what is expected in each category. Examples of categorical data include a set of outcomes from a discrete distribution with finite support, such as the multinomial (e.g., rolling a fair die); or from a discrete distribution with infinite support (e.g., Poisson, geometric), with the last bin being of infinite length; or arising from contingency tables.

The British polymath Karl Pearson (1857–1936), arguably the founder of mathematical statistics (though later eclipsed by Fisher), and father of Egon Pearson (of Neyman–Pearson fame), proposed use of the statistic

$$X_P^2 = \sum_{i=1}^{m} \frac{(O_i - E_i)^2}{E_i}, \tag{6.9}$$

where O_i and E_i respectively denote the observed and expected number of observations in the ith category, with the expected number being calculated under the null hypothesis of interest, and m is the number of categories. The form of (6.9) is used because its asymptotic (as the sample size n tends to infinity) distribution under the null hypothesis is χ_d^2, where the degrees of freedom, d, depend on the model and number of estimated parameters. The asymptotic distribution was derived by Pearson in 1900; see Stuart and Ord (1994, p. 520) or Ferguson (1996, Ch. 9) for the straightforward proof using modern notation. The use of the asymptotic distribution for finite samples is reasonably accurate in certain modeling contexts and for typical sample sizes, so that the test could be conducted in Pearson's time. This contrasts sharply with the Q-Q and related tests, requiring substantial computer power to obtain the correct pointwise significance levels. Discussions of the importance and history of the X_P^2 test are provided by the venerable statisticians Cox (2002) and Rao (2002) in a volume commemorating the centenary of Pearson's landmark paper.

The test is also applicable to continuous distributions, operationalized by **binning** the data into m groups. In this case, the choice of m is no longer obvious, and becomes a discrete **tuning parameter** of the test, with the quality of the asymptotic χ^2 distribution, and the power of the test, dependent on it. Moreover, the width of each bin has to be selected, adding $m + 1$ continuous tuning parameters. To avoid having to decide the (optimal) width of the bins, and their corresponding values of E_i, we can apply the probability integral transform (p.i.t., see Section I.7.4) using the purported distribution (most likely, with estimated parameters) to get $U_i = F(X_i; \widehat{\theta})$, with $U_i \overset{\text{i.i.d.}}{\sim} \text{Unif}(0, 1)$ if $F(\cdot; \widehat{\theta})$ is the true c.d.f. of the X_i. (This idea is a special case of what is called the **Rosenblatt transformation**, from Rosenblatt, 1952). Then we can apply any test that tests the uniform hypothesis; in this case, the **equiprobable Pearson test** with each bin having the same length and same probability, $p_j = 1/m$. The asymptotic distribution of X_P^2 is χ^2_{m-1-k}, where k is the dimension of θ (the number of estimated parameters) and the additional degree of freedom subtracted occurs because of the linear constraint $n = \sum_{i=1}^{m} O_i$.

To get an idea of what the U_i look like when F is both correctly and incorrectly specified, Figure 6.22 shows histograms of 1000 values of the U_i, based on $m = 30$ bins, taking F to be the normal c.d.f. with mean and variance parameters estimated from the data. The top left panel uses $X_i \overset{\text{i.i.d.}}{\sim} N(0, 1)$, and so the resulting histogram is, unexpectedly, that of (perfectly) uniform data. We say "perfectly" because the data were generated as $F^{-1}(V_i)$, where F is

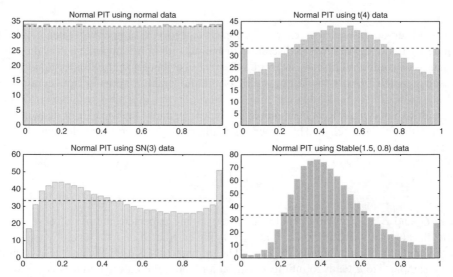

Figure 6.22 *Histograms of 1000 U_i values from the p.i.t., using 30 bins and F the normal c.d.f. with mean and variance parameters estimated from the data.*

the standard normal c.d.f., and, instead of taking the V_i to be a collection of i.i.d. Unif(0, 1) r.v.s (which would have generated a truly random sample of normal r.v.s), we use an equally spaced grid of values on (0, 1). Thus, the histogram is the prototype or typical histogram to be expected with a truly random sample. It deviates slightly from being perfectly uniform because the parameters of the normal distribution are estimated, and also because of the discrete nature of the histogram – in this case, the expected number of U_i in each bin is 1000/30 (as indicated by the dashed lines), and not integral.

The upper right panel of Figure 6.22 shows what happens when we take the X_i to be Student's $t(4)$ r.v.s, but wrongly assuming normality. They were computed as before using the grid of V_i values, so the resulting histogram is prototypical. Its shape is in accordance with the fact that the Student's t distribution has more mass near the center and in the tails compared to the normal distribution (which, again, was used in the p.i.t.) and is symmetric.

The lower left panel uses observations from a standard skew normal distribution with asymmetry parameter $\lambda = 3$. Finally, the bottom right panel uses a standard stable Paretian distribution with parameters $\alpha = 1.5$ and $\beta = 0.8$, thus exhibiting both fatter tails than the normal, and asymmetry. Indeed, its histogram has features that are prominent in both the Student's t and skew normal case.

Observe that there is a finite number of combinations of the O_i that can occur, the total number being $\binom{n+m-1}{n}$, from (I.2.1). For example, with $n = 20$ and $m = 3$, this number is 231. Also in this case, the X_P^2 statistic (6.9) easily simplifies to $(3/20)(\sum_1^3 O_i^2) - 20$, which is 1/10 for (any permutation of) $\mathbf{O} = [6, 7, 7]$; 4/10 for $\mathbf{O} = [6, 6, 8]$; 7/10 for $\mathbf{O} = [5, 7, 8]$; etc.

With this discrete support, we will not be able to get a cutoff value corresponding exactly to the 95% quantile. If we assume that the probability of an observation falling into one of the m bins is the same across all bins (true if the p.i.t. transformed data are precisely uniform distributed), then we can easily enumerate all possible cases and get the probability of each constellation via the multinomial distribution. For $n = 20$ and $m = 3$, the code to determine the exact probability that $X_P^2 = 1/10$ (this being the most likely value) is given in Listing 6.8. Running this results in 0.1145. (It would of course be quicker to just sum the three permutations of $\mathbf{O} = [6, 7, 7]$.)

The real use of the code would be to obtain the value

$$c_{0.05} = \min\{c \in S : \Pr(X_P^2 > c) \le 0.05\},$$

where S denotes the support of X_P^2; that is, to find the smallest of the 231 support values of X_P^2 such that the probability of exceeding it is less than or equal to 0.05. The problem with this idea for obtaining the cutoff value $c_{0.05}$ is that it does not take into account the fact that we use estimates of the two parameters of the normal distribution for the p.i.t., as needs to

```
1   ss=0;
2   for M1=0:20, for M2=0:(20−M1)
3     M3=20−M1−M2; M=[M1 M2 M3]; X2P=(3/20)*sum(M.^2)−20;
4      if X2P<0.101, p=C(20,M1)*C(20−M1,M2) / 3^20; ss=ss+p; end
5   end, end
```

Program Listing 6.8: Code for determining the exact probability that $X_P^2 = 1/10$. The function C is just the binomial coefficient, for example, $\binom{n}{k}$.

```
1   function [hyp X2P cutoff]=Pearsontestnormal(X,nbins,cutoff)
2   if nargin<3, cutoff=0.05; end
3   n=length(X); X=reshape(X,n,1);
4   edges=linspace(0,1,nbins+1); E=n/nbins * ones(nbins,1);
5   mu=mean(X); sig=std(X); U=normcdf(X,mu,sig);
6   O=histc(U,edges); O=O(1:end-1); X2P=sum((O-E).^2 ./ E);
7   if cutoff<1 % simulate to get the correct cutoff value
8       alpha=cutoff; sim=1e5; X2Pvec=zeros(sim,1);
9       for j=1:sim
10          if mod(j,1e4)==0, disp(['simulating, j=',int2str(j)]), end
11          R=randn(n,1); mu=mean(R); sig=std(R); U=normcdf(R,mu,sig);
12          O=histc(U,edges); O=O(1:end-1); X2Pvec(j)=sum((O-E).^2 ./ E);
13      end
14      cutoff=quantile(X2Pvec,1-alpha);
15  end
16  hyp=(X2P>cutoff);
```

Program Listing 6.9: The Pearson X_P^2 test, applied to binned, normal-p.i.t.-transformed data. The cutoff value can be passed to the function; we can use the $100(1 - \alpha)\%$ quantiles of χ^2_{m-3} (which are asymptotically justified, and where m is the number of bins), or obtained via simulation (which we have seen are more accurate). Note that, for $m = 3$, the χ^2_{m-3} quantile cannot be used, but simulation gives a value. If the passed value of cutoff is less than one, then it is taken to be the desired size of the test (default is 0.05 if not passed), and a simulation is then conducted to get the correct cutoff value (and also passed back).

be done when working with a composite null hypothesis. The results differ greatly in this case. Thus, as usual, the easiest way to get the correct cutoff values for the X_P^2 test is via simulation. This is implemented in the program in Listing 6.9.

Doing so in this case, with $n = 20$ and $m = 3$, results in the cutoff value of (precisely) 31/10, and yielding a test size of about 0.036, as can be seen from the first point in the upper left panel of Figure 6.23. For $m = 5$, we get a cutoff of 65/10. The simulation was done using $\alpha = 0.05$; for m, the number of bins, equal to $3, 5, 7, \ldots, 41$; and the four sample sizes $n = 20, 50, 100,$ and 1000. These cutoff values are then used to determine the power of the test; see below.

Simulation could be avoided by instead using the asymptotically correct cutoff values, obtained from the $100(1 - \alpha)\%$ quantiles of χ^2_{m-3} (valid only for $m > 3$). In this case, for $m = 5$, it is chi2inv(0.95,5-3) = 5.9915; this differs considerably from the value obtained via simulation. As n and m increase, the simulated and asymptotic values converge, as can be verified from Figure 6.23.

Remark.   Matlab has a built-in function chi2gof to perform the test (see its help file for details). It calculates the statistic differently, and, unfortunately the test does not have the correct size, as seen from Figure 6.23. For $n = 20$ the size is zero for all m, while for other sample sizes it is often above 0.05. It could be that, with the Matlab method, the power is higher than for our implementation (provided of course m is chosen so the size of the Matlab test is close to 0.05). It turns out to not be the case. As such, the Matlab method (as implemented in version 7.8) should not be used.   ∎

Figure 6.24 shows the power of the X_P^2 test for composite normality as a function of the number of bins, using the Student's t as the alternative, and having used the simulated

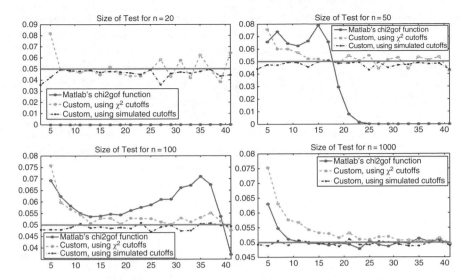

Figure 6.23 *Actual size of X_P^2 test as a function of the number of bins m, used as a composite test of normality (two unknown parameters), based on 100,000 replications, using the built-in Matlab function* `chi2gof` *(solid), and the custom implementation in Listing 6.9, using the asymptotically valid cutoff values from the χ_{m-1-2}^2 distribution (dashed), and cutoff values obtained via simulation (dash-dotted).*

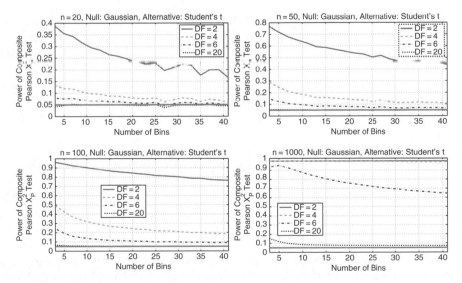

Figure 6.24 *The power of the X_P^2 test for normality, against Student's t alternatives with various degrees of freedom and for four sample sizes n, with nominal size $\alpha = 0.05$, using the method from Listing 6.9 with simulated cutoff values, and based on 1 million replications.*

cutoff values to ensure that the actual size of the test is close to the chosen nominal value of 0.05. It is immediately clear that the choice of three bins is optimal. Using this method, with its optimal choice of three bins, Figure 6.26(a) shows our usual power plot for the Student's t alternative; inspection shows that its power is comparable to that of the

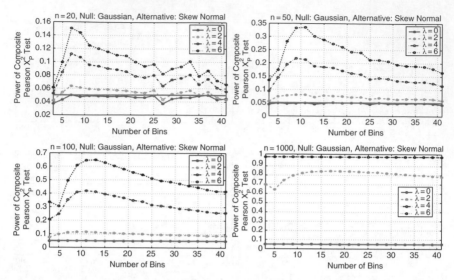

Figure 6.25   *Same as Figure 6.24 but using the skew normal as the alternative, with various asymmetry parameters λ and sample sizes n.*

Fowlkes-MP and KD tests, these being demonstrably lower than that of the AD and JB tests.

Figure 6.25 is the same as Figure 6.24, but using the skew normal as the alternative, with various asymmetry parameters λ and sample sizes n. Now, unlike with the Student's t alternative, the optimal number of bins depends on the sample size and the parameter under the alternative. Nevertheless, a compromise value of 11 bins appears reasonable. Figure 6.26(b) shows the usual power plot for the skew normal alternative, based on using 11 bins with the X_P^2 test. The power exceeds that of the AD test (recall that it did very poorly against this alternative) but is below that of the Fowlkes-MP and KD tests, and well below that of the JB, W^2, and MSP tests.

Based on the power for these two alternatives, the X_P^2 test for normality appears clearly inferior to most of the other tests we have seen. Moreover, it should be kept in mind that, in the X_P^2 power plots in Figure 6.26, we use the optimal number of bins for each of the two alternatives (three bins for Student's t, 11 for skew normal), thus giving it an advantage that the other tests do not have (they have no tuning parameters), lending further support against the use of X_P^2.

Remark.   The X_P^2 turns out to be a special case of what are called *smooth tests of goodness of fit*; see Rayner and Best (1990), Rayner and Rayner (1998), Rayner (2002), Rayner et al. (2009), and the references therein. Its use for a composite test of normality is discussed in Rayner et al. (2009, Sec. 6.2), with power results on their page 104.

Conveniently, they show a case with $n = 20$ and the Student's $t(4)$ and $t(6)$ alternatives, among others. Their stated power values are considerably better than those we obtained for the X_P^2 test, and are comparable to those of the JB test (about the same for $t(4)$; slightly lower for $t(6)$). However, as reported on their page 104, they used only 200 replications for their power study, so that the results are not very conclusive. For example, for a reported power of 0.30, this means 60 out of 200 trials were "successful" (rejected the null), and as

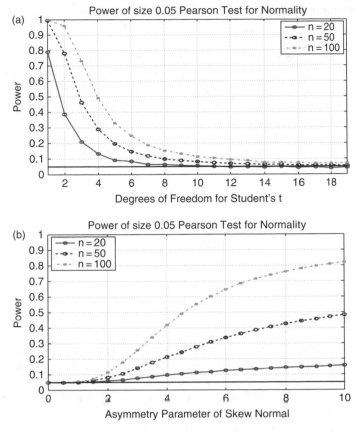

Figure 6.26 (a) Power of size 0.05 Pearson X_P^2 test for normality, based on 100,000 replications, using three bins (the optimal number, as indicated in Figure 6.24) and simulated cutoff values, for three different sample sizes, and Student's t alternative. (b) Same but using 11 bins (the compromise value from Figure 6.25) and the skew normal alternative.

these are just a set of i.i.d. Bernoulli trials, we can form a 95% c.i. for them, yielding (with the analytic method referred to at the end of Section 1.2, which yields shorter intervals than the bootstrap approach) $(0.24, 0.37)$. For a reported power of 0.17, the 95% c.i. is $(0.12, 0.23)$.   ∎

6.6 COMBINING TESTS AND POWER ENVELOPES

When two or more tests (which are not perfectly correlated) are available for testing the same null hypothesis, it might be possible to combine them such that a new test with higher power than its constituent components results. We consider two simple approaches for trying this in Section 6.6.1. We use the notation $A \prec B$ to indicate that test B has higher power than test A. Against a *specific* parametric distributional alternative that nests the null distribution, the likelihood ratio test will be the most powerful asymptotically, and thus, even with finite samples, can serve as a benchmark from which to judge other tests. Section 6.6.3 discusses this and the concept of the power envelope.

6.6.1 Combining Tests

6.6.1.1 Combining Tests: Method I

Let $T_{1,\alpha}$ and $T_{2,\alpha}$ be two tests of (nominal and actual) size α such that, for a given sample size and alternative hypothesis, $T_{1,\alpha} \prec T_{2,\alpha}$. Furthermore, assume initially that their outcomes (as realizations of Bernoulli r.v.s) are *independent* under the null. Our first attempt proceeds as follows. Let the combined test C_γ be the test of size γ that rejects the null when at least one of the $T_{i,\alpha}$ rejects. Size γ is determined as the probability that at least one $T_{i,\alpha}$ rejects the null, when the null is true. This is the complement of the event that they both do not reject, or $\gamma = 1 - (1 - \alpha)^2$. For $\alpha = 0.05$, this yields $\gamma = 0.0975$. (If we wish to have $\gamma = 0.05$, we take α to be the solution to

$$1 - (1 - \alpha)^2 = 0.05, \tag{6.10}$$

or $\alpha = 0.02532$.) Of interest is the possibility that C_γ has higher power than $T_{2,\gamma}$ (when both constituent tests are based on size γ). This will depend on the relative power properties of the T_i for sizes α and γ, though it seems plausible that, with a certain "uniformity" of the power curves, the most likely outcome will be $T_{1,\gamma} \prec C_\gamma \prec T_{2,\gamma}$.

To illustrate this with our battery of normality tests, we first need to assess whether any two of them exhibit pairwise independence under the null. Using the correlation as a measure of dependence, a simulation with 1 million replications was done for several of the tests (Q-Q was not used because it takes too long), using $n = 50$ and $\alpha = 0.05$, with the resulting correlation matrix shown in Table 6.3. As expected, the KD and Fowlkes-MP tests are perfectly correlated, yielding a sample correlation of 0.99 (the discrepancy from 1.00 resulting from having used a finite number of replications). The next highest correlation is, somewhat unsurprisingly, between W^2 and U^2. Also as expected, no two tests are uncorrelated, though the weakest correlation is rather small, 0.03, between AD and X_P^2. Alas, these are our worst tests against the skew normal alternative.

Let us choose MSP and JB, these having the relatively mild correlation of 0.23, and high power against the skew normal. We take test C to be the combined test based on MSP and JB. From the same simulation used to obtain the correlations, we obtain the empirical size of C to be $\gamma = 0.0860$. This is less than 0.0975 because the two tests are correlated. (If they were perfectly correlated, then γ would equal α.) We then run the code in Listing 6.10 to confirm the sizes of the three tests MSP, JB, and combined; we obtain 0.084, 0.086, and 0.086, respectively.

TABLE 6.3 Correlation between tests for normality under the null, using sample size $n = 50$, based on 1 million replications. For the X_P^2 test, nine bins were used

	KD	AD	W^2	U^2	MSP	F–MP	JB	X_P^2	TMG
KD	1.00								
AD	0.06	1.00							
W^2	0.59	0.10	1.00						
U^2	0.58	0.08	0.89	1.00					
MSP	0.44	0.17	0.43	0.41	1.00				
F–MP	0.99	0.06	0.59	0.58	0.44	1.00			
JB	0.14	0.71	0.21	0.18	0.23	0.14	1.00		
X_P^2	0.29	0.03	0.31	0.32	0.22	0.29	0.06	1.00	
TMG	0.56	0.09	0.78	0.76	0.43	0.56	0.18	0.30	1.00

```
1   alpha=0.05; gamma=0.0860; n=50; sim=1e5; hyp=zeros(sim,3);
2   for j=1:sim, y =  randn(n,1);
3     hyp(j,1)=MSPnormtest(y,gamma); hyp(j,2)=jbtest(y,gamma);
4     tst1=MSPnormtest(y,alpha); tst2=jbtest(y,alpha); hyp(j,3)=tst1|tst2;
5   end, mean(hyp) % actual power (or size under the null)
```

Program Listing 6.10: Code to confirm the sizes of the MSP, JB, and combined tests.

Running the code again, but using $t(4)$ data, shows the power of the MSP, JB and combined tests to be 0.42, 0.61, and 0.57, respectively, demonstrating that the combined test is unfortunately less powerful than JB. A similar finding occurred for other degrees of freedom. For the skew normal alternative with $\lambda = 8$, the powers are 0.83, 0.76, and 0.82, so that the combined test is nearly equal to, but not as powerful as, the MSP test.

6.6.1.2 Combining Tests: Method II

The following method is used in Section 9.5.4 in the context of testing the symmetric stable distribution. Here we provide more detail on how and why it works, and apply it to normality testing.

Let us first assume we have $k = 2$ independent tests of some null hypothesis. For a given data set, they yield p-values of, say, $p_1 = 0.11$ and $p_2 = 0.15$. While neither is below any of the traditional significance levels, the fact that two independent tests both have a relatively low p-value might still be evidence that the null hypothesis is questionable. Indeed, if we have k such independent tests, then, under the null, their p-values should be an i.i.d. sample of k values from a Unif$(0, 1)$ distribution. If they tend to cluster more towards zero, then this is certainly evidence that the null hypothesis may be false.

One might be tempted to look at the maximum of $p_1, \ldots, p_k$, whose distribution is easily computed. However, this is throwing information away, as easily seen by considering the two cases $p_1 = p_2 = p_3 = 0.06, p_4 = 0.4$, and $p_1 = p_2 = p_3 = 0.38, p_4 = 0.4$. A similar argument holds when considering just the minimum, or the minimum and maximum. To incorporate all the p-values, one idea would be to take their product, which, under the null, is their joint distribution. We use the log transformation (any monotonic transformation could be used), so we get the sum of logs of the p-values. A test based on this sum would then deliver a p-value commensurate with its distribution. This, like many good ideas in statistics, goes back to Ronald Fisher, and is known as Fisher's **combined probability test**. We will refer to it just as the **joint test**.

Recalling that if $U \sim$ Unif$(0, 1)$, then $-\log U \sim Exp(1)$ (Example I.7.7), that a sum of i.i.d. exponential r.v.s is gamma (Example II.2.3), and, finally, the relation between gamma and χ^2 r.v.s as given in, for example, (I.7.41), it follows that -2 times the sum of the log of the p-values follows a χ^2_{2k} distribution. The code in Listing 6.11 demonstrates the power increase obtained from the joint test.

The beta distribution is used to produce values of p between zero and one but such that they have a higher probability of being closer to zero (thus mimicking the behavior of p-values under an alternative hypothesis). Choosing the beta parameters as $a = b = 1$ results in the uniform distribution, useful for confirming the size of the test, while taking $a = 1$ and $b > 1$ will induce the desired effect of rendering the p-values closer to zero. Note that the joint test rejects the null when the product of the p-values, or the sum of their logs, is small. As we work with the negative of (twice) this sum, we reject for large values, which is why in the above code we use the right-tail area `1-chi2cdf(slALT,2*k)`.

```
1   k=2; sim=1e5; newpval=zeros(sim,1);
2   a=1; b=1; % a=b=1 corresponds to Unif(0,1), while b>1 yields values
3                   % that tend to be closer to zero, thus emulating the
4                   % behavior of p-values under some alternative hypothesis
5   for i=1:sim
6       pvals = betarnd(a,b,[k,1]); % iid draws from Beta(a,b)
7       sALT=-2*sum(log(pvals)); newpval(i) = 1-chi2cdf(sALT,2*k);
8   end
9   mean(newpval<0.05)
```

Program Listing 6.11: Demonstration of the concept of combining tests, using simulated beta random variables as the *p*-values.

Our application unfortunately involves tests that are not independent, so that the exact distribution of the sum of logs of the *p*-values is not χ^2_{2k}. So, before proceeding, we illustrate how the above exercise (but still assuming independence) could be done with simulation. Step one is to simulate $S_1, \ldots, S_B$, where the S_i are i.i.d., each being the sum of the log of k i.i.d. Unif(0, 1) r.v.s, and B is a big number, for which we use 1 million. In step two, we simulate $p_i \overset{\text{i.i.d.}}{\sim} \text{Beta}(a, b)$, and calculate the fraction of the S_i that are less than $\log p_1 + \cdots + \log p_k$. This yields the *p*-value of the joint test. The power of the joint test for a nominal size of 0.05 is obtained by repeating this sim times and calculating the fraction of times the *p*-value is less than 0.05.

For comparison purposes, when doing this, we also store, for each replication, $m_i = \min(p_1, \ldots, p_k)$. The power of using m_i is then given by the fraction of the m_i that are less than the appropriate cutoff value, which is calculated from (6.10) (there shown for $k = 2$).

Running the code in Listing 6.12 implements this for $k = 2$ independent tests. It outputs, for the *p*-value of the minimum and the *p*-value of the product method, 0.052 and 0.050 respectively, confirming that the size is correct. Running the second part of the code, but using $a = 1$ and $b = 3$ for the beta distribution, yields 0.14 and 0.22 respectively, showing that the product method does indeed lead to a test with higher power. Repeating this with $k = 4$ yields the correct size, and powers 0.14 and 0.39; for $k = 10$, we get 0.14 and 0.75.

Having established the required simulation framework, we can now proceed with tests that are not necessarily independent. With the ability to quickly and accurately approximate the *p*-values of the MSP and JB tests, we can easily determine the performance of the joint MSP+JB test. The first step is to compute $S_1, \ldots, S_B$, where the S_i are i.i.d., each being the sum of the log of the MSP and JB *p*-values, computed for a normal sample of length $n = 50$,

```
1   k=2; sim=1e6; sINULL=zeros(sim,1);
2   for i=1:sim, U=rand(k,1); sINULL(i) = sum(log(U)); end
3   sim=1e4; newpval=zeros(sim,1); minpval=zeros(sim,1); a=1; b=1;
4   for i=1:sim
5       pvals = betarnd(a,b,[k,1]); minpval(i)=min(pvals);
6       sALT=sum(log(pvals)); newpval(i) = mean(sINULL<sALT);
7   end
8   mcut = 1-0.95^(1/k); mean(minpval<mcut), mean(newpval<0.05)
```

Program Listing 6.12: Computes the *p*-value, `newpval`, based on the *p*-values of $k = 2$ independent tests (of the same null hypothesis and same data set), and the power of the resulting 5% test, as `mean(newpval<0.05)`.

and B is a big number; we use 100,000. Notice that we cannot use the (faster) method of generating the two p-values as independent uniforms, or the analytic method via the χ_4^2 distribution, because the MSP and JB tests (and, thus, their p-values) are not independent, as was seen in Table 6.3. Once these are computed, we can calculate the power of the joint test as follows. For a particular data set (of length 50) with MSP and JB p-values p_{MSP} and p_{JB}, respectively, the p-value of the joint test, $p_{\mathrm{MSP+JB}}$, is computed as the fraction of S_i that are less than $\log p_{\mathrm{MSP}} + \log p_{\mathrm{JB}}$. This is conducted for `sim` data sets, and the fraction of $p_{\mathrm{MSP+JB}}$ that are less than 0.05 is the power of the joint test.

Figure 6.27 shows the results of doing this, with `sim` $= 100,000$ replications, for the Student's t and skew normal alternatives. Overlaid in the plots are the power values of the MSP test and JB tests. The first order of business is to confirm is that the size of the joint test is correct. This is clear from the bottom panel for $\lambda = 0$. Regarding power, we see that the joint test has lower power than the JB for the Student's t alternative, but much higher than the MSP. For the skew normal alternative, the joint test has higher power than both MSP and JB tests, over the entire range of λ we considered, and substantially so for $3 \leq \lambda \leq 6$.

Our results are tempered, however, by the fact that the TMG test (Figure 6.21) still has higher power than this joint test. As mentioned, the reader is encouraged to augment the

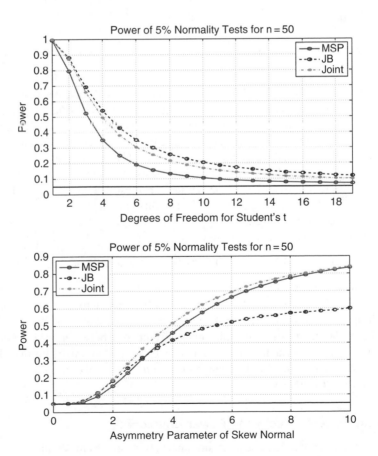

Figure 6.27 *Power of the MSP, JB, and joint tests using $n = 50$ and 100,000 replications.*

TMG test with p-values as we did above for the MSP and JB tests, and to attempt to create a joint test using TMG as one of the constituents such that the power of the joint test is higher.

Remark.   The Stengos and Wu (2010) test also delivers a quickly computed p-value, so that we can entertain combining it with the MSP and JB tests, in a similar way as described above, so that a total of three tests are used. Somewhat disappointingly, the resulting power for the skew normal was very close to that of the previous (two-factor) joint test, with only a very slight increase in power for smaller values of λ. For the power against Student's t, the three-factor test indeed has higher power than the two-factor test for all degrees of freedom considered, but still has power less than the JB test. As such, we do not report the graphs.   ∎

6.6.2 Power Comparisons for Testing Composite Normality

Having now seen several tests for the null hypothesis of composite normality, we can sum-marize their relative performance. Denote the Ghosh test by G, with obvious notation for the other tests, and let Joint refer to the joint test MSP+JB, introduced in Section 6.6.1.2 above. With respect to the Student's t alternative, the various tests we considered above can be ordered in terms of power as

$$X_P^2 \lessgtr \mathrm{KD} = \mathrm{F\text{-}MP} \prec \mathrm{MSP} \prec W^2 = U^2 \prec \mathrm{Q\text{-}Q} \prec \mathrm{AD} \lessgtr \mathrm{G} \lessgtr \mathrm{KL1} \lessgtr \mathrm{Joint} \prec \mathrm{JB},$$
$$(6.11)$$

where $A \prec B$ indicates that B has higher power than A; $A \lessgtr B$ indicates that A and B have similar power, with one exceeding the other under some conditions (sample size and alter-native), and vice versa; and $A = B$ means that the powers of A and B are, or appear to be, theoretically equal. It is important to keep in mind that (6.11) is based on a limited number of tests, and is intended to illustrate how one might display their ranking. For example, Gel et al. (2007) and Gel and Gastwirth (2008) study tests that exhibit higher power for some heavy-tailed alternatives than that of the Jarque–Bera test (see also the simulation results in Torabi et al., 2016, comparing 40 tests of normality, and also serving as a useful source for much of the recent literature).

For the skew normal alternative, we have

$$\mathrm{AD} \prec X_P^2 \prec \mathrm{KD} = \mathrm{F\text{-}MP} \prec U^2 \lessgtr \mathrm{Q\text{-}Q} \lessgtr \mathrm{JB} \lessgtr W^2 \lessgtr \mathrm{MSP} \prec \mathrm{Joint} \prec \mathrm{TMG}. \quad (6.12)$$

It is crucial to keep in mind that these results are based on having used just the three sample sizes $n = 20$, 50, and 100, and size $\alpha = 0.05$, except for the result for the Joint test, which only used $n = 50$ and $\alpha = 0.05$. The point here is not to make definitive com-parisons, but rather to illustrate the enormous discrepancy between orderings (6.11) and (6.12), as well as those pertaining to the various two-component mixed normal alternatives from Table 6.2, emphasizing the important point that *the power of a test can strongly depend on the alternative hypothesis.*

6.6.3 Most Powerful Tests and Power Envelopes

The fact that there is not a single test that dominates with respect to two alternatives implies that none of these tests is **most powerful** for testing composite normality. In order for a test to be most powerful, it has to exhibit, as the name suggests, the highest power *with respect to all alternatives*. It is unlikely that such a test exists. Indeed, one could presumably design a test that has higher power than all other known tests for a very specific alternative. Such a notion leads us to entertain the existence of a most powerful test *with respect to a specific alternative* (either composite or simple). In this case, a test indeed exists that is, under certain conditions, **asymptotically most powerful unbiased**. It is the (maximum) likelihood ratio test (l.r.t.), introduced in Section 3.3.2.

Consider the case of testing (composite) normality versus the alternative hypothesis of (composite) skew normal. The l.r.t. serves as an approximation to the **power envelope**, or the locus of points that gives, for a particular skew normal asymmetry parameter λ, the supremum of the power of all hypothesis tests for a given sample size n and test size α. The correct 5% cutoff value is obtained via simulation with the code in Listing 6.13.

We use 100,000 replications to get the correct cutoff, and, based on that cutoff value, 10,000 replications to determine the power under each alternative considered. The function SNestimation (not shown) parallels the tlikmax function in Listing 4.6, and computes the m.l.e. and the log-likelihood evaluated at the m.l.e. for the skew normal distribution and for data set data (and an initial vector, which we pass as the true parameter value), where the p.d.f. is given in (A.115). The function skewnormrnd is given in Listing 2.12.

Figure 6.28(b) overlays the power of the MSP test for normality and the power envelope from the l.r.t., with the skew normal as the specific alternative. We see that the l.r.t. has blatantly higher power. Keep in mind that this test was *designed* to have power against this specific alternative, and in many situations involving testing if a set of data comes from a particular distribution, it will not be realistic to specify a particular parametric alternative. *The main value of the l.r.t. power curves is that they serve as an envelope to indicate how close our general test (against unspecified alternatives) is to being most powerful against a particular alternative.*

Upon looking at the large increase in power, one might be inclined to believe that this l.r.t. (with its specific skew normal alternative) will still have high power as a general test for normality against a non-specific alternative. While this might hold for alternatives that are somehow "similar" to the skew normal, it will not be true in general. Indeed, as Thode (2002, p. 7) aptly warns in this context:

```
1  n=50; sim1=1e5; LRT=zeros(sim1,1);
2  for k=1:sim1, if mod(k,1e3)==0, k, end
3    lam=0; % set to 0 for the null hypothesis and to get the cutoff value
4    data=skewnormrnd(n,lam);
5    muhat=mean(data); sighat=std(data,1); % MLE for normal
6    normloglik = sum( log(normpdf(data,muhat,sighat)) );
7    [param,stderr,iters,loglik,Varcov] = SNestimation(data,[lam,0,1]);
8    LRT(k)=-2*(normloglik - loglik);
9  end
10 if lam==0, cut50=quantile(LRT,1-0.05), end % cutoff under the null
11 power = mean(LRT>cut50)    % use to get the power under the alternative
```

Program Listing 6.13: Simulates the l.r.t. for determining appropriate cutoff values for the test of normality versus skew normal.

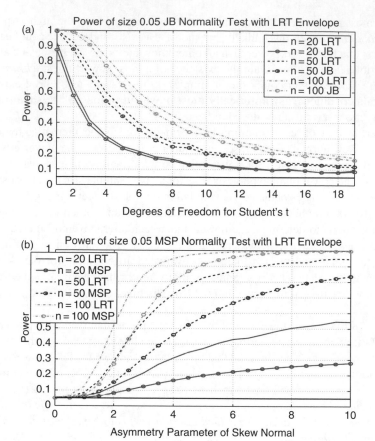

Figure 6.28   *(a) The power of the JB test (6.7) against the alternative of a Student's $t(v)$ (lines with circles; same power curves as given in Figure 6.18(a)), along with the power of the likelihood ratio test (using the Student's t as the specific alternative), based on 10,000 replications. (b) The power of the MSP test for normality against the alternative of skew normal (lines with circles; same power curves as in Figure 6.10(b)), along with the power of the likelihood ratio test (using the skew normal as the specific alternative), based on 10,000 replications.*

> Some tests, such as the likelihood ratio tests and most powerful location-
> and scale-invariant tests, were derived for detecting a specific alternative
> to normality. ... The disadvantages of these tests are that ... they may not
> be efficient as tests of normality if in fact neither the null nor the specified
> alternative hypotheses are correct.

This discussion does not imply that an l.r.t. will never have good power against other alternatives than the one it was designed for. As an example, Paolella (2016a) shows that an l.r.t. for the asymmetric stable Paretian distribution, using the noncentral Student's t as the alternative, has excellent power properties against a variety of other interesting alternative candidates. In the setting here, the reader is encouraged to investigate the power performance of the l.r.t. for composite normality with the skew normal as the alternative, for testing against the Student's t alternative, and vice versa.

Figure 6.28(a) shows the power of the JB test (6.7) against Student's $t(v)$, with the corresponding power envelope based on the l.r.t. using the Student's t as the specific alternative. In this case, particularly for smaller samples, the difference in power is not substantial, indicating that the JB test is close to being the most powerful test of normality for Student's t alternatives and probably among the most powerful tests for general fat-tailed alternatives.

6.7 DETAILS OF A FAILED ATTEMPT

This section can be skipped – it details an indulgence in a curiosity that does not lead to an effective test.

Ideas that do not work do not make their way into research papers (and usually also not into books). The point of illustrating this unsuccessful technique here (besides showing some nice graphics) is to add some further insight into the bootstrap, as well as to encourage the reader to experiment with ideas and to get used to the fact that research is challenging and that not all ideas will work. Problems 6.4 and 6.9 illustrate other "techniques" that also do not lead to tests with improved power.

With the ability to quickly approximate the p-value of the MSP test, and having spent most of Chapter 1 illustrating the usefulness of the bootstrap, one might consider what the bootstrap distribution of the MSP p-value looks like, and whether it is useful in some way. With respect to the first question, if we use the parametric bootstrap (generating i.i.d. normal random samples using the m.l.e. of the parameters μ and σ^2 fitted from the actual data set), then clearly, the resulting p-value distribution will be perfectly uniform (at least up to the level of the approximation used to calculate the MSP p-value), because the MSP test is for composite normality, and is thus invariant to the values of μ and σ^2. If, however, we use the nonparametric bootstrap, matters are not as clear.

One might initially think that a resample from the actual data is also normally distributed, but that is not the case: it is a discrete, multinomial distribution. We saw in Chapter 1 that the nonparametric and parametric bootstrap distributions are identical in the Bernoulli (or, more generally, multinomial) model, and that, for the geometric model, the parametric was preferred to the nonparametric, if in fact the true model is really geometric. In those cases, statistics from the sample were used to deliver a (consistent) point estimator. In our case here, we are not working with a consistent estimator of a model parameter, but rather using the MSP test statistic (6.6), whose behavior will differ if we use a multinomial distribution instead of a normal.

In each resample (except, possibly, the one in which all n sample values are taken), there will be repeats of values from the original sample, causing the mass function to have some "spikes" that otherwise have probability zero of occurring in an i.i.d. normal random sample. Figure 6.29 shows a histogram of a normal random sample of size $n = 50$, its MSP plot, and the same for a resample of that data set. The histograms are shown on the same scales; observe the pile-up of 18 values in the histogram around 0.25. The MSP test of the original data does not reject the null (its p-value is 0.25), but it does so for the resample (with a p-value of 0.00025). Thus, in general, we expect that the nonparametric bootstrap distribution of the MSP p-value will be highly asymmetric, with much more mass near zero. This is further confirmed in Figure 6.30, showing the actual p-value and the nonparametric bootstrap distribution for two normal data sets. The code used to generate those plots is given Listing 6.14.

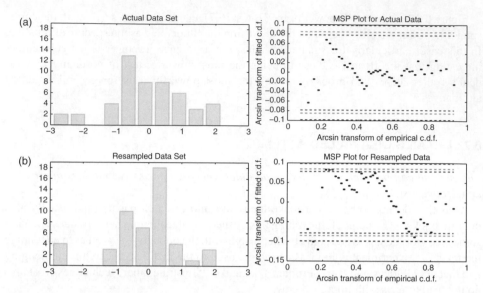

Figure 6.29 *(a) (left) histogram of an i.i.d. normal sample of size $n = 50$; (right) its MSP plot. (b) similar, but using a resample from the original data set.*

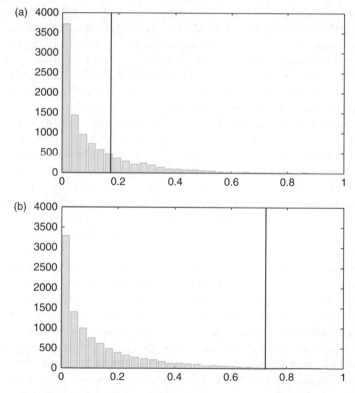

Figure 6.30 *(a) The nonparametric bootstrap distribution of the p-value based on a random normal sample of size $n = 50$, using the MSP test and $B = 1000$ resamples. Its p-value is marked with the vertical line. (b) Same, but for a different random normal sample.*

```
1  n=50; B=1e4; pb=zeros(B,1); x=randn(n,1);
2  [g1, g2, g3, pval0]=MSPnormtest(x,0.05);
3  for b=1:B
4      ind = unidrnd(n,[n,1]); xb=x(ind);
5      [g1, g2, g3, pval]=MSPnormtest(xb,0.05); pb(b)=pval;
6  end
7  hist(pb,30), ax=axis; line([pval0 pval0],[0 ax(4)])
```

Program Listing 6.14: Code used to generate Figure 6.30.

One might imagine that the fraction of bootstrap p-values less than, say, 0.05 will be a function of the p-value of the actual data set. In particular, denoting by p_0 the MSP p-value of the data, and by $p^*_{0.05}$ the fraction of bootstrap p-values under 0.05, we might expect that p_0 is the lower value, and $p^*_{0.05}$ the higher. This is indeed the case for the examples shown in Figure 6.30. Instead of just looking at two cases, let us consider 10,000. A scatterplot of p_0 and $p^*_{0.05}$ is shown in Figure 6.31(a), based on 10,000 replications, each using $B = 1000$ bootstrap resamples. As expected, $p^*_{0.05}$ tends to be higher for smaller values of p_0. To help quantify this, the overlaid lines are fitted values from a **quantile regression**. Instead of modeling the mean of the dependent variable as in a traditional regression, it models a specific quantile, for which we used 0.5 (the median), 0.01, 0.05, 0.95 and 0.99.[7] For example, the fitted median (for $n = 50$) is

$$F^{-1}_{p^*_{0.05}}(0.5) = \mathrm{median}(p^*_{0.05}) = 0.9841 + 0.8107 p_0 - 1.2456 p_0^{1/2} - 0.3527 p_0^2, \qquad (6.13)$$

and the 95% quantile is

$$F^{-1}_{p^*_{0.05}}(0.95) = 1.0435 + 0.4343 p_0 - 0.8843 p_0^{1/2} - 0.3171 p_0^2. \qquad (6.14)$$

Figure 6.31(b) is similar, with the points of the scatterplot having been generated under the alternative of a skew normal with $\lambda = 6$, *but the lines in the plot are the same as those in the top panel*. This allows us to more easily compare the behavior of the relationship between p_0 and $p^*_{0.05}$ under the null and alternatives. Observe how the p_0 under the alternative cluster much closer to zero. Also, and perhaps unexpectedly, the distribution of $p^*_{0.05}$ *conditional on p_0* does not differ appreciably from that under the null. This is unfortunate, because if it had, say, a longer right tail, then this difference could possibly have been exploited to develop a test with higher power. We can try nonetheless.

Consider the following: If $p_0 < 0.05$, then reject the null straight away, as usual. Otherwise, conduct the bootstrap to get $p_1, \ldots, p_B$, $B = 1000$, and calculate $p^*_{0.05}$ (the fraction of $p_1, \ldots, p_B$ below 0.05). Given p_0, compute the 95% quantile of $p^*_{0.05}$ under the null from (6.14), say (for ease of notation) $q_{0.95}$. We then reject the null hypothesis in this augmented MSP test if, as already stated, $p_0 < 0.05$, or if $p_0 > 0.05$ and $p^* > q_{0.95}$. Code to implement this is given in Listing 6.15. Running it with 10,000 replications under the null yields an actual size of 0.097 for the usual MSP test with nominal size 0.10, and an actual size of the augmented MSP test of 0.095. These actual sizes being very close, we can proceed to check the power. For the alternative of skew normal with $\lambda = 4$, the two tests have power values

---

[7] At this time, Matlab does not have routines for quantile regression in their statistics package, but function `rq_fnm` for Matlab is available, kindly posted on the web by Roger Koenker, a prominent researcher in quantile regression.

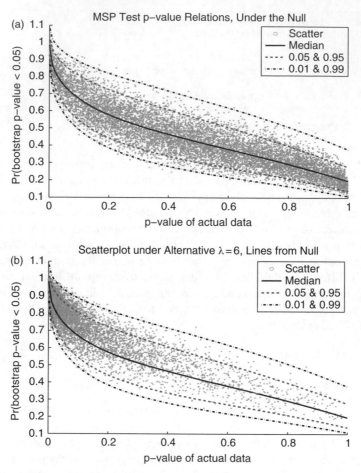

Figure 6.31 *(a) Scatterplot based on 10,000 replications, with x-axis showing the p-value p_0 from the MSP test, using a data set from the null, for $n = 50$, and y-axis showing the fraction of bootstrap p-values (B = 1000), based on that data set, that were less than 0.05. The lines were obtained from quantile regression using regressors a constant, p_0, $p_0^{1/2}$ and p_0^2. (b) Similar to the top panel, but the scatterplot corresponds to points obtained using a skew normal alternative with $\lambda = 6$, but the lines are the same as those in the top panel, that is, correspond to the quantiles under the null.*

0.58 and 0.52, respectively, demonstrating, as we suspected from the results in Figure 6.31, that the augmented test will not perform better.

Having a fast and accurate way to approximate the p-value, we can conduct the same exercise with the JB test as was done for the MSP test above; recall Figure 6.31. Figure 6.32 is similar, but based on the JB test (and only showing the median and 95% fitted quantile functions). The shape is very different compared to that for MSP, with $p_{0.05}^*$ being much closer to the value it would be using the parametric bootstrap – exactly 0.05. This is because the JB test uses the sample skewness and kurtosis, which are consistent estimators of their theoretical counterparts. From the bottom panel, we see that, for roughly $p_0 < 0.07$, $p_{0.05}^*$ under the alternative (the scatterplot points) is much higher than under the null (as indicated by the overlaid $F_{p_{0.05}^*}^{-1}(0.95)$ line).

```
1   bb95 =[1.044 0.434 −0.884 −0.317]'; bb99 =[1.091 0.433 −0.772 −0.375]';
2   sim=1e4; h=zeros(sim,1); hyp05=h; hyp10=h; hyp95=h; hyp99=h;
3   B=1e3; pb=zeros(B,1); n=50; lam=0;
4   for i=1:sim
5     x=rsn(n,0,1,lam);
6     [h05, g2, g3, pval05]=MSPnormtest(x,0.05); hyp05(i)=h05;
7     [h10, g2, g3, pval10]=MSPnormtest(x,0.10); hyp10(i)=h10;
8     pval0=pval05;
9     if pval0<0.05, hyp95(i)=1; hyp99(i)=1;
10    else
11      X=[1 pval0 pval0^(1/2) pval0^2]; hi95=X*bb95; hi99=X*bb99;
12      for b=1:B
13        ind = unidrnd(n,[n,1]); xb=x(ind);
14        [g1, g2, g3, pval]=MSPnormtest(xb,0.05); pb(b)=pval;
15      end
16      pstar = mean(pb<0.05); hyp95(i)=(pstar>hi95); hyp99(i)=(pstar>hi99);
17    end
18  end, powers = mean([hyp05 hyp10 hyp95 hyp99])
```

Program Listing 6.15: Computes size (or power if $\lambda \neq 0$, for the skew normal alternative) of the usual MSP tests with sizes 0.05 and 0.10, and also that of the augmented MSP test using the 95% and 99% cutoff values of $p^*_{0.05}$.

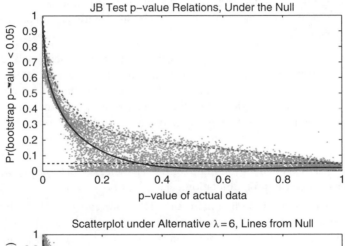

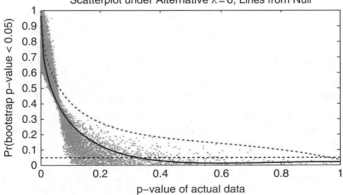

Figure 6.32 *Similar to Figure 6.31 but using the JB test, and only showing the median and 95% quantile fitted lines.*

Nevertheless, our hopes of obtaining a test with higher power are still dashed: Using the same method described above for the MSP test, we find the actual size of the augmented JB test to be $\alpha = 0.0948$ (based on 100,000 replications). Then, comparing the usual JB test with size $\alpha = 0.0948$ to the new augmented test using a skew normal alternative with $\lambda = 2$, we obtain (using 10,000 replications) power values 0.29 and 0.21, respectively. Similar discrepancies result for other values of λ.

6.8 PROBLEMS

6.1 Consider the data set 84, 86, 85, 82, 77, 76, 77, 80, 83, 81, 78, 78, 78, which are results from 1935 experiments to determine the acceleration due to gravity; see Davison and Hinkley (1997, p. 72) for the units, further description, and references.

   Davison and Hinkley (1997, p. 154) discuss the use of simulation in the context of the bootstrap to produce a normal Q-Q plot of this data set with a simultaneous significance level of 0.1. Make a normal Q-Q plot for this data (standardized, that is, subtract the mean and divide by the sample standard deviation) showing simultaneous null bands of sizes 0.05 and 0.10. Show that the third occurrence of the value 78 lies in the Q-Q plot between the 0.05 and 0.10 simultaneous null bands.[8]

6.2 Write a program that produces the mapping for Weibull data shown in Figure 6.4. Hint: To simulate and estimate the Weibull, use the built-in Matlab commands `wblrnd` and `wblfit`, respectively.

6.3 Recall that $f_{\text{Lap}}(x; 0, 1) = \exp(-|x|)/2$ is the location-zero, scale-one Laplace p.d.f. The Laplace distribution has been used to fit data in a wide variety of contexts (see, for example, Kotz et al., 2001, and the references therein), and so testing its goodness of fit is of importance. Various tests have been proposed in the literature; see, for example, Yen and Moore (1988), Puig and Stephens (2000), Noughabi and Arghami (2013), and the references therein.

   (a) Show that, if U_1 and U_2 are i.i.d. Unif(0, 1), then $L = \ln(U_1/U_2) \sim \text{Lap}(0, 1)$. We will need this result for simulation.

   (b) **(i)** Similar to the discussion in Section 2.3, write a program that, for a given sample size, calculates the cutoff values associated with the KD and AD statistics for testing whether a random sample is i.i.d. Laplace with location and scale parameters μ and σ. Also use the Cramér–von Mises statistic (2.30) and Watson's statistic (2.31). Hint: To estimate μ and σ from the data $X_1, \ldots, X_n$, use the maximum likelihood estimator, given by $\hat{\mu}_{\text{ML}} = \text{median}(X_1, \ldots, X_n)$ and $\hat{\sigma}_{\text{ML}} = n^{-1} \sum_{i=1}^{n} |X_i - \hat{\mu}|$.

   (ii) It turns out that, for Laplace, the AD test performs poorly. Instead of using the definition in (2.23), which takes the maximum, use the penultimate (second largest) value and denote the test by $\text{AD}^{[-1]}$. More generally, let $\text{AD}^{[-k]}$ be the resulting statistic defined in an obvious way. Figure 6.33(a)

---

[8] This data set and the associated normal Q-Q plot were also analyzed in Rosenkrantz (2000), who presents an analytical method for calculating $100(1 - \alpha)\%$ simultaneous null bands for the theoretical quantiles. These are, unfortunately, not valid for the more relevant empirical quantiles; see Webber (2001).

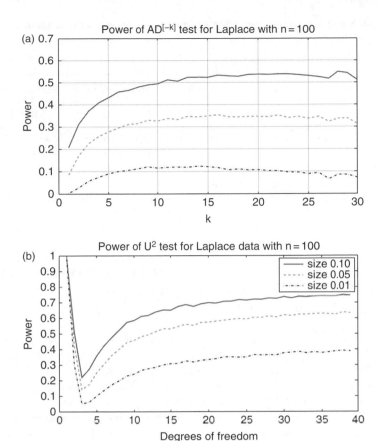

Figure 6.33 (a) Power of the Laplace $AD^{[-k]}$ test against the normal distribution as a function of k, for $n = 100$, for three different test sizes (see legend in the bottom panel). (b) Power of U^2 test for Laplace with $n = 100$ against Student's t alternative with v degrees of freedom, for three different test sizes.

shows the power of the Laplace $AD^{[-k]}$ test as a function of k, for $n = 100$, and three different test sizes. Replicate this. Note that the optimal k changes with respect to the size of the test. A good compromise value appears to be $k = 15$.

(iii) Compute the power of all these tests against Gaussian and Student's t alternatives. As a check, for $n = 100$ and $n = 200$, the powers against a Gaussian are given in Table 6.4.

The power values for KD, W^2, and U^2 for size 0.05 agree with those given in Puig and Stephens (2000, Table 6). (They did not use the AD statistic.) Indeed, as suggested by Durbin, U^2 is much more powerful than W^2, and has the highest power of all the statistics considered. Figure 6.33(b) shows the power of U^2 against Student's t with v degrees of freedom, as a function of v. Note that, as v increases, the power approaches the values given in Table 6.4.

(c) Similar to Figure 6.4, construct the mapping $s_{\mathrm{Lap}}(p, n)$ associated with pointwise and simultaneous significance levels for the null bands of the Laplace Q-Q plot for $n = 100$.

TABLE 6.4 The power of various tests for Laplace, against the Gaussian alternative, for $n = 100$ and $n = 200$

	n = 100				n = 200			
Size \ Test	KD	AD[-15]	W^2	U^2	KD	AD[-15]	W^2	U^2
0.10	0.545	0.542	0.590	0.795	0.836	0.827	0.918	0.983
0.05	0.391	0.346	0.403	0.690	0.708	0.677	0.820	0.965
0.01	0.161	0.117	0.120	0.439	0.397	0.337	0.468	0.877

Notice in Figure 6.4 that the curves showing the mapping are not perfectly smooth. This is, of course, the result of having used a finite number of replications in the simulation. Assuming the existence of a true mapping that is smooth (or at least first-order differentiable), and also that the observed values deviate randomly from the true mapping, we can obtain more accurate values by using a polynomial function that is fitted with, say, least squares, and results in an R^2 adequately close to one. (The increase in accuracy may not be necessary from a practical point of view, but the polynomial approximation is not only more elegant but also requires less storage.)

Using an intercept, linear, and quadratic term results in an excellent fit, as shown in Figure 6.34 for the Laplace case. To compare, for the values obtained in Figure 6.34, using the actual points and linear interpolation (Matlab's `interp1` gives a value of $p = 0.01320$ for $s_{Lap}^{-1}(0.05, 100)$, whereas using the fitted line gives $p = 0.01300$). Compute these values for your simulation.

Finally, observe also from Figure 6.34 that the pointwise values used were just shy of getting a simultaneous significance level of 0.10; if we *assume* that the function continues with the fitted model for a small distance outside of the range for which we obtained points, then we do not need to rerun the simulation, but can just extrapolate

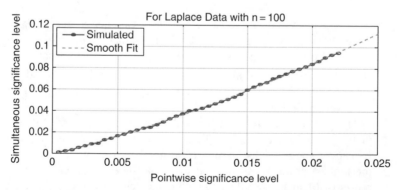

Figure 6.34 *The mapping between pointwise and simultaneous significance levels, for the Laplace Q-Q test using sample size n = 100, with the actual points obtained from the simulation (circles) and the regression line with intercept, linear, and quadratic term (dashed).*

the value of $s_{\text{Lap}}^{-1}(0.10, 100)$. We obtain 0.022790 for $s_{\text{Lap}}^{-1}(0.10, 100)$. Construct code to do this.

(d) Use the Laplace Q-Q plot as a test, and compute the power against normal and Student's t alternatives. For $n = 100$ and against normality, we get a power of 0.1470 corresponding to the test with size 0.05. This does not compare well with the previous tests.

(e) Let $X_i \sim \text{Lap}(0, s_i)$, with $f_{X_i}(x; s_i) = \exp\{-x/s_i\}/(2s_i)$, $i = 1, \ldots, n$, and define $S_n = \sum_{i=1}^{n} X_i$.

For a given set of scale terms $\mathbf{s} = (s_1, \ldots, s_n)$, write a program that computes and graphically compares the p.d.f. of S_n using the inversion formula and a second-order saddlepoint approximation (s.p.a.; see Appendix A for a brief review of the necessary formulas). For example, Figure 6.35(a) shows the s.p.a. and the

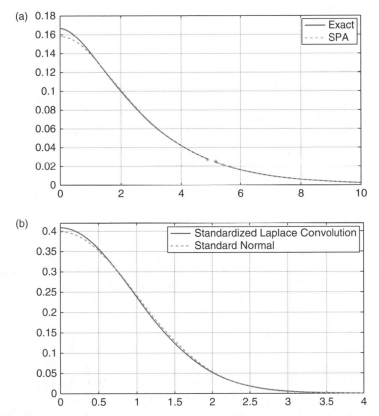

Figure 6.35 (a) The exact and second-order s.p.a. p.d.f. (on the half-line) of $X_1 + X_2$, where $X_i \stackrel{indep}{\sim} \text{Lap}(0, i)$, $i = 1, 2$. (b) The p.d.f. (on the half-line) of the standardized sum of 30 independent Laplace r.v.s, using (positive) random values as scale parameters, and the central limit theorem approximation.

true p.d.f. for $X_1 + X_2$, where $X_i \overset{\text{indep}}{\sim} \text{Lap}(0, i)$, $i = 1, 2$. As a check, for $n = 2$, the exact p.d.f. is

$$f_{S_2}(x; s_1, s_2) = \begin{cases} \dfrac{(1 + |x|/s)\exp\{-|x|/s\}}{4s}, & s = s_1 = s_2, \\[2ex] \dfrac{1}{2}\dfrac{s_1}{s_2}\dfrac{\exp\{-|x|/s_2\}/s_1 - \exp\{-|x|/s_1\}/s_2}{1 - (s_1/s_2)^2}, & s_1 \neq s_2, \end{cases}$$

as shown in Kotz et al. (2001, p. 38).[9]

Hint: For the saddlepoint, the c.g.f. $\mathbb{K}_{S_n}(t; \mathbf{s})$ is just

$$\mathbb{K}_{S_n}(t; \mathbf{s}) = -\sum_{i=1}^{n} \ln(1 - t^2 s_i^2), \quad -m^{-1} < t < m^{-1}, \quad m = \max s_i.$$

For the second-order s.p.a., we need $\mathbb{K}_{S_n}^{(3)}$ and $\mathbb{K}_{S_n}^{(4)}$ which, after some straightforward algebra, yields

$$\mathbb{K}_{S_n}^{(3)}(t; \mathbf{s}) = 4t \sum_{i=1}^{n} \frac{s_i^4(3 + t^2 s_i^2)}{(1 - t^2 s_i^2)^3}, \quad \mathbb{K}_{S_n}^{(4)}(t; \mathbf{s}) = 12 \sum_{i=1}^{n} \frac{s_i^4(1 + 6t^2 s_i^2 + t^4 s_i^4)}{(1 - t^2 s_i^2)^4}.$$

(f) Confirm that, as n increases, the central limit theorem is working, that is, the distribution of the sum approaches (after scaling) a standard normal. For example, Figure 6.35(b) shows the s.p.a. to the true p.d.f. (on the half line; it is symmetric) of the convolution of $n = 30$ independent Laplace r.v.s, each with a randomly chosen scale term, and the standard normal p.d.f.

6.4 Recall Section 6.2.4 for computing the pointwise to simultaneous mapping $s_{\text{Norm}}(p, n)$ and, in particular, its lengthy computation time. Consider the following faster method. Based on the fitted parameters $\hat{\theta} = (\hat{\mu}, \hat{\sigma}^2)'$, generate, sort, and store $s = 20{,}000$ normal random samples of length n, with parameters $\hat{\mu}$ and $\hat{\sigma}^2$, in an $s \times n$ matrix, say $\mathbf{Y}$. Then do exactly the same thing, storing the results in matrix $\mathbf{Z}$. Now, for a particular pointwise significance level p, (i) compute the quantiles for each of the i order statistics Y_i, using $\mathbf{Y}$, and (ii) for each of the data sets in $\mathbf{Z}$, compute the fraction of them such that at least one of the n points lies outside the null band corresponding to p. This last step is then repeated for several values of p to get the s-mapping.

Notice how a first set of data, $\mathbf{Y}$, is used to get the quantiles, and a second, independent, set of data, $\mathbf{Z}$, is used to assess the simultaneous coverage probability. Observe also that the method requires neither nested simulation nor parameter estimation, and so will be far faster than the method we used.

Confirm that this alternative method yields a value of 0.0029 for $s^{-1}(0.05, 50)$, differing from the correct value of 0.03816 by over a factor of 10. The top panel of Figure 6.36 shows the correct bounds, and indeed, all data points are within the null bands, as should happen on average 19 out of 20 times if we were to do this repeatedly, with normal data. The Q-Q plot of the data with the purported size-0.05 simultaneous null bands is shown in the bottom panel, based on the same data set as in the top panel.

---

[9] In their equation 2.2.23, each scale term s_i should be replaced by $1/s_i$. They are indeed using the usual scale formulation as seen in their equation 2.3.1, so it is a mistake.

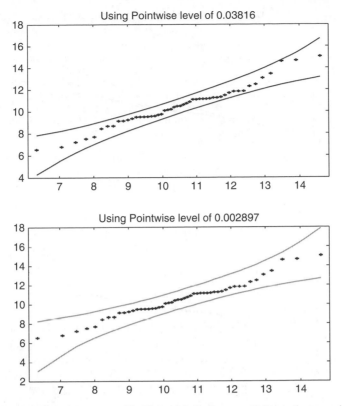

Figure 6.36 For the same data set used in Figure 6.2, the top panel shows the normal Q-Q plot using the correct pointwise significance level $p = s_{\text{Norm}}^{-1}(0.05, 50) = 0.03816$ to obtain a simultaneous one of 0.05. The bottom uses the value of p determined using the "fast but wrong" method.

The null bands are clearly too wide, and thus this alternative idea, while faster, does not work.

As a final illustration, Figure 6.37 shows the normal Q-Q plots using both methods, but with $n = 50$ observations from a Student's t distribution with 3 degrees of freedom. Based on the "fast but wrong" method, one would not have reason to reject the null hypothesis of normality. Thus, while the method is faster and might appear, at first blush, to be correct, it is not.

6.5 Design the program to produce the power plots in Figure 6.5.

6.6 Write the code to replicate Figure 6.18.

6.7 Based on the mostly admirable performance of the JB test for normality, a simple idea presents itself for testing *any* continuous distribution, for which we use the Laplace as an example. For $X_i \overset{\text{i.i.d.}}{\sim} \text{Lap}(\mu, \sigma)$, $i = 1, \dots, n$, we know from the p.i.t. that $U_i = F_{\text{Lap}}(X_i; \mu, \sigma) \overset{\text{i.i.d.}}{\sim} \text{Unif}(0, 1)$, so that $Z_i = F_N^{-1}(U_i) \overset{\text{i.i.d.}}{\sim} N(0, 1)$. So, with observed values $x_1, \dots, x_n$, we apply the JB test to $z_i = F_N^{-1}(F_{\text{Lap}}(x_i; \hat{\mu}, \hat{\sigma}))$. As μ and σ are unknown, we first need to find the nominal significance level corresponding to an actual significance level of, say, 0.05. This is done via simulation over a grid of nominal

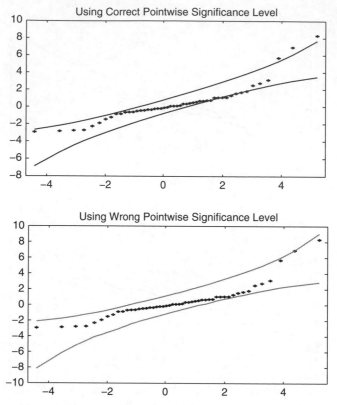

Figure 6.37   *Normal Q-Q plots with size 0.05 as in Figure 6.36, but using a random sample of 50 observations from a Student's t distribution with 3 degrees of freedom.*

significance levels; for $n = 100$, we get a nominal level of 0.06038 corresponding to an actual level of 0.05. Then, to assess the power against the Gaussian alternative, simulate the test but using normal data (with $n = 100$).

We obtain a power of 0.193, which is, disappointingly, much lower than that from the four tests shown in Table 6.4. However, the power improves considerably as the sample size increases: For $n = 200$, the nominal level is 0.0601 and the power is 0.8762, which is now better than the power of three of the four statistics in the table. For $n = 250$, the nominal level is 0.060 and the power is 0.970. The power values of W^2 and U^2 are respectively 0.918 and 0.990 in this case. Replicate these results.

6.8 Reproduce the two plots in Figure 6.28.

6.9 One obvious way of generating several tests that are definitely independent under the null is to split the (necessarily i.i.d.) sample up into subsets, and perform some (say, MSP) test on each of them. For example, with $n = 500$ and $k = 2$ equally sized subsets, we get the two p-values, p_1 and p_2, referring, respectively, to the first and second halves of the data. We reject the null at the 5% level if $-2(\log p_1 + \log p_2) > c$, where c is the 95% quantile of the χ^2_4 distribution.

With the fast approximation of the p-value of the MSP test for any sample size between 10 and 500, we can easily and quickly confirm that this test has the correct size. However, its power against a skew normal with $\lambda = 1.5$ is 0.40, whereas the power based on the test using the entire sample is 0.47. Similarly, with $k = 4$, the power is 0.32. Thus, as perhaps expected, it appears that we cannot extract more power out of the test by splitting the sample. Confirm these results.

7

Unbiased Point Estimation and Bias Reduction

In order to arrive at a distinct formulation of statistical problems, it is necessary to define the task which the statistician sets himself: briefly, and in its most concrete form, the object of statistical methods is the reduction of data. A quantity of data, which usually by its mere bulk is incapable of entering the mind, is to be replaced by relatively few quantities which shall adequately represent the whole, or which, in other words, shall contain as much as possible, ideally the whole, of the relevant information contained in the original data.

(R. A. Fisher, 1922, reproduced in Kotz and Johnson, 1992)

Chapter 3 emphasized the maximum likelihood estimator. It is intuitive, theoretically appealing, generally applicable, and usually possesses both good small- and large-sample properties. There are other concepts associated with estimation theory that are perhaps less intuitive, but serve to deepen our understanding of statistical inference, and sometimes allow for derivation of nonobvious estimators with better properties than the m.l.e.

7.1 SUFFICIENCY

7.1.1 Introduction

For certain sampling schemes, there exist functions of the data that contain just as much information about the unknown parameter θ as the full sample itself. Such a function, or statistic, if it exists, is referred to, appropriately, as a **sufficient statistic**. To illustrate, recall

Fundamental Statistical Inference: A Computational Approach, First Edition. Marc S. Paolella.
© 2018 John Wiley & Sons Ltd. Published 2018 by John Wiley & Sons Ltd.

the binomial model from Section 1.1.1, in which an i.i.d. Bernoulli sample $X_i \overset{\text{i.i.d.}}{\sim} \text{Bern}(p)$, $i = 1, \ldots, n$, is observed. The intuitive point estimator of p was $\hat{p} = S/n$, where S is the observed number of successes. This also coincides with $\hat{p}$ obtained via the method of moments and via maximum likelihood. One might then conjecture that S is sufficient for the sample $\mathbf{X} = (X_1, \ldots, X_n)$, meaning that the individual X_i are not important *per se*, only their sum $\sum_{i=1}^{n} X_i$ (and knowledge of n). This is indeed the case, the reason for which, however, needs to be made precise. To motivate a definition for a sufficient statistic, observe that (omitting the indicator functions for the X_i)

$$
f_{\mathbf{X}|S}(\mathbf{x} \mid s) = \frac{\Pr(\mathbf{X} = \mathbf{x}, \ S = s)}{\Pr(S = s)}
$$

$$
= \frac{p^{x_1}(1-p)^{1-x_1} \cdots p^{x_n}(1-p)^{1-x_n} \mathbb{I}_{\{s\}}\left(\sum_{i=1}^{n} x_i\right)}{\binom{n}{s} p^s (1-p)^{n-s}} = \frac{\mathbb{I}_{\{s\}}\left(\sum_{i=1}^{n} x_i\right)}{\binom{n}{s}},
$$

showing that the distribution of the X_i, conditional on S, does not depend on p for any $p \in (0, 1)$. This is sensible, because if S is sufficient for making inference about p, then, conditional on S, the distribution of $\mathbf{X}$ should contain no further information about p. This is taken to be the definition of a sufficient statistic.

> Let $\mathbf{S} = \mathbf{S}(\mathbf{X})$ be a function of $\mathbf{X}$, where the distribution of $\mathbf{X}$ is $f_{\mathbf{X}}(\cdot; \theta)$ for fixed parameter $\theta \in \Theta$. Then $\mathbf{S}$ is a **sufficient statistic** for the family $\{f_{\mathbf{X}}(\cdot; \theta), \theta \in \Theta\}$, if and only if the distribution of $\mathbf{X} \mid \mathbf{S}$ does not depend on θ.

The dimension of $\mathbf{S}$ need not be the same as that of parameter vector θ; see Example 7.12 and Problem 7.6 for two simple illustrations.

Example 7.1   *Let $X_i \overset{\text{i.i.d.}}{\sim} \text{Poi}(\lambda)$, $i = 1, \ldots, n$. We wish to show that $S(\mathbf{X}) = \sum_{i=1}^{n} X_i$ is a sufficient statistic. Using the fact that $S \sim \text{Poi}(n\lambda)$, we have (omitting the indicator functions)*

$$
\Pr(\mathbf{X} = \mathbf{x} \mid S = s) = \frac{\Pr(\mathbf{X} = \mathbf{x}, S = s)}{\Pr(S = s)} = \frac{\dfrac{e^{-n\lambda}\lambda^s}{x_1! \cdots x_n!}}{\dfrac{e^{-n\lambda}(n\lambda)^s}{s!}} = \frac{s! n^{-s}}{x_1! \cdots x_n!},
$$

which is not a function of λ.

Example 7.2   *Let $X_i \overset{\text{i.i.d.}}{\sim} \text{Exp}(\lambda)$ and define $S(\mathbf{X}) = \sum_{i=1}^{n} X_i \sim \text{Gam}(n, \lambda)$. From Example 1.9.3, the joint density of S and $Y_i = X_i$, $i = 2, \ldots, n$, is*

$$
f_{S,\mathbf{Y}}(s, \mathbf{y}) = \lambda^n e^{-\lambda s} \mathbb{I}_{(0,\infty)}(s) \mathbb{I}_{(0,s)}(t) \prod_{i=2}^{n} \mathbb{I}_{(0,\infty)}(y_i), \quad t = \sum_{i=2}^{n} y_i.
$$

It follows that

$$
f_{\mathbf{Y}|S}(\mathbf{y} \mid s) = \frac{f_{S,\mathbf{Y}}(s, \mathbf{y})}{f_S(s)} = (n-1)! s^{1-n} \mathbb{I}_{(0,s)}(t) \prod_{i=2}^{n} \mathbb{I}_{(0,\infty)}(y_i), \quad s > 0, \tag{7.1}
$$

which is not a function of λ. For example, with $n = 2$, this reduces to the uniform distribution $s^{-1}\mathbb{1}_{(0,s)}(y_2)$.[1] For $n = 3$, (7.1) reduces to

$$2s^{-2}\mathbb{1}_{(0,s)}(y_2 + y_3)\mathbb{1}_{(0,\infty)}(y_2)\mathbb{1}_{(0,\infty)}(y_3),$$

which was studied in Example I.8.2. Notice that the support of (Y_2, Y_3) is the triangle with coordinates $(0,0)$, $(0,s)$ and $(s,0)$, the area of which is $s^2/2$, so that the density is uniform.

If the distribution of $Y \mid S$ does not depend on λ, then $(S, Y) \mid S$ also does not. As $(X_1, \ldots, X_n)$ are related by a one-to-one transformation to S and Y, it follows that $(X_1, \ldots, X_n) \mid S$ cannot depend on λ. From this, it follows from the definition that S is sufficient.

Example 7.3 Let $X_i \overset{i.i.d.}{\sim} N(\mu, \sigma^2)$ and define

$$S(\mathbf{X}) = \sum_{i=1}^{n} X_i \sim N(n\mu, n\sigma^2).$$

Problem II.3.4 showed that $\mathbf{X} \mid S$ is multivariate normal with the n marginal distributions given by $(X_i \mid S = s) \sim N(s/n, (1 - 1/n)\sigma^2)$ and $\text{Cov}(X_i, X_j) = -n^{-1}$, $i \neq j$. As $f_{\mathbf{X}|S}$ does not depend on μ, it follows that S is sufficient for μ.

The sufficient statistic involved in the previous three examples, along with the initial illustration with the binomial distribution, were all the same, namely the sum of the X_i. This is certainly not always the case, not even when θ is a location parameter. For example, with i.i.d. Cauchy data with unknown location parameter μ, the mean is of no use for estimating μ.

Example 7.4 Let $X_i \overset{i.i.d.}{\sim} \text{Exp}(a, b)$, $i = 1, \ldots, n$, where

$$f_{X_i}(x; a, b) = b \exp(-b(x - a))\mathbb{1}_{(a,\infty)}(x). \tag{7.2}$$

Assume a is unknown but $b > 0$ is known. The p.d.f. of $Y(\mathbf{X}) = \min(X_i)$ is, from (A.177) with $F_X(x) = (1 - e^{-b(x-a)})\mathbb{1}_{(a,\infty)}(x)$,

$$f_Y(y) = n[1 - F_X(y)]^{n-1}f_X(y) = nbe^{-nb(y-a)}\mathbb{1}_{(a,\infty)}(y),$$

so that, as $a \leq y = \min(x_i)$, and defining $s = \sum_{i=1}^{n} x_i$,

$$f_{\mathbf{X}|Y}(\mathbf{x} \mid y) = \frac{f_{\mathbf{X},Y}(\mathbf{x}, y)}{f_Y(y)} = \frac{\mathbb{1}_{(a,\infty)}(y)\mathbb{1}_{[y,\infty)}(x_i)\mathbb{1}_{\{x_1,\ldots,x_n\}}(y)\prod_{i=1}^{n}be^{-b(x_i-a)}}{nbe^{-nb(y-a)}\mathbb{1}_{(a,\infty)}(y)}$$

$$= n^{-1}b^{n-1}\exp\{-b(s - ny)\}\prod_{i=1}^{n}\mathbb{1}_{[y,\infty)}(x_i)\mathbb{1}_{\{x_1,\ldots,x_n\}}(y).$$

As $f_{\mathbf{X}|Y}$ does not depend on a, Y is sufficient for a. See also Problem 3.3.

[1] It is interesting to note that, for the geometric distribution, being in a certain sense a discrete analog of the exponential as a waiting-time distribution, we have that, for $X_i \overset{i.i.d.}{\sim} \text{Geo}(p)$, $X_1 \mid (X_1 + X_2)$ is discrete uniform; see the examples after (A.78) or Example I.8.4.

7.1.2 Factorization

The previous examples illustrated how the direct use of the definition can be used to confirm that a statistic is sufficient. The following fact, referred to as the factorization theorem, is often more useful for determining whether a statistic is sufficient.[2]

> **Factorization theorem**: Let $f_{\mathbf{X}}(\mathbf{x}; \theta)$ denote the density or mass function of the random sample $\mathbf{X} = (X_1, \ldots, X_n)$. The statistics $\mathbf{S} = (S_1, \ldots, S_k)$ are sufficient for the family $\{f_{\mathbf{X}}(\mathbf{x}; \theta), \theta \in \Theta\}$ if and only if $f_{\mathbf{X}}$ can be algebraically factored as
>
> $$f_{\mathbf{X}}(\mathbf{x}; \theta) = g(\mathbf{S}, \; \theta) \cdot h(\mathbf{x}).$$
>
> More specifically, the S_i are sufficient if and only if $f_{\mathbf{X}}$ can be factored into two functions, one depending only on the S_i and θ, the other depending only on the X_i (and not θ).

The proof for the discrete case is quite straightforward, while to make the continuous case proof accessible, further assumptions are required (see below). A rigorous, general proof (requiring an understanding of measure theory) can be found in, for example, Lehmann and Casella (1998), Schervish (1995), and Shao (2003). Before showing the proofs, the technique is illustrated for several cases of interest.

Example 7.5 (Poisson, cont.)   *By expressing the joint density of the i.i.d. sample as*

$$f_{\mathbf{X}}(\mathbf{x}; \lambda) = e^{-n\lambda} \lambda^s \times \prod_{i=1}^{n} \frac{\mathbb{I}_{\{0,1,\ldots\}}(x_i)}{x_i!} = g(s, \lambda) \cdot h(\mathbf{x}), \quad s = \sum_{i=1}^{n} x_i,$$

it follows from the factorization theorem that S is a sufficient statistic.

Example 7.6 (Normal, cont.)   *It follows from the factorization theorem that* $S = \sum_{i=1}^{n} X_i$ *is sufficient for μ, as*

$$f_{\mathbf{X}}(\mathbf{x}; \mu) = \sigma^{-n}(2\pi)^{-n/2} \exp\left\{ -\frac{1}{2\sigma^2} \sum_{i=1}^{n} (x_i - \mu)^2 \right\}$$

$$= \sigma^{-n}(2\pi)^{-n/2} \exp\left\{ -\frac{1}{2\sigma^2} \left(\sum x_i^2 - 2\mu \sum x_i + n\mu^2 \right) \right\}$$

$$= \sigma^{-n}(2\pi)^{-n/2} \exp\left\{ -\frac{1}{2\sigma^2} \left(-2\mu \sum x_i + n\mu^2 \right) \right\} \exp\left\{ -\frac{1}{2\sigma^2} \left(\sum x_i^2 \right) \right\}$$

$$= g(s, \mu) \cdot h(\mathbf{x}),$$

where $h(\mathbf{x}) = \exp\left\{ -\left(\sum x_i^2 \right) / (2\sigma^2) \right\}$ does not depend on μ. This holds whether or not σ is a known or unknown parameter.

Next consider finding a sufficient statistic for σ^2. If μ is known, then writing

$$f_{\mathbf{X}}(\mathbf{x}; \sigma^2) = \sigma^{-n}(2\pi)^{-n/2} \exp\left\{ -\frac{1}{2\sigma^2} \sum_{i=1}^{n} (x_i - \mu)^2 \right\} = g(s', \sigma^2) \cdot h(\mathbf{x})$$

with $h(\mathbf{x}) = 1$ shows that $S' = \sum_{i=1}^{n} (X_i - \mu)^2$ is sufficient for σ^2.

---

[2] The theorem is often referred to as the Neyman factorization theorem or even the Fisher-Neyman factorization theorem, although a rigorous proof was first provided in 1949 by Halmos and Savage; see Casella and Berger (1990, p. 250).

If μ is also unknown, then there is no single statistic that is sufficient for σ^2. In this (more relevant) case, $\mathbf{S} = (S_1, S_2) = \left(\sum_{i=1}^n X_i, \sum_{i=1}^n X_i^2 \right)$ is sufficient for $\boldsymbol{\theta} = (\mu, \sigma^2)$ because

$$f_{\mathbf{X}}(\mathbf{x}; \mu, \sigma^2) = \sigma^{-n}(2\pi)^{-n/2} \exp \left\{ -\frac{1}{2\sigma^2} \left(\sum x_i^2 - 2\mu \sum x_i + n\mu^2 \right) \right\}$$

$$= g(s_1, s_2, \mu, \sigma^2) \cdot h(\mathbf{x}),$$

with $h(\mathbf{x}) = 1$.

It should be intuitively clear that one-to-one functions of the sufficient statistics $\mathbf{S}$ are also sufficient. With respect to the previous example, the statistics

$$(\bar{X}_n, S_n^2) = \left(n^{-1} \sum_{i=1}^n X_i, \ (n-1)^{-1} \sum_{i=1}^n (X_i - \bar{X})^2 \right)$$

are also jointly sufficient for μ and σ^2.

Example 7.7   *Let $\mathbf{Y}_i \overset{\text{i.i.d.}}{\sim} N(\boldsymbol{\mu}, \boldsymbol{\Sigma})$, $i = 1, \dots, n$, from the p-variate normal distribution (3.30), with $n > p$. It follows directly from (3.32) that*

$$\bar{\mathbf{Y}} = n^{-1} \sum_{i=1}^n \mathbf{Y}_i \quad \text{and} \quad \mathbf{S} = n^{-1} \sum_{i=1}^n (\mathbf{Y}_i - \bar{\mathbf{Y}})(\mathbf{Y}_i - \bar{\mathbf{Y}})'$$

are sufficient for $\boldsymbol{\mu}$ and $\boldsymbol{\Sigma}$. As shown in Example 3.8, $\widehat{\boldsymbol{\mu}}_{\text{ML}} = \bar{\mathbf{Y}}$ and $\widehat{\boldsymbol{\Sigma}}_{\text{ML}} = \mathbf{S}$.

Example 7.8 (Examples 7.2 and 7.4, cont.)   *With $X_i \overset{\text{i.i.d.}}{\sim} \text{Exp}(\lambda)$ and $S = \sum_{i=1}^n X_i$,*

$$f_{\mathbf{X}}(\mathbf{x}; \lambda) = \prod_{i=1}^n \lambda e^{-\lambda x_i} \mathbb{I}_{(0,\infty)}(x_i) = \lambda^n e^{-\lambda s} \times \prod_{i=1}^n \mathbb{I}_{(0,\infty)}(x_i) = g(s, \lambda) \cdot h(\mathbf{x}),$$

showing that $S = \sum_{i=1}^n X_i$ is sufficient for the (inverse) scale parameter λ. For the location family, let $X_i \overset{\text{i.i.d.}}{\sim} \text{Exp}(a, b)$, each with density (7.2), and b is known. Noting that

$$\prod_{i=1}^n \mathbb{I}_{(a,\infty)}(x_i) = \mathbb{I}_{(a,\infty)}(\min(x_i)),$$

we can write

$$f_{\mathbf{X}}(\mathbf{x}; a) = \prod_{i=1}^n b e^{-b(x_i - a)} \mathbb{I}_{(a,\infty)}(x_i) = e^{nba} \mathbb{I}_{(a,\infty)}(\min(x_i)) \times b^n e^{-bs}$$

$$= g(\min(x_i), a) \cdot h(\mathbf{x}),$$

confirming that $Y = \min(X_i)$ is sufficient for a. Similarly, if both a and b are unknown, we take $h(\mathbf{x}) = 1$ and

$$f_{\mathbf{X}}(\mathbf{x}; a, b) = e^{nba} \mathbb{I}_{(a,\infty)}(y) e^{-bs} b^n = g(y, s, \ a, b),$$

so that Y and S are jointly sufficient for a and b.

By emphasizing use of the maximum likelihood estimator, earlier chapters (should have) left the impression that the m.l.e. is an (if not the most) important estimator. It should come as no surprise that the m.l.e. is function of the sufficient statistics. This follows by expressing the likelihood as

$$\mathcal{L}(\theta; \mathbf{x}) = f_{\mathbf{X}}(\mathbf{x}; \theta) = g(\mathbf{S}; \theta) h(\mathbf{x}) \propto g(\mathbf{S}; \theta), \tag{7.3}$$

where $\mathbf{S}$ is a vector of sufficient statistics. (However, see Moore, 1971, and the discussion in Romano and Siegel, 1986, Sec. 8.13, for a clarification and the precise conditions under which this holds.)

Example 7.9 (Example 3.5, cont.)   *With* $X_i \overset{\text{i.i.d.}}{\sim} N(\mu, \sigma^2)$, $i = 1, \dots, n$, *and both* μ *and* σ *unknown, the m.l.e. is* $\hat{\mu}_{ML} = \bar{X}$, $\hat{\sigma}_{ML}^2 = n^{-1} \sum_{i=1}^{n} (X_i - \bar{X})^2$, *and is a function of the sufficient statistics* $\sum_{i=1}^{n} X_i$ *and* $\sum_{i=1}^{n} X_i^2$.

Example 7.10 (Example 7.8, cont.)   *The m.l.e. of* $\theta = (a, b)$ *is shown in Problem 3.3 to be given by* $\hat{a}_{ML} = \min(X_i)$, $\hat{b}_{ML} = 1/(\bar{X} - \hat{a}_{ML})$, *verifying that the m.l.e. is a function of the two sufficient statistics.*

Example 7.11   *Let* $X_1, \dots, X_n$ *be i.i.d. random variables sampled from the density*

$$f_X(x; \theta) = \theta x^{\theta-1} \mathbb{1}_{(0,1)}(x), \quad \theta > 0.$$

Then, with $s = x_1 x_2 \cdots x_n$,

$$f_{\mathbf{X}}(\mathbf{x}; \theta) = \theta^n s^{\theta-1} \times \prod_{i=1}^{n} \mathbb{1}_{(0,1)}(x_i) = g(s, \theta) \cdot h(\mathbf{x}),$$

it follows from the factorization theorem that $S(\mathbf{X}) = X_1 X_2 \cdots X_n$ *is sufficient for* θ. *Because one-to-one functions of sufficient statistics are also sufficient,* $\ln S = \sum_{i=1}^{n} \ln X_i$ *is also sufficient for* θ; *see also Problem 7.15(a).*

The contrapositive of the "$\Rightarrow$" part of the factorization theorem implies that, for a given statistic S, if the algebraic factorization of the density is not possible, then S is not sufficient. While this fact can be used, it is often more mathematically challenging to show it.

Example 7.12   *Let* $\mathbf{X} = (X_1, \dots, X_n)$ *be i.i.d. random variables from the location–scale Cauchy model* $\mathrm{Cau}(\mu, \sigma)$, *with joint density*

$$f_{\mathbf{X}}(\mathbf{x}) = \pi^{-n} \sigma^{-n} \prod_{i=1}^{n} (1 + \sigma^{-2}(x_i - \mu)^2)^{-1}.$$

As this does not (appear to) factor, there does not exist a set of sufficient statistics with dimension less than n.

The previous example for Cauchy data suggests that a similar result holds for the more general case of i.i.d. location–scale Student's t data with v degrees of freedom, where v is either known or unknown. In any case, Example 4.3 showed that, for known v and σ, the m.l.e. appears to be a complicated function of $\mathbf{X}$. Use of a trimmed mean, with the optimal trimming amount being a function of known v, led to an estimator nearly as good as the m.l.e., so that one might informally speak of the trimmed mean being "nearly sufficient" for μ, with v known, but it entails loss of information from $\mathbf{X}$.

The concept of partitioning the support of $\mathbf{X}$ into disjoint subsets is used in the proof of the factorization theorem and is also necessary for the subsequent discussion of minimally sufficient statistics. We state the notation here for the case where S is scalar. Let $S_{\mathbf{X}}$ be the support of $\mathbf{X}$, that is, the set of all values that $\mathbf{X}$ can assume. Similarly, let S_S be the support of sufficient statistic $S(\mathbf{X})$. Define, for each $s \in S_S$, the set $A_s = \{\mathbf{x} : \mathbf{x} \in S_{\mathbf{X}}, S(\mathbf{x}) = s\}$ so that $\bigcup_{s \in S_S} A_s = S_{\mathbf{X}}$, that is, the A_s form a disjoint partition of $S_{\mathbf{X}}$.

The following example, known as the **German tank problem** (see Wikipedia) helps to illustrate these definitions and the partition.

Example 7.13 *A city has one taxi company with a total of N cars, each uniquely labeled with a number, $1, \ldots, N$. Throughout the day, you observe a random sample of n taxis (with replacement) and wish to assess the size of N. The model is $X_i \overset{\text{i.i.d.}}{\sim} \text{DUnif}(N)$, $i = 1, \ldots, n$, with*

$$f_X(x; N) = N^{-1} \mathbb{1}_{\mathcal{M}}(x), \quad \mathcal{M} = \{1, 2, \ldots, N\}, \quad N \in \mathbb{N}.$$

Let $Y_n = \max(X_i)$ and $\mathbf{X} = (X_1, \ldots, X_n)$. The joint density can be expressed as

$$f_{\mathbf{X}}(\mathbf{x}; N) = \prod_{i=1}^n f_X(x_i; N) = \overbrace{N^{-n} \mathbb{1}_{\{1,2,\ldots,N\}}(y_n)}^{g(y_n, N)} \times \overbrace{\prod_{i=1}^n \mathbb{1}_{\{1,2,\ldots,y_n\}}(x_i)}^{h(\mathbf{x})},$$

or

$$f_{\mathbf{X}}(\mathbf{x}; N) = N^{-n} \mathbb{1}_{\{y_n, y_n+1, \ldots\}}(N) \times \prod_{i=1}^n \mathbb{1}_{\{1,2,\ldots\}}(x_i),$$

so that the factorization theorem implies that Y_n is a sufficient statistic. For this model, with $S_{\mathbf{X}} = \mathcal{M}^n$, $S_S = \mathcal{M}$, $S(\mathbf{X}) = Y_n = \max(X_i)$, and $A_s = \{\mathbf{x} : \mathbf{x} \in S_{\mathbf{X}}, \max(\mathbf{x}) = s\}$ for each $s \in S_S$, that is, A_s is the set of all n-vectors with positive integer elements such that the maximum is s, it contains s^n elements. Observe that $\bigcup_{s \in S_S} A_s = S_{\mathbf{X}}$, where the A_s are disjoint.

Proof of factorization theorem (discrete case). Let $\mathbf{X}$ be a discrete r.v. with p.m.f. $f_{\mathbf{X}}(\mathbf{x}; \theta) = \Pr_\theta(\mathbf{X} = \mathbf{x})$, and let $S(\mathbf{X})$ be a statistic that partitions the $\mathbf{X}$ such that

$$\Pr_\theta(\mathbf{S} = \mathbf{s}) = \sum_{\mathbf{x} \in A_s} \Pr_\theta(\mathbf{X} = \mathbf{x}) = \sum_{\mathbf{x} \in A_s} f_{\mathbf{X}}(\mathbf{x}; \theta),$$

where, for each $\mathbf{s} \in S_S$, $A_s = \{\mathbf{x} : \mathbf{x} \in S_{\mathbf{X}}, S(\mathbf{x}) = \mathbf{s}\}$, so that $\bigcup_{s \in S_S} A_s = S_{\mathbf{X}}$.

($\Leftarrow$) Assuming the factorization $f_{\mathbf{X}}(\mathbf{x}; \theta) = g(\mathbf{s}; \theta) \cdot h(\mathbf{x})$,

$$\Pr_\theta(\mathbf{S} = \mathbf{s}) = g(\mathbf{s}; \theta) \sum_{\mathbf{x} \in A_s} h(\mathbf{x}).$$

As $\Pr_\theta(\mathbf{X} = \mathbf{x}, \mathbf{S} = \mathbf{s}) = \Pr_\theta(\mathbf{X} = \mathbf{x})$,

$$f_{\mathbf{X}|\mathbf{S}}(\mathbf{x} \mid \mathbf{s}; \theta) = \frac{\Pr_\theta(\mathbf{X} = \mathbf{x})}{\Pr_\theta(\mathbf{S} = \mathbf{s})} = \frac{g(\mathbf{s}; \theta) \cdot h(\mathbf{x})}{g(\mathbf{s}; \theta) \sum_{\mathbf{x} \in A_s} h(\mathbf{x})} = \frac{h(\mathbf{x})}{\sum_{\mathbf{x} \in A_s} h(\mathbf{x})}$$

is independent of θ.

($\Rightarrow$) Let $\mathbf{S}$ be sufficient for θ so that, by definition, $p(\mathbf{x}, \mathbf{s}) := \Pr(\mathbf{X} = \mathbf{x} \mid \mathbf{S} = \mathbf{s})$ does not depend on θ. Then

$$\Pr_\theta(\mathbf{X} = \mathbf{x} \mid \mathbf{S} = \mathbf{s}) = \frac{\Pr_\theta(\mathbf{X} = \mathbf{x})}{\Pr_\theta(\mathbf{S} = \mathbf{s})} = p(\mathbf{x}, \mathbf{s})$$

if and only if $f_{\mathbf{X}}(\mathbf{x}; \theta) = \Pr_\theta(\mathbf{S} = \mathbf{s}) p(\mathbf{x}, \mathbf{s})$, so that setting $g(\mathbf{s}; \theta) = \Pr_\theta(\mathbf{S} = \mathbf{s})$ and $h(\mathbf{x}) = p(\mathbf{x}, \mathbf{s})$ yields the factorization $f_{\mathbf{X}}(\mathbf{x}; \theta) = g(\mathbf{s}; \theta) h(\mathbf{x})$. ∎

Proof of factorization theorem (continuous case). As in Hogg et al. (2014), let $\mathbf{r} : \mathbb{R}^n \to \mathbb{R}^n$ be a continuous bijection, mapping the support of $\mathbf{X} = (X_1, \dots, X_n)$ to the support of $(S_1, \dots, S_n)$, where $S_i = S_i(\mathbf{X})$, $i = 1, \dots, n$. Let $\mathbf{S} = \mathbf{S}(\mathbf{X}) = (S_1, \dots, S_j)$, for some $1 \le j \le n$, so that $\mathbf{X} = \mathbf{r}^{-1}(\mathbf{S}, S_{j+1}, \dots, S_n)$. Let $\mathbf{J}$ denote the Jacobian for the transformation of the function $\mathbf{r}$.

($\Leftarrow$) Assume the factorization $f_{\mathbf{X}}(\mathbf{x}; \theta) = g(\mathbf{S}; \theta) \cdot h(\mathbf{x})$, so that

$$f_{\mathbf{S}, S_{j+1}, \dots, S_n}(\mathbf{s}, s_{j+1}, \dots, s_n; \theta) = g(\mathbf{s}; \theta) \cdot h[\mathbf{r}^{-1}(\mathbf{s}, s_{j+1}, \dots, s_n)] |\det \mathbf{J}|$$

$$=: g(\mathbf{s}; \theta) \cdot h^*(\mathbf{s}, s_{j+1} \dots, s_n),$$

where h^* is so defined, and

$$f_{\mathbf{S}}(\mathbf{s}; \theta) = \int_{-\infty}^{\infty} \cdots \int_{-\infty}^{\infty} f_{\mathbf{S}, S_{j+1}, \dots, S_n}(\mathbf{s}, s_{j+1}, \dots, s_n; \theta) \, ds_{j+1} \cdots ds_n$$

$$= g(\mathbf{s}; \theta) \int_{-\infty}^{\infty} \cdots \int_{-\infty}^{\infty} h^*(\mathbf{s}, s_{j+1}, \dots, s_n) \, ds_{j+1} \cdots ds_n$$

$$=: g(\mathbf{s}; \theta) h^{**}(\mathbf{s}),$$

where h^{**} is so defined. Thus

$$f_{S_{j+1}, \dots, S_n \mid \mathbf{S}}(s_{j+1}, \dots, s_n \mid \mathbf{s}; \theta) = \frac{f_{\mathbf{S}, S_{j+1}, \dots, S_n}(\mathbf{s}, s_{j+1}, \dots, s_n; \theta)}{f_{\mathbf{S}}(\mathbf{s}; \theta)}$$

$$= \frac{g(\mathbf{s}; \theta) h^*(\mathbf{s}, s_{j+1}, \dots, s_n)}{g(\mathbf{s}; \theta) h^{**}(\mathbf{s})} = \frac{h^*(\mathbf{s}, s_{j+1}, \dots, s_n)}{h^{**}(\mathbf{s})}$$

is independent of θ, implying that the conditional distribution of $\mathbf{S}$ and $S_{j+1}, \dots, S_n$, given $\mathbf{S} = \mathbf{s}$, is also independent of θ. It follows from the assumed bijective correspondence between $(\mathbf{S}, S_{j+1}, \dots, S_n)$ and $\mathbf{X}$ that the conditional distribution of $\mathbf{X}$, given $\mathbf{S} = \mathbf{s}$, is independent of θ. (See Problem 7.2 for a simple illustration.)

($\Rightarrow$) Now let $\mathbf{S}$ be sufficient for θ. From the inverse transformation,

$$f_{\mathbf{X}}(\mathbf{x}; \theta) = f_{\mathbf{S}, S_{j+1}, \dots, S_n}(\mathbf{s}, s_{j+1}, \dots, s_n; \theta) |\det \mathbf{J}^{-1}|$$

$$= f_{S_{j+1}, \dots, S_n \mid \mathbf{S}}(s_{j+1}, \dots, s_n \mid \mathbf{s}) \times f_{\mathbf{S}}(\mathbf{s}; \theta) |\det \mathbf{J}^{-1}|,$$

as $f_{S_{j+1}, \dots, S_n \mid \mathbf{S}}$ does not depend on θ. Letting $h(\mathbf{x}) := f_{S_{j+1}, \dots, S_n \mid \mathbf{S}}(s_{j+1}, \dots, s_n \mid \mathbf{s}) |\det \mathbf{J}^{-1}|$, and $g(\mathbf{s}; \theta) = f_{\mathbf{S}}(\mathbf{s}; \theta)$, we can express $f(\mathbf{x}; \theta)$ as $f(\mathbf{x}; \theta) = g(\mathbf{s}; \theta) h(\mathbf{x})$, as was to be shown. ∎

7.1.3 Minimal Sufficiency

A set of sufficient statistics $\mathbf{S}$ contains all the information in the sample about the unknown parameter θ. Certainly then, taking $\mathbf{S}(\mathbf{X}) = \mathbf{X}$ is sufficient, but doing so provides no reduction of the data set. Indeed, the previous examples suggest that the dimension of **S** can be "close" to the dimension of θ, and in many cases they are equal. This behooves us to obtain the set of sufficient statistics that "reduces the data the most."

As a trivial example, the set $\mathbf{S}' = \left(\sum_{i=1}^{n} X_i, \ \sum_{i=1}^{n} X_i^2, \ \sum_{i=1}^{n} X_i^3 \right)$ is sufficient for (μ, σ^2) in the i.i.d. normal model, but could be reduced to $\mathbf{S} = \left(\sum_{i=1}^{n} X_i, \ \sum_{i=1}^{n} X_i^2 \right)$.[3] Slightly less trivial are the following two examples.

Example 7.14   *Let X_i, $i = 1, \ldots, n$, be observations from the uniform density $f_X(x; \theta) = (2\theta)^{-1} \mathbb{1}_{(-\theta,\theta)}(x)$. It seems intuitive that the two extremes of the sample, $Y_1 = \min(X_1, \ldots, X_n)$ and $Y_n = \max(X_1, \ldots, X_n)$, will provide all information about θ. This indeed follows from the factorization theorem, as*

$$f_{\mathbf{X}}(\mathbf{x}; \theta) = (2\theta)^{-n} \prod_{i=1}^{n} \mathbb{1}_{(-\theta,\theta)}(x_i) = (2\theta)^{-n} \mathbb{1}_{(-\theta,y_n)}(y_1) \mathbb{1}_{(y_1,\theta)}(y_n)$$

$$= g(Y_1, Y_n, \theta) \cdot h(\mathbf{x}),$$

with $h(\mathbf{x}) = 1$. However, a bit of reflection reveals that

$$f_{\mathbf{X}}(\mathbf{x}; \theta) = (2\theta)^{-n} \mathbb{1}_{(0,\theta)}(\max(|y_1|, |y_n|))$$

$$= (2\theta)^{-n} \mathbb{1}_{(0,\theta)}(\max(|x_1|, \ldots, |x_n|)),$$

that is, that $S = \max(|X_i|)$ is also sufficient for θ. The latter statistic is of lower dimension than $\mathbf{S}' = (Y_1, Y_n)$, and indeed appears to be "minimally sufficient," being of dimension 1.

Example 7.15   *From (I.8.12) or (II.3.12), the joint density of*

$$\begin{bmatrix} X_1 \\ Y_1 \end{bmatrix}, \ldots, \begin{bmatrix} X_n \\ Y_n \end{bmatrix} \overset{\text{i.i.d.}}{\sim} N_2 \left(\begin{bmatrix} 0 \\ 0 \end{bmatrix}, \begin{bmatrix} 1 & \rho \\ \rho & 1 \end{bmatrix} \right) \tag{7.4}$$

is, for $S_x = \sum_{i=1}^{n} x_i^2$, $S_y = \sum_{i=1}^{n} y_i^2$, $S_{xy} = \sum_{i=1}^{n} x_i y_i$, $\mathbf{X} = (X_1, \ldots, X_n)$ and $\mathbf{Y} = (Y_1, \ldots, Y_n)$,

$$f_{\mathbf{X},\mathbf{Y}}(\mathbf{x},\mathbf{y}) = K^n \exp \left\{ -\frac{1}{2(1-\rho)^2} \sum_{i=1}^{n} (x_i^2 - 2\rho x_i y_i + y_i^2) \right\}, \quad K = \frac{1}{2\pi \sqrt{1-\rho^2}}$$

$$= K^n \exp \left\{ -\frac{S_x - 2\rho S_{xy} + S_y}{2(1-\rho)^2} \right\} = g(\mathbf{S}, \rho) h(\mathbf{X}, \mathbf{Y}),$$

where $\mathbf{S} = (S_x, S_y, S_{xy})$, $h(\mathbf{X}, \mathbf{Y}) = 1$, and $\rho = \mathrm{Cov}(X, Y) = \mathrm{Corr}(X, Y)$.

From (3.25), the m.l.e. of $\mathrm{Corr}(X, Y)$ in the case where all five parameters are unknown is

$$\hat{\rho}_1 = \frac{\sum_{i=1}^{n} (X_i - \bar{X})(Y_i - \bar{Y})}{\sqrt{\sum_{i=1}^{n} (X_i - \bar{X})^2} \sqrt{\sum_{i=1}^{n} (Y_i - \bar{Y})^2}}.$$

---

[3] It is actually possible to reduce **S** down to a single real number, but the resulting function is not continuous; see Romano and Siegel (1986, Examples 7.1 and 7.2) for details. This serves to illustrate that the notion of "dimensionality" of a minimal sufficient statistic can be somewhat delicate.

Accounting here for the fact that μ_1 and μ_2 are known (and equal to zero), it makes sense to entertain the estimator

$$\hat{\rho}_2 = \frac{\sum_{i=1}^{n} X_i Y_i}{\sqrt{\sum_{i=1}^{n} X_i^2} \sqrt{\sum_{i=1}^{n} Y_i^2}} = \frac{S_{XY}}{\sqrt{S_X}\sqrt{S_Y}},$$

*which is a function of **S**. Finally, as σ_1 and σ_2 are also known (and equal to 1), we should consider $\hat{\rho}_3 = S_{XY}$. This is unbiased; see page II.88.*

One way of determining some of the properties of these estimators is via simulation. Figure 7.1 shows the bias and m.s.e. results of the three $\hat{\rho}_i$ as a function of ρ, for $\rho = 0, 0.02, 0.04, \ldots, 0.98$, based on 1 million replications and a sample size of $n = 10$. We see that $\hat{\rho}_1$ and $\hat{\rho}_2$ are biased for all $0 < \rho < 1$, $\hat{\rho}_1$ more so, as would be expected, while $\hat{\rho}_3$ is indeed unbiased. With respect to the (arguably more interesting) m.s.e., $\hat{\rho}_2$ is superior for all ρ, but $\hat{\rho}_1$ is close. As $|\rho| \to 1$, m.s.e.$(\hat{\rho}_1)$ and m.s.e.$(\hat{\rho}_2)$ decrease towards zero, while m.s.e.$(\hat{\rho}_3)$ increases. The reader is encouraged to confirm that, as the sample size n is increased, we obtain relatively similar results, with m.s.e. $(\hat{\rho}_1)$ and m.s.e. $(\hat{\rho}_2)$ becoming indistinguishable.

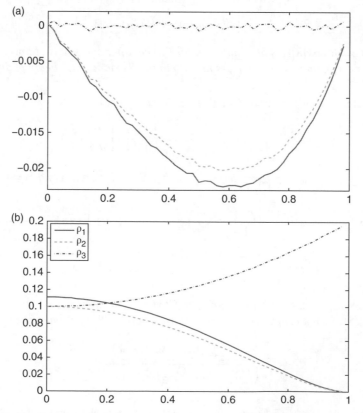

Figure 7.1   Bias (a) and m.s.e. (b) for estimators $\hat{\rho}_1$ (solid), $\hat{\rho}_2$ (dashed) and $\hat{\rho}_3$ (dash-dotted) for sample size $n = 10$ for the model in Example 7.15.

Similar to the illustration with geometric data in Section 1.1.2, we see that use of a biased statistic, $\hat{\rho}_2$, exhibits a lower m.s.e. for (in this case, all values of) ρ than an unbiased statistic, $\hat{\rho}_3$. See also Section 7.4 on estimating this correlation coefficient.

A minimally sufficient statistic should have the property that its partition $\{A_s\}$ of S_X is the "coarsest." That is, if $S(X)$ is minimally sufficient, then, for any other sufficient statistic $S^*(X)$ and $x, y \in S_X$, if $S^*(x) = S^*(y)$, then $S(x) = S(y)$. Observe that, in both the normal and uniform models shown above, the sufficient statistic of lower dimension can be expressed as a function of the one of higher dimension. Indeed, any sufficient statistic can be expressed as a function of the entire sample, X. In general, the coarser of two sufficient statistics will be a function of the other one. Thus, the following definition suggests itself.

> A sufficient statistic is said to be a **minimal sufficient statistic** if it is a function of all other sufficient statistics.

Contrary to the definition of sufficiency, the definition of a minimal sufficient statistic appears useless for actually determining if a set of statistics is minimal sufficient. The following theorem, due to Erich Leo Lehmann and Henry Scheffé, presented in two papers (1950 and 1955), does provide a useful method. Keep in mind the difference between S and S: the former denotes the support (as in S_X and S_S), while $S(X)$ (or $\mathbf{S}(X)$) denotes a (set of) sufficient statistic(s).

> **Lehmann and Scheffé Minimal Sufficieny:** For p.m.f. or p.d.f. f_X such that $f_X(x; \theta) > 0$ for $x \in S_X$ and $\theta \in \Theta$, if $\mathbf{S}(X)$ is such that, for $x, y \in S_X$, the ratio $f_X(x; \theta)/f_X(y; \theta)$ is constant with respect to $\theta \in \Theta \Leftrightarrow \mathbf{S}(x) = \mathbf{S}(y)$, then $\mathbf{S}(X)$ is minimal sufficient for θ.

Proof. Define sets $A_s = \{x : \mathbf{S}(x) = \mathbf{s}\}$ for each $\mathbf{s} \in S_S$. For each $\mathbf{s}$, fix a particular element of A_s, denoted $\mathbf{x}_s$, so that, for any $\mathbf{x} \in S_X$, $\mathbf{x}$ and $\mathbf{x}_{\mathbf{S}(x)}$ are in the same set $A_{\mathbf{S}(x)}$. Thus $\mathbf{S}(x) = \mathbf{S}(\mathbf{x}_{\mathbf{S}(x)})$. Then, under the $\Leftarrow$ assumption in the theorem, $h(x) := f_X(x; \theta)/f(\mathbf{x}_{\mathbf{S}(x)}; \theta)$ is well defined and constant with respect to $\theta \in \Theta$. Now letting $g(\mathbf{s}; \theta) := f(\mathbf{x}_s; \theta)$,

$$f_X(x; \theta) = \frac{f_X(\mathbf{x}_{\mathbf{S}(x)}; \theta) f_X(x; \theta)}{f_X(\mathbf{x}_{\mathbf{S}(x)}; \theta)} = g(\mathbf{s}; \theta) \cdot h(x),$$

showing via the factorization theorem that $\mathbf{S}(X)$ is sufficient for θ. We now need to establish that $\mathbf{S}(X)$ is *minimal* sufficient.

If $\mathbf{S}^*(X)$ is another sufficient statistic and $x, y \in S_X$ are such that $\mathbf{S}^*(x) = \mathbf{S}^*(y)$, then

$$\frac{f_X(x; \theta)}{f_X(y; \theta)} = \frac{g^*(\mathbf{S}^*(x); \theta) \times h^*(x)}{g^*(\mathbf{S}^*(y); \theta) \times h^*(y)} = \frac{h^*(x)}{h^*(y)},$$

where functions g^* and h^* exist via the factorization theorem. This ratio is well defined and constant with respect to $\theta \in \Theta$ so that, under the $\Rightarrow$ assumption, $\mathbf{S}(x) = \mathbf{S}(y)$. That is, for *any* sufficient statistic $\mathbf{S}^*$, $\mathbf{S}^*(x) = \mathbf{S}^*(y) \Rightarrow \mathbf{S}(x) = \mathbf{S}(y)$. Thus, $\mathbf{S}$ is a function of $\mathbf{S}^*$, which is the definition of minimal sufficiency. $\blacksquare$

Remark. A different approach to the proof, using an algebraic treatment and the concept of likelihood equivalence, instead of partitions, and written for a beginner audience in mathematical statistics, is developed in Sampson and Spencer (1976). $\blacksquare$

Example 7.16 (Normal, cont.) *For* $\mathbf{x}, \mathbf{y} \in S_{\mathbf{X}} = \mathbb{R}^n$, $\mu \in \mathbb{R}$, $\sigma > 0$,

$$\frac{f_{\mathbf{X}}(\mathbf{x}; \mu, \sigma^2)}{f_{\mathbf{X}}(\mathbf{y}; \mu, \sigma^2)} = \frac{\sigma^{-n}(2\pi)^{-n/2} \exp\left\{-\left(\sum x_i^2 - 2\mu \sum x_i + n\mu^2\right)/(2\sigma^2)\right\}}{\sigma^{-n}(2\pi)^{-n/2} \exp\left\{-\left(\sum y_i^2 - 2\mu \sum y_i + n\mu^2\right)/(2\sigma^2)\right\}} = e^Q,$$

where

$$Q = Q(\mathbf{x}, \mathbf{y}, \mu, \sigma^2) = -\frac{1}{2\sigma^2}\left(\sum x_i^2 - \sum y_i^2 - 2\mu \sum x_i + 2\mu \sum y_i\right). \tag{7.5}$$

For $\theta = \mu$ *(i.e.,* σ *known), the ratio or, equivalently,* Q, *is constant with respect to* μ *if and only if* $\sum x_i = \sum y_i$, *so that, from the Lehmann–Scheffé minimal sufficiency theorem,* $S(\mathbf{X}) = \sum X_i$ *is minimally sufficient for* μ.

For $\theta = (\mu, \sigma^2)$, Q *is constant with respect to* θ *if and only if* $\sum x_i^2 = \sum y_i^2$ *and* $\sum x_i = \sum y_i$, *so that* $S(\mathbf{X}) = \left(\sum X_i, \sum X_i^2\right)$ *is minimally sufficient for* θ.

Finally, consider the case for $\theta = \sigma^2$ *(and* μ *known). To see that* $\sum_{i=1}^n X_i^2$ *is not minimally sufficient for* σ^2, *note that* Q *in (7.5) is not constant with respect to* σ^2 *when only* $\sum x_i^2 = \sum y_i^2$, *unless* $\mu = 0$. *Thus,* $S(\mathbf{X}) = \left(\sum X_i, \sum X_i^2\right)$ *is also the minimal sufficient statistic for just* $\theta = \sigma^2$ *when* μ *is known but not equal to zero. This makes sense, because the m.l.e. of* σ^2 *is easily seen from (3.5) to be*

$$n^{-1} \sum_{i=1}^n (X_i - \mu)^2 = n^{-1}\left\{\sum_{i=1}^n X_i^2 + n\mu^2 - 2\mu \sum_{i=1}^n X_i\right\},$$

which reduces to $n^{-1} \sum_{i=1}^n X_i^2$ *when* μ *is known to equal zero, but requires both statistics in* $S(\mathbf{X}) = \left(\sum X_i, \sum X_i^2\right)$ *when* μ *is known but does not equal zero.*

Example 7.17 *Let* $X_i \overset{\text{i.i.d.}}{\sim} \text{Gam}(\alpha, \beta)$, $i = 1, \ldots, n$, *for* $\alpha, \beta > 0$, *with joint density of* $\mathbf{X} = (X_1, \ldots, X_n)$ *given by*

$$f_{\mathbf{X}}(\mathbf{x}; \alpha, \beta) = \prod_{i=1}^n \frac{\beta^\alpha}{\Gamma(\alpha)} x_i^{\alpha-1} e^{-\beta x_i} \mathbb{I}_{(0,\infty)}(x_i) \propto \left(\prod_{i=1}^n x_i\right)^{\alpha-1} \exp\left(-\beta \sum_{i=1}^n x_i\right).$$

Then, with $S_{\mathbf{x}} = \sum_{i=1}^n x_i$ *and* $P_{\mathbf{x}} = \prod_{i=1}^n x_i = \exp\left\{\sum_{i=1}^n \ln x_i\right\}$,

$$\frac{f_{\mathbf{X}}(\mathbf{x}; \alpha, \beta)}{f_{\mathbf{X}}(\mathbf{y}; \alpha, \beta)} = \frac{P_{\mathbf{x}}^{\alpha-1} \exp(-\beta S_{\mathbf{x}}) \prod_{i=1}^n \mathbb{I}_{(0,\infty)}(x_i)}{P_{\mathbf{y}}^{\alpha-1} \exp(-\beta S_{\mathbf{y}}) \prod_{i=1}^n \mathbb{I}_{(0,\infty)}(y_i)} \tag{7.6}$$

is constant with respect to $\theta = (\alpha, \beta)$ *if and only if* $S_{\mathbf{x}} = S_{\mathbf{y}}$ *and* $P_{\mathbf{x}} = P_{\mathbf{y}}$ *(and* $\mathbf{x}, \mathbf{y} \in S_{\mathbf{X}}$). *Thus,* $S(\mathbf{X}) = \left(\sum_{i=1}^n X_i, \prod_{i=1}^n X_i\right)$ *or, as a one-to-one function,* $\left(\sum_{i=1}^n X_i, \sum_{i=1}^n \ln X_i\right)$ *is minimal sufficient for* θ.

Now consider the cases where one of the two parameters is known. For $\theta = \alpha$, *we see from (7.6) that* $S(\mathbf{X}) = \sum_{i=1}^n \ln X_i$ *is minimal sufficient for* θ. *Recall from (7.3) that the m.l.e. is a function of the sufficient statistic. Indeed, the score function for* α *given in (3.22) in Example 3.6 only depends on the data via the sufficient statistic. Similarly, for* $\theta = \beta$, *from (7.6),* $S(\mathbf{X}) = \sum_{i=1}^n X_i$ *is minimal sufficient for* θ. *Again, as shown in Example 3.6,* $\hat{\beta}_{\text{ML}} = \alpha/\bar{X}$ *is just a function of* $\sum_{i=1}^n X_i$.

Example 7.18 *Let $X_i \overset{\text{i.i.d.}}{\sim}$ Beta(p, q), $i = 1, \ldots, n$, for $p, q > 0$. Then, for $\mathbf{X} = (X_1, \ldots, X_n)$,*

$$\frac{f_{\mathbf{X}}(\mathbf{x}; p, q)}{f_{\mathbf{X}}(\mathbf{y}; p, q)} = \frac{\left(\prod_{i=1}^n x_i\right)^{p-1}\left(\prod_{i=1}^n (1 - x_i)\right)^{q-1} \prod_{i=1}^n \mathbb{I}_{(0,1)}(x_i)}{\left(\prod_{i=1}^n y_i\right)^{p-1}\left(\prod_{i=1}^n (1 - y_i)\right)^{q-1} \prod_{i=1}^n \mathbb{I}_{(0,1)}(y_i)}$$

is constant with respect to $\theta = (p, q)$ if and only if $\mathbf{x}, \mathbf{y} \in S_{\mathbf{X}}$ and

$$\prod_{i=1}^n x_i = \prod_{i=1}^n y_i \quad \text{and} \quad \prod_{i=1}^n (1 - x_i) = \prod_{i=1}^n (1 - y_i),$$

showing that $\mathbf{S}(\mathbf{X}) = \left(\prod_{i=1}^n X_i, \prod_{i=1}^n (1 - X_i)\right)$ is minimal sufficient for θ.

Example 7.19 *Consider the inverse Gaussian distribution with density (I.7.54). In particular, let $X_i \overset{\text{i.i.d.}}{\sim} \text{IG}_2(\mu, \lambda)$, $i = 1, \ldots, n$, $\mu > 0$, $\lambda > 0$, with*

$$f_{\text{IG}_2}(x; \mu, \lambda) = \left(\frac{\lambda}{2\pi x^3}\right)^{1/2} \exp\left\{-\frac{\lambda}{2\mu^2 x}(x - \mu)^2\right\} \mathbb{I}_{(0,\infty)}(x).$$

(See also Section II.9.4.2.7.) The joint density of $\mathbf{X} = (X_1, \ldots, X_n)$ simplifies to

$$f_{\mathbf{X}}(\mathbf{x}; \mu, \lambda) = \left(\frac{\lambda}{2\pi}\right)^{n/2} \left(\prod_{i=1}^n x_i\right)^{-3/2}$$

$$\times \exp\left\{\frac{n\lambda}{\mu}\right\} \exp\left\{-\frac{\lambda}{2\mu}\left(\sum_{i=1}^n \frac{x_i}{\mu} + \sum_{i=1}^n \frac{\mu}{x_i}\right)\right\} \prod_{i=1}^n \mathbb{I}_{(0,\infty)}(x_i),$$

and the ratio $f_{\mathbf{X}}(\mathbf{x}; \mu, \lambda)/f_{\mathbf{X}}(\mathbf{y}; \mu, \lambda)$, given by

$$\frac{\left(\prod_{i=1}^n x_i\right)^{-3/2} \exp\left\{-\lambda\left(\mu^{-1}\sum_{i=1}^n x_i + \mu\sum_{i=1}^n x_i^{-1}\right)/(2\mu)\right\} \mathbb{I}_{(0,\infty)}(x_i)}{\left(\prod_{i=1}^n y_i\right)^{-3/2} \exp\left\{-\lambda\left(\mu^{-1}\sum_{i=1}^n y_i + \mu\sum_{i=1}^n y_i^{-1}\right)/(2\mu)\right\} \mathbb{I}_{(0,\infty)}(y_i)},$$

is constant with respect to $\theta = (\mu, \lambda)$ if and only if

$$\sum_{i=1}^n x_i = \sum_{i=1}^n y_i \quad \text{and} \quad \sum_{i=1}^n x_i^{-1} = \sum_{i=1}^n y_i^{-1}.$$

Notice it is not necessary that $\prod_{i=1}^n x_i = \prod_{i=1}^n y_i$. So, $\mathbf{S}(\mathbf{X}) = \left(\sum_{i=1}^n X_i, \sum_{i=1}^n X_i^{-1}\right)$ are the minimal sufficient statistics.

Example 7.20 *An important extension of the inverse Gaussian is the generalized inverse Gaussian (GIG), with density*

$$f_{\text{GIG}}(x; \lambda, \chi, \psi) \propto x^{\lambda-1} \exp\left(-\frac{1}{2}\left(\psi x + \frac{\chi}{x}\right)\right) \mathbb{I}_{(0,\infty)}(x),$$

for $\lambda \in \mathbb{R}$, $\chi, \psi > 0$; see Section II.9.4. An analysis similar to the IG case in the previous example shows that

$$\mathbf{S}(\mathbf{X}) = \left(\prod_{i=1}^n X_i, \sum_{i=1}^n X_i, \sum_{i=1}^n X_i^{-1}\right)$$

are the minimal sufficient statistics for $\theta = (\lambda, \chi, \psi)$.

Example 7.21 *For n i.i.d. observations $X_1, \ldots, X_n$ from a scale Laplace distribution* $\text{Lap}(0, \sigma)$, *it follows from the factorization theorem that $S = \sum_{i=1}^{n} |X_i|$ is sufficient because*

$$f_{\mathbf{X}}(\mathbf{x}; \sigma) = 2^{-n} \sigma^{-n} \exp\left(-\sigma^{-1} \sum_{i=1}^{n} |x_i|\right) = g(s, \sigma) \cdot h(\mathbf{x}),$$

where $h(\mathbf{x}) = 1$. Similarly, as

$$\frac{f_{\mathbf{X}}(\mathbf{x}; \sigma)}{f_{\mathbf{X}}(\mathbf{y}; \sigma)} = \exp\left(-\sigma^{-1} \sum_{i=1}^{n} |x_i| + \sigma^{-1} \sum_{i=1}^{n} |y_i|\right)$$

is constant with respect to σ if and only if $\sum_{i=1}^{n} |x_i| = \sum_{i=1}^{n} |y_i|$, S is minimally sufficient. For n i.i.d. observations from a location–scale Laplace distribution $\text{Lap}(\mu, \sigma)$, *it can be shown (see, for example, Lehmann and Casella, 1998) that all n order statistics are minimally sufficient. Thus, very little reduction of the data set is possible when a location parameter is introduced.*

The above six examples all have in common the fact that the corresponding densities (except the location Laplace) belong to the exponential family (3.51). In fact, if univariate density f_X is expressible as in (3.51), then so is $f_{\mathbf{X}}$ for $\mathbf{X} = (X_1, \ldots, X_n)$ with $X_i \overset{\text{i.i.d.}}{\sim} f_X$ and

$$f_{\mathbf{X}}(\mathbf{x}; \boldsymbol{\theta}) = (a(\boldsymbol{\theta}))^n \prod_{i=1}^{n} b(x_i) \exp\left\{\sum_{i=1}^{k} c_i(\boldsymbol{\theta}) \sum_{j=1}^{n} d_i(x_j)\right\}. \tag{7.7}$$

Thus, from the factorization theorem, $f_{\mathbf{X}}(\mathbf{x}; \boldsymbol{\theta}) = g(\mathbf{S}; \boldsymbol{\theta}) \cdot h(\mathbf{x})$ with $h(\mathbf{x}) = \prod_{i=1}^{n} b(x_i)$ and

$$\mathbf{S} = \left(\sum_{j=1}^{n} d_1(x_j), \ldots, \sum_{j=1}^{n} d_k(x_j)\right) \tag{7.8}$$

is sufficient. Furthermore, $\mathbf{S}$ is minimal sufficient because

$$\frac{f_{\mathbf{X}}(\mathbf{x}; \boldsymbol{\theta})}{f_{\mathbf{X}}(\mathbf{y}; \boldsymbol{\theta})} = \frac{\prod_{i=1}^{n} b(x_i) \, \exp\left\{\sum_{i=1}^{k}\left(c_i(\boldsymbol{\theta}) \sum_{j=1}^{n} d_i(x_j)\right)\right\}}{\prod_{i=1}^{n} b(y_i) \, \exp\left\{\sum_{i=1}^{k}\left(c_i(\boldsymbol{\theta}) \sum_{j=1}^{n} d_i(y_j)\right)\right\}}$$

is constant with respect to $\boldsymbol{\theta}$ if and only if $\sum_{j=1}^{n} d_i(x_j) = \sum_{j=1}^{n} d_i(y_j)$, $i = 1, \ldots, k$. Note that the "if" part of the if and only if statement ("sufficient") clearly still holds if the $c_i(\boldsymbol{\theta})$ are linearly dependent, but the "only if" part ("necessary") does not, that is, the $c_i(\boldsymbol{\theta})$ need to be linearly independent for minimal sufficiency of $\mathbf{S}$ to hold.

Armed with this general result, the results in the previous six examples could have been obtained more directly by expressing the joint density as (7.7). The reader is invited to do this for at least the gamma and beta cases above, as well as for the i.i.d. Bernoulli and i.i.d. geometric models. The next example shows the idea using the Poisson distribution.

Example 7.22 *The Poisson mass function is a member of the one-parameter exponential family:*

$$f_{\mathbf{X}}(\mathbf{x}; \lambda) = e^{-n\lambda}\left(\frac{1}{x_1! \cdots x_n!} \prod_{i=1}^{n} \mathbb{I}_{\{0,1,\ldots\}}(x_i)\right) \exp\left(\ln \lambda \sum_{i=1}^{n} x_i\right),$$

so that $S(\mathbf{X}) = \sum_{i=1}^{n} X_i$ is a minimal sufficient statistic for λ.

7.1.4 The Rao–Blackwell Theorem

The following theorem is due independently to Calyampudi Radhakrishna Rao, David Blackwell, and Andrey Nikolaevich Kolmogorov, from their papers in 1945, 1947, and 1950, respectively.

> Let $S(\mathbf{X})$ be a sufficient statistic for parameter vector $\theta \in \Theta$ and $T(\mathbf{X})$ an unbiased estimator of $\tau(\theta)$ with finite variance, for some function $\tau : \Theta \to \mathbb{R}$. Then $R = \mathbb{E}[T \mid \mathbf{S}]$ is (i) an estimator of $\tau(\theta)$, (ii) which is unbiased, and (iii) $\mathbb{V}(R) \le \mathbb{V}(T)$.

Before proving the result, we illustrate it with two examples.

Example 7.23 (Poisson, cont.) *It is easy to see that X_i is an unbiased estimator of λ. From the Rao–Blackwell theorem, we can do no worse with the estimator $\mathbb{E}[X_i \mid S]$, for $S = \sum_{j=1}^n X_j$. Letting $J_i = \{1, \dots, n\} \backslash i$ and using the independence of the X_i,*

$$\Pr(X_i = x \mid S = s) = \frac{\Pr(X_i = x, S = s)}{\Pr(S = s)} = \frac{\Pr(X_i = x)\Pr(\sum_{j \in J_i} X_j = s - x)}{\Pr(S = s)}$$

$$= \frac{e^{-\lambda}\lambda^x}{x!} \frac{e^{-(n-1)\lambda}[(n-1)\lambda]^{s-x}}{(s-x)!} \Big/ \frac{e^{-n\lambda}(n\lambda)^s}{s!}$$

$$= \binom{s}{x}\left(\frac{1}{n}\right)^x\left(\frac{n-1}{n}\right)^{s-x} \mathbb{I}_{\{0,1,\dots,s\}}(x),$$

or $X_i \mid S \sim \text{Bin}(\sum_{j=1}^n X_j, n^{-1})$, with expected value $n^{-1}\sum_{j=1}^n X_j = \bar{X}_n$.

Example 7.24 (Normal, cont.) *Let $X_i \overset{\text{i.i.d.}}{\sim} \text{N}(\mu, \sigma^2)$, assuming σ^2 known, and note that X_1 is unbiased for μ. The Rao–Blackwell theorem suggests use of $\mathbb{E}[X_1 \mid S]$, where $S = \sum_{i=1}^n X_i$ is the complete (see Section 7.2) and sufficient statistic for μ. From Example II.3.7, $X_1 \mid (S = s) \sim \text{N}(s/n, (n-1)\sigma^2/n)$, so that $\mathbb{E}[X_1 \mid S] = S/n = \bar{X}_n$. It should be kept in mind that the variance of $X_1 \mid S$ has nothing to do, per se, with the variance of $\bar{X}_n$.*

Proof of the Rao–Blackwell theorem.

(i) From the definition of sufficiency and the fact that T is a function of $\mathbf{X}$ (and not θ), the distribution of $T \mid \mathbf{S}$ is not a function of θ, and is thus an estimator.

(ii) From (A.88), $\mathbb{E}[R] = \mathbb{E}[\mathbb{E}[T \mid \mathbf{S}]] = \mathbb{E}[T] = \tau(\theta)$.

(iii) From (A.90), $\mathbb{V}(T) = \mathbb{V}(\mathbb{E}[T \mid \mathbf{S}]) + \mathbb{E}[\mathbb{V}(T \mid \mathbf{S})] \ge \mathbb{V}(R)$. ∎

Remark. If we did not know the conditional variance formula used in (iii) above, we could proceed as follows. Note that $\mathbb{V}(R) \le \mathbb{V}(T) \Leftrightarrow \mathbb{E}[R^2] \le \mathbb{E}[T^2]$, because $\mathbb{V}(R) = \mathbb{E}[R^2] - (\mathbb{E}[R])^2$ and $\mathbb{E}[R] = \mathbb{E}[T]$. To show $\mathbb{E}[R^2] \le \mathbb{E}[T^2]$, recall from Jensen's inequality (A.49) that $g(\mathbb{E}[X]) \le \mathbb{E}[g(X)]$ if $g(x)$ is convex. As $g(x) = x^2$ is convex, $(\mathbb{E}[T \mid \mathbf{S}])^2 \le \mathbb{E}[T^2 \mid \mathbf{S}]$ or

$$\mathbb{E}[R^2] = \mathbb{E}[(\mathbb{E}[T \mid \mathbf{S}])^2] \le \mathbb{E}[\mathbb{E}[T^2 \mid \mathbf{S}]] = \mathbb{E}[T^2].$$

What if we did not know Jensen's inequality? How about

$$\text{m.s.e.}(T) = \mathbb{E}[(T - \tau(\theta))^2] = \mathbb{E}[(T - R + R - \tau(\theta))^2]$$

$$= \mathbb{E}[(T-R)^2] + 2\mathbb{E}[(T-R)(R-\tau(\theta))] + \mathbb{E}[(R-\tau(\theta))^2]$$

$$= \mathbb{E}[(T-R)^2] + \text{m.s.e.}(R),$$

where the cross-term is zero (and explained below). Thus, m.s.e.$(R) \le$ m.s.e. (T). As T and R have the same expected value, it follows from (1.2) that $\mathbb{V}(R) \le \mathbb{V}(T)$.

For the cross-term, $\mathbb{E}[(T-R)(R-\tau(\theta))] = \mathbb{E}[(T-R)R]$ because $\tau(\theta)\mathbb{E}[T-R] = 0$. Then $\mathbb{E}[(T-R)R] = \mathbb{E}[\mathbb{E}[(T-R)R \mid \mathbf{S}]]$ and

$$\mathbb{E}[(T-R)R \mid \mathbf{S}] = R\mathbb{E}[(T-R) \mid \mathbf{S}] = R\{\mathbb{E}[T \mid \mathbf{S}] - \mathbb{E}[R \mid \mathbf{S}]\} = R\{R-R\} = 0,$$

as R is a function of $\mathbf{S}$.[4] ∎

The previous two examples were somewhat artificial in that the estimator $\bar{X}$ is both "obviously" better and suggests itself naturally as a candidate. The following example uses the Rao–Blackwell theorem to derive an estimator that is not so obvious.

Example 7.25 (Example 7.13, cont.) *Let* $X_i \overset{\text{i.i.d.}}{\sim} \text{DUnif}(N)$, $i = 1, \ldots, n$, *with* $Y_n = \max(X_i)$ *a sufficient statistic. As* $F_X(y) = (\lfloor y \rfloor / N)\mathbb{I}_{[1,N)}(y) + \mathbb{I}_{[N,\infty)}(y)$, *from (A.175)*,

$$F_{Y_n}(y) = \left(\frac{\lfloor y \rfloor}{N}\right)^n \mathbb{I}_{[1,N)}(y) + \mathbb{I}_{[N,\infty)}(y). \tag{7.9}$$

Differencing with $y \in \{1, 2, \ldots, N\}$ *yields*

$$\Pr(Y_n = y) = \Pr(Y_n \le y) - \Pr(Y_n \le y-1) = \frac{y^n - (y-1)^n}{N^n}\mathbb{I}_{\{1,2,\ldots,N\}}(y). \tag{7.10}$$

As $\mathbb{E}[X_i] = N^{-1}\sum_{i=1}^{N} i = (N+1)/2$, *applying the method of moments suggests* $2X_i - 1$, *for any* $i = 1, \ldots, n$, *is an unbiased estimator of* N. *Use* X_1, *so that* $T = 2X_1 - 1$. *Next, note that* $f_{X_1|Y_n}(x \mid y) = \Pr(X_1 = x \mid Y_n = y)$ *is given by*

$$\begin{cases} \dfrac{\Pr[X_1 = x \cap \max\{X_2, \ldots, X_n\} = y]}{\Pr(Y_n = y)}, & \text{if } x < y, \\ \dfrac{\Pr[X_1 = x \cap \max\{X_2, \ldots, X_n\} \le y]}{\Pr(Y_n = y)}, & \text{if } x = y, \end{cases}$$

and zero if $x > y$. *Recall that* X_1 *and* $\{X_2, \ldots, X_n\}$ *are independent. From the density,* $\Pr(X_1 = x) = N^{-1}$; *from (7.10),*

$$\Pr(\max\{X_2, \ldots, X_n\} = y) = \frac{y^{n-1} - (y-1)^{n-1}}{N^{n-1}};$$

and, from (7.9), $\Pr(\max\{X_2, \ldots, X_n\} \le y) = (y/N)^{n-1}$. *Putting these together yields*

$$f_{X_1|Y_n}(x, y) = \frac{1}{y^n - (y-1)^n} \begin{cases} y^{n-1} - (y-1)^{n-1}, & \text{if } x < y, \\ y^{n-1}, & \text{if } x = y, \end{cases}$$

assuming that $1 = \prod_{i=1}^{n} \mathbb{I}_{\{1,2,\ldots,N\}}(x_i)$.

---

[4] This being an occasional point of confusion, take the case in Example 7.24, with $R = \mathbb{E}[X_1 \mid S] = S/n$. Conditional on S, $\mathbb{E}[R \mid S] = \mathbb{E}[\bar{X} \mid S] = S/n = R$. On the contrary, unconditionally, $\mathbb{E}[\bar{X}] = \mu$.

Finally, applying the Rao–Blackwell theorem to derive a better estimator of N than $T = 2X_1 - 1$, say R, note that $R = \mathbb{E}[T \mid Y_n] = 2\mathbb{E}[X_1 \mid Y_n] - 1$ and

$$\mathbb{E}[X_1 \mid Y_n = y] = \sum_{x=1}^{y} x f_{X_1 \mid Y_n}(x, y) = \sum_{x=1}^{y-1} x f_{X_1 \mid Y_n}(x, y) + y f_{X_1 \mid Y_n}(y, y)$$

$$= \frac{y^{n-1} - (y-1)^{n-1}}{y^n - (y-1)^n} \sum_{x=1}^{y-1} x + y \frac{y^{n-1}}{y^n - (y-1)^n}$$

$$= \frac{(y-1)y}{2} \frac{y^{n-1} - (y-1)^{n-1}}{y^n - (y-1)^n} + \frac{y^n}{y^n - (y-1)^n}.$$

Straightforward algebra then yields

$$2\mathbb{E}[X_1 \mid Y_n = y] - 1 = \frac{y^{n+1} - (y-1)^{n+1}}{y^n - (y-1)^n}, \quad \text{or} \quad R = \frac{Y_n^{n+1} - (Y_n - 1)^{n+1}}{Y_n^n - (Y_n - 1)^n}. \tag{7.11}$$

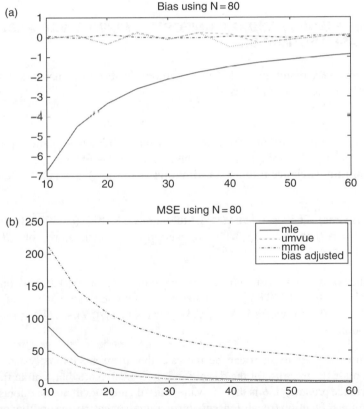

Figure 7.2 *Bias (a) and m.s.e. (b) as a function of sample size n for estimators of parameter $N = 80$ in the discrete uniform example, for the m.l.e. (solid), u.m.v.u.e. (dashed), m.m.e. (dash-dotted) and bias adjusted estimator (dotted). The m.s.e. of the u.m.v.u.e. and bias-adjusted estimator are graphically indistinguishable.*

Problem 7.16 confirms that the m.l.e. of N is $\hat{N}_{\mathrm{ML}} = Y_n = \max(X_i)$ and also derives an estimator based on the m.l.e., but with less bias, say $\hat{N}_{\mathrm{BA}}$, for "bias adjusted", given by $\hat{N}_{\mathrm{BA}} = (n+1)(Y_n - 1/2)/n$. Below we show that the estimator R in (7.11) is the uniform (meaning, for any $N \in \mathbb{N}$) minimum variance unbiased estimator (u.m.v.u.e.). Figure 7.2 compares the bias and m.s.e., obtained by simulation with 20,000 replications, for the four estimators, using $N = 80$ and a range of sample sizes n.

Except for the m.l.e., the estimators are not integers, and were rounded off. We see that the (rounded-off) u.m.v.u.e. R and the (rounded-off) bias-adjusted estimator are indeed unbiased, as is the m.m.e., but the latter has much higher m.s.e. than the other estimators. (If we do not round off, the bias for the three estimators appears to be exactly zero.) The u.m.v.u.e. and the bias-adjusted estimator have virtually the same m.s.e., which appears lower than that of the m.l.e. for all sample sizes.

Remark. The above discussion of the Rao–Blackwell theorem was restricted to the case for which $\tau(\theta)$ is scalar. The theorem can be extended to the multivariate case by showing that the difference between the mean squared error matrices of **T** and **R** is positive semi-definite.   ∎

7.2 COMPLETENESS AND THE UNIFORMLY MINIMUM VARIANCE UNBIASED ESTIMATOR

> An army may march great distances without distress, if it marches through country where the enemy is not.
>
> (Sun Tzu)

Here, interest centers on unbiased estimators that exhibit the smallest possible variance *uniformly*, meaning for all values of the parameter $\theta \in \Theta$. We refer to these as **uniformly minimum variance unbiased estimators** (u.m.v.u.e.), and take as definitions the following.

(**Univariate**) The estimator U^* is the u.m.v.u.e. of $\tau(\theta)$ if and only if $\mathbb{E}[U^*] = \tau(\theta)$ and $\mathbb{V}(U^*) \leq \mathbb{V}(U)$ for every unbiased estimator U of $\tau(\theta)$ and for all $\theta \in \Theta$.

(**Multivariate**) The estimator $\mathbf{U}^*$ is the u.m.v.u.e. of $\boldsymbol{\tau}(\theta)$ if and only if $\mathbb{E}[\mathbf{U}^*] = \boldsymbol{\tau}(\theta)$ and $\mathbb{V}(\mathbf{U}^*) \leq \mathbb{V}(\mathbf{U})$ for every unbiased estimator $\mathbf{U}$ of $\boldsymbol{\tau}(\theta)$ and for all $\theta \in \Theta$, where $\mathbb{V}(\mathbf{V}) \leq \mathbb{V}(\mathbf{W})$ means that $\mathbb{V}(\mathbf{W}) - \mathbb{V}(\mathbf{V})$ is positive semi-definite.

If an u.m.v.u.e. does exist, it can be the case that there are infinitely many unbiased estimators, making direct use of the definition impractical. Note that, unlike for the m.l.e. via its invariance property, if $\hat{\theta}$ is the u.m.v.u.e. of θ, then $\tau(\hat{\theta})$ will not be unbiased for $\tau(\theta)$ if τ is a nonlinear function (recall Jensen's inequality), so that $\tau(\hat{\theta})$ is not the u.m.v.u.e. for $\tau(\theta)$.

Before discussing the famous Lehmann–Scheffé theorem, the concept of completeness is required.

For the family of probability distributions $\mathcal{F} = \{f_S(\cdot; \theta); \ \theta \in \Theta\}$ of statistic S, $\mathcal{F}$ is **complete** if, for any function $Z : S_S \to \mathbb{R}$ satisfying $\mathbb{E}_\theta[Z(S)] = 0$ for all $\theta \in \Theta$, it is also the case that $Z(S) = 0$ with probability 1.

Recall that, in the discrete case, the condition "with probability 1" means that $Z(s) = 0$ for all values of $s \in S_S = \{s : \Pr(S = s) > 0\}$. In the continuous case, this means that $Z(s) = 0$ for $s \in S_S$ "almost everywhere," or such that the set of points in S_S such that $z \neq 0$ has measure zero; see page I.348 for a brief introduction to the latter concept. Observe that S is a statistic, and thus not a function of θ. Also, Z is a function of S but not of θ.

To help illustrate matters, let $m_1 : S_S \to \mathbb{R}$ and $m_2 : S_S \to \mathbb{R}$ be two different functions such that, for both $i = 1$ and $i = 2$, $\mathbb{E}[m_i(S)] = g(\theta)$ for some function $g : \Theta \to \mathbb{R}$. Then $D(S) = m_1(S) - m_2(S)$ has expected value zero. If statistic S is complete, then $D(S) = 0$ (and not just its expectation), and $m_1 = m_2$ with probability 1. This implies, for example, that if both $m_1(S)$ and $m_2(S)$ are unbiased estimators for $g(\theta)$ and S is a complete and sufficient statistic, then $m_1 = m_2$ (with probability one), showing that an unbiased estimator for $g(\theta)$ that is a function only of a complete and sufficient statistic is unique.

The following three examples are standard; see, for example, Rohatgi (1976, p. 345).

Example 7.26   Let $X_i \overset{\text{i.i.d.}}{\sim} N(\theta, \theta^2)$, $\theta \neq 0$, so that

$$f_X(x) = (2\pi\theta^2)^{-n/2} \exp\left\{ -\frac{1}{2\theta^2} \left(\sum x_i^2 - 2\theta \sum x_i + n\theta^2 \right) \right\}$$

and, from the factorization theorem with $h(x) = 1$, $S = (S_1, S_2) = \left(\sum_1^n X_i, \sum_1^n X_i^2 \right)$ is sufficient for θ. We have

$$\mathbb{E}[S_1^2] = (\mathbb{E}[S_1])^2 + \mathbb{V}(S_1) = \theta^2 n(n+1),$$

$$\mathbb{E}[S_2] = n(\mathbb{E}[X_i^2]) = n(\mathbb{V}(X_i) + (\mathbb{E}[X_i])^2) = 2n\theta^2.$$

With $Z(S) = 2S_1^2 - (n+1)S_2$, it follows that S is not complete because, for at least one $\theta \neq 0$ (and for all of them, in fact), $Z(S)$ has zero expectation but is not identically zero.

Example 7.27   Let $S \sim N(0, \theta)$, $\theta > 0$. With $Z(S) = S$, $\mathbb{E}[Z(S)] = \mathbb{E}[S] = 0$, but clearly $Z(S)$ is not zero with probability 1, so that S is not complete. As $S/\sqrt{\theta} \sim N(0, 1)$, we have $S^2/\theta \sim \chi^2(1)$. Define the statistic $T(S) := S^2 \overset{d}{=} \theta C$, where $C \sim \chi^2(1)$. Then

$$f_T(t) = \theta^{-1} f_C\left(\frac{t}{\theta} \right) = \frac{1}{\sqrt{2\pi\theta}} t^{-1/2} \exp\left(-\frac{t}{2\theta} \right) \mathbb{I}_{(0,\infty)}(t).$$

For T to be complete, it must be the case that, for all $\theta > 0$, and any function Z of T (but not of θ) such that

$$\mathbb{E}[Z(T)] = \int_0^\infty Z(t) t^{-1/2} \exp\left(-\frac{t}{2\theta} \right) dt = 0, \quad \forall \theta > 0,$$

it is also the case that $Z(T) = 0$, for (almost) all $T > 0$. This is the case, although more advanced results from analysis are required to prove it.

Example 7.28   *Let $X_i \overset{\text{i.i.d.}}{\sim} \text{Poi}(\lambda)$, $i = 1, \ldots, n$, and $S = \sum_{i=1}^{n} X_i \sim \text{Poi}(\theta)$, where $\theta = n\lambda$. For S to be complete, it needs to be shown that, for every function $Z : \mathbb{N} \to \mathbb{R}$, $\mathbb{E}[Z(S)] = 0$ for all $\theta > 0$ implies that $Z(k) = 0$, $k = 0, 1, 2, \ldots$. Observe that*

$$\mathbb{E}[Z(S)] = \sum_{s=0}^{\infty} Z(s) \frac{\theta^s e^{-\theta}}{s!} = e^{-\theta} \left(Z(0) + Z(1) \frac{\theta}{1!} + Z(2) \frac{\theta^2}{2!} + \cdots \right) = 0$$

implies $\sum_{k=0}^{\infty} Z(k) \theta^k / k! = 0$. What is then required is to show that, in order for this infinite series to converge to zero for all $\theta > 0$, each of the coefficients $Z(k)$ must also be zero. This is indeed the case, but we do not provide details.

The last two examples should have given the impression that confirmation of completeness could require some adeptness in analysis. Fortunately, a general case has been proven that encompasses many distributions of interest. This is the exponential family: if f_X is expressible as (7.7) and the range of $(c_1(\theta), \ldots, c_k(\theta))$ contains an open set in $\mathbb{R}^k$ of nonzero measure, then **S** in (7.8) is complete. Together with the results for exponential family discussed at the end of Section 7.1.3, this implies that **S** is a complete, minimally sufficient statistic.

A further result of interest is that, if sufficient statistic **S** is complete, then it is minimal sufficient. To outline a proof, let **S** be a complete, sufficient statistic and **M** a minimal sufficient statistic that, from the definition, can be expressed as **M(S)**. Let $Z(S) = S - \mathbb{E}[S \mid M]$, which is a function of **S** because **M** and, hence, **S** | **M**, is a function of **S**. As $\mathbb{E}[Z(S)] = \mathbb{E}[S] - \mathbb{E}[\mathbb{E}[S \mid M]] = \mathbf{0}$, and we are assuming that **S** is complete, it follows that $Z(S) = \mathbf{0}$, that is, $S = \mathbb{E}[S \mid M]$ (with probability 1). Thus, **S** is a function of **M**, and so, by definition, **S** is also minimally sufficient.

Remark.   Intuition behind the concept of completeness, and motivation for its name, are provided by Stigler (1972). Wackerly (1976) gives further discussion on the above result that, if sufficient statistic **S** is complete, then it is minimal sufficient.   ∎

We are now in a position to state the main result for the scalar parameter case.

> **Lehmann–Scheffé u.m.v.u.e.:** Let **S** be a complete, sufficient statistic. Then $R(S)$ is the u.m.v.u.e. of $\tau(\theta) = \mathbb{E}[R]$, provided $R : S_S \to \mathbb{R}$ and $\mathbb{E}[R]$ exists.

Proof. Let $T(\mathbf{X})$ be an unbiased estimator of $\tau(\theta)$ and $R^* = \mathbb{E}[T \mid S]$ so that, via the Rao–Blackwell theorem, $\mathbb{V}(R^*) \leq \mathbb{V}(T)$. Set $Z(S) = R^* - R$ so that $\mathbb{E}[Z(S)] = 0$. Completeness of **S** then implies that $Z(S) = 0$ so that $R = R^*$ with probability 1.   ∎

Recall the Poisson and normal cases in Examples 7.23 and 7.24, respectively. As both of these distributions belong to the exponential family, a complete, minimal sufficient statistic is available; in both of these cases it is $S = \sum_{i=1}^{n} X_i$. Thus, the application of the Rao–Blackwell theorem produces the u.m.v.u.e. For Example 7.25, showing that statistic Y_n is complete is straightforward; see Rohatgi (1976, p. 346) for details. Thus, R in (7.11) is the u.m.v.u.e.

In the multivariate setting, if $R_i(S)$ is the univariate u.m.v.u.e. of $\tau_i(\theta) = \mathbb{E}[R_i]$, $i = 1, \ldots, k$, then $\mathbf{R} = (R_1, \ldots, R_k)'$ is the u.m.v.u.e. of $\tau(\theta) = (\tau_1(\theta), \ldots, \tau_k(\theta))'$. In addition, for any vector $\mathbf{a} \in \mathbb{R}^k$, $\mathbf{a}'\mathbf{R}$ is the u.m.v.u.e. of $\mathbf{a}'\tau(\theta)$. See, for example, Mittelhammer (1996, pp. 406–407).

Example 7.29 (Normal, cont.)   *As $\bar{X}_n$ and S_n^2 are unbiased and functions of the complete and minimal sufficient statistics, they are the u.m.v.u.e.s for μ and σ^2.*

Example 7.30   *The m.l.e.s of the two parameters in the i.i.d. gamma model were given in Example 3.6 and are clearly functions of the sufficient statistics as given in Example 7.17. However, the m.l.e. is biased, with a first-order bias correction (i.e., less bias, but not unbiased) given by*

$$\hat{\alpha} = n^{-1}[(n-3)\hat{\alpha}_{\text{ML}} + 2/3], \quad \hat{\beta} = \hat{\alpha}/\bar{X}. \tag{7.12}$$

No u.m.v.u.e. is known for this model. See Bowman and Shenton (1988) and the references therein for further details relating to the gamma model, Section 7.4.1 below for the general technique, and Problem 7.19.

Remark.   The insistence on unbiasedness can give rise to absurd u.m.v.u.e.s; see, for example, Meeden (1987) and the references therein for examples. Recall the comments about use of unbiased estimators in Section 1.1.1. In particular, the Bayesian framework is more immune to such issues, and modern statistical inference, particularly for complicated, highly parameterized models, has moved away from unbiasedness and instead makes use of shrinkage estimation, empirical Bayes approaches, and model selection based on the lasso and related methods (see, for example, Hastie et al., 2015), as well as placing emphasis on what often really matters: the ability of a model to generate good forecasts.   ∎

7.3   AN EXAMPLE WITH I.I.D. GEOMETRIC DATA

We illustrate several of the concepts introduced here and in Chapter 3 with an application to the geometric model. Assume $X_1, \ldots, X_n \overset{\text{i.i.d.}}{\sim} \text{Geo}(\theta)$ with p.m.f. $\Pr(X_i = x) = \theta(1-\theta)^x \mathbb{I}_{\{0,1,\ldots\}}(x)$, $0 < \theta < 1$, and let $S = \sum_{i=1}^n X_i$. The joint density can be written

$$f_{\mathbf{X}}(\mathbf{x}; \theta) = \theta^n(1-\theta)^s \prod_{i=1}^n \mathbb{I}_{\{0,1,\ldots\}}(x_i) = g(s, \theta)h(\mathbf{x}), \tag{7.13}$$

so that, from the factorization theorem, S is sufficient. Even better, S is complete and minimally sufficient because

$$f_{\mathbf{X}}(\mathbf{x}; \theta) = \theta^n \prod_{i=1}^n \mathbb{I}_{\{0,1,\ldots\}}(x_i) \exp\left(\log(1-\theta)\sum_{i=1}^n x_i\right)$$

belongs to the exponential family. Of course, S is negative binomially distributed, with density

$$f_S(s; n, \theta) = \binom{n+s-1}{s}\theta^n(1-\theta)^s \mathbb{I}_{\{0,1,\ldots\}}(s). \tag{7.14}$$

As $\mathbb{E}[\bar{X}_n] = (1 - \theta)/\theta$, it follows from the Lehmann–Scheffé theorem that $\bar{X}_n$ is an u.m.v.u.e. of $(1 - \theta)/\theta$. From (7.13),

$$\ell(\theta) = \log f_X(x; \theta) = n \log \theta + (s) \log(1 - \theta) + \log \prod_{i=1}^{n} \mathbb{I}_{\{0,1,\dots\}}(x_i)$$

and

$$\dot{\ell} = \frac{n}{\theta} - \frac{1}{1 - \theta} s, \quad \ddot{\ell} = -\frac{n}{\theta^2} - \frac{s}{(1 - \theta)^2}, \quad \mathbb{E}[\ddot{\ell}] = -\frac{n}{\theta^2} - \frac{n\frac{1-\theta}{\theta}}{(1 - \theta)^2} = \frac{-n}{\theta^2(1 - \theta)} = -J.$$

With $\tau(\theta) = (1 - \theta)/\theta$ and $[\tau'(\theta)]^2 = \theta^{-4}$, the CRlb is $(1 - \theta)/(\theta^2 n)$, which coincides with $\mathrm{Var}(\bar{X}_n)$ (see Problem I.4.3), showing that $\bar{X}_n$ is the m.v.b.e. of $(1 - \theta)/\theta$. This also follows because (3.49) is fulfilled:

$$\frac{n}{\theta} - \frac{1}{1 - \theta} s = \dot{\ell}(\theta) = k(\theta)(U - \tau(\theta)) = \frac{n}{\theta - 1} \left(\frac{s}{n} - \frac{1 - \theta}{\theta} \right). \tag{7.15}$$

Now consider θ itself. The CRlb of θ is just $1/J = \theta^2(1 - \theta)/n$. The previous result for $\tau(\theta)$ and the fact that τ is nonlinear indicates that no m.v.b.e. exists for θ. Furthermore, the u.m.v.u.e. for θ is not $\tau^{-1}(\widehat{\tau(\theta)}) = (1 + \bar{X}_n)^{-1}$ because it is (necessarily) biased, but one might expect the latter to be close to the u.m.v.u.e., assuming one exists. Using the fact that $\theta = \Pr(X_1 = 0)$, the Rao–Blackwell theorem can be applied. Thus,

$$\hat{\theta} = \mathbb{E}[\mathbb{I}_{\{0\}}(X_1) \mid S = s] = \Pr(X_1 = 0 \mid S = s)$$

is the u.m.v.u.e. of $\tau(\theta) = \theta$ because S is complete and sufficient. But, from the i.i.d. assumption and (7.14),

$$\Pr(X_1 = 0 \mid S = s) = \frac{\Pr\left(X_1 = 0, \sum_{i=2}^{n} X_i = s\right)}{\Pr(S = s)} = \frac{\Pr(X_1 = 0)\Pr\left(\sum_{i=2}^{n} X_i = s\right)}{\Pr(S = s)},$$

so that

$$\hat{\theta} = \frac{\theta \binom{n+s-2}{s} \theta^{n-1}(1 - \theta)^s}{\binom{n+s-1}{s} \theta^n(1 - \theta)^s} = \frac{n - 1}{n + s - 1}, \quad n > 1, \tag{7.16}$$

is the u.m.v.u.e. Note that $\hat{\theta}$ approaches $(1 + \bar{X}_n)^{-1}$ as the sample size increases. From the unbiasedness of $\hat{\theta}$ and (7.14), we have, with $m = n - 1$, the identity

$$\theta = \mathbb{E}[\hat{\theta}] = \mathbb{E}\left[\frac{m}{m + S}\right] = \sum_{s=0}^{\infty} \frac{m}{m + s} \binom{m + s}{s} \theta^n(1 - \theta)^s, \quad m \geq 1,$$

or, equivalently,

$$\sum_{s=0}^{\infty} \binom{m + s - 1}{s} (1 - \theta)^s = \theta^{-m}, \quad m \geq 1, \tag{7.17}$$

which was directly proven in Example I.1.10.

Furthermore, with $K = m\,\theta^n(1-\theta)^{-m}$ and using (7.17),

$$\mathbb{E}[\hat{\theta}^2] = \sum_{s=0}^{\infty} \left(\frac{m}{m+s}\right)^2 \binom{m+s}{s} \theta^n(1-\theta)^s = K\sum_{s=0}^{\infty} \frac{(1-\theta)^{m+s}}{m+s}\binom{m+s-1}{s}$$

$$= K\sum_{s=0}^{\infty}\binom{m+s-1}{s}\int_0^{1-\theta} y^{m+s-1}\,\mathrm{d}y = K\int_0^{1-\theta} y^{m-1}\left\{\sum_{s=0}^{\infty}\binom{m+s-1}{s}y^s\right\}\mathrm{d}y$$

$$= K\int_0^{1-\theta} \frac{y^{m-1}}{(1-y)^m}\,\mathrm{d}y = KQ_m,$$

where Q_m is the so-defined integral. While Q_m can be computed via numerical integration, Problem 7.7 shows that it can be computed recursively as

$$Q_m = \frac{1}{m-1}\left(\frac{1-\theta}{\theta}\right)^{m-1} - Q_{m-1}, \qquad Q_1 = -\ln\theta, \tag{7.18}$$

this being computationally faster for m small. From Q_m, $\mathbb{E}[\hat{\theta}^2]$ and $\mathbb{V}(\hat{\theta})$ are easily computed. Figure 7.3 plots the variance of $\hat{\theta}$ and its CRlb as a function of θ for two values of n. As n grows, the variance of the u.m.v.u.e. approaches the CRlb.

We wish to derive the m.l.e.s of θ and $(1-\theta)/\theta$ and their asymptotic distributions. Equating $\dot{\ell}(\theta) = \mathrm{d}\ln f_{\mathbf{X}}(\mathbf{x},\theta)/\mathrm{d}\theta$ with zero and solving,

$$\frac{n}{\theta} - \frac{s}{1-\theta} = 0 \Rightarrow \hat{\theta}_{\mathrm{ML}} = \frac{n}{s+n} = (1+\bar{X}_n)^{-1}.$$

From the CRlb given above and the invariance property of the m.l.e. (and using the informal notation for the asymptotic distribution, such that sample size n appears in the distribution),

$$\hat{\theta}_{\mathrm{ML}} \overset{\mathrm{asy}}{\sim} \mathrm{N}\left(\theta, \frac{\theta^2(1-\theta)}{n}\right), \qquad \left(\widehat{\frac{1-\theta}{\theta}}\right)_{\mathrm{ML}} = \frac{1-\hat{\theta}_{\mathrm{ML}}}{\hat{\theta}_{\mathrm{ML}}} = \bar{X}_n \overset{\mathrm{asy}}{\sim} \mathrm{N}\left(\frac{1-\theta}{\theta}, \frac{1-\theta}{\theta^2 n}\right).$$

For $n = 1$, the u.m.v.u.e. for θ is invalid; it is zero for all $0 < \theta < 1$. In this case, the m.l.e. of θ is $1/(1+X_1)$. Its expected value is, with $y = 1 + x$,

$$g(\theta) = \mathbb{E}\left[\frac{1}{1+X}\right] = \theta\sum_{x=0}^{\infty}\frac{(1-\theta)^x}{1+x} = \frac{\theta}{1-\theta}\sum_{y=1}^{\infty}\frac{(1-\theta)^y}{y} = -\frac{\theta}{1-\theta}\ln(\theta),$$

using the Taylor series expansion $\sum_{y=1}^{\infty} t^y/y = -\ln(1-t)$. The discrepancy $\theta - g(\theta)$ is plotted in Figure 7.4. We see that the bias is worst around $\theta = 0.3$ and improves as θ approaches 0 or 1. It can be shown that no unbiased estimator for θ exists when $n = 1$.

In Section 1.1.2, we used geometric r.v.s with support $\{1, 2, \dots, \}$ instead of $\{0, 1, \dots, \}$ because we were interested in the number of failures and not the total number of trials. In that case, $\bar{X}_n$ is the u.m.v.u.e. for $\mathbb{E}[\bar{X}_n] = 1/\theta =: \psi(\theta)$. Also, $\psi^{-1}\widehat{(\psi(\theta))} = 1/\bar{X}_n$ is $\hat{p}_2$ in Section 1.1.2, which was shown (via simulation) to be biased. The u.m.v.u.e. of θ is, again from the Rao–Blackwell theorem but with $\theta = \Pr(X_1 = 1)$,

$$\frac{\Pr\left(X_1 = 1, \sum_{i=2}^{n} X_i = s-1\right)}{\Pr(S = s)} = \frac{\theta\binom{s-1-1}{n-1-1}\theta^{n-1}(1-\theta)^{(s-1)-(n-1)}}{\binom{s-1}{n-1}\theta^n(1-\theta)^{s-n}} = \frac{n-1}{s-1}, \tag{7.19}$$

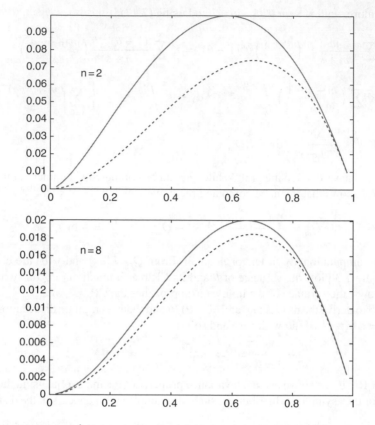

Figure 7.3   *Variance of $\hat{\theta}$ (the u.m.v.u.e. for θ) (solid) and the CRlb (dashed), as a function of θ.*

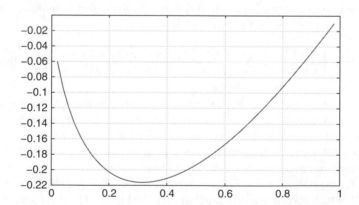

Figure 7.4   *Bias for the m.l.e. of the geometric distribution parameter for sample size 1.*

where the numerator follows from (1.3) with $x = s - 1$ and $r = n - 1$ and the independence of trials.[5] This estimator is $\hat{p}_3$ in Section 1.1.2, where it was shown (via simulation) to be unbiased, in agreement with result from the Rao–Blackwell theorem. Recall also that m.s.e.$(\hat{p}_2)$ was less than m.s.e.$(\hat{p}_3)$ for certain values of the parameter, demonstrating that an u.m.v.u.e. will not necessarily minimize the m.s.e.

7.4 METHODS OF BIAS REDUCTION

This section considers several methods to reduce (rather than eliminate) the bias of a consistent estimator. They are all related to a certain extent and are all, to differing degrees, numerically intensive. Only the univariate case is considered here, though the bias-function approach is readily generalized to the multivariate case; see MacKinnon and Smith (1998), as well as Kim (2016) and the references therein for more sophisticated procedures.

7.4.1 The Bias-Function Approach

Recall Example 7.30, in which a bias-adjusted estimator for the shape parameter of the gamma model was given. We present in this section how such estimators can be derived, doing so in the context of two examples. First consider the m.l.e. for σ^2 in the i.i.d. $N(\mu, \sigma^2)$ model with both parameters unknown. The expected value of $\hat{\sigma}^2_{\text{ML}} = n^{-1} \sum_{i=1}^{n} (X_i - \bar{X})^2$ takes the form of σ^2 times one minus some quantity, say Q, that is,

$$\mathbb{E}[\hat{\sigma}^2_{\text{ML}}] = \mathbb{E}\left[\frac{n-1}{n} S_n^2\right] = \frac{n-1}{n}\sigma^2 = \sigma^2\left(1 - \frac{1}{n}\right) = \sigma^2(1 - Q),$$

where $Q = Q(n) = 1/n$. The bias-corrected estimator

$$\frac{\sigma^2_{\text{ML}}}{1 - Q(n)} = \frac{\sigma^2_{\text{ML}}}{1 - 1/n} = \frac{n}{n-1}\sigma^2_{\text{ML}} = S_n^2 \qquad (7.20)$$

naturally suggests itself and, in this case, completely removes the bias.

The second example involves the bivariate normal distribution. The sample correlation coefficient $R = \hat{\rho}$ in (3.25) is the m.l.e. for ρ, and is biased, with expected value given in (3.28). In particular, from the approximation in (3.29),

$$\mathbb{E}[\hat{\rho}] = \rho\left(1 - \frac{1-\rho^2}{2n} + O(n^{-2})\right) \approx \rho(1 - Q). \qquad (7.21)$$

The expected value again takes the form of ρ times one minus the term Q. Unfortunately, in this case, $Q = Q(\rho, n) = (1 - \rho^2)/(2n)$ depends on the true value of ρ; if it did not, and ignoring the $O(n^{-2})$ terms, we could construct an unbiased estimator similar to the previous case for σ^2. What might be reasonable in such a case is the estimator $\hat{\rho}_{\text{adj}}$ given by the solution to the equation

$$\hat{\rho} = \hat{\rho}_{\text{adj}}(1 - Q) = \hat{\rho}_{\text{adj}}(1 - Q(\hat{\rho}_{\text{adj}}, n)) = \hat{\rho}_{\text{adj}}\left(1 - \frac{1 - \hat{\rho}^2_{\text{adj}}}{2n}\right), \qquad (7.22)$$

where $\hat{\rho}$ is the m.l.e. Observe that this parallels the construction in (7.20).

---

[5] This could have been directly obtained by noting that s in (7.16) is the total number of failures, so that $n + s$ is the total number of trials, which is what s represents in (7.19).

This can be algebraically solved so that $\hat{\rho}_{adj}$ is one of the solutions to a cubic equation. In general, a numeric root-finding procedure can be effectively used to compute the adjusted estimator. Rather obviously, because we have neglected the $O(n^{-2})$ term in this case, the resulting estimator will not be exactly unbiased. Less obvious is that, even if we were to somehow use the exact expression for $\mathbb{E}[\hat{\rho}]$, the resulting estimator will still not, in general, be exactly unbiased. The reason for this will be made clear below.

Let $\hat{\rho}$ be any consistent estimator of ρ, which we will refer to as the "initial estimator," and let $m(\rho) = \mathbb{E}[\hat{\rho}; \rho]$ denote the mean function of $\hat{\rho}$ when ρ is the true parameter, and m^{-1} : $(m(-1), m(1)) \rightarrow (-1, 1)$ denote its inverse. The values -1 and 1 are of course specific to the correlation coefficient and, in general, would be the left and right borders of the parameter space, which we assume to be an open interval. We further assume $m(\rho)$ is strictly monotone in ρ, in which case function m^{-1} is properly defined.[6] Then the bias-adjusted estimator of ρ, denoted $\hat{\rho}_{adj}$, is taken to be that value of ρ such that the initial estimator has a mean equal to the observed initial estimate. That is,

$$
\hat{\rho}_{adj} = \begin{cases} 1, & \text{if } \hat{\rho} \geq m(1), \\ m^{-1}(\hat{\rho}), & \text{if } m(-1) < \hat{\rho} < m(1), \\ -1, & \text{if } \hat{\rho} \leq m(-1). \end{cases} \tag{7.23}
$$

For $m(-1) < \hat{\rho} \leq m(1)$, and with $\hat{\rho}^O$ the observed value of initial estimator $\hat{\rho}$, we can write

$$
\hat{\rho}_{adj} = m^{-1}(\hat{\rho}) = \operatorname{argmin}_{\rho} |\mathbb{E}[\hat{\rho}; \rho, n] - \hat{\rho}^O|, \tag{7.24}
$$

which lends itself to computation with the use of, say, bisection, or more advanced numeric root-finding methods. As a very special case, observe that, if m takes the simple form $E = m(\rho) = \rho(1 - Q(n))$, then $\rho = m^{-1}(E) = E/(1 - Q(n))$ and

$$
\hat{\rho}_{adj} = m^{-1}(\hat{\rho}) = \frac{\hat{\rho}}{1 - Q(n)},
$$

as in (7.20).

Now consider the special case in (7.21). Ignoring the $O(n^{-2})$ term, (7.23) and the monotonicity of m together imply that $m(\hat{\rho}_{adj}) = \mathbb{E}[\hat{\rho}; \hat{\rho}_{adj}] = \hat{\rho}$, which leads precisely to the estimator in (7.22).

Figure 7.5 illustrates the procedure for the correlation coefficient using the exact mean (3.28). In this case, the observed value of $\hat{\rho}$, denoted $\hat{\rho}^O$, is equal to 0.5. Then $m^{-1}(\hat{\rho})$ is computed by locating that value on the x-axis such that the function m equals 0.5. This is seen to be very close to 0.6; computing it gives $\hat{\rho}_{adj} = 0.604$. In this case, the bias correction is quite large because the sample size was chosen extremely small ($n = 3$) in order to best illustrate the method. Use of the approximate expected value in (7.21) yields 0.564 which, again because of the extremely small sample size, differs considerably from 0.604. The programs to compute these are given below, after the other bias-adjusted estimators are discussed.

It is important to observe that, as

$$
\mathbb{E}[\hat{\rho}_{adj}] = \mathbb{E}[m^{-1}(\hat{\rho})] \neq m^{-1}(\mathbb{E}[\hat{\rho}]) = m^{-1}(m(\rho)) = \rho,
$$

[6] For the correlation coefficient $\hat{\rho}$, we will assume that $m(\rho)$ is strictly increasing for all $n > 2$ without actually checking it, although from (7.21) it is clear that, for n large enough, it will be the case. Figure 7.5 shows that $m(\rho)$ is increasing for $n = 3$.

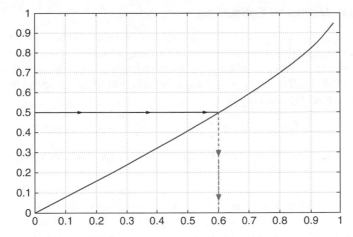

Figure 7.5 *Illustration of how the mean-adjusted estimator is determined. The graph shows the function $m(\rho) = \mathbb{E}[\hat{\rho}; \, \rho]$ for $n = 3$. If the observed value of $\hat{\rho}$, $\hat{\rho}^O$ is 0.5, then, as indicated with arrows in the figure, $\hat{\rho}_{adj} = m^{-1}(\hat{\rho}^O) \approx 0.6$.*

the bias-adjusted estimator will not be exactly unbiased, unless m^{-1} is linear. As Figure 7.5 shows, for the correlation coefficient, the function m^{-1} is approximately linear, so that the resulting estimator is also approximately unbiased. It is also important to realize that its use may not lead to a significant reduction in m.s.e. and could actually exhibit higher m.s.e.

Remark. Even if $\hat{\rho}_{adj}$ is exactly (or very close to) unbiased, it need not be precisely (or very close to) the u.m.v.u.e. If, however, the initial estimator used to construct $\hat{\rho}_{adj}$ is a function of the set of complete and sufficient statistics, then one might expect it to be close to the u.m.v.u.e. For the first example involving the estimators of σ^2, application of the bias-adjustment method does result in the u.m.v.u.e. because the m.l.e. is a function of the complete and sufficient statistics and because, in this simple case, the resulting estimator is exactly unbiased. (If, in this case, we applied the method to, say, $\hat{\sigma}^2_{Odd}$, defined to be the m.l.e. of the variance based on just observations $1, 3, 5, \ldots$, then, even though the bias-adjustment scheme in this case yields exact unbiasedness, it is not the u.m.v.u.e.)

For the correlation coefficient, $R = \hat{\rho}$ is a function of the set of complete and sufficient statistics, but, because of the nonlinearity of the mean function, $\hat{\rho}_{adj}$ is not exactly unbiased and, hence, not the u.m.v.u.e., though one would expect it to be close (because the mean function is "close" to linear). Interestingly enough, the u.m.v.u.e. of ρ does exist; Olkin and Pratt (1958) showed that it is given by

$$\hat{\rho}_{unb} = \hat{\rho} \; {}_2F_1\left(\frac{1}{2}, \frac{1}{2}, \frac{n-2}{2}, 1 - \hat{\rho}^2\right)$$

$$= \hat{\rho}\left(1 + \frac{1 - \hat{\rho}^2}{2(n-2)} + \frac{9(1 - \hat{\rho}^2)^2}{8n(n-2)} + O(n^{-3})\right). \tag{7.25}$$

We will see below via simulation that it is indeed unbiased for ρ, but has a higher m.s.e. than the m.l.e. over a large part of the parameter space.   ■

7.4.2 Median-Unbiased Estimation

An interesting alternative to pursuing a mean-unbiased estimator is to consider an estimator that is **median-unbiased**. An estimator $\hat{\theta}_{\text{med}}$ is median-unbiased for θ if, for each $\theta \in \Theta$, θ is a median of $\hat{\theta}_{\text{med}}$. In other words, $\hat{\theta}_{\text{med}}$ is median-unbiased if $\Pr(\hat{\theta}_{\text{med}} < \theta) = \Pr(\hat{\theta}_{\text{med}} > \theta)$. Lehmann (1959, p. 22) provides a formal definition in terms of absolute loss: $\hat{\theta}_{\text{med}}$ is median-unbiased for θ if and only if

$$\mathbb{E}_{\theta}[|\hat{\theta}_{\text{med}} - \theta|] \leq \mathbb{E}_{\theta}[|\hat{\theta}_{\text{med}} - \theta^*|], \quad \forall \, \theta, \theta^* \in \Theta,$$

where Θ is the parameter space of θ, and $\mathbb{E}_{\theta}$ denotes expectation when θ is the true parameter value.

While a mean-unbiased estimator is "correct on average", a median-unbiased estimator has a 50% chance of being too low and a 50% chance of being too high. This is arguably quite an appealing characteristic, perhaps even more so than mean-unbiasedness. For estimators whose density functions are rather asymmetric, the two forms of unbiasedness will give rise to point estimators with quite different small-sample properties.

The form of $\hat{\theta}_{\text{med}}$ is virtually the same as the mean-adjusted estimator considered previously, but with the function m (again monotone) denoting the median of the initial estimator $\hat{\theta}$ instead of the mean. That is, $m(\theta) = \text{Median}[\hat{\theta}; \theta, n]$, paralleling the definition of m in the mean-bias case above, and, as in Lehmann (1959, Sec. 3.5),

$$\hat{\theta}_{\text{med}} = m^{-1}(\hat{\theta}) = \operatorname{argmin}_{\theta}|\text{Median}[\hat{\theta}; \theta, n] - \hat{\theta}^{O}|, \tag{7.26}$$

where $\hat{\theta}^{O}$ is the observed value of the initial estimator. The computation of the median of $\hat{\theta}$ will require the ability to compute the inverse c.d.f. of $\hat{\theta}$. As an analytic expression for the inverse c.d.f. is rarely obtainable, calculating the median entails numerically solving the nonlinear equation $0.5 = F_{\hat{\theta}}(x; \theta)$ for x. The calculation of (7.26) itself, however, requires root searching, so that the entire process can be relatively time-consuming. A better computational alternative is to express (7.26) as $\text{Median}[\hat{\theta}; \hat{\theta}_{\text{med}}, n] = \hat{\theta}^{O}$ or

$$F_{\hat{\theta}}(\hat{\theta}^{O}; \hat{\theta}_{\text{med}}, n) = \int_{-\infty}^{\hat{\theta}^{O}} f_{\hat{\theta}}(x; \hat{\theta}_{\text{med}}) \, dx = \frac{1}{2}. \tag{7.27}$$

As (7.27) only requires one level of root searching, it will be considerably faster to compute.

It is noteworthy that, in comparison to the mean-adjusted estimator, $\hat{\theta}_{\text{med}}$ is *exactly* median-unbiased. This follows because the median function $m(\theta)$ is monotone. If it is monotone increasing,

$$\Pr(\hat{\theta}_{\text{med}} < \theta) = \Pr(m^{-1}(\hat{\theta}) < \theta) = \Pr(\hat{\theta} < m(\theta)) = \Pr(\hat{\theta} < \text{Median}[\hat{\theta}; \hat{\theta}_{\text{med}}, n]) = 0.5.$$

Just as for the mean-adjusted estimator, a median-unbiased estimator may not exhibit a smaller m.s.e. than the one to which the technique is applied. Problem 7.18 shows a case for which $\hat{\theta}_{\text{med}}$ will always be worse than the m.l.e. in terms of m.s.e.

Remark.   There are further interesting properties of median-unbiased estimators, including results that are analogous to the Rao–Blackwell and Lehmann–Scheffé theorems; a brief survey is provided by Read (1982). See also Ghosh and Sen (1989) on the relation of **Pitman closeness** and median-unbiased estimation.   ∎

```
1  function corrcoefsimDOIT(rho, n, sim)
2  [V,D]=eig([1 rho; rho 1]); C=V*sqrt(D)*V';
3  med=zeros(sim,1); mean1=med; mean2=med; themle=med;
4  olkinpratt=med; op2nd=med; themode=med; randn('state',1)
5  for i=1:sim, if mod(i,20)==0, i, end
6    r=randn(2,n); g=C*r; cc=corrcoef(g'); rhohat=cc(2,1); z=rhohat;
7    themle(i)=z; olkinpratt(i) = z * f_21(0.5,0.5,0.5*(n-2),1-z^2);
8    % op2nd(i) = z * (1+(1-z^2)/2/(n-2) + 9*(1-z^2)^2 / 8 / n / (n-2) );
9    [modu,medu,meanu,meanu2nd]=corrcoefmedunbiased(rhohat,n);
10   themode(i)=modu; med(i)=medu; mean1(i)=meanu; mean2(i)=meanu2nd;
11 end
12 str1 = ['save corrcoefstudyn',int2str(n),'rho',int2str(round(10*rho))];
13 str2 = [' themle olkinpratt themode med mean1 mean2'];
14 str = [str1 str2]; eval(str)
```

Program Listing 7.1: Program to simulate the various estimators of the correlation coefficient. We use the variable name `themle` instead of just `mle` because Matlab has alike-named function, `mle`, and likewise for `themode`.

Returning to the correlation coefficient example dealt with at length in the previous section, computation of the median-unbiased estimator requires the c.d.f. of $\hat{\rho}$. This can be calculated far faster by numerically integrating the p.d.f., using the Laplace approximation for the $_2F_1$ function from Butler and Wood (2002). Listing 7.1 shows the code used to simulate the various estimators; the program that actually computes $\hat{\theta}_{\mathrm{med}}$ and also the two mean-adjusted estimators (7.24) and (7.22) is shown in Listing 7.2.

For the correlation coefficient, the adjustment to $\hat{\rho}$ induced by $\hat{\rho}_{\mathrm{med}}$, computed from (7.26) based on the initial estimator $\hat{\rho}$, is the opposite of the adjustment induced by $\hat{\rho}_{\mathrm{adj}}$. For example, with $n = 3$ and $\hat{\rho}^O = 0.5$, we noted above that $\hat{\rho}_{\mathrm{adj}} = 0.604$. The median-unbiased estimator is $\hat{\rho}_{\mathrm{med}} = 0.343$. This may not have been expected; given that $|\hat{\rho}|$ is downward (mean)-biased for all ρ, one might have thought that $\hat{\rho}_{\mathrm{med}}$ would also correct upward, just by a different amount than $\hat{\rho}_{\mathrm{adj}}$. Similarly, for $n = 10$ and $\hat{\rho}^O = 0.5$, $\hat{\rho}_{\mathrm{med}} = 0.477$ and $\hat{\rho}_{\mathrm{adj}} = 0.522$, although the absolute amount of correction is less, owing to the larger sample size.

The performance of the various estimators is best depicted graphically; this is done below in a simulation study, after we introduce mode adjusted estimator. We will see that it offers a substantially different performance with respect to bias and m.s.e. for the correlation coefficient.

7.4.3 Mode-Adjusted Estimator

Note that the mean-adjusted estimator in (7.24) and the median-unbiased estimator in (7.26) are very similar in construction; they differ by the function m. As such, one might entertain use of other measures of central tendency, the other obvious one of which is the mode. As such, we define the **mode-adjusted estimator** as

$$\hat{\rho}_{\mathrm{mod}} = m^{-1}(\hat{\rho}) = \mathrm{argmin}_\rho |\mathrm{Mode}[\hat{\rho};\ \rho, n] - \hat{\rho}^O|, \tag{7.28}$$

as introduced in Broda et al. (2007). Its computation is also shown in the program in Listing 7.2. To assess the behavior of the estimator, Figure 7.6 shows the result of a simulation study for a grid of ρ-values from 0 to 0.9, and for two sample sizes, $n = 8$ and

```
1   function [modeu, medu, meanu, meanu2nd]=corrcoefmedunbiased(rhohat,n)
2
3   opt = optimset('Display','Off','tolFun',1e-6, 'tolX',1e-6);
4   modeu=fsolve(@(rho) here4(rho,rhohat,n),rhohat,opt); % mode unbiased
5   medu =fsolve(@(rho) here1(rho,rhohat,n),rhohat,opt); % median unbiased
6   meanu=fsolve(@(rho) here2(rho,rhohat,n),rhohat,opt);
7        % approx. mean unbiased, using exact expression for mean
8   meanu2nd=fsolve(@(cc) here3(cc,rhohat,n),rhohat,opt);
9        % app. mean unbiased, 2nd order exression from Hotelling (1953)
10
11  function d=here1(rho,rhohat,n), themedian=corrcoefinvcdf(rho,n);
12     d=(abs(themedian-rhohat));
13  function d=here2(rho,rhohat,n), themean = corrcoefmean(rho,n);
14     d=(abs(themean-rhohat));
15  function d=here3(c,r,n), d=r-c*(1-(1-c^2)/2/n);
16  function d=here4(rho,rhohat,n), themode=corrcoefmode(rho,n);
17     d=(abs(themode-rhohat));
18
19  function m=corrcoefmode(rho,n) % finds the mode of the density.
20  start=corrcoefmean(rho,n);      % Use expected value as start
21  opt = optimset('Display','Off','tolFun',1e-4, 'tolX',1e-4, ...
22       'LargeScale','On','Algorithm','active-set');
23  m=fmincon(@(r) minuscorrcoefpdf(r,rho,n),start,[],[],[],[],-1,1,[],opt);
24
25  function den=minuscorrcoefpdf(r,rho,n), den=-corrcoefpdf(r,rho,n);
```

Program Listing 7.2: For the correlation coefficient, computes the mode-adjusted estimator (7.28), median-unbiased (7.26), mean-adjusted (7.24), and (7.22). The function `corrcoefmean` is given in Listing 3.2 and the function `corrcoefinvcdf` is given in Listing 7.3. The function tofind the mode, `corrcoefmode`, uses Matlab's constrained minimization function `fmincon`. It supports quite general constraints; we only need to impose that the mode occurs for a value between -1 and 1.

```
1   function q=corrcoefinvcdf(rho,n,p)
2   if nargin<3, p=0.5; end
3   opt = optimset('Display','Off','tolFun',1e-6, 'tolX',1e-6);
4   q=fsolve(@(r) cdf(r,p,rho,n),rho,opt);
5
6   function v=cdf(r,p,rho,n), [~,F]=corrcoefpdf(r,rho,n); v= abs(F-p);
```

Program Listing 7.3: Computes quantiles of the correlation coefficient $\hat{\rho}$. The program `corrcoefpdf` is given in Listing 3.1.

$n = 16$. (The results are symmetric about zero and so we only need to discuss the results for $|\rho|$.) All the estimators discussed up to this point are compared with respect to mean bias, median bias, and m.s.e. We immediately see several things from the graphs. First, the u.m.v.u.e. is indeed mean-unbiased, but exhibits the highest m.s.e. for $|\rho| \leq 0.6$. Second, the median-unbiased estimator is indeed median-unbiased and has lower m.s.e. than the m.l.e. for $|\rho| \leq 0.6$. Third, $\hat{\rho}_{mod}$ has rather high bias, but lower m.s.e. than all the other estimators for $|\rho| \leq 0.5$, substantially so for ρ near zero. Finally, we see that both the bias and m.s.e. decrease when going from $n = 8$ to $n = 16$, and that the relative difference in m.s.e. of $\hat{\rho}_{mod}$ compared to the others also decreases as the sample size increases.

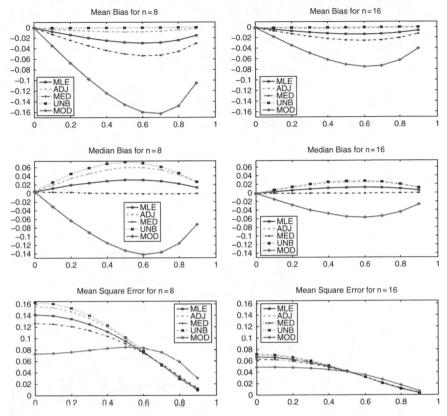

Figure 7.6 *Based on output from the program in Listing 7.1, this shows the mean and median bias, and the m.s.e., of the five estimators: the m.l.e. $\hat{\rho}$ given in (3.25) (denoted MLE in the legend); the mean-bias-adjusted estimator $\hat{\rho}_{adj}$ given in (7.24) (ADJ); the median-unbiased estimator $\hat{\rho}_{med}$ given in (7.26) (MED); the u.m.v.u.e. $\hat{\rho}_{unb}$ given in (7.25) (UNB); and the mode-adjusted estimator $\hat{\rho}_{mod}$ given in (7.28) (MOD), as a function of ρ, based on 10,000 replications, for $n = 8$ (left) and $n = 16$ (right).*

Given that $\hat{\rho}_{mod}$ is (apparently) unbiased at $\rho = 0$ and, for small sample sizes, exhibits a substantial reduction in m.s.e., it would be interesting to compare its density to those of the other estimators; this is shown in the left two panels of Figure 7.7. Indeed, the variance of $\hat{\rho}_{mod}$ is much smaller than that of the other estimators. The right panels show the densities for $\rho = 0.6$, for which $\hat{\rho}_{mod}$ has the highest median bias but for which the m.s.e.s are all very close.

It is clear from the graphs that, with respect to minimizing the m.s.e., if the true $|\rho|$ is less than 0.5, then one should use $\hat{\rho}_{mod}$; while for $|\rho| > 0.5$, any of the other estimators, the easiest of which is the m.l.e., could be used. This is the unfortunate, but not uncommon, situation that the best estimator (with respect to m.s.e.) depends on the true value of ρ. One could entertain a new estimator that, say, takes the value $\hat{\rho}_{mod}$ if $|\hat{\rho}_{med}|$ is less than 0.5; and $\hat{\rho}_{med}$ otherwise. Figure 7.8 is the same as Figure 7.6 but with an overlaid thicker line showing the mean bias, median bias, and m.s.e. of this estimator, obtained by simulation. The kink in the median-bias plot was to be expected as a result of the transition from $\hat{\rho}_{mod}$ to $\hat{\rho}_{med}$ around $\rho = 0.5$.

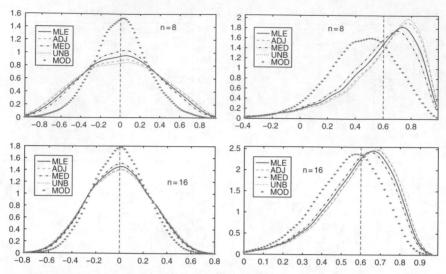

Figure 7.7 *Based on output from the program in Listing 7.1, this shows kernel density estimates of the five estimators: the m.l.e. $\hat{\rho}$ given in (3.25) (denoted MLE in the legend); the mean-bias-adjusted estimator $\hat{\rho}_{adj}$ given in (7.24) (ADJ); the median-unbiased estimator $\hat{\rho}_{med}$ given in (7.26) (MED); the u.m.v.u.e. $\hat{\rho}_{unb}$ given in (7.25) (UNB); and the mode-adjusted estimator $\hat{\rho}_{mod}$ given in (7.28) (MOD), for $\rho = 0$ (left) and $\rho = 0.6$ (right), based on 10,000 replications, for $n = 8$ (top) and $n = 16$ (bottom). The vertical dashed line indicates the true value of ρ.*

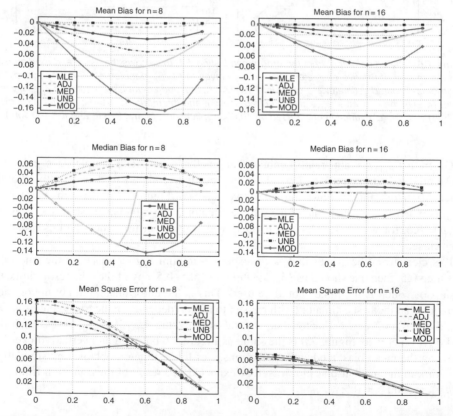

Figure 7.8 *Same as Figure 7.6 but with overlaid results, as the new, thicker line, corresponding to the properties of the estimator resulting from taking the value $\hat{\rho}_{mod}$ if $|\hat{\rho}_{med}|$ is less than 0.5, and $\hat{\rho}_{med}$ otherwise.*

With respect to m.s.e., the new estimator achieves a compromise between the small-ρ and large-ρ segments of the parameter space and is thus still quite far from the lower m.s.e. envelope, being nowhere overall superior. It is not clear that it should be preferred over use of $\hat{\rho}_{\text{mod}}$; the decision would involve the extent to which the researcher believes that, in the application at hand, $|\rho| < 0.5$, which is a subjective issue and touches upon the idea of using "prior information," or information not associated with the data set, for statistical inference.

Remarks

(a) It should be kept in mind that the above results are strictly valid when the data are precisely normally distributed – an assumption that can be violated in practice with real data. It is highly unlikely that the true distribution is known, and even if it were, the distribution of $\hat{\rho}$, unlike for the normal case in (3.26), will most likely be intractable. All is not lost: one could still use the above estimators and, via simulation with a set of plausible nonnormal distributions and/or use of the nonparametric bootstrap, assess their relative performance.

(b) The computation of $\hat{\rho}_{\text{mod}}$ as given in (7.28) is clearly the slowest of all the adjusted estimators considered because it involves nested optimization. The computation of all of the adjusted estimators, in particular $\hat{\rho}_{\text{mod}}$, can be sped up drastically by noting that, for a given sample size n, $\hat{\rho}_{\text{mod}}$ is a smooth function of $\hat{\rho}$ that can be approximated arbitrarily well by a polynomial in $\hat{\rho}$. (Note that the function is convex because $\hat{\rho}_{\text{mod}}$ adjusts towards zero.) By computing $\hat{\rho}_{\text{mod}}$ for a grid of $\hat{\rho}$-values and then fitting a polynomial by regression, we obtain, for $n = 8$,

$$\hat{\rho}_{\text{mod}} \approx k_0 + k_1\hat{\rho} + k_2\hat{\rho}^2 + k_3\hat{\rho}^3 + k_4\hat{\rho}^4, \quad \text{for } n = 8$$
$$= \text{sign}(\hat{\rho})(0.577|\hat{\rho}| + 0.276|\hat{\rho}|^2 - 0.440|\hat{\rho}|^3 + 0.583|\hat{\rho}|^4). \quad (7.29)$$

(There is no intercept term because $\hat{\rho}_{\text{mod}}$ is zero for $\hat{\rho} = 0$, and we use the symmetry property of the estimator.) This is obviously virtually instantaneously estimated and its use results in graphs that are indistinguishable from those shown above. It has the further benefit that it avoids the occasional numeric error associated with (7.28). Of course, it is specific to the chosen sample size, here $n = 8$, though this could be remedied by fitting a function using both $\hat{\rho}$ and n as regressors, or by constructing (7.29) for a grid of sample sizes n, and then approximating the resulting regression coefficients as functions of n, similar to what was done to obtain p-values for the MSP normality test in Section 6.4.3. The reader is encouraged to do this.

(c) The estimator $\hat{\rho}_{\text{mod}}$ adjusts $\hat{\rho}$ towards zero; as such, it is an example of a shrinkage estimator, similar to the one studied in Section 5.1.6. In our setting, we could attempt to approximate (7.29) with just the linear term, but for all n, expressing it as

$$\hat{\rho}_{\text{shr}} = k_1\hat{\rho} = \hat{\rho}\left(1 + \frac{c_1}{n} + \frac{c_2}{n^2}\right). \quad (7.30)$$

To estimate the values of c_1 and c_2, first estimate k_1 as was done in (7.29), for a grid of n values, that is, estimate $\hat{\rho}_{\text{mod}} \approx k_1\hat{\rho}$. (This is obviously very

crude, but still captures the shrinkage.) Then express the coefficients in
(7.30) as $n^2(k_1 - 1) = c_2 + nc_1$ and perform a regression to get the values
of c_1 and c_2. In this case, using a grid from $n = 8$ to $n = 60$, the values
$n^2(k_1 - 1)$ are nearly perfectly linear in n, and we obtain

$$\hat{\rho}_{\text{shr}} = \hat{\rho}\left(1 - \frac{1}{n} - \frac{3.6}{n^2}\right). \tag{7.31}$$

The estimator $\hat{\rho}_{\text{shr}}$ is clearly trivial to compute. Figure 7.9 shows the bias
and m.s.e. for (7.31), overlaid as was done above with the original set of
estimators. Unlike the other estimators, the bias continues to grow with
$|\rho|$, but the m.s.e. curve is, unlike the new estimator shown in Figure 7.8,
relatively good over the whole parameter range. Given its simplicity of
calculation, it is overall perhaps the best choice.

(d) The mean-, median- and mode-unbiased estimation procedures can be
applied in a variety of contexts. Broda et al. (2007) demonstrate their
use in the context of regression models with AR(1) disturbances using
near-exact (via saddlepoint approximations) small-sample distribution
theory. Via the median-unbiased estimator, a test for a **unit root** is
straightforward and does not require simulation or appeal to the asymptotics typically associated with unit root testing. This will be discussed at
length in Book IV.   ∎

7.4.4 The Jackknife

A simple (and rather ingenious) method for reducing the bias of a consistent estimator was
proposed by Quenouille (1956) and is referred to as the **jackknife**. (The name, however, is
attributed to John Tukey.) It can be viewed as an approximation of the bootstrap; see, for
example, Efron (1979) and Shao and Tu (1995).

Assume we have a sample of n observations $\mathbf{X} = (X_1, \ldots, X_n)$ and $\hat{\theta} = T(\mathbf{X})$ is an estimator of parameter θ. Let $\mathbf{X}_{(i)}$ denote the set of $n - 1$ observations resulting when X_i is not
included, and let $T_{(i)} = T(\mathbf{X}_{(i)})$, $i = 1, \ldots, n$. The **delete-one jackknife** estimator of θ based
on $\hat{\theta}$ is given by

$$\hat{\theta}^* = nT - (n - 1)\bar{T}_\bullet, \tag{7.32}$$

where $\bar{T}_\bullet = n^{-1}\sum_{i=1}^{n} T_{(i)}$ is the average of the $T_{(i)}$. (The delete-d case is discussed in Wu,
1986, and Shao and Tu, 1995).

To see what effect this procedure has on the bias, assume the expansion

$$\text{bias}(T) = \mathbb{E}[T] - \theta = \frac{a_1}{n} + \frac{a_2}{n^2} + \cdots \tag{7.33}$$

holds, for constants a_i that can depend on θ but not on n (see, for example, Bao, 2007, Bao
and Ullah, 2007, and the references therein). Then

$$\mathbb{E}[\hat{\theta}^*] = n\mathbb{E}[T] - (n - 1)\mathbb{E}[T_1]$$

$$= n\left(\theta + \frac{a_1}{n} + \frac{a_2}{n^2} + \frac{a_3}{n^3} + \cdots\right)$$

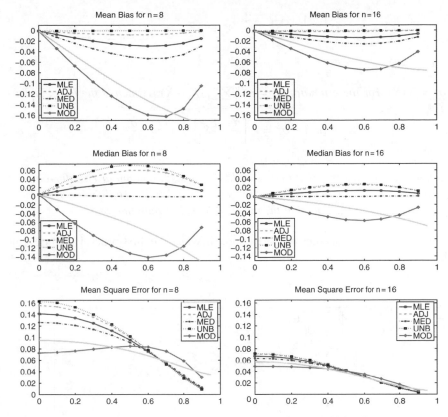

Figure 7.9 *Same as Figure 7.6 but with overlaid results, as the new, thicker line, corresponding to the estimator (7.31).*

$$- (n - 1)\left(\theta + \frac{a_1}{n - 1} + \frac{a_2}{(n - 1)^2} + \frac{a_3}{(n - 1)^3} + \cdots \right)$$

$$= \theta + a_2\left(\frac{1}{n} - \frac{1}{n - 1} \right) + a_3\left(\frac{1}{n^2} - \frac{1}{(n - 1)^2} \right) + \cdots$$

$$= \theta - \frac{a_2}{n(n - 1)} + O(n^{-3}),$$

showing that the first-order term a_1/n completely disappears and the second-order term is only slightly larger than a_2/n^2 in (7.33). If, for all θ, $\hat{\theta}$ itself is unbiased (so that $a_1 = a_2 = \cdots = 0$), then, clearly, $\hat{\theta}^*$ is also unbiased.

Example 7.31 *For $T = \bar{X}$, which is already unbiased for $\mu = \mathbb{E}[X]$ for all μ (assuming $|\mu| < \infty$), the jackknife procedure not is only unbiased, but also, in this case, yields precisely $\hat{\theta}^* = \bar{X}$. This is easily seen algebraically:*

$$\hat{\theta}^* = nT - (n - 1)\bar{T}_{\bullet} = n\bar{X} - (n - 1)\frac{1}{n}\sum_{i=1}^{n}\frac{1}{n - 1}\sum_{j=1, j\neq i}^{n}X_i$$

$$= n\bar{X} - (n-1)\frac{1}{n}\frac{1}{n-1}\sum_{i=1}^{n}\sum_{j=1, \, j\neq i}^{n} X_i = n\bar{X} - (n-1)\frac{1}{n}\frac{1}{n-1}\sum_{i=1}^{n}(n-1)X_i$$

$$= \bar{X}.$$

Let $\sigma^2 = \mathbb{V}(X)$. For the estimator $\hat{\sigma}^2 = n^{-1}\sum_{i=1}^{n}(X_i - \bar{X})^2$ (which is the m.l.e. under normality),

$$\mathbb{E}[\hat{\sigma}^2] - \sigma^2 = \mathbb{E}\left[\frac{n-1}{n}S_n^2\right] - \sigma^2 = -\frac{\sigma^2}{n},$$

so that, in (7.33), $a_1 = -\sigma^2$ and $a_i = 0$, $i \geq 2$.

The jackknife method applied to $\hat{\sigma}^2$ should thus result in an unbiased estimator. In fact, the procedure produces precisely S_n^2, as the reader should verify. Application of the jackknife to S_n^2 results again in S_n^2. Note that the m.s.e. of $\hat{\sigma}^2$ is less than that of S_n^2.

See Problem 7.13 for another simple example in which the (first-order) jackknife removes all the bias.

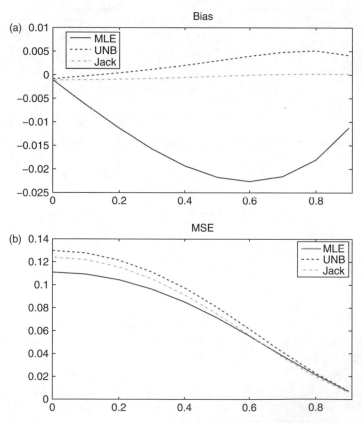

Figure 7.10 *The bias (a) and m.s.e. (b) of the m.l.e. $\hat{\rho}$ (solid), the jackknife estimator $\hat{\rho}^*$ (dashed), and the unbiased estimator $\hat{\rho}_{unb}$ given in (7.25) (dash-dotted), based on sample size $n = 10$ and 50,000 replications. The smoothness of the curves is obtained by using the same seed value when generating the data for each value of ρ, but this is almost irrelevant given the large number of replications.*

Example 7.32   *Let $X_i \overset{\text{i.i.d.}}{\sim} \text{Exp}(\lambda)$ and $S = \sum_{i=1}^n X_i$. The m.l.e. is $\hat{\lambda} = n/S = 1/\bar{X}$, which is biased. Problem 7.17 shows that the u.m.v.u.e. for λ is $U = (n-1)/S$. Thus,*

$$\mathbb{E}\left[\frac{n}{S}\right] - \lambda = \mathbb{E}\left[\frac{n}{n-1}\frac{n-1}{n}\frac{n}{S}\right] - \lambda = \mathbb{E}\left[\frac{n}{n-1}U\right] - \lambda = \frac{\lambda}{n-1} = \sum_{i=1}^{\infty}\frac{\lambda}{n^i}$$

and the jackknife estimator applied to the m.l.e. will not be exactly unbiased. Now let $\hat{\lambda}^*$ be the jackknife estimator applied to the u.m.v.u.e. U. It too must be unbiased, but it is not identical to U, so that $\mathbb{V}(\hat{\lambda}^*) \geq \mathbb{V}(U)$. Simulation quickly reveals some results: For $\lambda = 1$ and $n = 5$ (and based on 20,000 replications), the m.s.e. of U is 1.34, while that of $\hat{\lambda}^*$ is 1.54, showing that the jackknife procedure induces a considerable increase in variance for very small sample sizes. However, for $n = 7$, the relative difference in m.s.e. is much less, while for $n = 10$, the m.s.e. values are virtually the same.

Example 7.33 (Correlation coefficient, cont.)   *The m.l.e. $\hat{\rho}$ given in (3.25) is biased, unless $\rho = 0$. The bias and m.s.e. of $\hat{\rho}$ and $\hat{\rho}^*$ were computed for a grid of ρ-values $0, 0.1, \ldots, 0.9$ for $n = 10$; the results are shown in Figure 7.10. While the bias is indeed reduced for $\hat{\rho}^*$, its m.s.e. is higher for all values of ρ.*

In all the above examples, the jackknife estimator reduced (or eliminated) the bias, but increased the mean squared error. Cases exist for which the m.s.e. is also reduced; see e.g., Shao and Tu (1995, p. 66). Further details on the jackknife can be found in the textbook presentations of Efron and Tibshirani (1993, Ch. 11) and Shao and Tu (1995, Ch. 2).

7.5 PROBLEMS

7.1   Recall the Weibull distribution from (A.68). Let $X_i \overset{\text{i.i.d.}}{\sim} \text{Weib}(b, 0, s)$, for $i = 1, \ldots, n$, with density

$$f_{X_i}(x; b, s) = \frac{b}{s}\left(\frac{x}{s}\right)^{b-1}\exp\left\{-\left(\frac{x}{s}\right)^b\right\}\mathbb{I}_{(0,\infty)}(x).$$

Assuming that the shape parameter b is a known constant, find a complete sufficient statistic.

7.2   Let $X_1, X_2 \overset{\text{i.i.d.}}{\sim} N(\mu, \sigma^2)$, with $S = X_1 + X_2$ and $D = X_1 - X_2$. Show that (S, D) is a one-to-one transformation of (X_1, X_2); $f_{D|S}$ is independent of μ; and, from the definition of sufficiency, S is sufficient for μ.

7.3   Let **X** be an n-length i.i.d. sample from a population with density

$$f_X(x) = \theta x^{-2}\mathbb{I}_{(\theta,\infty)}(x), \quad \theta > 0.$$

   (a) Derive the c.d.f. of X.
   (b) Find a sufficient statistic, compute its density and verify that it integrates to 1.
   (c) What needs to be true in order for your sufficient statistic to be complete? Try to prove it by differentiating both sides of the condition.
   (d) Assuming completeness of your sufficient statistic, derive the u.m.v.u.e. of θ.
   (e) Compute the m.l.e. of θ.

7.4   Let $\mathbf{X}$ be an n-length i.i.d. sample from a population with density

$$f_X(x) = 2\theta^{-2}x\mathbb{1}_{(0,\theta)}(x), \quad \theta > 0.$$

(a) Find a sufficient statistic, its density and expected value.
(b) What needs to be true in order for your sufficient statistic to be complete?
(c) Assuming completeness of your sufficient statistic, derive the u.m.v.u.e. of θ.
(d) Compute the m.l.e. of θ.

7.5   Let $X_i \overset{\text{i.i.d.}}{\sim} N(\mu, \sigma^2)$, $i = 1, \ldots, n$, with μ known.
(a) Calculate the u.m.v.u.e. of σ^2 and call it Y.
(b) Calculate $\mathbb{V}(Y)$.
(c) Is Y also the m.v.b.e.?
(d) Let $T = n^{-1}\sum_{i=1}^{n}(X_i - \bar{X})^2$. Calculate the m.s.e. of T and compare it to that of Y. What do you conclude?

7.6   Let $X_1, \ldots, X_n \overset{\text{i.i.d.}}{\sim} \text{Unif}\left(\theta - \frac{1}{2}, \theta + \frac{1}{2}\right)$ for $\theta \in \mathbb{R}$.
(a) Calculate the mean of $X_{1:n}$ and $X_{n:n}$.
(b) Show that $(X_{1:n}, X_{n:n})$ is sufficient. (Observe that the dimension of the sufficient statistic is greater than the number of unknown parameters.)
(c) Show that $(X_{1:n}, X_{n:n})$ is not complete.
(d) Show X_1 is an unbiased estimator of θ.
(e) What estimator do you get by applying the Rao–Blackwell theorem to X_1?

7.7   Section 7.3 studied estimators for an i.i.d. geometric sample.
(a) Verify (7.18).
(b) Compute $\mathbb{V}(1/(1 + S))$ for $n = 2$ and $n = 3$ and verify that it is greater than the CRlb for $n = 2$.
(c) Verify that the u.m.v.u.e. for $\mathbb{V}(X_i) = (1 - \theta)/\theta^2$ is $(S^2 + nS)/(n(n + 1))$.

7.8   Assume $X_1, \ldots, X_m \overset{\text{i.i.d.}}{\sim} N(\mu_1, \sigma^2)$, $Y_1, \ldots, Y_n \overset{\text{i.i.d.}}{\sim} N(\mu_2, \sigma^2)$, with $X_i \perp Y_j$ for all i, j. We wish to estimate $\tau(\theta) = (\mu_1 + k\sigma^2)$ for k known, where $\theta = (\mu_1, \mu_2, \sigma^2)$. Define

$$s_{X,1} = \sum_{i=1}^{m} X_i, \quad s_{X,2} = \sum_{i=1}^{m} X_i^2, \quad \text{and} \quad s_{Y,1} = \sum_{i=1}^{n} Y_i, \quad s_{Y,2} = \sum_{i=1}^{n} Y_i^2. \qquad (7.34)$$

(a) Derive the u.m.v.u.e. of $\tau(\theta)$ based only on the sampled $\mathbf{X}$ vector; call it $T_{\mathbf{X}}^*$.
(b) Derive the m.l.e. of $\tau(\theta)$ and its asymptotic distribution.
(c) Compare the $T_{\mathbf{X}}^*$ and the asymptotic variance of $\hat{\tau}_{\text{ML}}$.
(d) Derive an u.m.v.u.e. of $\tau(\theta)$, say $T_{\mathbf{X},\mathbf{Y}}^*$, based on both sampled $\mathbf{X}$ and $\mathbf{Y}$ vectors in terms of the statistics defined in (7.34). Hint: Write out the likelihood, and find an unbiased estimate of σ^2 using sufficient statistics.

7.9   Assume $X_j \overset{\text{indep}}{\sim} N(\beta z_j, \sigma^2)$, $j = 1, \ldots, n$, where the z_j are known values, at least one of which is nonzero.
(a) Derive an u.m.v.u.e. of β.

(b) Consider the statistic

$$S_{\mathbf{z}}^2 := \frac{1}{n-1} \left[\sum_{i=1}^{n} x_i^2 - \frac{\left(\sum_{i=1}^{n} x_i z_i \right)^2}{\sum_{i=1}^{n} z_i^2} \right].$$

Compared to the usual definition of S_n^2, why might this be a reasonable estimator of σ^2?

(c) Calculate $\mathbb{E}[S_{\mathbf{z}}^2]$.

(d) Derive an u.m.v.u.e. of $\theta = (\beta, \sigma^2)'$.

7.10 Assume $X_1, \ldots, X_n \overset{\text{i.i.d.}}{\sim} \text{Poi}(\lambda)$ for $\lambda > 0$ and let $s = \sum_{i=1}^{n} x_i$.

(a) Compute the CRlb for $\tau(\lambda) = e^{-\lambda} = \Pr(X_i = 0)$.

(b) Compute $\mathbb{E}[Q_1]$ and $\text{Var}(Q_1)$, where $Q_1 = n^{-1} \sum_{i=1}^{n} \mathbb{I}_{\{0\}}(X_i)$.

(c) Show that the u.m.v.u.e. of $\tau(\lambda)$ is $Q_2 = \left(\frac{n-1}{n} \right)^s$ for $n > 1$.

(d) Apply the Rao–Blackwell theorem to $\mathbb{I}_{\{0\}}(X_1)$ (which is an unbiased estimator of $\tau(\lambda)$) to derive the estimator Q_2.

(e) Compute $\text{Var}(Q_2)$.

(f) Compare the CRlb, $\text{Var}(Q_1)$ and $\text{Var}(Q_2)$ by sketching a graph and consideration of Taylor series expansions.

(g) Prove the interesting result that $\mathbb{E}_\lambda[S_n^2 \mid \bar{X}_n] = \bar{X}_n$, where S_n^2 is the sample variance (5.32), that is,

$$S_n^2 = \frac{1}{n-1} \sum_{i=1}^{n} (X_i - \bar{X}_n)^2 = \frac{1}{n-1} \left(\sum_{i=1}^{n} X_i^2 - n\bar{X}_n^2 \right).$$

Hint: Use the distribution of $X_i \mid \sum_{j=1}^{n} X_j$.

7.11 Let $X_i, i = 1, \ldots, n$, be an i.i.d. sample from the density $f_X(x; \theta) = \theta^{-1} \mathbb{I}_{(0,\theta)}(x)$ for $\theta \in \mathbb{R}_{>0}$ with order statistics $Y_1 < Y_2 < \cdots < Y_n$.

(a) Is f_X a member of the exponential family?

(b) Derive the m.l.e. for θ and its c.d.f. and p.d.f.

(c) Determine if the m.l.e. for θ is

    (i) unbiased;

    (ii) asymptotically unbiased;

    (iii) m.s.e. consistent;

    (iv) weak consistent.

(d) Determine an u.m.v.u.e. for θ.

(e) Determine the estimator for θ of the form cY_n, where c is a constant that minimizes the m.s.e. among all estimators of the form cY_n.

(f) What is the limiting distribution of $T = n(\theta - Y_n)$?

7.12 Let $X_i \overset{\text{i.i.d.}}{\sim} \text{Gam}(r, \lambda), r, \lambda > 0$, with p.d.f.

$$f_{X_i}(x; r, \lambda) = \frac{\lambda^r}{\Gamma(r)} x^{r-1} e^{-\lambda x} \mathbb{I}_{(0,\infty)}(x),$$

and r a fixed, known constant.

(a) Calculate $\hat{\lambda}_{ML}$, the m.l.e. of λ.

(b) Confirm that $S = \sum X_i \sim \text{Gam}(nr, \lambda)$.

(c) Calculate $\mathbb{E}[\hat{\lambda}_{ML}]$.

(d) Show that $S = \sum_{i=1}^{n} X_i$ is a complete, minimal sufficient statistic for λ.

(e) Derive an u.m.v.u.e. for λ and its variance.

(f) Calculate the Cramér–Rao lower bound for the variance of unbiased estimators of λ und compare it with the variance of the u.m.v.u.e.

7.13 Let $X_i \overset{\text{i.i.d.}}{\sim} \text{Bern}(\theta)$.

(a) Calculate the m.l.e. for θ.

(b) Show that the m.l.e. for θ^2 is biased.

(c) For what value of θ does the m.l.e. exhibit the most bias?

(d) Now consider the (first-order) jackknife estimator

$$T^* = nT - \frac{n-1}{n} \sum_{i=1}^{n} T^{(i)},$$

with $T = T(X_1, \dots, X_n)$ and $T_{(i)} = T(X_1, \dots, X_{i-1}, X_{i+1}, \dots, X_n)$, $i = 1, \dots, n$, for $T = \hat{\theta}^2_{ML}$. Show that T^* is an unbiased estimator for θ^2.

(e) Show that, for $X_i \overset{\text{i.i.d.}}{\sim} \text{Bern}(p)$, the sample variance can be written as

$$S_n^2 = \frac{n}{n-1} \bar{X}(1 - \bar{X}).$$

(f) Show that $S = \sum_{i=1}^{n} X_i$ is a complete sufficient statistic.

(g) Using parts (e) and (f), give an u.m.v.u.e. for $\theta(1 - \theta)$.

(h) For $a, b \in \{0, 1\}$ and $n \geq s \geq 2$, show that

$$\Pr(X_1 = a;\ X_2 = b \mid S = s) = \binom{n-2}{s-a-b} \Big/ \binom{n}{s}.$$

(i) Derive the same u.m.v.u.e. from part (g) using the Rao–Blackwell theorem and noting that $X_1(1 - X_2)$ is an unbiased estimator of $\theta(1 - \theta)$.

(j) Derive the CRlb from $\tau(\theta) = \theta(1 - \theta)$. Do you think that the lower bound can be reached by any estimator?

(k) Derive an expression for $\mathbb{V}(S_n^2)$ and, for a few values of θ, plot the CRlb and $\mathbb{V}(S_n^2)$ for $n = 11$.

(l) Show that the u.m.v.u.e. for θ^2 is, for $S = \sum_{i=1}^{n} X_i$,

$$\frac{S}{n} \frac{S-1}{n-1}.$$

7.14 Let $X_i \overset{\text{i.i.d.}}{\sim} f_X(x) = a\theta^{-a} x^{a-1} \mathbb{1}_{(0,\theta)}(x)$, $i = 1, \dots, n$, for known $a > 0$ and let the order statistics be given by $0 < Y_1 < Y_2 < \dots < Y_n < \theta$.

(a) By computing both terms, demonstrate directly that (3.9) does not hold for this density. Explain why.

(b) Show that $Y_n = \max(X_i)$ is a sufficient statistic.

(c) Construct an unbiased estimate of θ using $Y = Y_n$.

(d) Derive the variance of this unbiased estimator, and compare to *what would be* the CRlb, if it existed.

7.15 Let $X_1, \ldots, X_n$ be i.i.d. random variables sampled from the density

$$f_X(x; \theta) = \theta x^{\theta-1} \mathbb{1}_{(0,1)}(x), \quad \theta > 0.$$

(a) Specify a complete and sufficient statistic for θ.

(b) Derive the ML estimator for θ and find its expected value. Hint: First derive the distribution of $Y = -\ln X$ and $S = -\sum_{i=1}^{n} \ln X_i$.

(c) Compute the asymptotic distribution of the m.l.e.

(d) Define $Z_i = \mathbb{1}_{(0,\phi)}(X_i)$, for a known value of ϕ such that $0 < \phi < 1$.

    (i) Derive $\hat{\theta}_{\mathrm{ML}}(Z)$, the m.l.e. of θ based only on the Z_i, $i = 1, \ldots, n$. It is useful to recall that, if $y = a^x$, then $dy/dx = (\log a)a^x$.

    (ii) Give an expression for the expected value of $\hat{\theta}_{\mathrm{ML}}(Z)$.

    (iii) Calculate the CRlb corresponding to estimators of θ based on the Z_i and, for several values of θ, find the optimal value of ϕ that leads to the CRlb being a minimum. Compare with the CRlb using the X_i.

(e) Specify the distribution function $F_X(x; \theta)$ for $x \in \mathbb{R}$.

(f) Derive the distribution of $Y_{n:n} = \max(X_i)$.

(g) Write an expression for the c.d.f. of $nY_{1:n}$. Then find the limiting density of $nY_{1:n}$ when $\theta = 1$.

(h) Calculate the rth raw moment of X.

(i) Find the asymptotic distribution of $\bar{X}_n$.

(j) Derive an expression for ξ_p in terms of p and θ, where ξ_p is the pth quantile, $0 < p < 1$.

(k) Find the asymptotic distribution of $Y_{np:n}$, where $Y_{i:n}$ is the ith order statistic and assume np is an integer.

(l) Compare the results for parts **(i)** and **(k)** with $p = 1/2$ and $\theta = 1$ (so that $X \sim \mathrm{Unif}(0,1)$).

(m) Determine value $p^* = p^*(\theta)$ such that, as $n \to \infty$, $\mathbb{E}[\bar{X}_n] = \mathbb{E}[Y_{np^*:n}]$.

(n) Using p^*, find an expression for $R_\theta(p)$, the ratio of the asymptotic variances of $\bar{X}_n$ and $Y_{np:n}$. Do not try to simplify it. Evaluate for values $\theta = 0.01$, 1, and 10.

(o) Show that, for finite λ and k,

$$e^\lambda = \lim_{y \to \infty} \left(1 + \frac{\lambda}{y} \right)^y = \lim_{y \to \infty} \left(1 + \frac{\lambda}{y+k} \right)^y. \tag{7.35}$$

Hint: For the latter, consider the substitution $\zeta = \theta + k$.

(p) What is $\lim_{\theta \to \infty} R_\theta(p^*)$?

(q) What is $\lim_{\theta \to 0} R_\theta(p^*)$? Hint: Use l'Hôpital's rule and $\lim_{x \searrow 0} x^x = 1$.

7.16 As in Example 7.25, let $X_i \overset{\text{i.i.d.}}{\sim} \text{DUnif}(N)$, $i = 1, \ldots, n$, with p.m.f.

$$f_X(x; N) = \frac{1}{N} \mathbb{1}_{\{1,2,\ldots,N\}}(x), \quad N \in \mathbb{N}.$$

Denote the order statistics of X_i by Y_i.

(a) Derive an m.m.e. $\hat{N}_{\text{MM}}$ of N; compute $\mathbb{E}[\hat{N}_{\text{MM}}]$ and $\mathbb{V}(\hat{N}_{\text{MM}})$.

(b) Let $p_{n,N} = \Pr(Y_n = N)$ be the probability that the largest possible value N is observed.

    (i) How does $p_{n,N}$ behave for $n \to \infty$?

    (ii) How does $p_{n,N}$ behave for $N \to \infty$?

    (iii) Sketch $p_{n,N}$ for $N = 10$ and $n = 5, 10, \ldots, 30$. How large does n have to be so that $p_{n,10} \geq 0.95$?

    (iv) How does $p_{N,N}$ behave for $N \to \infty$?

(c) Determine the m.l.e. $\hat{N}_{\text{ML}}$ for N.

(d) (Expected value of $\hat{N}_{\text{ML}}$)

    (i) Determine the exact expression for $\mathbb{E}[\hat{N}_{\text{ML}}]$ without trying to simplify it.

    (ii) The expected value $\mathbb{E}[\hat{N}_{\text{ML}}]$ for several values of n is given in Table 7.1. Calculate a "bias- adjusted" estimator, say $\hat{N}^*_{\text{ML}}$, for $\hat{N}_{\text{ML}}$ by ignoring all terms that are of order N^{-1} or smaller.

    (iii) Now give the $O(N^{-3})$ approximation to $\mathbb{E}[\hat{N}_{\text{ML}}]$ by ignoring terms of order N^{-3} and smaller.

TABLE 7.1 $\mathbb{E}[\hat{N}_{\text{ML}}]$ as a function of n, for Problem 7.16

n	$\mathbb{E}[\hat{N}_{\text{ML}}]$
1	$\frac{1}{2}N + \frac{1}{2}$
2	$\frac{2}{3}N + \frac{1}{2} - \frac{1}{6N}$
3	$\frac{3}{4}N + \frac{1}{2} - \frac{1}{4N}$
4	$\frac{4}{5}N + \frac{1}{2} - \frac{1}{3N} + \frac{1}{30N^3}$
5	$\frac{5}{6}N + \frac{1}{2} - \frac{5}{12N} + \frac{1}{12N^3}$
6	$\frac{6}{7}N + \frac{1}{2} - \frac{1}{2N} + \frac{1}{6N^3} - \frac{1}{42N^5}$
7	$\frac{7}{8}N + \frac{1}{2} - \frac{7}{12N} + \frac{7}{24N^3} - \frac{1}{12N^5}$
10	$\frac{10}{11}N + \frac{1}{2} - \frac{5}{6N} + \frac{1}{N^3} - \frac{1}{N^5} + \frac{1}{2N^7} - \frac{5}{66N^9}$
20	$\frac{20}{21}N + \frac{1}{2} - \frac{5}{3N} + \frac{19}{2N^3} - \frac{1292}{21N^5} + \frac{323}{N^7} + O(N^{-9})$

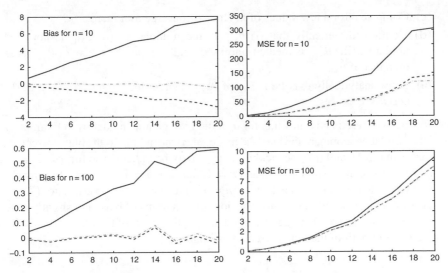

Figure 7.11 *For Problem 7.19, this shows the bias (left) and m.s.e. (right) of the m.l.e. $\hat{\alpha}$ (solid), the jackknife $\hat{\alpha}^*$ (dashed) and (7.12) (dash-dotted), as a function of α, based on 2000 replications, for $n = 10$ (top) and $n = 100$ (bottom).*

7.17 Let $X_i \overset{\text{i.i.d.}}{\sim} \text{Exp}(\lambda)$, $i = 1, \ldots, n$, each with density

$$f_X(x) = \lambda \exp(-\lambda x) \mathbb{I}_{(0,\infty)}(x),$$

and define $S = \sum_{i=1}^{n} X_i$.

(a) Show that S is complete and sufficient.

(b) Derive the u.m.v.u.e. of $\mathbb{E}[X] = \lambda^{-1}$ using the Lehmann–Schéffé theorem and also using the Rao–Blackwell theorem based in the unbiased estimate X_1.

(c) Calculate $\mathbb{E}[S^{-1}]$ and $\mathbb{V}(S^{-1})$.

(d) Derive the u.m.v.u.e. of $\mathbb{E}[X] = \lambda$ for $n > 2$ and calculate its variance.

(e) For given value $K > 0$, first verify that $\mathbb{I}_{(K,\infty)}(X_1)$ is unbiased for $e^{-K\lambda}$ for $n \geq 1$ and derive the u.m.v.u.e. of $e^{-K\lambda}$ using the Rao–Blackwell theorem.

(f) Using the previous result for the u.m.v.u.e. of $e^{-K\lambda}$, numerically compare its variance to the asymptotic variance of the m.l.e. as given in Example 3.11 for some values of n, λ and K.

7.18 Let $X_i \overset{\text{i.i.d.}}{\sim} \text{Exp}(\theta)$ with $f_X(x) = \theta^{-1} \exp(-x/\theta) \mathbb{I}_{(0,\infty)}(x)$.

(a) Write a program that computes $\hat{\theta}_{\text{med}}$ directly via (7.26).

(b) It turns out that $\hat{\theta}_{\text{med}}$ can be stated more explicitly for this model. Show that $\hat{\theta}_{\text{med}} = c\,\hat{\theta}_{\text{ML}}$, where c is given implicitly by

$$\frac{1}{2} = \int_0^{n/c} \frac{1}{\Gamma(n)} y^{n-1} \exp(-y)\, dy. \tag{7.36}$$

(c) Compute c for a few values of n. Does c converge as n increases?

(d) Confirm numerically that (i) $\hat{\theta}_{ML}$ is mean-unbiased, (ii) $\hat{\theta}_{ML}$ is not median-unbiased, (iii) $\hat{\theta}_{med}$ is median-unbiased, and (iv) that the m.s.e. of $\hat{\theta}_{med}$ is larger than that of $\hat{\theta}_{ML}$.

(e) Show analytically that the m.s.e. of $\hat{\theta}_{med}$ is larger than that of $\hat{\theta}_{ML}$ for all $\theta < \infty$ and $n < \infty$.

7.19 Recall that there is no u.m.v.u.e. for α in the i.i.d. gamma model from Example 7.17. Via simulation, compare the bias and m.s.e. of (i) the m.l.e., (ii) the bias-corrected estimator in (7.12), and (iii) the jackknife applied to the m.l.e. Do so for $\alpha = 2, 4, \ldots, 20$ and the two sample sizes $n = 10$ and $n = 100$. (Take $\beta = 1$.) The resulting graphs are shown in Figure 7.11. We see that, for $n = 10$, the bias-corrected estimator (7.12) outperforms the jackknife estimator considerably in terms of bias, and slightly in terms of m.s.e. Because of the large bias of the m.l.e., the m.s.e. of the m.l.e. is much higher than that of both of the bias-corrected estimators.

8

Analytic Interval Estimation

This chapter emphasizes analytically derived c.i.s. When applicable, they are usually easily computed functions of the data, and are faster than use of the bootstrap. Also, for a given data set, they are deterministic, in contrast to the stochastic nature of bootstrap c.i.s. In addition, analytic c.i.s tend to have actual coverage equal to, or at least closer to, the nominal, and also tend to be shorter than bootstrap c.i.s with the same actual coverage.

8.1 DEFINITIONS

Recall the introduction to confidence intervals in Section 1.2. Let $k = 1$ for simplicity. In almost all situations, $M(\mathbf{X})$ will be an interval: denoting the left and right endpoints by $\underline{\theta} = \underline{\theta}(\mathbf{X})$ and $\bar{\theta} = \bar{\theta}(\mathbf{X})$, respectively, $M(\mathbf{X}) = (\underline{\theta}, \bar{\theta})$ is referred to as a **confidence interval** (c.i.), for θ with **(nominal) confidence level** $1 - \alpha$ or, more commonly, a $100(1 - \alpha)\%$ c.i. for θ. It also makes sense to refer to a c.i. as an **interval estimator** of θ, as it draws attention to its purpose in comparison to that of a point estimator.

Example 8.1 *Let* $X_i \overset{\text{i.i.d.}}{\sim} \mathrm{N}(\mu, \sigma^2)$, σ^2 *known. For a c.i. of* μ *of the form* $(\underline{\mu}, \bar{\mu}) = (\bar{X} - c, \bar{X} + c)$, *simple manipulation gives*

$$1 - \alpha = \Pr(-d < Z < d) = \Pr\left(-d < \frac{\bar{X} - \mu}{\sigma / \sqrt{n}} < d\right) = \Pr(\bar{X} - c < \mu < \bar{X} + c), \quad (8.1)$$

where Z is so defined, $c = d\sigma / \sqrt{n}$ *and* $Z \sim N(0, 1)$. *Observe that Z is a function of* μ *and the data* $\mathbf{X} = (X_1, \dots, X_n)$, *but its* distribution *does not depend on* μ. *It is an example of a* **pivot**,

---

Fundamental Statistical Inference: A Computational Approach, First Edition. Marc S. Paolella.

discussed more below. The quantile d can be found by inverting the standard normal c.d.f., that is, $d = \Phi^{-1}(1 - \alpha/2)$. If $\alpha = 0.05$, then $d = 1.9600$ so that $(\underline{\mu}, \bar{\mu}) = \bar{X} \pm 1.96\, \sigma / \sqrt{n}$ is a 95% c.i. for μ.

Now assume more realistically that, in the previous example, σ^2 is unknown. Intervals of the form $(\bar{X} - c^*, \bar{X} + c^*)$ for any finite constant value c^* will not work, because the confidence coefficient $\inf_\theta \Pr(\mu \in \bar{X} \pm c^*) = 0$, seen by taking component σ^2 of $\theta = (\mu, \sigma^2)$ arbitrarily large. To remedy this, we let c^* depend on $\mathbf{X}$ by estimating σ^2. In particular,

$$1 - \alpha = \Pr(-d^* < T < d^*)$$

$$= \Pr\left(-d^* < \frac{\bar{X} - \mu}{S/\sqrt{n}} < d^*\right) = \Pr(\bar{X} - c^* < \mu < \bar{X} + c^*), \qquad (8.2)$$

where $T \sim t_{n-1}$ is a pivot, $S = \sqrt{(n-1)^{-1} \sum_{i=1}^{n} (X_i - \bar{X})^2}$ estimates σ, and $c^* = d^* S/\sqrt{n}$. The value d^* is the $1 - \alpha/2$ quantile of the t_{n-1} c.d.f., say $t_{n-1}^{-1}(1 - \alpha/2)$, which can be computed in virtually all statistical software packages.

Observe that the length of $\bar{X} \pm d\, \sigma /\sqrt{n}$ is deterministic, while that of $\bar{X} \pm d^* S /\sqrt{n}$ is random. Thus, we cannot state that the length of the c.i. of μ is larger when σ^2 is unknown. However, the *expected* length of $\bar{X} \pm d^* S /\sqrt{n}$ can be determined from (A.207), in particular, $\mathbb{E}[S] < \sigma$, which tends to reduce the length of the random interval. But this is counterbalanced by the fact that $t_{n-1}^{-1}(a) > \Phi^{-1}(a)$ for $a > 0.5$, that is, $d^* > d$. The expected lengths can be compared for any given n; Table 8.1 illustrates several such values for $\sigma^2 = 1$ and $\alpha = 0.05$, with the last column giving the ratio. The d^* length factor strongly outweighs the downward bias of S for (so it appears) all n.

To demonstrate, Figure 8.1 shows a histogram of the lengths of 1000 simulated intervals, $2d^* S n^{-1/2}$, for $\sigma^2 = 1$ and two values of n, with the inscribed line indicating the length of the interval with σ^2 known, that is, $2d\sigma n^{-1/2}$. Even for n as small as 5, there is still a considerable chance (about 0.30) that $d^* S < d\sigma$.

TABLE 8.1   Comparison of lengths of 95% c.i.s for μ in the normal model with sample size n

n	2	3	4	5	6	10	20	100
$2d\sigma n^{-1/2}$	2.77	2.26	1.96	1.75	1.60	1.24	0.877	0.392
$2d^* \mathbb{E}[S] n^{-1/2}$	4.86	3.26	2.56	2.16	1.90	1.37	0.921	0.396
$d/(d^* \mathbb{E}[S])$	0.571	0.695	0.766	0.811	0.842	0.904	0.952	0.990

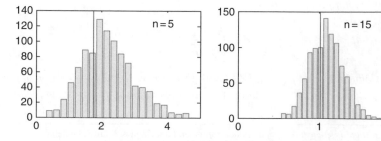

Figure 8.1 *Simulated lengths of 95% c.i.s for μ in the $\mathrm{N}(\mu, 1)$ model assuming σ^2 unknown.*

One might wonder why the more general $(\bar{X} - c_1, \bar{X} + c_2)$ was not used in the previous example, and also to what extent this c.i. is "optimal" in some sense. For the former, the reason comes from the symmetry of the normal and t distributions; in particular, starting with $(\bar{X} - c_1, \bar{X} + c_2)$ instead, a calculus exercise aimed at minimizing the length of the interval subject to the constraint that $1 - \alpha \leq \Pr(\bar{X} - c_1 < \mu < \bar{X} + c_2)$ reveals that the optimal values satisfy $c_1 = c_2$. This partly answers the second question: for given α, shorter intervals are preferred. We could have based the interval on another statistic instead of $\bar{X}$, such as X_1, which is also unbiased for μ and results in (8.1) for $n = 1$, which is clearly larger in length. Use of only one observation should obviously be inferior to use of n, but it is less clear why, for example, use of the median is also inferior to that of $\bar{X}$ when inference centers on μ. It turns out that use of a sufficient statistic often leads to intervals that are "best" in some sense.

Not all c.i.s need to be two-sided: if (1.4) holds with $M(\mathbf{x}) = (\underline{\theta}(\mathbf{x}), \infty)$, then $M(\mathbf{x})$ is referred to as a **lower confidence bound of** θ; similarly, $M(\mathbf{x})$ is an **upper confidence bound of** θ if $M(\mathbf{x}) = (-\infty, \bar{\theta}(\mathbf{x}))$. One-sided c.i.s are commonly used in conjunction with so-called **variance components**. In such cases, the lengths are infinite; for lower confidence bounds, comparisons can be conducted using $\theta - \underline{\theta}(\mathbf{X})$ or, if $\underline{\theta}(\mathbf{X})$ is random, $\mathbb{E}[\theta - \underline{\theta}(\mathbf{X})]$. Similarly, for upper confidence bounds, $\bar{\theta}(\mathbf{X}) - \theta$ or $\mathbb{E}[\bar{\theta}(\mathbf{X}) - \theta]$ is used.

In multi-parameter settings, there is often interest in a joint confidence region for θ. In certain situations, this is not difficult: A joint $100(1 - \alpha)\%$ confidence set for μ and σ^2 in the i.i.d. normal model is easy to form owing to the independence of $\bar{X}$ and S^2; see, for example, Shao (2003, pp. 130–131) for details. In general, however, things are not so easy. Bounds satisfying (1.4) can be constructed using the Bonferroni inequality, although these tend to be very conservative when k, the number of parameters, is large. In the important case of so-called **contrasts** in linear models, other methods are available that yield shorter intervals with the correct coverage probabilities.

8.2 PIVOTAL METHOD

The random variable $Q = Q(\mathbf{X}; \theta)$ is said to be a **pivotal quantity**, or just **pivot**, if its distribution does not depend on θ. It is not a statistic because it is a function of θ. The method consists of (i) determining a relevant pivotal quantity, (ii) obtaining values q_1 and q_2 such that $1 - \alpha = \Pr(q_1 \leq Q \leq q_2)$, and (iii) expressing the inequality $q_1 \leq Q \leq q_2$ as $\underline{\theta} \leq \theta \leq \bar{\theta}$, or "pivoting" it. This method was used in the normal example above: The random variable $T = T(\mathbf{X}; \mu) = (\bar{X} - \mu)/(S/\sqrt{n}) \sim t_{n-1}$ is a function of μ but whose distribution does not depend on $\theta = (\mu, \sigma^2)$. The event $-d^* < T < d^*$ was then pivoted to obtain the equivalent event $\bar{X} - c^* < \mu < \bar{X} + c^*$; the probability of the latter is determined by the choice of d^*.

8.2.1 Exact Pivots

Pivots with tractable distributions are often available when the parameter of interest is either a location or scale parameter. In the normal example above, note that μ is a location parameter. The next two examples are similar.

Example 8.2 *For the location exponential model $f_X(x) = \exp\{-(x - \theta)\}\mathbb{I}(x \geq \theta)$ with set of i.i.d. data $\mathbf{X} = (X_1, \dots, X_n)$, let $M = \min\mathbf{X} = \min\{X_1, \dots, X_n\}$. The log-likelihood is*

$$\ell(\theta; \mathbf{X}) = -\sum_{i=1}^{n}(X_i - \theta) = n(\theta - \bar{X}), \quad M \geq \theta,$$

with maximum occurring for $\hat{\theta}_{ML}$ as large as possible, but constrained by $\theta \le M$, so that $\hat{\theta}_{ML} = M$.[1] A simple calculation shows that M follows a location–scale exponential distribution with $f_M(m; n, \theta) = ne^{-n(m-\theta)}\mathbb{I}(m \ge \theta)$. The density of $Q = n(M - \theta)$ is, via transformation, seen to be $\mathrm{Exp}(1)$, and is not dependent on θ. Thus, with q_1 and q_2 such that

$$1 - \alpha = \Pr(q_1 < Q < q_2) = F_Q(q_2) - F_Q(q_1) = e^{-q_1} - e^{-q_2}, \tag{8.3}$$

inverting gives

$$1 - \alpha = \Pr\left(M - \frac{q_2}{n} < \theta < M - \frac{q_1}{n}\right).$$

Because we know that $\theta \le M$, it would seem reasonable to set $q_1 = 0$. This is in fact the optimal choice, seen as follows. Solving (8.3) for q_2 gives $q_2 = -\ln(e^{-q_1} - (1 - \alpha))$, so that the c.i. has length

$$L(q_1) = n^{-1}(q_2 - q_1) = n^{-1}(-\ln(e^{-q_1} - (1 - \alpha)) - q_1).$$

This can only be valid if q_1 is chosen such that at least $1 - \alpha$ probability mass remains to its right, that is, $F_Q(q_1) < \alpha$, or $q_1 < -\ln(1 - \alpha)$. (A quick sketch of the density makes this clear.) Because the density of Q is monotonically decreasing, an increase in q_1 induces a larger increase in q_2; as q_1 approaches $-\ln(1 - \alpha)$, $q_2 \to \infty$. Thus, taking q_1 as small as possible, meaning $q_1 = 0$, leads to the smallest L. Thus, the shortest $100(1 - \alpha)\%$ c.i. for θ based on the pivotal method is

$$(M + \ln(\alpha)/n, \ M). \tag{8.4}$$

For illustration, a plot of $L(q_1)$ versus q_1 for $\alpha = 0.05$ and $n = 1$ is shown in Figure 8.2.

*   *A more realistic case is the location–scale exponential model from Problem 3.3, and such that only the first, say, k order statistics are observed, as discussed in Example 3.3. Pivotal quantities for the two parameters in this setting and joint c.i.s are derived in Asgharzadeh and Abdi (2011). The distribution theory and statistical methodology employed there are straightforward and instructional, and the reader is encouraged to take a look.*

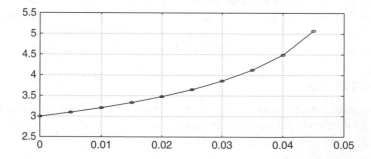

Figure 8.2 Length of the c.i. (8.4) for parameter θ of the location exponential model.

---

[1] As the m.l.e. is a function of the sufficient statistics, M must be sufficient, seen also from the factorization theorem, $f_X(x) = \exp\{-n\bar{x}\} \exp\{n\theta\}\mathbb{I}(m \ge \theta) = g(m; \theta) \cdot h(x)$.

Example 8.3 *Let $X_i \overset{indep}{\sim} \mathrm{Gam}(r_i, \lambda)$ with (for shape r)*

$$f_{X_i}(x; r, \lambda) = \frac{\lambda^r}{\Gamma(r)} x^{r-1} e^{-\lambda x} \mathbb{1}_{(0,\infty)}(x),$$

and such that the r_i are known constants. An important special case is with $r_i = 1$, so that $X_i \overset{i.i.d.}{\sim} \mathrm{Exp}(\lambda)$. With $r_\bullet = \sum_{i=1}^{n} r_i$, $S = \sum_{i=1}^{n} X_i \sim \mathrm{Gam}(r_\bullet, \lambda)$ and $Q = \lambda S \sim \mathrm{Gam}(r_\bullet, 1)$, choosing q_1 and q_2 to satisfy

$$1 - \alpha = \Pr(q_1 < Q < q_2) = \Pr\left(\frac{q_1}{S} \leq \lambda \leq \frac{q_2}{S}\right)$$

yields a $100(1 - \alpha)\%$ c.i. for λ. The length of the interval is $L = S^{-1}(q_2 - q_1)$; in order to minimize L, first differentiate both sides of the constraint $\int_{q_1}^{q_2} f_Q(q)\, dq = 1 - \alpha$ with respect to q_1 using Leibniz's rule to get

$$\frac{dq_2}{dq_1} f_Q(q_2) - f_Q(q_1) = 0 \quad \text{or} \quad \frac{dq_2}{dq_1} = \frac{f_Q(q_1)}{f_Q(q_2)}.$$

Setting dL/dq_1 equal to zero yields

$$\frac{dL}{dq_1} = S^{-1}\left(\frac{dq_2}{dq_1} - 1\right) = S^{-1}\left(\frac{f_Q(q_1)}{f_Q(q_2)} - 1\right) = 0,$$

or that $f_Q(q_1) = f_Q(q_2)$. (This result could also be obtained using the Lagrangian.) Values q_1 and q_2 subject to the constraints $f_Q(q_1) = f_Q(q_2)$ and $1 - \alpha = F_Q(q_2) - F_Q(q_1)$ can be numerically determined.

The gamma model is sometimes stated alternatively as $X_i \overset{indep}{\sim} \mathrm{Gam}(r_i, \theta)$, where

$$f_{X_i}(x; r, \theta) = \frac{1}{\theta^r \Gamma(r)} x^{r-1} e^{-x/\theta} \mathbb{1}_{(0,\infty)}(x),$$

r_i is known, and $\theta = \lambda^{-1}$ is a genuine scale parameter. Then $S = \sum X_i \sim \mathrm{Gam}(r_\bullet, \theta)$ and $Q = S/\theta \sim \mathrm{Gam}(r_\bullet, 1)$, so that

$$1 - \alpha = \Pr(q_1 < Q < q_2) = \Pr\left(\frac{S}{q_2} < \theta < \frac{S}{q_1}\right)$$

and $(s/q_2, s/q_1)$ is a $100(1 - \alpha)\%$ c.i. for θ.

Problem 8.1 shows that the constraints for minimizing the length $L^* = S(q_1^{-1} - q_2^{-1})$ are $F_Q(q_2) - F_Q(q_1) = 1 - \alpha$ and $q_1^2 f_Q(q_1) = q_2^2 f_Q(q_2)$.

The first result in the previous example regarding the minimum length of an interval whose length is proportional to $(q_2 - q_1)$ can be generalized: If f_Q is a unimodal p.d.f. and the interval (q_1, q_2) satisfies (i) $\int_{q_1}^{q_2} f_Q(q)\, dq = 1 - \alpha$, (ii) $f_Q(q_1) = f_Q(q_2) > 0$, and (iii) $q_1 \leq m \leq q_2$, where m is a mode of f_Q, then $[q_1, q_2]$ is the shortest interval satisfying (i). A proof can be found in Casella and Berger (1990, p. 431).

When a pivotal quantity can be found, the method is quite elegant and usually leads to "good" intervals. If one cannot be found, it is often the case that a pivotal quantity will be available *asymptotically*. This is considered next.

8.2.2 Asymptotic Pivots

Recall   from Section 3.1.4 that, for the k-length parameter vector $\theta = (\theta_1, \ldots, \theta_k)$, the asymptotic distribution of $\hat{\theta}_{ML}$ is (under certain regularity conditions on the density) given by

$$\hat{\theta}_{ML} \overset{asy}{\sim} N_k(\theta, \mathbf{J}^{-1}(\theta)), \tag{8.5}$$

where $\mathbf{J}(\theta) = -\mathbb{E}[\ddot{\ell}]$ is the information matrix. As $\mathbf{J}$ depends on the unknown parameter θ, $\mathbf{J}(\hat{\theta})$ is used in place of $\mathbf{J}(\theta)$ in practice. Let $\Sigma = \mathbf{J}(\hat{\theta})$ with (i,j)th element σ_{ij} and ith diagonal element $\sigma_{ii} = \sigma_i^2$. As $(\hat{\theta}_i - \theta_i)/\sigma_i \overset{asy}{\sim} N(0,1)$, it follows that an asymptotic $100(1-\alpha)\%$ c.i. for θ_i is, as in (3.45), the Wald interval

$$\hat{\theta}_i \pm d\sigma_i, \tag{8.6}$$

$i = 1, \ldots, k$, where $d = \Phi^{-1}(1 - \alpha/2)$. If interest centers, for example, on the difference between two parameters, say $\delta = \theta_1 - \theta_2$, then, with $\hat{\delta} = \hat{\theta}_1 - \hat{\theta}_2$ and $\hat{s}^2 = \mathbb{V}(\hat{\delta}) = \mathbb{V}(\hat{\theta}_1) + \mathbb{V}(\hat{\theta}_2) - 2\mathrm{Cov}(\hat{\theta}_1, \hat{\theta}_2)$, $(\hat{\delta} - \delta)/\hat{s} \overset{asy}{\sim} N(0,1)$. Thus, with $d = \Phi^{-1}(1 - \alpha/2)$, $1 - \alpha \approx \Pr(-d \le (\hat{\delta} - \delta)/\hat{s} \le d)$ or $1 - \alpha \approx \Pr(\hat{\delta} - d\hat{s} \le \delta \le \hat{\delta} + d\hat{s})$, that is, an asymptotic $100(1-\alpha)\%$ c.i. for δ is $\hat{\delta} \pm d\hat{s}$.

It is common practice to report the m.l.e. point estimators along with (typically 95%) approximate c.i.s based on (8.6) for all k parameters of a model. It must be emphasized that, while each individual interval indeed has the correct coverage probability (asymptotically), the probability that two or more, let alone all, of the intervals contain their respective θ_i is not equal to $1 - \alpha$. A joint $100(1-\alpha)\%$ c.i. for each θ_i can easily be constructed using Bonferroni's inequality (A.22). In particular, with events $A_i = \Pr(\theta_i \in (\hat{\theta}_i \pm d\sigma_i))$, $i = 1 \ldots, k$, it follows that $\Pr(\bigcap_{i=1}^k A_i) \ge 1 - \sum_{i=1}^k \Pr(\bar{A}_i) = 1 - k\alpha$. For example, to get simultaneous 95% c.i.s for θ_1, θ_2, and θ_3, take $\alpha = 0.05/3$, or $d = 2.394$. For $k = 1$, $d = 1.9600$.

Remark.   Notice how the Bonferroni method ignores possible correlation between the $\hat{\theta}_i$. For large k, such intervals will clearly be quite large and possibly of little inferential value. Improvements are possible; see, for example, Efron (1997). Joint c.i.s for model parameters with asymptotic correct overall actual coverage can be constructed via use of the bootstrap; see, for example, Davison and Hinkley (1997, p. 154) and Efron and Hastie (2016, Sec. 20.1).

The Bonferroni method can also be used for constructing intervals associated with point forecasts of, for example, time series, resulting in a set of **prediction intervals**; see, for example, Lütkepohl (1993, Sec. 2.2.3) and, for a bootstrap-based improvement, Wolf and Wunderli (2015).   ∎

Example 8.4   *Let $X_i \overset{i.i.d.}{\sim} \mathrm{Bern}(p)$, $i = 1, \ldots, n$. The m.l.e. of p is $\hat{p} = \bar{X}$, with variance $v^2 = pq/n$, where $q = 1 - p$. With $d = \Phi^{-1}(1 - \alpha/2)$, $1 - \alpha \approx \Pr(-d \le (\hat{p} - p)/\hat{v} \le d)$, where $\hat{v}^2 = \hat{p}(1 - \hat{p})/n$, so that an asymptotic $100(1-\alpha)\%$ c.i. for p is $\hat{p} \pm d\hat{v}$.*

This interval can also be used to ascertain the minimum required sample size so that $\hat{p}$ is within ϵ of the true p with (approximate) probability $1 - \alpha$, for some given $\epsilon > 0$. In particular, with p_0 an "educated guess" for p and $\hat{v}_0^2 = n^{-1}p_0(1 - p_0)$, solving $\epsilon = d\hat{v}_0$ for n gives $n = d^2\epsilon^{-2}p_0(1 - p_0)$, which would then be rounded up to the nearest integer. For example, with $\epsilon = 0.05$ and $p_0 = 0.3$, $n = 323$ for $\alpha = 0.05$ and $n = 558$ for $\alpha = 0.01$. To

numerically verify this, use the following Matlab code to confirm the result is 0.955 (use of
n = 322 results in 0.949):

```
1   vec=[]; n=323; p=0.3; epsilon=0.05;
2   for i=1:1e5, X=(rand(n,1) < p); phat=mean(X); vec=[vec phat]; end
3   mean(abs(vec-p) < epsilon)
```

It is easy to verify that $p(1-p)$ has its maximum at 0.5, so that taking $p_0 = 0.5$ will yield
a conservative estimate. In the previous cases, this yields $n = 385$ and $n = 664$ for the two
values of α, which are about 19% higher than n using $p_0 = 0.3$.

Example 8.5 *Let $X_i \overset{\text{i.i.d.}}{\sim} \text{Bern}(p_1)$, $i = 1, \ldots, n_1$, independent of $Y_i \overset{\text{i.i.d.}}{\sim} \text{Bern}(p_2)$, $i = 1,$*
$\ldots, n_2$. The m.l.e. of p_1 is $\hat{p}_1 = \bar{X}$, with variance $v_1^2 = p_1 q_1 / n_1$, where $q_1 = 1 - p_1$. Similar
expressions hold for p_2. With $\delta = p_1 - p_2$ and $\hat{\delta} = \hat{p}_1 - \hat{p}_2$, $(\hat{\delta} - \delta)/\hat{s} \overset{\text{asy}}{\sim} \text{N}(0, 1)$, where
$\hat{s} = \sqrt{\hat{v}_1^2 + \hat{v}_2^2}$ and $\hat{v}_i^2 = \hat{p}_i(1 - \hat{p}_i)/n_i$, $i = 1, 2$, an asymptotic $100(1 - \alpha)\%$ c.i. for the
difference $p_1 - p_2$ is $(\hat{p}_1 - \hat{p}_2) \pm d\hat{s}$, for $d = \Phi^{-1}(1 - \alpha/2)$.

Example 8.6 *Let $X_i \overset{\text{i.i.d.}}{\sim} \text{Poi}(\lambda_1)$, $i = 1, \ldots, n_1$, independent of $Y_i \overset{\text{i.i.d.}}{\sim} \text{Poi}(\lambda_2)$, $i = 1, \ldots, n_2$.*
An asymptotic $100(1 - \alpha)\%$ c.i. for $\lambda = \lambda_1 - \lambda_2$ is $\hat{\lambda} \pm d\hat{s}$, where $\hat{\lambda} = \hat{\lambda}_1 - \hat{\lambda}_2 = \bar{X} - \bar{Y}$ and
$\hat{s}^2 = \hat{\lambda}_1/n_1 + \hat{\lambda}_2/n_2$.

8.3 INTERVALS ASSOCIATED WITH NORMAL SAMPLES

While the distribution of an underlying population is often clearly of a particular type (such
as Bernoulli trials) or can, based on theoretical considerations, be assumed (such as Weibull
for measuring lifetimes), in many situations it is reasonable to assume normality, usually
via a central limit theorem argument. For this reason, c.i.s based on the normal assumption
are quite relevant and used repeatedly in practice. The use of pivots is the primary method
for deriving exact intervals, meaning that they have exactly the stated coverage. For some
quantities of interest, standard methods of obtaining an interval fail; for these, approximate
methods are subsequently discussed. We consider several cases in turn.

8.3.1 Single Sample

Let $X_i \overset{\text{i.i.d.}}{\sim} \text{N}(\mu, \sigma^2)$, $i = 1, \ldots, n$, with both μ and σ^2 unknown. The c.i. for μ was considered previously. A c.i. for σ^2 can be formed using the pivot $Q = (n-1)S^2/\sigma^2 \sim \chi^2_{n-1}$ from
(A.206), taking q_1 and q_2 such that $1 - \alpha = \text{Pr}(q_1 \leq Q \leq q_2)$ and inverting. This yields

$$1 - \alpha = \text{Pr}\left(\frac{(n-1)S^2}{q_2} \leq \sigma^2 \leq \frac{(n-1)S^2}{q_1} \right). \tag{8.7}$$

(Notice that taking square roots of the left and right bounds in (8.7) would give a c.i. for
σ as it is a monotone transformation.) For very large n, pivot Q, being a sum of $n-1$
χ^2_1 r.v.s, approaches a normal distribution, and thus is approximately symmetric, so that
the q_i can simply be chosen to give equal tail probabilities. For small to moderate n, it

```
1   function [q1, q2] = cisig2opt (n,alpha)
2   q1 = chi2inv(alpha/2,n-1); q2 = chi2inv(1-alpha/2,n-1); len=1/q1 - 1/q2;
3   equal=['Equal Tail: q1=', num2str(q1), ...
4           'q2=',num2str(q2), ' 1/q1 - 1/q2 = ', num2str(len)];
5   x0=[q1;q2]; % use these as starting values
6   x=fsolve(@fun,x0,optimset('Display','iter'),n,alpha);
7   q1=x(1); q2=x(2); len=1/q1 - 1/q2;
8   optim = ['Optimal: q1=',num2str(q1),' q2=',num2str(q2), ...
9           ' 1/q1 - 1/q2 = ',num2str(len)];
10  disp(equal), disp(optim)
11
12  function f = fun(x,n,alpha)
13  q1=x(1); q2=x(2);
14  f(1) = chi2cdf(q2,n-1) - chi2cdf(q1,n-1) - (1-alpha);
15  f(2) = q1^2 * chi2pdf(q1,n-1) - q2^2 * chi2pdf(q2,n-1);
```

Program Listing 8.1: Program to find the lower (q_1) and upper (q_2) quantiles of the χ^2 distribution that minimize the length of the usual $1 - \alpha$ confidence interval for σ^2.

pays to find q_1 and q_2 that minimize the c.i. length, $(n-1)S^2(q_1^{-1} - q_2^{-1})$. It is straight-forward to show (Problem 8.1) that the optimal values satisfy the two nonlinear equations $F_Q(q_2) - F_Q(q_1) = 1 - \alpha$ and $q_1^2 f_Q(q_1) = q_2^2 f_Q(q_2)$. These equations are easily solved using a high-level software package in which the p.d.f. and c.d.f. of the χ^2 distribution are available, along with programs for solving a set of nonlinear equations.

Table 8.2 illustrates the quantiles and lengths $l = q_1^{-1} - q_2^{-1}$ for the equal-tail and minimal-length 95% intervals, for several sample sizes n. For very small n, the difference is extreme, while even for moderate n, the difference can lead to substantially shorter intervals. The Matlab programs required to compute these are given below in Listing 8.1.

As an aside, if somehow μ is known, $\sum_{i=1}^{n} (X_i - \mu)^2/\sigma^2 \sim \chi_n^2$ so that $(T/q_2, T/q_1)$ is a $100(1 - \alpha)\%$ c.i. for σ^2, where $T = \sum_{i=1}^{n} (X_i - \mu)^2$ and $1 - \alpha = \Pr(q_1 \le \chi_n^2 \le q_2)$.

8.3.2 Paired Sample

Let

$$(X_i, Y_i) \stackrel{\text{i.i.d.}}{\sim} N_2 \left(\begin{bmatrix} \mu_1 \\ \mu_2 \end{bmatrix}', \begin{bmatrix} \sigma_1^2 & \rho\sigma_1\sigma_2 \\ \rho\sigma_1\sigma_2 & \sigma_2^2 \end{bmatrix} \right), \quad i = 1, \ldots, n,$$

with all parameters assumed unknown.

TABLE 8.2 Quantiles and lengths for the equal-tail (et) and minimal-length (min) 95% c.i.s for σ^2

	n	2	3	4	5	10	20	30	100
	q_1	0.0009821	0.05064	0.2158	0.4844	2.7004	8.9065	16.047	73.36
et	q_2	5.0239	7.378	9.3484	11.143	19.023	32.852	45.722	128.4
	l	1018	19.61	4.527	1.9746	0.3178	0.08184	0.04045	0.00584
	q_1	0.003932	0.1025	0.3513	0.7083	3.2836	9.899	17.271	75.15
min	q_2	24.35	21.48	20.74	21.063	26.077	38.33	50.58	132.2
	l	254.3	9.706	2.7988	1.3645	0.2662	0.07493	0.03813	0.00574

To construct a c.i. for $\mu_1 - \mu_2$, observe that $D_i = X_i - Y_i \overset{\text{i.i.d.}}{\sim} N(\mu_1 - \mu_2, v)$, where both $\mu_1 - \mu_2$ and $v = \sigma_1^2 + \sigma_2^2 - 2\rho\sigma_1\sigma_2$ are unknown. This is precisely the same situation as the one-sample case with c.i. (8.2) above, so that $\bar{D} \pm qS_D/\sqrt{n}$ forms a $100(1 - \alpha)\%$ c.i. for $\mu_1 - \mu_2$, where q is an appropriately chosen quantile from a t_{n-1} distribution and S_D is the sample standard deviation of the D_i.

Instead of the difference, interest occasionally centers on the ratio μ_2/μ_1. A clever method for deriving an exact c.i. for $\theta = \mu_2/\mu_1$ was given by Fieller (1954). Define $Z_i = Y_i - \theta X_i$ and note that

$$Z_i \overset{\text{i.i.d.}}{\sim} N(0, V), \quad V = \sigma_2^2 + \theta^2\sigma_1^2 - 2\theta\rho\sigma_1\sigma_2,$$

from which it follows that $\sqrt{n}\bar{Z}/S_Z \sim t_{n-1}$ is a pivotal quantity for θ, where $\bar{Z}$ and S_Z^2 are the sample mean and variance, respectively, of the Z_i, to be subsequently discussed. A confidence set with significance level $1 - \alpha$ is then $C = \{\theta : \bar{Z}^2/S_Z^2 \le k\}$ for $k = c^2/n$ and $c = t_{n-1}^{-1}(1 - \alpha/2)$. As $\bar{Z} = \bar{Y} - \theta\bar{X}$ and

$$(n - 1)S_Z^2 = \sum_{i=1}^{n}(Z_i - \bar{Z})^2 = \sum_{i=1}^{n}(Y_i - \bar{Y} - \theta(X_i - \bar{X}))^2$$

$$= \sum_{i=1}^{n}(Y_i - \bar{Y})^2 + \theta^2\sum_{i=1}^{n}(X_i - \bar{X})^2 - 2\theta\sum_{i=1}^{n}(X_i - \bar{X})(Y_i - \bar{Y})$$

$$= (n - 1)(S_Y^2 + \theta^2 S_X^2 - 2\theta S_{XY}),$$

C can be written as $\{\theta : P \le 0\}$, where $P = a_2\theta^2 - a_1\theta + a_0$, with

$$a_2 = \bar{X}^2 - k^2 S_X^2, \quad a_1 = 2(\bar{Y}\bar{X} - k^2 S_{XY}), \quad \text{and} \quad a_0 = \bar{Y}^2 - k^2 S_Y^2.$$

The solutions to $P = 0$ are given by $\theta_{+,-} = (a_1 \pm \sqrt{d})/2a_2$, where $d = a_1^2 - 4a_2a_0$. Several cases can occur, as follows: (i) If $a_2 > 0$ and $d > 0$, then P opens upward and has two distinct roots, so that C is an interval, that is, $(\theta_-, \theta_+) = (\underline{\theta}, \bar{\theta})$ is a $100(1 - \alpha)\%$ c.i. for θ. Less desirable cases include (ii) when $d > 0$ but $a_2 < 0$, in which case C is $(-\infty, \min(\theta_+, \theta_-)) \cup (\max(\theta_+, \theta_-), \infty)$; (iii) when $d < 0$ and $a_2 > 0$, in which case $C = \emptyset$, and (iv) when $d < 0$ and $a_2 < 0$, in which case $C = \mathbb{R}$. Observe also how the positivity of a_2 depends on the level of significance.

For numerical illustration, the method was simulated $s = 10,000$ times for parameters $\mu_1 = 2$, $\mu_2 = 3$, $\sigma_1^2 = \sigma_2^2 = 1$, $\rho = 0$, $n = 20$, and $\alpha = 0.05$ using program `fieller` as developed in Problem 8.6. All s trials resulted in case (i) above, that is, an interval, with empirical or actual coverage (how often the c.i. included 3/2) 0.9462. To determine if this is "close enough" to 0.95, we can build a c.i. for Bernoulli trials with $n = 10,000$ and $\hat{p} = 0.9462$ and check whether 0.95 is contained within it. Using the asymptotically valid interval developed in Example 8.5, the upper limit of the 95% c.i. of $\hat{p}$ is 0.9506, so that we "cannot reject the hypothesis" that the Fieller interval is exact. Repeating the exercise with $\mu_1 = \mu_2 = 1$, $\sigma_1^2 = 1$, $\sigma_2^2 = 3$, $\rho = 0.5$, and $n = 20$, the empirical coverage was 0.9481, well within the Bernoulli c.i. However, case (i) occurred 9864 times, case (ii) 78 times, and case (iv) 58 times. Notice that, for case (iv), coverage is automatically satisfied, but such an interval is of no value in practice.

It does not appear possible to use the pivotal method to derive an exact c.i. for ρ. The method developed in Section 8.4 below is applicable, however, and will be considered there.

8.3.3 Two Independent Samples

Let $X_i \overset{\text{i.i.d.}}{\sim} N(\mu_1, \sigma^2)$, $i = 1, \ldots, n_1$, and $Y_i \overset{\text{i.i.d.}}{\sim} N(\mu_2, \sigma^2)$, $i = 1, \ldots, n_2$ be two independent normal samples with the same variance $\sigma^2 > 0$, but possibly differing means. Intervals designed to assess the differences in the means of two independent normal samples arise often in applications. To develop a c.i. for $\mu_1 - \mu_2$, let

$$Z = \bar{X} - \bar{Y} \sim N(\mu_Z, \sigma_Z^2), \quad \text{where } \mu_Z = \mu_1 - \mu_2 \text{ and } \sigma_Z^2 = \sigma^2(n_1^{-1} + n_2^{-1}).$$

Then, with $C = (n_1 - 1)S_X^2/\sigma^2 + (n_2 - 1)S_Y^2/\sigma^2 \sim \chi^2_{n_1+n_2-2}$ independent of Z,

$$Q = \frac{(Z - \mu_Z)/\sigma_Z}{\sqrt{C/(n_1 + n_2 - 2)}} \sim t_{n_1+n_2-2}$$

is a pivotal quantity. By multiplying the numerator and denominator of Q by σ_Z and substituting $\sigma_Z^2 = \sigma^2(n_1^{-1} + n_2^{-1})$, it is easy to check that this can be expressed as

$$Q = \frac{Z - \mu_Z}{\sqrt{(n_1^{-1} + n_2^{-1})\,S_p^2}}, \quad \text{where } S_p^2 = \frac{(n_1 - 1)S_X^2 + (n_2 - 1)S_Y^2}{n_1 + n_2 - 2} \tag{8.8}$$

is the **pooled variance estimator** of σ^2. With q such that $1 - \alpha = \Pr(-q < Q < q)$, inverting Q gives the c.i. for μ_Z, the difference in means,

$$(\bar{X} - \bar{Y}) \pm q\sqrt{(n_1^{-1} + n_2^{-1})\,S_p^2}.$$

This interval is often constructed with particular attention paid to whether or not zero is included. If so, then one could surmise that there is no difference in the population means. Note though, that by taking α small enough, this conclusion can always be reached!

Of course, if σ_1^2 and σ_2^2 are known, then pivot

$$Q = \frac{\bar{X} - \bar{Y} - (\mu_1 - \mu_2)}{\sqrt{\sigma_1^2/n_1 + \sigma_2^2/n_2}} \sim N(0, 1) \tag{8.9}$$

can be used to form an interval in the usual way.

The assumption that σ^2 is the same for both populations might often be questionable. If $\sigma_X^2 \neq \sigma_Y^2$ instead, the above c.i. is no longer valid, with the derivation of an interval with the correct coverage probability being surprisingly quite complicated and the subject of ongoing research. The difficulty associated with inference on $\mu_Z = \mu_1 - \mu_2$ when $\sigma_X^2 \neq \sigma_Y^2$ is often referred to as the **Behrens–Fisher problem** after work conducted on it by Walter Behrens and Ronald Fisher in the 1930s. See Dudewicz and Mishra (1988, Sec. 9.9, 9.10, and p. 567), Weerahandi (1993, 1995), Kim and Cohen (1998), Stuart et al. (1999, pp. 138–148), and the references therein.

An obvious approximate solution is obtained by noting that

$$Q^* = \frac{\bar{X} - \bar{Y} - (\mu_1 - \mu_2)}{\sqrt{S_X^2 / n_1 + S_Y^2 / n_2}}, \tag{8.10}$$

as the analog of (8.9), is asymptotically standard normal and, thus, asymptotically pivotal. Better approximate methods that work well in practice will be considered in Sections 8.3.4 and 8.5 below.

A confidence interval for the ratio σ_1^2/σ_2^2 is easily obtained by noting that

$$Q = \frac{S_X^2/\sigma_1^2}{S_Y^2/\sigma_2^2} \sim F_{n_1-1,\, n_2-1}$$

is a pivot, so that, with $1 - \alpha = \Pr(q_1 \leq Q \leq q_2)$,

$$\left(\frac{S_X^2/S_Y^2}{q_2}, \frac{S_X^2/S_Y^2}{q_1} \right) \tag{8.11}$$

is a $100(1 - \alpha)\%$ c.i. for σ_1^2/σ_2^2. A confidence interval for the difference of two normal variances is sometimes desirable, but there exists no simple method of deriving one. Weerahandi (1995, Sec. 7.5) discusses an exact method for doing so; we consider an approximate method in Section 8.5 below.

Remark. One issue of great concern in applied work is the sensitivity of a proposed method to the necessary assumptions required for its use. The actual coverage probability of interval (8.11), for example, is known to be very sensitive to the underlying assumption of normality. Numerous composite normality tests are discussed in Chapter 6, so it seems that one can proceed with tests associated with (8.11) and (8.8) if a normality test fails to reject the null hypothesis. This procedure, however, changes the sampling theory and inference associated with such confidence intervals and tests. For example, its effect on use of (8.8) has been explicitly studied by Rochon et al. (2012). More generally, the idea of model estimation or further testing, after a decision is made based upon a preliminary test, is known as **pre-test estimation** or **pre-test testing**, respectively; see Giles and Giles (1993) and the references therein . ∎

We now turn to computation of approximate intervals for $\mu_1 - \mu_2$ that do not rely on the equal variance assumption, thus avoiding the complications associated with a pre-test of equal variances.

8.3.4 Welch's Method for $\mu_1 - \mu_2$ when $\sigma_1^2 \neq \sigma_2^2$

Preliminary tests of $\sigma_1^2 = \sigma_2^2$ seem to be a fruitless pastime.

(Rupert G. Miller Jr., 1997, p. 58)

In the earlier derivation with $\sigma_1^2 = \sigma_2^2 = \sigma^2$, a t-distributed pivot was easily obtained. Notice how the finite-sample distribution of Q^* in (8.10) depends on σ_1^2 and σ_2^2 and thus cannot be a pivot.

An approximate approach that is widely used and gives reasonable results for finite samples was proposed by Welch (1947) and involves approximating the distribution of $S_X^2/n_1 + S_Y^2/n_2$ with that of $w\chi_v^2/v$, a weighted χ^2 random variable, by equating the first two moments. That is, from (A.210),

$$\mathbb{E}\left[\frac{S_X^2}{n_1} + \frac{S_Y^2}{n_2}\right] = \frac{\sigma_1^2}{n_1} + \frac{\sigma_2^2}{n_2}, \quad \mathbb{V}\left(\frac{S_X^2}{n_1} + \frac{S_Y^2}{n_2}\right) = \frac{2\sigma_1^4}{(n_1 - 1)n_1^2} + \frac{2\sigma_2^4}{(n_2 - 1)n_2^2},$$

and

$$\mathbb{E}\left[\frac{w\chi_v^2}{v}\right] = \frac{w}{v}\mathbb{E}[\chi_v^2] = w, \quad \mathbb{V}\left(\frac{w\chi_v^2}{v}\right) = \frac{w^2}{v^2}\mathbb{V}(\chi_v^2) = \frac{2w^2}{v}.$$

It follows by equating terms that

$$w = \frac{\sigma_1^2}{n_1} + \frac{\sigma_2^2}{n_2}, \quad v = \frac{\left(\frac{\sigma_1^2}{n_1} + \frac{\sigma_2^2}{n_2}\right)^2}{\frac{\sigma_1^4}{(n_1-1)n_1^2} + \frac{\sigma_2^4}{(n_2-1)n_2^2}}, \quad \text{or} \quad \hat{v} = \frac{\left(\frac{S_X^2}{n_1} + \frac{S_Y^2}{n_2}\right)^2}{\frac{S_X^4}{(n_1-1)n_1^2} + \frac{S_Y^4}{(n_2-1)n_2^2}}, \qquad (8.12)$$

the latter being the "obvious" approximation to v. Then, as $[\bar{X} - \bar{Y} - (\mu_X - \mu_Y)]/\sqrt{w} \sim N(0, 1)$ and $[S_X^2/n_1 + S_Y^2/n_2]/w \overset{\text{appr}}{\sim} \chi_{\hat{v}}^2/\hat{v}$, we see that the asymptotic pivot (8.10) obeys $Q^* \overset{\text{appr}}{\sim} t_{\hat{v}}$, a Student's t distribution with $\hat{v}$ degrees of freedom.

The method involves two levels of approximation: the first being the weighted χ^2 distribution, the second being the "plug-in" version $\hat{v}$ of v. The performance of this and other approximations will be examined in Section 8.5.

8.3.5 Satterthwaite's Approximation

A simple procedure was derived by Satterthwaite (1946) to approximate the distribution of a positively weighted sum of k independent χ^2 r.v.s by that of a single weighted χ^2. It reduces to Welch's method above for $k = 2$.

Let $Y = \sum_{i=1}^{k} a_i C_i$, where $a_i > 0$ and $C_i \overset{\text{indep}}{\sim} \chi_{\delta_i}^2$. To approximate the distribution of Y by that of $C \sim \chi_d^2/d$, we first try equating $\mathbb{E}[Y] = \sum_{i=1}^{k} a_i \delta_i$ and $\mathbb{V}(Y) = 2\sum_{i=1}^{k} a_i^2 \delta_i$ to the corresponding quantities of C, namely $\mathbb{E}[C] = 1$ and $\mathbb{V}(C) = 2/d$. As only the variance of C involves d, this gives $d = (\sum_{i=1}^{k} a_i^2 \delta_i)^{-1}$. A different choice of d is obtained by incorporating the constraint on $\mathbb{E}[C]$ as follows. Observe first that equating $\mathbb{E}[C^2] = 2/d + 1$ with $\mathbb{E}[Y^2] = \mathbb{V}(Y) + \mathbb{E}[Y]^2$ yields

$$d = \frac{2}{2\sum_{i=1}^{k} a_i^2 \delta_i + \left(\sum_{i=1}^{k} a_i \delta_i\right)^2 - 1},$$

which could be negative. However, by writing

$$\mathbb{E}[Y^2] = \mathbb{E}[Y]^2 \left(\frac{\mathbb{V}(Y)}{\mathbb{E}[Y]^2} + 1\right)$$

and setting just the first $\mathbb{E}[Y]$ term on the right-hand side to $\mathbb{E}[C] = 1$, the solution to

$$\frac{2}{d} + 1 = \mathbb{E}[C^2] = \mathbb{E}[Y^2] = \left(\frac{\mathbb{V}(Y)}{\mathbb{E}[Y]^2} + 1\right)$$

is

$$d = \frac{2\mathbb{E}[Y]^2}{\mathbb{V}(Y)} = \frac{(\sum_{i=1}^{k} a_i \delta_i)^2}{\sum_{i=1}^{k} a_i^2 \delta_i}, \tag{8.13}$$

which is strictly positive. Using this value, χ_d^2/d offers an approximation to Y and can be used in any application involving positively weighted independent χ^2 r.v.s.

To see that this reduces to Welch's approximation, note that

$$\frac{S_X^2}{n_1} + \frac{S_Y^2}{n_2} = \frac{\sigma_1^2(n_1 - 1)S_X^2}{\sigma_1^2(n_1 - 1)n_1} + \frac{\sigma_2^2(n_2 - 1)S_Y^2}{\sigma_2^2(n_2 - 1)n_2} = a_1 C_1 + a_2 C_2$$

is a weighted average of two independent χ^2 r.v.s, where, with $S_1^2 = S_X^2$ and $S_2^2 = S_Y^2$ for the $k = 2$ case,

$$a_i = \frac{\sigma_i^2}{n_i(n_i - 1)} \quad \text{and} \quad C_i = (n_i - 1)\frac{S_i^2}{\sigma_i^2} \sim \chi_{n_i-1}^2.$$

Satterthwaite's approximation (8.13) applied to $a_1 C_1 + a_2 C_2$ yields

$$d = \frac{(a_1 \delta_1 + a_2 \delta_2)^2}{a_1^2 \delta_1 + a_2^2 \delta_2} = \frac{\left(\frac{\sigma_1^2}{n_1(n_1-1)}(n_1 - 1) + \frac{\sigma_2^2}{n_2(n_2-1)}(n_2 - 1) \right)^2}{\left(\frac{\sigma_1^2}{n_1(n_1-1)} \right)^2 (n_1 - 1) + \left(\frac{\sigma_2^2}{n_2(n_2-1)} \right)^2 (n_2 - 1)} = \frac{\left(\frac{\sigma_1^2}{n_1} + \frac{\sigma_2^2}{n_2} \right)^2}{\frac{\sigma_1^4}{(n_1-1)n_1^2} + \frac{\sigma_2^4}{(n_2-1)n_2^2}},$$

which is v given in (8.12).

Example 8.7 *Random effects models make great use of this approximation. The C_i are χ^2 r.v.s but whose expected value and variance depend on unknown quantities. This was the context Satterthwaite was interested in, and so he extended the approximation as follows. Write $\mathbb{V}(C_i) = 2\delta_i = 2\mathbb{E}[C_i]^2/\delta_i$, so that*

$$\mathbb{V}(Y) = \sum_{i=1}^{k} a_i^2 \mathbb{V}(C_i) = 2 \sum_{i=1}^{k} \frac{a_i^2 \mathbb{E}[C_i]^2}{\delta_i}.$$

Using this, (8.13) can be expressed as

$$d = \frac{2\mathbb{E}[Y]^2}{\mathbb{V}(Y)} = \frac{(\sum_{i=1}^{k} a_i \mathbb{E}[C_i])^2}{\sum_{i=1}^{k} a_i^2 \mathbb{E}[C_i]^2 / \delta_i}.$$

As C_i itself is certainly unbiased for $\mathbb{E}[C_i]$,

$$d' = \frac{(\sum_{i=1}^{k} a_i C_i)^2}{\sum_{i=1}^{k} a_i^2 C_i^2 / \delta_i}$$

can be used instead of d. Observe how $\chi_{d'}^2/d'$ involves two levels of approximation. A more advanced study in the context of random effects models, making use of saddlepoint approximations, is given in Butler and Paolella (2002).

8.4 CUMULATIVE DISTRIBUTION FUNCTION INVERSION

This is a straightforward method involving the lower and upper quantiles of a statistic and is particularly useful when a pivot cannot be found, such as in the Poisson, binomial, and negative binomial cases. It is referred to as the **statistical method** by Mood et al. (1974) and **guaranteeing an interval** by Casella and Berger (1990), both of which offer clear and detailed discussions of the method.

8.4.1 Continuous Case

Let the statistic T have a continuous c.d.f. $F_T(t; \theta)$ that is monotone in θ. For example, if $T \sim N(\mu, \sigma^2)$, then, for a given $t \in \mathbb{R}$, $F_T(t; \mu, \sigma^2)$ is monotone decreasing in μ, but is not monotone in σ^2; see Figure 8.3.

Note that, if $F_T(t; \theta)$ is monotone decreasing in θ, then, for $\theta_1 < \theta_2$, $F_T(t; \theta_1) - F_T(t; \theta_2) > 0$. Equivalently, for $\theta_1 < \theta_2$ and in terms of the survivor function $\bar{F}_T(t; \theta) = 1 - F_T(t; \theta)$,

$$\bar{F}_T(t; \theta_1) - \bar{F}_T(t; \theta_2) = F_T(t; \theta_2) - F_T(t; \theta_1) < 0,$$

that is, if $F_T(t; \theta)$ is monotone decreasing in θ, then $\bar{F}_T(t; \theta)$ is monotone increasing in θ.

With that in mind, the random interval $(\underline{\theta}(T), \bar{\theta}(T))$ is a $100(1 - \alpha)\%$ c.i. for θ if

- $F_T(t; \theta)$ is monotone *decreasing* in θ for all t, and values $\underline{\theta}(t), \bar{\theta}(t)$ satisfy

$$F_T(t; \bar{\theta}(t)) = \frac{\alpha}{2}, \quad F_T(t; \underline{\theta}(t)) = 1 - \frac{\alpha}{2}, \tag{8.14}$$

  or

- $F_T(t; \theta)$ is monotone *increasing* in θ for all t, and values $\underline{\theta}(t), \bar{\theta}(t)$ satisfy

$$F_T(t; \bar{\theta}(t)) = 1 - \frac{\alpha}{2}, \quad F_T(t; \underline{\theta}(t)) = \frac{\alpha}{2}. \tag{8.15}$$

To verify the former, first note that, because $F_T(t; \theta)$ is monotone decreasing in θ and $1 - \alpha/2 > \alpha/2$, $\underline{\theta}(t) < \bar{\theta}(t)$ and also that these values are unique. Using the right panel in Figure 8.3 as a representative case with $t = 0$ and assuming $\alpha = 0.05$, we have $\bar{\mu}(0) = 1.96$. This follows because, with $k = x - \bar{\mu}$,

$$\int_{-\infty}^{0} \frac{1}{\sqrt{2\pi}} e^{-(x-\bar{\mu})^2} \, dx = \int_{-\infty}^{-\bar{\mu}} \frac{1}{\sqrt{2\pi}} e^{-k^2} \, dk = \Phi(-\bar{\mu}; 0, 1),$$

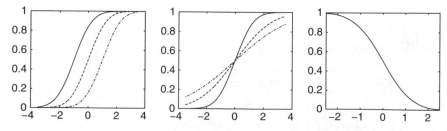

Figure 8.3 **Left:** The normal c.d.f. $\Phi(t; \mu, 1)$ versus t for $\mu = -1$ (solid), $\mu = 0$ (dashed) and $\mu = 1$ (dash-dotted). **Middle:** $\Phi(t; 0, \sigma^2)$ for $\sigma = 1$ (solid), $\sigma = 2$ (dashed) and $\sigma = 3$ (dash-dotted). **Right:** $\Phi(0; \mu, 1)$ versus μ.

or $-\bar{\mu} = -1.96$. Similarly, $\underline{\mu}(0) = -1.96$. It should be clear that

$$\theta > \bar{\theta}(t) \Leftrightarrow F_T(t; \theta) < \alpha/2 \quad \text{and} \quad \theta < \underline{\theta}(t) \Leftrightarrow F_T(t; \theta) > 1 - \alpha/2.$$

From this,

$$\Pr(\underline{\theta}(T) < \theta < \bar{\theta}(T)) = \Pr(\theta < \bar{\theta}(T)) - \Pr(\theta < \underline{\theta}(T))$$

$$= 1 - \Pr(F_T(T; \theta) < \alpha/2) - \Pr(F_T(T; \theta) > 1 - \alpha/2)$$

$$= 1 - \alpha/2 - \alpha/2,$$

where the last equality follows from the **probability integral transform**. The proof for $F_T(t; \theta)$ monotone increasing follows along similar lines.

Example 8.8 *Continuing with the i.i.d.* $N(\mu, \sigma^2)$ *model with* $T = \bar{X} \sim N(\mu, \sigma^2/n)$, *as* $F_T(t; \mu, \sigma^2)$ *is monotone decreasing in* μ, *(8.14) is applicable, yielding*

$$\frac{\alpha}{2} = F_{\bar{X}}(\bar{x}; \bar{\mu}(\bar{x})) = \Pr(\bar{X} < \bar{x}) = \Pr\left(Z < \frac{\bar{x} - \bar{\mu}(\bar{x})}{\sigma/\sqrt{n}}\right) = \Phi\left(\frac{\bar{x} - \bar{\mu}(\bar{x})}{\sigma/\sqrt{n}}\right),$$

or, applying Φ^{-1} *to both sides,* $\bar{\mu}(\bar{x}) = \bar{x} - \Phi^{-1}(\alpha/2)\sigma/\sqrt{n}$. *Similarly, for the upper bound,* $\underline{\mu}(\bar{x}) = \bar{x} - \Phi^{-1}(1 - \alpha/2)\sigma/\sqrt{n}$, *so that, with* $d = \Phi^{-1}(1 - \alpha/2) = -\Phi^{-1}(\alpha/2)$,

$$\left(\bar{X} - \frac{d\sigma}{\sqrt{n}}, \ \bar{X} + \frac{d\sigma}{\sqrt{n}}\right)$$

is a $100(1 - \alpha)\%$ *c.i. for* μ. *This is the same as the interval developed using the pivotal method.*

Example 8.9 (Example 8.2, cont.) *For the i.i.d. location exponential model with typical density* $f_X(x) = \exp\{-(x - \theta)\}\mathbb{I}(x \geq \theta)$ *with* $M = \min\{X_1, \ldots, X_n\}$, *the c.d.f.* $F_M(m; \theta) = (1 - \exp\{-n(m - \theta)\})\mathbb{I}(m \geq \theta)$ *is monotonically decreasing in* θ, *as pictured in the left panel of Figure 8.4, so that (8.14) is applicable.*

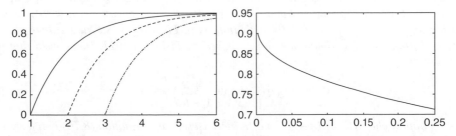

Figure 8.4 **Left**: *The c.d.f. of* M *in Example 8.9 for* $\theta = 1$ *(solid),* $\theta = 2$ *(dashed) and* $\theta = 3$ *(dash-dotted).* **Right**: *Ratio of pivotal method c.i. length to c.d.f. inversion method c.i. length versus* α.

Algebraically, $\bar{F}(m; \theta) = 1 - F(m; \theta)$ *must be monotonically increasing, so that, for* $\theta_1 < \theta_2$, *the ratio of* $\bar{F}(m; \theta_1)$ *and* $\bar{F}(m; \theta_2)$ *must be less than 1 for all m. This is satisfied, as*

$$\frac{\exp\{-n(m - \theta_1)\}}{\exp\{-n(m - \theta_2)\}} = \exp\{n(m - \theta_2) - n(m - \theta_1)\} = e^{-n(\theta_2 - \theta_1)} < e^0 = 1.$$

These give

$$\frac{\alpha}{2} = F_M\left(m; \bar{\theta}(m)\right) = (1 - \exp\{-n(m - \bar{\theta}(m))\}) \Rightarrow \bar{\theta}(m) = m + n^{-1} \ln\left(1 - \frac{\alpha}{2}\right)$$

and

$$1 - \frac{\alpha}{2} = F_M(m; \underline{\theta}(m)) = (1 - \exp\{-n(m - \underline{\theta}(m))\}) \Rightarrow \underline{\theta}(m) = m + n^{-1} \ln\left(\frac{\alpha}{2}\right),$$

yielding the $100(1 - \alpha)\%$ *c.i.*

$$(\underline{\theta}(m), \bar{\theta}(m)) = \left(M + n^{-1} \ln\left(\frac{\alpha}{2}\right), \ M + n^{-1} \ln\left(1 - \frac{\alpha}{2}\right)\right)$$

for θ. *The length of this interval,* $n^{-1}(\ln(1 - \alpha/2) - \ln(\alpha/2))$, *can be compared with that of the pivotal interval (8.4),* $-n^{-1} \ln(\alpha)$. *The right panel of Figure 8.4 plots their ratio*

$$\frac{-\ln(\alpha)}{\ln(1 - \alpha/2) - \ln(\alpha/2)}$$

versus α, *which is always less than (about) 0.9, confirming that the pivotal interval is significantly shorter. It is simple to algebraically confirm that the pivotal interval will always be longer:*

$$\ln\left(\frac{1}{\alpha}\right) \overset{?}{<} \ln\left(1 - \frac{\alpha}{2}\right) + \ln\left(\frac{2}{\alpha}\right) = \ln\left(\frac{2}{\alpha} - 1\right),$$

or

$$\ln\left(\frac{1}{\alpha}\right) - \ln\left(\frac{2}{\alpha} - 1\right) = \ln\left(\frac{1}{\alpha}\frac{\alpha}{2 - \alpha}\right) = \ln\left(\frac{1}{2 - \alpha}\right) \overset{?}{<} 0,$$

or $1/(2 - \alpha) \overset{?}{<} 1$, *which is true for* $0 \le \alpha \le 1$.

Example 8.10   *The c.d.f. inversion method is also useful for computing a c.i. for the sample correlation coefficient* $R = \hat{\rho}$. *Plotting the c.d.f. for numerous values of* ρ *(quickly accomplished via the program in Listing 3.1) confirms that* $F_R(r, \rho)$ *is monotone decreasing in* ρ, *so that (8.14) is applicable. Numerically solving the two equations in (8.14) is very similar to finding quantiles of the distribution as in Listing 7.3. Starting values can be obtained from the approximate method proposed by Fisher in 1921. He showed that the distribution of*

$$z(R) = \frac{1}{2} \ln\left(\frac{1 + R}{1 - R}\right) \tag{8.16}$$

rapidly approaches $N(z(\rho), (n - 3)^{-1})$ *as n increases. Transformation (8.16) is referred to as* **Fisher's variance-stabilizing transformation**. *While the true variance of z certainly depends on* ρ, *it is an amazing and quite useful fact that, for all* ρ, *the value* $v = 1/(n - 3)$ *is quite accurate. In particular, with* $d = \Phi^{-1}(1 - \alpha/2)$, $z \pm dv^{1/2}$ *is an approximate c.i. for*

```
1   function [lofish,hifish,lo,hi]=corrcoefCI(rhohat,n,alpha)
2   if nargin<3, alpha=0.05; end
3   r=rhohat; z=0.5*log((1+r)/(1-r)); d=norminv(1-alpha/2); sig=1/sqrt(n-3);
4   zlo = z-d*sig; zhi = z+d*sig;
5   lofish = (exp(2*zlo)-1) / (exp(2*zlo)+1); hifish = (exp(2*zhi)-1) /
6     (exp(2*zhi)+1);
7   lo=fsolve(@cdf,lofish,optimset('tolf',5e-8),1-alpha/2,r,n);
8   hi=fsolve(@cdf,hifish,optimset('tolf',5e-8),alpha/2,r,n);
9
10  function v=cdf(rho,a2,r,n), p=1; [~,F]=corrcoefpdf(r,rho,n); v= (abs(F-a2))^p;
11    % any p>0
```

Program Listing 8.2: Exact c.d.f.-inversion c.i.s for $\hat{\rho}$ and approximate ones via (8.16).

z. By inverting (8.16) via $R = (e^{2z} - 1)/(e^{2z} + 1)$ at these two endpoints, an interval for R is obtained. The program in Listing 8.2 can be used for the computations. The reader is encouraged to examine the accuracy of the Fisher approximation.

To compute the minimal sample size n such that the length of the equal-tails 95% c.i. is less than or equal to a certain amount, the value of ρ is required. In practice, the researcher can take this ρ to be an "educated guess," based perhaps on the results of similar studies. Taking $\rho = 0$ will lead to a conservative sample size. For example, using the code in Listing 8.2 with $\hat{\rho} = 0$, some trial and error reveals that $n = 384$ gives the interval 0 ± 0.100. For $\hat{\rho} = 0.5$ and $n = 384$, the interval is $(0.421, 0.571)$.

Example 8.11 The coefficient of variation (c.v.), introduced in Example 3.14, is another good case in point for the usefulness of the c.d.f. inversion method. It is given by σ/μ, where σ^2 is the variance and μ is the mean of the population which, for simplicity, we assume is positive. (It is sometimes also defined to be $\sigma/|\mu|$ or $100\sigma/\mu$ or $100\sigma/|\mu|$.) From Example 3.14, an approximate $100(1 - \alpha)$ c.i. for n observations is easily obtained: Letting $X_i \overset{i.i.d.}{\sim} N(\mu, \sigma^2)$, $i = 1, \dots, n$, and $z = \Phi^{-1}(\alpha/2)$, the interval is

$$\widehat{CV} \pm z\widehat{V}^{1/2}, \quad \text{where } \widehat{CV} = \frac{\hat{\sigma}}{\hat{\mu}} \text{ and } \widehat{V} = \frac{2\hat{\sigma}^4 + \hat{\sigma}^2\hat{\mu}^2}{2\hat{\mu}^4 n}, \tag{8.17}$$

with $\hat{\mu} = \bar{X}_n$ and $\hat{\sigma} = \sqrt{S_n^2}$ the usual estimators in the normal case.

To construct an exact c.i. assuming that the data are i.i.d. normal, use the fact that

$$\frac{\sqrt{n}\bar{X}}{\sigma} \sim N\left(\frac{\mu\sqrt{n}}{\sigma}, 1\right) \text{ is independent of } \frac{(n-1)S^2}{\sigma^2} \sim \chi^2_{n-1}$$

from (A.204) and (A.206), so that, similar to (A.208), the statistic

$$T = \frac{\sqrt{n}\bar{X}/\sigma}{S/\sigma} = \frac{\sqrt{n}\bar{X}}{S} = \frac{\sqrt{n}}{\widehat{CV}} \sim t_{n-1}(\sqrt{\tau}), \quad \tau = \frac{n}{CV^2},$$

that is, T is singly noncentral t with $n - 1$ degrees of freedom, and (numerator) noncentrality parameter $\sqrt{\tau}$; see Section A.14. Also, T^2 is singly noncentral F distributed:

$$F = T^2 = \frac{n\bar{X}^2}{S^2} \sim F_{1,n-1}(\tau),$$

and the c.d.f. of F is monotone decreasing in τ. Thus, a c.i. for τ is obtained by computing the two expressions in (8.14), giving, say,

$$1 - \alpha = \Pr\left(\underline{\tau} < \frac{n}{CV^2} < \bar{\tau}\right) = \Pr\left(\sqrt{\frac{n}{\bar{\tau}}} < CV < \sqrt{\frac{n}{\underline{\tau}}}\right).$$

For small sample sizes and large c.v. values, the upper end of the c.i. could be ∞, that is, $\underline{\tau} = 0$. This occurs because, in the right-hand side of (8.14), the noncentrality parameter $\underline{\tau}$ cannot go below zero, but (in a somewhat unfortunate notation) $F_F(F, 0) < 1 - \alpha/2$. For example, with $n = 5$, $\mu = 1$, $\sigma = 1.5$ and $\mathbf{x} = (3.6, 1.2, 1.5, -0.2, 1.5)'$, $\widehat{CV} = 0.894$, $F = n\bar{X}^2/S^2 = 6.254$ and $F_F(F, 0) = 0.935$. As $F_F(F, \eta) < F_F(F, 0)$ for all $\eta > 0$, the value $\underline{\tau}$ does not exist if $\alpha = 0.05$ (i.e., $1 - \alpha/2 = 0.975$). For the lower limit in this case, $\bar{\tau} = 24.94$, yielding $\sqrt{n / \bar{\tau}} = 0.448$.

For large sample sizes and small c.v.-values, n/CV^2 and, thus, its two c.i. endpoints $\underline{\tau}$ and $\bar{\tau}$ will necessarily be large. This results in numerous function evaluations for solving (8.14), and computing $F_F(F, \eta)$ is itself time-consuming and nontrivial. The otherwise prohibitive computation of (8.14) is, however, very fast when using the saddlepoint approximation to the noncentral F, developed in Section II.10.2. See Problem 8.7 for the program to compute the c.i.

While the speed advantage via the s.p.a. is convenient for a single calculation, its benefit becomes clear if the researcher wishes to conduct a simulation to determine, say, the average length of the c.i. for given parameters. This was done for $\alpha = 0.05$, parameters $\mu = 1$, $\sigma = 0.25$, $n = 10$, and 5000 replications. The empirical coverage, say C, was 0.9522, confirming the derivation.[2] The average length, say L, based on only those intervals with a finite upper limit, was 0.312. In this case, none of the 5000 replications had ∞ as the upper limit. For the same parameters but $n = 20$, $C = 0.9508$ and $L = 0.1874$; for $n = 40$, $C = 0.9486$ and $L = 0.124$; and for $n = 80$, $C = 0.9462$ and $L = 0.0847$.

These results can be compared to those obtained by using the asymptotic c.i. in (8.17). Based on the same 5000 replications for each sample size, the coverage obtained was 0.8894, 0.9136, 0.9332, and 0.9422 for $n = 10$, 20, 40, and 80. While the actual coverage improves as the sample size increases, we see that, even for $n = 80$, the coverage is still too low.

8.4.2 Discrete Case

The statement and proof of the method when the X_i are discrete r.v.s is almost identical to the continuous case (see, for example, Casella and Berger, 1990, pp. 420–421). The main difference is that, instead of taking $1 - F_X(x)$ to be $\Pr(X > x)$, as it is defined, we use $\Pr(X \geq x)$ in (8.14) and (8.15). The following example involving the binomial distribution illustrates the method; see also Dudewicz and Mishra (1988, Sec. 10.4).

Example 8.12   *For $X_i \stackrel{\text{i.i.d.}}{\sim} \text{Bern}(p)$, $S = \sum_{i=1}^{n} X_i \sim \text{Bin}(n, p)$ with*

$$F_S(s; n, p) = \sum_{i=1}^{s} \binom{n}{i} p^i (1 - p)^{n-i}.$$

---

[2] Note that, by construction, the 5000 variables taking on the values 0 (the true CV value of 0.25 is not in the interval) or 1 (which is in the interval) are i.i.d. Bernoulli, so that the asymptotic c.i. from Example 8.4 can be applied. The 95% c.i. is $0.9522 \pm 1.96 \times \sqrt{(0.9522)(1 - 0.9522)/5000}$ or $(0.946, 0.958)$, which includes 0.950.

```
1   function [lb,ub] = binomCI(n,s,alpha)
2   if s==n
3      lb=fzero(@(p) binopdf(n,n,p)-alpha,[1e-5 1-1e-5]); ub=1;
4   elseif s==0
5      lb=0; ub=fzero(@(p) binopdf(0,n,p)-alpha,[1e-5 1-1e-5]);
6   else
7      a=alpha; a1=a/2; a2=a1;              % equal-tails interval
8      tol = 1e-6; lo=0+tol; hi=1-tol; % range for the zero search
9      tol = 1e-5; opt=optimset('Display','None','TolX',tol);
10     if 1==2 % use the binomial cdf directly
11        ub = fzero(@(p) binomCI_(p,n,s   , a1),[lo,hi],opt);
12        lb = fzero(@(p) binomCI_(p,n,s-1,1-a2),[lo,hi],opt);
13     else    % use the incomplete beta function
14        ub = fzero(@(p) binomCIib(p,s+1,n-s   , a1),[lo,hi],opt);
15        lb = fzero(@(p) binomCIib(p,s  ,n-s+1,1-a2),[lo,hi],opt);
16     end
17   end
18
19   function z = binomCI_(p,n,x,cut), z = cut - binocdf(x,n,p);
20   function z = binomCIib(p,arg1,arg2,cut), z = cut - (1 - betainc(p,arg1,arg2));
```

Program Listing 8.3: Computes endpoints of $100(1 - \alpha)\%$ c.i. (`lb`, `ub`) for p for the binomial distribution for n trials and s observed successes. It uses the built-in Matlab functions `fzero`, `binocdf`, and `betainc` (incomplete beta function). The method discussed in Section 1.3 to deal with the case of $s = 0$ and $s = n$ is used here as well.

To algebraically determine if F_S is a decreasing function in p seems difficult. Using (A.178), however, it is easy to show that

$$1 - F_Y(p; k, n - k + 1) = F_S(k - 1; n, p),$$

where $Y \sim \text{Beta}(k, n - k + 1)$. Thus, for fixed n and k, F_S is decreasing in p if $F_Y(p)$ is increasing in p, which is clearly the case. Thus, we wish to solve

$$\alpha_1 = F_S(s; n, \bar{p}) = 1 - F_Y(\bar{p}; s + 1, n - s)$$

for the upper bound and

$$\alpha_2 = \Pr(S \geq s) = 1 - \Pr(S < s) = 1 - F_S(s - 1; n, \underline{p}) = F_Y(\underline{p}; s, n - s + 1)$$

for the lower bound, where $0 < \alpha_i < \alpha$ and $\alpha_1 + \alpha_2 = \alpha$, usually taken to be $\alpha_i = \alpha/2$. In order to efficiently calculate $\underline{p}$ and $\bar{p}$, a routine for finding a zero of a function is required, along with either the binomial or beta c.d.f. Listing 8.3 shows Matlab code to perform this.

A more compact expression can be obtained for $\underline{p}$ and $\bar{p}$ by using the relation between the F and beta distributions detailed in Problem I.7.20. In particular, if $F \sim F_{2u,2v}$, then

$$\frac{qF}{1 + qF} \sim \text{Beta}(u, v), \quad q = \frac{u}{v}.$$

For the upper bound, with $u = s + 1$, $v = n - s$ and $F \sim F_{2u,2v}$,

$$\alpha_1 = 1 - F_Y(\bar{p}; u, v) = 1 - \Pr\left(\frac{qF}{1 + qF} \leq \bar{p}\right) = 1 - \Pr\left(F \leq \frac{\bar{p}}{q - \bar{p}q}\right),$$

or, with c_1 the value such that $1 - \alpha_1 = \Pr(F \leq c_1)$, $\bar{p} = c_1 q / (1 + c_1 q)$. For the lower bound, let $u = s$, $v = n - s + 1$, $r = u/v$, and $F \sim F_{2u,2v}$, so that

$$\alpha_2 = F_Y(\underline{p}; s, n - s + 1) = \Pr\left(\frac{rF}{1 + rF} < \underline{p}\right) = \Pr\left(F < \frac{\underline{p}}{r - pr}\right),$$

or, with c_2 such that $\alpha_2 = \Pr(F \leq c_2)$, $\underline{p} = c_2 r / (1 + c_2 r)$. Thus, with cutoff values c_1 and c_2 defined above,

$$\left(\frac{c_2 r}{1 + c_2 r}, \frac{c_1 q}{1 + c_1 q}\right), \quad r = \frac{s}{n - s + 1}, \quad q = \frac{s + 1}{n - s}, \tag{8.18}$$

is a $100(1 - \alpha)\%$ c.i. for p. Of course, the quantiles of the F distribution have to be available to compute this. In Matlab, for example, these are obtained with the `finv` *function, which ultimately computes the quantiles by applying a root finder to the incomplete beta function! Thus, computationally speaking, nothing is saved by using it. It is, however, faster to program and makes the code somewhat more portable. The program to compute (8.18) is given in Listing 8.4.*

Example 8.13   *An interesting application of the Bernoulli parameter interval is the following. Let $X_i \overset{\text{i.i.d.}}{\sim} \text{Poi}(\lambda_1)$, $i = 1, \ldots, n_1$, independent of $Y_i \overset{\text{i.i.d.}}{\sim} \text{Poi}(\lambda_2)$, $i = 1, \ldots, n_2$. An asymptotically valid interval for $\lambda_1 - \lambda_2$ was developed in Example 8.6. Interest might, however, center on the ratio $\rho = \lambda_1/\lambda_2$ instead. An exact interval for ρ is easily obtained if we are willing to condition on the sum of all the observations. As $S_1 = \sum_{i=1}^{n_1} X_i \sim \text{Poi}(n_1 \lambda_1)$ and $S_2 = \sum_{i=1}^{n_2} Y_i \sim \text{Poi}(n_2 \lambda_2)$, we know from (A.79) that*

$$S_1 \mid (S_1 + S_2) \sim \text{Bin}(n, p),$$

where $n = s_1 + s_2$ is the observed sum of all the r.v.s and $p = n_1 \lambda_1/(n_1 \lambda_1 + n_2 \lambda_2)$. If we are willing to condition our inference of ρ on the observed sum n, then an exact c.i. for p is available from Example 8.12. Denoting the lower and upper c.i. limits in (8.18) by $\underline{p}$ and $\bar{p}$, respectively, then

$$1 - \alpha = \Pr(\underline{p} \leq p \leq \bar{p}) = \Pr\left(\underline{p} \leq \frac{n_1 \lambda_1}{n_1 \lambda_1 + n_2 \lambda_2} \leq \bar{p}\right) = \Pr\left(\underline{p} \leq \left(1 + \frac{n_2 \lambda_2}{n_1 \lambda_1}\right)^{-1} \leq \bar{p}\right)$$

$$= \Pr\left(\underline{\rho} \leq \frac{\lambda_1}{\lambda_2} \leq \bar{\rho}\right),$$

```
1   function [lb,ub] = binomCIF(n,s,alpha)
2   a=alpha; a1=a/2; a2=a1;
3   r=s/(n-s+1);                    q=(s+1)/(n-s);
4   c2 = finv(a1,2*s,2*(n-s+1)); c1 = finv(1-a2,2*(s+1),2*(n-s));
5   lb = c2*r/(1+c2*r);            ub = c1*q/(1+c1*q);
```

Program Listing 8.4: Same function as program `binomCI` above, but uses the `finv` function to find the quantiles of the *F* distribution so that we need not conduct the root finding ourselves.

```
1  function [lb,ub] = poisratioCI(n1,sum1,n2,sum2,alpha)
2  [lb0,ub0] = binomCIF(sum1+sum2,sum1,alpha);  f = n1/n2;
3  lb = 1 / ( f*(1/lb0 - 1) );  ub = 1 / ( f*(1/ub0 - 1) );
```

Program Listing 8.5: Exact c.i. for λ_1/λ_2 conditioned on $\sum_{i=1}^{n_1} X_i + \sum_{i=1}^{n_2} Y_i$. Calls binomCIF in Listing 8.4.

where

$$\underline{\rho} = \left(\frac{n_1}{n_2} \left(\underline{p}^{-1} - 1 \right) \right)^{-1} \quad \text{and} \quad \bar{\rho} = \left(\frac{n_1}{n_2} \left(\bar{p}^{-1} - 1 \right) \right)^{-1},$$

being careful to distinguish ρ from p. The method from Example 8.12 (and the associated programs) for computing $\underline{p}$ and $\bar{p}$ require values of n and s. That for n is $s_1 + s_2$; for s, which represents the sum of i.i.d. Bernoulli trials, we use n times $\hat{p}$, or

$$n\hat{p} = (s_1 + s_2) \frac{n_1 \hat{\lambda}_1}{n_1 \hat{\lambda}_1 + n_2 \hat{\lambda}_2} = (s_1 + s_2) \frac{s_1}{s_1 + s_2} = s_1.$$

Listing 8.5 shows the code to compute the interval.

Because of the conditioning argument, it is no longer clear that the coverage is exactly $1 - \alpha$. For several parameter constellations, we consider the performance of the interval, evaluated by the proportion of 20,000 simulated intervals that contain the true ratio. Table 8.3 contains the results. The last two columns give the (asymptotic) 95% c.i. on the coverage, recalling that, by construction, it is itself a realization of a binomial experiment with $n = 20,000$. We see that the actual coverage is too high, though not by much, and improves as n_1 and/or n_2 increase, and as λ_1/λ_2 decreases. Further simulation would be required before making general conclusions, however.

The next two examples further illustrate the c.d.f. inversion method and are similar to the binomial case.

Example 8.14   *Let $X_i \overset{i.i.d.}{\sim} \mathrm{Poi}(\lambda)$, with $S = \sum_{i=1}^{n} X_i \sim \mathrm{Poi}(n\lambda)$. Recalling (A.66), with $G \sim \mathrm{Gam}(s, \lambda)$ and $S \sim \mathrm{Poi}(n\lambda)$,*

$$\Pr(S \geq s; n\lambda) = \Pr(G \leq n; s, \lambda) = F_G(n; s, \lambda),$$

TABLE 8.3 Accuracy of the 95% c.i. of $\rho = \lambda_1/\lambda_2$ that conditions on the observed sum. EC is the empirical coverage proportion, with lo and hi denoting the endpoints of the asymptotic 95% c.i. for EC

n_1	λ_1	n_2	λ_2	EC	lo	hi
5	3	7	4	0.9660	0.9635	0.9685
10	3	14	4	0.9610	0.9583	0.9637
15	3	21	4	0.9608	0.9581	0.9635
5	3	7	8	0.9630	0.9604	0.9657
10	3	14	8	0.9596	0.9569	0.9623
15	3	21	8	0.9553	0.9524	0.9581

so that F_S is strictly decreasing in its parameter $n\lambda$. Consider first the lower bound. We wish to solve

$$\frac{\alpha}{2} = \Pr(S \geq s; n\lambda) = \Pr(G \leq n; s, \lambda).$$

This is easily done using a computer, although one has to check how the program treats the gamma scale parameter. Instead, this can be expressed in terms of a χ^2 c.d.f., which has only one parameter and is implemented uniformly in software. Recalling the relation between χ^2 and gamma, observe that $2\lambda G \sim \text{Gam}(s, 1/2)$ or $2\lambda G \sim \chi^2_{2s}$ so that

$$\Pr(G \leq n; \ s, \lambda) = \Pr(2\lambda G \leq 2\lambda n; \ s, \lambda) = \Pr(C \leq 2\lambda n),$$

where $C \sim \chi^2_{2s}$. That is, we need to solve $\alpha/2 = F_C(2\lambda n; \ 2s)$, giving $2\lambda n = c_1$, where c_1 is such that $\Pr(C \leq c_1) = \alpha/2$, or $\underline{\lambda} = c_1/2n$. For the upper bound, we solve

$$\frac{\alpha}{2} = \Pr(S \leq s; n\lambda) = 1 - \Pr(S > s; n\lambda) = 1 - \Pr(S \geq s + 1; n\lambda)$$

$$= 1 - \Pr(G' \leq n; s + 1, \lambda),$$

where $G' \sim \text{Gam}(s + 1, \lambda)$, or

$$1 - \frac{\alpha}{2} = \Pr(G' \leq n; s + 1, \lambda) = \Pr(2\lambda G' \leq 2\lambda n; \ s + 1, \lambda) = \Pr(C' \leq 2\lambda n),$$

where $C' \sim \chi^2_{2(s+1)}$. Letting c_2 be such that $\Pr(C' \leq c_2) = 1 - \alpha/2$, $\bar{\lambda} = c_2/2n$, giving the $100(1 - \alpha)\%$ c.i. for λ,

$$\left(\frac{c_1}{2n}, \frac{c_2}{2n} \right).$$

Listing 8.6 shows a simple program for computing this using an equal-tails interval. For example, if $n = 10$ and $s = 6$, a 90% equal-tails c.i. for λ is $(0.2613, 1.184)$. The program in Listing 8.7 computes the shortest $100(1 - \alpha)\%$ c.i. For the previous values, the shortest interval is $(0.2100, 1.098)$, which occurs for $\alpha_1 = 0.0204$ (and $\alpha_2 = 0.0796$). Figure 8.5 plots the interval length versus α_1.

Remark. For improved intervals in several discrete cases, see Blaker (2000). ∎

8.5 APPLICATION OF THE NONPARAMETRIC BOOTSTRAP

The Behrens–Fisher problem mentioned in Section 8.3.3 offers a situation for which the nonparametric bootstrap could be of use, as discussed in Section 1.3. This is a two-sample

```
1  function [lb,ub] = PoisClequaltail(n,s,alpha)
2  a=alpha; a1=a/2; a2=a1; % equal-tails interval
3  lb = chi2inv(a1,2*s)/(2*n); ub = chi2inv(1-a2,2*s+2)/(2*n);
```

Program Listing 8.6: Computes an equal-tails c.i. for Poisson parameter λ based on n i.i.d. Poisson observations with observed sum s.

```
1   function [a1,a2,len,lb,ub] = PoisCI(alpha,n,s)
2   tol = 1e-5; opt=optimset('Display','None','TolX',tol);
3   a1 = fminbnd(@PoisCI_,tol,alpha-tol,opt,n,s,alpha); a2 = alpha-a1;
4   [len,lb,ub] = PoisCI_(a1,n,s,alpha);
5
6   function [len,lb,ub] = PoisCI_(a1,n,s,alpha)
7   a2 = alpha-a1; lb=chi2inv(a1,2*s)/(2*n); ub=chi2inv(1-a2,2*s+2)/(2*n);
8     len=ub-lb;
```

Program Listing 8.7: Same as function `PoisCIequaltail` but computes shortest c.i. for λ.

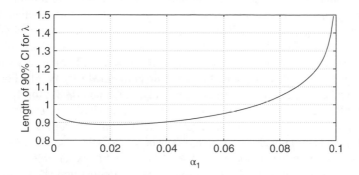

Figure 8.5   *The length of the 90% c.i. of λ in Example 8.14, with $n = 10$ and $s = 6$, as a function of α_1.*

problem and the sampling is conducted as one might imagine: A random sample of size n_1 is drawn, with replacement, from the first vector of observations and, independently, a random sample of size n_2 is drawn from the second vector of observations. The means are computed and their difference is recorded. This is done B times. The bootstrap percentile c.i. is then formed by taking the appropriate quantiles of the simulated distribution.

The performance of this method can be compared with that of Welch's approximate method discussed in Section 8.3.4. Taking $n_1 = n_2 = 5$, $\mu_1 = \mu_2 = 0$, $\sigma_1^2 = \sigma_2^2 = 1$, $B = 1000$ bootstrap replications, $\alpha = 0.05$, and 1000 simulated trials, we find that the actual coverage of the Welch method is 0.977 and that of the bootstrap is 0.887; thus, both have incorrect empirical coverage, with Welch being too conservative and the bootstrap being too liberal. For the same constellation of parameters but with $n_1 = n_2 = 10$, both methods improve, with respective empirical coverage values 0.964 and 0.922. Doubling the sample size again to 20 yields 0.944 and 0.923. Lastly, we take $n_1 = 10$, $n_2 = 20$, $\sigma_1^2 = 1$, and $\sigma_2^2 = 3$. This gives coverage values of 0.962 and 0.935. Both methods appear to perform reasonably well and also somewhat consistently: Welch is too conservative and the bootstrap is too liberal.

What if a c.i. for the difference in means of two independent samples were now desired but the assumption of normality is no longer tenable? One would presume that the Welch method would suffer somewhat, the extent of which depending on the deviation from normality. The nonparametric bootstrap, however, makes no use whatsoever of the normality assumption; only that the means of the populations are finite. To illustrate, let the data follow a Laplace distribution instead of normal. For $n_1 = n_2 = 10$, location parameters

$\mu_1 = \mu_2 = 0$ and scale parameters $\sigma_1 = \sigma_2 = 1$, the Welch c.i. had an empirical coverage of 0.967, while that bootstrap gave 0.918. Interestingly enough, these numbers are quite close to the values obtained in the normal case. The Welch method, however, is less justified on theoretical grounds, contrary to the bootstrap, but in some cases it might be possible to analytically determine how a method designed for normality will perform under a specific nonnormal distribution. Of course, simulation is the fastest and easiest way to determine actual performance.

Now consider the same experiment ($\mu_1 = \mu_2 = 0$, $\sigma_1 = \sigma_2 = 1$) but using Student's t data with three degrees of freedom. The empirical coverage values were 0.983 and 0.941. While the performance of Welch's method begins to break down for very fat-tailed distributions, that of the bootstrap actually improves. A situation almost destined for failure would be the use of the Welch method with nonnormal, possibly asymmetric stable Paretian data. As the variance does not exist, the S^2 statistic is no longer a scaled chi-square and Q^* in (8.10) is most likely quite far from being approximately t distributed. This was examined using a stable tail index of $\alpha = 1.5$ and asymmetry parameter $\beta = 0.95$. For the very small sample sizes used previously, both Welch and the bootstrap performed very poorly. For $n_1 = n_2 = 50$, the empirical coverage values were, respectively, 0.995 and 0.987; for $n_1 = n_2 = 100$, they were 0.994 and 0.984; for $n_1 = n_2 = 500$, they were 0.976 and 0.950; and for $n_1 = n_2 = 1000$, they were 0.974 and 0.941. The bootstrap begins to work well as the sample size increases, while it is not clear if Welch is converging to 0.95. Use of the Welch method in this context appears to be *ad hoc*, while use of the bootstrap is sensible theoretically.

Recall the application of the c.d.f. inversion method for constructing c.i.s for the coefficient of variation at the end of Section 8.4.1. While certainly straightforward, the derivation nevertheless required knowledge of the noncentral F distribution and the ability to compute it. The nonparametric bootstrap could offer an "easy way out." Using the same set of simulated series, bootstrap percentile c.i.s based on 2000 bootstrap replications were computed. Their actual coverage values, however, were not good. For $n = 10, 20, 40$, and 80, they were 0.7910, 0.8582, 0.9026, and 0.9240. As would be expected, the performance increases as n increases, but is disappointingly poor for very small sample sizes.

The sample correlation coefficient R is another good example of a statistic whose distribution is difficult to obtain and approaches normality rather slowly, owing to its high skewness and finite support. It is thus a good candidate for bootstrapping. Of course, in this case, we happen to have both a highly accurate approximation via Fisher's variance-stabilizing transformation and even the exact c.d.f. of R. As such, we can, via a simulation exercise, ascertain not only the actual coverage probability of a bootstrap interval (for a given ρ and n), but can also compare the *lengths* of the exact and bootstrap intervals.

We performed precisely such a comparison, conducted as follows: For a given n and ρ, sim $= 1000$ bivariate normal data sets were generated and, for each, (i) the exact c.i. via the c.d.f. inversion method, (ii) the approximate c.i. via Fisher, and (iii) a bootstrap percentile c.i. based on $B = 1000$ bootstrap replications were computed (each with $\alpha = 0.05$). The results: For $n = 10$ and $\rho = 0$, the exact c.i. had actual (or empirical) coverage 0.95, Fisher had 0.94 and the bootstrap had 0.91. It would appear that the Fisher interval, while extremely close to 0.95, was a bit too short, while the bootstrap interval appears (relatively speaking) much too short. However, just the opposite occurred: The average length of the exact interval was 1.12, that of Fisher was 1.16, and that of the bootstrap was 1.20. This means that the bootstrap interval (for this n and ρ) not only "misses" the true ρ more than it should, but also gives rise to longer intervals than necessary.

For $n = 10$ and $\rho = 0.3$, the exact, Fisher, and bootstrap actual coverage values were 0.94, 0.94, and 0.91, with respective average lengths 1.06, 1.10, and 1.13. For $n = 10$ and $\rho = 0.7$, the coverage values were 0.95, 0.95, and 0.91, with lengths 0.750, 0.754, and 0.750. It thus appears that, as $|\rho|$ increases, the coverage of the bootstrap c.i. stays about the same, but the length, relative to the exact interval, becomes shorter. Indeed, for $n = 10$ and $\rho = 0.95$, the coverage values were 0.96, 0.95, and 0.91, while the lengths were 0.214, 0.202, and 0.190. To verify that the performance of the bootstrap c.i. in terms of actual coverage increases with larger n, a run with $n = 50$ and $\rho = 0.3$ was conducted: the empirical coverage values were 0.96, 0.96, and 0.94, and lengths 0.498, 0.501, and 0.494. (Notice that about five times as many data points were necessary to halve the average interval length.)

8.6 PROBLEMS

8.1 Recall how the length of the c.i. developed in Example 8.3 was minimized.
   (a) Using a similar derivation, verify the constraint $q_1^2 f_Q(q_1) = q_2^2 f_Q(q_2)$ which arose for the minimal c.i. length of σ^2 in the i.i.d. $N(\mu, \sigma^2)$ model.
   (b) Derive the same condition as in (a), but using the method of Lagrange multipliers.

8.2 Let $X_i \overset{\text{i.i.d.}}{\sim} \text{Weib}(b, 0, s)$ for $i = 1, \dots, n$ with density

$$f_{X_i}(x; b, s) = \frac{b}{s}\left(\frac{x}{s}\right)^{b-1} \exp\left\{-\left(\frac{x}{s}\right)^b\right\} \mathbb{I}_{(0,\infty)}(x).$$

   Assume that the shape parameter b is a known constant. Derive (i) the density of $Z_i = s^{-b}X_i^b$, (ii) the density of $Q = \sum_{i=1}^{n} Z_i$, (iii) a $100(1-\alpha)\%$ c.i. for s, and (iv) the conditions for which the c.i. has minimal length.

8.3 In sampling from a $N(\mu, \sigma^2)$ distribution with μ and σ unknown, how large a sample must be drawn to make the probability 0.95 that a 90% confidence interval for μ will have length less than $\sigma/5$?   (Mood et al., 1974, p. 399(16))

8.4 Assume $X_1, \dots, X_n \overset{\text{i.i.d.}}{\sim} f_{X_i}(x) = \theta(1+x)^{-(1+\theta)}\mathbb{I}_{(0,\infty)}(x)$. Calculate (i) the c.d.f. $F_{X_i}(x)$, (ii) the density of $Y_i = \ln(1+X_i)$, (iii) the density of $Q = \theta\sum_{i=1}^{n} Y_i$. (iv) Derive a c.i. for θ.

8.5 Let $X_i \overset{\text{i.i.d.}}{\sim} \text{Unif}(0, \theta)$, $i = 1, \dots, n$.
   (a) Construct a $100(1-\alpha)\%$ minimal length c.i. for θ using a pivot.
   (b) Construct a $100(1-\alpha)\%$ minimal length c.i. for θ using the c.d.f. inversion method of Section 8.4.

8.6 Construct a Matlab program that simulates bivariate normal samples and computes the means ratio interval suggested by Fieller (1954) in Section 8.3.2 above. It should also keep track of the number of occurrences of each of the four outcome possibilities.

8.7 Write programs to compute (i) the exact c.i. for the coefficient of variation and (ii) the asymptotic c.i. and the bootstrap c.i. The latter program should also plot the normal density and a kernel density estimate of the bootstrap distribution.

Part III

Additional Topics

9

Inference in a Heavy-Tailed Context

The normal distribution arises in many stochastic processes involving large numbers of independent variables, and certainly the market place should fulfill this condition, at least.

(M. Osborne, 1959, p. 151)

The tails of the distributions of price changes are in fact so extraordinarily long that the sample second moments typically vary in an erratic fashion.

(Benoit Mandelbrot, 1963, p. 395)

Non-existence of moments is sometimes seen as a disadvantage of heavy-tailed distributions, but the difficulty is the statistical world's obsession with moments rather than with heavy-tailed distributions per se.

(Chris Jones, 2007, p. 63)

This chapter illustrates further inferential methods less commonly addressed in introductory accounts of statistical inference, though they are of great relevance, if not essential, in many contexts. The topics revolve around inference (parameter estimation and testing) for univariate heavy-tailed data and distributions. Section 9.1 illustrates a basic way of attempting to assess the maximally existing moment of the underlying distribution governing a set of data, and also how *not* to do it. This leads into the notion of so-called **tail estimation** in Section 9.2, or assessing the behavior of the underlying distribution only in the extreme tails. Sections 9.3 and 9.4 discuss fast estimation methods for the distributional parameters

Fundamental Statistical Inference: A Computational Approach, First Edition. Marc S. Paolella.
© 2018 John Wiley & Sons Ltd. Published 2018 by John Wiley & Sons Ltd.

associated with the location–scale noncentral Student's t (NCT) and asymmetric stable, respectively, both of which have densities (and thus likelihoods) that are numerically slow to evaluate. Finally, Section 9.5 details various methods for testing the univariate (symmetric and asymmetric) stable Paretian distribution hypothesis.

9.1 ESTIMATING THE MAXIMALLY EXISTING MOMENT

The maximally existing moment of a distribution is typically characterized by stating its supremum, or lowest upper bound, say $\alpha \in \mathbb{R}_{>0}$, such that, if X follows that distribution, then $\mathbb{E}[|X|^r]$ exists for $0 \leq r < \alpha$, and does not exist for $r \geq \alpha$. Our goal is to estimate α based on a set of data. This turns out to be a rather difficult task, a fact that should not be too surprising, recalling that the maximally existing moment is determined by the tail behavior of the distribution and, by definition, in a finite set of data, there will only be a relatively small number of tail observations. For a set of (i.i.d.) data drawn from any distribution, certainly any sample moment can be numerically computed (as the usual plug-in estimator). However, it is very important to emphasize:

> If the distribution does not possess finite absolute moments of all orders, and a sample moment is computed such that the associated theoretical moment counterpart does not exist, then the sample value is meaningless in the sense that, even as the sample size tends to infinity, there is no law of large numbers at work, and the computed value will never converge.

For example, if the data are i.i.d. non-Gaussian stable Paretian, then the traditional measures of asymmetry and heavy-tails vis sample skewness $\hat{\mu}_3$ (as the scale-standardized third central moment) and kurtosis $\hat{\mu}_4$ (as the fourth), respectively, are not valid, because their theoretical counterparts do not exist. As such, while an empirically computed sample kurtosis will be large, the law of large numbers is not applicable, and it will not converge as the sample size increases. This is not the case with Student's t data with degrees of freedom, v, larger than four. In that case, the sample kurtosis is informative for estimation of v. For example, Singh (1988) proposed an estimator of v, assuming $v > 4$, as $\hat{v} = 2(2\hat{\mu}_4 - 3)(\hat{\mu}_4 - 3)^{-1}$.

We begin by illustrating a simple graphical method that is exploratory in nature, as opposed to a formal estimation or testing paradigm. Assume we have an i.i.d. sample $X_1, \ldots, X_n$ from a symmetric distribution about zero that possesses finite moments up to, but not including, order α. For a fixed $r > 0$, consider plotting $\hat{\mu}'_r(Y_1, \ldots, Y_m)$, versus m, where m ranges from, say, $\lfloor 0.01n \rfloor$ to n in increments of $\lfloor 0.01n \rfloor$ and $Y_i = |X_i|$. Based on the aforementioned fact regarding convergence and divergence of the sample moment, we would expect that, if $r < \alpha$, then the plot will tend to oscillate around a constant value (μ'_r) and appear to converge to it as m increases. If, instead, $r \geq \alpha$, then the plot should appear more erratic, and tend to increase. This is because, as m increases and relatively large values of $|X_i|$ enter into the set, $\hat{\mu}'_r(Y_1, \ldots, Y_m)$ will experience a jump. We will call such graphs **moment plots**. Of course, inference from such plots is rather subjective.

To illustrate, let $X_1, \ldots, X_{2000}$ be an i.i.d. sample from a Student's t distribution with $v = 4$ degrees of freedom. Figure 9.1 shows the moment plots for several values of r. Indeed, for $r = 2$ and $r = 3$, the plots appear to converge, while for $r = 5, 6, 7$, the "increasing staircase" is visible. In theory, the $r = 4$ graph should also diverge for this data set, but that is not so

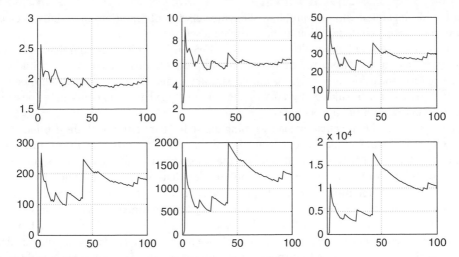

Figure 9.1   *Moment plots for r = 2, 3, 4 (top, from left to right) and r = 5, 6, 7 (bottom, from left to right) for 2000 simulated i.i.d. Student's t_4 realizations.*

clear from the plot. As fourth moments are directly on the border, we would expect this only to become clear as the sample size becomes very large. Notice, finally, that fractional values of r could also be used, although it is doubtful whether one can differentiate between the behavior of plots with, say, $r = 3.9$ and $r = 4.1$.

Example 9.1   *The maximally existing moment of daily stock returns continues to be debated. Consider the NASDAQ stock returns from Example 4.4. The distribution of the returns is clearly heavy-tailed and mildly asymmetric. Figure 9.2 shows the moment plots for r = 2, 3, and 4. They suggest that third and higher moments may not exist, and even the existence of the variance might be questioned.*

Before leaving this brief section, we discuss how *not* to estimate the tail index α when the parametric form of the distribution governing the data is not known. If we *assume* a parametric distribution whose supremum of the maximally existing moment is a known function of the shape parameter(s) (such as v in the Student's t, vd in the GAt from (A.125), and α for the stable Paretian), then the maximally existing moment can be estimated. Notice

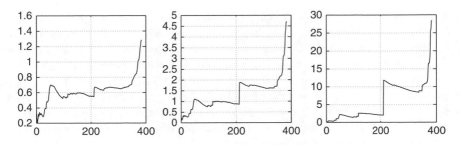

Figure 9.2   *Moment plots for r = 2, 3, 4 (from left to right) for the NASDAQ return series.*

this avoids the aforementioned problem of having very few observations in the tails, because we fit a heavy-tailed distribution to the entire data set.

There are two problems with this approach. First, it is virtually certain that, with real data (such as financial asset returns), they are not precisely distributed as either Student's t or stable Paretian (or any other common parametric distribution), so that the model is misspecified and the resulting inference on the maximally existing moment can be (possibly highly) biased, even in large samples.

The second problem is that, when assuming the stable Paretian model, the estimated tail index for data with sample kurtosis considerably higher than 3 (that of the normal) will, almost certainly, be less than 2, this being the upper limit of the parameter space of α. Thus, with leptokurtic data (even from a distribution that possesses absolute moments of all orders, such as the mixed normal, Laplace, NIG or, more generally, the proper generalized hyperbolic), we are *forced* to conclude that the second moment does not exist, which is unlikely for many real data sets that exhibit excess kurtosis. If $\hat{\alpha} = 2$, we are forced to conclude (or if a confidence interval of α includes 2, we are behooved to entertain) that the data are Gaussian, thus possessing finite positive absolute moments of all order. This would seem to suggest the use of the Student's t model, for which the parameter space of v is $(0, \infty)$, thus not excluding infinite variance, but still allowing for a maximally existing moment. Unfortunately, the problem goes both ways: For a finite set of data from a non-Gaussian stable Paretian distribution but with existing mean (i.e., with tail index $1 < \alpha < 2$), estimation of a Student's t model can deliver a value of $\hat{v}$ greater than 2.

This is best illustrated with simulation. The top panel of Figure 9.3 shows the histogram of 1000 estimated values of v for the location–scale Student's t model (such that the three parameters were estimated jointly, by maximum likelihood), whereby each of the 1000 data sets are actually 2020 realizations from a $S_{1.6,0}(0, 1)$ model, that is, symmetric stable Paretian with $\alpha = 1.6$, location 0 and scale 1. (We used a sample size of 2020 because of the example with the 30 stock return series below, and the value 1.6 because it is the average value of $\hat{\alpha}$ for those stocks.) We see that *not a single data set* had a $\hat{v}$ below 2.0, and the mean of the $\hat{v}$ is 2.65. Conversely, the bottom panel shows the histogram of 1000 estimated values of α for the location–scale symmetric stable model, whereby each of the data sets is actually 2020 realizations from a Student's t with $v = 2.75$ degrees of freedom (and location 0, scale 1). The value 2.75 was chosen because it is the average value of $\hat{v}$ for the 30 stocks considered below. All values of $\hat{\alpha}$ are well below 2.

For the NASDAQ returns from Example 9.1, $\hat{v} = 2.38$ under the Student's t model, and $\hat{\alpha} = 1.47$ under the (asymmetric) stable Paretian. Based on the above simulation results, one simply cannot conclude from these fits whether or not the second moment exists.

Example 9.2   *We estimate both Student's t and stable Paretian models for the daily returns on the 30 stocks composing the Dow Jones Industrial Average index from June 2001 to March 2009 (each series has 2020 observations). This period captures the large economic downturn and banking crisis starting in August 2007, with the accompanying high volatility and extreme stock market price changes. Figure 9.4 shows the results in the form of two overlaid histograms of the 30 values. Even the Student's t model has $\hat{v} < 2$ for six out of the 30 series (the lowest being $\hat{v} = 1.410$, for Bank of America, which also yielded the lowest $\hat{\alpha}$-value of 1.193; the highest was $\hat{v} = 3.928$ for Wal-Mart), while, by construction, the values of $\hat{\alpha}$ are all (quite far) below 2.*

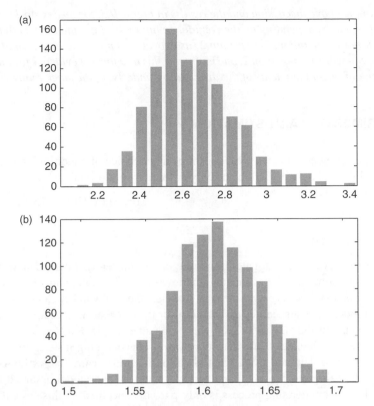

Figure 9.3 *(a) Estimated values of the degrees-of-freedom parameter for the Student's t distribution, but for data sets that are symmetric stable with tail index 1.6. (b) Estimated values of tail index α for the symmetric stable distribution, but the data are Student's t with $\upsilon = 2.75$ degrees of freedom.*

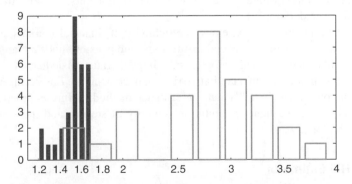

Figure 9.4 *Thin, solid lines correspond to the 30 estimated tail index values $\hat{\alpha}_{ML}$ for the location–scale asymmetric stable Paretian model, while the thick, empty boxes correspond to the 30 estimated degrees of freedom values $\hat{\upsilon}$ for the (symmetric) location–scale Student's t model.*

In light of the above discussion and the results in Figure 9.3, for the six stocks with $\hat{v} < 2$ (and possibly others, depending on the confidence intervals of v), and the fact that not only is the estimated stable tail index less than 2 (as expected, if not necessary), but also the tail index of the Student's t is less than 2, one might use this as a form of evidence for questioning the existence of second moments of the distribution underlying the stock returns.

9.2   A PRIMER ON TAIL ESTIMATION

Here, as elsewhere in statistics, questions as to choice of conditioning events are subtle, and answers will be heavily colored by philosophical outlook.

(Bruce M. Hill, 1975, p. 1164)

9.2.1   Introduction

Throughout this text, we investigate fitting one or several relevant candidate distributions to i.i.d. data by estimating the associated unknown location, scale, and shape parameters. These are examples of *(fully) parameterized* models, others of which include regression and time series models with a specified distribution for the error term. In Chapters 2 and 6, we emphasize the use of the empirical c.d.f., this being an example of a *(fully) nonparametric* model. This section deviates from these two forms, concentrating instead only on the tail behavior of a set of data and using what is called a **semi-parametric estimator**. As its name suggests, it implies that the parametric structure of the model assumed to govern the underlying data-generating process is only partially specified; in this case, it is for the tail behavior. Focusing on only the tail of a distribution is of interest when one is concerned about extreme events, and also serves as a way of assessing the maximally existing moment associated with the underlying process. The latter is relevant, among other reasons, because most statistical methods assume existence of particular moments, and thus the validity of their use needs to be determined.

With respect to the study of extremes, the behavior of financial markets and the risks faced by insurance companies, financial institutions, and pension funds are among the primary examples of concern to many people, and a large literature is dedicated to their study, given their central importance in quantitative risk management; see Embrechts et al. (1997). However, numerous scientific fields make use of such methods: a further practical example includes the changing weather, precipitation, severity of storms, and increased flooding associated with global warming.

9.2.2   The Hill Estimator

This section introduces a simple and fundamental estimator associated with tail estimation, serving to initiate some basic concepts, and requires only techniques from probability and statistical theory that we have already seen. More generally, such methods fall into the category of so-called **extreme value theory** (EVT). Textbook introductions to EVT, with discussions of numerous applications, include Embrechts et al. (1997), Coles (2001), Beirlant et al. (2004), Finkenstädt and Rootzén (2004), de Haan and Ferreira (2006), and Reiss and Thomas (2007); see also Longin (2017) for numerous applications of EVT in finance. The paper by Gilli and Këllezi (2006) is a highly readable overview of basic methodology applied to financial risk.

Let $X, X_t \overset{\text{i.i.d.}}{\sim} \text{Par}(\alpha, x_0)$, $t = 1, \ldots, T$, with c.d.f. (A.75) and order statistics for the X_t denoted by $X_{1:T} < X_{2:T} < \ldots < X_{T:T}$. For some $\kappa \in \{1, 2, \ldots, T\}$, let $x_0 = X_{\kappa:T}$, and observe that, from (A.76), $\{\log(X) - \log(x_0)\} \sim \text{Exp}(\alpha)$. Our goal is to somehow use this fact to elicit information about α from a set of data.

Recall Rényi's representation (A.183). Using it, with $j = T - h + 1$, we can write

$$(T - (j - 1))(\log(X_{j:T}) - \log(X_{j-1:T})) = h \cdot (\log(X_{T-h+1:T}) - \log(X_{T-h:T})) \overset{\text{i.i.d.}}{\sim} \text{Exp}(\alpha),$$

$h = 1, \ldots, T - \kappa$. Next, recall Example 3.3, showing that, for $X_i \overset{\text{i.i.d.}}{\sim} \text{Exp}(\alpha)$, $i = 1, \ldots, n$, the m.l.e. of α is $1/\bar{X}_n$. Thus, for given κ, with $R_h := \log(X_{T-h+1:T}) - \log(X_{T-h:T})$,

$$\frac{1}{\hat{\alpha}_{\text{ML}}} = \frac{1}{T - \kappa} \sum_{h=1}^{T-\kappa} h R_h = \frac{1}{T - \kappa} \sum_{h=1}^{T-\kappa} \sum_{m=1}^{h} R_h = \frac{1}{T - \kappa} \sum_{m=1}^{T-\kappa} \sum_{h=m}^{T-\kappa} R_h$$

$$= \frac{1}{T - \kappa} \sum_{m=1}^{T-\kappa} \log X_{T-m+1:T} - \log X_{k:T} = \frac{1}{k} \sum_{j=1}^{k} \log X_{T-j+1:T} - \log X_{T-k:T}, \quad (9.1)$$

where the third equality comes from a simple rearrangement of the sum; the fourth equality follows because of the "telescope canceling" of terms (both of which the reader should quickly verify); and the last inequality just sets $k = T - \kappa$. Estimator (9.1) is referred to as the **Hill estimator**, after Hill (1975). The associated program is given in Listing 9.1.

As $S = \sum_{h=1}^{T-\kappa} h R_h \sim \text{Gam}(T - \kappa, \alpha)$ (recall Examples I.93, I.9.11, and II.2.3), we have, from Problem I.7.9 (as the reader should confirm),

$$\mathbb{E}[\hat{\alpha}_{\text{ML}}] = (T - \kappa)\mathbb{E}[S^{-1}] = \alpha \frac{T - \kappa}{T - \kappa - 1} \approx \alpha$$

for large T, and

$$\mathbb{V}(S^{-1}) = \frac{\alpha^2}{(T - \kappa - 1)(T - \kappa - 2)} - \left(\frac{\alpha}{T - \kappa - 1}\right)^2 = \frac{\alpha^2}{(T - \kappa - 1)^2(T - \kappa - 2)},$$

so that

$$\mathbb{V}(\hat{\alpha}_{\text{ML}}) = \frac{(T - \kappa)^2 \alpha^2}{(T - \kappa - 1)^2(T - \kappa - 2)} \approx \frac{\alpha^2}{T - \kappa},$$

```
1  function [hill,hstd]=hill(x,krange)
2  use=abs(x(x~=0)); use=reshape(use,length(use),1);
3  y=sort(use,1,'ascend'); % traditional order statistics
4  n=length(y); lny=log(y); hillmat=zeros(1,length(krange)); vmat=hillmat;
5  for loop=1:length(krange)
6    k=krange(loop); a=1 / ( mean(lny( (n+1-k):n )) - lny(n-k) );
7    hillmat(loop)= a; vmat(loop)=a*a/k;
8  end
9  hill=hillmat'; hstd=sqrt(vmat');
```

Program Listing 9.1: Computes the Hill estimator (9.7) and its approximate standard error.

for large T. Further developments, and proof that $(T - \kappa)^{1/2}(\hat{\alpha}_{\mathrm{ML}}^{-1} - \alpha^{-1}) \overset{\text{asy}}{\sim} \mathrm{N}(0, \alpha^{-2})$, can be found in Hall (1982), Davis and Resnick (1984), Haeusler and Teugels (1985), Csörgoő and Mason (1985), Goldie and Smith (1987), and de Haan and Ferreira (2006, Sec. 3.2).

While (9.1) is often used (see, for example, Haeusler and Teugels, 1985; Mason and Turova, 1994; Reiss and Thomas, 2007, Eq. 5.3), the estimator will sometimes appear in a different form, because here we used the traditional definition of order statistics, with $X_{1:T} < X_{2:T} < \ldots < X_{T:T}$, as opposed to reversing them, as is common in the literature on tail estimation (see, for example, Hill, 1975; McNeil et al., 2015, and the references therein). Embrechts et al. (1997, pp. 330–336), Beirlant et al. (2004, Sec. 4.2.1), and de Haan and Ferreira (2006, Sec. 3.2) provide several different ways of deriving the Hill estimator. Further developments, a large literature overview, simulation studies, and a historical account going back to the nineteenth and early twentieth centuries can be found in Csörgő and Viharos (1998).

For genuine i.i.d. Pareto data, the Hill estimator is accurate for a wide range of k, as demonstrated below, and enjoys the usual properties of the m.l.e., namely consistency and asymptotic normality. However, its main purpose is to provide an estimate of the tail index for distributions such that the right tail is asymptotically Pareto, such as the Student's t and stable Paretian. While it is the cornerstone of semi-parametric inference for the tail index, we will see that it does not perform well with practical (and even large) sample sizes, and numerous variations and improvements exist. In particular, when the entire distribution is not Pareto, the choice of tuning parameter k, indicating "where the tail starts," becomes critical. For a given sample size T, large values of k would seem to be preferred because one is further into the right tail, but the dearth of the ensuing number of tail observations results in high inaccuracy of the estimator. This is a classic example of a **bias–variance tradeoff**, and attempts at choosing k to minimize the mean squared error generally require knowledge of the true distribution or assumptions that cannot be verified.

As an illustration, consider computing the Hill estimator for a simulated set of i.i.d. data $X_1, \ldots, X_T$ and plotting it as a function of tuning parameter k, known as a Hill plot. To do this for X_i Pareto with c.d.f. $F_X(x) = 1 - x^{-\alpha}$, $x > 1$, the probability integral transform implies that we can take $X_i = P_i^{-1/\alpha}$ for $P_i \sim \mathrm{Unif}(0, 1)$, $i = 1, \ldots, T$. Figure 9.5(a) illustrates a Hill plot for Pareto and symmetric stable data, both with the same tail index $\alpha = 1.3$, as well as Student's t data with degrees of freedom $v = 3$. Figure 9.5(b) is similar, but using $\alpha = 1.7$ and $v = 6$, all based on $T = 10,000$ observations. While the Hill estimator is highly accurate for the Pareto data over a wide range of k, it never "stabilizes" for stable Paretian or Student's t data. Its near uselessness in these and other contexts gives rise to the term "Hill horror plot" (see Embrechts et al., 1997, pp. 194, 270, 343).

Calculating the graphs in Figure 9.5 but having multiplied the simulated data by scaling factor $\sigma > 0$ numerically confirms that the Hill estimator is scale-invariant. This is also obvious from term R_h in (9.1). However, it is not location-invariant. A modification of the Hill estimator that is both location- and scale-invariant and involves jointly estimating a location term is developed in Aban and Meerschaert (2001). Alternatively, one could just center the data by subtracting a robust estimator of the location, such as the median, as also suggested in Fofack and Nolan (1999).

Several authors have developed methods for the selection of the optimal k based on bias or m.s.e. reduction, such as via use of the bootstrap; see Danielsson et al. (2001), Clementi et al. (2006), Brilhante et al. (2013), and the reference therein. A different way, designed

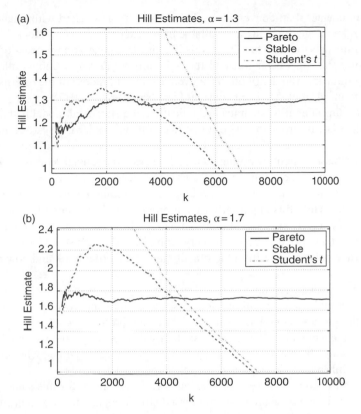

Figure 9.5 *Hill estimates as a function of k, known as Hill plots, for simulated Pareto, symmetric stable Paretian, and Student's t, with tail index for Pareto and stable $\alpha = 1.3$, and Student's t degrees of freedom $v = 3$ (a) and $\alpha = 1.7$, $v = 6$ (b), based on sample size $T = 10,000$.*

instead to decide which part of the sample is most informative, or "belongs to the tail," is developed in Nguyen and Samorodnitsky (2012), where further references on the selection of k can be found. The Hill estimator can also be robustified (see Vandewalle et al., 2007; Beran and Schell, 2012; and the references therein) and extended to the multivariate case (see Dominicy et al., 2017).

9.2.3 Use with Stable Paretian Data

With regard to use of the stable Paretian distribution, several papers, such as Loretan and Phillips (1994), provide evidence of the existence of second moments in financial returns data, thus disqualifying the stable model. The problem with all such attempts at inference in this regard was mentioned above: Determination of the maximally existing moment of the distribution underlying a given set of data based on (necessarily) a finite number of observations is notoriously difficult in general, and certainly so via Hill-type estimators; see McCulloch (1997b), Mittnik et al. (1998), Weron (2001), Heyde and Kou (2004), and the references therein for critiques of studies drawing inference from such methods.

One might counter that, conveniently, the sample size associated with financial returns data can be rather large, by using all available historical price data on a stock. The problem with this argument is that it is hardly tenable that the data-generating process of the returns on a particular stock has been constant throughout time. Putting it more colloquially, do you actually think that the distribution (or even just the tail index) of the returns associated with some financial stock has been constant from, say, 1967 to 2017?

Another argument against such naive application of a statistical model is the concept of **survivorship bias**: In order to get so much data, one is required to restrict the choice of stocks to those that, *a posteriori*, did not go bankrupt, or survived, the whole period. One could counter this argument by noting that, with respect to financial returns data, one could fix the calendar time (to, say, one or two years), and increase the sample size by increasing the frequency of the observed price process, as is common now with high-frequency stock price data. While this is partially true, a problem with this line of reasoning is that, as the frequency increases (beyond, approximately, 5 minutes, for highly liquid stocks), the data-generating process becomes very complicated because of so-called **market microstructure noise** arising from the functioning of the order book and how markets are cleared.

Another, yet weaker, argument occasionally brought forth regarding the existence of variance is that, ultimately, in all applications, the actual support of the random variable underlying the phenomenon under study is bounded (and thus must have existence of all positive absolute moments). While this is certainly theoretically true, we should note that, as stated eloquently by Nolan (1999, p. 2), "Of course the same people who argue that the population is inherently bounded and therefore must have a finite variance routinely use the normal distribution – with unbounded support – as a model for this same population. The variance is but one measure of spread for a distribution, and it is not appropriate for all problems."

Another issue is that financial returns data (certainly at the daily or higher frequency level) are blatantly not i.i.d., as already seen in Figure 4.7, so that the process is obviously not i.i.d. stable. It could be *conditionally* stable, if the model is such that the scale term is allowed to evolve over time, but this then needs to be accounted for appropriately; see Mittnik et al. (2000, Sec. IV.12.5) and the references therein for further discussion of this issue.

The Hill estimator certainly can be applied to the returns (or any non-i.i.d. data set) to assess the tail index of the *unconditional* data, though this violates the assumptions used in its derivation, complicating further its usage outside of the genuine i.i.d. Pareto (not Paretian!) framework. Discussions of the pitfalls, and some solutions, associated with its use and choice of k are discussed in Mittnik et al. (1998), Pictet et al. (1998), and the references therein. The applicability of Hill and smoothed Hill estimators to the filtered residuals of GARCH-type models is developed in Kim and Lee (2016).

In light of the poor performance of the Hill estimator in assessing the tail index, Adler (1997) comically states that "Overall, it seems that the time may have come to relegate Hill-like estimators to the *Annals of Not-Terribly-Useful Ideas*." Despite its weaknesses, the Hill estimator is far from ready to be relegated to the dustbin of statistical history. As indicated by the small sample of references above, progress continues, and, as discussed in Section 9.4.2 below, it forms the basis of an accurate estimator of the stable tail index.

9.3 NONCENTRAL STUDENT'S t ESTIMATION

9.3.1 Introduction

Recall from Section A.14 and Problem A.18 that if $X \sim N(\gamma, 1)$ independent of $Y \sim \chi^2(k)$, then $T = X/\sqrt{Y/k}$ is singly noncentral t (NCT) with degrees of freedom $k \in \mathbb{R}_{>0}$ and non-centrality parameter $\gamma \in \mathbb{R}$, written $T \sim t'(k, \gamma)$. If $k > 1$, then

$$\mathbb{E}[T] = \gamma \left(\frac{k}{2}\right)^{1/2} \frac{\Gamma(k/2 - 1/2)}{\Gamma(k/2)}, \tag{9.2}$$

as detailed in Section II.10.4.3. A location parameter μ and scale $c > 0$ can be incorporated in the usual manner. The distribution is important, as it arises in power calculations (and thus sample size determination) in analysis of variance and linear regression models, in which case k is an integer. Moreover, as an asymmetric version of the Student's t, it can serve as (the error term in) a model for heavy-tailed, asymmetric data, most notably in empirical finance; see, for example, Harvey and Siddique (1999, 2000). This usage is somewhat *ad hoc*, in the sense that there is no theoretical justification for its use as the error term *per se*, unlike the use of the stable distribution, though it captures very well precisely the observed features inherently required.[1]

Another benefit is that the NCT is closed under addition, in the sense that the sum of univariate margins of the Kshirsagar (1961) multivariate NCT (MVNCT), is (univariate) NCT. This happens because the MVNCT is a location–scale continuous mixture of normals. Thus, if a set of financial asset returns is modeled using the MVNCT, then the portfolio distribution (a weighted sum of the univariate margins) is analytically tractable, as used by Jondeau (2016); see also Genton (2004) and Paolella and Polak (2015a).

The p.d.f. of the NCT at a single point x can be expressed as an integral expression, as given in (A.339), and also as an infinite sum, given by

$$f_T(x; k, \gamma) = \frac{\Gamma((k+1)/2)}{(\pi k)^{1/2}\Gamma(k/2)} \left(\frac{k}{k+x^2}\right)^{\frac{k+1}{2}} \times \exp\left\{-\frac{\gamma^2}{2}\right\} \times \sum_{i=0}^{\infty} g_i(x; k, \gamma), \tag{9.3}$$

where

$$g_i(x; k, \gamma) = \frac{2^{i/2}\Gamma((k+1+i)/2)}{i!\,\Gamma((k+1)/2)} \left(\frac{x\gamma}{\sqrt{k+x^2}}\right)^i; \tag{9.4}$$

see Section II.10.4.1.1 for derivation. This can be evaluated in a vectorized fashion for calculating the likelihood of a set of data, yielding a substantial speed increase, and is used in Matlab's built-in implementation.[2]

---

[1] This latter claim is not completely true: The NCT is a continuous mixture of normals, as is the generalized hyperbolic (see, for example, Chapter II.9) and so can be seen as generalizations of a discrete mixture of normals – which is a foundational model for asset returns; see Jondeau et al. (2007, Sec. 3.3, 3.4), Andersen et al. (2010), and the references therein.

[2] Unfortunately, it is faulty for some values. For example, in Matlab version 10, calling `nctpdf(0.01,3,4)` returns the wrong value, though this is fixed in version 14. However, even in version 14, `nctpdf(1e-15,3,1)` should result in virtually the same answer as `nctpdf(0,3,1)`, but those values differ already in the first decimal place.

As an alternative to use of exact expressions for the p.d.f. and c.d.f., Broda and Paolella (2007) provide a saddlepoint approximation (s.p.a.) with a closed-form solution to the saddlepoint equation (obviating root searching and potential numerical problems) that is easily vectorized, resulting in a substantial increase in speed, with accuracy of about three significant digits. (Note three digits, and not decimal places; the s.p.a. in general exhibits *relative* accuracy; see Butler, 2007, and the references therein.) It is renormalized (via numeric integration), and thus integrates to 1. As such, one can interpret its use as fitting an alternative distribution to the data (which happens to virtually coincide with the NCT) with *no loss of accuracy*, as the exact NCT model (or stable, or GAt, etc.) is anyway not the true data-generating process in applied contexts such as fitting financial asset returns data.

The speed benefit associated with use of the s.p.a. is of particular importance for applications in which an estimator of $\theta = (k, \gamma, \mu, c)'$ needs to be computed many times, such as in simulation studies, bootstrap calculations, or through a large number of moving windows of time series data in backtesting exercises. In particular, fast estimation becomes crucial for implementing methods of portfolio optimization requiring a large number of repeated estimations of the NCT; see Paolella and Polak (2018a).

We now present two additional methods to speed up estimation of the NCT distribution parameters.

9.3.2 Direct Density Approximation

As in Krause and Paolella (2014), we wish to develop an approximation to (9.3) that is much faster to evaluate, without appreciable loss in accuracy. We refer to this as (for want of a better name) the direct density approximation (d.d.a.). Observe in (9.3) how the first term (before the $\times$) corresponds to the usual (central) Student's t, while the last term, with the infinite sum, is only relevant in the noncentral case.

Let x be a point in the support of T, let $\varepsilon > 0$ be a small threshold value (e.g., the machine precision), and let $f_T^{\gamma=0}(x)$ be the first two terms in (9.3). The idea is to evaluate the third term in (9.3) only if $f_T^{\gamma=0}(x) > \varepsilon$. By construction, this approximation involves an error in the outer tail area, for which f_T will anyway evaluate to a value close to zero if extreme cases of noncentrality are neglected, this being a reasonable assumption in the context of modeling financial asset returns. As such, the approximation error will be negligible and depends on the choice of ε. Using machine precision for ε, the approximation is nearly exact for $|\gamma| < 1$.

Evaluation of infinite sums always involves specifying "when to stop summing." For (9.3), we have (i) $g_0 = 1$; (ii) g_i is oscillating when $x\gamma$ is negative; (iii) $g_i \to 0$ as $i \to \infty$; and (iv) the series $\{|g_i|\}$ has a global maximum. While (iii) is obvious as the sum is convergent, we can confirm this by letting

$$\nabla_i = 2^{i/2}\Gamma((k+1+i)/2)\left(\frac{x\gamma}{\sqrt{k+x^2}}\right)^i \quad \text{and} \quad \Delta_i = i!\Gamma((k+1)/2)$$

denote the numerator and denominator, respectively, of g_i as functions of i. Observe that the denominator exhibits a higher growth rate than the numerator. That is, Δ_i outweighs ∇_i as i increases, and $g_i \to 0$ in the limit as $i \to \infty$.

For (iv), analogously to (ii), we look at $|g_i|$ and consider the absolute value of ∇_i and Δ_i. It is easy to see that $|\Delta_i|$ is a monotonically increasing function in i, while $|\nabla_i|$ can be monotonically decreasing if $|x\gamma| < 1$, or is increasing if $|x\gamma| \geq 1$. From the monotonicity of $|\Delta_i|$ and $|\nabla_i|$, it follows that $|g_i|$ either takes its maximum at $i = 0$ if $|\nabla_i|$ is decreasing, or starts with $|g_0| = 1$, takes a maximum for some i dependent on x, $k,0$ and γ, and then declines towards zero if $|\nabla_i|$ is increasing. In both cases, $|g_i|$ has a global maximum. We can truncate the infinite sum at $i = i^\star$ with $h_{i\star} \leq \varepsilon$, where $\varepsilon > 0$ is an absolute threshold. Alternatively, the sum can be truncated at the first summand that does not significantly contribute to the sum, that is, at index $i = i^\star$ with $g_{i\star} / \sum_{j=0}^{i^\star-1} g_j \leq \varepsilon$.

Very large values of g_i, such as triggered by large values of k, can break the numerical limitations of the underlying finite arithmetic architecture when the resulting sum, $g_i + \sum_{j=1}^{i-1} g_j$, becomes sufficiently large. To address this, use the identity

$$\log\left(\exp\{a\} + \sum_j \exp\{b_j\}\right) = a + \log\left(1 + \sum_j \exp\{b_j - a\}\right), \quad a, b_j \in \mathbb{R},$$

and reformulate the infinite sum as

$$s_{i+1} = s_i + \log(1 + \exp\{\log g_i - s_i\}), \tag{9.5}$$

where $s_0 = \log g_0$ and

$$\log g_i(x; k, \gamma) = \frac{i \log 2}{2} \log \Gamma\left(\frac{k+1+i}{2}\right) - \log \Gamma(i+1) - \log \Gamma\left(\frac{k+1}{2}\right)$$
$$+ i \log(x\gamma) - \frac{i \log(k+x^2)}{2}.$$

That is, $s_i = \log \sum_{j=0}^j g_j$, and the infinite sum is computed based on $\log(g_i/s_i)$ instead of g_i. The advantage of using (9.5) is improved numerical robustness.

The resulting approximation can be seen as $\hat{f}_T^{\text{NCT}} = f_T^{\text{T}} + g_T^{\text{NCT}}\mathbb{I}\{f_T^{\text{T}} \geq \varepsilon\}$, where f_T^{T} denotes the p.d.f. of the central case, and g_T^{NCT} refers to the infinite sum computed based on (9.5). The program in Listing 9.2 computes the log density for a vector of points.[3]

9.3.3 Quantile-Based Table Lookup Estimation

Recall the discussion of the quantile least squares estimator in Section 5.2.3, and its potential for use with table lookup. This was implemented for estimating the two shape parameters of the NCT in Krause and Paolella (2014), along with use of the method in Example 4.3 for the location term. The scale needs to be known: While this is unrealistic in general, the estimator, hereafter denoted $\hat{\theta}_Q^{|c} = (\hat{k}_Q, \hat{\gamma}_Q, \hat{\mu}_Q)'$, was designed for use with financial returns time series data, requiring modeling of the time-varying scale term. The latter is accomplished via use of a GARCH-type model, resulting in the filtered innovations having scale term 1.

The data are then location- and scale-transformed and the two shape parameters are determined by table lookup. In particular, for every pair in a tight grid of k- and γ-values, the set of corresponding quantiles $\{Q_1, \ldots, Q_m\}$ is computed once (with each Q_i obtained by numerically inverting the NCT c.d.f.) and stored. While this takes many hours of computing,

---

[3] Because of the related nature of the MVNCT construction, the same logic used for the d.d.a. also applies in the multivariate case, thus enabling straightforward maximum likelihood estimation. This will be detailed in a future volume.

```
1   function pdfln = stdnctpdfln_j(x, nu, gam)
2   vn2 = (nu + 1) / 2; rho = x.^2;
3   pdfln = gammaln(vn2) - 1/2*log(pi*nu) - gammaln(nu/2) - vn2*log1p(rho/nu);
4   if (all(gam == 0)), return, end
5   idx = (pdfln >= -37); % -36.841 = log(1e-16)
6   if (any(idx))
7     gcg = gam.^2; pdfln = pdfln - 0.5*gcg; xcg = x .* gam;
8     term = 0.5*log(2) + log(xcg) - 0.5*log(max(realmin,nu+rho));
9     term(term == -inf) = log(realmin); term(term == +inf) = log(realmax);
10    maxiter = 1e4; k = 0;
11    logterms = gammaln((nu+1+k)/2) - gammaln(k+1) - gammaln(vn2) + k*term;
12    fractions = real(exp(logterms)); logsumk = log(fractions);
13    while (k < maxiter)
14      k = k + 1;
15      logterms = gammaln((nu+1+k)/2) - gammaln(k+1) - gammaln(vn2) + k*term(idx);
16      fractions = real(exp(logterms-logsumk(idx)));
17      logsumk(idx) = logsumk(idx) + log1p(fractions);
18      idx(idx) = (abs(fractions) > 1e-4); if (all(idx == false)), break, end
19    end
20    pdfln = real(pdfln+logsumk);
21  end
```

Program Listing 9.2: The direct density approximation to the NCT(v, γ), computing the log density. x is the vector of points at which to evaluate the log density; nu and gam are the degrees of freedom and noncentrality parameters, respectively.

once finished, the storage of the coefficients (on a drive, but particularly in memory) is, on modern personal computers, trivial. For example, with $m = 21$, we choose quantiles corresponding to probabilities $\{0.01, 0.05, 0.10, 0.15, \ldots, 0.95, 0.99\}$. With the completed table, parameter estimation is conducted by finding in the table that pair $\{v, \gamma\}$ for which

$$\sum_{i=1}^{m} w_i(\hat{Q}_i - Q_i)^2 \tag{9.6}$$

is minimized, where $\hat{Q}_i$ refers to the sample counterpart of Q_i, and $(w_1, \ldots, w_m)'$ is the vector of weights obtained using the asymptotic distribution (A.189).

There are several tuning parameters to be chosen: the number m of quantiles, which quantiles, the size (granularity) of the lookup table, and whether to use weighting in (9.6) or not, as it might not yield sufficiently large improvements in accuracy compared to the additional required computation time. Krause and Paolella (2014) considered use of $m = 6, 11, 21, 41$, three grid sizes for k and γ (resulting in tables with 3621, 14,241, and 56,481 entries), and presence and absence of weighting in (9.6). In the most elaborate table, use of $T = 250$ resulted in $\hat{\theta}_Q^{|c}$ (slightly) outperforming the m.l.e. in terms of m.s.e. for k and γ. For speed and memory reasons, in the results that follow, we use the variant with $m = 21$, no weighting in (9.6), and based on the table with 14,241 (k, γ) pairs.

9.3.4 Comparison of NCT Estimators

In addition to the exact NCT density, we have the saddlepoint and direct approximations to the p.d.f., all of which can be used for computing the m.l.e. $\hat{\theta}^{|c}$. The resulting estimators can be compared to $\hat{\theta}_Q^{|c}$ from the quantile-based table lookup estimator.

Regarding estimation time for the three parameters in $\widehat{\theta}^{|c}$, based on a sample size of $T = 500$, use of the m.l.e. with exact density using Matlab's built-in vectorized method (though see footnote 2) and a convergence tolerance of 0.00001 takes (on a 2.5 GHz laptop) 0.89 seconds; m.l.e. with s.p.a. 0.22 seconds; m.l.e. with d.d.a. 0.15 seconds; and the table lookup estimator $\widehat{\theta}_Q^{|c}$ 0.028 seconds. Thus, the latter is about 32 times faster than use of the vectorized NCT density for maximum likelihood estimation, while the s.p.a. and direct approximations offer a substantial decrease in estimation time. As mentioned in Section 9.4.3, the benefit of being able to evaluate the p.d.f. is that non-i.i.d. models can also be estimated with maximum likelihood, whereas, in the stable Paretian case, the McCulloch and empirical c.f.-based estimators are only suitable for i.i.d. data, and likewise for the NCT with the table lookup method.

Consider comparing the small-sample behavior of the various estimators. The top two rows of Figure 9.6 show, for the two sample sizes $T = 500$ and $T = 1000$, kernel density estimates of the estimated parameters based on (the same) 10,000 replications and use of $k = 3, \gamma = -0.4, \mu = 0$ (and known $c = 1$). It is important to mention that the m.l.e.s based on all three methods of computing the p.d.f. estimate only $\theta^{|c}$, that is, the scale is assumed known to be 1 and is not estimated, so that comparisons with the table lookup method are fair. The bottom two rows are similar, but with $k = 6$.

In both parameter settings, and for each of the three parameters, moving from $T = 500$ to $T = 1000$ results in higher precision, as expected. The empirical distributions of the m.l.e. based on the three computation methods are virtually identical, as well as unbiased and Gaussian, as would be expected for these sample sizes. Thus, based on their relative computation times, the d.d.a. is recommended for contexts in which the asymmetry is not particularly extreme, as for financial returns data. The s.p.a. is almost as fast, and can be used for a wider range of γ in the univariate case, while for moderate degrees of freedom (say, 5 and higher), as is typical for conditional models accounting for the time-varying scale term, the table lookup method is fastest and quite reasonable in performance.

Comparing the $k = 3$ to $k = 6$ cases, the shape of the kernel density of $\widehat{v}_Q$ changes from highly discrete to much smoother. This is due to the granularity of the vector of v-values used in the table construction. The m.l.e. is clearly more accurate than the quantile-based table lookup estimator for γ and μ, and, particularly for the $v = 3$ case, $\widehat{\mu}_Q$ is mildly biased.

Remarks

(a) The reader is encouraged to compare the quality (in terms of length and empirical coverage) of confidence intervals resulting from use of the asymptotic distribution of the m.l.e., as well as that of parametric and nonparametric bootstrap c.i.s using the m.l.e. and the table lookup estimator. For example, based on $\widehat{v}_Q$ and the nonparametric bootstrap, for $T = 500$, true $v = 6$ and $\gamma = -0.4$, $B = 1000$ bootstrap replications, and 1000 simulations, the 90% nominal c.i. resulted in an actual coverage of 0.88.

(b) Problem 10.3, in the chapter on indirect inference, asks the reader to develop a different estimator of the four parameters of the location–scale NCT. Its performance can be compared to use of the m.l.e. as in Figure 9.6.   ∎

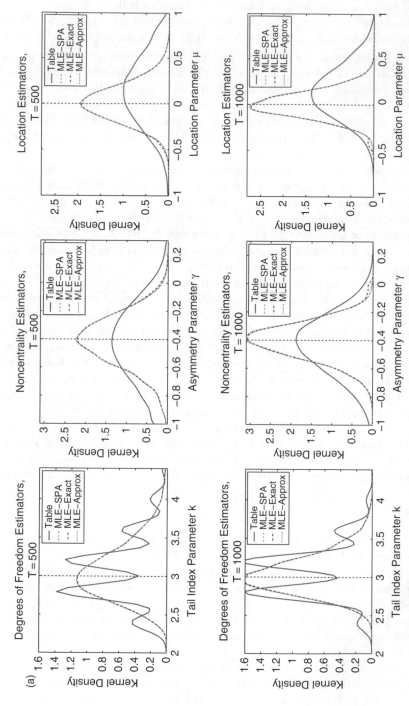

Figure 9.6 *Comparison of four estimators of the NCT distribution based on 10,000 replications, two sample sizes, and two parameter constellations. (a) Correspond to $k = 3$, $\gamma = -0.4$, $\mu = 0$; (b) to $k = 6$, $\gamma = -0.4$, $\mu = 0$. True parameter values are indicated by vertical dashed lines. The m.l.e.-based distributions are optically almost indistinguishable.*

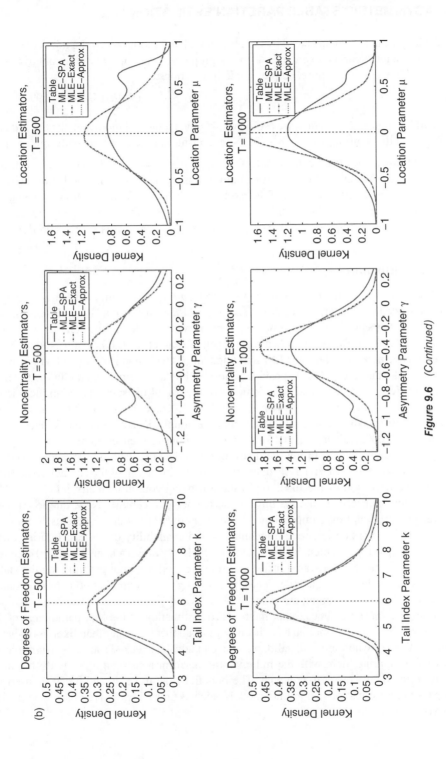

Figure 9.6 *(Continued)*

9.4 ASYMMETRIC STABLE PARETIAN ESTIMATION

> Many of the problems faced by the Hill and related estimators of the tail decay parameter α can be overcome if one is prepared to adopt a more parametric model and assume, for example, stable innovations.
>
> (Robert J. Adler, 1997)

Recall Example 5.6, in which we estimated, via the method of moments, the tail index of the $S_{\alpha,0}(0, 1)$ distribution (for $1 < \alpha < 2$). Now consider the case in which all four parameters are unknown. We observe the sample $X_t \overset{\text{i.i.d.}}{\sim} S_{\alpha,\beta}(\mu, c), t = 1, \ldots, T$, and wish to estimate $\theta = (\alpha, \beta, \mu, c)'$. We present three ways of doing this; two very fast methods not requiring the likelihood (and thus evaluation of the p.d.f.), and the m.l.e. Before doing so, we begin in Section 9.4.1 with some background remarks, and in 9.4.2 with an estimator designed only for estimating the tail index α for $1 < \alpha \le 2$, such that the mean μ is known and β is zero.

9.4.1 Introduction

Interest continues in estimating and using the stable Paretian distribution in a large variety of contexts, since the pioneering work of Mandelbrot (1963) and Fama (1963, 1965a,b), who investigated its theoretical and practical applicability for modeling financial asset returns. McCulloch (1997a), Rachev and Mittnik (2000), Borak et al. (2005), and Nolan (2018) offer extensive accounts of the stable distribution and its wide applicability in finance, while Samorodnitsky and Taqqu (1994) provide a more technical development including the multivariate setting. Extensions and complements to the use of the stable Paretian include the tempered stable (see, for example, Kim et al., 2011; Küchler and Tappe, 2013; and the references therein) and the geometric stable (see, for example, Kozubowski and Rachev, 1999; Kozubowski, 2000; Halvarsson, 2013; and the numerous references therein).

In the 1960s and 1970s, fitting the distribution to data (as well as easy access to daily stock prices) was far from trivial, with direct numerical inversion of the characteristic function and maximization of the likelihood function being computationally infeasible. This changed approximately around the turn of the twenty-first century. Now, due to (i) inexpensive, powerful and ever-improving computer hardware; (ii) developments in numeric optimization, vectorized integration techniques and availability of the fast Fourier transform (FFT); (iii) their implementation in high-level software packages; and (iv) free and commercially available software for fast and reliable methods of approximating the stable p.d.f. (as well as the c.d.f., quantiles, etc.), performing maximum likelihood has become a routine, five-line programming exercise.

In addition, since the 1960s, other methods for estimation of the four parameters of the location–scale stable distribution have been developed that are faster than likelihood-based methods. These are not only still valid, but also can now be checked routinely in simulation exercises comparing them with the m.l.e.; some perform nearly on a par with the m.l.e. and are dramatically faster to compute. The benefit of the m.l.e. is that it can be used for joint parameter estimation in non-i.i.d. models, such as time series and regression, and other contexts.

9.4.2 The Hint Estimator

Mittnik and Paolella (1999) propose a scale-invariant estimator of α designed explicitly for tail index α for location-zero symmetric alpha stable (SαS) data, referred to as the Hill-intercept (or Hint) estimator, and denoted by $\hat{\alpha}_{\text{Hint}}$. It turns out that it exhibits excellent statistical properties, and is based on the Hill estimator. It is valid for (at least) $1 \le \alpha \le 2$ and makes use of the empirical observation that the Hill estimator is very nearly linear as a function of k for stable Paretian data over a large range of $1 \le \alpha \le 2$, tuning parameter k, and sample size T; recall Figure 9.5. It was found that both the intercept and slope of this linear approximation can be used to derive estimates of α. Using the intercept, the estimator takes the simple form

$$\hat{\alpha}_{\text{Hint}} = \hat{\alpha}_{\text{Hint}}(\mathbf{X}) = -0.8110 - 0.3079\,\hat{b} + 2.0278\,\hat{b}^{0.5}, \tag{9.7}$$

where $\hat{b}$ is the intercept in the simple linear regression of $\hat{\alpha}_{\text{Hill}}(\mathbf{k}; \text{abs}(\mathbf{X}))$ on $\mathbf{k}/1000$; the elements of $\mathbf{k}$ are such that $0.2T \le k \le 0.8T$ in steps of $\max\{\lfloor T/100 \rfloor, 1\}$; and the absolute value of the data are used because we assume symmetry.

In addition to its trivial computation, even in samples as small as $T = 50$, the estimator is essentially unbiased for $\alpha \in [1, 2]$ and almost exactly normally distributed. In comparison, the McCulloch (1986) estimator (discussed below in Section 9.4.4, and which is also based on order statistics) exhibits higher sample variance, downward bias as α approaches 2 and, even with sample sizes in excess of 5000, is not normally distributed. Furthermore, for given sample size, the variance of $\hat{\alpha}_{\text{Hint}}$ is practically constant across α, reaching its maximum for $\alpha = 1.5$. For sample sizes $50 < T < 10{,}000$, this is given approximately by

$$\widehat{\text{SE}}(\hat{\alpha}_{\text{Hint}}) \approx 0.0322 - 0.00205T_* + 0.02273T_*^{-1} - 0.0008352T_*^{-2}, \tag{9.8}$$

where $T_* = T/1000$. Finally, simulations show that, for sample sizes $50 < T < 10{,}000$, the m.l.e. performs only slightly better in terms of mean squared error.

To illustrate its performance and small-sample properties, Figure 9.7 shows boxplots across various α and for two sample sizes, based on 100 replications, comparing it to the McCulloch estimator and the m.l.e., where the latter estimates the three parameters location, scale, and α (and not β). We see that the Hint estimator has nearly the same sampling variance as the m.l.e., but, as it is not constrained to lie in $(1, 2)$, its sampling distribution is nearly Gaussian also for α close to 1 or 2, enabling easy construction of accurate confidence intervals even in the case of only $T = 100$. For the larger sample case of $T = 1000$, observe the much larger variation of the m.l.e. compared to Hint in the tails, as seen in the boxplots indicated by plus signs outside of the whiskers.

Remarks

(a) Keep in mind that the McCulloch estimator is for all four parameters of the location–scale asymmetric stable, whereas the Hint estimator (9.7) is only applicable in the symmetric case and known location parameter (though it is scale-invariant). The intrigued reader could attempt to use the left- and right-tail data points separately, and develop an estimator similar to Hint but such that it delivers point estimates (and ideally approximate standard

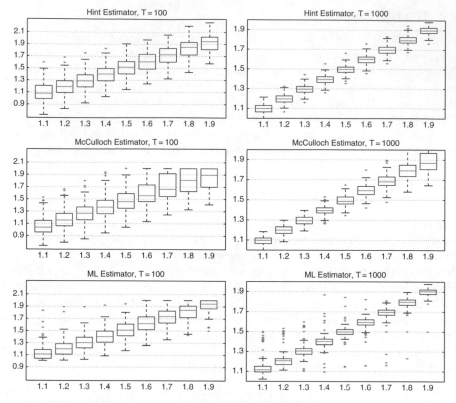

Figure 9.7 *Performance comparison via boxplots of the Hint, McCulloch, and ML estimators of tail index α for i.i.d. symmetric stable Paretian data based on sample size T.*

errors) for both α and asymmetry parameter β, possibly in conjunction with the simple method in Problem 5.2.

(b) As both the McCulloch and Hint estimators of α are based on order statistics, but are very different in nature, we will see that, when the data are *not* stable Paretian, they behave very differently. This will be used to form the basis of a test for SαS data in Section 9.5 below. ∎

9.4.3 Maximum Likelihood Estimation

Once the stable p.d.f. is computationally available, maximum likelihood estimation is straightforward. It can be conducted accurately and efficiently by use of standard gradient/Hessian-based optimization routines, as discussed in Section 4.3. The m.l.e. $\hat{\boldsymbol{\theta}}_{\mathrm{ML}}$ will often be the best choice in terms of accuracy as the sample size grows, owing to its asymptotic properties; see, for example, DuMouchel (1973, 1975). Another benefit of having the density is that the m.l.e. can be computed for the parameters of non-i.i.d. models in essentially the same way as under a Gaussian assumption. Examples include linear and nonlinear regression (see McCulloch, 1998; Tsonias, 2000; Nolan and Revah, 2013; Hallin et al., 2013), ARMA time series (see Mikosch et al., 1995; Adler et al., 1998;

Lin and McLeod, 2008), and GARCH-type models (see Mittnik et al., 2002; Mittnik and Paolella, 2003; Broda et al., 2013; and the references therein).

Sections A.16 and II.1.3.3 discuss how the density can be computed for a set of x-values, using vectorized integration of a real expression for the p.d.f., and via the FFT, respectively. Despite the ease with which we can now compute the m.l.e. of the stable Paretian distribution, there are still some caveats with maximum likelihood estimation in this context. Evaluation of the density function to very high accuracy is still relatively time-consuming compared to other estimation methods, as it requires numeric integration of some sort, or evaluation of infinite sums, or specification of tuning parameters for the FFT that dictate its accuracy.

The fastest way to directly evaluate the density (also in the asymmetric case) with very high accuracy appears to be via the use of the routine `stablepdf` from John Nolan's stable toolbox for Matlab,[4] this being based on numeric integration of a real expression as in Section A.16, programmed in a low-level language optimized for speed. A single estimation with $n = 5000$ data points takes about 4 minutes. Much faster, and with almost the same accuracy (for $1.2 < \alpha \le 2$), is to use the FFT and linear interpolation approach to compute the density, as described in Section II.1.3.3. Using this method, the same run takes about 35 seconds. A much faster way still is to use a spline approximation to the density, as implemented in Nolan's toolbox with routine `stableqkpdf` (the qk for "quick"). Estimation with this routine requires only half a second, and the resulting parameter estimates are the same as those using "exact" density evaluation to about two or three significant digits.

The price to pay for use of the quick routine is the accuracy of the m.l.e., and also the behavior of the distribution of the estimator. Some small-sample distributions of $\hat{\alpha}_{\mathrm{ML}}$, obtained via simulation, using the FFT and the spline approximation from Nolan are shown later in Figure 9.11, in the context of a comparison with other estimators. The accuracy of $\hat{\alpha}_{\mathrm{ML}}$ based on the spline approximation to the p.d.f. is noticeably lower than with use of the FFT, and, as the true value of α moves closer towards 2, its distribution starts to deviate substantially from the expected Gaussian bell shape. This peculiar behavior stems from the lack of smoothness (because of the knots associated with use of splines) and accuracy associated with the piecewise polynomial approximations to the stable distribution. As such, one might wish to use the non-gradient-based heuristic optimization algorithms discussed in Section 4.4, as they are insensitive to discontinuities in the objective function. They do, however, tend to require more function evaluations, thus (mildly) tempering the speed benefit.

9.4.4 The McCulloch Estimator

Perhaps the most remarkable estimator is that of McCulloch (1986). The method involves the use of five particular order statistics and some simple computations involving interpolation from several, relatively small, two-dimensional tables. As such, it is trivial to compute. While the estimator $\hat{\theta}_{\mathrm{McC}}$ is consistent (to the extent that the granularity of the mapping to α and β of (9.9) below increases with the sample size) and applicable for $0.6 < \alpha \le 2$ and $-1 \le \beta \le 1$, its performance is not comparable to the (now fully accessible) m.l.e., and so it is less used nowadays, but was a breakthrough in the early 1980s, given the lack of computing power and software.

---

[4] An internet search with the obvious keywords, and/or Robust Analysis Inc., immediately leads to it.

Let $Q(p) = F_X^{-1}(p)$ denote the quantile function of $X \sim S_{\alpha,\beta}(\mu, \sigma)$ at p, $0 < p < 1$, and define

$$v_\alpha = \frac{Q(0.95) - Q(0.05)}{Q(0.75) - Q(0.25)} \quad \text{and} \quad v_\beta = \frac{Q(0.95) + Q(0.05) - 2Q(0.5)}{Q(0.95) - Q(0.05)}. \tag{9.9}$$

The functions v_α and v_β are invariant to location μ and scale σ. These functions are tabulated for a grid of α- and β-values, and by replacing theoretical quantiles with (linearly adjusted) sample counterparts from the data $\mathbf{X}$, those two functions can be inverted to obtain estimators $\hat{\alpha}_{\text{McC}} = \hat{\alpha}_{\text{McC}}(\mathbf{X})$ and $\hat{\beta}_{\text{McC}} = \hat{\beta}_{\text{McC}}(\mathbf{X})$. Similar clever analysis in McCulloch (1986) yields expressions for the location and scale terms. For computation of the McCulloch estimator, we use the publicly available Matlab implementation `stablecull` from Szymon Borak and Rafał Weron.

To assess its accuracy, we compare the small-sample distribution of the parameter estimates by simulation, using the m.l.e. as a benchmark. Figure 9.8 shows boxplots for two sample sizes and a particular constellation of parameters, $\alpha = 1.5$, $\beta = -0.4$, $\mu = 0$, and $c = 1/2$ (these being relatively typical for daily financial return series). The difference in performance of the two estimators is only slight, with the McCulloch estimator of μ for $n = 50$ having lower variance and less skewness than the m.l.e. Figure 9.9 shows the m.s.e. for $\hat{\alpha}$, over a grid of α-values, for both estimators. We see that the m.s.e. of the m.l.e. is two to three times lower than that of McCulloch, for both sample sizes. Still, it is fair to say that, particularly when accounting for its simplicity and the trivial amount of computer resources required for its computation, the McCulloch estimator compares reasonably well to the m.l.e., based on the results in Figure 9.8.

Example 9.3 *Consider comparing estimators of α for the symmetric case, with $X_i \overset{\text{i.i.d.}}{\sim} S_{\alpha,0}(\mu, c)$, $i = 1, \ldots, n$. The McCulloch estimator $\hat{\theta}_{\text{McC}}$, by design, estimates all four parameters and (without modification of the method) cannot incorporate the additional*

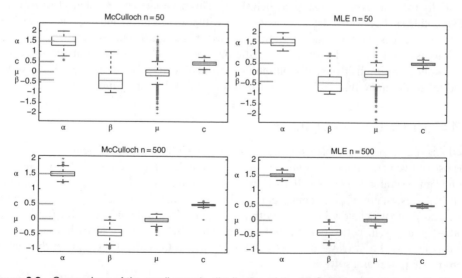

Figure 9.8 *Comparison of the small-sample distribution of the McCulloch and maximum likelihood estimators of the parameters of the $S_{\alpha,\beta}(\mu, c)$ model for an i.i.d. data set with $n = 50$ and $n = 500$, based on values $\alpha = 1.5$, $\beta = -0.4$, $\mu = 0$, and $c = 0.5$.*

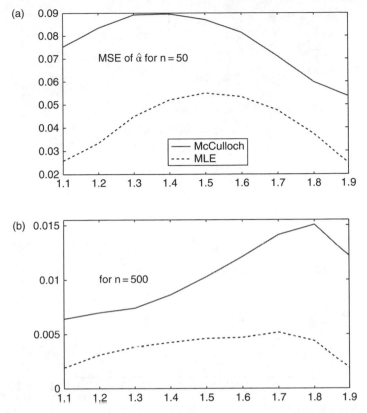

Figure 9.9 *Mean squared error of $\hat{\alpha}$ for the McCulloch estimator (solid) and m.l.e. (dashed) for $n = 50$ (a) and $n = 500$ (b), for the i.i.d. model with n observations and $S_{\alpha,-0.4}(0, 0.5)$ distribution. For both McCulloch and the m.l.e., all four parameters are assumed unknown and are estimated.*

information that β, μ, and c are known. The Hint estimator (9.7), by design, can only estimate α, for $1 \leq \alpha \leq 2$, and relies on β being zero and μ being known (and subtracted from the data). It is, however, scale-invariant, so knowledge of c is not necessary. Observe that the m.l.e. has the best of both worlds: if β, μ, and c are known, then the likelihood can be maximized only over α (with correspondingly higher speed and accuracy, compared to having to estimate all four parameters).

We wish to assess via simulation the small-sample properties of $\hat{\alpha}_{ML}$, $\hat{\alpha}_{McC}$, and $\hat{\alpha}_{MM}$ from Example 5.6. Figure 9.10 shows the results. By comparing Figures 9.9 and 9.10, we can see how much improvement in m.s.e. for the m.l.e. there is when β, μ, and c are known (notice we also changed β from -0.4 to zero, but this has little effect on the m.l.e. $\hat{\alpha}_{ML}$). The decrease is relatively substantial for $n = 50$ but virtually zero for $n = 500$.

Finally, in a comparison with other estimators, the top left panel of Figure 9.11 shows the distribution of $\hat{\alpha}_{McC}$ based on 10,000 replications and three values of α (the remaining parameters used were $\beta = 0$, $\mu = 0$, $c = 1$, and their estimates are not reported). It is noteworthy that it takes about 2 seconds to perform 10,000 evaluations of the McCulloch estimator with $n = 500$.

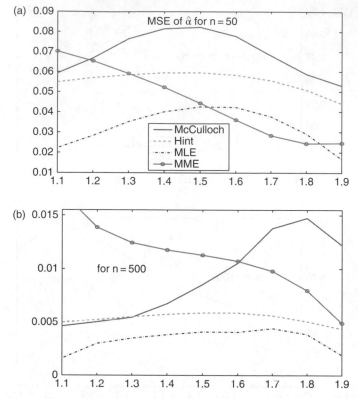

Figure 9.10   *Mean squared error of $\hat{\alpha}$ for the McCulloch estimator (solid), the Hint estimator (9.7) (dashed), the m.l.e. (dash-dotted), and the method of moments estimator $\hat{\alpha}_{MM}$ from Example 5.6 (circles), for $n = 50$ (a) and $n = 500$ (b), for the i.i.d. model with n observations and $S_{\alpha,0}(0,1)$ distribution. For the m.l.e., maximization was done only with respect to α; parameters β, μ and c were fixed at their known values.*

9.4.5   The Empirical Characteristic Function Estimator

The c.f. of the stable distribution is easily evaluated, whereas repeatedly evaluating the density for a large number of points, as associated with maximum likelihood estimation, was for most of the twentieth century computationally infeasible. As such, the use of the empirical characteristic function (e.c.f.) naturally suggested itself. It delivers a point estimator $\hat{\theta}_{CF}$ that minimizes a suitable distance measure between the theoretical and empirical characteristic functions, as discussed in Section 5.2.6.

Several variations of this method for computing the four parameters of the location–scale asymmetric stable Paretian distribution have been proposed, including Press (1972), Paulson et al. (1975), Koutrouvelis (1980), and Kogon and Williams (1998). Simulations in Misiorek and Weron (2004) comparing the latter two show that the performance of both is very similar, with that of Kogon and Williams (1998) performing somewhat better near $\alpha = 1$ and $\beta \neq 0$. Matlab programs for both of them (`stablereg.m` and `stableregkw.m`) are publicly available from Szymon Borak and Rafał Weron. The method of estimation is conceptually straightforward; the reader is encouraged to consult

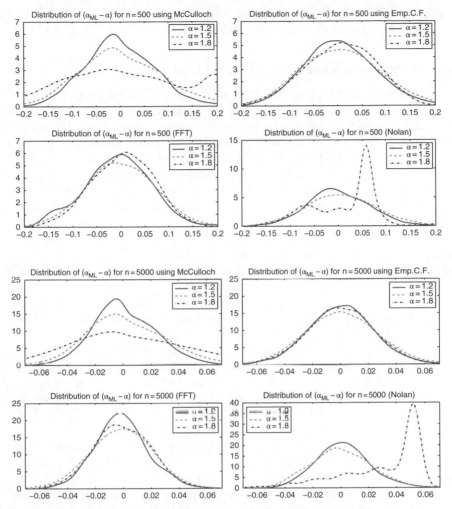

Figure 9.11 *First row: Kernel density, based on 10,000 replications, of the McCulloch estimator (left) and the Kogon and Williams (1998) empirical c.f. estimator (right) of α, for sample size n = 500. **Second row***: Same, for the m.l.e. of α, but based on only 1000 replications, using the FFT method to calculate the stable density (and, thus, the log-likelihood) (left) and the fast spline-approximation routine for the stable density provided in Nolan's toolbox (function* `stableqkpdf`*) (right). **Third and fourth rows**: The bottom four panels are the same as the top four, but using n = 5000 observations.*

Kogon and Williams (1998) to understand their regression-based method in conjunction with the e.c.f., and its relation to previous uses of the e.c.f. for estimation.

The top right panel of Figure 9.11 shows the small-sample performance of the Kogon and Williams (1998) estimator (also available in Nolan's toolbox; call `stablefit(y,3,1)`). This method, like that of McCulloch, is also extraordinarily fast, requiring about 4 seconds for 10,000 estimations. We see that it compares rather well with the m.l.e. computed via the FFT, both of which result in nearly unbiased estimates and sampling distributions resembling the Gaussian, much more so than McCulloch and use of the fast spline approximation to the density provided by Nolan. The reader is encouraged to replicate Figure 9.10 and augment it with the results from the e.c.f. estimator.

9.4.6 Testing for Symmetry in the Stable Model

If the conjecture of a stable distribution (as an i.i.d. model, the error term in a regression model, or the innovations sequence in a time series, etc.) is tenable, then the distribution is almost surely not perfectly symmetric. Indeed, the point $\beta = 0$ has measure zero, and, with enough data, an accurate point estimate of β could be determined. This appears to lead to the conclusion that one should also incorporate β into the model. As a counter to this, recall the discussion in Section 3.3: It stands to reason that, as the sample size increases, other features of the true data-generating process could be modeled, while, for a particular sample size, the goal is to find a suitable model that captures the primary features of the data, without over-fitting. As such, it is plausible to assess whether parameter β is "close enough" to zero to deem it unnecessary, and this can be conducted, for example, by confidence intervals for β, or use of the likelihood ratio test from Section 3.3.2.

The first way we consider for assessing the tenability of $\beta = 0$ is to use the m.l.e. and look at the approximate confidence intervals based on its asymptotic distribution and the approximate standard errors obtained from the numeric optimization of the likelihood. (Higher accuracy is of course obtained via use of the bootstrap, though we are working here with large sample sizes, so that Wald intervals are adequate.)

Recall Example 3.9, which made use of the 1945 daily returns on each of the 30 stocks comprising the Dow Jones Industrial Average (DJIA) stock market index from June 2001 to March 2009. Figure 9.12(a) shows, for each of the 30 DJIA stock return series, the individual 90% confidence intervals for β as straight lines, $\hat{\beta}_{ML} \pm 1.645\widehat{std}(\hat{\beta}_{ML})$, with the stars and triangles representing the 95% and 99% interval endpoints, respectively. To the extent that (i) the model is correctly specified (it most surely is not), (ii) the intervals are accurate (they are reasonably so, given the sample size), and (iii) the 30 time series are independent (they are anything but), we would *expect* that three (10%) of the 90% intervals will not contain zero under the null hypothesis that $\beta = 0$. There are four such series, 5, 6, 8 and 12, with the intervals for 8 and 12 (corresponding to Chevron and Exxon Mobil Corporations, respectively) deviating considerably from zero. As an aside, for Chevron, $\hat{\beta}_{McC} = -0.43$, but the m.l.e. in this case is the less extreme value of $\hat{\beta}_{ML} = -0.35$. For the other 29 stocks, $\hat{\beta}_{McC}$ and $\hat{\beta}_{ML}$ were much closer.

The second way is to look at the likelihood ratio test statistics, computed from the $S_{\alpha,\beta}(\mu, c)$ and $S_{\alpha,0}(\mu, c)$ i.i.d. models. Figure 9.12(b) plots the test statistic for each of the 30 assets, along with lines indicating the 90%, 95%, and 99% asymptotically valid cutoff levels (from the χ^2_1 distribution). The inference is virtually the same as obtained from the confidence intervals, because of the large sample size and the accuracy of the asymptotic distributions employed. In general, with smaller samples, c.i.s should be computed with the bootstrap, and the appropriate small-sample cutoff values of the l.r.t. obtained via simulation.

Finally, we proceed in a different way that sheds some further insight over and above just looking at the individual confidence bands or l.r.t.s. We compare the boxplot of the 30 $\hat{\beta}$-values obtained from the McCulloch estimator to boxplots of the distribution of $\hat{\beta}$ arising when $\beta = 0$, obtained via simulation of i.i.d. stable data, using the McCulloch estimator (because of speed). The first boxplot, on the far left in Figure 9.13, shows the distribution

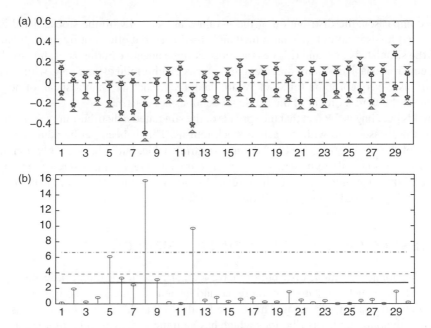

Figure 9.12 (a) The 90%, 95%, and 99% Wald confidence intervals for β for each of the 30 DJIA stock return series, obtained from having estimated the four-parameter location–scale asymmetric stable distribution. (b) Likelihood ratio test statistics and associated 90%, 95%, and 99% cutoff values.

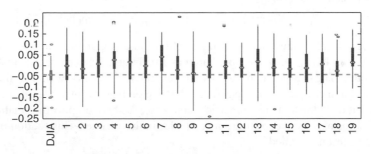

Figure 9.13 The first boxplot represents the 30 estimated stable Paretian asymmetry parameters, $\hat{\beta}$, for the 30 daily return series on the Dow Jones Industrial Average index, using the McCulloch estimator. The dashed line illustrates their median. Each of the other 19 boxplots is based on the 30 values, $\hat{\beta}_{McC}(1), \ldots, \hat{\beta}_{McC}(30)$, the ith of which was estimated from a simulated data set of 2020 i.i.d. $S_{1.6,0}(0,1)$ values.

of the 30 values of $\hat{\beta}$ from the stock returns. (The plot was truncated at the bottom: the "outlier" corresponding to Chevron, with $\hat{\beta} = -0.43$, is not shown.) To determine whether such a boxplot could have arisen by chance if all the stocks were from i.i.d. symmetric stable distributions, we generated a set of 30 series, each of length 2020, from a $S_{1.6,0}(0,1)$ model (the 1.6 being the average $\hat{\alpha}$ from the 30 DJIA stocks) and plotted the boxplot of its 30 estimated β values. This was done 19 times, with the resulting boxplots also shown in Figure 9.13.

By comparison, the distribution of the $\hat{\beta}$-values associated with the stocks listed in the DJIA is (i) less spread out, (ii) has a median value lower, but often not by much, than that of all the simulated ones, and (iii) has the most extreme negative outlier (not shown). Thus, it appears reasonable to claim that the asset returns exhibit mild asymmetry, with a longer left tail, as is very common for stock return data. (Note that we cannot speak of negative skewness: the third moment of the stable model does not exist.)

What is puzzling is the fact that the spreads of the simulated $\hat{\beta}$ distributions are wider than that of the $\hat{\beta}$ associated with the actual stock returns. This is because the data-generating process of the stock returns is not i.i.d. (asymmetric) stable, but rather strongly dominated by a time-varying scale term. The reader is encouraged to repeat this exercise, but replacing the i.i.d. stable model for the stock returns (and the 19 simulated sets of data) with a so-called $S_{\alpha,\beta}$-GARCH model; see Paolella (2016b).

9.5 TESTING THE STABLE PARETIAN DISTRIBUTION

> The nonnormal stable distributions have ordinarily been given very little attention in the field of statistical inference and indeed are surely not so important as the normal distributions. The reason for this is that the normal distributions are the only stable distributions which have a finite variance, and infinite variance does seem inappropriate in many statistical contexts. In fact, it seems to be widely felt that infinite variance is inappropriate in almost any context, because an empirical distribution with literally infinite variance seems almost a contradiction in terms.
>
> (William H. DuMouchel, 1973, p. 949)

Recall our feeble, humiliating, contemptible and reputation-ruining attempt in Section 2.6 to test the symmetric stable Paretian distribution. Its purpose was instructional, and the methodology used there works in general (and works well for the Laplace distribution), but there are better ideas that capitalize on features of the stable distribution. Also, we want to address the asymmetric stable case, further complicating matters. To do so requires some of the concepts introduced so far in this chapter. This section presents an additional five testing methods.

We will see that one method emerges as quite powerful and very easy to compute, not requiring evaluation of the p.d.f., c.d.f., quantiles, or c.f. of the stable distribution, while another, based on a likelihood ratio type of test, appears the most powerful against a range of alternatives, but does require computing the m.l.e. of both the symmetric stable and usual Student's t in the symmetric case, and the asymmetric stable and the NCT if asymmetry is deemed relevant. However, we saw above that, for both the stable and NCT distributions, computing the m.l.e. is no longer a limiting factor, via use of spline approximations, or vectorized density evaluations, or the FFT, for the stable; and via the saddlepoint or the direct density approximation for the NCT.

9.5.1 Test Based on the Empirical Characteristic Function

Given the tractability of the stable c.f., a natural test is to reject the null of stability for a large enough discrepancy between the theoretical and empirical c.f.s. This was pursued in the symmetric stable case by Koutrouvelis and Meintanis (1999), Meintanis (2005), and Matsui and Takemura (2008). For data vector $\mathbf{X} = (X_1, \ldots, X_T)$, the e.c.f. is given by

$$\phi_T(t; \mathbf{X}) = \phi_T(t; \mathbf{X}; \hat{\mu}, \hat{\sigma}) = \frac{1}{T} \sum_{j=1}^{T} \exp(itY_j), \quad Y_j = \frac{X_j - \hat{\mu}}{\hat{\sigma}}, \tag{9.10}$$

where consistent estimators of the location μ and scale σ of the symmetric stable are used. As the c.f. of each Y_j, with estimated α, is $\phi(t; \hat{\alpha}) = \exp(-|t|^{\hat{\alpha}})$, a natural test statistic for assessing the validity of the assumed distribution is

$$D_{T,\kappa} = D_{T,\kappa}(\mathbf{X}) = T \int_{-\infty}^{\infty} |\phi_T(t; \mathbf{X}) - \phi(t; \hat{\alpha})|^2 w(t) \, dt, \quad w(t) = \exp(-\kappa|t|), \quad \kappa > 0. \tag{9.11}$$

Matsui and Takemura (2008) derive the asymptotic distribution of (9.11) under the null and, via simulation for power assessment, recommend choosing $\kappa = 5$.

Delivering the test statistic is clearly easy, but it is meaningless without reference to the null distribution. Via simulation for a set of sample sizes, one could obtain the appropriate cutoff values for a test at the 10%, 5%, and 1% levels, as was done in Section 2.4 for the KD and AD statistics. What is more useful is to (quickly) deliver an accurate approximation to the p-value. Observe that this is more general than just the binary result at the three usual test levels, and it allows for the construction of combined tests; see Section 9.5.4 below. Thus, instead of modeling the three relevant quantiles of the distribution under the null, our goal is to model *the entire density function*, so that p-values can be elicited.

This is done by fitting a flexible distribution to values of the test statistic obtained by simulation, for a variety of sample sizes T and tail index values α, and storing the fitted *parameters* of the chosen flexible distribution. For an actual data set to be tested, the p-value is then computed for each stored distribution in the table (this being fast, because the distribution is chosen such that, in addition to being flexible, its c.d.f. is quickly evaluated), and then interpolation is used based on the actual sample size and estimate of α. This technique is general, and is used for the some of the other tests developed below. As such, we provide some details on how it is accomplished.

For test (9.11), we proceed as follows:

(i) Simulate $n = 10,000$ realizations of (9.11) under the null, for each element in a grid of sample sizes T and tail indexes α, for which we use $T = 250, 500, 750, 1000, 1500, 2000, \ldots, 10,000$, and $\alpha = 1.05, 1.1, 1.2, \ldots, 1.9, 1.99$. We use the McCulloch (1986) estimator (being consistent and instantly computed) to obtain $\hat{\mu}$ and $\hat{\sigma}$, as required for the location–scale transformation in (9.11), though the e.c.f. estimator from Section 9.4.5 can also be used.

(ii) Fit a flexible parametric distribution (such that the c.d.f. is easily evaluated) to the (log of) the n values for each (T, α) combination in step 1, and store the estimated parameters.

For (ii), we use a two-component mixture of generalized asymmetric t distributions (MixGAt), recalling the GAt as discussed in Example 4.9. As its c.d.f. admits a closed-form expression, the p-value corresponding to a particular test statistic can be quickly computed. With five parameters (including location and scale), the GAt is a rather flexible distribution.

However, it was found to fit the simulated data not as well as desired. Greater accuracy can be obtained by using a two-component mixture of GAt, with mixing parameters $0 < \lambda_1 < 1$ and $\lambda_2 = 1 - \lambda_1$. This 11-parameter construction is extraordinarily flexible, and was found to fit the simulated test statistics essentially perfectly, as confirmed by kernel density plots. Its p.d.f. and c.d.f. are just weighted sums of GAt p.d.f.s and c.d.f.s respectively, so that evaluation of the MixGAt c.d.f. is no more involved than that of the GAt.

Observe that steps (i) and (ii) of this simulation exercise need to be conducted only once. Then, for a particular data set **X** of interest:

(iii) Compute $\mathbf{Y} = (Y_1, \ldots, Y_T)$, where $Y_j = (X_j - \hat{\mu})/\hat{\sigma}$, and $\hat{\mu}$ and $\hat{\sigma}$ are based on the McCulloch estimator.

(iv) Compute the statistic $D_{T,\kappa}(\mathbf{X})$ from (9.11), using $\hat{\alpha}_{\text{Hint}}$ as the estimator for the tail index, given its higher accuracy than the McCulloch estimator and its low computation cost.

(v) Build a matrix of p-values, say $\mathbf{P}_D$, based on the parametric MixGAt approximations, for each entry in the (T, α) grid, where each p-value is the right tail area from $D_{T,\kappa}(\mathbf{X})$ (recall that we wish to reject for large values of (9.11)) of the fitted MixGAt distribution.

(vi) Use bivariate interpolation (as implemented in Matlab's `interp2` function, for example) with table $\mathbf{P}_D$ to deliver the p-value, based on the actual length T and estimated tail index $\hat{\alpha}_{\text{Hint}}$ of the data set of interest. (Of course, to enhance speed, if T coincides with one of the entries used in the grid, only that row of the grid is computed, and univariate interpolation is used.)

The actual sizes of the test were checked, for the usual three nominal sizes of 10%, 5%, and 1%, first using the true value of α instead of the estimator $\hat{\alpha}_{\text{Hint}}$, and sample sizes that were used in the computation of the grid (and using a different set of seed values for the generation of the stable data sets), so that interpolation of p-values was not required. As expected, the actual sizes were very close to their nominal counterparts, confirming that the method of construction of the p-value table, via use of the parametric approximation MixGAt, is effective.

Of course, of real interest are the actual sizes when an estimator is used for α, in which case interpolation into the matrix of p-values is required. We find that the results are not satisfactory for small sample sizes (e.g., $T = 250$) and values of $\alpha < 1.6$. The reason is that the tail cutoff values corresponding to the three sizes change nonlinearly as a function of α, so that, even though the distribution of $\hat{\alpha}_{\text{Hint}}$ is symmetric about the true α, the actual size will not agree with the nominal size. This problem can be addressed by using a much denser grid of α-values; however, such a scheme would entail an enormous amount of simulation. We propose a second method.

Let $P_{T,\alpha}(\hat{\alpha})$ denote the random variable associated with the p-value of SαS data $\mathbf{X} = (X_1, \ldots, X_T)$ with tail index α, based on the procedure given by steps (iii)–(vi) above, using the estimator $\hat{\alpha} = \hat{\alpha}_{\text{Hint}} \in [1, 2]$. Then, for a fixed T and nominal cutoff probability $c_{\text{nom}} \in \{0.10, 0.05, 0.01\}$, the actual size c_{act} as a function of c_{nom} is

$$c_{\text{act}}(c_{\text{nom}}) = \Pr(P_{T,\alpha}(\hat{\alpha}) \leq c_{\text{nom}}) = \mathbb{E}[\mathbb{I}\{P_{T,\alpha}(\hat{\alpha}) \leq c_{\text{nom}}\}] = \int_1^2 \mathbb{I}\{P_{T,\alpha}(a) \leq c_{\text{nom}}\} f_{\hat{\alpha}}(a)\, da,$$

$$(9.12)$$

where $\mathbb{I}\{A\}$ is the indicator function of event A and $f_{\hat{\alpha}}$ denotes the density of $\hat{\alpha}_{\text{Hint}}$. As an example, imagine that, for a fixed T and true $\alpha = 1.55$, $\hat{\alpha}_{\text{Hint}}$ only takes on values in $(1.45, 1.64)$, and that nearest-neighbor interpolation is used to determine the p-value from the matrix of computed p-values $\mathbf{P}_D$. Thus, given the symmetry of $f_{\hat{\alpha}}$, only the tabulated p-values corresponding to $\alpha = 1.5$ and $\alpha = 1.6$ are used, with equal probability for a given data set, so that the resulting p-value is, on average, the average of those two values. The nonlinearity of the tabulated p-values as a function of α is what causes $c_{\text{act}}(c_{\text{nom}}) \neq c_{\text{nom}}$. The argument of course still holds for $\hat{\alpha}_{\text{Hint}} \in [1, 2]$ and use of other methods of interpolation, such as linear or cubic spline.

To resolve this, observe that the mean value theorem for integrals (see, for example, (I.A.58)) implies that there exists a constant $\alpha_k = \alpha_k(T, \alpha, c_{\text{nom}}) \in [1, 2]$ such that the integral in (9.12) equals

$$\mathbb{I}\{P_{T,\alpha}(\alpha_k) \leq c_{\text{nom}}\} \int_1^2 f_{\hat{\alpha}}(a)\, \mathrm{d}a = \mathbb{I}\{P_{T,\alpha}(\alpha_k) \leq c_{\text{nom}}\}, \tag{9.13}$$

i.e., $\exists \alpha_k$ such that $c_{\text{act}} = \Pr(P_{T,\alpha}(\alpha_k) \leq c_{\text{nom}}) = c_{\text{nom}}$.

To operationalize this, we let $m = m(T, \alpha, c_{\text{nom}}) = \alpha_k(T, \alpha, c_{\text{nom}})/\hat{\alpha}_{\text{Hint}}$, and find via simulation the values of m for a small set of T and a grid of α-values. For interpolation into the vector of p-values, we use linear interpolation (as opposed to nearest neighbor or cubic spline), because it has the effect of making the values of m closest to unity. The values obtained, based on 10,000 replications, are plotted in Figure 9.14. We see that almost all are below 1.0, very much so for $\alpha < 1.6$, confirming the nonlinearity of the p-values as α changes. As expected, as the sample size moves from $T = 250$ to 1000, the values of m are all closer to unity.

Having determined this mapping, for a given data set $\mathbf{X}$, linear interpolation into the vector of values of m is used, based on $\hat{\alpha}_{\text{Hint}}(\mathbf{X})$, for each of the three c_{nom}-values. The value of α used to index the table of p-values is taken to be

$$\alpha_k = \alpha_k(\mathbf{X}, c_{\text{nom}}) = m(T, \hat{\alpha}_{\text{Hint}}(\mathbf{X}), c_{\text{nom}}) \times \hat{\alpha}_{\text{Hint}}(\mathbf{X}), \tag{9.14}$$

for each of the three values of c_{nom}. Observe that, once the values of m are computed, this method entails no more computation time than the original test and the interpolation to get the p-value. We expect the use of this method to be helpful for values $1 < \alpha < 1.6$; while, given that the true m is near unity for $1.6 < \alpha < 2$, and the fact that the estimator of α is stochastic (in this case, $\hat{\alpha}_{\text{Hint}}$), performance could worsen as α moves towards 2.

This is precisely the case, as demonstrated in Table 9.1, which shows the actual levels corresponding to levels of 5% and 1% (similar conclusions hold for the less interesting case of 10% and are not shown), using the procedure developed above for $T = 250$ and $T = 1000$ (shown in italics) and otherwise, using the standard method, without augmentation by factor m. The results become very good as the sample size T increases, because $\hat{\alpha}_{\text{Hint}}$ is consistent for α and the use of the MixGAt approximation (for any sample size and α) is very accurate.

After all this work, the power of this test turns out to be rather disappointing, as will be shown in Section 9.5.6 below, in comparison to the other tests developed here.

9.5.2 Summability Test and Modification

Recall from (A.298) that if the data are i.i.d. stable, then the value of the tail index α should not change when the data are summed. In contrast, for many alternatives outside the stable

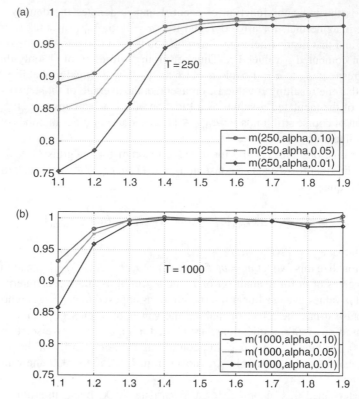

Figure 9.14   *Values of $m(T, \alpha, c_{nom})$, as a function of stable tail index α, based on 10,000 replications, for the $D_{T,\kappa}(\mathbf{X})$ test (9.11) for the two sample sizes $T = 250$ (a) and $T = 1000$ (b).*

TABLE 9.1   Actual sizes of the nominal 5% and 1% $D_{T,\kappa}(\mathbf{X})$ test (9.11) for data of length T with tail index α. The entries in italics (for $T = 250$ and 1000) make use of the adjustment procedure via multiplicative factor $m(T, \alpha, c_{nom})$. The remaining entries do not use the adjustment procedure

Nominal			0.05					0.01		
$T \backslash \alpha$	1.20	1.35	1.55	1.75	1.90	1.20	1.35	1.55	1.75	1.90
250	*0.086*	*0.062*	*0.049*	*0.043*	*0.017*	*0.0373*	*0.0184*	*0.0096*	*0.0070*	*0.0024*
250	0.129	0.093	0.060	0.052	0.050	0.0643	0.0290	0.0136	0.0136	0.0143
1000	*0.059*	*0.048*	*0.047*	*0.050*	*0.022*	*0.0181*	*0.0082*	*0.0092*	*0.0096*	*0.0030*
1000	0.075	0.050	0.048	0.054	0.051	0.0267	0.0095	0.0096	0.0120	0.0109
2500	0.050	0.043	0.049	0.051	0.051	0.0103	0.0075	0.0076	0.0108	0.0114
5000	0.045	0.042	0.049	0.052	0.052	0.0087	0.0065	0.0074	0.0103	0.0112

class, the classic central limit theorem (A.160) will be at work, and the values of α will tend to increase towards 2 as the data are summed. These facts can be used to form a test, as considered informally by Fama and Roll (1971), Lau and Lau (1993), and formally by Paolella (2001), with the latter yielding a correct-sized test and delivering the binary results at the three usual test significance levels. It was designed for i.i.d. location-zero SαS data with tail index $1 < \alpha \leq 2$, but is scale-invariant. As such, an estimate of μ is required. We suggest use of the McCulloch (1986) estimator, for speed and simplicity reasons, and then proceeding with $\mathbf{X} = \mathbf{Y} - \hat{\mu}$. The estimator $\hat{\alpha}_{\text{Hint}}$ is used in place of the theoretical value of α required below.

Denote by s the level of aggregation applied to data vector $\mathbf{X} = (X_1, \ldots, X_T)$ – that is, for $s = 1$, the entire data vector is used; for $s = 2$, the data are reduced to $\mathbf{X}_{(2)} = (X_1 + X_2, X_3 + X_4, X_5 + X_6, \ldots)$; for $s = 3$, $\mathbf{X}_{(3)} = (X_1 + X_2 + X_3, X_4 + X_5 + X_6, \ldots)$, etc. – and let $\hat{\alpha}(s) = \hat{\alpha}_{\text{Hint}}(s)$ denote the estimate of α based on the Hint estimator for the given level of aggregation s. For sample size T, the aggregation values used are $s = 1, 2, \ldots, [T/100]$, so that the last $\hat{\alpha}(s)$ is based on at least 100 observations. Under the null hypothesis of (i.i.d., symmetric) stable data the $\hat{\alpha}(s)$ should be constant, while for non-stable i.i.d. data they are expected to increase.

Figure 9.15 illustrates the graphical output of the method, using i.i.d. stable and Student's t data. For the former, the point estimates of α do not vary much with respect to the aggregation value s, while those for the latter tend to increase.

Thus, we consider estimating a simple linear trend model, as a linear regression of $\hat{\alpha}(s)$ on a constant and s, with the slope coefficient denoted $\hat{b}$. By using the Hill-intercept estimator with its desirable properties discussed in Section 9.4.2, each $\hat{\alpha}(s)$ can be treated

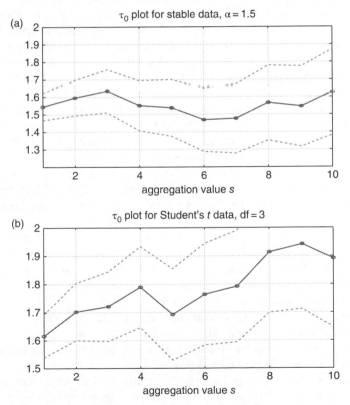

Figure 9.15 *Plots associated with the τ_0 summability test, based on $T = 2000$, using (a) symmetric stable data with $\alpha = 1.5$, (b) Student's t with three degrees of freedom.*

as a realization from a normal distribution with known variance, so that weighted least squares can be used to compute $\hat{b}$, where the weights are inversely proportional to the standard error of $\hat{\alpha}(s)$, in (9.8), delivered with the Hill-intercept estimator. Consider using the *studentized* test statistic given by $\tau_0 = \tau_0(\mathbf{X}) = \hat{b}/\widehat{\mathrm{SE}}(b)$. Paolella (2001) determined simple functions of T and α to compute the cutoff values for sizes 0.01, 0.05, and 0.10, associated with the right tail of its distribution under the null hypothesis.

This was augmented in Paolella (2016a) in two ways. The first issue to realize is that the test statistic τ_0 (and thus the hypothesis test outcome) is not invariant to permutations of the data (though observe that $\hat{\alpha}_{\mathrm{Hint}}$ is invariant). As the data are purported to be i.i.d., the ordering should not play a role. To alleviate this issue, we take

$$\tau_B = \tau_B(\mathbf{X}) = B^{-1} \sum_{i=1}^{B} \tau_0(\mathbf{X}^{[i]}), \tag{9.15}$$

where $\mathbf{X}^{[i]}$ denotes a random permutation of the data such that all the X_i appear once (i.e., sampling without replacement). Notice that, unless all theoretically possible permutations are used (or, taking $B = \infty$ if they are randomly drawn), the procedure outlined will still return different test statistics for the same data set (unless the set of seed values for the random permutations is held constant at some arbitrary choice). While this feature is still undesirable, it cannot be avoided with finite B and random permutations.

The second augmentation is to deliver an approximate p-value of the τ_B test statistic, instead of only the test outcomes at the usual three levels of significance. This is now described in detail, using $B = 20$. Determination of the p-value of the τ_{20} test is done in the same way as for the $D_{T,\kappa}(\mathbf{X})$ test (9.11) via the MixGAt approximation to the distribution of the test statistic under the null, for a grid of values over various T and α. (As this test is only applicable for sample sizes realistically larger than $T = 500$, there is less need to use the multiplicative adjustment technique from (9.14), and thus we omit it.) The statistic $\tau_B(\mathbf{X})$ is then computed from location-adjusted $\mathbf{X}$ (and B can be chosen much larger for a single data set to ensure a nearly unique test statistic), and, for each element in the (T, α) grid, the associated p-value is computed, where it is given as the right tail area greater than the test statistic (i.e., we use a one-sided test, rejecting for large τ_B, as we expect the slope coefficient $\hat{b}$ to be positive under the alternative hypothesis). Then, in the same way as for (9.11), bivariate interpolation is used to deliver the p-value, according to the actual sample size T and $\hat{\alpha}_{\mathrm{Hint}}(\mathbf{X})$.

With respect to the null distribution, it was found that, as $\alpha \to 2$, the shape of the p.d.f. (from kernel density plots based on 10,000 replications) becomes very nonstandard, exhibiting spiky behavior, bimodality, and very extreme asymmetry, so that the quality of the MixGAt fits begin to deteriorate. This explains why the actual size of the test, investigated next, is poor for small sample sizes and $\alpha = 1.9$.

Analogously to the test (9.11), we wish to inspect the actual size of the $\tau_B(\mathbf{X})$ test (9.15) for a set of T and α. For each replication, we use $B = 20$. Table 9.2 shows the results for nominal sizes 5% and 1% (numbers for 10% were computed, and were qualitatively similar to the 5% and 1% cases). The actual and nominal sizes coincide very well for values of α close to 1.5 and/or T large. This is due to the limitations of the interpolation method when $\hat{\alpha}$ is near the border of $(1, 2)$, exacerbated for smaller T because of the higher variance of the estimator. The power of the test turns out to be quite good for a variety of (but not all) alternatives. This will be illustrated below in Section 9.5.6.

TABLE 9.2   Actual sizes of the nominal 5% and 1% $\tau_{20}(X)$ test (9.15) for data of length T with tail index α

Nominal	0.05					0.01				
$T \backslash \alpha$	1.20	1.35	1.55	1.75	1.90	1.20	1.35	1.55	1.75	1.90
500	0.071	0.060	0.054	0.046	0.033	0.0238	0.0132	0.0100	0.0046	0.0040
750	0.060	0.056	0.052	0.044	0.036	0.0132	0.0095	0.0104	0.0089	0.0070
1000	0.060	0.055	0.053	0.045	0.035	0.0123	0.0111	0.0102	0.0074	0.0028
1500	0.058	0.055	0.055	0.047	0.033	0.0104	0.0098	0.0096	0.0081	0.0025
2500	0.055	0.050	0.051	0.053	0.039	0.0112	0.0101	0.0109	0.0083	0.0032
5000	0.053	0.051	0.050	0.054	0.050	0.0102	0.0091	0.0099	0.0103	0.0095
10000	0.048	0.048	0.052	0.053	0.048	0.0099	0.0115	0.0108	0.0096	0.0064

9.5.3   ALHADI: The α-Hat Discrepancy Test

The motivation for this test statistic comes from the structure of the McCulloch (1986) and Hint estimators for α. Observe that the McCulloch estimator uses only the order statistics associated with quantiles 0.95, 0.05, 0.75, and 0.25, whereas Hint uses a much larger set, in such a way as to measure the tail behavior, via the Hill estimator. One might speculate that, when applying this finding to data that are not stable Paretian, these estimates could differ substantially. This is indeed the case for several alternatives of interest. This idea of assessing the appropriateness of the stable assumption by comparing different consistent estimators of the tail index is not new; its possibility was mentioned by, for example, Nolan (1999, p. 9). Our goal is to formalize the process and develop a test that is (as nearly as possible) size-correct and, with it, a quickly computed p-value.

Observe that this idea is potentially generally applicable for distribution testing when there are different (ideally consistent) estimators for one or more of the shape parameters (and not location and scale), as opposed to the summability method from Section 9.5.2, which accesses a specific characteristic of the stable distribution.

Recall Figure 9.7, comparing the small-sample distribution of the Hint, McCulloch, and maximum likelihood estimators of tail index α for i.i.d. symmetric stable Paretian data. Figure 9.16 is similar, but using four values of α and the two sample sizes $T = 250$ and $T = 2500$ (and based on 1000 replications). The symmetry of $\hat{\alpha}_{\text{Hint}}$, as well as its unbiasedness and a variance nearly the same as that of the m.l.e., is apparent. It is to be compared with Figure 9.17, which shows the same plot, but using Student's t data with v degrees of freedom, $v = 1.5, 2.5, 3.5, 4.5$, for the sample size $T = 25,000$.[5] The discrepancy between $\hat{\alpha}_{\text{Hint}}$ and $\hat{\alpha}_{\text{McC}}$ is readily apparent, and indicates that their asymptotic values converge to constants – and these constants are different when the data are Student's t, for $1 < v < \infty$ (recall that $v = 1$ corresponds to stable with $\alpha = 1$, while as $v \to \infty$, it approaches the normal, or stable with $\alpha = 2$).

The behavior of the m.l.e. is noteworthy. Not only is $\hat{\alpha}_{\text{McC}} < \hat{\alpha}_{\text{ML}} < \hat{\alpha}_{\text{Hint}}$ as $T \to \infty$, but as $v \to 1$, $\hat{\alpha}_{\text{McC}}$ and $\hat{\alpha}_{\text{ML}}$ are close, and appear to converge as $T \to \infty$, while as $v \to \infty$, $\hat{\alpha}_{\text{ML}}$ "moves towards" $\hat{\alpha}_{\text{Hint}}$, and they appear to converge. Of course, all three estimators will converge as $v \to 1$ (to $\alpha = 1$) and as $v \to \infty$ (to $\alpha = 2$), though we see that, for $1 <$

---

[5] This exercise was conducted for various sample sizes and the results are qualitatively exactly the same. We use this rather extreme sample size just to make the point optically very clear.

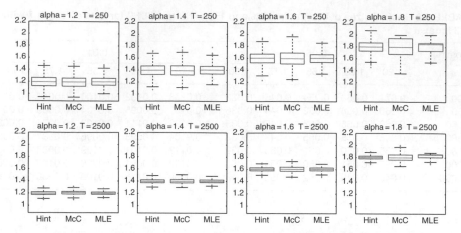

Figure 9.16 Boxplots of $\hat{\alpha}_{\text{Hint}}$, $\hat{\alpha}_{\text{McC}}$, and $\hat{\alpha}_{\text{ML}}$ based on 1000 simulated symmetric stable data sets, each of length T and for tail index α.

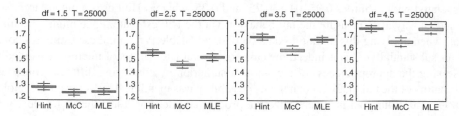

Figure 9.17 Similar to Figure 9.16 but based on simulated Student's t data with υ degrees of freedom (here denoted by df), and using $T = 25{,}000$. MLE refers to the maximum likelihood estimator of stable tail index α.

$\upsilon < \infty$, the behavior of $\hat{\alpha}_{\text{McC}}$ and $\hat{\alpha}_{\text{Hint}}$ always differ. (The reader is encouraged to repeat this exercise, also using the empirical c.f. estimator for α.)

Thus, given that the discrepancy is largest between the Hint and McCulloch estimators when using Student's t data (and other alternatives; see below), the test statistic given by

$$A = A(\mathbf{X}) = \hat{\alpha}_{\text{Hint}}(\mathbf{X}) - \hat{\alpha}_{\text{McC}}(\mathbf{X}) \tag{9.16}$$

suggests itself for testing the i.i.d. SαS null hypothesis. As in Paolella (2016a), we call this the α-hat discrepancy test (ALHADI) or the stable Paretian order statistics test (SPORT). Under the null, $\mathbb{E}[A(\mathbf{X})] \approx 0$, recalling that $\hat{\alpha}_{\text{Hint}}$ and $\hat{\alpha}_{\text{McC}}$ are both consistent, with the former being practically unbiased even in small samples. We note that, while it is almost surely the case that there exists some function of the data $\mathbf{X}$ that yields a test with higher power than ALHADI for a range of viable alternatives, it is far from obvious how to find such a statistic. For a *specific* alternative, the likelihood ratio can be used; see Section 9.5.5 below.

For the Student's t alternative, we see from Figure 9.17 that $\mathbb{E}[A(\mathbf{X})] > 0$. This will not be the case for all alternatives. We consider four further distributions that are suitable for modeling leptokurtic data, and which exhibit different tail behaviors. The first is the two-component Gaussian mixture from (5.1), that is, for $X \sim \text{MixN}(\mu_1, \sigma_1, \mu_2, \sigma_2, \lambda)$, the p.d.f. of X is $f_X(x) = \lambda f_N(x; \mu_1, \sigma_1) + (1 - \lambda)f_N(x; \mu_2, \sigma_2)$, where $0 < \lambda < 1$ and f_N denotes the normal p.d.f. The use of the mixed normal for modeling financial asset returns has a substantial history; see Haas et al. (2004a, 2013), Paolella (2015a), and the references therein. The next is a two-component SαS mixture: for $X \sim \text{MixS}(\alpha_1, \mu_1, \sigma_1, \alpha_2, \mu_2, \sigma_2, \lambda)$ its p.d.f. is $f_X(x) = \lambda f_S(x; \alpha_1, \mu_1, \sigma_1) + (1 - \lambda)f_S(x; \alpha_2, \mu_2, \sigma_2)$, where $0 < \lambda < 1$ and f_S denotes the SαS p.d.f. See Broda et al. (2013) and the references therein for the use of stable mixtures for modeling asset returns. Clearly, the mixed normal has short tails, while the mixed stable has power tails.

The third case we consider is the normal inverse Gaussian (NIG) distribution, as detailed in Section II.9.5.2.7. Its p.d.f. (allowing for the unfortunate convention of using α and β as shape parameters, as in the stable Paretian) is

$$f_{\text{NIG}}(x; \alpha, \beta, \delta, \mu) = e^{\delta \sqrt{\alpha^2 - \beta^2}} \frac{\alpha \delta}{\pi \sqrt{\delta^2 + (x - \mu)^2}} K_1(\alpha \sqrt{\delta^2 + (x - \mu)^2}) e^{\beta(x - \mu)}, \qquad (9.17)$$

for $\alpha > 0$, $\beta \in (-\alpha, \alpha)$, $\delta > 0$ and $\mu \in \mathbb{R}$, where K_ν denotes the modified Bessel function of the third kind with index ν, given in (A.19). The limiting tail behavior is $f_{\text{NIG}}(x; \alpha, \beta) \propto |x|^{-3/2} e^{(\mp \alpha + \beta)x}$, as $x \to \pm\infty$, which is referred to as "semi-heavy", as discussed in Chapter II.9. See Broda and Paolella (2009) and the references therein for its use in modeling financial returns. As $\alpha \to 0$, and recalling $0 \le |\beta| < \alpha$, the tail behavior approaches $(2\pi)^{-1/2} x^{-3/2}$. This limiting case is the Lévy distribution (A.305), which coincides with the stable distribution for (now referring to the stable Paretian parameters) $\alpha = 1/2$ and $\beta = 1$.

The fourth case is the GAt distribution. Clearly, from Figure 9.17, for the special case of Student's t, the test has power, but we are interested in parameter constellations that best mimic the stable Paretian. To determine such a set of parameters, we take a large simulated sample of SαS data with $\alpha = 1.5$, and estimate the GAt; this yielded $\hat{d} = 2.9073$, $\hat{\nu} = 0.6429$, $\hat{c} = 1.705$, and $\theta = 1$, $\mu = 0$. The implied supremum of the maximally existing moment is $\hat{\nu}\hat{d} = 1.87$. Simulation of GAt realizations is conducted via the probability integral transform and inverting the c.d.f., as output in program Listing 4.8.

Figure 9.18 shows the behavior of the three estimators under these four alternatives, based on $T = 25{,}000$. For the mixed normal, we use $\mu_1 = \mu_2 = 0$, $\sigma_1 = 1$, $\sigma_2 = 4$, and $\lambda = 1/2$; for the mixed stable, $\mu_1 = \mu_2 = 0$, $\sigma_1 = \sigma_2 = 1$, $\alpha_1 = 1.2$, $\alpha_2 = 3 - \alpha_1 = 1.8$, and $\lambda = 1/2$; and for the NIG, $\alpha = 0.6$, $\beta = 0$, $\mu = 0$ and $\delta = 1$. We again see strong separation between $\hat{\alpha}_{\text{Hint}}$ and $\hat{\alpha}_{\text{McC}}$ for the mixed normal and NIG cases, while the separation in the mixed stable case depends on the value of α_1: as α_1 moves from 1.5 to 1.0, the separation increases.

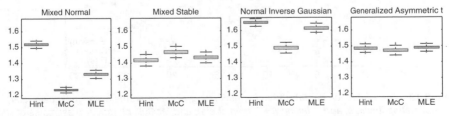

Figure 9.18 Boxplots of $\hat{\alpha}_{\text{Hint}}$, $\hat{\alpha}_{\text{McC}}$, and $\hat{\alpha}_{\text{ML}}$ under four non-stable-Paretian distributional assumptions, based on 1000 replications, each of length $T = 25{,}000$.

For the GА*t* alternative, it is remarkable that the estimated values of stable tail index α are close to 1.5 (recall how the GА*t* parameters were selected), and the separation is rather weak. In sample sizes of 5000 or less, the separation is barely apparent. As such, the ALHADI test (9.16) will have virtually no power against some GА*t* alternatives. We will see below that only the τ_{20} and likelihood ratio tests have some power, though relatively low, compared to other alternatives, when fixing $d = 2.9073$ and varying the parameter v (the location and scale parameters do not play a role; and we set θ to its value corresponding to symmetry, $\theta = 1$). Thus, the GА*t* with $d \approx 2.9$ serves as a challenging alternative distribution for future developments of tests for stability.

For the mixed stable alternative, $\mathbb{E}[\hat{\alpha}_{\text{Hint}}(\mathbf{X})] < \mathbb{E}[\hat{\alpha}_{\text{McC}}(\mathbf{X})]$. Our test will return a *p*-value, so that one can reject for either case, and the user can decide between a one- and two-sided test. In our comparisons later, we will always use a one-sided test, rejecting for large values of (9.16), and not consider the mixed stable case, though we observe that (9.16) does have some power against this alternative.

Example 9.4   *For each of the 30 DJIA daily stock return series as were used in Example 9.2, Figure 9.19 shows four different estimates of stable tail index α: Hint; the m.l.e. estimating all four parameters as unknown; the m.l.e. estimating just α, μ, and c, taking $\beta = 0$; and McCulloch . The lines indicate the $\pm 2\widehat{\text{std}}(\hat{\alpha}_{\text{ML}})$ interval using the four-parameter m.l.e.*

Observe that the two m.l.e. values, one assuming symmetry (taking $\beta = 0$) and the other estimating β along with the other three parameters, are nearly identical. While this is not evidence that $\beta = 0$ per se (recall the analysis in Section 9.4.6), it is useful because it indicates that the amount of asymmetry in each series is small enough such that the point estimate of α under the (possibly misspecified symmetric) model is virtually unaffected. For most of the 30 series, the estimated standard errors of $\hat{\alpha}$ for the two m.l.e.s are also the same, to about two significant digits.

Next, note that $\hat{\alpha}_{\text{McC}} < \hat{\alpha}_{\text{ML}} < \hat{\alpha}_{\text{Hint}}$ for almost all the assets (asset 8, Chevron Corp., being an exception), exactly as in Figure 9.17, and for many of the series they differ substantially. Without invoking a formal testing paradigm via computing the p-value of the ALHADI test (to be done below), we can be skeptical that the return series are i.i.d. stable Paretian. This should come as no surprise: as discussed above, the returns exhibit strong volatility clustering, and are thus anyway far from being i.i.d. *This is not a useless exercise, however: it can be conducted on the residuals , or the filtered innovation sequence based*

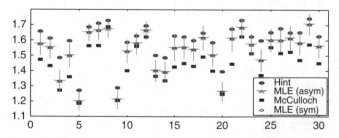

Figure 9.19   *The Hint (thick circle); the m.l.e. estimating all four parameters as unknown (star); the m.l.e. estimating just α, μ, and c, taking $\beta = 0$ (thin circle); and McCulloch (square) estimates of stable tail index α for each of the 30 DJIA daily stock return series. The lines indicate the interval of $\pm 2\widehat{\text{std}}(\hat{\alpha}_{\text{ML}})$ using the four-parameter m.l.e.*

on estimated parameters, of the so-called stable Paretian asymmetric power autoregressive conditional heteroskedasticity model, or $S_{\alpha,\beta}$-APARCH, applied to the return series. This model will be discussed at length in Book IV. To the extent that the model is "reasonably specified" and the residuals are approximately i.i.d., it makes sense to apply tests for stability to them.

Continuing the graphical analysis, Figure 9.20 compares the distribution of the ALHADI test statistic, as the difference between $\hat{\alpha}_{\mathrm{Hint}}(\mathbf{X})$ and $\hat{\alpha}_{\mathrm{McC}}(\mathbf{X})$ (top), and the difference between $\hat{\alpha}_{\mathrm{ML}}(\mathbf{X})$ and $\hat{\alpha}_{\mathrm{McC}}(\mathbf{X})$ (bottom), based on (i) 2000 replications of simulated $S_{\alpha,\beta}(\mu,c)$ data with parameters chosen as the m.l.e. using the fourth DJIA component, AT&T (left); and (ii) the nonparametric bootstrap using $B = 2000$ replications, applied to the returns of AT&T (right). From the left panels, we see that, under the null hypothesis of stable data, the distributions of the two differences are both centered around zero, close to Gaussian, and have approximately the same variance. The right panels indicate that, for this choice of real data (which is surely not i.i.d. stable Paretian), the ALHADI difference is substantially larger than that of $\hat{\alpha}_{\mathrm{ML}}(\mathbf{X}) - \hat{\alpha}_{\mathrm{McC}}(\mathbf{X})$, in line with the results of Figures 9.17 and 9.18, so that, as a test statistic, ALHADI will have much higher power.

Instead of showing the histograms in Figure 9.20 corresponding to a single asset (AT&T), one could reduce each of the four histograms to a boxplot, and plot them for all 30 assets. This is done in Figure 9.21. The top two panels of Figure 9.21 correspond to the ALHADI test statistic: For each of the 30 DJIA return series (the first 15 in the left panel; the last 15 in the right panel), there are two thin box plots. The first corresponds to the distribution of the ALHADI test statistic based on simulation of $S_{\alpha,\beta}(\mu,c)$ data, with the parameters

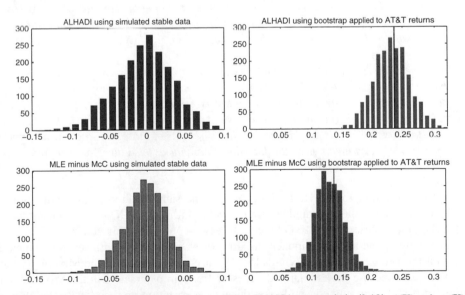

Figure 9.20 **Top left**: Simulated distribution of the ALHADI test statistic (9.16), $A(\mathbf{X}) = \hat{\alpha}_{\mathrm{Hint}}(\mathbf{X}) - \hat{\alpha}_{\mathrm{McC}}(\mathbf{X})$, using 2000 series of i.i.d. $S_{\alpha,\beta}(\mu,c)$ data of length $T = 2020$, where the parameter vector $(\alpha,\beta,\mu,c)'$ is the m.l.e. of the daily returns of the AT&T closing stock price, this being the fourth component of the DJIA index. **Top right**: The nonparametric bootstrap distribution of $A(\mathbf{X})$, using $B = 2000$ bootstrap draws from the AT&T return series. The thin vertical line shows the actual value of $A(\mathbf{X})$ for the AT&T returns. **Bottom**: Similar, but using $\hat{\alpha}_{\mathrm{ML}} - \hat{\alpha}_{\mathrm{McC}}$ instead of the ALHADI test statistic.

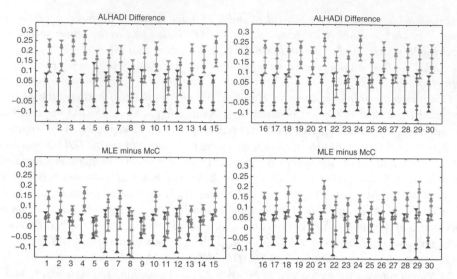

Figure 9.21 **Top**: *The ALHADI test statistic for each of the 30 DJIA return series: For each, the left boxplot corresponds to the distribution of the ALHADI test statistic based on simulation of $S_{\alpha,\beta}(\mu, c)$ data using 2000 replications, with the parameters being the m.l.e. for that asset; the second (usually higher up in the graphic) to the distribution of the ALHADI test statistic using the nonparametric bootstrap with B = 2000 replications, applied to the return series.* **Bottom**: *Same, but based on the difference $\hat{\alpha}_{ML}(\mathbf{X}) - \hat{\alpha}_{McC}(\mathbf{X})$.*

being the m.l.e. for that asset. The second corresponds to the distribution of the ALHADI test statistic using the nonparametric bootstrap applied to the return series. The bottom two panels are similar, but based on the difference $\hat{\alpha}_{\mathrm{ML}}(\mathbf{X}) - \hat{\alpha}_{\mathrm{McC}}(\mathbf{X})$ instead of the ALHADI difference.

Most assets are such that the distributions based on simulated stable data, and the non-parametric bootstrap of the actual returns, are highly separated, indicating rejection of stability. Notice that they are nearly identical for asset 8 (Chevron Corp.), and overlap substantially for asset 22 (McDonald's Corp.), suggesting that, with respect to the ALHADI test, the stability hypothesis for the unconditional returns of these two stocks would not be rejected. However, as mentioned, the fact that financial asset returns are far from being i.i.d. tempers such conclusions.

For the p-value of the ALHADI test, we use the same method of approximation as employed to calculate a quick and accurate p-value for the $D_{T,\kappa}(\mathbf{X})$ test (9.11) and the τ_B test (9.15), but instead of the MixGAt, we use a two-component mixed normal (5.1), with five parameters, and a two-component mixture of skew normals, with seven parameters, recalling the skew normal density (A.115). We use these instead of the MixGAt (which also fits well) because the estimated degrees-of-freedom parameters in the MixGAt tend to be very large, and numeric problems during optimization arose occasionally, as it is over-parameterized and not suited for modeling near-Gaussian data. Figure 9.22 shows four cases, each based on 100,000 replications. For these cases, the matches between the kernel density and the fitted MixGAt, mixed normal, and mixture of skew normals are all very good. It appears that the extra flexibility from the mixture of skew normals is not necessary,

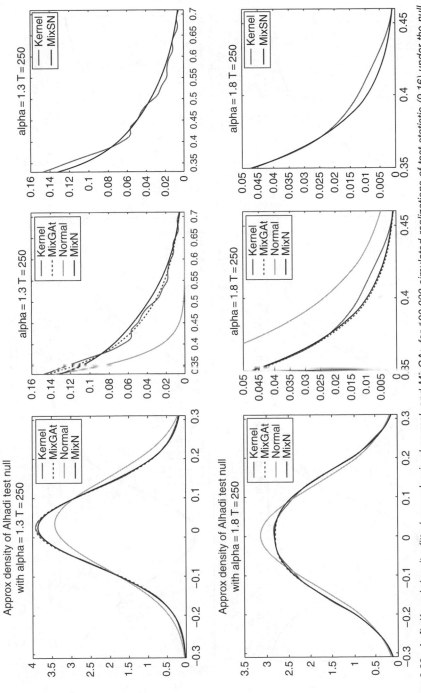

Figure 9.22 **Left**: *Kernel density, fitted normal, mixed normal, and MixGAt, for 100,000 simulated realizations of test statistic (9.16) under the null, for the two indicated sample sizes T and two values of α.* **Middle**: *Same as left, but magnified view of right tail.* **Right**: *Same as middle, but showing the fit of the two-component mixture of skew normals (MixSN).*

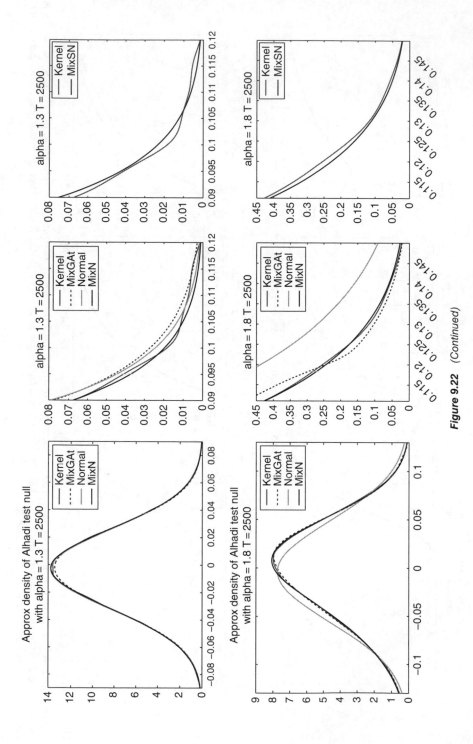

Figure 9.22 *(Continued)*

TABLE 9.3 **Actual sizes of the nominal 5% and 1% A(X) test (9.16) for data of length *T* with tail index α. The rows labeled 250* and 5000* indicate the use of the true value of α instead of using $\hat{\alpha}_{\text{Hint}}$ and linear interpolation into the constructed table of p-values**

Nominal	0.05					0.01				
$T \backslash \alpha$	1.20	1.35	1.55	1.75	1.90	1.20	1.35	1.55	1.75	1.90
250	0.0963	0.0617	0.0434	0.0431	0.0549	0.0436	0.0203	0.0075	0.0063	0.0091
250*	0.0532	0.0501	0.0525	0.0477	0.0504	0.0095	0.0096	0.0101	0.0098	0.0090
500	0.0837	0.0501	0.0441	0.0455	0.0551	0.0342	0.0120	0.0068	0.0068	0.0102
1000	0.0665	0.0479	0.0459	0.0473	0.0556	0.0211	0.0097	0.0073	0.0075	0.0109
2500	0.0501	0.0438	0.0461	0.0472	0.0555	0.0122	0.0076	0.0076	0.0081	0.0105
5000	0.0468	0.0453	0.0484	0.0495	0.0546	0.0093	0.0100	0.0088	0.0089	0.0118
5000*	0.0502	0.0502	0.0496	0.0501	0.0535	0.0092	0.0110	0.0098	0.0098	0.0097

so we use the mixed normal as the parametric choice for modeling the distribution of the test statistic under the null.

As with the previous two tests, the p-values for each entry in a grid of T- and α-values are computed (but based on the mixed normal parametric approximation) and bivariate spline interpolation is applied, using the actual T and $\hat{\alpha}_{\text{Hint}}$ in place of the true α. Anticipating the same issue that arose with the $D_{T,\kappa}(\mathbf{X})$ test, we use a tighter grid of α-values for the p-value interpolation construction, $1.05, 1.10, 1.15, \ldots, 1.95, 1.99$, though we will see that this is still not adequate for obtaining the desired accuracy.

The actual sizes of the ALHADI test (using our usual method of p-value interpolation, but not using the multiplicative modification method in (9.14)) are shown in Table 9.3, for several T and α, for nominal sizes 5% and 1% (the values for 10% were also recorded, but are qualitatively similar and not shown). The accuracy of the actual size compared to the nominal is reasonably good for $1.3 < \alpha < 1.95$ and $T \geq 500$, though poor for $\alpha < 1.4$ and $T = 250$. The row labeled 250* shows the performance when the true value of α is used (instead of $\hat{\alpha}_{\text{Hint}}$), so that interpolation is not required. The performance for $\alpha = 1.2$ and $\alpha = 1.35$, with $T = 250$, is then much better, showing that the mixed normal parametric approximation is adequate, and that the problem has to do with the fact that we do not entertain values of $\hat{\alpha}_{\text{Hint}} < 1.05$. The row indicated by 5000* is similar, and shows that the effect of p-value interpolation is still apparent (though less consequential) even for substantial sample sizes.

We could attempt to deploy the same modification method in (9.14) for, say, sample sizes $250 \leq T \leq 500$ and $1.05 \leq \hat{\alpha} \leq 1.4$, though it would be more effective to use a finer grid of α-values and values of $\alpha < 1$. We omit this because, in our testing applications involving financial data, $\hat{\alpha}$ tends to be well above 1.4, and sample sizes of at least 500 are anyway required to get reasonable power.

9.5.4 Joint Test Procedure

With the ability to quickly and accurately approximate the p-values of the three general tests, any two of them, or all three, can be combined to form a joint test. The constituent tests are not independent, so that the usual χ^2 distributional result is not applicable, and simulation is required. This procedure is discussed in more detail in Section 6.6.1.2 in the context of testing normality.

In the context here, we are interested in combining the ALHADI and τ_{20} tests, resulting in what we will refer to as the $A + \tau_{20}$ test. To obtain a p-value of this joint test, we first compute $S_1, \ldots, S_B$, where the S_i are i.i.d., each being the sum of the logs of the two p-values under the null of SαS, for a grid of α-values, say $\boldsymbol{\alpha} = (1.05, 1.1, 1.15, \ldots, 1.95, 1.99)$, and B a large number, for which we use 10,000. Once these are computed, we can quickly calculate the power of the joint test for a particular data set $\mathbf{X}$ as follows: Having used the sum of logs (and not their negative), for each α-value, the fraction of the S_i that are less than the sum of the logs of p-values corresponding to the constituent tests applied to $\mathbf{X}$ is recorded, and this is stored as a vector, say $\mathbf{p}_C$. Then the p-value of the joint test, p_C, is formed via interpolation into the vectors $\boldsymbol{\alpha}$ and $\mathbf{p}_C$, using $\hat{\alpha}_{\text{Hint}}(\mathbf{X})$. This can be done for numerous simulated data sets, and the fraction of the p_C values that is less than, say, 0.05, is the actual size, or power, of the joint test with nominal size 0.05.

Note that, for a single particular data set, given the ability to quickly compute the ALHADI and τ_{20} p-values, the joint test p-value is easily computed via simulation based on the actual sample size of the data set and the associated estimate of α. The only reason for simulating it over a grid of α-values (for a fixed sample size) is so that it can be used for a large number of data sets (of the same length), as required in our size and power simulations, and in empirical applications.

We will see below in Section 9.5.6 that the combined test based on τ_{20} and ALHADI is indeed – and often quite substantially – superior in terms of power compared to its two constituent component tests for nearly all distributions and parameter constellations considered. The power of the $D_{T,5}$ test (9.11) based on the characteristic function is, unfortunately, rather low for most alternatives of interest. When combining it with the ALHADI test, matters do not improve.

9.5.5 Likelihood Ratio Tests

Recall from Section 2.4 that a composite distributional test is such that the null hypothesis consists of a family of distributions indexed by an unspecified set of parameters θ, as opposed to the null being a fully specified distribution. Our interest throughout this section is in composite tests for stability. The empirical c.f. test (9.11), the summability test (9.15), and the ALHADI test (9.16) are all composite, and are also such that there is no specific distributional alternative specified. We now consider a composite test but such that an explicit composite alternative is specified, via a likelihood ratio.

A natural candidate as the alternative is the (location–scale) Student's t, this also being a heavy-tailed distribution with, like the SαS, one shape parameter that determines the tail index. In this case, the likelihood ratio test (l.r.t.) would not be of the classic form, in which there is a model that nests both alternatives, but rather a comparison of two nonnested models, as mentioned in Remark (c) in Section 3.3.2. The l.r.t. can exhibit very high power against a specific (composite) alternative, and as such, serves to give an upper bound on the power of a test, with respect to that alternative. Interestingly, we will see below that this l.r.t. also has excellent power properties against other alternatives besides the Student's t.

This method can be operationalized to conduct a size-0.05 test of the null of composite SαS versus the alternative of composite Student's t, for a particular data set $\mathbf{X}$ of length T, by performing the following steps:

(1) Estimate the parameters of the SαS distribution, say $\widehat{\boldsymbol{\theta}}_0 = (\hat{\alpha}, \hat{\sigma}, \hat{\mu})$, using the m.l.e., with associated log-likelihood denoted by $\ell_{\text{S}\alpha\text{S}}(\widehat{\boldsymbol{\theta}}_0; \mathbf{X})$.

(2) Estimate the parameters of the location–scale Student's t distribution and compute the associated log-likelihood $\ell_t(\cdot; \mathbf{X})$.

(3) Compute the ratio

$$\mathrm{LR}_0(\mathbf{X}) = 2 \times (\ell_t(\cdot; \mathbf{X}) - \ell_{\mathrm{S}\alpha\mathrm{S}}(\widehat{\boldsymbol{\theta}}_0; \mathbf{X})). \tag{9.18}$$

(4) For $i = 1, \ldots, s_1$,

(a) simulate $\mathbf{X}_{(i)}$, consisting of T i.i.d. SαS realizations with parameter vector $\widehat{\boldsymbol{\theta}}_0$.

(b) similar to steps 1–3, compute the ratio $\mathrm{LR}_i(\mathbf{X}_{(i)}) = 2 \times (\ell_t(\cdot; \mathbf{X}_{(i)}) - \ell_{\mathrm{S}\alpha\mathrm{S}} (\widehat{\boldsymbol{\theta}}_i; \mathbf{X}_{(i)}))$.

(5) Reject the SαS null hypothesis in favor of the Student's t alternative if LR_0 is equal to or exceeds the 95% empirical quantile of $(\mathrm{LR}_1, \ldots, \mathrm{LR}_{s_1})$.

However, for the large number of simulations required to obtain the empirical size and power, this parametric bootstrap procedure will be too slow. As such, we proceed similarly to the development of the previous tests: For a grid of α-values for a fixed T (or a over two-dimensional grid of T- and α-values), for each element in the grid, simulate (9.18) s_1 times, based on SαS data, and record the empirical 95% quantile of the s_1 likelihood ratio values. This yields a set of actual cutoff values $c_{\mathrm{act}}(0.05; T, \alpha)$. Then, for a particular data set of length T, and based on an estimator of α, use interpolation into the grid based on the actual T and $\hat{\alpha}$ to approximate the appropriate cutoff value corresponding to a 5% level test. For three sample sizes $T = 500$, $T = 1000$ and $T = 2500$, and based on a grid of values $\alpha = 1.05, 1.1, 1.15, \ldots, 1.95, 1.99$, the construction of the grid is relatively fast (using $s_1 = 20{,}000$), assuming fast maximum likelihood estimation of the SαS model, for which we use Nolan's spline approximation to the density, in conjunction with the Hessian-based general multivariate optimization routines discussed in Section 4.3.1.

Once functions $c_{\mathrm{act}}(0.05; T, \alpha)$ are available, the application of the test is very fast. Simulations (based on 10,000 realizations and different seeds than those used to determine the cutoffs) confirm that the test is size-correct to two decimal places for SαS data with $1.2 \leq \alpha \leq 1.9$. For example, for $T = 1000$, the average size was 0.0516 (where the true value of α is *not* used, but rather estimated from the data, as detailed above, and used for interpolating into the grid of simulated cutoff values to get the approximate cutoff value corresponding to a 5% level test).

To help envision the test, Figure 9.23(a) shows a kernel density estimate based on a sample of $T = 1000$ simulated SαS data with $\alpha = 1.6$, along with the m.l.e.-fitted location–scale SαS and Student's t densities. The fitted SαS is indeed slightly closer to the kernel density, though the Student's t fit is nearly as good. Figure 9.23(b) shows as a histogram the l.r.t. values, based on $T = 1000$, and the (negatives of the) interpolated cutoff values of the test (making them negative just for ease of graphic illustration), based on 1000 replications, using SαS data. The cutoff values are all rather close, but not equal because the m.l.e. of α is used for interpolation into the grid of cutoff values. By construction, they cluster around the 95% quantile of the simulated l.r.t. values.

9.5.6 Size and Power of the Symmetric Stable Tests

We first illustrate the actual sizes obtained, for several values of stable tail index α. Table 9.4 contains the results, all of which are based on $s = 10{,}000$ replications and such that, in each

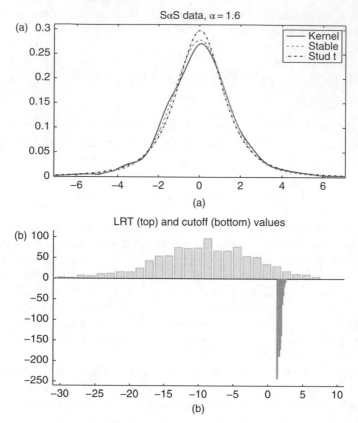

Figure 9.23 *SαS and Student's t fitted to SαS data (a) and 1000 l.r.t. and cutoff values (made negative, for graphic illustration) under the null (b).*

column, the same s data sets were used. The last row of the table shows the average of the $\hat{\alpha}_{\text{Hint}}$ estimates over the s samples. The most accurate tests with respect to a nominal size of 5% are $A + \tau_{20}$ and the l.r.t. – and conveniently so, as these will be seen to have the highest power.

To assess the power of the four general (nonspecific alternative) tests for the symmetric case, and the l.r.t., we will use the Student's t and the various alternatives discussed in Figure 9.18. We restrict ourselves to (i) varying only one of the parameters of the alternative distribution; (ii) use of the nominal size of 5%, as it the most common in practice; and (iii) use of three sample sizes, $T = 500$, $T = 1000$, and $T = 2500$. All reported powers are based on use of $s = 10{,}000$ replications, and such that, in each column of the forthcoming tables, the same s data sets were used.

Beginning with the Student's t alternative, Table 9.5 shows the power of the four general alternative tests and the l.r.t., for six values of the degrees-of-freedom parameter, $v = 1.5, 2, 3, 4, 5, 6$. The last row of the table shows the average of the $\hat{\alpha}_{\text{Hint}}$ estimates over the $s = 10{,}000$ replications. For all tests, the power decreases as $v \to 1$ and $v \to \infty$ as expected. In all cases, the $D_{T,5}$ test has the lowest power, and demonstrably so. Using the notation $a \succ b$ to indicate that test a has higher power than b, and $a \vee b = \max(\text{power}(a), \text{power}(b))$,

TABLE 9.4 Actual sizes of tests, for i.i.d. symmetric stable data with tail index α = 1.3, 1.45, 1.6, 1.75, 1.9, for sample size T, using the nominal size of 5%

Sample size		T = 500					T = 1000					T = 2500				
Test	Eq.	1.3	1.45	1.60	1.75	1.90	1.3	1.45	1.60	1.75	1.90	1.3	1.45	1.60	1.75	1.90
$D_{T,5}$	(9.11)	0.068	0.053	0.050	0.053	0.050	0.050	0.048	0.050	0.050	0.046	0.045	0.046	0.050	0.051	0.051
τ_{20}	(9.15)	0.069	0.060	0.055	0.045	0.035	0.056	0.056	0.051	0.049	0.038	0.053	0.050	0.051	0.050	0.036
A	(9.16)	0.058	0.045	0.042	0.046	0.055	0.050	0.046	0.043	0.046	0.055	0.049	0.045	0.046	0.048	0.055
$A + \tau_{20}$		0.051	0.052	0.051	0.050	0.054	0.050	0.052	0.051	0.050	0.050	0.050	0.050	0.050	0.051	0.050
LRT	(9.18)	0.052	0.050	0.049	0.053	0.051	0.051	0.050	0.050	0.049	0.053	0.049	0.048	0.048	0.052	0.056
Average $\hat{\alpha}_{Hint}$		1.30	1.45	1.60	1.75	1.90	1.30	1.45	1.60	1.75	1.90	1.30	1.45	1.60	1.75	1.90

TABLE 9.5 Power against the Student's t alternative, for degrees-of-freedom values ν = 1.5, 2, 3, 4, 5, 6 and sample size T, using the nominal size of 5%

Sample size		T = 500						T = 1000						T = 2500					
Test	Eq.	1.5	2	3	4	5	6	1.5	2	3	4	5	6	1.5	2	3	4	5	6
$D_{T,5}$	(9.11)	0.12	0.10	0.10	0.11	0.10	0.10	0.09	0.12	0.14	0.13	0.12	0.11	0.09	0.16	0.29	0.28	0.23	0.20
τ_{20}	(9.15)	0.18	0.29	0.42	0.42	0.35	0.29	0.22	0.46	0.72	0.74	0.68	0.59	0.28	0.65	0.95	0.98	0.98	0.97
A	(9.16)	0.18	0.30	0.36	0.32	0.27	0.23	0.30	0.54	0.61	0.54	0.44	0.38	0.59	0.90	0.94	0.88	0.79	0.69
$A + \tau_{20}$	(9.16)	0.17	0.36	0.54	0.52	0.45	0.37	0.29	0.61	0.82	0.82	0.74	0.66	0.59	0.93	0.99	0.99	0.99	0.97
LRT	(9.18)	0.27	0.46	0.61	0.57	0.53	0.50	0.44	0.71	0.89	0.88	0.85	0.81	0.77	0.97	1.00	1.00	1.00	0.99
Average $\hat{\alpha}_{\text{Hint}}$		1.28	1.46	1.64	1.73	1.79	1.82	1.28	1.46	1.64	1.73	1.79	1.82	1.29	1.45	1.64	1.73	1.79	1.82

we see that, for $v = 1.5$ and $v = 2$, $A \succ \tau_{20}$, while for $v \geq 3$, $\tau_{20} \succ A$. However, in both cases, we have $A + \tau_{20} \succ (A \vee \tau_{20})$ (except at $v = 1.5$, in which case the power values of $A + \tau_{20}$ and A differ at most by 0.01, but this is most likely due to A having slightly too liberal size; see Table 9.4), with the power of the combined test often substantially higher than $A \vee \tau_{20}$.

For all three sample sizes and for all v, the l.r.t. dominates, as must be the case asymptotically for the Student's t alternative, by construction. Interestingly, for $v = 4$, the power values of $A + \tau_{20}$ and l.r.t. are rather close (0.52 and 0.57, respectively, for $T = 500$; and 0.82 and 0.88 for $T = 1000$). This is relevant, because $v = 4$ is considered to be the most typical and best "default choice" degrees-of-freedom value for financial asset returns data; see Platen and Heath (2006, p. 90).

Table 9.6 shows the power for a range of mixed normal alternatives at the 5% nominal testing level. The choice of this alternative for investigation was fortuitous, as we have a case (the only one) for which the power of the $D_{T.5}$ test is strong, and such that, for a segment of the parameter space (depending on sample size T), $D_{T.5} \succ \tau_{20}$. Observe that the power of τ_{20} begins to decrease as a function of σ_2 (with the starting point depending on the sample size). It decreases so much that, at some point, the test is biased for level 5%, that is, the power is less than the size under the alternative. Closer inspection indicates that the τ_{20} test can still have some value for this alternative: when conducted at the 10% level, the power values for $T = 2500$ and $\sigma_2 = 2, 3, 4, 5, 6$ are 1.0, 1.0, 0.96, 0.76, and 0.54, respectively. Nevertheless, as with the Student's t case, $A + \tau_{20} \succ (A \vee \tau_{20})$ and $A + \tau_{20} \succ D_{T.5}$.

What is perhaps unexpected is the high power of the l.r.t. test, given that it has been designed for a specific (composite) alternative of Student's t. To help envision the test in this case, Figure 9.24 is similar to Figure 9.23, but using simulated mixed normal data with $\sigma_2 = 3$ (and again based on $T = 1000$). The fitted SαS and Student's t distributions are very close, and both differ significantly from the mixed normal density, so that one might have expected the l.r.t. test statistic to be close to 1, and the test to have low power. Yet, as revealed by the bottom panel, this is not the case. This occurs because the Student's t can accommodate the mixed normal shape slightly (but enough for a test) better than the stable. As seen in Table 9.6, except for the $\sigma_2 = 6$ case, it has power nearly equivalent to $A + \tau_{20}$. The reason for the breakdown as σ_2 increases is that the m.l.e. of α tends towards 1 as σ_2 increases, as seen in the last line of the table, but the l.r.t. was only calibrated for $1.05 \leq \alpha \leq 1.99$. As such, the test will not be applicable as σ_2 grows.

Next, consider the NIG alternative, with p.d.f. (9.17). Table 9.7 is similar to the previous two tables, showing the power as the NIG shape parameter α varies. As expected, for all tests, the power decreases as the NIG shape parameter α increases. As in the Student's t case, the $D_{T.5}$ test performs relatively poorly. For the other three nonspecific tests, we have a uniformity result across all used sample sizes and values of the NIG shape parameter: $A + \tau_{20} \succ \tau_{20} \succ A$, with the joint test $A + \tau_{20}$ having substantially higher power than τ_{20} for the smaller sample sizes. Between $A + \tau_{20}$ and the l.r.t., the results are not uniform, with respect to either shape parameter α or sample size, except that power increases for both as the sample size increases, as must be the case for consistent tests.

Recall that, for the GAt distribution when calibrated to stable Paretian data, the ALHADI test is nearly powerless, as shown in Figure 9.18. We consider this case, varying parameter v, with power shown in Table 9.8. Indeed, the ALHADI and $D_{T.5}$ tests perform poorly, and such that the tests are biased, with some power values lower than the size. The τ_{20}

TABLE 9.6 Power against the mixed normal alternative, with p.d.f. $f_X(x) = (1/2)f_N(x; 0, 1) + (1/2)f_N(x; 0, \sigma_2)$, for second component scale values $\sigma_2 = 2, 3, 4, 5, 6$, using the nominal size of 5%

Sample size		$T = 500$					$T = 1000$					$T = 2500$				
Test	Eq.	2	3	4	5	6	2	3	4	5	6	2	3	4	5	6
$D_{T,5}$	(9.11)	0.14	0.41	0.74	0.90	0.93	0.20	0.81	0.98	1.00	1.00	0.51	1.00	1.00	1.00	1.00
τ_{20}	(9.15)	0.52	0.84	0.67	0.52	0.41	0.87	0.95	0.72	0.46	0.32	0.99	0.77	0.09	0.01	0.01
A	(9.16)	0.49	0.98	1.00	1.00	1.00	0.75	1.00	1.00	1.00	1.00	0.98	1.00	1.00	1.00	1.00
$A + \tau_{20}$		0.71	1.00	1.00	1.00	1.00	0.95	1.00	1.00	1.00	1.00	1.00	1.00	1.00	1.00	1.00
LRT	(9.18)	0.90	0.97	1.00	0.99	0.79	0.99	0.97	1.00	1.00	0.87	0.99	0.99	1.00	1.00	0.96
Average $\hat{\alpha}_{\text{Hint}}$		1.80	1.61	1.52	1.48	1.46	1.80	1.61	1.52	1.48	1.46	1.80	1.61	1.52	1.48	1.46
Average $\hat{\alpha}_{\text{ML}}$		1.85	1.55	1.34	1.19	1.08	1.85	1.55	1.34	1.19	1.08	1.85	1.55	1.33	1.19	1.08

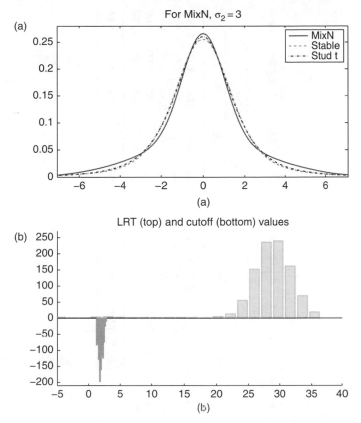

Figure 9.24 *SαS and Student's t fitted to mixed normal data (a) and 1000 l.r.t. and cutoff values under the mixed normal alternative (b).*

test performs the best overall, dominating all other tests in most cases. The τ_{20} and l.r.t. tests have their maximal power (for the values considered) at $v = 1$, which corresponds to a maximally existing moment of $vd \approx 2.9$. As $\theta = 1$ (symmetric case), the GAt reduces to the Cauchy distribution for $v = 1/d \approx 0.34$, and converges to an exponential-tail distribution as $v \to \infty$; this explains the decrease in power as v decreases or increases away from 1.0. Unlike in the Student's t, NIG and mixed normal cases, the l.r.t. test is not the most powerful, but performs well overall. As v decreases, the m.l.e. of the stable parameter α decreases, and the l.r.t. test will not be applicable, for the same reason as mentioned for the mixed normal case. This explains why the l.r.t. is biased and has nearly zero power for $v = 0.4$.

The last alternative we consider is (a particular form of) the mixed stable distribution, with p.d.f. $f_X(x) = (1/2)f_S(x; \alpha_1, 0, 1) + (1/2)f_S(x; 3 - \alpha_1, 0, 1)$, where f_S denotes the SαS p.d.f. The tabulated power values are not shown. A two-sided ALHADI test would have power, but, based on Figure 9.18, the power will be very low for this alternative. The τ_{20} test is not expected to have power against this alternative, as sums of the data do not submit to the Gaussian central limit theorem. Indeed, this is the case: except for $\alpha_1 = 1.5$ (which reduces to SαS with tail index 1.5), the power of τ_{20} for $\alpha_1 > 1.5$ is below the nominal size. This is also the case with the $D_{T,5}$ test.

TABLE 9.7 Power against the NIG alternative, with p.d.f. (9.17), using $\beta = 0$, $\mu = 0$, $\delta = 1$, and shape values α = 0.3, 0.6, 0.9, 1.2, 1.5, for the nominal size of 5%

Sample size		T = 500					T = 1000					T = 2500				
Test	Eq.	0.3	0.6	0.9	1.2	1.5	0.3	0.6	0.9	1.2	1.5	0.3	0.6	0.9	1.2	1.5
$D_{T,5}$	(9.11)	0.20	0.19	0.16	0.14	0.12	0.36	0.33	0.26	0.20	0.17	0.85	0.82	0.68	0.52	0.41
τ_{20}	(9.15)	0.83	0.75	0.65	0.55	0.45	0.99	0.98	0.95	0.88	0.80	1.00	1.00	1.00	1.00	1.00
A	(9.16)	0.78	0.69	0.57	0.46	0.40	0.98	0.93	0.84	0.74	0.63	1.00	1.00	0.99	0.98	0.94
$A + \tau_{20}$		0.94	0.89	0.81	0.69	0.61	1.00	1.00	0.98	0.94	0.88	1.00	1.00	1.00	1.00	1.00
LRT	(9.18)	0.98	0.75	0.87	0.85	0.80	1.00	0.87	0.96	0.95	0.95	1.00	0.97	0.99	0.98	0.97
Average $\hat{\alpha}_{\text{Hint}}$		1.53	1.65	1.72	1.76	1.79	1.53	1.65	1.72	1.76	1.79	1.53	1.65	1.72	1.76	1.79

TABLE 9.8 Power against the GAt alternative using $d = 2.9073$ and $\vartheta = 1$ (and $\mu = 0$, $c = 1.7$, though the location and scale terms are irrelevant for power considerations), and shape values $\nu = 0.4, 0.7, 1.0, 1.3, 1.6$, for nominal size of 5%

Sample size		T = 500					T = 1000					T = 2500				
Test	Eq.	0.4	0.7	1.0	1.3	1.6	0.4	0.7	1.0	1.3	1.6	0.4	0.7	1.0	1.3	1.6
$D_{T,5}$	(9.11)	0.13	0.06	0.07	0.05	0.05	0.10	0.06	0.06	0.05	0.04	0.04	0.06	0.07	0.06	0.04
τ_{20}	(9.15)	0.07	0.21	0.23	0.15	0.08	0.07	0.33	0.40	0.26	0.14	0.07	0.52	0.79	0.70	0.41
A	(9.16)	0.09	0.07	0.05	0.03	0.02	0.07	0.09	0.06	0.02	0.01	0.03	0.13	0.06	0.01	0.00
$A + \tau_{20}$		0.06	0.16	0.15	0.09	0.04	0.04	0.24	0.26	0.12	0.05	0.03	0.42	0.54	0.34	0.10
LRT	(9.18)	0.02	0.15	0.20	0.11	0.05	0.01	0.21	0.36	0.22	0.15	0.00	0.43	0.73	0.61	0.64
Average $\hat{\alpha}_{Hint}$		1.14	1.54	1.73	1.84	1.90	1.14	1.54	1.73	1.84	1.90	1.14	1.54	1.73	1.84	1.90

Remark.   Recall that, in the mixed normal case, the l.r.t. has high power because the Student's t fits the mixed normal adequately better than the $S\alpha S$. One might conjecture that the stable will fit better than the Student's t and the l.r.t. test would have power in its left tail for distributions that are "closer" to the stable, such as the mixed stable, as well as the tempered and geometric stable, as mentioned in Section 9.4.1.

This conjecture is partially confirmed in Figure 9.25, which is similar to Figures 9.23 and 9.24, but for mixed stable data with $\alpha_1 = 1.7$ (and again based on $T = 1000$). (For this data-generating process, the average estimated stable tail index over the 1000 simulations was 1.47, which is very close to the average of 1.7 and 1.3. This average being well over 1.2, it also confirms that the l.r.t. test can be computed in this case.) Comparing Figures 9.24(b) and 9.25(b), it is clear that the l.r.t. will have power against the mixed stable alternative, but the power is nowhere near as strong as the l.r.t. against the mixed normal, as seen from Figure 9.24(b).

Based on this observation, one is behooved to use a two-sided l.r.t. test, and compute the 2.5% and 97.5% quantiles instead of only the 95% quantile, if such "stable-like" alternatives are of interest. The mixed stable case we use here being somewhat of a toy example, the reader is encouraged to pursue the development of testing procedures for stable versus the more relevant cases of tempered and geometric stable.   ∎

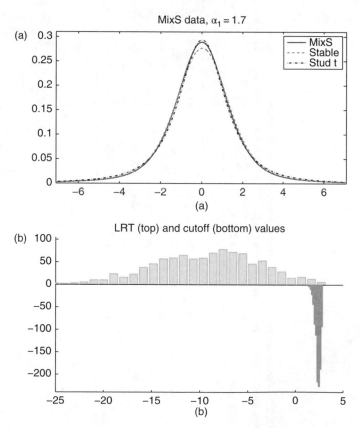

Figure 9.25   *SαS and Student's t fitted to mixed stable data (a) and 1000 l.r.t. and cutoff values under the mixed stable alternative (b).*

9.5.7 Extension to Testing the Asymmetric Stable Paretian Case

9.5.7.1 Ignoring the Asymmetry

Let $X_t \sim S_{\alpha,\beta}(\mu, \sigma)$ be i.i.d. copies, $t = 1, \ldots, T$, each with c.d.f. $F_X(\cdot;\ \alpha, \beta, \mu, \sigma)$. First consider what happens to the actual size of the τ_{20} and ALHADI tests when asymmetry is ignored. This is explored via simulation using 10,000 replications, based on the nominal size of 5%, for the single (and reasonably typical for financial asset returns) value of $\alpha = 1.6$, three sample sizes, and values of $\beta = 0, \pm 0.2, \pm 0.4, \pm 0.6, \pm 0.8$.

The results are shown in Table 9.9. Both tests appear to have actual size symmetric in β, and such that they become too liberal as $|\beta| \to 1$, much more so for ALHADI. The actual size of the ALHADI test not only breaks down as $|\beta|$ moves away from zero, but gets worse as the sample size increases; whereas the size of the τ_{20} test is still somewhat reasonable across sample sizes and for moderate β, and its actual size even improves (mildly) for positive β as the sample size grows, but worsens for negative β. Thus, for data sets that only exhibit a mild amount of asymmetry, the τ_{20} test can still be used, though this conclusion is limited, as we studied only the $\alpha = 1.6$ case. As generation of stable random variates and computation of the τ_{20} test are very fast, one can always use simulation, based on the m.l.e. of α and β, for a particular data set, to assess the actual size for a given nominal size.

9.5.7.2 Use of CDF and Inverse CDF Transform

Our simple idea to deal with the asymmetric case is to apply the aforementioned tests for SαS data to $\tilde{\mathbf{X}} = (\tilde{X}_1, \ldots, \tilde{X}_T)$, where

$$\tilde{X}_t = F_X^{-1}(P_t;\ \hat{\alpha}_{\mathrm{ML}}, 0, 0, 1), \quad P_t = F_X((X - \hat{\mu}_{\mathrm{ML}})/\hat{\sigma}_{\mathrm{ML}};\ \hat{\alpha}_{\mathrm{ML}}, \hat{\beta}_{\mathrm{ML}}, 0, 1), \quad (9.19)$$

$t = 1, \ldots, T$, that is, the data are transformed by the inverse c.d.f. based on the m.l.e. to produce SαS data. This procedure is numerically feasible using the fast m.l.e. c.d.f., and inverse c.d.f. routines provided in Nolan's toolbox, and clearly asymptotically valid, though in finite samples, and in light of the relative difficulty of estimating β accurately, it can result in size distortions, depending on the nature of the test.

TABLE 9.9 Actual sizes of the τ_{20} and ALHADI nominal 5% tests, as designed for symmetric stable data but applied to asymmetric stable data, using 10,000 replications, and based on $\alpha = 1.6$ and $\beta = 0, \pm 0.2, \pm 0.4, \pm 0.6, \pm 0.8$, using sample size T, and ignoring the asymmetry

Test	Eq.	0.8	0.6	0.4	0.2	0	−0.2	−0.4	−0.6	−0.8
					$T = 500$					
τ_{20}	(9.15)	0.14	0.10	0.075	0.061	0.066	0.070	0.083	0.098	0.14
A	(9.16)	0.39	0.23	0.12	0.047	0.040	0.067	0.13	0.24	0.40
					$T = 1000$					
τ_{20}	(9.15)	0.13	0.093	0.069	0.057	0.051	0.057	0.072	0.094	0.13
A	(9.16)	0.55	0.32	0.15	0.065	0.043	0.068	0.15	0.32	0.55
					$T = 2{,}500$					
τ_{20}	(9.15)	0.12	0.082	0.059	0.048	0.047	0.061	0.082	0.11	0.18
A	(9.16)	0.82	0.49	0.20	0.073	0.049	0.087	0.23	0.51	0.83

TABLE 9.10 Similar to Table 9.9, showing actual size for a nominal size of 5%, again based on sample size *T* and using 10,000 replications, but accounting for asymmetry by having applied transform (9.19). Also shown are the actual sizes of the combined test $A + \tau_{20}$ and l.r.t.

Test	Eq.	0.8	0.6	0.4	0.2	0	−0.2	−0.4	−0.6	−0.8
						T = 500				
τ_{20}	(9.15), (9.19)	0.034	0.043	0.050	0.051	0.052	0.051	0.049	0.045	0.032
A	(9.16), (9.19)	0.029	0.031	0.034	0.035	0.035	0.034	0.035	0.031	0.032
$A + \tau_{20}$	(9.19)	0.028	0.035	0.042	0.042	0.044	0.043	0.042	0.036	0.029
LRT	(9.18), (9.19)	0.031	0.042	0.050	0.052	0.052	0.052	0.050	0.043	0.031
						T = 1000				
τ_{20}	(9.15), (9.19)	0.034	0.048	0.050	0.052	0.050	0.053	0.052	0.049	0.034
A	(9.16), (9.19)	0.034	0.038	0.038	0.039	0.039	0.037	0.038	0.039	0.032
$A + \tau_{20}$	(9.19)	0.031	0.043	0.046	0.047	0.047	0.047	0.046	0.043	0.029
LRT	(9.18), (9.19)	0.031	0.045	0.050	0.051	0.052	0.052	0.050	0.044	0.029
						T = 2,500				
τ_{20}	(9.15), (9.19)	0.034	0.049	0.050	0.051	0.051	0.050	0.049	0.046	0.031
A	(9.16), (9.19)	0.029	0.040	0.042	0.043	0.042	0.044	0.042	0.042	0.030
$A + \tau_{20}$	(9.19)	0.029	0.044	0.048	0.049	0.048	0.047	0.047	0.047	0.030
LRT	(9.18), (9.19)	0.029	0.045	0.051	0.051	0.052	0.052	0.052	0.048	0.030

Table 9.10 is similar to Table 9.9, but applying transform (9.19), and also showing the results for the combined test $A + \tau_{20}$ and l.r.t. As conjectured, the actual sizes of all the tests improve as the sample size increases, and also, for a given sample size, decrease away from 0.05 as $|\beta|$ increases. The actual sizes of τ_{20} and the l.r.t. are quite reasonable for $|\beta| \leq 0.6$, as is that of the combined test for $|\beta| \leq 0.4$.

We now turn to the power of these tests for two asymmetric alternatives. We use the asymmetric NIG distribution (9.17), with $\alpha = 1.2$ and $\beta = 0, -0.3, -0.6, -0.9$; and the non-central Student's t with $v = 4$ degrees of freedom and noncentrality (asymmetry) parameters $\gamma = 0, -0.6, -1.2, -1.8$, these values having been chosen to capture mild, strong, and very strong asymmetry, relative to financial returns data. Table 9.11 shows the results based on 1000 replications. We see that, for the NCT and NIG alternatives, the power of all the tests weakens as the asymmetry grows in magnitude. We have, for all three distributional alternatives considered, the ordering with respect to power among the first four tests as LRT > $A + \tau_{20}$ > τ_{20} > A. The fifth test, (9.21), will be discussed below.

We also inspect the power against another leptokurtic, asymmetric alternative, namely the inverse hyperbolic sine (IHS) distribution, attributed to Johnson (1949) and used for empirical financial applications by Brooks et al. (2005) and, most notably, Choi and Nam (2008). A random variable Y follows an IHS distribution, denoted $Y \sim \text{IHS}(\lambda, \theta)$, if $\sinh^{-1}(Y) \sim N(\lambda, \theta^2)$ or, with $Z \sim N(0, 1)$, $Y = \sinh(\lambda + \theta Z)$. As $\sinh^{-1}(\cdot)$ is a nondecreasing function of its argument, the p.d.f. of Y is straightforwardly obtained and given by

$$f_Y(y; \lambda, \theta) = \frac{1}{\sqrt{2\pi(y^2 + 1)\theta^2}} \exp\left\{ -\frac{(\sinh^{-1}(y) - \lambda)^2}{2\theta^2} \right\}. \tag{9.20}$$

The IHS distribution is very convenient for financial applications, as the quantiles can be explicitly determined. In particular, using

$$w = \frac{1}{\theta}(\sinh^{-1}(x) - \lambda) = \frac{1}{\theta}(\ln(x + \sqrt{1+x^2}) - \lambda), \quad dw = \frac{dx}{\theta\sqrt{1+x^2}},$$

yields

$$F_Y(y; \lambda, \theta) = \int_{-\infty}^{\frac{1}{\theta}(\ln(y+\sqrt{1+y^2})-\lambda)} f_Z(w)\,dw = \Phi\left(\frac{1}{\theta}(\sinh^{-1}(y) - \lambda)\right),$$

where Φ is the standard normal c.d.f., and the quantile function is

$$\Pr(Y \le y_q) = \Pr(\sinh(\lambda + \theta Z) \le y_q) \Rightarrow F_Y^{-1}(q) = \sinh(\lambda + \theta\Phi^{-1}(q)).$$

For a power comparison, we consider only one parameter set, obtained by fitting the location–scale IHS to a large sample of simulated $S_{1.7,-0.3}(0, 1)$ data, these being typical values for financial returns data, yielding IHS shape parameters $\lambda = -0.17$ and $\theta = 0.82$. Figure 9.26 shows the four-parameter stable and four-parameter fitted IHS distribution (along with a kernel density estimate of simulated values from the latter, confirming the method of simulation).

The rightmost column of Table 9.11 shows the power of the various tests against the IHS distribution. In this case, the l.r.t. dominates with a power of 0.99, followed by the combined test $A + \tau_{20}$, with power 0.95.

In light of the excellent performance of the l.r.t. for several alternatives, one naturally considers the development of joint (ideally two-sided, recalling the performance of the l.r.t. with the mixed stable) tests based on asymmetric-transformed ALHADI, τ_{20}, and possibly several likelihood ratio tests for a variety of judiciously chosen composite alternatives. This could lead to a very powerful test for sensible alternatives to the asymmetric stable Paretian distribution for parameter constellations typical in finance and other applications. The interested reader is encouraged to pursue this.

TABLE 9.11 For $T = 1000$ and nominal size 5%, power values against asymmetric alternatives of the τ_{20}, ALHADI, combined $A + \tau_{20}$, and l.r.t. (9.18) tests, using transform (9.19); and, in the last row, l.r.t. (9.21). The left panels show the power for the noncentral Student's t based on $v = 4$ degrees of freedom and noncentrality (asymmetry) parameters $\gamma = 0, -0.6, -1.2, -1.8$. The center panels use the asymmetric NIG (9.17) with NIG shape parameters $\alpha = 1.2$ and $\beta = 0, -0.3, -0.6, -0.9$. The rightmost column is for the IHS distribution (9.20) with $\lambda = -0.17$ and $\theta = 0.82$

Alternative		Noncentral t				Asymmetric NIG				IHS
Test	Eq.	0.0	−0.6	−1.2	−1.8	0.0	−0.3	−0.6	−0.9	
τ_{20}	(9.15), (9.19)	0.71	0.70	0.40	0.13	0.89	0.81	0.63	0.40	0.91
A	(9.16), (9.19)	0.50	0.42	0.20	0.11	0.72	0.65	0.43	0.27	0.75
$A + \tau_{20}$	(9.19)	0.79	0.73	0.42	0.17	0.94	0.90	0.69	0.44	0.95
LRT	(9.18), (9.19)	0.90	0.84	0.44	0.17	0.99	0.94	0.78	0.51	0.99
LRT	(9.21)	0.92	0.92	0.85	0.29	1.00	0.98	0.88	0.28	0.99

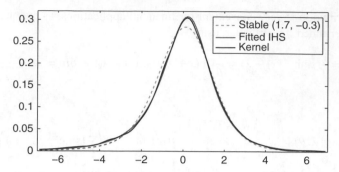

Figure 9.26 *The $S_{1.7,-0.3}(0,1)$ density (dashed) and the best-fitting location–scale IHS density (yielding $\lambda = -0.17$ and $\theta = 0.82$), based on the m.l.e. from a sample of size $T = 50{,}000$ of $S_{1.7,-0.3}(0,1)$ data; and a kernel density estimate of simulated IHS realizations based on the four parameters of the fitted distribution.*

9.5.7.3 LRT for Asymmetric Stable versus Noncentral Student's *t*

The composite l.r.t. can be extended to the asymmetric stable case. This requires applying the procedure discussed in Section 9.5.5 to a two-dimensional grid of α and β values for a fixed sample size, for which we use $T = 1000$ (or a three-dimensional grid also using various sample sizes), and forming the likelihood ratio based on the likelihood of the fitted location–scale asymmetric stable and a particular asymmetric fat-tailed alternative. For this, we use the NCT, generalizing the Student's *t* as we used in the symmetric l.r.t. case, so that

$$\mathrm{LR}_0(\mathbf{X}) = 2 \times (\ell_{\mathrm{NCT}}(\cdot\,;\mathbf{X}) - \ell_{S_{\alpha,\beta}}(\widehat{\boldsymbol{\theta}}_0;\mathbf{X})), \qquad (9.21)$$

where now $\widehat{\boldsymbol{\theta}}_0 = (\hat{\alpha}, \hat{\beta}, \hat{\sigma}, \hat{\mu})$. For the stable m.l.e., as speed will be crucial for doing size and power studies, we use the canned method in Nolan's toolbox, based on a spline approximation to the stable density. The m.l.e. of the NCT can be computed by using the saddlepoint approximation to the density, as discussed in Section 9.3.

The cutoff values for $T = 1000$ corresponding to (9.21) and a nominal size of 5% were computed for each element in the two-dimensional grid based on $\alpha = 1.05,$ $1.1, 1.15, \ldots, 1.95, 1.99$ and $\beta = -0.9, -0.85, -0.8, \ldots, 0.85, 0.9$, as discussed in Section 9.5.5, based on 1000 replications. The actual size of this procedure, when using estimates of α and β and bivariate interpolation into the grid of stored cutoff values, is quite reasonable for $1.5 \leq \alpha \leq 1.9$ and zero to moderate asymmetry of either sign, as shown in Figure 9.27 (the average of the plotted points is 0.0504).

However, as the asymmetry decreases towards -1 and increases towards 1, the actual size becomes more variable and is in almost all cases too low. (The inaccuracy outside this range arises because of the limitations of the spline density approximation to the stable p.d.f. Greater accuracy could be achieved for larger $|\beta|$ in the cutoff table by using the FFT approach for computing the density, when the former reports potential problems, though this will incur quite some numeric cost. To develop a cutoff grid with accuracy for $0.6 < |\beta| \leq 1$ and $1 < \alpha < 1.4$, even slower methods of computing the density would be required, and the computation times would become ominous.) As such, application of the test against the NCT, NIG, and IHS for the parameters used in Table 9.11 is valid, though as the asymmetry increases, the power will be artificially somewhat lower, given the lower actual size.

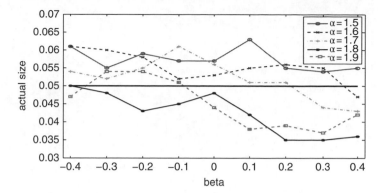

Figure 9.27 *Actual size of the nominal 5% level l.r.t. (9.21), for sample size T = 1000, using a ratio of noncentral Student's t and asymmetric stable Paretian.*

The resulting power values, listed in the last line of Table 9.11, indicate that the method is overall superior to use of the symmetric l.r.t. (9.18) with transformation (9.19). The power values are greater for the NCT, as should be expected, with a near doubling of the power for the $\gamma = -1.2$ case. For the NIG, the asymmetric l.r.t. delivers much lower power for the extreme asymmetry case of NIG parameter $\beta = -0.9$, though with the previous comments about the actual size in mind, the reported power in this case is not accurate, and would be higher if the cutoff values for high asymmetry were determined with greater accuracy.

10

The Method of Indirect Inference

This chapter can be seen as an extension of Section 5.2, useful when the only aspect of the data-generating process available to the researcher is that it can be simulated. It is not immediately obvious what kinds of models would be entertained that do not have an accessible p.d.f., m.g.f., c.f., or quantiles, rendering maximum likelihood and the methods of Section 5.2 inapplicable, though it turns out that several classes of models useful for inference, particularly in economics, are of precisely this form.

As its name makes clear, the indirect inference method (IIM), for computing the point estimator of θ, say $\widehat{\theta}_{\text{IIM}}$, is indirect. For the method to be operational, it presupposes that the assumed data-generating process can be (ideally quickly and easily) simulated, for a given parameter vector $\theta \in \Theta$, but it is not necessary to be able to evaluate its likelihood or characteristic function.

Section 10.1 introduces the IIM, explains when it is applicable, and outlines the basic theory of how and why it works. This is followed by three detailed cases that serve to exemplify it, illustrate its mechanics, and assess its performance.

10.1 INTRODUCTION

Denote the "true" (or supposed) data-generating process by $g(\theta)$. The indirect inference method consists of the following two steps. Firstly, we fit (usually via maximum likelihood, but any consistent method can be used) an incorrect but more easily estimated model to the data $Y_1, \ldots, Y_n$, referred to as the **auxiliary model**, denoted by $g_a(\gamma)$, with parameter vector $\gamma \in \Gamma$, to get $\widehat{\gamma}_{\text{data}}$. Secondly, $\widehat{\theta}$ is obtained by repeatedly simulating from $g(\theta)$ a set of data

Fundamental Statistical Inference: A Computational Approach, First Edition. Marc S. Paolella.
© 2018 John Wiley & Sons Ltd. Published 2018 by John Wiley & Sons Ltd.

$X_1, \ldots, X_{n^*}$ (where the choice of n^* is discussed below), based on a candidate value θ, and computing the m.l.e. of γ under model g_a, say $\widehat{\gamma}_{\text{sim}}(\theta)$, until $\widehat{\gamma}_{\text{data}}$ and $\widehat{\gamma}_{\text{sim}}(\theta)$ are close. Formally,

$$\widehat{\theta}_{\text{IIM}} = \arg\min_{\theta} \|\widehat{\gamma}_{\text{data}} - \widehat{\gamma}_{\text{sim}}(\theta)\|. \tag{10.1}$$

The method is valuable because there are models of interest for which a computable expression of the likelihood associated with g is problematic, computationally prohibitive, or does not exist. A simple example is the stable Paretian model, for which a closed-form expression for the p.d.f. and, thus, the likelihood, is not directly available. (In this case, several effective methods exist for computing the p.d.f. (see Section A.16), while other estimation techniques are also available (see Section 9.4).) Various models arising in economics, such as simultaneous equations, dynamic stochastic general equilibrium (DSGE), discrete choice, and discrete- and continuous-time stochastic volatility models, are significantly more complicated, and IIM is among the most viable and important methods of estimation and inference.[1]

While the idea embodied in (10.1) should appear intuitive, the technical apparatus underlying it is nontrivial; it was first developed by Smith (1993) and Gourieroux et al. (1993). There are several variations, such as the so-called efficient method of moments (EMM); see Gallant and Tauchen (1996).

It should be clear that the choice of auxiliary model will be decisive for the success of the method. There needs to exist a continuous mapping between θ and γ such that, for every $\theta \in \Theta$ (except possibly at the border of the parameter space), there exists a $\gamma \in \Gamma$ that "corresponds" to θ in the sense that $g_a(\gamma)$ is the best approximation to $g(\theta)$. This mapping is referred to as the **binding function**. We have actually already seen an example of a binding function in Figure 2.14(a), showing the estimated value of the tail index of a stable Paretian distribution, but based on Student's t data for a range of degrees-of-freedom values. One of the examples below illustrates the method for estimating the parameters of the i.i.d. stable Paretian model.

Above, in the description of the method, we did not explicitly state the sample size n^* to be drawn from g. Ideally, $n^* = \infty$, so that the true binding function is "traced out" while θ is being searched for, though not only is this theoretically impossible, but also, from a practical point of view, the larger n^* is taken to be, the longer the estimation process will take. Thus, as with all nontrivial modeling applications, a tradeoff between estimation quality and computation time needs to be made.

It is also imperative to realize that, for any (necessarily finite) value of n^*, the same *seed value* needs to be used for generating samples from g. The reason is that n^* is finite. For a value $\theta_0 \in \Theta$, with the corresponding value of the auxiliary parameter γ_0 based on the simulated data set $X_{0,1}, \ldots, X_{0,n^*}$, it must be the case that, for $\theta_1 \in \Theta$ "close" to θ_0, the resulting value of γ_1 based on the sample $X_{1,1}, \ldots, X_{1,n^*}$ is "close" to γ_0, that is, we need the binding function to be continuous. If the seed value is not kept constant, then each $X_{1,i}$ will not be a perturbation of $X_{0,i}$, and γ_1 will not be close to γ_0. This also implies that the

---

[1] A partial list of contributions includes Smith (1993), Gourieroux et al. (1993), Engle and Lee (1996), Gallant and Tauchen (1996), Gallant et al. (1997), Andersen and Lund (2003), Monfardini (1998), Andersen et al. (1999), Pastorello et al. (2000), Carrasco and Florens (2002), Ahn et al. (2002), Billio and Monfort (2003), Calzolari et al. (2004), Heggland and Frigessi (2004), Dridi et al. (2007), Sentana et al. (2008), and Gallant and McCulloch (2009).

choice of seed value will influence the final result, though as $n^* \rightarrow \infty$, its effect diminishes. Below, we will also investigate the effect on the estimator of using different seed values, and consider how to "integrate its effect out" by averaging.

10.2 APPLICATION TO THE LAPLACE DISTRIBUTION

We first demonstrate the method using a simple model to illustrate the mechanics. Consider an i.i.d. sequence of $n = 40$ Laplace r.v.s with location parameter μ and scale parameter σ, so that $\theta = (\mu, \sigma)'$. For simulating Laplace r.v.s, Problem 6.3 shows that if U_1 and U_2 are i.i.d. Unif(0, 1), then $L = \ln(U_1/U_2) \sim \text{Lap}(0, 1)$. The auxiliary model is taken to be an i.i.d. sequence of n^* normal r.v.s, for which the m.l.e. of its location and scale parameters $\gamma = (\mu_{\text{aux}}, \sigma_{\text{aux}})'$ is given in Example 3.5 and trivial to compute. In this case, the likelihood is accessible, so that we can use the m.l.e. as a benchmark to judge the quality of $\hat{\theta}_{\text{IIM}}$. In fact, the m.l.e. for data $X_1, \ldots, X_n$ is expressible in closed form as $\hat{\mu}_{\text{ML}} = \text{median}(X_1, \ldots, X_n)$ and $\hat{\sigma}_{\text{ML}} = n^{-1} \sum_{i=1}^{n} |X_i - \hat{\mu}_{\text{ML}}|$. In what follows, we monitor only the performance of the estimator of the scale parameter.

The program in Listing 10.1 implements the method, showing two ways to compue $\hat{\theta}_{\text{IIM}}$. The first way just uses a grid of σ-values and then (linear) interpolation to obtain $\hat{\sigma}_{\text{IIM}}$. This method is applicable when the parameter vector is one-dimensional, and helps to explicitly visualize the connection between θ and γ. In fact, the resulting sct of values can then be easily plotted to depict the binding function, shown in Figure 10.1(a), along with inscribed arrows indicating how $\hat{\sigma}_{\text{IIM}}$ is obtained from $\hat{\sigma}_{\text{aux}}$. We see that the binding function is linear, and it is easy to see why. The variance of $L \sim \text{Lap}(\mu, \sigma)$ is most easily calculated as follows: taking $\mu = 0$ and $\sigma = 1$ without loss of generality, we have that $\mathbb{V}(L) = \mathbb{E}[L^2] = \mathbb{E}[X^2]$, where $X \sim \text{Exp}(1)$, but $\mathbb{E}[X^2] = \int_0^\infty x^{3-1} \exp(-x) \, dx = \Gamma(3) = 2$, implying that the variance of $L \sim \text{Lap}(\mu, \sigma)$ is $2\sigma^2$. Thus, as the m.l.e. of the scale term under normality is just the square root of the sample variance (but using divisor n instead of $n - 1$), the binding function between σ and σ_{aux} is linear, with slope $\sqrt{2}$, achieved when taking $n^* = \infty$.

The second way to compute the IIM is more general, and makes use of a black-box method of minimization to search for $\hat{\theta}_{\text{IIM}}$. It can be used as a general template program for IIM. Figure 10.1(b) shows the mean squared error of $\hat{\sigma}_{\text{ML}}$, and $\hat{\sigma}_{\text{IIM}}$ for $n^* = 40, 200$, and ∞, where the latter is available from the analytic relation between σ and σ_{aux}. We confirm that the estimator improves as n^* increases, and also ascertain that, even for $n^* = \infty$, the m.l.e. is superior.

10.3 APPLICATION TO RANDOMIZED RESPONSE

10.3.1 Introduction

The **randomized response technique** is a way of designing survey questions about sensitive topics to enhance the chances of getting correct responses and avoiding **evasive answer bias**. For example, if the survey involves asking people if they cheat on their taxes, cheat on their spouses, have lied about worker disability, take drugs, engage in criminal activity or behaviors deemed socially reprehensible, etc., then the "guilty" respondent might

```
1   function [sigmahatIIM,sigmahatMLE] = iim_play(data,nstar)
2   % call as follows for demonstration:
3   %    n=20; sigma=1; % the parameter of interest; here, the Laplace scale
4   %    data = sigma * ( log(rand(1,n)./rand(1,n)) ); % true Laplace data
5   %    [sigmahatIIM,sigmahatMLE] = iim_play(data,nstar)
6   n=length(data); if nargin<2, nstar=1e5; end
7   muhat=median(data); sigmahatMLE = mean(abs(data-muhat)); % MLE
8   % IIM using a grid of values and interpolation and plots the binding function
9   param_a = sqrt(var(data,1)); % MLE under the auxiliary model
10  sigmagrd = 0.1:0.1:2; % we know in this case the function is linear
11                       % so the grid coarseness is not an issue.
12  glen=length(sigmagrd); param_a_grd=zeros(glen,1);
13  % For the simulations, we want to use the same seed value across
14  %    different values of sigma, but as sigma is a scale parameter,
15  %    we can do this more quickly as follows:
16  X= ( log(rand(1,nstar)./rand(1,nstar)) );
17  if 1==1
18    for g=1:glen
19      use=sigmagrd(g)*X; % now just change the scale
20      param_a_grd(g)=sqrt(var(use,1)); % its MLE under aux
21    end
22    plot(sigmagrd,param_a_grd,'r-','linewidth',2), set(gca,'fontsize',16)
23    xlabel('Parameter of Interest (Laplace scale term)')
24    ylabel('Parameter of Auxiliary Model (normal scale term)')
25    title('Binding Function for Indirect Inference')
26    sigmahatIIM0 = interp1(param_a_grd,sigmagrd,param_a);
27    h1=arrow([0 param_a],[sigmahatIIM0 param_a]);
28    h2=arrow([sigmahatIIM0 param_a],[sigmahatIIM0 0]);
29    set(h1,'linewidth',2,'edgecolor','b','facecolor','b')
30    set(h2,'linewidth',2,'edgecolor','b','facecolor','b')
31  end
32  % The more general method, using a minimization algorithm
33  maxiter=200; tol=1e-5; MaxFunEvals=400;
34  opts=optimset('Display','none','Maxiter',maxiter,'TolFun',tol,...
35               'TolX',tol,'MaxFunEvals',MaxFunEvals,'LargeScale','Off');
36  initvec=sigmahatMLE; % this is of course not realistic in general!
37  bound.lo=1e-4; bound.hi=10; bound.which=1;
38  pout=fminunc(@(param) AuxFunc(param,X,param_a,bound), ...
39               einschrk(initvec,bound),opts);
40  sigmahatIIM=einschrk(pout,bound,1);
41
42  function disc=AuxFunc(param,X,param_a,bound)
43  paramvec=einschrk(real(param),bound,999);   sigma=paramvec(1);
44  use=sigma*X; candidate=sqrt(var(use,1)); % its MLE under aux
45  disc = (candidate - param_a).^2;
```

Program Listing 10.1: Indirect inference for estimating σ, the scale term of the Laplace distribution (with known mean zero) based on n i.i.d. observations, using the scale term of a normal distribution as the auxiliary parameter and model. `data` is the n-length vector of Laplace observations, and `nstar` is the sample size when simulating the target (Laplace) model. The function `arrow` can be downloaded for free from the Mathworks website.

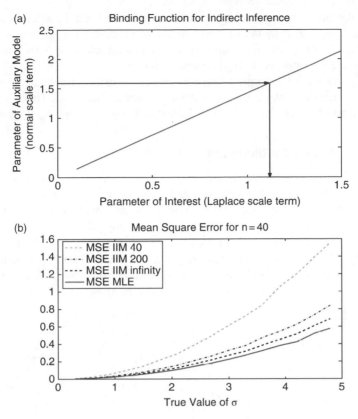

Figure 10.1 *(a) The binding function between the desired parameter (scale term of the Laplace) and the parameter of the auxiliary model (scale term of the normal) for a random data set with $n = 40$ observations, and using $n^* = 100,000$. For this data set, the estimated value of the scale term under the auxiliary (normal) model is $\hat{\sigma}_{aux} = 1.590$, yielding the estimate of the scale term under the desired model of $\hat{\sigma}_{IIM} = 1.120$. Observe that $1.590/\sqrt{2} = 1.124 \approx 1.120$, which would be exact if we used $n^* = \infty$. The m.l.e. for the data set was $\hat{\sigma}_{ML} = 1.1818$. (b) The mean squared error of $\hat{\sigma}_{ML}$ and $\hat{\sigma}_{IIM}$, based on 10,000 replications, for $n = 40$ observations, as a function of the true scale parameter σ. For $\hat{\sigma}_{IIM}$, we use two values of n^* and the value $n^* = \infty$, for which $\hat{\sigma}_{IIM} = \hat{\sigma}_{aux}/\sqrt{2}$.*

well lie about his or her transgressions out of embarrassment, fear of nonanonymity, legal action, and stigmatization. The seminal work on this is Warner (1965), followed by a large academic literature, including the textbook survey by Chaudhuri (2010) and the overview emphasizing estimation methods in Blair et al. (2015).

There are several ways of designing such questions, such as the **forced response**, which involves asking the respondent to roll a fair, six-sided die, say, taking care not reveal the outcome to the interviewer, and if the outcome is a 1, then answer the actual question "no"; if a 6, answer "yes"; and otherwise answer the question truthfully. Another setup is the **unrelated question design**, again making use of some randomization device whose outcome is not revealed to the interviewer, say again the fair die, and such that if the die outcome is 1 or 2, the respondent answers an innocuous question, such as "Does the month in which you were born start with a J?", and otherwise answers the real question of interest.

Consider the latter setup, such that the question of interest and the artificial question are of the yes/no type. Let θ be the true probability of interest; let θ_A be the probability of answering "yes" to the artificial question, with θ_A known, such as 1/4 in the case of months starting with a J (as opposed to it also being unknown, such as "Do you like green-tea ice cream?"); and let π be the probability, determined from the randomization device, of obtaining the question of interest, as opposed to the artificial question. Then the probability of a "yes" response, by the law of total probability (A.30), is $\lambda = \pi\theta + (1 - \pi)\theta_A$.

10.3.2   Estimation via Indirect Inference

The indirect inference paradigm is not only applicable here, but also very straightforward. It appears to have been overlooked as an alternative estimation methodology. Let $X_1, \ldots, X_n$ be the i.i.d. sequence of Bernoulli responses, with outcome 1 indicating a "yes" response. Then, unconditionally, that is, not knowing which question was answered, $f_{X_i}(x) = \lambda^x(1 - \lambda)^{1-x}\mathbb{1}_{\{0,1\}}(x)$. A natural choice of auxiliary model is the binomial, with $(S_X \mid \theta) = \sum_{i=1}^{n} X_i \sim \text{Bin}(n, \lambda)$, so that $\hat{\gamma}_{\text{data}} = S_X/n$. Figure 10.2 shows the binding function for a particular constellation of parameters and three values of π, from which it is clear it is linear. Thus, estimation via IIM can be done by computing the binding

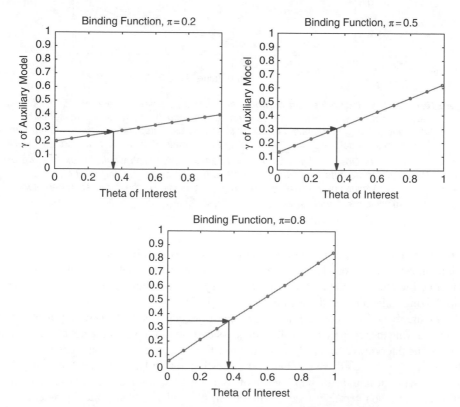

Figure 10.2   *The binding function in the randomized response setup, based on $\theta = 0.4$, $\theta_A = 0.25$, and three values of π.*

```
1    function [thetahatIIM,data] = RandRespIIM(data, truetheta)
2    % IIM for the Randomized Response Method
3    % data is either the bernoulli sequence of answers, or the seed value to
4    %   generate the data set, along with the 2nd argument: The true theta
5    global nstar pie theta
6    pie=0.5; thetaA=0.25; nstar=1e5;
7    if length(data)==1
8      seed=data; n=1e3; data=RRsim(n,truetheta,seed);
9    end
10   if isempty(data), n=1e3; data=RRsim(n,truetheta,1); end
11   gMLE=mean(data); % MLE under aux model g, Bin(n, theta)
12   if 1==1 % IIM using a grid of values.
13     thetagrd=[0.01 0.5 0.99];
14     glen=length(thetagrd); param_grd=zeros(glen,1);
15     for g=1:glen  % mean(Xsim) is the MLE under aux model
16       tt=thetagrd(g); Xsim=RRsim(nstar,tt); param_grd(g)=mean(Xsim);
17     end
18     thetahatIIM0 = interp1(param_grd,thetagrd,gMLE);
19     if 1==2 % plots the binding function
20       figure,
21       plot(thetagrd,param_grd,'r-o','linewidth',2), set(gca,'fontsize',16)
22       xlabel('Theta of Interest'), ylabel('\gamma of Auxiliary Model')
23       title('Binding Function')
24       h1=arrow([0 gMLE],[thetahatIIM0 gMLE]);
25       h2=arrow([thetahatIIM0 gMLE],[thetahatIIM0 0]);
26       set(h1,'linewidth',2,'edgecolor','b','facecolor','b')
27       set(h2,'linewidth',2,'edgecolor','b','facecolor','b')
28     end
29     thetahatIIM = thetahatIIM0;
30     return
31   end
32
33   % Use a minimization algorithm to get the estimate
34   maxiter=6; tol=1e-3;
35   opts=optimset('Display','iter','Maxiter',maxiter,'TolFun',tol,'TolX',tol);
36   bound.lo=1e-4; bound.hi=1-1e-4; bound.which=1; inittheta=0.6;
37   pout=fminsearch(@(param) AuxFunc(param,gMLE,bound), ...
38              einschrk(inittheta,bound),opts);
39   thetahatIIM=einschrk(pout,bound,1);
40
41   function X=RRsim(n,truetheta,seed)
42   global pie theta
43   if nargin<3, seed=1; end
44   stream = RandStream('mt19937ar','Seed',seed);
45   RandStream.setDefaultStream(stream);
46   X=zeros(n,1);
47   for i=1:n
48     usetrue=binornd(1,pie,1,1);
49     X(i) = binornd(1,usetrue*truetheta+(1-usetrue)*thetaA,1,1);
50   end
51
52   function disc=AuxFunc(param,gMLE,bound)
53   global nstar
54   thetahat=einschrk(real(param),bound,999);
55   X=RRsim(nstar,thetahat); candidate=mean(X);
56   disc = (candidate - gMLE).^2;
```

Program Listing 10.2: Simulates and then estimates the randomized response model via the IIM.

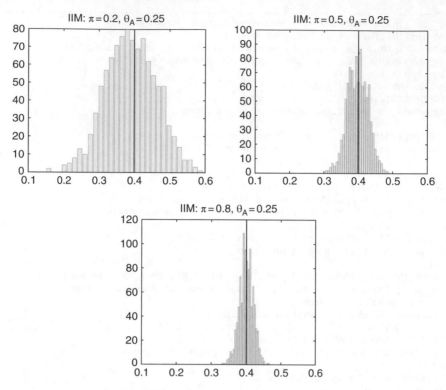

Figure 10.3   *Histograms of $\hat{\theta}_{IIM}$ based on a sample size of $n = 1000$, 1000 replications, true $\theta = 0.4$ (as indicated by the vertical line), $\theta_A = 0.25$, and three values of π.*

function based on two different values of θ (we use three: 0.01, 0.5, and 0.99), and taking $n^* = 100,000$, followed by linear interpolation.

As π decreases, so does the "effective sample size" of respondents to the actual question of interest, so the variance of $\widehat{\theta}_{IIM}$ is expected to increase. This is also revealed in Figure 10.2, showing that the slope of the binding function decreases as π decreases, inducing a wider variation in $\widehat{\theta}_{IIM}$ to occur for a given range of γ.

The program in Listing 10.2 implements this, and also includes code for the more general use of computing (10.1).[2] Figure 10.3 shows the behavior of $\widehat{\theta}_{IIM}$ based on a sample size of $n = 1000$, $\theta = 0.4$, $\theta_A = 0.25$, and the same values of π as used in Figure 10.2. The IIM estimator appears unbiased and (unsurprisingly for this large sample size) Gaussian, though we see that, as π decreases, the variance of $\widehat{\theta}_{IIM}$ increases. Use of other values of θ yields similar performance.

The reader is encouraged to develop the EM algorithm required for maximum likelihood estimation in this context, and compare the performance of $\widehat{\theta}_{IIM}$ and $\widehat{\theta}_{ML}$; see Blair et al. (2015) for details.

---

[2] The latter method, despite using the same seed value for the random draws, did not perform well with `fminunc`, while use of the non-gradient-based `fminsearch` works well, albeit more slowly than use of the simple linear interpolation of the binding function, and returns the same value of $\widehat{\theta}_{IIM}$ as determined by the use of interpolation from the linear binding function, even with starting values far away from the truth.

10.4 APPLICATION TO THE STABLE PARETIAN DISTRIBUTION

The next illustration of the method is less trivial, using the IIM to estimate the four parameters $\theta = (\alpha, \beta, \mu, c)'$ of the location–scale asymmetric stable. Given that the stable distribution has Pareto-type (power) tails as $|y| \to \infty$, it might seem sensible to use one of the various asymmetric Student's t distributions as the auxiliary model, as these exhibit similar tail behavior. Several such candidates were presented in Book II; here, we will first use the Fernández and Steel (1998) asymmetric extension of the Student's t (with location and scale parameters), which is a special case of the GAt distribution (A.125) with $d = 2$. This was also the choice of auxiliary model used in Garcia et al. (2009) and Lombardi and Calzolari (2008), who independently studied its application for estimating the i.i.d. univariate stable model. See Lombardi and Veredas (2009) for its use for estimating the parameters of the multivariate elliptical stable distribution.

The first requirement to implement indirect inference is a program to draw a random sample from the $S_{\alpha, \beta}(\mu, c)$ distribution, for a specified seed value. This is shown in Listing A.6, implementing (A.307).

We conducted a simulation study based on 100 replications, for two sample sizes ($n = 100$ and $n = 1000$) for the i.i.d. stable model with $\alpha = 1.6$, $\beta = 0$, location $\mu = 0$ and scale $c = 1$, using four values of n^*, namely $n^* = 100, 250, 500, 1000$, and the constant seed value (recall the discussion about the necessity of this at the end of Section 10.1) of 1. For each, the initial value is taken to be c.f.-based estimator $\widehat{\theta}_{CF}$ discussed in Section 9.4. This arguably gives IIM an unfair advantage because these starting values are excellent (and arguably render the use of IIM unnecessary) and the function to be minimized may not be smooth or unimodal.

Consider the first issue, with smoothness of the objective function, h. Each point of h is itself the result of a likelihood maximization (of the asymmetric Student's t), and while we used a tolerance of 10^{-8} for these, apparently h is still not smooth enough for use with the BFGS algorithm. In particular, in the majority of cases associated with this simulation exercise, BFGS did not move from the starting value – and several starting values were tried.

Thus, use of $\widehat{\theta}_{CF}$ as an initial estimate and blind application of BFGS would erroneously produce essentially the same performance results as those from $\widehat{\theta}_{CF}$. The non-gradient-based optimization methods, such as CMAES, as given in Listing 4.13, did not suffer from this problem. As such, we use, and recommend use of, CMAES.

Turning to the second issue of nonunimodality, if the objective function h has multiple minima, then using $\widehat{\theta}_{CF}$ as starting values will bring it immediately close to the global minimum with high probability. In the far more realistic case in which such a convenient initial estimate is not available, one could compute the IIM estimator for each of, say, 10 starting values, formed from a grid over a range of α-values, and choose the best one, thus significantly increasing the probability of obtaining the global minimum. This, of course, would take far longer than using an accurate starting value. Moreover, the idea of using a grid was stated for only one of the four parameters, whereas in principle, a four-dimensional grid should be entertained, for which the curse of dimensionality quickly becomes present.

A better idea would be to use the grid of, say, 10 α-values and, for each, fix α, and estimate (via IIM if that is all that is available) the remaining three parameters, this being analogous to computing the profile likelihood. Then the best fit of those 10 is chosen and used as a

```
1   function iimstablesim(n,auxmodel,seedval)
2   sim=100; alpha=1.6;
3   filename = ['IIMs',int2str(seedval),'n',int2str(n),'aux', ...
4                 int2str(auxmodel),'alpha',int2str(10*alpha)]
5   if ~exist(strcat(filename,'.mat'),'file')
6     nstarvec=[100 250 500 1000];
7     for j=1:length(nstarvec)
8       nstar=nstarvec(j);
9       eval(['a',int2str(nstar),'=zeros(sim,1);'])
10      for i=1:sim, i\_and\_nstar = [i nstar]
11        data=stabgen (n,alpha,0,1,0,i+1);
12        paramIIM = iimstable(data,auxmodel,nstar,[],[],seedval);
13        eval(['a',int2str(nstar),'(i)=paramIIM(1);'])
14      end
15      save(filename)
16    end
17    aNL1=zeros(sim,1); aNL3=zeros(sim,1);
18    for i=1:sim
19      data=stabgen (n,alpha,0,1,0,i+1);
20      param = stablefit(data,1,1); aNL1(i)=param(1);
21      param = stablefit(data,3,1); aNL3(i)=param(1);
22    end
23    save(filename)
```

Program Listing 10.3: Simulates indirect inference for the stable Paretian model with a fixed tail index α (set in the program); continued in Listing 10.4. Input n is sample size; set auxmodel to 3 for GAt, 5 for NIG, 6 for GAt with $d = 2$ (and thus the asymmetric Student's t) and 7 for generalized logistic. The function iimstable is given in Listing 10.5; the function stablefit is part of John Nolan's Stable toolbox, and estimates the four parameters of the univariate stable distribution.

```
1   else
2     load(filename)
3     if verLessThan('matlab', '7.8')
4       lab={'n*=100','n*=250','n*=500','n*=1000','MLE','ch.f.'};
5     else
6       lab={'$n^*=100$','$n^*=250$','$n^*=500$','$n^*=1000$','MLE','ch.f.'};
7     end
8     figure, boxplot([a100 a250 a500 a1000 aNL1 aNL3],'labels',lab)
9     set(gca,'fontsize',16), ylim([1 2]), ylabel('')
10    if ~verLessThan('matlab', '7.8')
11      h = get(get(gca,'Children'),'Children');
12      ht = findobj(h,'type','text'); set(ht,'fontsize',18,'Rotation',30)
13      set(ht,'Interpreter','latex','horizontalalignment','right')
14    end
15    line([0.5 6.5],[alpha alpha],'linestyle','--','color','g','linewidth',2)
16    str=['IIM of \alpha (true = ',num2str(alpha),', n=',int2str(n),')'];
17    if auxmodel==3, str=strcat(str,' Using GAt');
18    elseif auxmodel==5, str=strcat(str,' Using NIG');
19    elseif auxmodel==6, str=strcat(str,' Using Skew t');
20    elseif auxmodel==7, str=strcat(str,' Using GenLog');
21    end
22    str=strcat(str,' as Aux'); title(str)
23  end
```

Program Listing 10.4: Continuation of the program iimstablesim in Listing 10.3.

starting value, with all four parameters allowed to vary. This idea was used to avoid inferior local maxima for GARCH models in Paolella and Polak (2015a).

The programs required for computing the IIM estimator in this context, and conducting the simulation study, are given in Listings 10.3–10.6. Only the estimates for $\hat{\alpha}_{\text{IIM}}$ are shown, as boxplots, in the top two panels of Figure 10.4, along with the corresponding results based on $\hat{\theta}_{\text{ML}}$ and $\hat{\theta}_{\text{CF}}$. We see that, for this choice of seed value (1) and this choice of true parameter α (1.6), there is a bit of a downward bias of $\hat{\alpha}_{\text{IIM}}$ with $n^* = 250$ for both sample sizes $n = 100$ and $n = 1000$, but otherwise the method is nearly unbiased and it performs quite admirably relative to the m.l.e. with respect to the spread of the distribution.

```
1   function paramIIM = iimstable(data,auxmodel,nstar,dval,initvec,seed)
2
3   if nargin<6, seed=1; end, if nargin<5, initvec=[]; end
4   if nargin<4, dval-[]; end, if nargin<3, nstar=1e3; end
5   if isempty(dval), dval=2; end
6
7   data=reshape(data,length(data),1);
8   if isempty(initvec)
9     NOL = stablefit(data,3,1); initvec=[NOL(1) NOL(2) NOL(4) NOL(3)];
10  end
11  %%%%%%%%%       alpha    beta    mu    scale
12  bound.lo=    [0.2000    -1    -1    0.01];
13  bound.hi=    [1.9999     1     2     100];
14  bound.which=[1           1     0       1];
15
16  maxiter=300; tol=1e-5; MaxFunEvals=9000;
17  opts=optimset('Display' , 'Iter','MaxIter',maxiter,'TolFun',tol,
18            'TolX',tol,'MaxFunEvals',MaxFunEvals,'LargeScale','Off');
19  if       auxmodel==3, param_a = GAtestimation(data);
20  elseif auxmodel==5, param_a = NIGestimation(data);
21  elseif auxmodel==6, param_a = GAtestimation(data,[],[],dval);
22  elseif auxmodel==7, param_a = GLestimation(data);
23  end
24
25  if 1==2
26    pout=fminsearch(@(param) AuxFunc(param,param_a,bound,nstar, ...
27          auxmodel,dval,seed), einschrk(initvec,bound),opts);
28  elseif 1==2
29    pout=fminunc(@(param) AuxFunc(param,param_a,bound,nstar, ...
30          auxmodel,dval,seed), einschrk(initvec,bound),opts);
31  elseif 1==1
32    pout=fmincmaes(@(param) AuxFunc(param,param_a,bound,nstar, ...
33          auxmodel,dval,seed), einschrk(initvec,bound),opts);
34  end
35  paramIIM=einschrk(pout,bound,1);
```

Program Listing 10.5: Performs indirect inference of the stable Paretian model, returning a vector of estimates for α, β, location μ and scale c. Input `nstar` is the sample size of the draws. See Listing 10.3 for detail on the auxiliary models, and Listing 10.6 for the function `AuxFunc`. The program assumes we have functions `NIGestimation` and `GLestimation`, which are trivially constructed based on function `GAtestimation`, given in Listing 4.9.

```
1   function disc=AuxFunc(param,param_a,bound,nstar,auxmodel,dval,seed)
2   paramvec=einschrk(real(param),bound,999);
3   alpha=paramvec(1); beta=paramvec(2);
4   mu=paramvec(3); scale=paramvec(4);
5
6   use=stabgen (nstar,alpha,beta,scale,mu,seed);
7   use=reshape(use,length(use),1);
8   if auxmodel==3
9      candidate=GAtestimation(use,[],param_a');
10  elseif auxmodel==5
11     candidate=NIGestimation(use,param_a');
12  elseif auxmodel==6
13     candidate=GAtestimation(use,[],param_a',dval);
14  elseif auxmodel==7
15     candidate=GLestimation(use,param_a');
16  end
17  disc = sum( (candidate – param_a).^2 );
```

Program Listing 10.6: Continuation of Listing 10.5.

Before turning to the performance of the method using other auxiliary models, we mention that, with respect to computation time, the IIM estimates for the 100 replications in the simulation study required just over 3 hours, whereas for the 100 replications (not each, but for all 100) the m.l.e. based on the spline approximation to the stable density in Nolan's toolbox required about 1 second, and the characteristic function-based estimator required about 0.02 seconds.

The remaining panels of Figure 10.4 are similar to the top panels, but based on different auxiliary models. Of course, the same 100 sets of data were used, as was the same seed value (1). The first uses the GAt distribution, which nests the previous auxiliary model used; the second uses the normal inverse Gaussian (NIG) distribution (9.17); while the third uses the generalized logistic (GL) distribution (see Problems II.7.5 and II.7.6). While, roughly speaking, the bias of $\hat{\alpha}_{\text{IIM}}$ does appear to decrease as n^* increases, that is about the only good news; the performance of the method is drastically worse than when based on the original asymmetric Student's t model, thus showing that the choice of auxiliary model will be critical for the success of the method.

As the GAt model nests the asymmetric t, it might have been expected to perform similarly to the use of the latter. It is, indeed, the best of the three other auxiliary models considered, but still considerably worse than the asymmetric t, indicating that too much flexibility of the auxiliary model can be detrimental. Recall from Section II.9.6.2 that the NIG distribution has so-called semi-heavy tails, while the GL is a thin-tailed distribution. These results lend some evidence to the conclusion that (i) the auxiliary model g_a ideally has the same number of parameters as the assumed model g; (ii) each of the parameters of g_a can be associated with a similarly interpretable parameter of g (here, the degrees of freedom and tail index α, etc.), and certain properties of g and g_a coincide, such as the tail behavior (here, both the asymmetric t and stable Paretian have power tails). Based on these results with respect to the choice of auxiliary model, we subsequently only consider use of the asymmetric t. See also Problem 10.1.

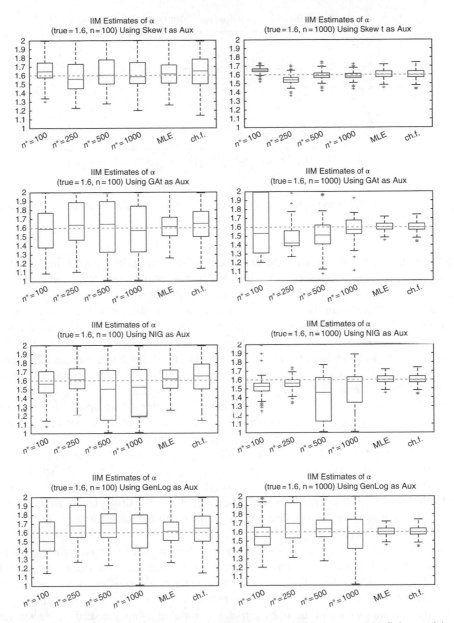

Figure 10.4 *Boxplots of $\hat{\alpha}_{IIM}$ based on 100 replications for sample sizes $n = 100$ (left panels) and $n = 1000$ (right panels) for four different auxiliary models and four values of n^*, along with $\hat{\alpha}_{ML}$ and $\hat{\alpha}_{CF}$, with the latter used as a starting value.*

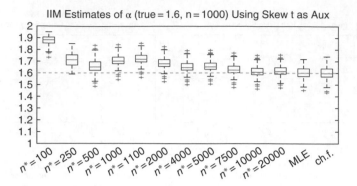

Figure 10.5   Similar to the top right panel of Figure 10.4, but using a different seed value (105 instead of 1) and more values of n^*.

Recall the importance of using the same seed value during the computation of the IIM estimator. This implies that the results are conditional on that seed value, and will change when the seed value is changed. To investigate this, the results in the top right panel of Figure 10.4 (pertaining to the use of the asymmetric t distribution as the auxiliary model, with $n = 1000$) were recomputed using the same 100 samples, but a different seed value (105), with the results shown in Figure 10.5. Now the results are less impressive for $n^* \leq 1000$, with a significant bias. It indeed appears that, as n^* increases, the bias disappears, and along the way there is a certain damping oscillatory behavior, so that it is not the case that the performance monotonically increases as n^* increases.

Based on this finding, and the fact that there is no prior information (i.e., no natural preferences or aversions) for the choice of seed value, it would seem logical to conclude that the best choice would be to average the results computed over many seed values. This is justified if, for each seed value s, we can decompose the IIM estimator for a given seed value, $\widehat{\theta}_{\text{IIM}}(s)$, as $\widehat{\theta}_{\text{IIM}}(s) = \widehat{\theta}_{\text{IIM}} + b(s)$, where $\widehat{\theta}_{\text{IIM}}$ represents the "true" IIM estimator (for a given auxiliary model and choice of n^*) and $b(s)$ is a bias term resulting from the particular seed value, with expectation zero and finite variance. From the strong law of large numbers, if we average over many $\widehat{\theta}_{\text{IIM}}(s)$, then the bias of the resulting estimator due to the $b(s)$ term should be close to zero.

Observe, crucially, that this does *not* mean that if we average over enough $\widehat{\theta}_{\text{IIM}}(s)$, then the resulting IIM estimator is unbiased; it only means that we numerically approach the value of $\widehat{\theta}_{\text{IIM}}$, which is devoid of bias from the $b(s)$ term, but itself might still be biased in finite samples. The price to pay for removing the effect of $b(s)$ is estimation time: We saw above that the simulation of the IIM estimator, computed using one seed value, was quite time-consuming, and this should now be repeated as many times as feasible, using different seed values. We content ourselves with using a set of 10 seed values $(102, 103, \ldots, 111)$, a single value of n^*, and also investigate the behavior not just at the value of $\alpha = 1.6$, but over a grid of 10 α-values.

Figure 10.6 shows the results, for the two sample sizes $n = 100$ (left) and $n = 1000$ (right), and the fixed value of $n^* = 1000$. To facilitate comparison, the boxplots now show the difference $\widehat{\alpha}_{\text{IIM}} - \alpha$. The first and second rows are based on seed values 1 and 105, respectively, while the third row is based on the average of the estimates resulting from

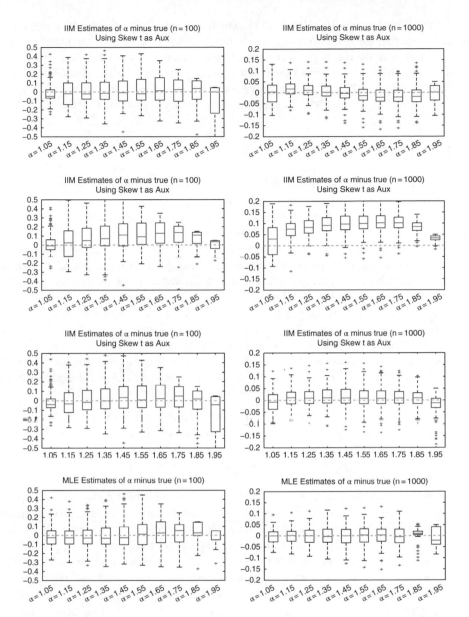

Figure 10.6 *Boxplots of $\hat{\alpha}_{IIM} - \alpha$, over a grid of α-values, based on the asymmetric t auxiliary model (denoted by Skew t) and $n^* = 1000$, for two sample sizes, $n = 100$ (left) and $n = 1000$ (right). The panels in the first row are based on the seed value 1; the panels in the second row are based on the seed value 105; those in the third row are based on averaging the results over the 10 seed values $102, 103, \dots, 111$; and the last row shows the performance of the m.l.e.*

using the 10 seed values 102,103, ... , 111. The fourth row shows the results based on the m.l.e. Comparing the first two rows, we see the remarkable difference the particular seed value makes, with the use of the second one (the arbitrarily chosen 105) clearly inferior to the first one. The third row shows the result of averaging over 10 seed values. We see that, particularly for $n = 1000$, it is virtually as good as the use of seed value 1 (top row), and is, with the exception of $\alpha = 1.95$, of nearly the same quality as exhibited by the m.l.e.

10.5 PROBLEMS

10.1 This exercise provides further evidence that the quality of the IIM estimator is highly sensitive to the choice of auxiliary model. In particular, we show how the quality changes as a function of a fixed parameter associated the auxiliary model.

   We saw that the IIM estimator of the stable Paretian model was best when using the skewed t model, which is a special case of the GAt with $d = 2$. While it was also superior to the GAt when d was estimated as a free parameter, it is unlikely that the optimal value of d is precisely 2. Investigate the performance for several values of d using a grid of α-values similar to those in Figure 10.6, and based on averaging the IIM estimator over 10 (or more) seed values. The results based on using six values of d besides 2.0, and a sample size of $n = 1000$, are shown in Figure 10.7, where each boxplot is comparable to the one in the third row, second column, of Figure 10.6. We see that, as d decreases, the quality for α near 2 decreases, while the quality for the other values of α is not affected. Reproduce these results, and additionally study the quality of the estimates of the other estimated stable parameters.

10.2 Compute the IIM estimator for g being the Student's t distribution (with location and scale parameters μ and σ, respectively) with unknown degrees of freedom v and using as the auxiliary model g_a a k-component mixture of normals, with parameters μ_i, σ_i and λ_i, $i = 1, \dots , k$, but such that the means are constrained to be equal, $\mu_1 = \mu_2 = \dots = \mu_k$.

   To illustrate for $k = 2$, Figure 10.8(a) shows part of the binding function, computed based on $n = 100,000$ observations. It shows just the ratio of the larger to the smaller variance, $\hat{\sigma}_2 / \hat{\sigma}_1$, which we expect to be relevant for v, and $\hat{\lambda}_1$, the estimated weight of the first component, as a function of v. Reproduce this graph.

   Indeed, as v increases, $\hat{\sigma}_2 / \hat{\sigma}_1$ monotonically decreases towards 1, though the behavior of $\hat{\lambda}_1$ was not as obvious to predict. As v decreases, the weight of the first component moves towards 1, while the ratio $\hat{\sigma}_2 / \hat{\sigma}_1$ begins to approach an asymptote.[3]

   Figure 10.8(b,c) shows the result of a simulation study, comparing the IIM estimator to the m.l.e. for parameter v. We see that the m.l.e. is better for each of the four values of v considered in terms of variance, and also that both estimators tend to be

---

[3] Presumably, with $k = 2$, it makes sense to consider only values of v larger than 2, otherwise it is doubtful that the scale parameters σ_i can be estimated, given the meaning of their squares. Nevertheless, keeping in mind that the Student's t is an infinite mixture of normals, for any value $v > 0$, it is possibly the case that, for a given value of v, say v^*, as long as the sample size used to determine the binding function, n, tends to infinity, there exists a value $k^* = k^*(v^*)$ such that, for all $k \geq k^*$, the binding function is well defined and continuous in v on some open ball within the allowed parameter range, and the IIM estimator is computable.

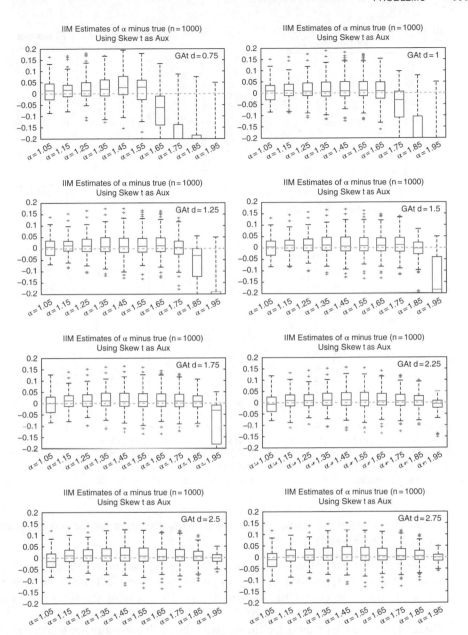

Figure 10.7 *Boxplots of $\hat{\alpha}_{IIM} - \alpha$ over a grid of α-values, based on the GAt auxiliary model and $n^* = 1000$, for the single sample size $n = 1000$, from averaging the results over the 10 seed values $102, 103, \ldots, 111$, and using different fixed values of the parameter d in the GAt distribution. The box-plot based on $d = 2$ is in Figure 10.6, third row, second column.*

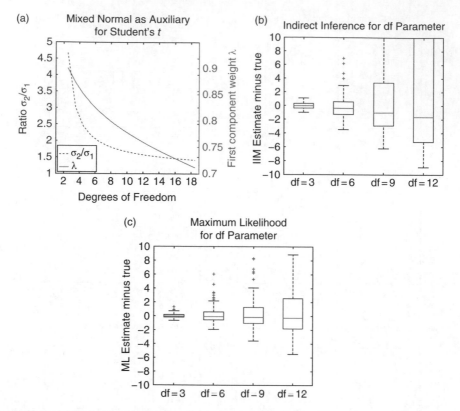

Figure 10.8 *(a) Part of the binding function between g, the Student's t, and g_a, the two-component mixed normal. (b and c) Boxplots of the estimated parameter v (degrees of freedom) of the Student's t distribution, minus the true value, for indirect inference (b) and maximum likelihood (c) based on $n = 1000$ observations and 100 replications. The IIM values are based on use of a two-component mixed normal distribution with equal-mean restriction ($\mu_1 = \mu_2$) and $n^* = 10,000$.*

upward biased as v increases, but the bias is markedly larger for the IIM estimator. Reproduce these results.

10.3 Recall from Section 9.3 that computing the m.l.e. of the noncentral Student's t (NCT) using the density expression (A.339) (even possibly vectorized) is rather slow. Serving as an ironic converse of the study in Section 10.4, develop the IIM for the four parameters of the location–scale NCT using the location–scale asymmetric stable Paretian as the auxiliary model, estimated with the c.f.-based estimator $\hat{\theta}_{CF}$ discussed in Section 9.4. Observe that $\hat{\theta}_{CF}$ is nearly instantaneously computed and is nearly as efficient as the m.l.e.

Appendix A

Review of Fundamental Concepts in Probability Theory

Anyone who conducts an argument by appealing to authority is not using his intelligence; he is just using his memory.

(Leonardo da Vinci)

In addition to serving as a review, this chapter collects various formulas referred to throughout the book, helping to keep this volume reasonably self-contained. Throughout, occasional reference to equations and sections in Paolella (2006, 2007) will be made. These books are denoted by the capital Roman numerals I and II, respectively. Mastery of the primary concepts contained in Book I (or in any book at a comparable level) is essential, while some elements from Book II will be occasionally required, such as moment generating functions, characteristic functions, the multivariate normal distribution, the basics of saddlepoint approximations, order statistics, the stable Paretian distribution, and concepts related to inequalities, convergence, and asymptotic theory. These topics are summarized here for convenience and can be referred to as needed, but do not necessarily need to be studied in order to gain an appreciation of, and some competence with, the major concepts associated with statistical inference.

Along with some results in the exercises, new material given here that is not found in Books I and II includes a detailed section on expected shortfall, more computer code for the stable Paretian distribution, and proofs of the if-and-only-if version of Helly–Bray and the continuous mapping theorem (as used for proving the asymptotic normality of the m.l.e.).

The exercises in this review appendix require solid knowledge of basic probability theory and calculus, and serve as a testing ground for the skills required for this book. Some are challenging; many are useful; all have solutions provided. The student is advised to try them first before inspecting the solutions, and to self-assess those elements of fundamental probability or calculus requiring review.

Fundamental Statistical Inference: A Computational Approach, First Edition. Marc S. Paolella.
© 2018 John Wiley & Sons Ltd. Published 2018 by John Wiley & Sons Ltd.

A.1 COMBINATORICS AND SPECIAL FUNCTIONS

The number of ways in which n distinguishable objects can be ordered is given by

$$n(n-1)(n-2)\cdots 2\cdot 1 = n!,$$

pronounced "n factorial," and such that $0! \equiv 1$. The number of ways in which k objects can be chosen from n, $0 \le k \le n$, when order is irrelevant is the binomial coefficient

$$\frac{n(n-1)\cdots(n-k+1)}{k!} = \frac{n!}{(n-k)!\,k!} =: \binom{n}{k}, \tag{A.1}$$

pronounced "n choose k." The results

$$\binom{n}{k} = \binom{n-1}{k} + \binom{n-1}{k-1}, \quad 1 \le k < n,$$

and its generalization

$$\binom{n}{k} = \sum_{i=0}^{k} \binom{n-i-1}{k-i}, \quad 1 \le k < n,$$

are fundamental identities, along with, for $\binom{a}{b} \equiv 0$ if $b > a$,

$$\binom{N+M}{k} = \sum_{i=0}^{k} \binom{N}{i}\binom{M}{k-i}.$$

As a generalization of (A.1), if a set of n distinct objects is to be divided into k distinct groups whereby the size of each group is n_i, $i = 1, \dots, k$, and $\sum_{i=1}^{k} n_i = n$, then the number of possible divisions is given by

$$\binom{n}{n_1, n_2, \dots, n_k} := \binom{n}{n_1}\binom{n-n_1}{n_2}\binom{n-n_1-n_2}{n_3}\cdots\binom{n_k}{n_k} = \frac{n!}{n_1!\,n_2!\cdots n_k!}. \tag{A.2}$$

Another useful generalization of (A.1) is, for n and k positive integers,

$$\binom{-n}{k} = (-1)^k \binom{n+k-1}{k}. \tag{A.3}$$

Example   *Use of (A.3) and the Taylor series expansion of $f(x;t) = (1+x)^t$, $t \in \mathbb{R}$, $|x| < 1$, given by*

$$f(x;t) = \sum_{j=0}^{\infty} \binom{t}{j} x^j, \quad |x| < 1,$$

leads to, for $n \in \mathbb{N}_{>0}$,

$$g(x;n) = (1-x)^{-n} = \sum_{j=0}^{\infty} \binom{-n}{j}(-x)^j = \sum_{j=0}^{\infty} \binom{n+j-1}{j} x^j, \quad |x| < 1.$$

For n a positive integer, the binomial theorem is

$$(x+y)^n = \sum_{i=0}^{n} \binom{n}{i} x^i y^{n-i}. \tag{A.4}$$

The multinomial theorem generalizes (A.4) and, for $n_{\bullet} = \sum_{i=1}^{r} n_i$, is given by

$$\left(\sum_{i=1}^{r} x_i \right)^n = \sum_{\mathbf{n}:\ n_{\bullet}=n,\ n_i \geq 0} \binom{n}{n_1, \ldots, n_r} \prod_{i=1}^{r} x_i^{n_i}, \quad r, n \in \mathbb{N},$$

where the combinatoric is defined in (A.2), and $\mathbf{n}$ denotes the vector $(n_1, \ldots, n_r)$.

The gamma function is defined as

$$\Gamma(a) := \int_0^{\infty} x^{a-1} e^{-x} \, dx, \quad a \in \mathbb{R}_{>0}, \quad \Gamma(0) := 1, \tag{A.5}$$

and has the fundamental property that

$$\Gamma(a) = (a-1)\Gamma(a-1), \quad a \in \mathbb{R}_{>1}. \tag{A.6}$$

In particular, $\Gamma(n) = (n-1)!$ for $n \in \mathbb{N}_{\geq 2}$. Useful facts include

$$\Gamma(a) = 2 \int_0^{\infty} u^{2a-1} e^{-u^2} \, du, \tag{A.7}$$

$\Gamma(1/2) = \sqrt{\pi}$, and Legendre's duplication formula for gamma functions,

$$\Gamma(2a) = \frac{2^{2a-1}}{\sqrt{\pi}} \Gamma(a)\Gamma\left(a + \frac{1}{2}\right). \tag{A.8}$$

The incomplete gamma function is given by

$$\Gamma_x(a) = \int_0^x t^{a-1} e^{-t} \, dt, \quad a \in \mathbb{R}_{>0}, \quad x \in \mathbb{R}_{\geq 0}. \tag{A.9}$$

The beta function is defined as

$$B(a,b) := \int_0^1 x^{a-1} (1-x)^{b-1} \, dx, \quad a, b \in \mathbb{R}_{>0}, \tag{A.10}$$

and is related to the gamma function by the famous identity

$$B(a,b) = \frac{\Gamma(a)\Gamma(b)}{\Gamma(a+b)}, \tag{A.11}$$

three proofs of which are mentioned in Section I.1.5.2, and another given below in Problem A.17.2.

Example *To compute*

$$I = I(a,b) = \int_0^s x^a (s-x)^b \, dx, \quad s \in (0,1), \quad a, b > 0,$$

use $u = 1 - x/s$ (so that $x = (1 - u)s$ and $dx = -s\,du$) to get

$$I = \int_0^s x^a (s - x)^b \, dx = s^{a+b+1} B(b + 1, a + 1), \tag{A.12}$$

so that the integral can be expressed in closed form, in terms of the beta function.

The incomplete beta function is given by

$$B_x(p, q) = \int_0^x t^{p-1}(1 - t)^{q-1} \, dt, \quad x \in \mathbb{R} \cap [0, 1], \tag{A.13}$$

and the incomplete beta ratio is $B_x(p, q)/B(p, q)$, denoted $\bar{B}_x(p, q)$. One method for computing the incomplete beta function uses the relation

$$\sum_{j=k}^n \binom{n}{j} p^j (1 - p)^{n-j} = \frac{\Gamma(n + 1)}{\Gamma(k)\Gamma(n - k + 1)} \int_0^p x^{k-1}(1 - x)^{n-k} \, dx, \tag{A.14}$$

for $0 \le p \le 1$ and $k = 1, 2, \ldots$; see, for example, Example II.6.5 or (A.178) below.

Two low-order generalized hypergeometric functions of interest are

$$_1F_1(a, b; z) = \sum_{n=0}^\infty \frac{a^{[n]}}{b^{[n]}} \frac{z^n}{n!} \quad \text{and} \quad _2F_1(a, b; c; z) = \sum_{n=0}^\infty \frac{a^{[n]} b^{[n]}}{c^{[n]}} \frac{z^n}{n!}, \tag{A.15}$$

where

$$a^{[j]} = \begin{cases} a(a + 1) \cdots (a + j - 1), & \text{if } j \ge 1, \\ 1, & \text{if } j = 0. \end{cases}$$

Also, as integral expressions,

$$_1F_1(a, b; z) = \frac{1}{B(a, b - a)} \int_0^1 y^{a-1}(1 - y)^{b-a-1} e^{zy} \, dy \tag{A.16}$$

and

$$_2F_1(a, b; c; z) = \frac{1}{B(a, c - a)} \int_0^1 y^{a-1}(1 - y)^{c-a-1}(1 - zy)^{-b} \, dy, \tag{A.17}$$

where, for $_1F_1$, it is necessary that $a > 0$ and $b - a > 0$, while for $_2F_1$, $a > 0$, $c - a > 0$ and $z < 1$ must hold.

The (real) digamma function is given by

$$\psi(s) = \frac{d}{ds} \ln \Gamma(s) = \int_0^\infty \left[\frac{e^{-t}}{t} - \frac{e^{-zt}}{1 - e^{-t}} \right] dt, \quad s \in \mathbb{R}_{>0}, \tag{A.18}$$

with higher-order derivatives denoted by

$$\psi^{(n)}(s) = \frac{d^n}{ds^n} \psi(s) = \frac{d^{n+1}}{ds^{n+1}} \ln \Gamma(s) = (-1)^{n+1} \int_0^\infty \frac{t^n e^{-zt}}{1 - e^{-t}} \, dt, \quad n = 1, 2, \ldots \, .$$

The modified Bessel function of the third kind is given by

$$K_z(x) = \frac{1}{2} \int_0^\infty u^{z-1} \exp\left[-\frac{x}{2} \left(\frac{1}{u} + u \right) \right] du, \quad z \in \mathbb{R}, \quad x \in \mathbb{R}_{>0}, \tag{A.19}$$

and it satisfies

$$-2K_z'(x) = K_{z-1}(x) + K_{z+1}(x), \quad z \in \mathbb{R}, \ x \in \mathbb{R}_{>0}. \tag{A.20}$$

An introduction is provided in Section II.9.2. In Matlab, $K_z(x)$ can be computed with the built-in function `besselk(z,x)`.

A.2 BASIC PROBABILITY AND CONDITIONING

A field $\mathcal{A}$ is a collection of subsets of (nonempty) Ω such that (i) $\Omega \in \mathcal{A}$, (ii) $A \in \mathcal{A} \Rightarrow A^c \in \mathcal{A}$, and (iii) $A, B \in \mathcal{A} \Rightarrow A \cup B \in \mathcal{A}$. A probability measure is a set function that assigns a real number $\Pr(A)$ to each event $A \in \mathcal{A}$ such that $\Pr(A) \geq 0$, $\Pr(\Omega) = 1$, and, for a countable infinite sequence of mutually exclusive events A_i,

$$\Pr\left(\bigcup_{i=1}^{\infty} A_i\right) = \sum_{i=1}^{\infty} \Pr(A_i). \tag{A.21}$$

Requirement (A.21) is known as countable additivity. If $A_i \cap A_j = \emptyset$, $i \neq j$, and $A_{n+1} = A_{n+2} = \cdots = \emptyset$, then $\Pr\left(\bigcup_{i=1}^{n} A_i\right) = \sum_{i=1}^{n} \Pr(A_i)$, which is referred to as finite additivity. The triplet $\{\Omega, \mathcal{A}, \Pr(\cdot)\}$ refers to the probability space with sample space Ω, collection of measurable events $\mathcal{A}$, and probability measure $\Pr(\cdot)$.

In addition to the basic properties (i) $\Pr(\emptyset) = 0$, (ii) if $A \subset B$, then $\Pr(A) \leq \Pr(B)$, (iii) $\Pr(A) \leq 1$, (iv) $\Pr(A^c) = 1 - \Pr(A)$, (v) $\Pr\left(\bigcup_{i=1}^{\infty} A_i\right) \leq \sum_{i=1}^{\infty} \Pr(A_i)$, and (vi) $\Pr(A_1 \cup A_2) = \Pr(A_1) + \Pr(A_2) - \Pr(A_1 A_2)$, we have Bonferroni's inequality,

$$\Pr\left(\bigcap_{i=1}^{n} A_i\right) \geq \sum_{i=1}^{n} \Pr(A_i) - (n - 1) = 1 - \sum_{i=1}^{n} \Pr(\bar{A}_i), \tag{A.22}$$

Poincaré's theorem or the inclusion–exclusion principle,

$$\Pr\left(\bigcup_{i=1}^{n} A_i\right) = \sum_{i=1}^{n} (-1)^{i+1} S_i, \qquad S_j = \sum_{i_1 < \cdots < i_j} \Pr(A_{i_1} \cdots A_{i_j}); \tag{A.23}$$

the sieve or de Moivre–Jordan theorem, where, for events $B_1, \ldots, B_n$, the probability that exactly m of the B_i occur, $m = 0, 1, \ldots, n$, is given by

$$p_{m,n} = \sum_{i=m}^{n} (-1)^{i-m} \binom{i}{m} S_i, \tag{A.24}$$

with $S_j = \sum_{i_1 < \cdots < i_j} \Pr(B_{i_1} \cdots B_{i_j})$ and $S_0 = 1$, and the probability that at least m of the B_i occur is given by

$$P_{m,n} = \sum_{i=m}^{n} (-1)^{i-m} \binom{i-1}{i-m} S_i; \tag{A.25}$$

and Bonferroni's general inequality, which says that $p_{m,n}$ in (A.24) is bounded by

$$\sum_{i=0}^{2k-1} (-1)^i \binom{m+i}{i} S_{m+i} \leq p_{m,n} \leq \sum_{i=0}^{2k} (-1)^i \binom{m+i}{i} S_{m+i}, \tag{A.26}$$

for any $k = 1, 2, \ldots, \lfloor (n - m)/2 \rfloor$.

The problem of coincidences states that, if n objects, labeled $1, \ldots, n$, are randomly arranged in a row, then the probability that the positions of exactly m of them coincide with their label, $0 \le m \le n$, is given by

$$c_{m,n} = \frac{1}{m!} \left(\frac{1}{2!} - \frac{1}{3!} + \frac{1}{4!} - \cdots + \frac{(-1)^{n-m}}{(n-m)!} \right). \tag{A.27}$$

If $\Pr(B) > 0$, then the probability of A given B is

$$\Pr(A \mid B) = \frac{\Pr(AB)}{\Pr(B)}. \tag{A.28}$$

Two events A and B are independent if $\Pr(A \mid B) = \Pr(A)$ and $\Pr(B \mid A) = \Pr(B)$. Equivalently, if events A and B are independent, then $\Pr(AB) = \Pr(A)\Pr(B)$, which is referred to as pairwise independence. In general, events A_i, $i = 1, \ldots, n$, are mutually independent if and only if, for every collection $A_{i_1}, A_{i_2}, \ldots, A_{i_m}$, $1 \le m \le n$,

$$\Pr(A_{i_1} A_{i_2} \cdots A_{i_m}) = \prod_{j=1}^{m} \Pr(A_{i_j}). \tag{A.29}$$

For events $A, B_1, \ldots, B_n$, the law of total probability is

$$\Pr(A) = \sum_{i=1}^{n} \Pr(A \mid B_i)\Pr(B_i), \tag{A.30}$$

while Bayes' rule is given by

$$\Pr(B \mid A) = \frac{\Pr(A \mid B)\Pr(B)}{\sum_{i=1}^{n} \Pr(A \mid B_i)\Pr(B_i)}. \tag{A.31}$$

A.3 UNIVARIATE RANDOM VARIABLES

Given a probability space $\{\Omega, \mathcal{A}, \Pr(\cdot)\}$, X is a univariate random variable relative to the collection of measurable events $\mathcal{A}$ if and only if it is a function with domain Ω and range the real number line and such that, for every $x \in \mathbb{R}$, $\{\omega \in \Omega \mid X(\omega) \le x\} \in \mathcal{A}$.

The cumulative distribution function (c.d.f.) of random variable (r.v.) X is denoted $F_X(\cdot)$ or just $F(\cdot)$, and defined to be $\Pr(X \le x)$ for some point $x \in \mathbb{R}$. It has the properties (i) $0 \le F(x) \le 1$ for all $x \in \mathbb{R}$, (ii) F is nondecreasing, (iii) F is right continuous, and (iv) $\lim_{x \to -\infty} F(x) = 0$ and $\lim_{x \to \infty} F(x) = 1$.

An r.v. is discrete if it takes on either a finite or countably infinite number of values. The probability mass function (p.m.f.) of the discrete r.v. X is given by $f_X(x) = f(x) = \Pr(X = x)$. If the c.d.f. of r.v. X is absolutely continuous and differentiable, then X is a continuous r.v. and the function $f_X(x) = dF_X(x)/dx$ is the probability density function (p.d.f.) of X.

The support S of the r.v. X is the subset $x \in \mathbb{R}$ such that $f_X(x) > 0$. If there exists a number $B > 0$ such that $|x| < B$ for all $x \in S$, then X is said to have bounded support. Defining

$$\int_{-\infty}^{\infty} g(x)\, dF_X(x) = \begin{cases} \int_S g(x) f_X(x)\, dx, & \text{if } X \text{ is continuous,} \\ \sum_{i \in S} g(x_i) f_X(x_i), & \text{if } X \text{ is discrete,} \end{cases} \tag{A.32}$$

we have $\int_{-\infty}^{\infty} dF_X(x) = 1$.

Example   *A discrete uniform r.v., denoted $X \sim \text{DUnif}(\theta)$, has mass function*

$$f_X(x) = \theta^{-1} \mathbb{1}_{\{1,2,\ldots,\theta\}}(x) \tag{A.33}$$

and c.d.f.

$$F_X(t) = \min\left[\max\left(0, \frac{\lfloor t \rfloor}{\theta}\right), 1\right] = \begin{cases} 0, & \text{if } t < 1, \\ \lfloor t \rfloor / \theta, & \text{if } 1 \le t \le \theta, \\ 1, & \text{if } t > \theta, \end{cases} \tag{A.34}$$

where $\lfloor \cdot \rfloor$ is the floor function.

The quantile ξ_p of a continuous r.v. X is defined to be that value such that $F_X(\xi_p) = p$ for given $0 < p < 1$. The median m is a special case, with $F_X(m) = 0.5$.

A family of distributions indexed by parameter vector $\theta = (\theta_1, \ldots, \theta_k)$ belongs to the exponential family if it can be algebraically expressed as

$$f(x; \theta) = a(\theta)b(x) \exp\left\{ \sum_{i=1}^{k} c_i(\theta)d_i(x) \right\}, \tag{A.35}$$

where $a(\theta) \ge 0$ and $c_i(\theta)$ are real-valued functions of θ but not x; and $b(x) \ge 0$ and $d_i(x)$ are real-valued functions of x but not θ. A wealth of information on exponential families can be found in Brown (1986).

If X is truncated on the left at a and on the right at b, $a < b$, then the density of the truncated random variable is given by

$$\frac{f_X(x)}{F_X(b) - F_X(a)} \mathbb{1}_{(a,b)}(x). \tag{A.36}$$

We now turn to the basic discrete sampling schemes. These involve randomly choosing elements either with or without replacement, and such that either a fixed number of draws, say n, are conducted, or sampling is continued until a specified number of objects from each class are obtained. These give rise to several fundamental types of random variables, notably the Bernoulli, binomial, hypergeometric, geometric, and negative binomial, as reviewed next.

A Bernoulli r.v. X has support $\{0, 1\}$ and takes on the value 1 ("success") with probability p or 0 ("failure") with probability $1 - p$. The p.m.f. can be written as

$$\Pr(X = x) = f_X(x) = p^x(1 - p)^{1-x} \mathbb{1}_{\{0,1\}}(x). \tag{A.37}$$

Let X be the sum of n independently and identically distributed (i.i.d.) Bernoulli r.v.s, that is, X is the number of successes obtained from n draws with replacement. Then X is binomially distributed, and we write $X \sim \text{Bin}(n, p)$ with p.m.f.

$$f_X(x) = \binom{n}{x} p^x(1 - p)^{n-x} \mathbb{1}_{\{0,1,\ldots,n\}}(x). \tag{A.38}$$

If an urn contains N white and M black balls, and n balls are randomly withdrawn without replacement, then $X =$ *the number of white balls drawn* is a random variable with a hypergeometric distribution, written $X \sim \mathrm{HGeo}(N, M, n)$, with

$$f_X(x) = f_{\mathrm{HGeo}}(x; N, M, n) = \frac{\binom{N}{x}\binom{M}{n-x}}{\binom{N+M}{n}} \mathbb{I}_{\{\max(0, n-M), 1, \dots, \min(n, N)\}}(x). \tag{A.39}$$

Now let sampling be conducted with replacement and until a success is obtained. Let the r.v. X denote the ensuing number of failures until the success occurs. Then $X \sim \mathrm{Geo}(p)$ is a geometric r.v. with density

$$f_X(x) = f_{\mathrm{Geo}}(x; p) = p(1 - p)^x \mathbb{I}_{\{0, 1, \dots\}}(x) \tag{A.40}$$

if the success is not counted, or

$$f_X(x) = p(1 - p)^{x-1} \mathbb{I}_{\{1, 2, \dots\}}(x) \tag{A.41}$$

if it is counted. If X is the number of failures until r successes occur, $r = 1, 2, \dots$, then X is said to follow a negative binomial distribution, or $X \sim \mathrm{NBin}(r, p)$, with p.m.f.

$$f_X(x; r, p) = \binom{r + x - 1}{x} p^r (1 - p)^x \mathbb{I}_{\{0, 1, \dots\}}(x). \tag{A.42}$$

The last sampling scheme we mention is when trials continue until k successes are obtained and sampling is without replacement, giving rise to what is called an inverse hypergeometric random variable. If an urn contains w white and b black balls, then the probability that a total of x balls need to be drawn to get k white balls, $1 \leq k \leq w$, is given by

$$f_X(x; k, w, b) = \binom{x - 1}{k - 1} \frac{\binom{w+b-x}{w-k}}{\binom{w+b}{w}} \mathbb{I}_{\{k, k+1, \dots, b+k\}}(x), \quad 1 \leq k \leq w. \tag{A.43}$$

A limiting distribution of the r.v.s associated with these basic random sampling schemes is the Poisson. For $\lambda > 0$, the mass function of $X \sim \mathrm{Poi}(\lambda)$ is

$$f_X(x; \lambda) = \frac{e^{-\lambda} \lambda^x}{x!} \mathbb{I}_{\{0, 1, \dots\}}(x). \tag{A.44}$$

Moments of random variables play a significant role in statistical inference. The expected value of the function $g(X)$, where X is a random variable, is defined to be

$$\mathbb{E}[g(X)] = \int_{S_X} g(x) \, dF_X(x), \tag{A.45}$$

using the notation in (A.32) above. The expected value of X is $\mathbb{E}[X]$ and, more generally, the rth raw moment of X is

$$\mu_r' = \mu_r'(X) = \mathbb{E}[X^r] = \int_S x^r \, dF_X(x), \tag{A.46}$$

while the rth central moment of X is

$$\mu_r = \mu_r(X) = \mathbb{E}[(X - \mu)^r] = \int_S (X - \mu)^r \, dF_X(x), \tag{A.47}$$

where $\mu = \mathbb{E}[X]$. The variance of X is

$$\mu_2 = \mathbb{V}(X) = \int_S (x - \mu)^2 \, \mathrm{d}F_X(x) = \mu_2' - \mu^2, \tag{A.48}$$

often denoted σ^2; and the standard deviation of X is $\sigma = \mu_2^{1/2}$. The skewness is given by $\mu_3/\mu_2^{3/2}$ and the kurtosis is μ_4/μ_2^2. These measures, like the variance, are invariant to location changes, while the skewness and kurtosis are also invariant to scale changes in X.

Remark. The modern use of kurtosis as a measure of the "thickness of the tails" is well established, though it was named from the Greek for "humped shape," along with the terms platykurtic ("broad" kurtosis) and leptokurtic ("narrow" kurtosis), for describing the center of the distribution. Its origins lie with Karl Pearson and his system of "frequency curves"; see Fiori and Zenga (2009) for a clear discussion of its history and development and relationship to Pearson's work in biological evolution, while a very enjoyable and informative account of Pearson's life, polymath career, and influence on the development of mathematical statistics is given in Magnello (2009).

The sample kurtosis is reported in uncountable empirical finance papers as evidence (not that any more is needed) that, say, monthly, daily, or higher-frequency stock returns are leptokurtic and non-Gaussian, and usually presented within a page-long table of various sample statistics for a variety of stocks (that nobody reads anyway). Besides its redundancy for displaying the well-known heavier-tailed nature of financial asset returns, it is arguably incorrect. The empirical statistic only has meaning if its theoretical counterpart exists, and there is ample evidence that it does not; see Chapter 9 and the numerous references therein for discussion, and Fiori and Beltrami (2014) on its inappropriateness for this and other reasons, and presentation of robust estimators of leptokurtosis better suited for studying financial asset returns data.

There are other interpretations of kurtosis; see Moors (1986) and, particularly, DeCarlo (1997) for overviews. ∎

Jensen's inequality states that, for any r.v. X with finite mean,

$$\mathbb{E}[g(X)] \geq g(\mathbb{E}[X]), \text{ for } g(\cdot) \text{ convex}; \quad \mathbb{E}[g(X)] \leq g(\mathbb{E}[X]), \text{ for } g(\cdot) \text{ concave.} \tag{A.49}$$

Example *Let $g(x) = x^{-1}$, $x > 0$. As $g''(x) = 2/x^3 \geq 0$, g is convex, and, assuming X is a positive random variable with existing first moment μ, the first statement in (A.49) implies that, if $\mathbb{E}[X^{-1}]$ exists, then $\mathbb{E}[X^{-1}] \geq \mu^{-1}$.*

A.4 MULTIVARIATE RANDOM VARIABLES

Extending the univariate case, the n-variate vector function

$$\mathbf{X} = (X_1, X_2, \ldots, X_n) = (X_1(\omega), X_2(\omega), \ldots, X_n(\omega)) = \mathbf{X}(\omega)$$

is defined to be a (multivariate or vector) random variable relative to the collection of events $\mathcal{A}$ from the probability space $\{\Omega, \mathcal{A}, \Pr(\cdot)\}$ if and only if it is a function with domain Ω and range (possibly a subset of) $\mathbb{R}^n$ and such that

$$\forall \mathbf{x} = (x_1, \ldots, x_n) \in \mathbb{R}^n, \quad \{\omega \in \Omega \mid X_i(\omega) \leq x_i, i = 1, \ldots, n\} \in \mathcal{A}.$$

The joint cumulative distribution function of $\mathbf{X}$ is denoted by $F_{\mathbf{X}}(\cdot)$ and defined to be the function with domain $\mathbb{R}^n$ and range $[0, 1]$ given by

$$F_{\mathbf{X}}(\mathbf{x}) = \Pr(\mathbf{X} \le \mathbf{x}) := \Pr(-\infty < X_i \le x_i, \ i = 1, \dots, n), \quad \forall \mathbf{x} \in \mathbb{R}^n,$$

where vector inequalities are defined to operate elementwise on the components.

The marginal densities of the bivariate r.v.s X and Y with joint p.d.f. $f_{X,Y}$ are given by

$$f_X(x) = \int_{-\infty}^{\infty} f_{X,Y}(x, y) \, dy, \quad f_Y(y) = \int_{-\infty}^{\infty} f_{X,Y}(x, y) \, dx.$$

More generally, for the n-length vector random variable $\mathbf{X}$ with p.m.f. or p.d.f. $f_{\mathbf{X}}$, the marginal density of the subset $\mathbf{X}_m := (X_{i_1}, \dots, X_{i_m})$, $1 \le m \le n$, is obtained by integrating out (in the continuous case) or summing out (in the discrete case) all of the remaining X_j. That is, with $\mathbf{X}_{\bar{m}} := \{X_j : j \in \{1, \dots, n\} \setminus \{i_1, \dots, i_m\}\}$,

$$f_{\mathbf{X}_{\bar{m}}}(\mathbf{x}) = \int_{\mathbf{x} \in \mathbb{R}^{n-m}} dF_{\mathbf{X}_{\bar{m}}}(\mathbf{x}).$$

The expected value of a function $g(\mathbf{X})$ for $g : \mathbb{R}^n \to \mathbb{R}$, with respect to the n-length vector random variable $\mathbf{X}$ with p.m.f. or p.d.f. $f_{\mathbf{X}}$, is defined by

$$\mathbb{E}[g(\mathbf{X})] = \int_{\mathbf{x} \in \mathbb{R}^n} g(\mathbf{x}) \, dF_{\mathbf{X}}(\mathbf{x}). \tag{A.50}$$

Often only a subset of the $\mathbf{X}$ are used in g, say $g(\mathbf{X}_m) = g(X_1, \dots, X_m)$, $m < n$, in which case $X_{m+1}, \dots, X_n$ get integrated out in (A.50) so that

$$\mathbb{E}[g(\mathbf{X}_m)] = \int_{\mathbf{x} \in \mathbb{R}^m} g(\mathbf{x}) \, dF_{\mathbf{X}_m}(\mathbf{x}), \tag{A.51}$$

where $F_{\mathbf{X}_m}$ denotes the marginal c.d.f. of $(X_1, \dots, X_m)$.

Important cases include the mean $\mu_i = \mathbb{E}[X_i]$ and variance $\sigma_i^2 = \mathbb{E}[(X_i - \mu_i)^2]$ of the individual components of $\mathbf{X}$. A generalization of the variance is the covariance, defined for any two X_i, if it exists, as

$$\sigma_{ij} := \operatorname{Cov}(X_i, X_j) = \mathbb{E}[(X_i - \mu_i)(X_j - \mu_j)] = \mathbb{E}[X_i X_j] - \mu_i \mu_j, \tag{A.52}$$

where $\mu_i = \mathbb{E}[X_i]$. The correlation of two r.v.s X_i and X_j, if both of their second moments exist, is defined to be

$$\operatorname{Corr}(X_i, X_j) = \frac{\operatorname{Cov}(X_i, X_j)}{\sqrt{\mathbb{V}(X_i)\mathbb{V}(X_j)}} = \frac{\sigma_{ij}}{\sigma_i \sigma_j}, \tag{A.53}$$

and is bound between -1 and 1. This follows by squaring the left and right sides of the Cauchy–Schwarz inequality

$$|\mathbb{E}[UV]| \le \mathbb{E}[|UV|] \le +\sqrt{\mathbb{E}[U^2]\mathbb{E}[V^2]}, \tag{A.54}$$

for any two r.v.s U and V with existing second moments.

The r.v.s $X_1, \dots, X_n$ are independent if and only if their joint density can be factored as

$$f_{\mathbf{X}}(\mathbf{x}) = \prod_{i=1}^{n} f_{X_i}(x_i). \tag{A.55}$$

If the n components of r.v. $\mathbf{X} = (X_1, \dots, X_n)$ are independent and the function $g(\mathbf{X})$ can be partitioned as, say, $g_1(X_1)g_2(X_2)\cdots g_n(X_n)$, then

$$\mathbb{E}_{\mathbf{X}}[g(\mathbf{X})] = \prod_{i=1}^{n} \int_{x_i \in \mathbb{R}} g_i(x_i)\, dF_{X_i}(x_i) = \prod_{i=1}^{n} \mathbb{E}_{X_i}[g_i(X_i)], \tag{A.56}$$

where the notation $\mathbb{E}_Y$ denotes taking the expectation with respect to the distribution of r.v. Y. Note that, if X_i and X_j are independent, then, from (A.56) for $i \neq j$,

$$\operatorname{Cov}(X_i, X_j) = \mathbb{E}[X_i - \mu_i]\, \mathbb{E}[X_j - \mu_j] = 0, \tag{A.57}$$

if the expectation exists.

Let $Y = \sum_{i=1}^{n} X_i$, where the X_i are random variables. Then

$$\mathbb{E}[Y] = \sum_{i=1}^{n} \mathbb{E}[X_i], \tag{A.58}$$

if the expected value exists for each X_i. If the variance for each X_i also exists, then

$$\mathbb{V}(Y) = \sum_{i=1}^{n} \mathbb{V}(X_i) + \sum_i \sum_{i \neq j} \operatorname{Cov}(X_i, X_j), \tag{A.59}$$

with special case

$$\mathbb{V}(X_i + X_j) = \mathbb{V}(X_i) + \mathbb{V}(X_j) + 2\operatorname{Cov}(X_i, X_j). \tag{A.60}$$

Extending (A.58) and (A.59) to the weighted sum case with $X = \sum_{i=1}^{n} a_i X_i$, for fixed constants $a_1, \dots, a_n$, and finite $\mu_i = \mathbb{E}[X_i]$, yields $\mathbb{E}[X] = \sum_{i=1}^{n} a_i \mu_i$ and

$$\mathbb{V}(X) = \sum_{i=1}^{n} a_i^2 \mathbb{V}(X_i) + \sum_i \sum_{i \neq j} a_i a_j \operatorname{Cov}(X_i, X_j). \tag{A.61}$$

For example, with $n = 2$, $a_1 = 1$, and $a_2 = -1$, $\mathbb{V}(X_1 + X_2) = \mathbb{V}(X_1) + \mathbb{V}(X_2) - 2\operatorname{Cov}(X_i, X_j)$, or, for general indices i and j,

$$\mathbb{V}(X_i \pm X_j) = \mathbb{V}(X_i) + \mathbb{V}(X_j) \pm 2\operatorname{Cov}(X_i, X_j), \tag{A.62}$$

generalizing (A.60).

The covariance between two r.v.s $X = \sum_{i=1}^{n} a_i X_i$ and $Y = \sum_{i=1}^{m} b_i Y_i$ is, if it exists,

$$\operatorname{Cov}(X, Y) = \sum_{i=1}^{n} \sum_{j=1}^{m} a_i b_j \operatorname{Cov}(X_i, Y_j), \tag{A.63}$$

of which (A.61) is a special case, as $\mathbb{V}(X) = \operatorname{Cov}(X, X)$.

A.5 CONTINUOUS UNIVARIATE RANDOM VARIABLES

A handful of continuous univariate r.v.s are omnipresent in applied statistical analysis. These include the gamma (and special cases exponential and χ^2), with

$$f_{\text{Gam}}(x; \alpha, \beta) = \frac{\beta^{\alpha}}{\Gamma(\alpha)} x^{\alpha-1} \exp\{-\beta x\} \mathbb{I}_{(0,\infty)}(x), \tag{A.64}$$

or, as a member of the exponential family (A.35),

$$\underbrace{\frac{\beta^{\alpha}}{\Gamma(\alpha)}}_{a(\alpha,\beta)} \underbrace{\mathbb{I}_{(0,\infty)}(x)}_{b(x)} \exp\{\underbrace{(\alpha-1)}_{c_1(\alpha,\beta)} \underbrace{\ln x}_{d_1(x)}\} \exp\{\underbrace{-\beta}_{c_2(\alpha,\beta)} \underbrace{x}_{d_2(x)}\}. \tag{A.65}$$

There exists an important relationship between gamma and Poisson r.v.s. Let $X \sim$ Gam(α, λ) with $\alpha \in \mathbb{N}$ and $Y \sim$ Poi$(\cdot)$. Then

$$F_X(t; \alpha, \lambda) = \Pr(Y \geq \alpha; \lambda t). \tag{A.66}$$

Other important r.v.s include the beta (with special case uniform),

$$f_{\text{Beta}}(x; p, q) = \frac{1}{B(p,q)} x^{p-1} (1-x)^{q-1} \mathbb{I}_{[0,1]}(x)$$

$$= \underbrace{\frac{1}{B(p,q)}}_{a(p,q)} \underbrace{\mathbb{I}_{[0,1]}(x)}_{b(x)} \exp\{\underbrace{(p-1)}_{c_1(p,q)} \underbrace{\ln x}_{d_1(x)}\} \exp\{\underbrace{(q-1)}_{c_2(p,q)} \underbrace{\ln(1-x)}_{d1(x)}\};$$

the Laplace, with

$$f_{\text{Lap}}(x; 0, 1) = \frac{\exp\{-|x|\}}{2}, \qquad F_{\text{Lap}}(x; 0, 1) = \frac{1}{2} \begin{cases} e^x, & \text{if } x \leq 0, \\ 2 - e^{-x}, & \text{if } x > 0; \end{cases} \tag{A.67}$$

the Weibull, with

$$f_{\text{Weib}}(x; \beta, x_0, \sigma) = \frac{\beta}{\sigma} \left(\frac{x - x_0}{\sigma}\right)^{\beta-1} \exp\left\{-\left(\frac{x - x_0}{\sigma}\right)^{\beta}\right\} \mathbb{I}_{(x_0,\infty)}(x), \tag{A.68}$$

and

$$F_{\text{Weib}}(x; \beta, x_0, \sigma) = 1 - \exp\left\{-\left(\frac{x - x_0}{\sigma}\right)^{\beta}\right\} \mathbb{I}_{(x_0,\infty)}(x);$$

the Cauchy, with

$$f_{\text{C}}(x; 0, 1) = \frac{1}{\pi} \cdot \frac{1}{1 + x^2}; \tag{A.69}$$

and the normal, given below. Finally, the Student's t_n, with

$$f_t(x; n) = \frac{\Gamma\left(\frac{n+1}{2}\right) n^{\frac{n}{2}}}{\sqrt{\pi}\, \Gamma\left(\frac{n}{2}\right)} (n + x^2)^{-\frac{n+1}{2}} = \frac{n^{-\frac{1}{2}}}{B\left(\frac{n}{2}, \frac{1}{2}\right)} (1 + x^2/n)^{-\frac{n+1}{2}}; \tag{A.70}$$

and Fisher's $F(n_1, n_2)$, with

$$f_F(x; n_1, n_2) = \frac{n}{B\left(\frac{n_1}{2}, \frac{n_2}{2}\right)} \frac{(nx)^{n_1/2-1}}{(1+nx)^{(n_1+n_2)/2}}, \quad n = \frac{n_1}{n_2}, \tag{A.71}$$

are of great importance in the distribution theory associated with Gaussian models.

If X is a continuous random variable with p.d.f. $f_X(x)$, then the linearly transformed random variable $Y = \sigma X + \mu$, $\sigma > 0$, has density

$$f_Y(y) = \frac{1}{\sigma} f_X\left(\frac{y-\mu}{\sigma}\right). \tag{A.72}$$

The distributions of X and Y are said to be members of the same location–scale family, with location parameter μ and scale parameter σ.

Example *If Z is a normal r.v. with location 0 and scale 1, then*

$$f_Z(z; 0, 1) = \frac{1}{\sqrt{2\pi}} \exp\left\{-\frac{1}{2}z^2\right\},$$

while, with $\sigma > 0$, the density of $X = \sigma Z + \mu \sim N(\mu, \sigma^2)$ is

$$f_X(x; \mu, \sigma) = \frac{1}{\sqrt{2\pi}\sigma} \exp\left\{-\frac{1}{2}\left(\frac{x-\mu}{\sigma}\right)^2\right\}.$$

The central moments of X are $\mathbb{E}\left[(X - \mu)^k\right] = 0$, $k = 1, 3, 5, \ldots$, and

$$\mathbb{E}[(X - \mu)^{2r}] = \sigma^{2r}\mathbb{E}[Z^{2r}] = (2\sigma^2)^r \pi^{-1/2}\Gamma\left(r + \frac{1}{2}\right), \quad r = 1, 2, \ldots, \tag{A.73}$$

which reduces to $3\sigma^4$ for $r = 2$.

The kernel of a p.d.f. or p.m.f. is the part of it that involves only the variables associated with the r.v.s of interest. The remaining quantities just form the constant of integration.

If X is a continuous random variable with p.d.f. f_X, g is a continuous differentiable function with domain contained in the range of X, and $dg/dx \neq 0$ for all $x \in S_X$, then f_Y, the p.d.f. of $Y = g(X)$, can be calculated by

$$f_Y(y) = f_X(x) \left|\frac{dx}{dy}\right|, \tag{A.74}$$

where $x = g^{-1}(y)$ is the inverse function of Y.

Example *The Pareto distribution is given in terms of its c.d.f. as*

$$F_X(x) = \left[1 - \left(\frac{x_0}{x}\right)^\alpha\right] \mathbb{I}_{(x_0, \infty)}(x), \tag{A.75}$$

yielding p.d.f. $f_X(x; \alpha, x_0) = \alpha x_0^\alpha x^{-(\alpha+1)} \mathbb{I}_{(x_0, \infty)}(x)$. We are interested in $Y = \log(X/x_0)$. From (A.74), with $x = x_0 \exp(y)$, $dx/dy = x_0 \exp(y)$, and $\mathbb{I}_{(x_0, \infty)}(x) = \mathbb{I}_{(x_0, \infty)}(x_0 \exp(y)) = \mathbb{I}_{(0, \infty)}(y)$,

$$f_Y(y; \alpha) = f_X(x; \alpha, x_0) \quad |x_0 \exp(y)| = \alpha \exp(-\alpha y) \mathbb{I}_{(0, \infty)}(y), \tag{A.76}$$

so that $Y \sim \text{Exp}(\alpha)$.

Transformation (A.74) is the univariate version of the more general multivariate result (A.139) discussed in Section A.9 below.

A.6 CONDITIONAL RANDOM VARIABLES

Let X and Y be discrete r.v.s. with joint mass function $f_{X,Y}$. Let event $B = \{(x,y) : y = y_0\}$. If event $A = \{(x,y) : x \le x_0\}$, then the conditional c.d.f. of X given $Y = y_0$ is given by

$$\Pr(A \mid B) = \frac{\Pr(X \le x, Y = y_0)}{\Pr(Y = y_0)} = \sum_{i=-\infty}^{x} \frac{f_{X,Y}(i, y_0)}{f_Y(y_0)} =: F_{X|Y=y_0}(x \mid y_0),$$

and, likewise, if A is the event $\{(x,y) : x = x_0\}$, then the conditional p.m.f. of X given $Y = y_0$ is

$$\Pr(A \mid B) = \frac{\Pr(X = x, Y = y_0)}{\Pr(Y = y_0)} = \frac{f_{X,Y}(x, y_0)}{f_Y(y_0)} =: f_{X|Y}(x \mid y_0). \tag{A.77}$$

As an important example, let X_1 and X_2 be independently distributed discrete r.v.s. and consider the conditional distribution of X_1 given that $S = X_1 + X_2$ is some particular value s. From (A.77) with $X_1 = X$ and $S = Y$ and the fact that X_1 and X_2 are independent,

$$\Pr(X_1 = x \mid S = s) = \frac{\Pr(X_1 = x)\Pr(X_2 = s - x)}{\Pr(S = s)}. \tag{A.78}$$

As special cases, for $X_i \overset{\text{i.i.d.}}{\sim} \text{Bin}(n,p)$, $X_1 + X_2 \sim \text{Bin}(2n,p)$ and $X_1 \mid (X_1 + X_2)$ follows a hypergeometric distribution with p.m.f. $\binom{n}{x}\binom{n}{s-x} / \binom{2n}{s}$. For $X_i \overset{\text{i.i.d.}}{\sim} \text{Geo}(p)$ with density (A.40), $X_1 + X_2 \sim \text{NBin}(r = 2, p)$ and $X_1 \mid (X_1 + X_2)$ follows a discrete uniform distribution with p.m.f. $(1 + s)^{-1} \mathbb{1}_{(0,1,\dots,s)}(x)$. For $X_i \overset{\text{indep}}{\sim} \text{Poi}(\lambda_i)$, $X_1 + X_2 \sim \text{Poi}(\lambda_1 + \lambda_2)$ and

$$X_1 \mid (X_1 + X_2) \sim \text{Bin}(s,p), \tag{A.79}$$

where $p = \lambda_1/(\lambda_1 + \lambda_2)$.

Now let X and Y be continuous r.v.s. with joint density function $f_{X,Y}$. The conditional p.d.f. and conditional c.d.f. of X given y can be motivated by use of the mean value theorem for integrals (see Section I.8.2.2), and are defined to be

$$f_{X|Y}(x \mid y) := \frac{f_{X,Y}(x, y)}{f_Y(y)}, \quad F_{X|Y}(x \mid y) := \int_{-\infty}^{x} f_{X|Y}(t \mid y)\, dt = \frac{\int_{-\infty}^{x} f_{X,Y}(t, y)\, dt}{f_Y(y)}. \tag{A.80}$$

A set of r.v.s are mutually independent if their marginal and conditional distributions coincide. For two r.v.s, this implies

$$f_{X,Y}(x, y) = f_{X|Y}(x \mid y)f_Y(y) = f_X(x)f_Y(y). \tag{A.81}$$

From the c.d.f. expression in (A.80),

$$F_{X|Y}(x \mid y)f_Y(y) = \int_{-\infty}^{x} f_{X,Y}(t, y)\, dt,$$

and, integrating both sides,

$$\int_{-\infty}^{y} F_{X|Y}(x \mid w) f_Y(w) \, dw = \int_{-\infty}^{y} \int_{-\infty}^{x} f_{X,Y}(t, w) \, dt \, dw = F_{X,Y}(x, y),$$

that is, the joint bivariate c.d.f. can be expressed as a function of a single integral. Similarly,

$$F_{X,Y}(x, y) = \int_{-\infty}^{x} F_{Y|X}(y \mid t) f_X(t) \, dt.$$

In words, the joint c.d.f. of X and Y can be interpreted as a weighted average of the conditional c.d.f. of Y given X, weighted by the density of X.

If X and Y are continuous random variables and event $A = \{X < aY\}$, then, conditioning on Y,

$$\Pr(A) = \Pr(X < aY) = \int_{-\infty}^{\infty} \Pr(X < aY \mid Y = y) f_Y(y) \, dy = \int_{-\infty}^{\infty} F_{X|Y}(ay) f_Y(y) \, dy. \quad \text{(A.82)}$$

Multiplying both sides of the conditional p.d.f. in (A.80) with $f_Y(y)$ and integrating with respect to y gives an expression for the marginal of X,

$$f_X(x) = \int_{-\infty}^{\infty} f_{X|Y}(x \mid y) \, dF_Y(y), \quad \text{(A.83)}$$

which can be interpreted as a weighted average of the conditional density of X given Y, weighted by density Y, analogous to (A.30) in the discrete case.

Letting $\mathbf{Y} = (X_{m+1}, \ldots, X_n)$ so that $\mathbf{X} = (\mathbf{X}_m, \mathbf{Y})$, the expected value of the function $g(\mathbf{X}_m)$ conditional on $\mathbf{Y}$ is given by

$$\mathbb{E}[g(\mathbf{X}_m) \mid \mathbf{Y} = \mathbf{y}] = \int_{\mathbf{x} \in \mathbb{R}^m} g(\mathbf{x}) \, dF_{\mathbf{X}_m|\mathbf{Y}}(\mathbf{x} \mid \mathbf{y}). \quad \text{(A.84)}$$

From (A.84), $\mathbb{E}[g(\mathbf{X}_m) \mid \mathbf{Y} = \mathbf{y}]$ is a function of $\mathbf{y}$, so that the expectation $\mathbb{E}[g(\mathbf{X}_m) \mid \mathbf{Y}]$ with respect to $\mathbf{Y}$ can be computed. An important special case of (A.84) for univariate random variables X and Y with bivariate density $f_{X,Y}$ is $\mathbb{E}[X \mid Y = y]$, the conditional expectation of X given $Y = y$, given by

$$\mathbb{E}[X \mid Y = y] = \int_{-\infty}^{\infty} x f_{X|Y}(x \mid y) \, dx. \quad \text{(A.85)}$$

There are two interpretations of $\mathbb{E}[X \mid Y = y]$. The first is as a *univariate function of* y, that is, as $g : S \subset \mathbb{R} \to \mathbb{R}$, say, with $g(y) := \mathbb{E}[X \mid Y = y]$ and S is the support of Y. The second interpretation, sometimes emphasized by using the shorter notation $\mathbb{E}[X \mid Y]$, is as a *random variable*. This follows because Y is an r.v., and, from the first interpretation, $\mathbb{E}[X \mid Y]$ is a function of Y. Thus, one could compute its expectation.

These two interpretations of course hold in the more general case of (A.84). For S the support of $\mathbf{Y}$ and $\mathbf{y} \in S$, $\mathbb{E}[g(\mathbf{X}_m) \mid \mathbf{Y} = \mathbf{y}]$ is a function of $\mathbf{y}$. As such, it also makes sense to treat $\mathbb{E}[g(\mathbf{X}_m) \mid \mathbf{Y}]$ as a random variable, and we could take, say, its expectation. As an important special case, let X and Y be continuous univariate random variables with joint density $f_{X,Y}$. Then, subscripting the expectation operators for clarity,

$$\mathbb{E}_Y \mathbb{E}_{X|Y}[g(X) \mid Y] = \int_{-\infty}^{\infty} f_Y(y) \int_{-\infty}^{\infty} g(x) \frac{f_{X,Y}(x, y)}{f_Y(y)} \, dx \, dy$$

$$= \int_{-\infty}^{\infty} \int_{-\infty}^{\infty} g(x) f_{X,Y}(x, y) \, dx \, dy$$

$$= \int_{-\infty}^{\infty} g(x) \int_{-\infty}^{\infty} f_{X,Y}(x, y) \, dy \, dx$$

$$= \int_{-\infty}^{\infty} g(x) f_X(x) \, dx = \mathbb{E}_X[g(X)]. \tag{A.86}$$

The same result holds if Y is discrete, in which case we can write

$$\mathbb{E}_X[g(X)] = \mathbb{E}_Y \mathbb{E}_{X|Y}[g(X) \mid Y] = \sum_{y=-\infty}^{\infty} \mathbb{E}_{X|Y}[g(X) \mid Y = y] \Pr(Y = y). \tag{A.87}$$

We write both results (A.86) and (A.87) as

$$\mathbb{E}\mathbb{E}[g(X) \mid Y] = \mathbb{E}[g(X)], \tag{A.88}$$

referred to as the law of the iterated expectation, or tower property of expectation.

Analogous to the conditional expectation, the conditional variance of X given Y is obtained from (A.84) with $g(X, Y) = (X - \mathbb{E}[X \mid Y])^2$, that is,

$$\text{Var}(X \mid Y) = \mathbb{E}[(X - \mathbb{E}[X \mid Y])^2 \mid Y] = \mathbb{E}[X^2 \mid Y] - (\mathbb{E}[X \mid Y])^2, \tag{A.89}$$

which leads to the conditional variance formula

$$\mathbb{V}(X) = \mathbb{E}[\mathbb{V}(X \mid Y)] + \mathbb{V}(\mathbb{E}[X \mid Y]). \tag{A.90}$$

A.7   GENERATING FUNCTIONS AND INVERSION FORMULAS

The moment generating function (m.g.f.) of the random variable X is the function $\mathbb{M}_X :$ $\mathbb{R} \to \mathbb{X}_{\geq 0}$ (where $\mathbb{X}$ denotes the extended real line) given by $\mathbb{M}_X(t) = \mathbb{E}[e^{tX}]$. It exists if it is finite on a neighborhood of zero, that is, if there is an $h > 0$ such that, for all $t \in (-h, h)$, $\mathbb{M}_X(t) < \infty$. If it exists, then the largest (open) interval $\mathcal{I}$ around zero such that $\mathbb{M}_X(t) < \infty$ for $t \in \mathcal{I}$ is referred to as the convergence strip (of the m.g.f. of) X.

If the m.g.f. of r.v. X exists, then, as detailed in Section II.1.1.1, the limit operators of derivative and integral can be interchanged, so that

$$\mathbb{M}_X^{(j)}(t) = \frac{d^j}{dt^j} \mathbb{E}[e^{tX}] = \mathbb{E}\left[\frac{d^j}{dt^j} e^{tX}\right] = \mathbb{E}[X^j e^{tX}],$$

and $\mu_j' = \mathbb{E}[X^j] = \mathbb{M}_X^{(j)}(0)$, $j = 1, 2, \ldots$. When it exists, the m.g.f. uniquely determines, or characterizes, the distribution, that is, for a given m.g.f., there is a unique corresponding c.d.f. (up to sets of measure zero). This fact is useful when the m.g.f. of a r.v. is known, but not its p.d.f. or c.d.f.

If $\mathbb{M}_Z(t)$ is the m.g.f. of r.v. Z and $X = \mu + \sigma Z$ for $\sigma > 0$, then

$$\mathbb{M}_X(t) = \mathbb{E}[e^{tX}] = \mathbb{E}[e^{t(\mu + \sigma Z)}] = e^{t\mu} \mathbb{M}_Z(t\sigma). \tag{A.91}$$

If X_i is a sequence of r.v.s with corresponding m.g.f.s $\mathbb{M}_i(t)$ and the latter converge to a function $\mathbb{M}(t)$ for all t in an open neighborhood containing zero, then the distribution of X_i converges to F_X, the c.d.f. corresponding to $\mathbb{M}(t)$; see (A.283) below. We write this as $X_i \overset{\text{asy}}{\sim} F_X$ or, more commonly, if F_X is, say, the normal c.d.f., we write $X_i \overset{\text{asy}}{\sim} N(\cdot, \cdot)$. If F_X is continuous, it need not be the case that the F_{X_i} are continuous.

The cumulant generating function (c.g.f.) is $\mathbb{K}_X(t) = \log \mathbb{M}_X(t)$, where the terms κ_i in the series expansion $\mathbb{K}_X(t) = \sum_{r=0}^{\infty} \kappa_r t^r / r!$ are referred to as the cumulants of X. As the ith derivative of $\mathbb{K}_X(t)$ evaluated at $t = 0$ is κ_i, a good exercise for the reader is to show that

$$\kappa_1 = \mu, \quad \kappa_2 = \mu_2, \quad \kappa_3 = \mu_3 \quad \kappa_4 = \mu_4 - 3\mu_2^2. \tag{A.92}$$

Example *It is straightforward to show (see Problem I.7.17 and Example II.1.3) that the m.g.f. and c.g.f. of $X \sim N(\mu, \sigma^2)$ are given by*

$$\mathbb{M}_X(t) = \exp\left\{ \mu t + \frac{1}{2}\sigma^2 t^2 \right\}, \quad \mathbb{K}_X(t) = \mu t + \frac{1}{2}\sigma^2 t^2. \tag{A.93}$$

Thus,

$$\mathbb{K}_X'(t) = \mu + \sigma^2 t, \quad \mathbb{E}[X] = \mathbb{K}_X'(0) = \mu, \quad \mathbb{K}_X''(t) = \sigma^2, \quad \mathbb{V}(X) = \mathbb{K}_X''(0) = \sigma^2,$$

and $\mathbb{K}_X^{(i)}(t) = 0$, $i \geq 3$, so that $\mu_3 = 0$ and $\mu_4 = \kappa_4 + 3\mu_2^2 = 3\sigma^4$. As such, X has skewness $\mu_3 / \mu_2^{3/2} = 0$ and kurtosis $\mu_4 / \mu_2^2 = 3$.

The m.g.f. of vector $\mathbf{X} = (X_1, \ldots, X_d)'$ is given by $\mathbb{M}_{\mathbf{X}}(\mathbf{t}) = \mathbb{E}[e^{\mathbf{t}'\mathbf{X}}]$, where $\mathbf{t} = (t_1, \ldots, t_d)'$. As in the univariate case, this characterizes the distribution of $\mathbf{X}$ and, thus, all the (univariate and multivariate) marginals as well. In particular, observe that

$$\mathbb{M}_{\mathbf{X}}((0, \ldots, 0, t_i, 0, \ldots, 0)') = \mathbb{E}[e^{t_i X_i}] = \mathbb{M}_{X_i}(t_i), \quad i = 1, \ldots, d, \tag{A.94}$$

so knowledge of $\mathbb{M}_{\mathbf{X}}$ implies knowledge of $\mathbb{M}_{X_i}$, $i = 1, \ldots, d$, similarly to knowledge of $f_{\mathbf{X}}$ implying knowledge of all the d univariate marginal p.d.f.s f_{X_i}, but knowing all the f_{X_i} (or all the $\mathbb{M}_{X_i}$) does not convey knowledge of $f_{\mathbf{X}}$ (or $\mathbb{M}_{\mathbf{X}}$).

For r.v. $\mathbf{Z} = (Z_1, \ldots, Z_d)'$ with m.g.f. $\mathbb{M}_{\mathbf{Z}}$, let $\mathbf{X} = \boldsymbol{\mu} + \boldsymbol{\Sigma}^{1/2}\mathbf{Z}$, for vector $\boldsymbol{\mu} = (\mu_1, \ldots, \mu_d)' \in \mathbb{R}^d$ and $d \times d$ positive definite matrix $\boldsymbol{\Sigma}$ with typical entry denoted by σ_{ij} and diagonal elements denoted by σ_j^2, $j = 1, \ldots, d$. Then the extension of (A.91) to the multivariate case takes the form

$$\mathbb{M}_{\mathbf{X}}(\mathbf{t}) = e^{\mathbf{t}'\boldsymbol{\mu}} \mathbb{M}_{\mathbf{Z}}(\boldsymbol{\Sigma}^{1/2}\mathbf{t}). \tag{A.95}$$

The characteristic function (c.f.) of univariate random variable X is $\varphi_X(t) = \mathbb{E}[e^{itX}]$, where i is the imaginary unit. Using the notation from (A.32) and Euler's formula (II.1.31),

$$\varphi_X(t) = \int_{-\infty}^{\infty} e^{itx} \, dF_X(x) = \int_{-\infty}^{\infty} \cos(tx) \, dF_X(x) + i \int_{-\infty}^{\infty} \sin(tx) \, dF_X(x)$$
$$= \mathbb{E}[\cos(tX)] + i\mathbb{E}[\sin(tX)], \tag{A.96}$$

which exists for all random variables, though in some cases obtaining an analytic expression can be difficult. The uniqueness theorem states that a distribution is uniquely determined by its c.f., that is, if random variables X and Y have c.d.f.s F_X and F_Y, and c.f.s φ_X and φ_Y, respectively, and $\varphi_X(t) = \varphi_Y(t)$ for all $t \in \mathbb{R}$, then $F_X = F_Y$ "almost everywhere", meaning, they can differ only on a set of measure zero; see also (A.263) below. The c.f. of random vector $\mathbf{X} = (X_1, \ldots, X_d)'$ is given by $\varphi_{\mathbf{X}}(\mathbf{t}) = \mathbb{E}[e^{i\mathbf{t}'\mathbf{X}}]$, where $\mathbf{t} = (t_1, \ldots, t_d)'$, similar to the vector m.g.f.

Section II.1.2.4 provides a discussion on when one can simply take $\varphi_X(t) = \mathbb{M}_X(it)$. It is only possible when the m.g.f. exists, and even then, observe that this relation is not obvious because of the definition $\mathbb{M}_X : \mathbb{R} \to \mathbb{X}_{\geq 0}$. It is the case for essentially all of the primary distributions used in applied statistical inference. The normal distribution is such a case: for $X \sim N(\mu, \sigma^2)$, from (A.93), we immediately obtain

$$\varphi_X(t) = \mathbb{E}[e^{itX}] = \exp\left\{ \mu it - \frac{1}{2}\sigma^2 t^2 \right\}, \tag{A.97}$$

without the need for explicit complex integration. Sometimes extra work may be required to operationalize $\varphi_X(t) = \mathbb{M}_X(it)$; see Problem A.19(a) for an example.

Let X be a univariate discrete random variable with support $\{x_j\}_{j=1}^{\infty}$, probability mass function f_X, and characteristic function φ_X. The inversion formula states that, for $j \geq 1$,

$$f_X(x_j) = \frac{1}{2\pi} \int_{-\pi}^{\pi} e^{-itx_j} \varphi_X(t) \, dt. \tag{A.98}$$

Similarly, if X is a univariate continuous random variable with p.d.f. f_X and c.f. φ_X such that $\int_{-\infty}^{\infty} |\varphi_X(t)| \, dt < \infty$, then

$$f_X(x) = \frac{1}{2\pi} \int_{-\infty}^{\infty} e^{-itx} \varphi_X(t) \, dt. \tag{A.99}$$

Given the theoretical and practical importance of (A.99), we include the outline of its proof. For $T > 0$,

$$\int_{-T}^{T} e^{-itx} \varphi_X(t) \, dt = \int_{-T}^{T} e^{-itx} \int_{-\infty}^{\infty} e^{ity} f_X(y) \, dy \, dt$$

$$= \int_{-\infty}^{\infty} f_X(y) \int_{-T}^{T} e^{it(y-x)} \, dt \, dy = \int_{-\infty}^{\infty} f_X(y) \frac{2 \sin T(y-x)}{y-x} \, dy. \tag{A.100}$$

Then, with $A = T(y-x)$ and $dy = dA/T$,

$$\int_{-\infty}^{\infty} e^{-itx} \varphi_X(t) \, dt = 2 \lim_{T \to \infty} \int_{-\infty}^{\infty} f_X(y) \frac{\sin T(y-x)}{y-x} \, dy$$

$$= 2 \lim_{T \to \infty} \int_{-\infty}^{\infty} f_X\left(x + \frac{A}{T}\right) \frac{\sin A}{A} \, dA = 2\pi f_X(x), \tag{A.101}$$

because

$$\int_0^{\infty} \frac{\sin x}{x} \, dx = \frac{\pi}{2}, \tag{A.102}$$

as shown in Example I.A.30. A more rigorous proof adds justification for the interchanging of limit operations in (A.100) and (A.101).

More suitable for numeric work are the counterparts to (A.98) and (A.99), given by

$$f_X(x_j) = \frac{1}{\pi} \int_0^{\pi} \text{Re}[e^{-itx_j} \varphi_X(t)] \, dt, \quad f_X(x) = \frac{1}{\pi} \int_0^{\infty} \text{Re}[e^{-itx} \varphi_X(t)] \, dt, \tag{A.103}$$

respectively. The integrand in (A.103) can be transformed such that the range of integration is over $(0, 1)$. The substitution $u = 1/(1 + t)$ leads, in the continuous case, to

$$f_X(x) = \frac{1}{\pi} \int_0^1 h\left(\frac{1-u}{u}\right) u^{-2} \, du, \quad h(t) = \text{Re}[e^{-itx}\varphi_X(t)]. \tag{A.104}$$

For c.f. φ_X of r.v. X with c.d.f. F_X, if $F_X(x)$ is continuous at the two points $a \pm h$, $h > 0$, then

$$F_X(a + h) - F_X(a - h) = \lim_{T \to \infty} \frac{1}{\pi} \int_{-T}^{T} \frac{\sin(ht)}{t} e^{-ita} \varphi_X(t) \, dt. \tag{A.105}$$

More useful for computation is the expression from Gil-Peleaz (1951),

$$F_X(x) = \frac{1}{2} + \frac{1}{2\pi} \int_0^\infty \frac{e^{itx}\varphi_X(-t) - e^{-itx}\varphi_X(t)}{it} \, dt. \tag{A.106}$$

Proofs of (A.105) and (A.106) are provided in Section II.1.2.6. For computation, we suggest the use of

$$F_X(x) = \frac{1}{2} - \frac{1}{\pi} \int_0^\infty g(t) \, dt, \quad g(t) = \frac{\text{Im}(z)}{t}, \quad z = z(t) = e^{-itx}\varphi_X(t), \tag{A.107}$$

and

$$F_X(x) = \frac{1}{2} - \frac{1}{\pi} \int_0^1 g\left(\frac{1-u}{u}\right) u^{-2} \, du. \tag{A.108}$$

A.8 VALUE AT RISK AND EXPECTED SHORTFALL

Expected shortfall, the risk measure that the Basel Accords contemplate as an eventual substitute for VaR, encounters its own theoretical troubles. Notwithstanding the virtues of subadditivity and coherence, expected shortfall does not represent a platonically ideal risk measure. Its principle problem is that it cannot be reliably backtested in the sense that forecasts of expected shortfall cannot be verified through comparison with historical observations. This is the primary respect in which VaR holds a regulatory advantage vis-á-vis expected shortfall as a measure of risk.

Indeed, nearly the entire class of spectral risk measures, of which expected shortfall is a special case, is not elicitable. Whatever efforts we undertake to surmise the true shape and size of the tails of market-based loss, distributions are just that, informed guesses in the face of incurable leptokurtic blindness.

(Chen, 2016, p. 310)

The value at risk (VaR) and expected shortfall (ES) are examples of **tail risk measures** used in empirical finance and quantitative (financial) risk management (QRM). They have become the dominant risk measures in QRM, with ES growing in importance, being a so-called **coherent** risk measure that obeys the property of **subadditivity**; see Embrechts and Wang (2015) and the references therein.

Motivated by (A.28), for a random variable X, the expected value of the measurable function $g(X)$, given that $X \leq c$, is

$$\mathbb{E}[g(X) \mid X \leq c] = \frac{\int_{-\infty}^{c} g(x)\, dF_X(x)}{F_X(c)}, \tag{A.109}$$

if the integral exists. This can be generalized to conditioning on any measurable event of X with nonzero measure, though the use of $X \leq c$ is common, as it is associated with the ES, given by $\mathbb{E}[X \mid X \leq c]$. In particular, from (A.109) and assuming the r.v. X is continuous with finite expected value, the ξ-level ES of X, denoted $\mathrm{ES}(X, \xi)$, can be expressed as the tail conditional expectation

$$\mathrm{ES}(X, \xi) = \frac{1}{\xi} \int_{-\infty}^{q_{X,\xi}} u\, f_X(u)\, du = \mathbb{E}[X \mid X \leq q_{X,\xi}], \quad \xi \in (0,1), \tag{A.110}$$

where $q_{X,\xi}$ is the ξ-quantile of X and is such that $\mathrm{VaR}(X, \xi) = q_{X,\xi}$ is the ξ-level value at risk corresponding to one unit of investment. In some presentations, VaR and ES are the negatives of the definitions above, so that the risk measures are positive numbers.

The ES can be expressed alternatively as follows. Let $c = q_{X,\xi} < 0$ be the ξ-quantile of X, let $Y = X - c$, so that $f_Y(y) = f_X(y + c)$, and, for any r.v. Z, define $Z^- = Z\mathbb{I}\{Z \leq 0\}$. Then

$$\mathbb{E}[(X - c)^-] = \mathbb{E}[Y^-] = \int_{-\infty}^{0} y f_Y(y)\, dy + \int_{0}^{\infty} 0 \cdot f_Y(y)\, dy = \int_{-\infty}^{0} y f_X(y + c)\, dy$$

$$= \int_{-\infty}^{c} (x - c) f_X(x)\, dx = \int_{-\infty}^{c} x f_X(x)\, dx - c F_X(c)$$

$$= F_X(c)\{\mathrm{ES}(X; \xi) - c\}. \tag{A.111}$$

Remarks

(a) The ES (sometimes referred to as tail VaR or conditional VaR) is a moment (in this case, the first moment) of a random variable conditional on that random variable exceeding a certain threshold. The variance, as well as cross-moments involving two random variables (such as for correlation), are also used in practice. The collections and books by Dempster (2002), Szegö (2004), Dowd (2005), Christoffersen (2011), and McNeil et al. (2015) provide highly useful accounts of the use (and misuse) of VaR and ES in empirical finance and QRM.

(b) In the language of extreme value theory (see the textbook references in Section 9.2), $\Pr(X > c)$ (the right tail is usually used) is often denoted as the **exceedance probability**. Let $X^+ = X\mathbb{I}\{X > 0\}$. The **stop-loss premium**, as commonly used in insurance, is given by $\mathbb{E}[(X - c)^+]$, for some value c in the right tail, and, similar to (A.111), the reader is encouraged to confirm that

$$\mathbb{E}[(X - c)^+] = \Pr(X > c)\{\mathrm{ES}(-X; \xi) - c\}. \tag{A.112}$$

The stop-loss premium also plays an important role in the pricing of collateralized debt obligations (CDOs) and options on realized variance.

(c) In biostatistics and survival analysis, $\Pr(X > c)$ is referred to as the **survival function**. The ES is related to the **conditional expected future lifetime** in survival analysis. If T is a nonnegative random variable that denotes the survival time (of a living creature, or possibly an industrial product such as a light bulb), the c.d.f. of T, given survival until time t_0, is

$$F_T(t \mid t_0) := \Pr(T \le t_0 + t \mid T > t_0) = \frac{\Pr(t_0 < T \le t_0 + t)}{\Pr(T > t_0)}$$

$$= \frac{F_T(t_0 + t) - F_T(t_0)}{1 - F_T(t_0)},$$

with (in the continuous case) density

$$f_T(t \mid t_0) = \frac{d}{dt}F_T(t \mid t_0) = \frac{f_T(t_0 + t)}{1 - F_T(t_0)} = \frac{f_T(t_0 + t)}{S_T(t_0)},$$

where $S_T(t) = 1 - F_T(t)$ is the survivor function. If $\mathbb{E}[T]$ exists, then the conditional expected future lifetime is

$$\int_0^\infty t f_T(t \mid t_0)\, dt = \frac{1}{S_T(t_0)}\int_0^\infty t f_T(t_0 + t)\, dt = \frac{1}{S(t_0)}\int_{t_0}^\infty S(t)\, dt,$$

$$\tag{A.113}$$

where the second equality is obtained from integration by parts applied to the last term: with $u = S_T(t)$ and $dv = dt$, $du = -f_T(t)\, dt$, $v = t$,

$$\int_{t_0}^\infty S_T(t)\, dt = [1 - F_T(t)]t \Big|_{t_0}^\infty + \int_{t_0}^\infty t f_T(t)\, dt \qquad (\text{now let } h = t - t_0)$$

$$= -t_0 S_T(t_0) + \int_0^\infty (h + t_0) f_T(h + t_0)\, dh$$

$$= -t_0 S_T(t_0) + \int_0^\infty h f_T(h + t_0)\, dh + t_0 \int_0^\infty f_T(h + t_0)\, dh$$

$$= \int_0^\infty t f_T(t + t_0)\, dt,$$

and $\lim_{t \to \infty} t(1 - F_T(t)) = 0$ in the second equality follows from the same argument as used below in (A.133). Expression (A.113) can also be compared to (A.132).

(d) From the definition of ES, accurate calculation of the predictive VaR quantile is required. For an overview of successful methods when applied to daily financial returns data, and methods for testing the adequacy of out-of-sample VaR forecasts, see Manganelli and Engle (2004), Haas (2005), Kuester et al. (2006), Hartz et al. (2006), Haas (2009), Francioni and Herzog (2012), Haas et al. (2013), Santos et al. (2013), Krause and Paolella (2014), Abad et al. (2014), Pelletier and Wei (2016), Slim et al. (2016), and the numerous references therein.

(e) The use of the bootstrap for calculating confidence intervals associated with the VaR and ES is recommended. Christoffersen and Gonçalves

(2005) and Gao and Song (2008) investigate the use of the nonparametric bootstrap, while the parametric bootstrap is demonstrated in Pritsker (1997) and (also with the use of weighted likelihood) Broda and Paolella (2011). In both cases, a GARCH-type model can be used to model the time-varying scale term; this is discussed in detail in Book IV.

Such c.i.s are valuable because they help quantify the uncertainty arising from the estimated model parameters. Moreover, imagine two portfolios with (approximately) the same point estimate of the VaR, say, but such that the lower endpoint of the associated c.i. based on the first portfolio is substantially less than that of the second portfolio. This implies that the first one is riskier, and should be granted a higher capital reserve.

It is important to keep in mind that, when using the nonparametric bootstrap, the choice of data window size will heavily influence the outcome, though it has the usual benefit associated with nonparametric inference of "letting the data speak for themselves," while with the parametric bootstrap, both point and interval estimates of VaR and ES will depend on the parametric model assumed. This is especially the case for the ES, and depends to a large extent on the tail behavior of the assumed distribution, for example, a thin-tailed one, such as the mixed normal; a semi-heavy-tailed one, such as the normal inverse Gaussian; a heavy-tailed one that allows the existence of any positive moment, such as the (noncentral) Student's t, and a heavy-tailed one with a maximal possible moment, such as the non-Gaussian (asymmetric) stable Paretian.

The outcomes of both bootstrap methods are obviously dependent on the choice of nominal size of the confidence interval, and, as in other statistical contexts, the best choice is not obvious. Given modern computing power, we recommend computing 90% nominal c.i.s based on the nonparametric bootstrap using a variety of data window sizes, as well as based on the parametric bootstrap (also for a variety of data window sizes), using the aforementioned distributions, and comparing them. As with all nontrivial statistical analyses, some subjective expert knowledge will be required to decide what method to favor. If a very conservative estimate is desired, the use of the parametric bootstrap with the (asymmetric) stable Paretian distribution makes sense.

(f) For an account of, and some resolutions to, backtesting and so-called *elicitability* of VaR and ES for risk management, see Gneiting et al. (2007), Gneiting (2011), Embrechts and Hofert (2014), Bellini and Bignozzi (2015), Davis (2016), Kou and Peng (2016), Roccioletti (2016), Du and Escanciano (2017), and the references therein.

(g) Broda and Paolella (2011) detail the computation of ES for various distributions popular in finance, while Nadarajah et al. (2013) give an overview of ES and a strong literature review on estimation methods. Formulas for the ES associated with a portfolio (weighted sums of margins) from an elliptic distribution have been derived by Landsman and Valdez (2003), Kamdem (2005), and Dobrev et al. (2017).

(h) More theoretical accounts of ES, issues related to discrete distributions, and applications to portfolio theory can be found in Pflug (2000), Acerbi

(2002, 2004), Acerbi and Tasche (2002), Rockafellar and Uryasev (2000, 2002), and Rockafellar et al. (2006a,b, 2007). Some drawbacks and limitations of ES ("unexpected shortfalls of expected shortfall") are discussed in Koch-Medina and Munari (2016) and Asimit and Li (2016). So-called **distortion risk measures** generalize ES, and are studied in Kusuoka (2001) and Tsukahara (2009, 2014).

(i) The saddlepoint approximation (s.p.a.) is discussed below in Section A.11, and provides highly accurate approximations to the p.d.f. and c.d.f. based on the m.g.f. Martin (2006) has shown that the integral in $ES(X; \xi)$ can be approximated as

$$\int_{-\infty}^{q} x f_X(x) \, \mathrm{d}x \approx \mu_X F_X(q) - f_X(q) \frac{q - \mu_X}{\hat{s}}, \qquad (A.114)$$

where $\mu_X = \mathbb{E}[X]$ and $\hat{s}$ is the saddlepoint. The quantities $f_X(q)$ and $F_X(q)$ can of course be replaced by their s.p.a. counterparts.

More accurate s.p.a.s for expected shortfall than (A.114), also applicable in the highly relevant case for which there is no m.g.f. (such as the noncentral t, generalized exponential distribution, and stable Paretian) are developed in Broda and Paolella (2010) and Broda et al. (2017), with an example using the stable Paretian shown below in Section A.16. In a related vein, and with applications to risk management, Kim and Kim (2017) consider s.p.a.s to expressions of the form $\mathbb{E}[X \mid \mathbf{Y} = \mathbf{a}]$ and $\mathbb{E}[X \mid \mathbf{Y} \geq \mathbf{a}]$, where X and $\mathbf{Y}$ are continuous univariate and multivariate random variables respectively, such that they possess a joint moment generating function.

Saddlepoint approximations to (A.112) are developed in Antonov et al. (2005), Yang et al. (2006), Huang and Oosterlee (2011), and Zheng and Kwok (2014); see the discussion in Broda et al. (2017) for the relation of these s.p.a.s to those of the ES.

A problem related to calculation of the stop-loss premium is approximating $\mathbb{E}[(\exp\{X\} - c)^+]$, as required in option pricing. This can be reduced to calculating a tail probability through an exponential change of measure. This has been exploited for constructing saddlepoint approximations in Rogers and Zane (1999), Carr and Madan (2009), Glasserman and Kim (2009), among others.

(j) VaR and ES are but two of numerous measures of financial risk, many of which, like VaR and ES, emphasize so-called downside risk, or the left tail of the portfolio returns distribution, as opposed to, say, the variance; see Cogneau and Hübner (2009a,b) for a presentation of just over 100 of them. Links between ES and so-called investment prudence, temperance, higher-order risk attitudes, and loss aversion are discussed in Eeckhoudt et al. (2016). ∎

Example *Azzalini (1985, 1986) proposed and studied an asymmetric generalization of the normal, referred to as the skew normal, with density*

$$f_{\mathrm{SN}}(z; \lambda) = 2\phi(z)\Phi(\lambda z), \qquad (A.115)$$

for some $\lambda \in \mathbb{R}$, where ϕ and Φ are the standard normal p.d.f. and c.d.f., respectively. The c.d.f. is

$$F_{\text{SN}}(z; \lambda) = 2 \int_{-\infty}^{z} \phi(t)\Phi(\gamma t) \, dt = 2 \int_{-\infty}^{z} \int_{-\infty}^{\gamma t} \phi(t)\phi(s) \, ds \, dt, \qquad (A.116)$$

so that canned routines, bivariate integration, or simulation for computation of the bivariate normal c.d.f. can be used for its evaluation; see, for example, Section II.3.4. The m.g.f. of a standard skew normal random variable X is

$$\mathbb{M}_X(t) = 2 \exp\{t^2/2\}\Phi(t\delta), \quad \delta = \frac{\lambda}{\sqrt{1 + \lambda^2}}, \qquad (A.117)$$

yielding

$$\mathbb{E}[X] = \delta \sqrt{\frac{2}{\pi}} \quad \text{and} \quad \mathbb{V}(X) = 1 - \frac{2\delta^2}{\pi}; \qquad (A.118)$$

see Problem A.18(c). For higher-order moments, see Henze (1986), Martínez et al. (2008), and Haas (2012).

An appealing and useful property of the SN distribution not shared by other more ad hoc *methods for introducing skewness into a normal density is that if $X \sim \text{SN}(\lambda)$, then $X^2 \sim \chi_1^2$. While this was proven directly in Azzalini (1985), it easily follows from a more general and interesting result, as noted in Gupta et al. (2004), due to Roberts and Geisser (1966):*

$W^2 \sim \chi_1^2$ if and only if the p.d.f. of W has the form
$$f(w) = h(w) \exp(-w^2/2), \text{ where } h(w) + h(-w) = \sqrt{2/\pi}.$$

Further distributional aspects of skew normal random variables with density (A.115) are discussed below in Problems A.19 and A.20. In addition to other fields such as biostatistics, the SN has found extensive use in finance. See, for example, Adcock and Shutes (2005), Christodoulakis and Peel (2009), Harvey et al. (2010), Augustyniak and Boudreault (2012), Adcock et al. (2015); and, for mixtures of them, see, for example, Lin et al. (2007), Lin (2009), Haas (2010), Haas and Paolella (2012), and Bernardi (2013). The generalization to the skew Student t is considered in Problem A.22, with applications discussed in Adcock (2010, 2014). The multivariate skew normal is developed in Azzalini and Dalla Valle (1996) and Azzalini and Capitanio (1999).

Figure A.1(a) shows the true ES (see Problem A.19(a); or numeric integration based on the definition of ES can be used) and its s.p.a. based on (A.114), the standard skew normal distribution, with $\xi = 0.01$. The approximation is exact for the normal case ($\lambda = 0$), is accurate for $\lambda < 0$, but worsens as λ increases from zero, though its accuracy is clearly still very high. Figure A.1(b) shows its relative percentage error, as well as that of a second-order s.p.a. developed in Broda et al. (2017), which is seen to be much more accurate. The reader is encouraged to reproduce the results in the top panel.

Example   *Let $R \sim N(0, 1)$ with p.d.f. ϕ and c.d.f. Φ. For fixed $c < 0$, let $u = -r^2/2$. Then*

$$\mathbb{E}[R \mid R \leq c] = \frac{1}{\Phi(c)} \int_{-\infty}^{c} r\phi(r) \, dr = \frac{1}{\Phi(c)} \frac{1}{\sqrt{2\pi}} \int_{-\infty}^{c} r \exp\left\{-\frac{1}{2}r^2\right\} dr$$

$$= \frac{1}{\Phi(c)} \frac{1}{\sqrt{2\pi}} \left(-\exp\left\{-\frac{1}{2}c^2\right\}\right) = -\frac{\phi(c)}{\Phi(c)}. \qquad (A.119)$$

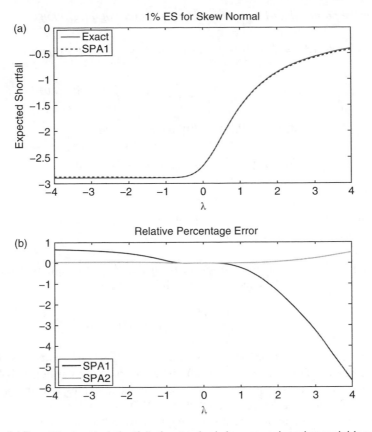

Figure A.1 *(a) True 1% expected shortfall of a standard skew normal random variable as a function of asymmetry parameter λ (solid) and its s.p.a. based on (A.114) (dashed). (b) The relative percentage error of the s.p.a. based on (A.114) (denoted SPA1) and that of the less accurate of two second-order s.p.a.s (denoted SPA2) developed in Broda et al. (2017).*

This calculation reveals the interesting result that $\int_{-\infty}^{c} r\phi(r)\,dr = -\phi(c)$, or $c\phi(c) = -\phi'(c)$, the former quickly confirmed in Matlab with the following code:

```
1  G = @(r) r.*normpdf(r); c=-1; format long, -quadl(G,-20,c), normpdf(c)
```

This result can be used to show what is referred to as Stein's lemma. *For $Z \sim N(0,1)$ and differentiable function $h : \mathbb{R} \to \mathbb{R}$ such that $\mathbb{E}[h'(Z)] < \infty$ and $|h(0)| < \infty$,*

$$\mathbb{E}[Zh(Z)] = \mathbb{E}[h'(Z)]. \tag{A.120}$$

To prove this, first observe that, with $c < 0$ and $u = -r$,

$$\phi(c) = -\int_{-\infty}^{c} r\phi(r)\,dr = \int_{-c}^{\infty} u\phi(u)\,du = \left[\int_{c}^{\infty} u\phi(u)\,du + \int_{c}^{-c} u\phi(u)\,du\right]$$

$$= \int_{c}^{\infty} u\phi(u)\,du.$$

Then, for the positive half-line, and switching the order of integration,

$$\int_0^\infty h'(z)\phi(z)\,\mathrm{d}z = \int_0^\infty h'(z)\int_z^\infty u\phi(u)\,\mathrm{d}u\,\mathrm{d}z = \int_0^\infty \int_0^u h'(z)u\phi(u)\,\mathrm{d}z\,\mathrm{d}u$$

$$= \int_0^\infty [h(u) - h(0)]u\phi(u)\,\mathrm{d}u.$$

Likewise, $\int_{-\infty}^0 h'(z)\phi(z)\,\mathrm{d}z = \int_{-\infty}^0 [h(u) - h(0)]u\phi(u)\,\mathrm{d}u$, so that

$$\mathbb{E}[h'(Z)] = \int_{-\infty}^\infty h(u)u\phi(u)\,\mathrm{d}u = \mathbb{E}[Zh(Z)],$$

as $\int_{-\infty}^\infty h(0)u\phi(u)\,\mathrm{d}u = 0$.

Stein's lemma is also useful when working with the capital asset pricing model; see Cochrane (2001, p. 164) and Panier (1998, Sec. 4.5).

In Problem A.17 below, we show the general result that if, for continuous random variable X with c.f. φ_X, the m.g.f. exists and is such that $\varphi_X(s) = \mathbb{M}_X(is)$, then

$$\int_{-\infty}^q xf_X(x)\,\mathrm{d}x = -\frac{1}{2\pi i}\int_{c-i\infty}^{c+i\infty} \exp\{\mathbb{K}_X(s) - qs\}\mathbb{K}_X'(s)\frac{\mathrm{d}s}{s}. \tag{A.121}$$

It can be evaluated with computing software that supports complex numbers (Julia, Matlab, Python, R, etc.). This expression is useful when the p.d.f. of X is not available, but its m.g.f. is, such as for sums of independent r.v.s. Alternative expressions that also assume the existence of an m.g.f. are given in Kim et al. (2010).

A more general formulation in terms of the c.f. is given by Broda (2011). Importantly, it does not require the existence of the m.g.f., such as with a (sum of independent) stable Paretian r.v.s; see, for example, Broda et al. (2013) for an application. Let $F = F_X$ denote the distribution function of X. For $n \in \{0, 1, 2, \dots\}$, define $G_n(x) \equiv \int_{-\infty}^x x^n\,\mathrm{d}F(x)$, so that $F(x) \equiv G_0(x)$, and observe that, at every point of continuity of F (and hence G_n),

$$\mathbb{E}[X^n \mid X \le x] = \frac{G_n(x)}{F(x)}.$$

Broda (2011) proves that, if the nth moment of X, $n \in \{0, 1, 2, \dots\}$, is finite, and $F(x)$ is continuous at x, then

$$G_n(x) = \frac{\varphi^{(n)}(0)}{2i^n} - \frac{1}{\pi}\int_0^\infty \mathrm{Im}\left[\frac{e^{-itx}\varphi^{(n)}(t)}{i^n t}\right]\mathrm{d}t, \tag{A.122}$$

where $\varphi^{(n)}(t)$ is the nth derivative of the c.f. of X.

Example   *Let $R \sim N(0, 1)$ with characteristic function φ. From (A.97), (A.122), Euler's formula $\exp(it) = \cos(t) + i\sin(t)$, and the basic fact (which the reader should verify) that, for $z = a + bi \ne 0$,*

$$\frac{1}{z} = \frac{a}{a^2 + b^2} - \frac{b}{a^2 + b^2}i, \tag{A.123}$$

or, in particular, $1/i = -i$ (trivially confirmed by multiplying both sides by i), we have

$$\varphi(t) = e^{-t^2/2}, \quad \varphi^{(1)}(t) = -te^{-\frac{1}{2}t^2}, \quad \varphi^{(1)}(0) = 0,$$

and

$$G_1(x) = -\frac{1}{\pi} \int_0^\infty \text{Im}\left[-\frac{[\cos(tx) + i\sin(tx)]\exp(-t^2/2)}{i}\right] dt$$

$$= -\frac{1}{\pi} \int_0^\infty \cos(tx) \exp\left(-\frac{1}{2}t^2\right) dt = -\frac{1}{\sqrt{2\pi}} \exp\left(-\frac{1}{2}x^2\right) = -\phi(x),$$

where the second integral can be found in Gradshteyn and Ryzhik (2007, p. 488, Sec. 3.896, Eq. 4) (and is built in to symbolic computing packages such as Maple). This answer of course agrees with (A.119).

We can use simulation to obtain the ES. It works because of the (weak or strong) law of large numbers; see Section A.15.3 below. In particular, for some $\xi \in (0,1)$ and $q = q_{Z,\xi}$ the ξ-quantile of Z, recall that

$$\text{ES}(Z; \xi) = \mathbb{E}[Z \mid Z < q] = \frac{1}{\xi} \int_{-\infty}^q z f_Z(z) \, dz.$$

The integral in the ES formula can be written as $\mathbb{E}[g(Z)]$, with $g(Z) = Z\mathbb{I}(Z < q)$. So, defining $Y_i = Z_i \mathbb{I}(Z_i < q)$ and $\bar{Y}_n = n^{-1} \sum_{i=1}^n Y_i$, the weak law of large numbers confirms that

$$\frac{1}{\xi} \bar{Y}_n \xrightarrow{p} \text{ES}(Z; \xi).$$

As in the previous example, let $Z_i \overset{\text{i.i.d.}}{\sim} N(0,1)$, $i = 1, \ldots, n$. The empirical ES can be computed in Matlab as follows:

```
xi=0.05; q=norminv(xi); Z=randn(1e6,1); I=(Z<q); mean(Z.*I)/xi
```

An alternative way of computing this is to approximate $\mathbb{E}[Z \mid Z < q]$ directly by taking the average of just those values of Z which are less than q. In Matlab,

```
xi=0.05; q=norminv(xi); Z=randn(1e6,1); W=Z(Z<q); mean(W)
```

The two methods are asymptotically equivalent, and are numerically identical if we take the first one to be

```
I=(Z<q); mean(Z.*I)/mean(I)
```

This method of simulation and empirical determination of the ES is advantageous if the distribution of the Z_i is not known, in which case q is determined by the `quantile` function in Matlab. In particular, assuming the set of i.i.d. data is contained in the vector **P**,

```
VaR=quantile(P,xi); Plo=P(P<=VaR); ES=mean(Plo);
```

yields the empirical VaR and ES.

Example   *Let ϕ_v and Φ_v respectively denote the p.d.f. and c.d.f. of the Student's t distribution (A.70) with v degrees of freedom. Assume $v > 1$, and let $K = v^{-1/2} / B(v/2, 1/2)$ and $u = 1 + r^2/v$. Then*

$$
\mathbb{E}[R \mid R < c] = \frac{1}{\Phi_v(c)} \int_{-\infty}^{c} r\phi_v(r)\, dr = \frac{K}{\Phi_v(c)} \int_{-\infty}^{c} r(1 + r^2/v)^{-(v+1)/2}\, dr
$$

$$
= \frac{K}{\Phi_v(c)} \frac{v}{2} \int_{\infty}^{1+c^2/v} u^{-(v+1)/2}\, du
$$

$$
= -\frac{K}{\Phi_v(c)} \frac{v}{1-v} u^{1-(v+1)/2} \big|_{1+c^2/v}^{\infty} = \frac{K}{\Phi_v(c)} \frac{v}{1-v} (1 + c^2/v)^{1-(v+1)/2}
$$

$$
= -\frac{\phi_v(c)}{\Phi_v(c)} \times \left[\frac{v + c^2}{v - 1}\right], \tag{A.124}
$$

from which it is clear that, as $v \to \infty$, the expression approaches that based on the normal distribution in (A.119), and also why we require first moments to exist, that is, $v > 1$. Using

```
1  N=10^6; df=3; xi=0.01; q=tinv(xi,df); X=trnd(df,N,1); use=X(X<q);
2  ES = -tpdf(q,df) / tcdf(q,df) * (df+q^2)/(df-1);
3  empiricalES_and_trueES = [mean(use) ES]
```

we can confirm the result via simulation.

Example   *The generalized asymmetric t (GAt) distribution generalizes the Student's t in two ways. Its p.d.f. with location 0 and scale 1 is given by*

$$
f_{\mathrm{GAt}}(z; d, v, \theta) = C_{d,v,\theta} \times
\begin{cases}
\left(1 + \dfrac{(-z \cdot \theta)^d}{v}\right)^{-\left(v + \frac{1}{d}\right)}, & \text{if } z < 0, \\[3mm]
\left(1 + \dfrac{(z/\theta)^d}{v}\right)^{-\left(v + \frac{1}{d}\right)}, & \text{if } z \geq 0,
\end{cases}
\tag{A.125}
$$

$d, v, \theta \in \mathbb{R}_{>0}$, *where* $C_{d,v,\theta}^{-1} = (\theta^{-1} + \theta)\, d^{-1} v^{1/d} B(1/d, v)$. *The c.d.f. of $Z \sim \mathrm{GA}\, t(d, v, \theta)$ is*

$$
F_Z(z; d, v, \theta) =
\begin{cases}
\dfrac{\bar{B}_L(v, 1/d)}{1 + \theta^2}, & \text{if } z \leq 0, \\[3mm]
\dfrac{\bar{B}_U(1/d, v)}{1 + \theta^{-2}} + (1 + \theta^2)^{-1}, & \text{if } z > 0,
\end{cases}
\tag{A.126}
$$

where $\bar{B}$ is the incomplete beta ratio,

$$
L = \frac{v}{v + (-z\theta)^d}, \quad \text{and} \quad U = \frac{(z/\theta)^d}{v + (z/\theta)^d}.
$$

   Of the three shape parameters, d measures the peakedness of the density, with values near 1 indicative of Laplace-type behavior, while values near 2 indicate a peak similar to that of the Gaussian and Student's t distributions. The parameter v indicates the tail thickness, and *is analogous to the degrees-of-freedom parameter in the Student't t, except that moments of order vd and higher do not exist, so that, if $d = 2$, then we would double the value of v to make it comparable to the degrees-of-freedom parameter in the Student's t distribution. As $v \to \infty$, the GAt approaches the (possibly asymmetric) generalized exponential distribution which for $d = 2$ is normal, and for $d = 1$ is Laplace. The third parameter, θ, controls the asymmetry, with values less than 1 indicating negative skewness.*

The necessary constraints during estimation of the five parameters (location, scale, and three shape parameters) are $\hat{d} > 0$, $\hat{v} > 0$, $\hat{\theta} > 0$, and scale $\hat{c} > 0$. Parameter estimation is discussed in Example 4.9, with the p.d.f. and c.d.f. of the GAt computed in the function in Listing 4.8.

The rth moment for integer r such that $0 \le r < vd$ is

$$\mathbb{E}[Z^r] = \frac{I_1 + I_2}{K^{-1}} = \frac{(-1)^r \theta^{-(r+1)} + \theta^{r+1}}{\theta^{-1} + \theta} \frac{B((r+1)/d, v - r/d)}{B(1/d, v)} v^{r/d},$$

i.e., the mean is

$$\mathbb{E}[Z] = \frac{\theta^2 - \theta^{-2}}{\theta^{-1} + \theta} \frac{B(2/d, v - 1/d)}{B(1/d, v)} v^{1/d} \qquad (A.127)$$

when vd > 1, and the variance is computed in the obvious way. For $Z \sim \text{GAt}(d, v, \theta)$, Problem II.7.7(d) shows that $S_{r,Z}(c) = \mathbb{E}[Z^r \mid Z < c]$ for $c < 0$ is given by

$$S_{r,Z}(c) = (-1)^r v^{r/d} \frac{(1 + \theta^2)}{(\theta^r + \theta^{r+2})} \frac{B_L(v - r/d, (r+1)/d)}{B_L(v, 1/d)}, \qquad L = \frac{v}{v + (-c\theta)^d}, \qquad (A.128)$$

from which the ES can be computed.

Example *Problem A.13 derives the ES for the so-called asymmetric double Weibull distribution, whose asymmetry and tail behavior make it suitable for modeling (particularly the tails of) financial asset returns.*

Example *The discrete mixture of normals distribution is considered at length in Chapter 5. Let $X \sim \text{MixN}(\boldsymbol{\mu}, \boldsymbol{\sigma}, \boldsymbol{\lambda})$, with $f_{\text{MixN}}(x; \boldsymbol{\mu}, \boldsymbol{\sigma}, \boldsymbol{\lambda}) = \sum_{c=1}^{k} \lambda_c \phi(x; \mu_c, \sigma_c^2)$. The ξ-quantile of X, $q_{X,\xi}$, can be found numerically by solving*

$$\xi - F_X(q_{X,\xi}; \boldsymbol{\mu}, \boldsymbol{\sigma}, \boldsymbol{\lambda}) = 0,$$

where F_X is just a weighted sum of the normal c.d.f.s. The ES can be computed directly from the definition, using numeric integration (and replacing $-\infty$ with, say, -100). This is easy to implement and fast to compute. However, a bit of algebra shows that the ES can be expressed in other forms that are more convenient for numerical calculation and also interpretation.

Let $X_j \sim \text{N}(\mu_j, \sigma_j^2)$ be the jth component in the mixture with p.d.f. $f_{X_j}(x; \mu_j, \sigma_j^2) = \phi(x; \mu_j, \sigma_j^2)$. Now (i) use the fact that, if $Z \sim \text{N}(0, 1)$ and $X_j = \mu_j + \sigma_j Z$, then $f_{X_j}(x) = \sigma_j^{-1} f_Z(z)$, where $z = (x - \mu_j)/\sigma_j$; (ii) substitute $z = (x - \mu_j)/\sigma_j$; and (iii) recall $\int_{-\infty}^{c} z f_Z(z) \, dz = -\phi(c)$ to get

$$\text{ES}(X; \xi) = \frac{1}{\xi} \int_{-\infty}^{q_{X,\xi}} x f_X(x) \, dx = \frac{1}{\xi} \sum_{j=1}^{k} \lambda_j \int_{-\infty}^{q_{X,\xi}} x f_{X_j}(x; \mu_j, \sigma_j^2) \, dx$$

$$= \frac{1}{\xi} \sum_{j=1}^{k} \lambda_j \int_{-\infty}^{q_{X,\xi}} x \, \sigma_j^{-1} f_Z \left(\frac{x - \mu_j}{\sigma_j} \right) \, dx$$

$$= \frac{1}{\xi} \sum_{j=1}^{k} \lambda_j \int_{-\infty}^{\frac{q_{X,\xi} - \mu_j}{\sigma_j}} (\sigma_j z + \mu_j) \sigma_j^{-1} f_Z(z) \, \sigma_j \, dz$$

$$= \frac{1}{\xi} \sum_{j=1}^{k} \lambda_j \left[-\sigma_j \phi \left(\frac{q_{X,\xi} - \mu_j}{\sigma_j} \right) + \mu_j \Phi \left(\frac{q_{X,\xi} - \mu_j}{\sigma_j} \right) \right],$$

which is easily calculated numerically. Further, letting $c_j := (q_{X,\xi} - \mu_j)/\sigma_j$ and factoring out $\Phi(c_j)$ gives

$$\text{ES}(X; \xi) = \frac{1}{\xi} \sum_{j=1}^{k} \lambda_j \Phi(c_j) \left[\mu_j - \sigma_j \frac{\phi(c_j)}{\Phi(c_j)} \right] = \sum_{j=1}^{k} \frac{\lambda_j \Phi(c_j)}{\xi} \left[\mu_j - \sigma_j \frac{\phi(c_j)}{\Phi(c_j)} \right],$$

which has the appearance of a weighted sum of the component ES values, but notice that $\mu_j - \sigma_j \phi(c_j)/\Phi(c_j)$ is not $\text{ES}(X_j; \xi)$ because $c_j = (q_{X,\xi} - \mu_j)/\sigma_j \neq (q_{X_j,\xi} - \mu_j)/\sigma_j = q_{Z,\xi} = \Phi^{-1}(\xi)$. Thus, the ES of a discrete mixture distribution is not a mixture (with the same mixture weights) of ES of the components. We could write $\text{ES}(X; \xi) = \sum_{j=1}^{k} \omega_j \text{ES}(X_j; \xi)$ for

$$\omega_j := \frac{\lambda_j \Phi(c_j)}{\xi} \frac{\mu_j - \sigma_j \phi(c_j)/\Phi(c_j)}{\mu_j - \sigma_j \phi(q_{Z,\xi})/\Phi(q_{Z,\xi})},$$

and let $\omega_j^* = \omega_j / \sum_{j=1}^{k} \omega_j$, and the ω_j^* can be interpreted as the fraction of the ES attributed to component j.

Example A program for computing the ES for the noncentral t distribution is given in Section A.14, while Section A.16 details its computation for the (symmetric and asymmetric) stable Paretian distribution.

Let Z be a location-zero, scale-one random variable, and let $Y = \sigma Z + \mu$ for $\sigma > 0$. An important result is that

$$\text{ES}(Y; \xi) = \mu + \sigma \text{ES}(Z; \xi), \tag{A.129}$$

that is, ES preserves location–scale transformations. To see this, first note that

$$\Pr(Z \leq q_{Z,\xi}) = \xi \quad \Leftrightarrow \quad \Pr(\sigma Z + \mu \leq \sigma q_{Z,\xi} + \mu) = \xi \quad \Leftrightarrow \quad q_{Y,\xi} = \sigma q_{Z,\xi} + \mu. \tag{A.130}$$

Then

$$\text{ES}(Y; \xi) = \mathbb{E}[Y \mid Y \leq q_{Y,\xi}]$$

$$= \mathbb{E}[\sigma Z + \mu \mid \sigma Z + \mu \leq \sigma q_{Z,\xi} + \mu]$$

$$= \sigma \mathbb{E}[Z \mid Z \leq q_{Z,\xi}] + \mu = \sigma \text{ES}(Z; \xi) + \mu.$$

Let Q_X be the quantile function of continuous r.v. X, that is, $Q_X : (0, 1) \rightarrow \mathbb{R}$ with $p \mapsto F_X^{-1}(p)$. Then $\text{ES}(X; \xi)$ can be expressed as

$$\text{ES}(X; \xi) = \frac{1}{\xi} \int_0^\xi Q_X(p) \, dp. \tag{A.131}$$

```
1  xi=0.01; c=norminv(xi);
2  ES1 = -normpdf(c)/normcdf(c), ES2 = quadl(@norminv, 1e-7, xi, 1e-7, 0) / xi
```

Program Listing A.1: Code to verify (A.131).

This is easily seen by letting $u = Q_X(p)$, so that $p = F_X(u)$ and $dp = f_X(u)\,du$. Then, with $q_\xi = Q_X(\xi)$,

$$\int_0^\xi Q_X(p)\,dp = \int_{-\infty}^{q_\xi} u\,f_X(u)\,du.$$

This is a common way to express ES because a weighting function (called the risk spectrum or risk-aversion function) can be incorporated into the integral in (A.131) to form the so-called spectral risk measure. To verify this in Matlab, we use the N(0, 1) case and run the code in Listing A.1.

Another useful result, easily obtained by integration by parts, is

$$\text{ES}(R;\xi) = q_{R,\xi} - \frac{1}{\xi}\int_{-\infty}^{q_{R,\xi}} F_R(r)\,dr. \tag{A.132}$$

Recalling that $q_{R,\xi}$ is the ξ-level VaR, this shows that, in absolute terms, $\text{ES}(R;\xi)$ will be more extreme than the VaR. For the integral in (A.132), with $u = F_R(r)$ and $dv = dr$,

$$\frac{1}{\xi}\int_{-\infty}^{q_{R,\xi}} F_R(r)\,dr = \frac{1}{\xi}rF_R(r)\Big|_{-\infty}^{q_{R,\xi}} - \frac{1}{\xi}\int_{-\infty}^{q_{R,\xi}} rf_R(r)\,dr \overset{?}{=} q_{R,\xi} - \text{ES}(R;\xi).$$

The result then follows if we can show that $rF_R(r)|_{-\infty}^{q_{R,\xi}}$ is $q_{R,\xi} \times \xi$, that is, it $\lim_{r\to-\infty} rF_R(r) = 0$. To this end, let X be a continuous random variable with finite expected value. We wish to show that $\lim_{x\to-\infty} xF_X(x) = 0$ and, if X is non-positive, then $\mathbb{E}[X] = -\int_{-\infty}^0 F_X(x)\,dx$. To show $\lim_{x\to-\infty} xF_X(x) = 0$, note that, as $0 \le F_X(x) \le 1$ for all x,

$$0 \ge \lim_{x\to-\infty} xF_X(x) = \lim_{x\to-\infty} x\int_{-\infty}^x f_X(t)\,dt$$

$$= \lim_{x\to-\infty}\int_{-\infty}^x xf_X(t)\,dt \overset{x>t}{\ge} \lim_{x\to-\infty}\int_{-\infty}^x tf_X(t)\,dt = 0, \tag{A.133}$$

where the last equality follows because $\mathbb{E}[X]$ exists. Thus, $\lim_{x\to-\infty} xF_X(x)$ is bounded above and below by zero and, thus, is zero.

Another way is to note that, if $t < x < 0$, then $|x| < |t|$, so that

$$0 \le |xF_X(x)| = |x|\int_{-\infty}^x f_X(t)\,dt$$

$$\le \int_{-\infty}^x |t|f_X(t)\,dt = \int_{-\infty}^x (-t)f_X(t)\,dt = -\int_{-\infty}^x tf_X(t)\,dt,$$

and taking limits shows that $0 \le \lim_{x\to-\infty}|xF_X(x)| \le -\lim_{x\to-\infty}\int_{-\infty}^x tf_X(t)\,dt = 0$, where, as before, the last equality follows because we assumed that $\mathbb{E}[X]$ exists. Thus, $\lim_{x\to-\infty}|xF_X(x)| = 0$, which implies $\lim_{x\to-\infty} xF_X(x) = 0$.

If it exists, the nth-order lower partial moment with respect to reference point c is

$$\text{LPM}_{n,c}(X) = \int_{-\infty}^{c} (c-x)^n f_X(x)\,dx, \quad n \in \mathbb{N}. \tag{A.134}$$

This is another tail risk measure and is related to ES. The LPM can be computed by numeric integration, though in some cases closed-form solutions will exist. Applying the binomial theorem to $(c-x)^n$, we can write

$$\text{LPM}_{n,c}(X) = \sum_{h=0}^{n} K_{h,c} T_{h,c}(X), \tag{A.135}$$

where $K_{h,c} = K_{h,c}(n) = \binom{n}{h} c^{n-h}(-1)^h$ and $T_{h,c}(X) = \int_{-\infty}^{c} x^h f_X(x)\,dx$.

Example *For $Z \sim N(0,1)$ and $c < 0$, calculation shows (let $u = z^2/2$ for $z < 0$) that, for $h \in \mathbb{N}$,*

$$T_{h,c}(Z) = \frac{(-1)^h 2^{h/2-1}}{\sqrt{\pi}} \left[\Gamma\left(\frac{h+1}{2}\right) - \Gamma_{c^2/2}\left(\frac{h+1}{2}\right) \right], \tag{A.136}$$

where $\Gamma_x(a)$ is the incomplete gamma function. Note $T_{0,c}(Z) = \Phi(c)$ and $T_{1,c}(Z) = -\phi(c)$. For $X \sim t_v$, substitute $u = 1 + x^2/v$ for $x < 0$ and then $x = (u-1)/u$, for $h < v$, to get

$$T_{h,c}(X;v) = \frac{(-1)^h v^{h/2}}{2B\left(\frac{v}{2},\frac{1}{2}\right)} \left[B\left(\frac{h+1}{2}, \frac{v-h}{2}\right) - B_w\left(\frac{h+1}{2}, \frac{v-h}{2}\right) \right], \tag{A.137}$$

where $w = (c^2/v)/(1 + c^2/v)$ and B_w is the incomplete beta function. In particular,

$$T_{0,c}(X;v) = F_X(c;v) = \Phi_v(c) \quad \text{and} \quad T_{1,c}(X;v) = \phi_v(c)\frac{v+c^2}{1-v},$$

as shown in (A.124).

Similar to simulation of ES, for the LPM in (A.134), we have $g(Z) = (c-Z)^n \mathbb{I}(Z < c)$, and $\bar{Y}_n \xrightarrow{p} \text{LPM}_{n,c}(Z)$ for $Y_i = (c-Z_i)^n \mathbb{I}(Z_i < c)$, computed for $n = 2$ as follows:

```
xi=0.05; q=norminv(xi); Z=randn(1e6,1); I=(Z<q); Y=(q-Z).^2; mean(Y.*I)
```

Or we can compute the conditional version as

$$\frac{\Pr(Z < c)}{\Pr(Z < c)} \int_{-\infty}^{c} (c-z)^n f_Z(z)\,dz = \Pr(Z < c)\mathbb{E}[(c-Z)^n \mid Z < c]$$

using, for $n = 2$,

```
xi=0.05; q=norminv(xi); Z=randn(1e6,1); W=Z(Z<q); mean((q-W).^2)*xi
```

Example *If $X \sim \text{MixN}(\boldsymbol{\mu}, \boldsymbol{\sigma}, \boldsymbol{\lambda})$ with k components, then $\text{LPM}_{n,c}(X)$ can be computed from (A.135) and (A.136), where $T_{h,c}(X)$ is, similarly to the derivation of ES for mixtures, given by*

$$T_{h,c}(X) = \sum_{j=1}^{k} \lambda_j \sum_{m=0}^{h} \binom{h}{m} \sigma_j^m \mu_j^{h-m} T_{m,(c-\mu_j)/\sigma_j}(Z), \tag{A.138}$$

for $Z \sim \text{N}(0, 1)$. An important special case is for $k = 1$, which corresponds to $X \sim \text{N}(\mu, \sigma^2)$. One can also construct a mixture of Student's t distributions analogous to the mixed normal. Letting $\mathbf{v}$ denote the vector of k degrees-of-freedom parameters, we write $X \sim \text{MixT}(\mathbf{v}, \boldsymbol{\mu}, \boldsymbol{\sigma}, \boldsymbol{\lambda})$. For $X \sim \text{MixT}(\mathbf{v}, \boldsymbol{\mu}, \boldsymbol{\sigma}, \boldsymbol{\lambda})$ and $h < \min v_j$, a result similar to that for the mixed normal case holds by replacing $T_{m,(c-\mu_j)/\sigma_j}(Z)$ in (A.138) with $T_{m,(c-\mu_j)/\sigma_j}(Y; v_j)$ from (A.137), where $Y \sim t_{v_j}$.

The $\text{LPM}_{n,c}(X)$ can also be easily computed with simulation (though is far slower and less accurate), and serves as a check that (A.138) is programmed correctly. The reader is encouraged to perform these calculations.

A.9 JACOBIAN TRANSFORMATIONS

Let $\mathbf{X} = (X_1, \ldots, X_n)$ be an n-dimensional continuous r.v. and $\mathbf{g} = (g_1(\mathbf{x}), \ldots, g_n(\mathbf{x}))$ a continuous bijection that maps $S_{\mathbf{X}} \subset \mathbb{R}^n$, the support of $\mathbf{X}$, onto $S_{\mathbf{Y}} \subset \mathbb{R}^n$. Then the p.d.f. of $\mathbf{Y} = (Y_1, \ldots, Y_n) = \mathbf{g}(\mathbf{X})$ is given by

$$f_{\mathbf{Y}}(\mathbf{y}) = f_{\mathbf{X}}(\mathbf{x})|\det \mathbf{J}|, \tag{A.139}$$

where $\mathbf{x} = \mathbf{g}^{-1}(\mathbf{y}) = (g_1^{-1}(\mathbf{y}), \ldots, g_n^{-1}(\mathbf{y}))$ and

$$\mathbf{J} = \begin{pmatrix} \dfrac{\partial x_1}{\partial y_1} & \dfrac{\partial x_1}{\partial y_2} & \cdots & \dfrac{\partial x_1}{\partial y_n} \\[2mm] \dfrac{\partial x_2}{\partial y_1} & \dfrac{\partial x_2}{\partial y_2} & & \dfrac{\partial x_2}{\partial y_n} \\[2mm] \vdots & \vdots & \ddots & \vdots \\[2mm] \dfrac{\partial x_n}{\partial y_1} & \dfrac{\partial x_n}{\partial y_2} & \cdots & \dfrac{\partial x_n}{\partial y_n} \end{pmatrix} = \begin{pmatrix} \dfrac{\partial g_1^{-1}(\mathbf{y})}{\partial y_1} & \dfrac{\partial g_1^{-1}(\mathbf{y})}{\partial y_2} & \cdots & \dfrac{\partial g_1^{-1}(\mathbf{y})}{\partial y_n} \\[2mm] \dfrac{\partial g_2^{-1}(\mathbf{y})}{\partial y_1} & \dfrac{\partial g_2^{-1}(\mathbf{y})}{\partial y_2} & & \dfrac{\partial g_2^{-1}(\mathbf{y})}{\partial y_n} \\[2mm] \vdots & \vdots & \ddots & \vdots \\[2mm] \dfrac{\partial g_n^{-1}(\mathbf{y})}{\partial y_1} & \dfrac{\partial g_n^{-1}(\mathbf{y})}{\partial y_2} & \cdots & \dfrac{\partial g_n^{-1}(\mathbf{y})}{\partial y_n} \end{pmatrix} \tag{A.140}$$

is the Jacobian associated with $\mathbf{g}$.

Via transformation (A.139), it is straightforward to show that, for $G \sim N(0, 1)$ independent of $C \sim \chi_n^2$, $T = G/\sqrt{C/n} \sim t_n$, that is, T is a Student's t r.v. with n degrees of freedom. Similarly, for $X_i \overset{\text{indep}}{\sim} \chi_{n_i}^2$, $i = 1, 2$, $Y = (X_1/n_1)/(X_2/n_2) \sim F(n_1, n_2)$, that is, Y is an F r.v. with n_1 and n_2 degrees of freedom.

It is sometimes necessary to partition $S_{\mathbf{X}}$ into disjoint subsets, say $S_{\mathbf{X}}^{(1)}, \ldots, S_{\mathbf{X}}^{(d)}$, in order for $\mathbf{g}$ to be piecewise invertible. Letting $\mathbf{g}_i = (g_{1i}(\mathbf{x}), \ldots, g_{ni}(\mathbf{x}))$ be continuous bijections from $S_{\mathbf{X}}^{(i)}$ to (a subset of) $\mathbb{R}^n$, the general transformation can be written as

$$f_{\mathbf{Y}}(\mathbf{y}) = \sum_{i=1}^{d} f_{\mathbf{X}}(\mathbf{x}_i)|\det \mathbf{J}_i|, \tag{A.141}$$

where $\mathbf{x}_i = (g_{1i}^{-1}(\mathbf{y}), \dots, g_{ni}^{-1}(\mathbf{y}))$ and

$$\mathbf{J}_i = \begin{pmatrix} \dfrac{\partial g_{1i}^{-1}(\mathbf{y})}{\partial y_1} & \dfrac{\partial g_{1i}^{-1}(\mathbf{y})}{\partial y_2} & \cdots & \dfrac{\partial g_{1i}^{-1}(\mathbf{y})}{\partial y_n} \\[2mm] \dfrac{\partial g_{2i}^{-1}(\mathbf{y})}{\partial y_1} & \dfrac{\partial g_{2i}^{-1}(\mathbf{y})}{\partial y_2} & & \dfrac{\partial g_{2i}^{-1}(\mathbf{y})}{\partial y_n} \\[2mm] \vdots & \vdots & \ddots & \vdots \\[2mm] \dfrac{\partial g_{ni}^{-1}(\mathbf{y})}{\partial y_1} & \dfrac{\partial g_{ni}^{-1}(\mathbf{y})}{\partial y_2} & \cdots & \dfrac{\partial g_{ni}^{-1}(\mathbf{y})}{\partial y_n} \end{pmatrix}.$$

Example   *Let X_1 and X_2 be continuous r.v.s with joint p.d.f. f_{X_1,X_2}. To derive an expression for the density of $S = X_1^2 + X_2^2$, start by letting $S = X_1^2 + X_2^2$, $Z = X_1^2$, $X_1 = \pm\sqrt{Z}$ and $X_2 = \pm\sqrt{S - Z}$. Then, considering each of the four possible sign configurations on the x_i,*

$$f_{S,Z}(s,z) = |\mathbf{J}| f_{X_1,X_2}(x_1, x_2) \mathbb{1}_{(-\infty,0)}(x_1) \mathbb{1}_{(-\infty,0)}(x_2) + \cdots$$

with

$$\mathbf{J} = \begin{pmatrix} \partial x_1/\partial z & \partial x_1/\partial s \\ \partial x_2/\partial z & \partial x_2/\partial s \end{pmatrix} = \begin{pmatrix} \pm\frac{1}{2}z^{-1/2} & \cdot \\ 0 & \pm\frac{1}{2}(s-z)^{-1/2} \end{pmatrix},$$

and, as z and $s - z$ are both positive, all $|\mathbf{J}|$ are the same:

$$|\mathbf{J}| = \frac{1}{4}z^{-1/2}(s-z)^{-1/2}.$$

Thus

$$f_{S,Z}(s,z) = \frac{1}{4}z^{-1/2}(s-z)^{-1/2} \times [f_{X_1,X_2}(-\sqrt{z}, -\sqrt{s-z}) + \cdots] \mathbb{1}_{(0,s)}(z), \tag{A.142}$$

where the term in brackets has the form $f(-,-) + f(-,+) + f(+,-) + f(+,+)$ and

$$f_S(s) = \int_0^s f_{S,Z}(s,z)\, \mathrm{d}z.$$

For the special case of X_1 and X_2 being i.i.d. standard normal r.v.s, $f_{X_1,X_2}(x_1,x_2) = \exp\left\{-\frac{1}{2}(x_1^2 + x_2^2)\right\}/(2\pi)$, so that all four terms in (A.142) are the same, yielding

$$f_{S,Z}(s,z) = \frac{1}{4}z^{-1/2}(s-z)^{-1/2}\frac{4}{2\pi}e^{-\frac{1}{2}(z+(s-z))} = \frac{1}{2\pi}z^{-1/2}(s-z)^{-1/2}e^{-\frac{1}{2}s}\mathbb{1}_{(0,s)}(z).$$

From (A.12),

$$f_S(s) = \frac{1}{2\pi}e^{-\frac{1}{2}s}\int_0^s z^{-1/2}(s-z)^{-1/2}\, \mathrm{d}z = \frac{1}{2\pi}e^{-\frac{1}{2}s}B\left(\frac{1}{2}, \frac{1}{2}\right) = \frac{1}{2}e^{-\frac{1}{2}s}\mathbb{1}_{(0,\infty)}(s),$$

so that $S \sim \mathrm{Exp}(1/2)$ and also $S \sim \chi_2^2$.

A.10 SUMS AND OTHER FUNCTIONS

Let X_i, $i = 1, \dots, n$, be independent r.v.s, each of which possesses an m.g.f., and define $Y = \sum_{i=1}^n a_i X_i$. Then

$$\mathbb{M}_Y(s) = \mathbb{E}[e^{sY}] = \mathbb{E}[e^{sa_1 X_1}] \cdots \mathbb{E}[e^{sa_n X_n}] = \prod_{i=1}^n \mathbb{M}_{X_i}(sa_i), \qquad (A.143)$$

with a similar result holding for the c.f. This is useful for demonstrating that particular random variables are (or are not) closed under addition. For example, the normal and χ^2, and, under certain conditions on their parameters, also the binomial, negative binomial, and gamma are closed under addition.

Example For $X \sim \text{Gam}(\alpha, \beta)$, the reader should quickly confirm that its m.g.f. is given by $\mathbb{M}_X(s) = (\beta/(\beta - s))^\alpha$ for $s < \beta$. If $X_i \overset{\text{indep}}{\sim} \text{Gam}(\alpha_i, \beta)$, then $\sum_i X_i \sim \text{Gam}\left(\sum_i \alpha_i, \beta\right)$, as seen from (A.143). We mention two important special cases: if $X_i \overset{\text{i.i.d.}}{\sim} \text{Exp}(\lambda)$, then $\sum_i^n X_i \sim \text{Gam}(n, \lambda)$; and if $Y_i \overset{\text{indep}}{\sim} \chi^2(\nu_i)$, then $\sum_i^n Y_i \sim \chi^2\left(\sum_i^n \nu_i\right)$.

Example If $Z_i \overset{\text{i.i.d.}}{\sim} N(0, 1)$, $i = 1, \dots, n$, then $\sum_{i=1}^n Z_i^2 \sim \chi^2(n)$. More generally,

$$\text{if } X_i \overset{\text{i.i.d.}}{\sim} N(\mu_i, \sigma_i^2), \quad \text{then} \quad \sum_{i=1}^n \left(\frac{X_i - \mu_i}{\sigma_i}\right)^2 \sim \chi^2(n), \qquad (A.144)$$

an important result that the reader should be able to easily prove.

If the m.g.f. does not exist, then the c.f. (A.96) can be used.

Example Example II.1.22 shows that, for c.f. $\varphi_X(t) = \exp\{-|t|\}$, the p.d.f. of X is (A.69), that is, X is standard Cauchy. Let $X_i \overset{\text{i.i.d.}}{\sim} \text{Cau}(0, 1)$. Then $S = n^{-1} \sum_{i=1}^n X_i$ has c.f.

$$\varphi_S(t) = \mathbb{E}[e^{itS}] = \mathbb{E}[e^{itn^{-1}X_1}] \cdots \mathbb{E}[e^{itn^{-1}X_n}] = [\varphi_{X_1}(t/n)]^n = \exp\{-|t|\}, \qquad (A.145)$$

so that $S \sim \text{Cau}(0, 1)$. This serves as one of the primary examples for demonstrating that the law of large numbers does not give the result of convergence to the expected value of X, which in this case does not exist.

If X and Y are jointly distributed continuous random variables with density $f_{X,Y}(x, y)$, then the densities of $S = X + Y$, $D = X - Y$, $P = XY$ and $R = X/Y$ can be respectively expressed as

$$f_S(s) = \int_{-\infty}^\infty f_{X,Y}(x, s - x)\, dx = \int_{-\infty}^\infty f_{X,Y}(s - y, y)\, dy, \qquad (A.146)$$

$$f_D(d) = \int_{-\infty}^\infty f_{X,Y}(x, x - d)\, dx = \int_{-\infty}^\infty f_{X,Y}(d + y, y)\, dy, \qquad (A.147)$$

$$f_P(p) = \int_{-\infty}^\infty \frac{1}{|x|} f_{X,Y}\left(x, \frac{p}{x}\right) dx = \int_{-\infty}^\infty \frac{1}{|y|} f_{X,Y}\left(\frac{p}{y}, y\right) dy, \qquad (A.148)$$

$$f_R(r) = \int_{-\infty}^{\infty} \frac{|x|}{r^2} f_{X,Y}\left(x, \frac{x}{r}\right) dx = \int_{-\infty}^{\infty} |y| f_{X,Y}(ry, y)\, dy. \tag{A.149}$$

That for the sum, (A.146), is referred to as the *convolution* of X and Y.

Example *The reader can confirm with (A.149) that, for $Z_i \overset{\text{i.i.d.}}{\sim} N(0,1)$,*

$$C_1 = Z_1/Z_2 \sim \mathrm{Cau}(0,1), \quad \text{independent of} \quad C_2 = Z_3/Z_4 \sim \mathrm{Cau}(0,1), \tag{A.150}$$

each with density (A.69). Of interest is the distribution of the product P and ratio R of two independent standard Cauchy r.v.s. Using their distributional relationship to the normal, observe that

$$R := \frac{C_1}{C_2} = \frac{Z_1}{Z_2}\frac{Z_4}{Z_3} = C_1 C_3 =: P, \quad \text{where } C_3 = \frac{Z_4}{Z_3} \perp C_1,$$

so that R and P have the same distribution. The reader is encouraged to show that

$$f_R(r) = f_P(r) = \frac{\ln(r^2)}{\pi^2(r^2 - 1)}. \tag{A.151}$$

In the scaled case, if $C_1 \sim \mathrm{Cau}(0,s)$ independent of $C_2 \sim \mathrm{Cau}(0,s)$, then, clearly, $f_R(r)$ does not change, but $f_P(r)$ depends on s. The reader can derive the p.d.f. for R and P when $C_1 \sim \mathrm{Cau}(0,s_1)$ independent of $C_2 \sim \mathrm{Cau}(0,s_2)$.

Remark.   Results (A.145) and (A.150) can be expressed as follows. Let $\mathbf{X} = (X_1, \ldots, X_n)'$ and $\mathbf{Y} = (Y_1, \ldots, Y_n)'$ be i.i.d. $N(\mathbf{0}, \mathbf{I}_n)$ and $W_i = 1/n$, $i = 1, \ldots, n$. Then

$$S = \sum_{i=1}^{n} W_i \frac{X_i}{Y_i} \sim \mathrm{Cau}(0,1). \tag{A.152}$$

Remarkably, Pillai and Meng (2016) have shown that (A.152) holds when (i) the W_i are random, such that $W_i \geq 0$, $\sum_{i=1}^{n} W_i = 1$, and independent of $\mathbf{X}$ and $\mathbf{Y}$; and (ii) $\mathbf{X}$ and $\mathbf{Y}$ are i.i.d. $N(\mathbf{0}, \boldsymbol{\Sigma})$, with $\boldsymbol{\Sigma}$ being an arbitrary covariance matrix with positive diagonals.

Furthermore, they show that, for $\mathbf{V} = (W_1/X_1, \ldots, W_n/X_n)'$,

$$\mathbf{V}'\boldsymbol{\Sigma}\mathbf{V} \sim \mathrm{L\acute{e}vy}(0,1), \tag{A.153}$$

where the Lévy distribution is given in (A.305), and is notably the distribution of Z^{-2} for $Z \sim N(0,1)$. Statistical applications of results (A.152) and (A.153) are discussed in Drton and Xiao (2016). ∎

Example *Let $X_0 \sim N(0,1)$ independent of $Y_0 \sim \mathrm{Lap}(0,1)$. For a known value of c, with $0 \leq c \leq 1$, define*

$$Z = cX_0 + (1-c)Y_0 = X + Y,$$

where $X = cX_0 \sim N(0, c^2)$ and, with $k = 1 - c$, $Y = kY_0 \sim \mathrm{Lap}(0, k)$. We write $Z \sim \mathrm{NLap}(c)$. The exact density of Z for $c \in (0,1)$ is obtained from the convolution formula (A.146) as

$$f_Z(z) = \int_{-\infty}^{\infty} f_X(z - y) f_Y(y)\, dy,$$

which needs to be numerically evaluated. Algebraic manipulation shows that f_Z can also be written as

$$f_Z(z) = \frac{1}{2(1-c)} \left(\exp\left\{ \frac{s^2 - z^2}{2c^2} \right\} \Phi\left(-\frac{s}{c}\right) + \exp\left\{ \frac{d^2 - z^2}{2c^2} \right\} \bar{\Phi}\left(-\frac{d}{c}\right) \right), \quad (A.154)$$

which is faster to evaluate, but becomes numerically problematic as $c \to 1$. See pages II.81–83 for further details, and Haas et al. (2006) on combining the Laplace and normal distributions.

Let the continuous function $g(\mathbf{x}) := g(x_1, x_2)$ be defined on an open neighborhood of $\mathbf{x}^0 = (x_1^0, x_2^0)$ with $\dot{g}_i(\mathbf{x}) = (\partial g/\partial x_i)(\mathbf{x})$, $\ddot{g}_{ij}(\mathbf{x}) = (\partial^2 g/\partial x_i \partial x_j)(\mathbf{x})$ and $\triangle_i = (x_i - x_i^0)$, $i = 1, 2$. Let $\mathbf{X} = (X_1, X_2)$ be a bivariate r.v. with mean $\mathbf{x}^0 = (\mu_{X_1}, \mu_{X_2})$. Then

$$\mathbb{E}[g(\mathbf{X})] \approx g(\mathbf{x}^0) + \frac{1}{2}\ddot{g}_{11}(\mathbf{x}^0)\mathbb{V}(X_1) + \frac{1}{2}\ddot{g}_{22}(\mathbf{x}^0)\mathbb{V}(X_2) + \ddot{g}_{12}(\mathbf{x}^0)\mathrm{Cov}\,(X_1, X_2) \quad (A.155)$$

and

$$\mathbb{V}(g(\mathbf{X})) \approx \dot{g}_1^2(\mathbf{x}^0)\mathbb{V}(X_1) + \dot{g}_2^2(\mathbf{x}^0)\mathbb{V}(X_2) + 2\dot{g}_1(\mathbf{x}^0)\dot{g}_2(\mathbf{x}^0)\mathrm{Cov}\,(X_1, X_2). \quad (A.156)$$

In particular,

$$\mathbb{E}\left[\frac{X}{Y}\right] \approx \frac{\mu_X}{\mu_Y} + \frac{\mu_X}{\mu_Y^3}\sigma_Y^2 - \frac{c}{\mu_Y^2} \quad (A.157)$$

and

$$\mathbb{V}\left(\frac{X}{Y}\right) \approx \mu_Y^{-2}\sigma_X^2 + \mu_X^2\mu_Y^{-4}\sigma_Y^2 - 2\mu_X\mu_Y^{-3}c = \left(\frac{\mu_X}{\mu_Y}\right)^2 \left(\frac{\sigma_X^2}{\mu_X^2} + \frac{\sigma_Y^2}{\mu_Y^2} - \frac{2\,c}{\mu_X\mu_Y}\right), \quad (A.158)$$

where $\sigma_X^2 = \mathbb{V}(X)$, $\sigma_Y^2 = \mathbb{V}(Y)$ and $c = \mathrm{Cov}\,(X, Y)$. Also,

$$\mathbb{E}\left[\frac{1}{Y}\right] \approx \frac{1}{\mu_Y} + \frac{\sigma_Y^2}{\mu_Y^3} \quad \text{and} \quad \mathbb{V}\left(\frac{1}{Y}\right) \approx \frac{\sigma_Y^2}{\mu_Y^4}. \quad (A.159)$$

The simplest form of the central limit theorem states that, for a sequence of i.i.d. r.v.s $X_1, X_2, \ldots$, with finite mean and variance, denoted $X_i \overset{\text{i.i.d.}}{\sim} (\mu, \sigma^2)$,

$$S_n = n^{1/2}\frac{\bar{X}_n - \mu}{\sigma} = \frac{\bar{X}_n - \mu}{\sigma/\sqrt{n}} \overset{\text{asy}}{\sim} N(0, 1). \quad (A.160)$$

A straightforward proof of (A.160) assuming existence of the moment generating function is as follows (Tardiff, 1981). Let $Z_i = (X_i - \mu)/\sigma$ with common m.g.f. $\mathbb{M}_Z(t)$ that exists for t in a neighborhood of zero, and observe that $S_n = n^{-1/2}\sum_{i=1}^n Z_i$. From (A.91), $\mathbb{M}_{S_n}(t) = (\mathbb{M}_Z(tn^{-1/2}))^n$. With $\mathbb{K}_{S_n}(t) = \log \mathbb{M}_{S_n}(t)$, we wish to show that $\lim_{n\to\infty}\mathbb{K}_{S_n}(t) = t^2/2$. As $\mathbb{K}_{S_n}(t) = n\mathbb{K}_Z(tn^{-1/2})$,

$$\lim_{n\to\infty}\mathbb{K}_{S_n}(t) = \lim_{n\to\infty}\frac{\mathbb{K}_Z(tn^{-1/2})}{1/n} = \lim_{n\to\infty}\frac{\mathbb{K}_Z'(tn^{-1/2})t(-1/2)n^{-3/2}}{-n^{-2}}$$

$$= \frac{t}{2} \lim_{n \to \infty} \frac{\mathbb{K}'_Z(tn^{-1/2})}{1/n^{1/2}} = \frac{t}{2} \lim_{n \to \infty} \frac{\mathbb{K}''_Z(tn^{-1/2})t(-1/2)n^{-3/2}}{(-1/2)n^{-3/2}}$$

$$= \frac{t^2}{2} \lim_{n \to \infty} \mathbb{K}''_Z(tn^{-1/2}),$$

applying l'Hôpital's rule twice. Assuming the interchange of limit and integral and using (A.92),

$$\lim_{n \to \infty} \mathbb{K}''_Z(tn^{-1/2}) = \mathbb{K}''_Z(\lim_{n \to \infty} tn^{-1/2}) = \mathbb{K}''_Z(0) = \mathbb{V}(Z) = 1,$$

so that $\lim_{n \to \infty} \mathbb{K}_{S_n}(t) = t^2/2$.

A.11 SADDLEPOINT APPROXIMATIONS

For small sample sizes, the normal approximation to a random variable X via the central limit theorem (A.160) will often not be accurate. If the c.g.f. of X, $\mathbb{K}_X$, is available, then the saddlepoint approximation (s.p.a.) to the p.d.f. and c.d.f. of X at a value x in the (interior of the) support of X can be computed. The s.p.a. can be thought of as an approximation to the inversion formulas. Book-length presentations of s.p.a. methods and original references are provided by Jensen (1995) and Butler (2007). The following saddlepoint p.d.f. and c.d.f. approximations are very basic, yet very accurate and highly useful for our purposes.

The (first-order) saddlepoint density approximation to f is

$$\hat{f}_X(x) = \frac{1}{\sqrt{2\pi\, \mathbb{K}''_X(\hat{s})}} \exp\{\mathbb{K}_X(\hat{s}) - x\hat{s}\}, \quad x = \mathbb{K}'_X(\hat{s}), \qquad (A.161)$$

where $\hat{s} = \hat{s}(x)$ is the solution to the saddlepoint equation and is referred to as the saddlepoint at x. An elementary derivation of (A.161) is given in Section II.5.1.1, while more advanced derivations that also lend themselves to computing higher-order terms are given in Butler (2007).

The density s.p.a. (A.161) will not, in general, integrate to 1, although it will usually not be far off. It will often be possible to renormalize it, that is,

$$\bar{\hat{f}}_X(x) = \frac{\hat{f}_X(x)}{\int \hat{f}_X(x)\, \mathrm{d}x}, \qquad (A.162)$$

which is a proper density. The s.p.a. to the "almost c.d.f." for the random variable X is (note that it is $X < x$ and not $X \leq x$, which is relevant in the discrete case)

$$\hat{F}_X(x^-) = \Pr(X < x) = \Phi(\hat{w}) + \phi(\hat{w})\left\{\frac{1}{\hat{w}} - \frac{1}{\hat{u}}\right\}, \quad x \neq \mathbb{E}[X], \qquad (A.163)$$

where Φ and ϕ are the c.d.f. and p.d.f. of the standard normal distribution, respectively,

$$\hat{w} = \mathrm{sgn}(\hat{s})\sqrt{2\hat{s}x - 2\mathbb{K}(\hat{s})} \quad \text{and} \quad \hat{u} = \begin{cases} \hat{s}\sqrt{\mathbb{K}''(\hat{s})}, & \text{if } X \text{ is continuous,} \\ (1 - \mathrm{e}^{-\hat{s}})\sqrt{\mathbb{K}''(\hat{s})}, & \text{if } X \text{ is discrete.} \end{cases}$$

$$(A.164)$$

At the mean,

$$\widehat{F}_X(\mathbb{E}[X]) = \frac{1}{2} + \frac{\mathbb{K}'''(0)}{6\sqrt{2\pi}\mathbb{K}''(0)^{3/2}}. \tag{A.165}$$

The second-order s.p.a. approximations are given by

$$\tilde{f}(x) = \hat{f}(x)\left(1 + \frac{\hat{\kappa}_4}{8} - \frac{5}{24}\hat{\kappa}_3^2\right), \tag{A.166}$$

where $\hat{\kappa}_i = \mathbb{K}^{(i)}(\hat{s})/[\mathbb{K}''(\hat{s})]^{i/2}$, and

$$\tilde{F}(x) = \widehat{F}(x) - \phi(\hat{w})\left\{\hat{u}^{-1}\left(\frac{\hat{\kappa}_4}{8} - \frac{5}{24}\hat{\kappa}_3^2\right) - \hat{u}^{-3} - \frac{\hat{\kappa}_3}{2\hat{u}^2} + \hat{w}^{-3}\right\} \tag{A.167}$$

for $x \neq \mathbb{E}[X]$.

Example   *Let $Z \sim \text{NLap}(c)$, that is, $Z = X + Y$, whereby $X \sim \text{N}(0, c^2)$ independent of $Y \sim \text{Lap}(0, k)$, $k = 1 - c$, $0 \leq c \leq 1$, with exact density (A.154). The m.g.f. of X is $\exp(c^2 s^2/2)$ and the m.g.f. of Y is $\mathbb{M}_Y(s) = (1 - s^2 k^2)^{-1}$ for $|s| < 1/k$. It follows that*

$$\mathbb{K}_{X+Y}(s) = \frac{1}{2}c^2 s^2 - \ln(1 - s^2 k^2), \quad |s| < 1/k,$$

so that, for $0 < c < 1$, the solution to the saddlepoint equation

$$\mathbb{K}_Z'(s) = c^2 s + \frac{2sk^2}{1 - s^2 k^2} = z$$

involves finding the root of the cubic $s^3 c^2 k^2 - s^2 k^2 z - s(2k^2 + c^2) + z = 0$ that satisfies $|s| < 1/k$. The first-order s.p.a. density is then given by (A.161), with

$$\mathbb{K}_Z''(s) = c^2 + 2k^2 \frac{s^2 k^2 + 1}{(1 - s^2 k^2)^2}.$$

As

$$\mathbb{K}_Z^{(3)}(s) = 4sk^4 \frac{3 + s^2 k^2}{(1 - s^2 k^2)^3}, \qquad \mathbb{K}_Z^{(4)}(s) = 12k^4 \frac{1 + 6s^2 k^2 + s^4 k^4}{(1 - s^2 k^2)^4},$$

the second-order s.p.a. approximation can also be used.

Example   *Let $\lambda \in \mathbb{R}$, $\omega > 0$, $-1 < \rho < 1$, $\mu \in \mathbb{R}$ and $\sigma > 0$. Then r.v. X follows a generalized hyperbolic (GHyp) density, written $X \sim \text{GHyp}(\lambda, \omega, \rho, \sigma, \mu)$, if its density is given by*

$$f_{\text{GHyp}}(x; \lambda, \omega, \rho, \sigma, \mu) = \frac{\omega^\lambda \bar{y}^{\lambda - 1/2}}{\sqrt{2\pi}\bar{\alpha}^{\lambda - 1/2}\sigma K_\lambda(\omega)}K_{\lambda - 1/2}(\bar{\alpha}\bar{y})\exp\{\rho\bar{\alpha}z\}, \tag{A.168}$$

where $z = (x - \mu)/\sigma$, $\bar{\alpha} \equiv \omega(1 - \rho^2)^{-1/2}$, $\bar{y} = \sqrt{1 + z^2}$, and $K_\nu(x)$ is the modified Bessel function of the third kind with index ν, given in (A.19). The parameters of the GHyp have the following interpretation: μ and σ are genuine location and scale parameters, respectively, while λ, ω and ρ are location- and scale-invariant. The parameter ω controls the tail thickness, and ρ is a measure of the skewness. Derivations of the GHyp as a continuous *normal mixture distribution, and its moments and m.g.f., are given in Chapter II.9, along with alternative density expressions and a detailed discussion of its numerous special and limiting cases. Two important special cases that still allow for leptokurtosis and asymmetry, and possess an m.g.f., include the normal inverse Gaussian (NIG), obtained by letting $\lambda = -1/2$, and the hyperbolic, for $\lambda = 1$.*

The expected value and variance of the GHyp distribution are given by $\mathbb{E}[X] = \mu + \sigma\rho(1-\rho^2)^{-1/2}k_1(\omega)$ and $\mathbb{V}(X) = \sigma^2[\omega^{-1}k_1(\omega) + \rho^2(1-\rho^2)^{-1}k_2(\omega)]$, respectively, where $k_1(\omega) = K_{\lambda+1}(\omega)/K_\lambda(\omega)$ and $k_2(\omega) = [K_\lambda(\omega)K_{\lambda+2}(\omega) - K_{\lambda+1}(\omega)^2]/K_\lambda(\omega)^2$. For certain applications, it is useful to standardize the GHyp to have zero mean and unit variance. We call the resulting distribution the standard generalized hyperbolic, with p.d.f. given by

$$f_{\text{SGH}}(x, \lambda, \omega, \rho) = f_{\text{GHyp}}(x, \lambda, \omega, \rho, \hat{\sigma}, \hat{\mu}), \tag{A.169}$$

where $\hat{\sigma} = [\omega^{-1}k_1(\omega) + \rho^2(1-\rho^2)^{-1}k_2(\omega)]^{-1/2}$ and $\hat{\mu} = -\rho(1-\rho^2)^{-1/2}\hat{\sigma}k_1(\omega)$.

Let $X \sim \text{GHyp}(\lambda, \omega, \rho, \sigma, \mu)$ with density (A.168). With $\beta = \omega\sigma^{-1}\rho(1-\rho^2)^{-1/2}$ and $\psi = \omega^2\sigma^{-2}$, the m.g.f. of X is given by

$$\mathbb{M}_X(t) = e^{\mu t} \frac{K_\lambda\left(\omega\sqrt{1 - \frac{2\beta t + t^2}{\psi}}\right)}{K_\lambda(\omega)\left(1 - \frac{2\beta t + t^2}{\psi}\right)^{\lambda/2}}, \tag{A.170}$$

with convergence strip given by those values of t such that

$$1 - \frac{2\beta t + t^2}{\psi} > 0 \overset{\psi > 0}{\Rightarrow} t^2 + 2\beta t - \psi < 0.$$

The solutions of $t^2 + 2\beta t - \psi = 0$ are $t = -\beta \pm \sqrt{\beta^2 + \psi}$, so that the convergence strip is

$$-\beta - \sqrt{\beta^2 + \psi} < t < -\beta + \sqrt{\beta^2 + \psi}. \tag{A.171}$$

The c.g.f. corresponding to (A.170) is

$$\mathbb{K}_X(t) = \mu t + \ln K_\lambda(\omega Q) - \ln K_\lambda(\omega) - \lambda \ln(Q), \tag{A.172}$$

where $Q = Q(t) := \sqrt{1 - (2\beta t + t^2)/\psi}$. It is straightforward to see that

$$\frac{dQ(t)}{dt} = \frac{1}{2}\left(1 - \frac{2\beta t + t^2}{\psi}\right)^{-1/2}\left(-\frac{2\beta + 2t}{\psi}\right) = -\frac{\beta + t}{Q\psi},$$

so, via (A.20) and some simplification,

$$\mathbb{K}'_X(t) = \mu + \frac{\beta + t}{Q\psi}\left(\frac{\omega}{2}\frac{K_{\lambda-1}(\omega Q) + K_{\lambda+1}(\omega Q)}{K_\lambda(\omega Q)} + \frac{\lambda}{Q}\right).$$

Numerically solving $\mathbb{K}'_X(t) = x$ in the range (A.171) gives the unique saddlepoint $\hat{t}$. For the second derivative, with

$$A(t) := \frac{\omega}{2}\frac{K_{\lambda-1}(\omega Q) + K_{\lambda+1}(\omega Q)}{K_\lambda(\omega Q)} + \frac{\lambda}{Q},$$

we can write

$$\mathbb{K}''_X(t) = \frac{\beta + t}{Q\psi} \times \frac{dA(t)}{dt} + A(t) \times \frac{d}{dt}\left(\frac{\beta + t}{Q\psi}\right),$$

with

$$\frac{d}{dt}\left(\frac{\beta + t}{Q\psi}\right) = \frac{Q\psi + (\beta + t)^2/Q}{(Q\psi)^2} = \frac{1}{Q\psi}\left(1 + \frac{(\beta + t)^2}{Q^2\psi}\right).$$

Next, with $N(t) := K_{\lambda-1}(\omega Q) + K_{\lambda+1}(\omega Q)$,

$$\frac{dA(t)}{dt} = \frac{\omega}{2}\left(\frac{K_\lambda(\omega Q) \times \frac{d}{dt}N(t) - N(t) \times \frac{d}{dt}K_\lambda(\omega Q)}{K^2_\lambda(\omega Q)}\right) + \frac{\lambda(\beta + t)}{Q^3\psi},$$

and, using (A.20),

$$\frac{d}{dt}K_\lambda(\omega Q) = \frac{\omega}{2}(K_{\lambda-1}(\omega Q) + K_{\lambda+1}(\omega Q))\left(\frac{\beta + t}{Q\psi}\right) = \frac{\omega}{2}\left(\frac{\beta + t}{Q\psi}\right)N(t)$$

and

$$\frac{d}{dt}N(t) = \frac{\omega}{2}\left(\frac{\beta + t}{Q\psi}\right)(K_{\lambda-2}(\omega Q) + 2K_\lambda(\omega Q) + K_{\lambda+2}(\omega Q)).$$

That is,

$$\mathbb{K}''_X(t) = \frac{\beta + t}{Q\psi} \times P_1 + \left[\frac{\omega}{2}\frac{K_{\lambda-1}(\omega Q) + K_{\lambda+1}(\omega Q)}{K_\lambda(\omega Q)} + \frac{\lambda}{Q}\right] \times \left[\frac{1}{Q\psi}\left(1 + \frac{(\beta + t)^2}{Q^2\psi}\right)\right],$$

where P_1 *and* P_2 *are given by*

$$P_1 = \frac{\omega}{2}P_2 + \frac{\lambda(\beta + t)}{Q^3\psi},$$

and

$$K^2_\lambda(\omega Q)P_2 = K_\lambda(\omega Q) \times \left[\frac{\omega}{2}\left(\frac{\beta + t}{Q\psi}\right)(K_{\lambda-2}(\omega Q) + 2K_\lambda(\omega Q) + K_{\lambda+2}(\omega Q))\right]$$

$$-[K_{\lambda-1}(\omega Q) + K_{\lambda+1}(\omega Q)] \times \left[\frac{\omega}{2}(K_{\lambda-1}(\omega Q) + K_{\lambda+1}(\omega Q))\left(\frac{\beta + t}{Q\psi}\right)\right].$$

With these expressions, the first-order s.p.a. to the p.d.f. and c.d.f. can be computed.

Now consider the case of working with weighted sums of independent GHyp random variables. Let $X_i \overset{\text{indep}}{\sim} GHyp(\lambda_i, \omega_i, \rho_i, \sigma_i, \mu_i)$ *with density (A.168). Let* $\beta_i = \omega_i\sigma_i^{-1}\rho_i(1 - \rho_i^2)^{-1/2}$ *and* $\psi_i = \omega_i^2\sigma_i^{-2}$, $i = 1, \ldots, d$. *The c.g.f. of* $S = \sum_{i=1}^d a_iX_i$, $a_i \neq 0$, *is* $\mathbb{K}_S(t) = \sum_{i=1}^d \mathbb{K}_{X_i}(a_it)$, *where the c.g.f. of each* X_i *is given in (A.172). Similarly, the m.g.f. is the product of the individual m.g.f.s as given in (A.170). The singularities of the c.g.f. lie on both sides of the origin at the points* $(-\beta_i \pm \sqrt{\beta_i^2 + \psi_i})/a_i$.

In the special case of the NIG distribution, matters simplify considerably. Specifically,

$$\mathbb{K}_X(t) = \mu t - \omega Q + \omega, \quad \mathbb{K}'_X(t) = \mu + \omega\frac{\beta + t}{Q\psi}, \quad \mathbb{K}''_X = \frac{\omega}{Q\psi} + \frac{\omega(\beta + t)^2}{Q^3\psi^2},$$

and the saddlepoint is now given explicitly as

$$\hat{t} = z\frac{\bar{\alpha}}{\bar{y}\sigma} - \beta,$$

where z, $\bar{y}$, and $\bar{\alpha}$ are as in (A.168), $\mathbb{K}''_X(\hat{t}) = \bar{y}^3\sigma^2\bar{\alpha}$, and $\hat{w} = \text{sgn}(\hat{t})\sqrt{2(\bar{y}\bar{\alpha} - z\rho\bar{\alpha} - \omega)}$. Also,

$$\hat{\kappa}_3 = 3\frac{z}{\sqrt{\bar{y}\bar{\alpha}}} \quad \text{and} \quad \hat{\kappa}_4 = 3\frac{1 + 5z^2}{\bar{y}\bar{\alpha}}.$$

This saddlepoint approximation was used in Broda and Paolella (2009) in the context of non-Gaussian portfolio optimization based on an independent components analysis decomposition.

A.12 ORDER STATISTICS

The order statistics of a random i.i.d. sample X_i, $i = 1, \ldots, n$, are the n values arranged in ascending order and denoted $X_{1:n} \leq X_{2:n} \leq \cdots \leq X_{n:n}$ or $X_{(1)} \leq X_{(2)} \leq \cdots \leq X_{(n)}$ or $Y_1 \leq Y_2 \leq \cdots \leq Y_n$.

The ith order statistic of i.i.d. sample X_i, $i = 1, \ldots, n$, from distribution $F = F_X$ (and density $f = f_X$) has c.d.f.

$$F_{Y_i}(y) = \Pr(Y_i \leq y) = \sum_{j=i}^{n} \binom{n}{j} [F(y)]^j [1 - F(y)]^{n-j}. \tag{A.173}$$

Special cases of interest are the sample minimum

$$F_{Y_1}(y) = 1 - [1 - F(y)]^n, \tag{A.174}$$

and sample maximum

$$F_{Y_n}(y) = [F(y)]^n. \tag{A.175}$$

The p.d.f. of Y_i is

$$f_{Y_i}(y) = \frac{n!}{(i-1)!(n-i)!} F(y)^{i-1} [1 - F(y)]^{n-i} f(y), \tag{A.176}$$

with special cases

$$f_{Y_1}(y) = n[1 - F(y)]^{n-1} f(y) \quad \text{and} \quad f_{Y_n}(y) = n[F(y)]^{n-1} f(y), \tag{A.177}$$

which also follow from differentiating (A.174) and (A.175).

Example For $X_i \overset{\text{i.i.d.}}{\sim} \text{Unif}(0, 1)$, it is straightforward to verify that $Y_i \sim \text{Beta}(i, n - i + 1)$, where Y_i denotes the ith order statistic. Hence, the c.d.f. of Y_i can be expressed using (A.173) or integrating (A.176), that is, for $0 \leq y \leq 1$,

$$F_{Y_i}(y) = \sum_{j=i}^{n} \binom{n}{j} y^j (1 - y)^{n-j} = \frac{n!}{(i-1)!(n-i)!} \int_0^y x^{i-1}(1 - x)^{n-i} \, dx, \tag{A.178}$$

which gives rise to an interesting identity as well as a computation method for evaluating the incomplete beta function (A.13).

In the bivariate case, for $x < y$,

$$F_{Y_i, Y_j}(x, y) = \sum_{a=j}^{n} \sum_{b=i}^{a} \frac{n!}{b!(a-b)!(n-a)!} [F(x)]^b [F(y) - F(x)]^{a-b} [1 - F(y)]^{n-a} \quad \text{(A.179)}$$

and

$$f_{Y_i, Y_j}(x, y) = K F(x)^{i-1} [F(y) - F(x)]^{j-i-1} [1 - F(y)]^{n-j} f(x) f(y) \, \mathbb{I}_{(x,\infty)}(y), \quad \text{(A.180)}$$

where $K = n!/(i-1)!(j-i-1)!(n-j)!$.

Taking $i = 1$ and $j = 2$ in (A.180) gives the joint density of the first two order statistics as

$$f_{Y_1, Y_2}(x, y) = \frac{n!}{(n-2)!} [1 - F(y)]^{n-2} f(x) f(y) \mathbb{I}_{(x,\infty)}(y),$$

and generalizing this to the first k order statistics gives

$$f_{Y_1, \dots, Y_k}(y_1, \dots, y_k) = \frac{n!}{(n-k)!} [1 - F(y_k)]^{n-k} \prod_{i=1}^{k} f(y_i), \quad y_1 < y_2 < \cdots < y_k. \quad \text{(A.181)}$$

Taking $k = n$ in (A.181) gives the p.d.f. of the whole sample of order statistics,

$$f_{Y_1, \dots, Y_n}(y_1, y_2, \dots, y_n) = n! \prod_{i=1}^{n} f(y_i), \quad y_1 < y_2 < \cdots < y_n. \quad \text{(A.182)}$$

Example   *Let $X_i \overset{\text{i.i.d.}}{\sim} \text{Exp}(\lambda)$, $i = 1, \dots, n$, with $\mathbb{E}[X] = \lambda^{-1}$ and $\mathbb{V}(X) = \lambda^{-2}$. Let Y_i be the ith order statistic, and define $D_0 = Y_1, D_1 = Y_2 - Y_1, D_2 = Y_3 - Y_2, \dots, D_{n-1} = Y_n - Y_{n-1}$. As shown in Example II.6.16, via (A.182) and the Jacobian transformation (A.139),*

$$D_j \overset{\text{indep}}{\sim} \text{Exp}(\lambda(n-j)) \quad \text{or} \quad (n-j)D_j \overset{\text{indep}}{\sim} \text{Exp}(\lambda), \quad j = 0, \dots, n-1, \quad \text{(A.183)}$$

so that

$$Y_i = \sum_{j=0}^{i-1} D_j \overset{d}{=} \frac{1}{\lambda} \sum_{k=1}^{i} \frac{Z_k}{n-k+1}, \quad Z_k \overset{\text{i.i.d.}}{\sim} \text{Exp}(1), \quad i = 1, \dots, n, \quad \text{(A.184)}$$

As reported in Galambos and Kotz (1978, p. 3), this result appears to date back to Sukhatme (1937), and was rediscovered by Malmquist (1950), Epstein and Sobel (1953), and Rényi (1953), and is often called, after the latter author, Rényi's representation.

Solving $p = F_X(\xi_p) = 1 - e^{-\lambda \xi_p}$ gives the closed-form solution $\xi_p = -\lambda^{-1} \ln(1 - p)$. For $j = 1, \dots, n$,

$$\mathbb{E}[Y_j] = \sum_{i=0}^{j-1} \mathbb{E}[D_i] = \frac{1}{\lambda} \left[\frac{1}{n} + \frac{1}{(n-1)} + \cdots + \frac{1}{(n-j+1)} \right] \quad \text{(A.185)}$$

and, because of the independence of the D_i,

$$\mathbb{V}(Y_j) = \sum_{i=0}^{j-1} \mathbb{V}(D_i) = \frac{1}{\lambda^2} \left[\frac{1}{n^2} + \frac{1}{(n-1)^2} + \cdots + \frac{1}{(n-j+1)^2} \right]. \quad \text{(A.186)}$$

From (A.194) given below, and the independence of the D_i,

$$\text{Cov}\,(Y_i, Y_j) = \sum_{p=0}^{i-1}\sum_{q=0}^{j-1} \text{Cov}\,(D_p, D_q) = \sum_{p=0}^{U-1} \mathbb{V}(D_p) = \frac{1}{\lambda^2}\sum_{p=0}^{U-1} \frac{1}{(n-p)^2}, \qquad \text{(A.187)}$$

where $U = \min(i,j)$.

Let X_i, $i = 1, \dots, n$, be an i.i.d. sample from a continuous distribution with p.d.f. f and c.d.f. F, and denote the order statistics by $Y_1, \dots, Y_n$. The sample range is defined to be $R = Y_n - Y_1$, and the sample midrange is defined to be $T = (Y_1 + Y_n)/2$. The joint distribution of the sample range and sample midrange is given by

$$f_{R,T}(r,t) = n(n-1)\left[F\left(t+\frac{r}{2}\right) - F\left(t-\frac{r}{2}\right)\right]^{n-2} f\left(t-\frac{r}{2}\right) f\left(t+\frac{r}{2}\right)\mathbb{1}_{(0,\infty)}(r). \quad \text{(A.188)}$$

Let $Y_1, \dots, Y_n$ denote the order statistics of the i.i.d. sample $X_1, \dots, X_n$, where each X_i has p.d.f. and c.d.f. f_X and F_X, respectively. Then, as $n \to \infty$,

$$Y_{\lfloor np \rfloor} \overset{\text{app}}{\sim} \mathrm{N}\left(F_X^{-1}(p), \frac{p(1-p)}{n\{f_X[F_X^{-1}(p)]\}^2}\right), \qquad p \in (0,1); \qquad \text{(A.189)}$$

see, for example, Reiss (1989) and Ferguson (1996, Ch. 13). For a fixed n, this asymptotic approximation tends to be relatively accurate for the center order statistics, but suffers as p in (A.189) approaches 0 or 1.

A wealth of further information on order statistics can be found in Reiss (1989).

A.13 THE MULTIVARIATE NORMAL DISTRIBUTION

Let $\mathbf{X} = (X_1, \dots, X_n)'$ be a vector random variable such that $\mathbb{E}[X_i] = \mu_i$, $\mathbb{V}(X_i) = \sigma_i^2$, $i = 1, \dots, n$, and $\text{Cov}\,(X_i, X_j) = \sigma_{ij}$. Then

$$\mathbb{E}[\mathbf{X}] := \mathbb{E}[(X_1, \dots, X_n)'] = (\mu_1, \dots, \mu_n)',$$

usually denoted by $\boldsymbol{\mu}_{\mathbf{X}}$ or just $\boldsymbol{\mu}$; and

$$\mathbb{V}(\mathbf{X}) := \mathbb{E}[(\mathbf{X} - \boldsymbol{\mu}_{\mathbf{X}})(\mathbf{X} - \boldsymbol{\mu}_{\mathbf{X}})'] = \begin{bmatrix} \sigma_1^2 & \sigma_{12} & \cdots & \sigma_{1n} \\ \sigma_{21} & \sigma_2^2 & & \sigma_{2n} \\ \vdots & & \ddots & \vdots \\ \sigma_{n1} & \sigma_{n2} & & \sigma_n^2 \end{bmatrix}, \qquad \text{(A.190)}$$

which is symmetric and often denoted by $\boldsymbol{\Sigma}_{\mathbf{X}}$ or just $\boldsymbol{\Sigma}$. A particular element of $\boldsymbol{\Sigma}$ is given by

$$\sigma_{ij} = \mathbb{E}[(X_i - \mu_i)(X_j - \mu_j)]. \qquad \text{(A.191)}$$

For a real $n \times n$ matrix $\mathbf{A}$ and $n \times 1$ real column vector $\mathbf{b}$, $\mathbb{E}[\mathbf{AX} + \mathbf{b}] = \mathbf{A}\boldsymbol{\mu}_{\mathbf{X}} + \mathbf{b}$, and

$$\mathbb{V}(\mathbf{AX} + \mathbf{b}) = \mathbf{A}\boldsymbol{\Sigma}\mathbf{A}'. \qquad \text{(A.192)}$$

If $\mathbf{a} = (a_1, a_2, \dots, a_n)' \in \mathbb{R}^n$, then (A.192) reduces to

$$\mathbb{V}(\mathbf{a}'\mathbf{X}) = \mathbf{a}'\boldsymbol{\Sigma}\mathbf{a} = \sum_{i=1}^n a_i^2 \mathbb{V}(X_i) + \sum_{i \neq j}\sum a_i a_j \text{Cov}\,(X_i, X_j). \qquad \text{(A.193)}$$

Also,

$$\text{Cov}(\mathbf{AX}, \mathbf{BY}) = \mathbb{E}[\mathbf{A}(\mathbf{X} - \boldsymbol{\mu}_\mathbf{X})(\mathbf{Y} - \boldsymbol{\mu}_\mathbf{Y})'\mathbf{B}'] = \mathbf{A}\boldsymbol{\Sigma}_{\mathbf{X},\mathbf{Y}}\mathbf{B}',$$

with important special case

$$\text{Cov}(\mathbf{a}'\mathbf{X}, \mathbf{b}'\mathbf{Y}) = \sum_{i=1}^{n}\sum_{j=1}^{m} a_i b_j \text{Cov}(X_i, Y_j) \tag{A.194}$$

for vectors $\mathbf{a} = (a_1, a_2, \dots, a_n)' \in \mathbb{R}^n$ and $\mathbf{b} = (b_1, b_2, \dots, b_m)' \in \mathbb{R}^m$, as in (A.63). If $X_1, \dots, X_n \overset{\text{indep}}{\sim} N(\mu_i, \sigma_i^2)$, then their joint density is

$$f_\mathbf{X}(\mathbf{x}) = \frac{1}{\sqrt{(2\pi)^n \prod_{i=1}^{n} \sigma_i^2}} \exp\left\{ -\frac{1}{2} \sum_{i=1}^{n} \left(\frac{x_i - \mu_i}{\sigma_i} \right)^2 \right\}. \tag{A.195}$$

More generally, $\mathbf{Y}$ is an n-variate multivariate normal r.v. if its density is given by

$$f_\mathbf{Y}(\mathbf{y}; \boldsymbol{\mu}, \boldsymbol{\Sigma}) = \frac{1}{|\boldsymbol{\Sigma}|^{1/2}(2\pi)^{n/2}} \exp\left\{ -\frac{1}{2}((\mathbf{y} - \boldsymbol{\mu})'\boldsymbol{\Sigma}^{-1}(\mathbf{y} - \boldsymbol{\mu})) \right\}, \tag{A.196}$$

written $\mathbf{Y} \sim N(\boldsymbol{\mu}, \boldsymbol{\Sigma})$, where $\boldsymbol{\mu} = (\mu_1, \dots, \mu_n)' \in \mathbb{R}^n$ and $\boldsymbol{\Sigma} > 0$ with (i, j)th element σ_{ij}, $\sigma_i^2 := \sigma_{ii}$. The following are some important facts:

(1) $\mathbb{E}[\mathbf{Y}] = \boldsymbol{\mu}$, $\mathbb{V}(\mathbf{Y}) = \boldsymbol{\Sigma}$, and the parameters $\boldsymbol{\mu}$ and $\boldsymbol{\Sigma}$ completely determine the distribution.

(2) All $2^n - 2$ marginals are normally distributed with mean and variance given appropriately from $\boldsymbol{\mu}$ and $\boldsymbol{\Sigma}$.

(3) An important special case is the bivariate normal,

$$\begin{pmatrix} Y_1 \\ Y_2 \end{pmatrix} \sim N\left(\begin{pmatrix} \mu_1 \\ \mu_2 \end{pmatrix}, \begin{pmatrix} \sigma_1^2 & \rho\sigma_1\sigma_2 \\ \rho\sigma_1\sigma_2 & \sigma_2^2 \end{pmatrix} \right), \tag{A.197}$$

where, from (A.53), $\text{Corr}(Y_1, Y_2) = \rho$. Its density is

$$f_{Y_1, Y_2}(x, y) = K \exp\left\{ -\frac{X^2 - 2\rho XY + Y^2}{2(1 - \rho^2)} \right\}, \tag{A.198}$$

where

$$K = \frac{1}{2\pi\sigma_1\sigma_2(1 - \rho^2)^{1/2}}, \qquad X = \frac{x - \mu_1}{\sigma_1}, \qquad Y = \frac{y - \mu_2}{\sigma_2}$$

and the marginal distributions are $Y_i \sim N(\mu_i, \sigma_i^2)$, $i = 1, 2$.

(4) If Y_i and Y_j are jointly normally distributed, then they are independent if and only if $\text{Cov}(Y_i, Y_j) = 0$.

(5) For nonoverlapping subsets $\mathbf{Y}_{(i)}$ and $\mathbf{Y}_{(j)}$ of $\mathbf{Y}$, the conditional distribution of $\mathbf{Y}_{(i)} \mid \mathbf{Y}_{(j)}$ is also normally distributed. The general case is given in (A.203). In the bivariate normal case,

$$Y_1 \mid Y_2 \sim N(\mu_1 + \rho\sigma_1\sigma_2^{-1}(y_2 - \mu_2), \ \sigma_1^2(1 - \rho^2)) \tag{A.199}$$

and

$$Y_2 \mid Y_1 \sim \mathrm{N}(\mu_2 + \rho\sigma_2\sigma_1^{-1}(y_1 - \mu_1),\ \sigma_2^2(1 - \rho^2)).$$

(6) The linear combination $L = \mathbf{a}'\mathbf{Y} = \sum_{i=1}^{n} a_i Y_i$ is normally distributed with mean $\mathbb{E}[L] = \sum_{i=1}^{n} a_i \mu_i = \mathbf{a}'\boldsymbol{\mu}$ and variance $\mathbb{V}(L)$ from (A.193). More generally, we have the set of linear combinations

$$\mathbf{L} = (L_1, \dots, L_m)' = \mathbf{A}\mathbf{Y} \sim \mathrm{N}(\mathbf{A}\boldsymbol{\mu}, \mathbf{A}\boldsymbol{\Sigma}\mathbf{A}'), \tag{A.200}$$

using (A.192).

Example   Let $X_i \overset{\text{i.i.d.}}{\sim} N(\mu, \sigma^2)$, $i = 1, \dots, n$, $S = \sum_{i=1}^{n} X_i$, and $L_1 = X_1 = \mathbf{a}_1'\mathbf{X}$, $L_2 = \sum_{i=1}^{n} X_i = \mathbf{a}_2'\mathbf{X}$, where $\mathbf{a}_1 = (1, 0, \dots, 0)'$ and $\mathbf{a}_2 = (1, 1, \dots, 1)'$. From property 6,

$$\begin{bmatrix} L_1 \\ L_2 \end{bmatrix} \sim \mathrm{N}\left(\begin{bmatrix} \mu \\ \sum_{i=1}^{n} \mu \end{bmatrix}, \begin{bmatrix} 1 & 0 & \vdots & 0 \\ 1 & 1 & \vdots & 1 \end{bmatrix} \boldsymbol{\Sigma} \begin{bmatrix} 1 & 1 \\ 0 & 1 \\ \vdots & \vdots \\ 0 & 1 \end{bmatrix} \right) = \mathrm{N}\left(\begin{bmatrix} \mu \\ n\mu \end{bmatrix}, \begin{bmatrix} \sigma^2 & \sigma^2 \\ \sigma^2 & n\sigma^2 \end{bmatrix} \right),$$

with $\rho = n^{-1/2}$. From property 5, $(X_1 \mid S = s) \sim \mathrm{N}(s/n,\ \sigma^2(1 - n^{-1}))$.

Let $\mathbf{Y} \sim \mathrm{N}_n(\boldsymbol{\mu}, \boldsymbol{\Sigma})$, where $\boldsymbol{\Sigma} > 0$. Then $\mathbf{Z} = \boldsymbol{\Sigma}^{-\frac{1}{2}}(\mathbf{Y} - \boldsymbol{\mu}) \sim \mathrm{N}_n(\mathbf{0}, \mathbf{I})$. It follows from (A.144) that $\mathbf{Z}'\mathbf{Z} \sim \chi^2(n)$, that is, for $\mathbf{Y} \sim \mathrm{N}_n(\boldsymbol{\mu}, \boldsymbol{\Sigma})$, the *quadratic form*

$$(\mathbf{Y} - \boldsymbol{\mu})'\boldsymbol{\Sigma}^{-1}(\mathbf{Y} - \boldsymbol{\mu}) \sim \chi^2(n). \tag{A.201}$$

We will have much more to say about quadratic forms in Book IV.

Similarly, let $\mathbf{Z} \sim \mathrm{N}_n(\mathbf{0}, \mathbf{I})$. Then $\mathbf{Y} = \boldsymbol{\mu} + \boldsymbol{\Sigma}^{1/2}\mathbf{Z} \sim \mathrm{N}(\boldsymbol{\mu}, \boldsymbol{\Sigma})$, where $\boldsymbol{\Sigma}^{1/2}$ is the symmetric square root of $\boldsymbol{\Sigma}$ as obtained via the spectral decomposition. (The Cholesky decomposition of $\boldsymbol{\Sigma}$ can be used as well; the resulting distribution theory is the same.) Based on this and the fact that $\mathbf{Z}$ is easy to simulate, $\mathbf{Y}$ can be straightforwardly simulated as well.

The m.g.f. of $\mathbf{Y}$ is

$$\mathbb{M}_{\mathbf{Y}}(\mathbf{t}) = \exp\{\mathbf{t}'\boldsymbol{\mu} + \mathbf{t}'\boldsymbol{\Sigma}\mathbf{t}/2\}. \tag{A.202}$$

Let $\mathbf{Y} = (Y_1, \dots, Y_n)' \sim \mathrm{N}(\boldsymbol{\mu}, \boldsymbol{\Sigma})$ and consider the partition $\mathbf{Y} = (\mathbf{Y}_{(1)}', \mathbf{Y}_{(2)}')'$, where $\mathbf{Y}_{(1)} = (Y_1, \dots, Y_p)'$ and $\mathbf{Y}_{(2)} = (Y_{p+1}, \dots, Y_n)'$ for $1 \le p < n$, with $\boldsymbol{\mu}$ and $\boldsymbol{\Sigma}$ partitioned accordingly such that $\mathbb{E}[\mathbf{Y}_{(i)}] = \boldsymbol{\mu}_{(i)}$, $\mathbb{V}(\mathbf{Y}_{(i)}) = \boldsymbol{\Sigma}_{ii}$, $i = 1, 2$, and $\mathrm{Cov}\,(\mathbf{Y}_{(1)}, \mathbf{Y}_{(2)}) = \boldsymbol{\Sigma}_{12}$, that is, $\boldsymbol{\mu} = (\boldsymbol{\mu}_{(1)}', \boldsymbol{\mu}_{(2)}')'$ and

$$\boldsymbol{\Sigma} = \begin{bmatrix} \boldsymbol{\Sigma}_{11} & \vdots & \boldsymbol{\Sigma}_{12} \\ \cdots\cdots\cdots\cdots \\ \boldsymbol{\Sigma}_{21} & \vdots & \boldsymbol{\Sigma}_{22} \end{bmatrix}, \quad \boldsymbol{\Sigma}_{21} = \boldsymbol{\Sigma}_{12}'.$$

If $\boldsymbol{\Sigma}_{22} > 0$ (which is true if $\boldsymbol{\Sigma} > 0$), then the conditional distribution of $\mathbf{Y}_{(1)}$ given $\mathbf{Y}_{(2)}$ is

$$(\mathbf{Y}_{(1)} \mid \mathbf{Y}_{(2)} = \mathbf{y}_{(2)}) \sim \mathrm{N}(\boldsymbol{\mu}_{(1)} + \boldsymbol{\Sigma}_{12}\boldsymbol{\Sigma}_{22}^{-1}(\mathbf{y}_{(2)} - \boldsymbol{\mu}_{(2)}),\ \boldsymbol{\Sigma}_{11} - \boldsymbol{\Sigma}_{12}\boldsymbol{\Sigma}_{22}^{-1}\boldsymbol{\Sigma}_{21}), \tag{A.203}$$

generalizing (A.199).

If $X_i \overset{\text{i.i.d.}}{\sim} N(\mu, \sigma^2)$, then the statistics

$$\bar{X} = \bar{X}_n = n^{-1} \sum_{i=1}^{n} X_i \quad \text{and} \quad S^2 = S_n^2(X) = (n-1)^{-1} \sum_{i=1}^{n} (X_i - \bar{X})^2 \qquad \text{(A.204)}$$

are independent, as shown via use of the m.g.f. in Section II.3.7. The marginal distributions are

$$\bar{X} \sim N\left(\mu, \frac{\sigma^2}{n}\right), \qquad \text{(A.205)}$$

$$B = \frac{(n-1)S^2}{\sigma^2} \sim \chi^2_{n-1}, \qquad \text{(A.206)}$$

and

$$\mathbb{E}[S] = K\sigma, \qquad K = \frac{\sqrt{2}}{\sqrt{n-1}} \frac{\Gamma\left(\frac{n}{2}\right)}{\Gamma\left(\frac{n-1}{2}\right)}. \qquad \text{(A.207)}$$

Furthermore, with $Z := (\bar{X} - \mu)/(\sigma/\sqrt{n})$ standard normal, the random variable

$$\frac{Z}{\sqrt{B/(n-1)}} = \frac{(\bar{X} - \mu)/(\sigma/\sqrt{n})}{\sqrt{\sigma^{-2} \sum_{i=1}^{n} (X_i - \bar{X})^2/(n-1)}} = \frac{\bar{X} - \mu}{S_n/\sqrt{n}} \qquad \text{(A.208)}$$

has a Student's t distribution with $n-1$ degrees of freedom (and does not depend on σ).

As S^2 in (A.204) is a statistic, it also has a variance. A straightforward but tedious calculation shows that

$$\mathbb{V}(S_n^2) = n^{-1}\left(\mu_4 - \frac{n-3}{n-1}\sigma^4\right), \qquad \text{(A.209)}$$

where σ^2 is the population variance. For i.i.d. normal r.v.s., $\mu_4 = 3\sigma^4$, so that

$$\mathbb{V}(S_n^2) = n^{-1}\left(3\sigma^4 - \frac{n-3}{n-1}\sigma^4\right) = \frac{2\sigma^4}{n-1}. \qquad \text{(A.210)}$$

Observe that (A.210) follows directly from (A.206), and that $\mathbb{V}(C) = 2(n-1)$, where $C \sim \chi^2_{n-1}$.

A.14 NONCENTRAL DISTRIBUTIONS

Let $X_i \overset{\text{indep}}{\sim} N(\mu_i, 1)$, $\mu_i \in \mathbb{R}$, $i = 1, \ldots, n$. The random variable $X = \sum_{i=1}^{n} X_i^2 \sim \chi^2(n, \theta)$, where $\theta = \boldsymbol{\mu}'\boldsymbol{\mu} = \sum_{i=1}^{n} \mu_i^2$, and one says that X follows a noncentral χ^2 distribution with n degrees of freedom and noncentrality parameter θ. If all the μ_i are zero, then this reduces to the usual, central χ_n^2.

The p.d.f. of X is derived in Section II.10.1.1, and shown to be

$$f_X(x) = \sum_{i=0}^{\infty} \omega_{i,\theta} \, g_{n+2i}(x), \qquad \text{(A.211)}$$

where g_v denotes the central χ_v^2 density and the $\omega_{i,\theta} = e^{-\theta/2}(\theta/2)^i / i!$ are weights corresponding to a Poisson distribution. The c.d.f. of X is

$$\Pr(X \leq x) = \sum_{i=0}^{\infty} \omega_{i,\theta} \, G_{n+2i}(x), \tag{A.212}$$

where G_v is the c.d.f. of a χ_v^2 random variable. Also,

$$\mathbb{E}[X] = n + \theta, \tag{A.213}$$

$\mathbb{V}(X) = 2n + 4\theta$, and

$$\mathbb{M}_X(t) = (1 - 2t)^{-n/2} \exp\left\{ \frac{t\theta}{1 - 2t} \right\}, \quad t < 1/2, \tag{A.214}$$

as was shown in two ways in Problem II.10.6.

Let $X_i \overset{\text{indep}}{\sim} \chi^2(n_i, \theta_i)$, $i = 1, 2$, and define $F = (X_1/n_1)/(X_2/n_2)$. The random variable F is said to follow a doubly noncentral F distribution, denoted $F \sim \mathrm{F}(n_1, n_2, \theta_1, \theta_2)$. With $\omega_{i,\theta} = e^{-\theta/2}(\theta/2)^i / i!$, Problem II.10.8 shows that

$$f_F(x) = \sum_{i=0}^{\infty} \sum_{j=0}^{\infty} \omega_{i,\theta_1} \omega_{j,\theta_2} \frac{n_1^{n_1/2+i}}{n_2^{-n_2/2-j}} \frac{x^{n_1/2+i-1}(xn_1 + n_2)^{-(n_1+n_2)/2-i-j}}{B(i + n_1/2, j + n_2/2)}. \tag{A.215}$$

If $\theta_2 = 0$, this reduces to the singly noncentral F distribution,

$$f_F(x) = \sum_{i=0}^{\infty} \omega_{i,\theta_1} \frac{n_1^{n_1/2+i}}{n_2^{-n_2/2}} \frac{x^{n_1/2+i-1}(xn_1 + n_2)^{-(n_1+n_2)/2-i}}{B(i + n_1/2, n_2/2)}. \tag{A.216}$$

If $X \sim \mathrm{F}(n_1, n_2, \theta_1, 0)$, then

$$\mathbb{E}[X] = \frac{n_2}{n_1} \frac{n_1 + \theta_1}{n_2 - 2}, \quad n_2 > 2, \tag{A.217}$$

and

$$\mathbb{V}(X) = 2\frac{n_2^2}{n_1^2} \frac{(n_1 + \theta_1)^2 + (n_1 + 2\theta_1)(n_2 - 2)}{(n_2 - 2)^2(n_2 - 4)}, \quad n_2 > 4. \tag{A.218}$$

With $X \sim \mathrm{N}(\mu, 1)$ independent of $Y \sim \chi^2(k, \theta)$, $T = X/\sqrt{Y/k}$ follows a doubly noncentral t distribution with k degrees of freedom, numerator noncentrality parameter μ and denominator noncentrality parameter θ. If $\theta = 0$, then T is singly noncentral t with noncentrality parameter μ.

Section II.10.4.1.1 shows that the c.d.f. of the singly noncentral t can be expressed as

$$F_T(t; k, \mu) = \frac{2^{-k/2+1}k^{k/2}}{\Gamma(k/2)} \int_0^{\infty} \Phi(tz; \mu, 1) \, z^{k-1} \exp\left\{ -\frac{1}{2}kz^2 \right\} \, dz, \tag{A.219}$$

where

$$\Phi(tz; \mu, 1) = (2\pi)^{-1/2} \int_{-\infty}^{tz} \exp\left\{ -\frac{1}{2}(x - \mu)^2 \right\} \, dx = \Phi(tz - \mu; 0, 1) \equiv \Phi(tz - \mu).$$

```
1  function [ES, VaR] = nctES(xi,v,theta)
2  howfar = nctinv(1e-8,v,theta); % how far into the left tail to integrate
3  VaR = nctinv(xi,v,theta); % matlab routine for the quantile
4  I = quadl(@int,howfar,VaR,1e-6,[],v,theta); ES = I/xi;
5
6  function I = int(u,v,theta), pdf = nctpdf(u,v,theta); I = u .* pdf;
```

Program Listing A.2: For a given tail probability $\mathtt{xi}$, computes the ξ-quantile and ES for the singly noncentral t with v degrees of freedom and noncentrality θ.

We will have much more to say about the singly noncentral t in Section 9.3. Its p.d.f., c.d.f., and quantile function are conveniently programmed already in Matlab (though see the footnote in Section 9.3.1), so that computation of the quantiles and expected shortfall is very easy; see the program in Listing A.2.

A.15 INEQUALITIES AND CONVERGENCE

A.15.1 Inequalities for Random Variables

There are various random variable inequalities of great utility, including Jensen (A.49), and Cauchy–Schwarz (A.54). For others, it is useful to define the set indicating the existence of absolute moments, say

$$L_r = \{\text{r.v.s } X : \mathbb{E}[|X|^r] < \infty\}. \tag{A.220}$$

Further, let the k norm of r.v. $X \subset L_k$ be $\|X\|_k = \Gamma[|X|^k]^{1/k}$ for $k \geq 1$.
For r.v.s $U, V \in L_1$, the triangle inequality states that

$$\mathbb{E}[|U + V|] \leq \mathbb{E}[|U|] + \mathbb{E}[|V|], \tag{A.221}$$

while for $U, V \in L_r$ for $r > 0$, it is straightforward to show that

$$\mathbb{E}[|U + V|^r] \leq \mathbb{E}[(|U| + |V|)^r] \leq 2^r(\mathbb{E}[|U|^r] + \mathbb{E}[|V|^r]). \tag{A.222}$$

This can be sharpened to

$$\mathbb{E}[|U + V|^r] \leq c_r(\mathbb{E}[|U|^r] + \mathbb{E}[|V|^r]), \quad c_r = \begin{cases} 1, & \text{if } 0 < r \leq 1, \\ 2^{r-1}, & \text{if } r \geq 1 \end{cases} \tag{A.223}$$

(see, for example, Gut, 2005, p. 127), so that (A.221) is a special case.
Hölder's inequality generalizes Cauchy–Schwarz to

$$\|UV\|_1 \leq \|U\|_p \|V\|_q, \quad p, q > 1, \quad p^{-1} + q^{-1} = 1, \quad U \in L_p, \quad V \in L_q. \tag{A.224}$$

From this, it is easy to prove Lyapunov's inequality,

$$\|X\|_r \leq \|X\|_s, \quad 1 \leq r \leq s, \quad X \in L_s. \tag{A.225}$$

Minkowski's inequality generalizes (A.221) to

$$\|U + V\|_p \leq \|U\|_p + \|V\|_p, \quad p \geq 1, \quad U, V \in L_p. \tag{A.226}$$

Markov's inequality states that, if $X \in L_r$ for some $r > 0$, then, for all $a > 0$,

$$\Pr(|X| \geq a) \leq \frac{\mathbb{E}[|X|^r]}{a^r}. \qquad (A.227)$$

The most common special case is, for all $a > 0$,

$$\Pr(X \geq a) \leq \frac{\mathbb{E}[X]}{a}, \quad X \in L_1, \quad \Pr(X > 0) = 1, \qquad (A.228)$$

seen from

$$\mathbb{E}[X] = \int_0^\infty x \, dF_X = \int_0^a x \, dF_X + \int_a^\infty x \, dF_X$$

$$\geq \int_a^\infty x \, dF_X \geq \int_a^\infty a \, dF_X = a \int_a^\infty dF_X = a \Pr(X \geq a).$$

Chernoff's inequality states that

$$\Pr(X \geq c) \leq \inf_{t>0} \mathbb{E}[e^{t(X-c)}], \quad c > 0, \qquad (A.229)$$

which is used to derive the Chernoff bound: for X_i i.i.d., $i = 1, \ldots, n$,

$$\Pr(\bar{X}_n \geq c) \leq \inf_{t>0} \exp\left(n \log \mathbb{M}\left(\frac{t}{n}\right) - tc \right), \qquad (A.230)$$

where $\bar{X}_n = n^{-1} \sum_{i=1}^n X_i$, and $\mathbb{M}$ is the m.g.f. of the X_i.

Chebyshev's inequality states that, for $X \in L_2$ with mean μ and variance σ^2, and for any $b > 0$,

$$\Pr(|X - \mu| \geq b) \leq \frac{\sigma^2}{b^2}. \qquad (A.231)$$

Example   Let $X \in L_2$ with $\mathbb{E}[X] = \mu$ and $\mathbb{V}(X) = \sigma^2$. For some $a > 0$, as $\{X - \mu > a\} \Rightarrow \{|X - \mu| > a\}$, that is, $\Pr(X - \mu > a) \leq \Pr(|X - \mu| > a)$, Chebyshev's inequality (A.231) implies $\Pr(X - \mu > a) \leq \Pr(X - \mu \geq a) \leq \sigma^2/a^2$. However, this bound can be sharpened to

$$\Pr(X > \mu + a) \leq \frac{\sigma^2}{\sigma^2 + a^2} \quad , \quad \Pr(X < \mu - a) \leq \frac{\sigma^2}{\sigma^2 + a^2}, \qquad (A.232)$$

which is known as the one-sided Chebyshev, or Cantelli's inequality. To see this, first let $\mu = 0$, so that $\mathbb{E}[X] = 0$, or

$$-a = \int_{-\infty}^\infty (x - a) \, dF_X \geq \int_{-\infty}^\infty (x - a)\mathbb{I}_{(-\infty,a)}(x) \, dF_X = \mathbb{E}[(X - a)\mathbb{I}_{(-\infty,a)}(X)],$$

or, multiplying by -1, squaring, and applying (the squares of both sides of) the Cauchy–Schwarz inequality (A.54),

$$a^2 \leq (\mathbb{E}[(a - X)\mathbb{I}_{(-\infty,a)}(X)])^2 \leq \mathbb{E}[(a - X)^2]\mathbb{E}[\mathbb{I}_{(-\infty,a)}^2(X)].$$

As $\mathbb{E}[\mathbb{I}_{(-\infty,a)}^2(X)] = \mathbb{E}[\mathbb{I}_{(-\infty,a)}(X)] = F_X(a)$, expanding each term, $\mathbb{E}[(a - X)^2] = a^2 + \sigma^2$, and $a^2 \leq (a^2 + \sigma^2)F_X(a)$, or

$$\Pr(X > a) \leq \frac{\sigma^2}{a^2 + \sigma^2}. \qquad (A.233)$$

Assume $\mu \neq 0$ and observe that $X - \mu$ and $\mu - X$ have mean zero, so that both statements in (A.232) follow from (A.233).

Chebyshev's order inequality states that, for discrete r.v. X and nondecreasing real functions f and g,

$$\mathbb{E}[f(X)]\mathbb{E}[g(X)] \le \mathbb{E}[f(X)g(X)]. \tag{A.234}$$

In what follows, we use the notation $X_i \overset{\text{indep}}{\sim} (0, \sigma_i^2)$ to indicate that the X_i are independent r.v.s in L_2, each with mean zero and variance $\sigma_i^2 < \infty$. For $X_i \overset{\text{indep}}{\sim} (0, \sigma_i^2)$ and $n \in \mathbb{N}$, let $S_n = \sum_{i=1}^n X_i$, so that $\mathbb{E}[S_n] = 0$, $\mathbb{V}(S_n) = \sum_{i=1}^n \sigma_i^2$. Then Chebyshev's inequality (A.231) implies $\Pr(|S_n| \ge a) \le \mathbb{V}(S_n)/a^2$, that is,

$$\Pr(|X_1 + \cdots + X_n| \ge a) \le \frac{1}{a^2} \sum_{i=1}^n \sigma_i^2.$$

However, it turns out that this bound applies to the larger set

$$A_{a,n} := \bigcup_{j=1}^n \{|S_j| \ge a\} = \{\max_{1 \le j \le n} |S_j| \ge a\}, \tag{A.235}$$

instead of just $\{|S_n| \ge a\}$, and leads to Kolmogorov's inequality: Let $X_i \overset{\text{indep}}{\sim} (0, \sigma_i^2)$, $S_j := \sum_{i=1}^j X_i$, and $A_{a,n}$ as in (A.235). For any $a > 0$ and $n \in \mathbb{N}$,

$$\Pr(A_{a,n}) \le \frac{1}{a^2} \sum_{i=1}^n \sigma_i^2. \tag{A.236}$$

If there exists a c such that $\Pr(|X_k| \le c) = 1$ for each k, then

$$\Pr(A_{a,n}) \ge 1 - \frac{(c+a)^2}{\sum_{i=1}^n \sigma_i^2}, \tag{A.237}$$

sometimes referred to as the "other" Kolmogorov inequality (Gut, 2005, p. 123).

A.15.2 Convergence of Sequences of Sets

Let Ω denote the sample space and let $\{A_n \in \Omega, \ n \in \mathbb{N}\}$ be an infinite sequence of subsets of Ω, which we abbreviate to just $\{A_n\}$. The union and intersection of $\{A_n\}$ are given by

$$\bigcup_{n=1}^\infty A_n = \{\omega : \omega \in A_n \text{ for some } n \in \mathbb{N}\}, \quad \bigcap_{n=1}^\infty A_n = \{\omega : \omega \in A_n \text{ for all } n \in \mathbb{N}\},$$

respectively. Sequence $\{A_n\}$ is monotone increasing if $A_1 \subset A_2 \subset \cdots$, monotone decreasing if $A_1 \supset A_2 \supset \cdots$, and monotone if it is either monotone increasing or monotone decreasing. For sets $A, B \subset \Omega$,

$$A \subset B \Rightarrow \Pr(A) \le \Pr(B), \tag{A.238}$$

and, for sequence of sets $\{A_n\}$,

$$\Pr\left(\bigcup_{n=1}^\infty A_n\right) \le \sum_{n=1}^\infty \Pr(A_n), \tag{A.239}$$

which is Boole's inequality, or the property of countable subadditivity.

As a simple illustration, let $X_1, \ldots, X_n$ be i.i.d. r.v.s from a continuous distribution with p.d.f. f_X, and let $E_{ij} = \mathbb{I}\{X_i = X_j\}$. Then, from (A.51) and (A.55), $\Pr(E_{ij})$ is, for $i \neq j$,

$$\Pr(E_{ij}) = \mathbb{E}[E_{ij}] = \iint\limits_{\mathbb{R}^2 \,:\, x=y} f_X(x) f_X(y) \, dx \, dy = 0, \quad i \neq j,$$

as the set $x = y$ has measure zero in $\mathbb{R}^2$, and (A.239) implies, for $A = \{1 \leq i, j \leq n, \ i \neq j\}$,

$$\Pr\left(\bigcup_{i,j \in A} E_{ij}\right) \leq \sum_{i,j \in A} \Pr(E_{ij}) = 0, \tag{A.240}$$

that is, all the X_i are different w.p. 1.

Another useful and easily verified fact is that

$$A \subset B \quad \Leftrightarrow \quad B^c \subset A^c \tag{A.241}$$

or, when combined with (A.238),

$$A \subset B \quad \Rightarrow \quad \Pr(B^c) \leq \Pr(A^c). \tag{A.242}$$

As detailed in Section I.2.3.3, the continuity property of $\Pr(\cdot)$ states that if $A_1, A_2, \ldots$ is a monotone sequence of events, then

$$\lim_{n \to \infty} \Pr(A_n) = \Pr\left(\lim_{n \to \infty} A_n\right). \tag{A.243}$$

If $\{A_n\}$ is a monotone increasing sequence, then $\lim_{n \to \infty} A_n = A := \bigcup_{n=1}^{\infty} A_n$. This is commonly written as $A_n \uparrow A$. Similarly, if the A_i are monotone decreasing, then $\lim_{n \to \infty} A_n = A := \bigcap_{n=1}^{\infty} A_n$, written $A_n \downarrow A$.

The question arises as to the limits of sets that are not monotone. Let $\{A_n\}$ be an arbitrary (not necessarily monotone) sequence of sets. Analogous to the limit of a deterministic sequence of real numbers, the *limit supremum* (or *limit superior*) of $\{A_n\}$, and the *limit infimum* (or *limit inferior*) of $\{A_n\}$, are denoted and defined as

$$A^* = \limsup_{i \to \infty} A_i = \bigcap_{k=1}^{\infty} \bigcup_{n=k}^{\infty} A_n, \quad A_* = \liminf_{i \to \infty} A_i = \bigcup_{k=1}^{\infty} \bigcap_{n=k}^{\infty} A_n. \tag{A.244}$$

To better interpret what A^* contains, observe that, for an $\omega \in \Omega$, if $\omega \in A^*$, then $\omega \in \bigcup_{n=k}^{\infty} A_n$ for *every* k. In other words, for any k, no matter how large, there exists an $n \geq k$ with $\omega \in A_n$. This means that $\omega \in A_n$ for infinitely many values of n. Likewise, if $\omega \in \Omega$ belongs to A_*, then it belongs to $\bigcap_{n=k}^{\infty} A_n$ for *some* k, that is, there exists a k such that $\omega \in A_n$ for all $n \geq k$. Thus, definitions (A.244) are equivalent to

$$A^* = \{\omega : \omega \in A_n \text{ for infinitely many } n \in \mathbb{N}\},$$
$$A_* = \{\omega : \omega \in A_n \text{ for all but finitely many } n \in \mathbb{N}\}, \tag{A.245}$$

and are thus sometimes abbreviated as $A^* = \{A_n \text{ i.o.}\}$ and $A_* = \{A_n \text{ ult.}\}$, where "i.o." stands for "infinitely often" and "ult." stands for "ultimately."

As a definition, the sequence $\{A_n\}$ converges to A, written $A_n \to A$, if and only if $A = A^* = A_*$, that is,

$$A_n \to A \quad \Leftrightarrow \quad A = \limsup A_n = \liminf A_n. \tag{A.246}$$

For the sequence of events $\{A_n\}$, De Morgan's law states that

$$\left(\bigcup_{n=1}^{\infty} A_n\right)^c = \bigcap_{n=1}^{\infty} A_n^c \quad \text{and} \quad \left(\bigcap_{n=1}^{\infty} A_n\right)^c = \bigcup_{n=1}^{\infty} A_n^c.$$

With $B_k = \bigcup_{n=k}^{\infty} A_n$, these imply $B_k^c = \bigcap_{n=k}^{\infty} A_n^c$ and, thus,

$$(A^*)^c = \left(\bigcap_{k=1}^{\infty} \bigcup_{n=k}^{\infty} A_n\right)^c = \left(\bigcap_{k=1}^{\infty} B_k\right)^c = \bigcup_{k=1}^{\infty} B_k^c = \bigcup_{k=1}^{\infty} \bigcap_{n=k}^{\infty} A_n^c, \tag{A.247}$$

and, similarly, $(A_*)^c = \bigcap_{k=1}^{\infty} \bigcup_{n=k}^{\infty} A_n^c$.

For $\{A_n\}$ an arbitrary sequence of sets, and with $B_k := \bigcup_{n=k}^{\infty} A_n$, $k = 1, 2, \ldots$, $\{B_k\}$ is a monotone decreasing sequence of events, so that

$$\bigcup_{n=k}^{\infty} A_n = B_k \downarrow \bigcap_{k=1}^{\infty} B_k = \bigcap_{k=1}^{\infty} \bigcup_{n=k}^{\infty} A_n = A^*.$$

That is, as $k \to \infty$, $\bigcup_{n=k}^{\infty} A_n \downarrow A^*$, so that, from (A.243),

$$\Pr(A^*) = \lim_{k \to \infty} \Pr\left(\bigcup_{n=k}^{\infty} A_n\right). \tag{A.248}$$

Similarly, with $B_k := \bigcap_{n=k}^{\infty} A_n$ a monotone increasing sequence of events,

$$\bigcap_{n=k}^{\infty} A_n = B_k \uparrow \bigcup_{k=1}^{\infty} B_k = \bigcup_{k-1}^{\infty} \bigcap_{n=k}^{\infty} A_n = A_*,$$

that is, as $k \to \infty$, $\bigcap_{n=k}^{\infty} A_n \uparrow A_*$ and

$$\Pr(A_*) = \lim_{k \to \infty} \Pr\left(\bigcap_{n=k}^{\infty} A_n\right). \tag{A.249}$$

If $\{A_n\}$ is a sequence of events that is not necessarily monotone, then

$$\Pr(A_*) \leq \liminf_n \Pr(A_n) \quad \text{and} \quad \limsup_n \Pr(A_n) \leq \Pr(A^*). \tag{A.250}$$

For the former, let $B_k = \bigcap_{n=k}^{\infty} A_n$. As $B_k \subset A_k$ for each k, (A.238) implies that $\Pr(B_k) \leq \Pr(A_k)$ for each k, and, as B_k is a monotone increasing sequence, $B_k \uparrow \bigcup_{k=1}^{\infty} B_k$, and $\bigcup_{k=1}^{\infty} B_k = \bigcup_{k=1}^{\infty} \bigcap_{n=k}^{\infty} A_n = \liminf_n A_n$. Then, from (A.249),

$$\Pr(\liminf A_k) = \lim_{k \to \infty} \Pr(B_k) \leq \liminf \Pr(A_k).$$

The last inequality is true because (i) if sequences b_k and a_k are such that $b_k \leq a_k$ for all k, then $\lim_{k \to \infty} b_k \leq \lim_{k \to \infty} a_k$, and (ii) while $\lim_{k \to \infty} a_k$ may not exist, $\liminf_k a_k$ always does, so that $\lim_{k \to \infty} b_k \leq \liminf_{k \to \infty} a_k$. The second inequality in (A.250) is similar: let $B_k = \bigcup_{n=k}^{\infty} A_k$, so $A_k \subset B_k$ and $B_k \downarrow \limsup_n A_n$. Then, from (A.248) and the aforementioned facts on real sequences, $\Pr(\limsup A_n) = \lim_{k \to \infty} \Pr(B_k) \geq \limsup \Pr(A_k)$.

We can now show the fundamental result that extends the convergence result for monotone sequences. Let $\{A_n\}$ be a sequence of events which is not necessarily monotone. We wish to show that

$$\text{if } A_n \to A, \text{ then } \lim_{n \to \infty} \Pr(A_n) \text{ exists, and } \lim_{n \to \infty} \Pr(A_n) = \Pr(A).$$

First recall that if s_n is a deterministic sequence of real numbers, then $U = \lim \sup s_n$ and $L = \lim \inf s_n$ exist, and $\lim s_n$ exists if and only if $U = L$, in which case $\lim s_n = U = L$. From (A.84), for any $\epsilon > 0$, $\exists N_U \in \mathbb{N}$ such that, for all $n \geq N_U$, $s_n < U + \epsilon$. Likewise, $\exists N_L \in \mathbb{N}$ such that, for all $n \geq N_L$, $s_n > L - \epsilon$. Thus, for all $n \geq \max(N_U, N_L)$, $L - \epsilon < s_n < U + \epsilon$, and as $\epsilon > 0$ is arbitrary, it must be the case that $L \leq U$. In particular, if A_n is a sequence of events, and $s_n = \Pr(A_n)$, then $\lim \inf_n \Pr(A_n) \leq \lim \sup_n \Pr(A_n)$. Then, from (A.250),

$$\Pr(\lim \inf_n A_n) \leq \lim \inf_n \Pr(A_n) \leq \lim \sup_n \Pr(A_n) \leq \Pr(\lim \sup_n A_n). \tag{A.251}$$

From the assumption that $A_n \to A$ and definition (A.246), we know that $A = \lim_n A_n = \lim \inf_n A_n = \lim \sup_n A_n$, so that (A.251) implies

$$\Pr(A) \leq \lim \inf_n \Pr(A_n) \leq \lim \sup_n \Pr(A_n) \leq \Pr(A),$$

that is, $p := \lim \inf_n \Pr(A_n) = \lim \sup_n \Pr(A_n)$. Thus, $\lim_n \Pr(A_n)$ exists and $\lim_n \Pr(A_n) = p$. Again from (A.251), we have $\Pr(A) \leq p \leq \Pr(A)$, or $\lim_n \Pr(A_n) = \Pr(A)$, as was to be shown.

The two standard *Borel–Cantelli lemmas*, named in recognition of work by Émile Borel and Francesco Cantelli around 1909, are also fundamental results. They are as follows. First, for a sequence $\{A_n\}$ of arbitrary events,

$$\sum_{n=1}^{\infty} \Pr(A_n) < \infty \quad \Rightarrow \quad \Pr(A_n \text{ i.o.}) = 0. \tag{A.252}$$

Second, for a sequence $\{A_n\}$ of *independent* events,

$$\sum_{n=1}^{\infty} \Pr(A_n) = \infty \quad \Rightarrow \quad \Pr(A_n \text{ i.o.}) = 1. \tag{A.253}$$

To prove (A.252), use (A.248), (A.239), and the Cauchy criterion for convergent sums (I.A.86) to get

$$\Pr(A_n \text{ i.o.}) = \lim_{k \to \infty} \Pr\left(\bigcup_{n=k}^{\infty} A_n\right) \leq \lim_{k \to \infty} \sum_{n=k}^{\infty} \Pr(A_n) = 0.$$

To prove (A.253), use (A.247) and (A.249) to get

$$\Pr(A_n \text{ i.o.}) = 1 - \Pr\left(\bigcup_{k=1}^{\infty} \bigcap_{n=k}^{\infty} A_n^c\right) = 1 - \lim_{k \to \infty} \Pr\left(\bigcap_{n=k}^{\infty} A_n^c\right).$$

As the A_n are independent, so are the events A_n^c. Thus,

$$\Pr(A_n \text{ i.o.}) = 1 - \lim_{k \to \infty} \prod_{n=k}^{\infty} \Pr(A_n^c) = 1 - \lim_{k \to \infty} \prod_{n=k}^{\infty} [1 - \Pr(A_n)].$$

As $1 - x \leq e^{-x}$ for $x \geq 0$,[1]

$$\Pr(A_n \text{ i.o.}) \geq 1 - \lim_{k \to \infty} \exp \left\{ - \sum_{n=k}^{\infty} \Pr(A_n) \right\} = 1 - 0 = 1,$$

because $\sum_{n=1}^{\infty} \Pr(A_n) = \infty$ implies that, for any $k \in \mathbb{N}$, $\sum_{n=k}^{\infty} \Pr(A_n) = \infty$.

By imposing independence, the two lemmas can be combined to give a so-called *zero–one law*: For a sequence $\{A_n\}$ of independent events, $\Pr(A_n \text{ i.o.}) = 0$ when $\sum_{n=1}^{\infty} \Pr(A_n)$ is finite, and $\Pr(A_n \text{ i.o.}) = 1$ otherwise. This, implies, for example, that if one shows $\Pr(A_n \text{ i.o.}) < 1$, then $\Pr(A_n \text{ i.o.}) = 0$.

As an example of the first lemma, let X_n be a sequence of r.v.s with $\Pr(X_n = 0) = n^{-2}$, $n \geq 1$. Then, from the well-known result (see Example II.1.26) that $\sum_{n=1}^{\infty} \Pr(X_n = 0) = \pi^2/6 < \infty$, (A.245), and the first lemma, the probability of the event $\{X_n = 0\}$ occurring for infinitely many n is zero.

A.15.3 Convergence of Sequences of Random Variables

Let X and Y be r.v.s defined on the same probability space $\{\mathbb{R}, \mathcal{B}, \Pr(\cdot)\}$. If $\Pr(X \in A) = \Pr(Y \in A)$ for all $A \in \mathcal{B}$, then X and Y are said to be *equal in distribution*, written $X \overset{d}{=} Y$. If the set $\{\omega : X(\omega) \neq Y(\omega)\}$ is an event in $\mathcal{B}$ having probability zero (the null event), then X and Y are said to be *equal almost surely* or *almost surely equal*, written $X \overset{a.s.}{=} Y$. To emphasize the difference, let X and Y be i.i.d. standard normal. They have the same distribution, but, as they are independent, they are equal with probability 0.

The sequence of (univariate) random variables $\{X_n\}$ is said to converge in probability to the random variable X (possibly degenerate) if and only if, for all $\epsilon > 0$,

$$\lim_{n \to \infty} \Pr(|X_n - X| > \epsilon) = 0 \quad \text{or, equivalently,} \quad \lim_{n \to \infty} \Pr(|X_n - X| < \epsilon) = 1, \quad (\text{A.254})$$

for all $\epsilon > 0$, and we write $X_n \overset{p}{\to} X$. Observe $X_n \overset{p}{\to} X$ if and only if, for all $\epsilon > 0$ and $\delta > 0$, $\exists N \in \mathbb{N}$ such that $\Pr(|X_n - X| > \epsilon) < \delta$ for all $n \geq N$. We write "Assume $X_n \overset{p}{\to} X$" to mean "Let $\{X_n\}$ be a sequence of r.v.s that converges in probability to X." This is also expressed as the probability limit of X_n is X, and written plim $X_n = X$.

Let $\{X_n\}$ be a sequence of uncorrelated r.v.s in L_2, each with mean μ and variance σ^2, and let $\bar{X}_n = n^{-1} \sum_{i=1}^{n} X_i$, the average of the first n elements of the sequence. The weak law of large numbers (for uncorrelated r.v.s with the same finite first and second moments) states that

$$\bar{X}_n \overset{p}{\to} \mu. \quad (\text{A.255})$$

To prove this, as $\bar{X}_n$ has mean μ and variance σ^2/n, it follows immediately from Chebyshev's inequality (A.231) that, for any $\epsilon > 0$, $\Pr(|\bar{X}_n - \mu| \geq \epsilon) \leq \sigma^2/(n\epsilon^2)$, so that, in the limit, from definition (A.254), $\bar{X}_n \overset{p}{\to} \mu$.

Let $c, k \in \mathbb{R}$ and assume $X_n \overset{p}{\to} X$. If $c = 0$, then it is immediate from (A.254) that $(cX_n + k) \overset{p}{\to} cX + k$, while for $c \neq 0$, observe that, for any $\epsilon > 0$,

$$\lim_{n \to \infty} \Pr(|(cX_n + k) - (cX + k)| \geq \epsilon) = \lim_{n \to \infty} \Pr(|X_n - X| \geq \epsilon/|c|) = 0, \quad (\text{A.256})$$

that is, for any $c, k \in \mathbb{R}$, $(cX_n + k) \overset{p}{\to} cX + k$.

---

[1] To see this, with $f(x) = e^{-x}$ and $g(x) = 1 - x$, $f(0) = g(0) = 1$, and $g'(x) \leq f'(x)$ because $x \geq 0 \Leftrightarrow 0 \geq -x \Leftrightarrow 1 \geq e^{-x} \Leftrightarrow -1 \leq -e^{-x} \Leftrightarrow g'(x) \leq f'(x)$.

Assume $X_n \xrightarrow{p} a$, let $A \subset \mathbb{R}$, and let $g : A \to \mathbb{R}$ be a function continuous at point a with $a \in A$. We wish to confirm that $g(X_n) \xrightarrow{p} g(a)$. Recall that g is continuous at a if, for a given $\epsilon > 0$, $\exists \delta > 0$ (with δ being a function of a and ϵ) such that, if $|x - a| < \delta$ and $x \in A$, then $|g(x) - g(a)| < \epsilon$. The contrapositive of this is: if g is continuous at a, then, for a given $\epsilon > 0$, $\exists \delta > 0$ such that if $|g(x) - g(a)| \geq \epsilon$, then $\{|x - a| \geq \delta\}$. This implies that (recalling that r.v. X_n is a function of $\omega \in \Omega$)

$$\{\omega : |g(X_n(\omega)) - g(a)| \geq \epsilon\} \subset \{\omega : |X_n(\omega) - a| \geq \delta\}. \tag{A.257}$$

From (A.238), this implies that, for a given $\epsilon > 0$, $\exists \delta > 0$ such that

$$\Pr\{|g(X_n) - g(a)| \geq \epsilon\} \leq \Pr\{|X_n - a| \geq \delta\}.$$

The right-hand-side probability tends to zero for all δ, including the one corresponding to the choice of ϵ, so that

$$\lim_{n \to \infty} \Pr\{|g(X_n) - g(a)| \geq \epsilon\} = 0, \quad \text{i.e., } g(X_n) \xrightarrow{p} g(a). \tag{A.258}$$

Generalizing (A.256) to the nonlinear case, for g continuous (written $g \in C^0$),

$$X_n \xrightarrow{p} X, \; g \in C^0 \; \Rightarrow \; g(X_n) \xrightarrow{p} g(X), \tag{A.259}$$

as proven in Example II.4.11.

Assume $X_n \xrightarrow{p} X$, $Y_n \xrightarrow{p} Y$, and $\epsilon > 0$. Let $d_X = X_n - X$, $d_Y = Y_n - Y$, $S_n = X_n + Y_n$, and $S = X + Y$. From the triangle inequality,

$$\{|S_n - S| > \epsilon\} = \{|d_X + d_Y| > \epsilon\} \subset \{|d_X| + |d_Y| > \epsilon\} =: C. \tag{A.260}$$

With $A = \{|d_X| > \epsilon/2\}$, $B = \{|d_Y| > \epsilon/2\}$, Figure A.2 confirms that $C \subset \{A \cup B\}$ (Problem II.4.11 proves this algebraically), in which case (A.238) implies

$$\Pr(|S_n - S| > \epsilon) \leq \Pr(C) \leq \Pr(A \cup B) \leq \Pr(A) + \Pr(B) \to 0,$$

so that

$$X_n \xrightarrow{p} X, \; Y_n \xrightarrow{p} Y \; \Rightarrow \; X_n + Y_n \xrightarrow{p} X + Y. \tag{A.261}$$

Combining (A.261) and (A.256), we see that convergence in probability is closed under linear transformations, that is, if $X_n \xrightarrow{p} X$ and $Y_n \xrightarrow{p} Y$, then, for constants $a, b \in \mathbb{R}$, $aX_n + bY_n \xrightarrow{p} aX + bY$. More generally, from (A.259), if $X_n \xrightarrow{p} X$, $Y_n \xrightarrow{p} Y$, and $g, h \in C^0$, then $g(X_n) + h(Y_n) \xrightarrow{p} g(X) + g(Y)$. For example,

$$X_n Y_n = \frac{1}{2}X_n^2 + \frac{1}{2}Y_n^2 - \frac{1}{2}(X_n + Y_n)^2 \xrightarrow{p} \frac{1}{2}X^2 + \frac{1}{2}Y^2 - \frac{1}{2}(X + Y)^2 = XY,$$

that is, if $X_n \xrightarrow{p} X$ and $Y_n \xrightarrow{p} Y$, then $X_n Y_n \xrightarrow{p} XY$.

The concept of convergence in probability is easily extended to sequences of multivariate r.v.s. In particular, the sequence $\{\mathbf{X}_n\}$ of k-dimensional r.v.s converges in probability to the k-dimensional r.v. $\mathbf{X}$ if and only if

$$\lim_{n \to \infty} \Pr(\|\mathbf{X}_n - \mathbf{X}\| > \epsilon) = 0, \tag{A.262}$$

and we write $\mathbf{X}_n \xrightarrow{p} \mathbf{X}$.

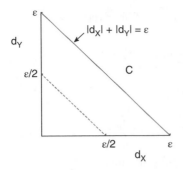 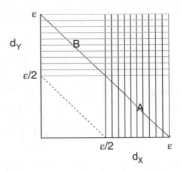

Figure A.2 *The vertical axis is d_Y, and the horizontal axis is d_X. This graphically verifies that $C \subset \{A \cup B\}$, where C is the region above the line $|d_X| + |d_Y| = \epsilon$ (left plot), B is the region indicated by horizontal lines, and A is the region indicated by vertical lines.*

The sequence $\{X_n\}$ is said to converge almost surely, or almost everywhere, or with probability 1 to the random variable X if and only if

$$\Pr(\omega : \lim_{n\to\infty} X_n(\omega) = X(\omega)) = 1, \qquad (A.263)$$

and we write $\lim_{n\to\infty} X_n = X$ a.s., or $X_n \overset{a.s.}{\to} X$. Observe how this definition differs from (A.254) for convergence in probability.

Almost sure convergence is similar to pointwise convergence of functions, but does not impose $\lim_{n\to\infty} X_n(\omega) = X(\omega)$ for *all* $\omega \in \Omega$, but rather only on a set of ω with probability 1. In particular, $X_n \overset{a.s.}{\to} X$ if and only if there exists a null event $E \in \mathcal{A}$ (often termed the *exception set*) with $\Pr(E) = 0$ and, for all $\omega \in E^c$, $\lim_{n\to\infty} X_n(\omega) = X(\omega)$. Observe that the definition allows E to be empty.

Example *Let ω be a random number drawn from the probability space $\{\Omega, \mathcal{A}, \Pr(\cdot)\}$ with $\Omega = [0, 1]$, $\mathcal{A}$ the Borel σ-field given by the collection of intervals $[a, b]$, $a \leq b$, $a, b \in [0, 1]$, and $\Pr(\cdot)$ uniform, that is, for event $A = [a, b] \in \mathcal{A}$, $\Pr(A) = b - a$. (In short, let ω be a uniformly distributed r.v. on the interval $[0, 1]$.) Let $X_n(\omega) = n\mathbb{1}_{[0,1/n]}(\omega)$ and let $E = \{0\}$ be the exception set, with $\Pr(E) = 0$ and $E^c = (0, 1]$. Then*

$$\forall \omega \in E^c, \ \lim_{n\to\infty} X_n(\omega) \to 0, \text{ but as } X_n(0) = n, \ \lim_{n\to\infty} X_n(\omega) \nrightarrow 0 \quad \forall \omega \in \Omega. \qquad (A.264)$$

Thus, the sequence $\{X_n\}$ converges almost surely to 0. See Resnick (1999, p. 168) for further discussion and a less trivial example.

Example II.4.14 proves that (i) X in (A.263) is unique up to a set of measure zero, that is, if $X_n \overset{a.s.}{\to} X$ and $X_n \overset{a.s.}{\to} Y$, then $\Pr(X = Y) = 1$; (ii) almost sure convergence is preserved under addition, that is, if $X_n \overset{a.s.}{\to} X$ and $Y_n \overset{a.s.}{\to} Y$, then $X_n + Y_n \overset{a.s.}{\to} X + Y$; and (iii) almost sure convergence is preserved under continuous transformation, that is,

$$\text{if } X_n \overset{a.s.}{\to} X \text{ and } g \in C^0, \text{ then } g(X_n) \overset{a.s.}{\to} g(X). \qquad (A.265)$$

To help exemplify the difference between almost sure convergence and convergence in probability, we use an equivalent statement of (A.263), as proven, for example, in Gut

(2005, Sec. 5.1.2). The sequence $X_n \overset{a.s.}{\to} X$ if and only if, for every $\epsilon > 0$,

$$\lim_{m \to \infty} \Pr\left(\bigcup_{n=m}^{\infty} \{ |X_n - X| > \epsilon \} \right) = \lim_{m \to \infty} \Pr(\sup_{n \geq m} |X_n - X| > \epsilon) = 0, \qquad \text{(A.266)}$$

or, equivalently,

$$\lim_{m \to \infty} \Pr\left(\bigcap_{n \geq m} \{ |X_n - X| \leq \epsilon \} \right) = 1. \qquad \text{(A.267)}$$

Using (A.248) with $A_n = |X_n - X| > \epsilon$, (A.266) states that

$$X_n \overset{a.s.}{\to} X \quad \Leftrightarrow \quad \Pr(A^*) = \lim_{m \to \infty} \Pr\left(\bigcup_{n=m}^{\infty} A_n \right) = 0, \qquad \text{(A.268)}$$

while (A.249) implies that

$$X_n \overset{a.s.}{\to} X \quad \Leftrightarrow \quad \Pr((A^c)_*) = \lim_{m \to \infty} \Pr\left(\bigcap_{n=m}^{\infty} A_n^c \right) = 1.$$

Informally speaking, the latter statement, for example, says that, with probability 1, an ω occurs such that, for any $\epsilon > 0$, $|X_n - X| \leq \epsilon$ for all n sufficiently large.

Based on (A.266), almost sure convergence can also be expressed by saying that $X_n \overset{a.s.}{\to} X$ if and only if, for all $\epsilon > 0$ and for all $\delta \in (0, 1)$,

$$\exists N \in \mathbb{N} \text{ such that, } \forall m \geq N, \quad \Pr\left(\bigcap_{n \geq m} \{ |X_n - X| < \epsilon \} \right) > 1 - \delta.$$

This is easily used to show that $X_n \overset{a.s.}{\to} X$ implies $X_n \overset{p}{\to} X$: Let $X_n \overset{a.s.}{\to} X$ and define $A_n :=$ $\{ |X_n - X| > \epsilon \}$. Clearly, for all $n \in \mathbb{N}$, $A_n \subset \bigcup_{k=n}^{\infty} A_k$. Using (A.238) on, and taking limits of, the latter expression, and using (A.266), we have

$$\lim_{n \to \infty} \Pr(|X_n - X| > \epsilon) = \lim_{n \to \infty} \Pr(A_n) \leq \lim_{n \to \infty} \Pr\left(\bigcup_{k=n}^{\infty} A_k \right) = 0,$$

that is,

$$X_n \overset{a.s.}{\to} X \quad \Rightarrow \quad X_n \overset{p}{\to} X. \qquad \text{(A.269)}$$

Examples II.4.15 and II.4.16 show that the converse of (A.269) is not true in general. The sequence $\{ X_n \}$ is said to converge completely to r.v. X if and only if, for all $\epsilon > 0$,

$$\sum_{n=1}^{\infty} \Pr(|X_n - X| > \epsilon) < \infty; \qquad \text{(A.270)}$$

we write $X_n \overset{c.c.}{\to} X$. This is a rather strong form of convergence. Indeed, Problem II.4.10 proves that

$$X_n \overset{c.c.}{\to} X \quad \Rightarrow \quad X_n \overset{a.s.}{\to} X. \qquad \text{(A.271)}$$

Let $\{X_n\}$ be a sequence of i.i.d. r.v.s in L_4 with expected value μ and $K := \mathbb{E}[X_1^4]$, and let $S_r = \sum_{i=1}^{r} X_i$ and $\bar{X}_r = r^{-1} S_r$. The strong law of large numbers states that

$$\bar{X}_n \overset{a.s.}{\to} \mu; \tag{A.272}$$

see Example II.4.17 for proof.

The sequence $\{X_n\}$ in L_r is said to converge in r-mean to $X \in L_r$ if and only if

$$\lim_{n\to\infty} \mathbb{E}[|X_n - X|^r] = 0. \tag{A.273}$$

In this case, we write $X_n \overset{L_r}{\to} X$, with other popular notation being $X_n \overset{L^r}{\to} X$ or just $X_n \overset{r}{\to} X$. Convergence in r-mean is also written *convergence in* L_r. In the common case when $r = 2$, one speaks of *mean square convergence* or *convergence in quadratic mean* and sometimes writes $X_n \overset{q.m.}{\to} X$. Problem II.4.9 shows that (i) r.v. X in (A.273) is unique up to a set of measure zero, that is, if $X_n \overset{r}{\to} X$ and $X_n \overset{r}{\to} Y$, then $\Pr(X = Y) = 1$; and (ii) if $X_n \overset{r}{\to} X$ and $Y_n \overset{r}{\to} Y$, then $X_n + Y_n \overset{r}{\to} X + Y$.

As an example, for $\{X_n\}$ an i.i.d. sequence of random variables in L_2 with $\mathbb{E}[X_n] = \mu$, $\mathbb{V}(X_n) = \sigma^2$, and $\bar{X} = n^{-1} S_n$, $S_n = \sum_{i=1}^{n} X_i$,

$$\lim_{n\to\infty} \mathbb{E}[|\bar{X}_n - \mu|^2] = \lim_{n\to\infty} \mathbb{V}(\bar{X}_n) = \lim_{n\to\infty} \frac{\sigma^2}{n} = 0,$$

so that $\bar{X} \overset{L_2}{\to} \mu$. In this case, we also have $\bar{X} \overset{p}{\to} \mu$ and $\bar{X} \overset{a.s.}{\to} \mu$, from the weak and strong laws of large numbers, respectively.

That $X_n \overset{L_1}{\to} X \Rightarrow X_n \overset{p}{\to} X$ follows from Markov's inequality (A.228), that is, for any $\epsilon > 0$,

$$\Pr(|X_n - X| \geq \epsilon) \leq \frac{\mathbb{E}[|X_n - X|]}{\epsilon}.$$

Section II.4.3.3 shows that

$$X_n \overset{L_r}{\to} X \Rightarrow X_n \overset{p}{\to} X, \quad r > 0, \tag{A.274}$$

$$X_n \overset{L_s}{\to} X \Rightarrow X_n \overset{L_r}{\to} X, \quad s \geq r \geq 1, \tag{A.275}$$

$$X_n \overset{L_r}{\to} X \Rightarrow \mathbb{E}[|X_n|^r] \to \mathbb{E}[|X|^r], \quad r > 0, \tag{A.276}$$

$$X_n \overset{L_2}{\to} X, \quad Y_n \overset{L_2}{\to} Y \Rightarrow X_n Y_n \overset{L_1}{\to} XY, \tag{A.277}$$

and provides two examples confirming that, in general, $X_n \overset{a.s.}{\to} X \not\Rightarrow X_n \overset{L_r}{\to} X$. Also, $X_n \overset{p}{\to} X \not\Rightarrow X_n \overset{L_r}{\to} X$ and $X_n \overset{L_r}{\to} X \not\Rightarrow X_n \overset{a.s.}{\to} X$.

For a given c.d.f. F, let $C(F) = \{x : F(x)$ is continuous at $x\}$. The sequence $\{X_n\}$ is said to converge in distribution to X if and only if

$$\lim_{n\to\infty} F_{X_n}(x) = F_X(x) \quad \forall x \in C(F_X), \tag{A.278}$$

and we write $X_n \overset{d}{\to} X$. Convergence in distribution is the weakest form of convergence. It can also be written as convergence in law or weak convergence.

Similar to the other types of convergence, if $X_n \xrightarrow{d} X$, then X is unique. Suppose that $X_n \xrightarrow{d} X$ and $X_n \xrightarrow{d} Y$. Then, for an $x \in C(F_X) \cap C(F_Y)$, the triangle inequality (A.221) implies that

$$|F_X(x) - F_Y(x)| \le |F_X(x) - F_{X_n}(x)| + |F_{X_n}(x) - F_Y(x)|,$$

and in the limit, the right-hand side goes to zero. Examples II.4.23 and II.4.24 show that $X_n \xrightarrow{d} X \not\Rightarrow X_n \xrightarrow{L_r} X$ and $X_n \xrightarrow{p} X \Rightarrow X_n \xrightarrow{d} X$.

Similarly to (A.262) for convergence in probability, the concept of convergence in distribution is easily extended in a natural way to sequences of multivariate r.v.s. The sequence $\{\mathbf{X}_n\}$ of k-dimensional r.v.s, with distribution functions $F_{\mathbf{X}_n}$, converges in distribution to the k-dimensional r.v. $\mathbf{X}$ with distribution $F_{\mathbf{X}}$ if

$$\lim_{n\to\infty} F_{\mathbf{X}_n}(\mathbf{x}) = F_{\mathbf{X}}(\mathbf{x}) \quad \forall \mathbf{x} \in C(F_{\mathbf{X}}), \tag{A.279}$$

and we write $\mathbf{X}_n \xrightarrow{d} \mathbf{X}$.

Some relationships between the various methods of convergence are summarized in the following diagram, for $\{X_n\}$ a sequence of r.v.s, X an r.v., and constant $c \in \mathbb{R}$.

The most important ones to memorize are between almost sure convergence, convergence in probability, and convergence in distribution. Some further important, well-known results regarding convergence in distribution are as follows:

(1) If $X_n \xrightarrow{d} X$, $Y_n \xrightarrow{d} Y$, $X \perp Y$ (X is independent of Y) and $X_n \perp Y_n$ for all n, then $X_n + Y_n \xrightarrow{d} X + Y$.

(2) (Helly–Bray) Let $X_n \xrightarrow{d} X$. If $g \in C^0[a,b]$ with $a,b \in C(F_X)$, then $\mathbb{E}[g(X_n)] \to \mathbb{E}[g(X)]$.

(3) (Extended Helly–Bray) Let $X_n \xrightarrow{d} X$. If $g \in C^0$ and bounded, then $\mathbb{E}[g(X_n)] \to \mathbb{E}[g(X)]$.

(4) (Continuous mapping theorem)

$$\text{Let } X_n \xrightarrow{d} X. \text{ If function } f \in C^0, \text{ then } f(X_n) \xrightarrow{d} f(X). \tag{A.280}$$

(5) (Scheffé's lemma) Let $X, X_1, X_2, \ldots$ be continuous r.v.s with respective p.d.f.s $f_X, f_{X_1}, f_{X_2}, \ldots$, and such that $f_{X_n} \to f_X$ for "almost all" x (for all $x \in \mathbb{R}$ except possibly on a set of measure zero). Then $X_n \xrightarrow{d} X$.

(6) (Slutsky's theorem) Let $X_n \xrightarrow{d} X$ and $Y_n \xrightarrow{p} a$ for some $a \in \mathbb{R}$. Then, for continuous $f : \mathbb{R}^2 \to \mathbb{R}$,

$$f(X_n, Y_n) \xrightarrow{d} f(X, a). \tag{A.281}$$

In particular, $X_n \pm Y_n \xrightarrow{d} X \pm a$, $X_n \cdot Y_n \xrightarrow{d} X \cdot a$, and, for $a \ne 0$, $X_n/Y_n \xrightarrow{d} X/a$.

(7) (Continuity theorem for c.f.) For r.v.s $X, X_1, X_2, \ldots$ with respective characteristic functions $\varphi_X, \varphi_{X_1}, \varphi_{X_2}, \ldots$,

$$\lim_{n \to \infty} \varphi_{X_n}(t) = \varphi_X(t) \ \forall t \quad \Leftrightarrow \quad X_n \overset{d}{\to} X. \tag{A.282}$$

Moreover, if the c.f.s of the X_n converge to some function φ for all $t \in \mathbb{R}$, which is continuous at zero, then there exists an r.v. X with c.f. φ such that $X_n \overset{d}{\to} X$.

(8) (Continuity Theorem for m.g.f.) Let X_n be a sequence of r.v.s such that the corresponding m.g.f.s $\mathbb{M}_{X_n}(t)$ exist for $|t| < h$, for some $h > 0$, and for all $n \in \mathbb{N}$. If X is a random variable whose m.g.f. $\mathbb{M}_X(t)$ exists for $|t| \le h_1 < h$ for some $h_1 > 0$, then

$$\lim_{n \to \infty} \mathbb{M}_{X_n}(t) = \mathbb{M}_X(t) \text{ for } |t| < h_1 \quad \Rightarrow \quad X_n \overset{d}{\to} X. \tag{A.283}$$

(9) (Cramér–Wold device) Let $\mathbf{X}$ and $\{\mathbf{X}_n\}$ be k-dimensional r.v.s. Then

$$\mathbf{X}_n \overset{d}{\to} \mathbf{X} \quad \Leftrightarrow \quad \mathbf{t}'\mathbf{X}_n \overset{d}{\to} \mathbf{t}'\mathbf{X}, \quad \forall \mathbf{t} \in \mathbb{R}^k. \tag{A.284}$$

Proofs can be found in many books. Gut (2005) conveniently has them all, shown respectively on pages 247, 222, 223, 246, 227, 249, 238, 242, and 246, though we will show proofs and give other useful references for some of the results now. For example, a very accessible and detailed proof of Slutsky's theorem is given in Ferguson (1996, pp. 39–41). Observe that the multivariate version of Slutsky's theorem is related to Cramér-Wold, in one direction, as follows: if for k-dimensional r.v.s $\mathbf{X}_n \overset{d}{\to} \mathbf{X}$ and $\mathbf{Y}_n \overset{p}{\to} \mathbf{t}$, then $\mathbf{Y}'_n\mathbf{X}_n \overset{d}{\to} \mathbf{t}'\mathbf{X}$.

We first prove the stronger statement of the Helly–Bray result 3 above, as an if-and-only-if condition.

Theorem A.1 Let $X_1, X_2, \ldots$ denote a sequence of r.v.s with support S. Then, for r.v. X such that $\Pr(X \in S) = 1$, and g a bounded, continuous function on S,

$$X_n \overset{d}{\to} X \quad \Leftrightarrow \quad \mathbb{E}[g(X_n)] \to \mathbb{E}[g(X)]. \tag{A.285}$$

A very concise proof for the d-dimensional case can be found in Ferguson (1996, p. 13). We present a proof for the univariate case following Severini (2005, p. 325), which, while longer, uses more basic principles and is thus highly instructive at this level. Once established, the proof of the continuous mapping theorem is very short and easy.

Proof. ($\Rightarrow$) Suppose $X_n \overset{d}{\to} X$ as $n \to \infty$ and let F denote the c.d.f. of X. Consider two cases: case 1 assumes $X, X_1, X_2, \ldots$ are bounded; case 2 removes this restriction.

Case 1. Let $M \in \mathbb{R}_{>0}$ be such that, with probability one, $|X_n| \le M$, $n = 1, 2, \ldots$, and $|X| \le M$. Assume without loss of generality that M is a continuity point of F. For a function $g : \mathbb{R} \to \mathbb{R}$ and an $\epsilon > 0$, let $x_1, x_2, \ldots, x_m$ be continuity points of F such that $-M = x_0 < x_1 < \cdots < x_{m-1} < x_m < x_{m+1} = M$ and $\max_{1 \le i \le m} \sup_{x_i \le x < x_{i+1}} |g(x) - g(x_i)| \le \epsilon$. Observe this is possible, because g is bounded and continuous. Define $g_m(x) = g(x_i)$ for $x_i \le x < x_{i+1}$, and set $g_m(x) = 0$ for $x < x_0$ and $x \ge x_{m+1}$. Then, as $n \to \infty$, we have

$$\int_{-M}^{M} g_m(x) \, dF_n(x)$$

$$= \int_{x_0}^{x_1} g_m(x)\,dF_n(x) + \int_{x_1}^{x_2} g_m(x)\,dF_n(x) + \cdots + \int_{x_m}^{x_{m+1}} g_m(x)\,dF_n(x)$$

$$= g(x_0)\int_{x_0}^{x_1} dF_n(x) + g(x_1)\int_{x_1}^{x_2} dF_n(x) + \cdots + g(x_m)\int_{x_m}^{x_{m+1}} dF_n(x)$$

$$= g(x_0)[F_n(x_1) - F_n(x_0)] + \cdots + g(x_m)[F_n(x_{m+1}) - F_n(x_m)]$$

$$= \sum_{i=0}^{m} g(x_i)[F_n(x_{i+1}) - F_n(x_i)]$$

$$\to \sum_{i=0}^{m} g(x_i)[F(x_{i+1}) - F(x_i)] = \int_{-M}^{M} g_m(x)\,dF(x). \tag{A.286}$$

Hence, we can find an N such that, for all $n \geq N$, both

$$\left|\int_{-M}^{M} g_m(x)A(x)\right| \leq \epsilon \quad \text{and} \quad \left|\int_{-M}^{M} A(x)\right| \leq \epsilon \tag{A.287}$$

are satisfied, where we define

$$A(x) := dF_n(x) - dF(x) \tag{A.288}$$

for convenience. Observe that the latter inequality in (A.287) is satisfied because we assume $X_n \xrightarrow{d} X$. Alternatively, note that this is just (A.286) with $g_m(x) \equiv 1$ for all x, this being bounded and continuous.

From the triangle inequality, $|g(x)| \leq |g(x) - g_m(x)| + |g_m(x)|$, or

$$\left|\int_{-M}^{M} g(x)A(x)\right| \leq \left|\int_{-M}^{M} [g(x) - g_m(x)]A(x)\right| + \left|\int_{-M}^{M} g_m(x)A(x)\right|$$

$$\leq \int_{-M}^{M} |g(x) - g_m(x)|\,|A(x)| + \epsilon$$

$$\leq \max_{1 \leq i \leq m} \sup_{x_i \leq x < x_{i+1}} |g(x) - g_m(x)| \int_{-M}^{M} |A(x)| + \epsilon$$

$$\leq \epsilon^2 + \epsilon. \tag{A.289}$$

Because ϵ is arbitrary, it follows that $\lim_{n\to\infty} \mathbb{E}[g(X_n)] = \mathbb{E}[g(X)]$.

Case 2. For the general case in which the $X_1, X_2, \ldots$ and X are not necessarily bounded, let $0 < \epsilon < 1$ be arbitrary and let M and $-M$ denote continuity points of F such that

$$\int_{-M}^{M} dF(x) = F(M) - F(-M) \geq 1 - \epsilon. \tag{A.290}$$

It is useful to let $\overline{\int}^M = \int_{-\infty}^{-M} + \int_{M}^{\infty}$, so that (A.290) can be expressed as $\overline{\int}^M dF(x) < \epsilon$. As both F_n and F are c.d.f.s, and $F_n(x) \to F(x)$ for all x, we can find an N and an M such that, for $n \geq N$, both $\overline{\int}^M dF(x) < \epsilon$ and $\overline{\int}^M dF_n(x) < \epsilon$. Then, as g is bounded,

$$\overline{\int}^M g(x)A(x) \leq \sup_x |g(x)| \left|\overline{\int}^M A(x)\right|$$

$$\leq \sup_x |g(x)| \left[\left| \overline{\int^M} dF_n(x) \right| + \left| \overline{\int^M} dF(x) \right| \right] \leq \sup_x |g(x)| 2\epsilon.$$

Thus, from this and (A.289),

$$\left| \int_{-\infty}^{\infty} g(x)A(x) \right| \leq \left| \overline{\int^M} g(x)A(x) \right| + \left| \int_{-M}^{M} g(x)A(x) \right| \leq \sup_x |g(x)| 2\epsilon + \epsilon^2 + \epsilon,$$

showing that $\left| \int_{-\infty}^{\infty} g(x)[dF_n(x) - dF(x)] \right| \to 0$ as $n \to \infty$.

($\Leftarrow$) Now suppose that $\mathbb{E}[g(X_n)]$ converges to $\mathbb{E}[g(X)]$ for all real-valued, bounded, continuous functions g. Define

$$h(x) = \begin{cases} 1, & \text{if } x < 0, \\ 1 - x, & \text{if } 0 \leq x \leq 1, \\ 0, & \text{if } x > 1, \end{cases} \tag{A.291}$$

and for any $t > 0$, define $h_t(x) = h(tx)$. This is plotted in Figure A.3.

Note that, for fixed t, h_t is bounded and continuous, so that, for all $t > 0$,

$$\lim_{n \to \infty} \mathbb{E}[h_t(X_n)] = \mathbb{E}[h_t(X)].$$

For fixed x, it is easy to see from a plot (take $y = u - x$) that, for all u and t,

$$\mathbb{I}(u \leq x) \leq h_t(u - x) \leq \mathbb{I}(u \leq x + 1/t).$$

Hence, from the first inequality,

$$F_n(x) = \int_{-\infty}^{x} dF_n(u) = \int_{-\infty}^{\infty} \mathbb{I}(u \leq x) dF_n(u) \leq \int_{-\infty}^{\infty} h_t(u - x) dF_n(u) = \mathbb{E}[h_t(X_n - x)].$$

As the limit of $\mathbb{E}[h_t(X_n)]$ exists, for any $t > 0$,

$$\limsup_{n \to \infty} F_n(x) \leq \lim_{n \to \infty} \mathbb{E}[h_t(X_n - x)] = \mathbb{E}[h_t(X - x)],$$

and, as $0 \leq tX - tx \leq 1 \Leftrightarrow 0 \leq X - x \leq 1/t \Leftrightarrow x \leq X \leq x + 1/t$,

$$\mathbb{E}[h_t(X - x)] = \mathbb{E}[h(tX - tx)] = \Pr(tX - tx < 0) + \int_x^{x+1/t} [1 - (tu - tx)] dF(u)$$

$$= \Pr(X < x) + \Pr(x \leq X \leq x + 1/t) - t \int_x^{x+1/t} \underbrace{(u - x)}_{>0} dF(u)$$

$$\leq F(x + 1/t), \quad t > 0.$$

It follows that, if F is continuous at x,

$$\limsup_{n \to \infty} F_n(x) \leq F(x). \tag{A.292}$$

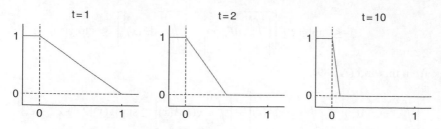

Figure A.3   *The function $h_t(x) = h(tx)$, where h is given in (A.291).*

Similarly, for fixed x, $\mathbb{I}(u \leq x - 1/t) \leq h_t(u - x + 1/t) \leq \mathbb{I}(u \leq x)$ for all u and t. Hence, using the second inequality,

$$F_n(x) = \int_{-\infty}^{\infty} \mathbb{I}(u \leq x)\, dF_n(u) \geq \int_{-\infty}^{\infty} h_t(u - x + 1/t)\, dF_n(u) = \mathbb{E}[h_t(X_n - x + 1/t)].$$

For any value of $t > 0$, and because the limit of $\mathbb{E}[h_t(X_n)]$ exists,

$$\liminf_{n \to \infty} F_n(x) \geq \lim_{n \to \infty} \mathbb{E}[h_t(X_n - x + 1/t)] = \mathbb{E}[h_t(X - x + 1/t)]$$

and, as

$$h_t(X - x + 1/t) = h(tX - tx + 1) = \begin{cases} 1, & \text{if } tX - tx + 1 < 0, \\ 1 - (tX - tx + 1), & \text{if } 0 \leq tX - tx + 1 \leq 1, \\ 0, & \text{if } tX - tx + 1 > 1, \end{cases}$$

$$= \begin{cases} 1, & \text{if } X < x - 1/t, \\ t(x - X), & \text{if } x - 1/t \leq X \leq x, \\ 0, & \text{if } X > x, \end{cases}$$

$$\mathbb{E}[h_t(X - x + 1/t)] = \Pr(X < x - 1/t) + t\int_{x-1/t}^{x} \underbrace{(x - u)\, dF(u)}_{>0} \geq F(x - 1/t), \quad t > 0.$$

That is, $\liminf_{n \to \infty} F_n(x) \geq F(x - 1/t)$ for $t > 0$, thus, if F is continuous at x,

$$\liminf_{n \to \infty} F_n(x) \geq F(x). \tag{A.293}$$

From (A.292) and (A.293), $\lim_{n \to \infty} F_n(x) = F(x)$ at all continuity points of F.   ∎

We now use (A.285) to prove the continuous mapping theorem (sometimes also called Slutsky's theorem). Let $X_n \xrightarrow{d} X$, let $f, g : \mathbb{R} \to \mathbb{R}$ with $f, g \in C^0$, additionally with g bounded, and let $h(x) = g(f(x)) \in C^0$. Then, the $\Rightarrow$ of (A.285) implies $\mathbb{E}[h(X_n)] \to \mathbb{E}[h(X)]$, that is, $\mathbb{E}[g(f(X_n))] \to \mathbb{E}[g(f(X))]$, and the $\Leftarrow$ of (A.285) implies $f(X_n) \xrightarrow{d} f(X)$.

A.16 THE STABLE PARETIAN DISTRIBUTION

The stable distribution exhibits heavy tails and possible asymmetry, making it a useful candidate for modeling a variety of observed processes that exhibit such characteristics. While this is the case for many such distributions, only the stable possesses certain theoretical aspects that can help justify it as the potentially "correct" distribution in certain modeling cases. One is the generalized central limit theorem, which, as its name suggests, generalizes Gaussian-based central limit theorems such as (A.160). In particular, it relaxes the finite-mean and finite-variance constraints, allowing the sequence of r.v.s in the sum to be nearly any set of i.i.d. (or possibly weakly dependent) r.v.s.

By nesting the c.f.s associated with the Cauchy and normal distributions, we can write

$$\varphi_X(t; \alpha) = \exp\{-|t|^\alpha\}, \quad 0 < \alpha \leq 2. \tag{A.294}$$

This is the symmetric stable Paretian distribution with tail index α, or α-stable or SαS distribution for short. Note that, as $\varphi_X(t; \alpha)$ is real, f_X must be symmetric about zero. It can be shown that the p.d.f. is unimodal; see, for example, Lukacs (1970, Sec. 5.10). This also holds for the asymmetric case below. For $\alpha = 2$, $\varphi_X(t; \alpha) = \exp\{-t^2\}$, which is the same as $\exp\{-t^2\sigma^2/2\}$ with $\sigma^2 = 2$, that is, as $\alpha \to 2$, $X \xrightarrow{d} N(0, 2)$ (and not a standard normal). Location and scale parameters are incorporated as usual by setting $Y = cX + \mu$, $c > 0$, and using $f_Y(y; \alpha, \mu, c) = c^{-1}f_X((y - \mu)/c)$. We will write $S_\alpha(\mu, c)$.

The density at a given point x can be computed via the inversion formula (A.99) or, simplifying it,

$$f_X(x) = \frac{1}{\pi} \int_0^\infty \cos(tx)e^{-t^\alpha} \, dt. \tag{A.295}$$

When evaluating the density at a large number of points, it is numerically much faster to use the FFT method, as discussed in Sections II.1.3.3 and II.8.1. Alternatively, we can use the density expression from Zolotarev (1986),

$$f_X(x; \alpha, \mu, \sigma) = \frac{1}{\sigma} \begin{cases} \frac{\alpha}{\pi|\alpha-1|z} \int_0^{\pi/2} V(y; \alpha, z) \exp\{-V(y; \alpha, z)\} dy, & \text{if } z > 0, \\ \Gamma(1 + \frac{1}{\alpha})/\pi, & \text{if } z = 0, \\ f_X(-x; \alpha, \mu, \sigma), & \text{if } z < 0, \end{cases} \tag{A.296}$$

where $z = (x - \mu)/\sigma$ and

$$V(y; \alpha, z) = \left(\frac{z \cos(y)}{\sin(\alpha y)}\right)^{\alpha/(\alpha-1)} \left(\frac{\cos((\alpha - 1)y)}{\cos(y)}\right).$$

Other real expressions exist, but the benefit of (A.296), as discussed by Nolan (1997), is its suitability for numeric computation. This can be "vectorized," yielding a very large decrease in computation time compared to elementwise evaluation. Replacing all mathematical operators by their elementwise counterparts, and using the vectorized variant of the adaptive Simpson quadrature from Lyness (1969), we evaluate

$$\oint_{\mathbf{a}}^{\mathbf{b}} f(\mathbf{x})d\mathbf{x} = \begin{cases} (\mathbf{q}_1 + \mathbf{q}_2)/2, & \text{if } \|\mathbf{q}_2 - \mathbf{q}_1\|_\infty \leq 10^{-6}, \\ \oint_{\mathbf{a}}^{\mathbf{c}} f(\mathbf{x})d\mathbf{x} + \oint_{\mathbf{c}}^{\mathbf{b}} f(\mathbf{x})d\mathbf{x}, & \text{otherwise}, \end{cases} \tag{A.297}$$

where $\oint$ denotes the (recursive) Simpson integral, $\|\mathbf{x}\|_\infty = \max(|x_1|, \ldots, |x_n|)$ is the sup norm for vector $\mathbf{x} \in \mathbb{R}^n$,

$$\mathbf{q}_1 = \frac{\mathbf{b} - \mathbf{a}}{6}(f(\mathbf{a}) + 4f(\mathbf{c}) + f(\mathbf{b})), \quad \mathbf{q}_2 = \frac{\mathbf{b} - \mathbf{a}}{12}(f(\mathbf{a}) + 4f(\mathbf{d}) + 2f(\mathbf{c}) + 4f(\mathbf{e}) + f(\mathbf{b})),$$

with $\mathbf{c} = (\mathbf{a} + \mathbf{b})/2$, $\mathbf{d} = (\mathbf{a} + \mathbf{c})/2$ and $\mathbf{e} = (\mathbf{c} + \mathbf{b})/2$. The resulting routine is about 40 times faster than repeated use of the scalar implementation, but equally robust and accurate. See below for the Matlab code for computing the p.d.f. in the asymmetric case, as well as direct inversion of the characteristic function.

One of the most important properties of the stable distribution is summability or stability: with $X_i \overset{\text{indep}}{\sim} S_\alpha(\mu_i, c_i)$ and $S = \sum_{i=1}^n X_i$,

$$\varphi_S(t) = \prod_{i=1}^n \varphi_{X_i}(t) = \exp(i\mu_1 t - c_1^\alpha |t|^\alpha) \cdots \exp(i\mu_n t - c_n^\alpha |t|^\alpha) = \exp(i\mu t - c^\alpha |t|^\alpha),$$

that is,

$$S \sim S_\alpha(\mu, c), \quad \mu = \sum_{t=1}^n \mu_i, \quad c = (c_1^\alpha + \cdots + c_n^\alpha)^{1/\alpha}. \tag{A.298}$$

The word "Paretian" in the name reflects the fact that the asymptotic tail behavior of the S_α distribution is the same as that of the Pareto distribution, that is, S_α has power tails, for $0 < \alpha < 2$. In particular, it can be shown that, for $X \sim S_\alpha(0, 1)$, $0 < \alpha < 2$, as $x \to \infty$,

$$\bar{F}_X(x) = \Pr(X > x) \approx k(\alpha)x^{-\alpha}, \quad k(\alpha) = \pi^{-1} \sin(\pi\alpha/2)\Gamma(\alpha), \tag{A.299}$$

where $a \approx b$ means that a/b converges to 1 as $x \to \infty$. Informally, differentiating the limiting value of $1 - \bar{F}_X(x)$ gives the asymptotic density in the right tail,

$$f_X(x) \approx \alpha k(\alpha)x^{-\alpha-1}. \tag{A.300}$$

Expressions for the asymptotic left-tail behavior follow from the symmetry of the density about zero, $f_X(x) = f_X(-x)$ and $F_X(x) = \bar{F}_X(-x)$. It follows that, for $0 < \alpha < 2$, the (fractional absolute) moments of $X \sim S_\alpha$ of order α and higher do not exist, that is, $\mathbb{E}[|X|^r]$ is finite for $0 \leq r < \alpha$, and infinite otherwise. For $\alpha = 2$, all positive moments exist. If $\alpha > 1$, then the mean exists,[2] and, from the symmetry of the density, it is clearly zero. The variance does not exist unless $\alpha = 2$.

Remark.   The fact that moments of $X \sim S_\alpha$ of order α and higher do not exist for $0 < \alpha < 2$, while for $\alpha = 2$, all positive moments exist, implies quite different tail behavior than

---

[2] Distributions with $\alpha \leq 1$ might appear to be of only academic interest, but there exist data sets, comprising for example file sizes of data downloaded from the World Wide Web, whose tails are so heavy that the mean does not exist; see Resnick and Rootzén (2000) and the references therein.

the Student's t distribution, which also nests the Cauchy and, as the degrees-of-freedom parameter $v \to \infty$, the normal. For the t, moments of order v and higher do not exist, and as $v \to \infty$, the tail behavior gradually moves from power tails to exponential tails. For the stable distribution, there is a "knife-edge" change from $\alpha < 2$ (power tails) to $\alpha = 2$ (normal, with exponential tails). The transition is still smooth, however, in the sense that the p.d.f. of $X \sim S_\alpha$ for $\alpha = 2 - \epsilon, \epsilon > 0$, can be made arbitrarily close to that of the normal p.d.f. This is a consequence of the continuity of the c.f. and the fact that, like parameter $v \in \mathbb{R}_{>0}$, the set of numbers $[2 - \epsilon, 2]$ is uncountably infinite. ∎

The more general case allows for asymmetry. We write $X \sim S_{\alpha,\beta}(\mu, c)$ for X a location–scale asymmetric stable Paretian, with tail index α, $0 < \alpha \leq 2$, asymmetry parameter β, $-1 \leq \beta \leq 1$, and c.f.

$$\ln \varphi_X(t) = \begin{cases} -c^\alpha |t|^\alpha \left[1 - i\beta \text{sgn}(t) \tan \dfrac{\pi\alpha}{2} \right] + i\mu t, & \text{if } \alpha \neq 1, \\ -c|t| \left[1 + i\beta \dfrac{2}{\pi} \text{sgn}(t) \ln |t| \right] + i\mu t, & \text{if } \alpha = 1. \end{cases} \tag{A.301}$$

As $\alpha \to 2$, the effect of β diminishes because $\tan(\pi\alpha/2) \to 0$; and when $\alpha = 2$, $\tan(\pi) = 0$, and β has no effect. Thus, there is no "skewed normal" distribution within the stable family.

The programs in Listings A.3 and A.4 compute the p.d.f. (and, in the former, optionally also the c.d.f.) using direct numerical inversion of the characteristic function, and numerical

```
1   function [f,F] = asymstab(xvec,a,b)
2   % pdf and, optionally, cdf of the asymmetric stable. See also asymstabpdf.m
3   % Nolan's routine: Call stablepdf(xvec,[a,b],1)
4
5   if nargin<3, b=0; end
6   bordertol=1e-8; lo=bordertol; hi=1-bordertol; tol=1e-7;
7   xl=length(xvec); F=zeros(xl,1); f=F;
8   for loop=1:length(xvec)
9     x=xvec(loop); f(loop)= quadl(@fff,lo,hi,tol,[],x,a,b,1) / pi;
10    if nargout>1
11      F(loop)=0.5-(1/pi)* quadl(@fff,lo,hi,tol,[],x,a,b,0);
12    end
13  end;
14
15  function I = fff(uvec,x,a,b,dopdf)
16  subs = 1; I = zeros(size(uvec));
17  for ii=1:length(uvec)
18    u=uvec(ii);
19    if subs==1, t =(1-u)/u; else t =u/(1-u); end
20    if a==1, cf = exp( -abs(t)*( 1 + 1i*b*(2/pi)*sign(t) * log(t)) );
21    else cf = exp( - ((abs(t))^a) *( 1 - 1i*b*sign(t) * tan(pi*a/2)) );
22    end
23    z   = exp(-1i*t*x).* cf; if dopdf==1, g=real(z); else g=imag(z)./t; end
24    if subs==1, I(ii)=g*u^(-2); else I(ii)=g*(1-u)^(-2); end
25  end
```

Program Listing A.3: Computes the p.d.f. and, optionally, the c.d.f. of the asymmetric stable Paretian distribution based on the inversion formulas (A.104) and (A.108).

```
1   function f = asymstabpdf(xvec, a, b, plotintegrand)
2   % pdf of the asymmetric stable. See also asymstab.m
3   % Set plotintegrand to 1, and xvec a scalar, to plot the integrand.
4
5   if nargin<4, plotintegrand=0; end
6   if a==1, error('not ready yet'), end
7   xl = length(xvec); f = zeros(xl, 1); n = length(xvec);
8   for loop = 1:n
9     x = xvec(loop);
10    if (x < 0), f(loop) = stab(-x, a, -b); else f(loop) = stab(x, a, b); end
11  end
12  if plotintegrand && xl==1 % show a plot of the integrand
13    if x<0, x=-x; b=-b; end
14    t0 = (1 / a) * atan(b * tan((pi * a) / 2));
15    t=-t0:0.001:(pi/2); l=integrand2(t, x, a, t0); plot(t,l,'r-')
16  end
17
18  function pdf = stab(x, a, b)
19  t0 = (1 / a) * atan(b * tan((pi * a) / 2));
20  if (x == 0) % Borak et al. (2005)
21    pdf = gamma(1+(1/a)) * cos(t0)/pi/(1 + (b*tan((pi*a)/2))^2)^(1/(2*a));
22  else
23    tol = 1e-9; display = 0;
24    integ = quadv(@integrand, -t0, pi/2, tol, display, abs(x), a, t0);
25    pdf = a * x^(1 / (a-1)) / pi / abs(a-1) * integ;
26  end
27
28  function I = integrand(t, x, a, t0)
29  ct = cos(t); s = t0 + t;
30  v = (cos(a*t0))^(1/(a-1))*(ct/sin(a*s))^(a/(a-1))*(cos(a*s - t)/ct);
31  term = -x^(a / (a-1)); I = v .* exp(term * v);
32
33  function I = integrand2(t, x, a, t0)
34  ct = cos(t); s = t0 + t;
35  v = (cos(a*t0))^(1/(a-1))*(ct/sin(a*s)).^(a/(a-1)) .* (cos(a*s - t) ./ ct);
36  term = -x.^(a / (a-1)); I = v .* exp(term * v);
```

Program Listing A.4: Computes the asymmetric stable Paretian p.d.f. via numeric integration based on a real, finite integral expression.

integration of the real, definite integral form of the density, respectively. Each method takes about 2 seconds to compute the p.d.f. on a grid of 600 points (on a typical desktop PC at the time of writing). Figure A.4 shows example output of the two functions. The code used to generate the graphs is given in Listing A.5.

A much faster method for computing the p.d.f. (and c.d.f.) is to use commercially available routines based on spline approximations. We use that from John Nolan, available for Matlab; see his website. For the stable p.d.f., he provides two routines, one, called stablepdf, that delivers the density to a very high accuracy, and one, called stable-qkpdf, that is based on spline approximations and is less accurate, but virtually as fast as, say, evaluating the normal distribution in Matlab. The latter typically agrees with the former to about five decimal places, and is thus accurate enough for use with maximum likelihood estimation. Similarly, it contains routines stablecdf and stableqkcdf for the c.d.f. In all of these routines, the type of parameterization of the stable distribution needs

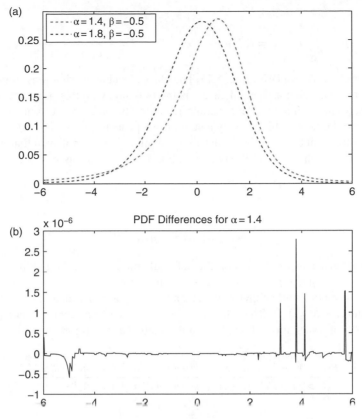

Figure A.4 *(a) Asymmetric stable p.d.f. for $\beta = -0.5$ and two values of α. (b) Discrepancy between the two computation methods using the $\alpha = 1.4$ case.*

```
1  xvec=-6:0.02:6; b=-0.5;
2  a=1.8; f18=asymstab(xvec,a,b); f28=asymstabpdf(xvec,a,b);
3  a=1.4; f14=asymstab(xvec,a,b); f24=asymstabpdf(xvec,a,b);
4  figure, set(gca,'fontsize',16)
5  plot(xvec,f14,'r-',xvec,f18,'b-','linewidth',2)
6  legend('\alpha = 1.4, \beta = -0.5', ...
7          '\alpha = 1.8, \beta = -0.5', 'Location','NorthWest')
8  hold on, plot(xvec,f24,'g:',xvec,f28,'g:'), hold off, ylim([0,0.3])
9  figure, set(gca,'fontsize',16)
10 plot(xvec,f14-f24,'k-'), title('PDF Differences for \alpha = 1.4')
```

Program Listing A.5: Generates the graphics in Figure A.4.

to be specified; we always use type 1, which corresponds to the characteristic function in (A.301).

The sum of independent stable r.v.s, each with the same tail index α, also follows a stable distribution. In particular,

$$\text{if } X_i \overset{\text{indep}}{\sim} S_{\alpha,\beta_i}(\mu_i, c_i), \quad \text{then } S = \sum_{i=1}^n X_i \sim S_{\alpha,\beta}(\mu, c),$$

where

$$\mu = \sum_{t=1}^{n} \mu_i, \quad c = (c_1^\alpha + \cdots + c_n^\alpha)^{1/\alpha}, \quad \beta = \frac{\beta_1 c_1^\alpha + \cdots + \beta_n c_n^\alpha}{c_1^\alpha + \cdots + c_n^\alpha};$$

see, for example, Lukacs (1970, Sec. 5.7). The class of stable distributions (thus including the Cauchy and Gaussian as special cases) is the only one with this property. (Recall, for example, that sums of independent gamma r.v.s are also gamma, but only if their scale parameters are the same, which is why a sum of independent χ^2 r.v.s is χ^2.)

The asymptotic tail behavior is similar to the symmetric case, but such that parameter β dictates the "relative heights" of the two tails. For $X \sim S_{\alpha,\beta}(0, 1)$, as $x \to \infty$,

$$\bar{F}_X(x) = \Pr(X > x) \approx k(\alpha)(1 + \beta)\, x^{-\alpha} \tag{A.302}$$

and

$$f_X(x) \approx \alpha k(\alpha)(1 + \beta)\, x^{-\alpha-1}, \tag{A.303}$$

where $k(\alpha)$ is given in (A.299). For the left tail, the symmetry relations $f_X(x; \alpha, \beta) = f_X(-x; \alpha, -\beta)$ and $F_X(x; \alpha, \beta) = \bar{F}_X(-x; \alpha, -\beta)$ can be used.

From the asymptotic tail behavior, we can ascribe a particular meaning to parameter β. Let $\alpha < 2$ and $X \sim S_{\alpha,\beta}(0, 1)$. Then, from (A.302) and the symmetry relation $F_X(x; \alpha, \beta) = \bar{F}_X(-x; \alpha, -\beta)$, a calculation devoid of any mathematical rigor suggests (correctly), as $x \to \infty$,

$$\frac{P(X > x) - P(X < -x)}{P(X > x) + P(X < -x)} \approx \frac{k(\alpha)(1 + \beta)x^{-\alpha} - k(\alpha)(1 - \beta)x^{-\alpha}}{k(\alpha)(1 + \beta)x^{-\alpha} + k(\alpha)(1 - \beta)x^{-\alpha}}$$

$$= \frac{2k(\alpha)\beta x^{-\alpha}}{2k(\alpha)x^{-\alpha}} = \beta, \tag{A.304}$$

that is, β measures the asymptotic difference in the two tail masses, scaled by the sum of the two tail areas.

There is one more special case of the stable distribution for which a closed-form density expression is available, namely when $\alpha = 1/2$ and $\beta = 1$, referred to as the Lévy, or Smirnov, or inverse χ^2 distribution. If $X \sim S_{1/2,1}(0, 1)$, then

$$f_X(x) = (2\pi)^{-1/2} x^{-3/2} e^{-1/(2x)} \mathbb{I}_{(0,\infty)}(x). \tag{A.305}$$

It arises, for example, in the context of hitting times for Brownian motion (see, for example, Feller, 1971, pp. 52, 173), and also appeared above in (A.153). It is a limiting case of the inverse Gaussian distribution, and is the distribution of Z^{-2} for $Z \sim N(0, 1)$ (see Example I.7.15). Two applications of l'Hôpital's rule yield that $f_X(x) \downarrow 0$ as $x \downarrow 0$. The tail approximation (A.303) simplifies in this case to $(2\pi)^{-1/2} x^{-3/2}$, as is obvious from (A.305).

As in the symmetric case, it is clear from (A.303) that moments of $X \sim S_{\alpha,\beta}$ of order α and higher do not exist when $\alpha < 2$. When $1 < \alpha \leq 2$, the mean of $X \sim S_{\alpha,\beta}(\mu, c)$ is μ, irrespective of β.

An expression for $\mathbb{E}[|X|^r]$, $-1 < r < \alpha$, is available. Two (lengthy) methods of derivation are given in Section II.8.3, showing that

$$\mathbb{E}[|X|^r] = \kappa^{-1}\Gamma\left(1 - \frac{r}{\alpha}\right)(1 + \tau^2)^{r/2\alpha}\cos\left(\frac{r}{\alpha}\arctan\tau\right), \quad -1 < r < \alpha, \tag{A.306}$$

```
1   function x=stabgen (nobs,a,b,c,d,seed)
2   if nargin<3, b=0; end, if nargin<4, c=1; end
3   if nargin<5, d=0; end, if nargin<6, seed=rand; end
4   z=nobs;
5   rand('twister',seed), V=unifrnd(-pi/2,pi/2,1,z);
6   rand('twister',seed+42), W=exprnd(1,1,z);
7   if a==1
8     x=(2/pi)*(((pi/2)+b*V).*tan(V)-b*log((W.*cos(V))./((pi/2)+b*V)));
9     x=c*x+d-(2/pi)*d*log(d)*c*b;
10  else
11    Cab=atan(b*tan(pi*a/2))/(a); Sab=(1+b^2*(tan((pi*a)/2))^2)^(1/(2*a));
12    A=(sin(a*(V+Cab)))./((cos(V)).^(1/a));
13    B0=(cos(V-a*(V+Cab)))./W;  B=(abs(B0)).^((1-a)/a);
14    x=Sab*A.*(B.*sign(B0)); x=c*x+d;
15  end
```

Program Listing A.6: Generates a random sample of size nobs from the $S_{a,b}(d,c)$ distribution.

where, noting that, for ϵ not a negative integer, $\Gamma(-1+\epsilon) = \Gamma(\epsilon)/(-1+\epsilon)$,

$$\tau = \beta \tan(\pi\alpha/2), \quad \kappa = \begin{cases} \Gamma(1-r)\cos(\pi r/2), & \text{if } r \neq 1, \\ \pi/2, & \text{if } r = 1. \end{cases}$$

A method for simulating stable Paretian r.v.s was given by Chambers et al. (1976). Let $U \sim \text{Unif}(-\pi/2, \pi/2)$ independent of $E \sim \text{Exp}(1)$. Then $Z \sim S_{\alpha,\beta}(0,1)$, where

$$Z = \begin{cases} \dfrac{\sin \alpha(\theta + U)}{(\cos \alpha\theta \cos U)^{1/\alpha}} \left[\dfrac{\cos (u\theta + (u-1)U)}{E} \right]^{(1-\alpha)/\alpha}, & \text{if } \alpha \neq 1, \\[4mm] \dfrac{2}{\pi} \left[\left(\dfrac{\pi}{2} + \beta U \right) \tan U - \beta \ln \left(\dfrac{(\pi/2)E \cos U}{(\pi/2) + \beta U} \right) \right], & \text{if } \alpha = 1, \end{cases} \quad \text{(A.307)}$$

and $\theta = \arctan(\beta \tan(\pi\alpha/2))/\alpha$ for $\alpha \neq 1$. Incorporation of location and scale changes is done as usual when $\alpha \neq 1$, that is, $X = \mu + cZ \sim S_{\alpha,\beta}(\mu, c)$, but for $\alpha = 1$, $X = \mu + cZ + (2\beta c \ln c)/\pi \sim S_{1,\beta}(\mu, c)$ (note that the latter term is zero for $\beta = 0$ or $c = 1$). This is implemented in the program in Listing A.6.

Let $X_1, X_2, \dots$ be an i.i.d. sequence of r.v.s (with or without finite means and variances). The generalized central limit theorem states that there exist real constants $a_n > 0$ and b_n, and a nondegenerate r.v. Z such that, as $n \to \infty$,

$$a_n \sum_{j=1}^{n} X_j - b_n \overset{d}{\to} Z, \quad \text{(A.308)}$$

if and only if $Z \sim S_{\alpha,\beta}(\mu, c)$. In the standard theorem (A.160), $a_n = n^{-1/2}$. The term a_n in (A.308) is of the form $n^{-1/\alpha}$, generalizing the $\alpha = 2$ case.

Recall the discussion of expected shortfall (ES) in Section A.8. The stable Paretian for $\alpha < 2$ does not possess an m.g.f., so that (A.121) cannot be used for computing the ES, though the more general (A.122) is applicable for $1 < \alpha < 2$. This is implemented in Listing A.9 below.

Stoyanov et al. (2006) also derive a (definite) integral expression specifically for the stable Paretian distribution that does not explicitly involve the stable density to compute the ES. For $\alpha > 1$ (in which case the mean exists, as required for the ES), $S \sim S_\alpha(0, 1)$, $0 < \xi < 1$, and $q_{S,\xi} = F_S^{-1}(\xi)$,

$$\text{ES}(S; \xi) = \frac{1}{\xi}\text{Stoy}(q_{S,\xi}, \alpha), \qquad (A.309)$$

where the tail component $\int_{-\infty}^{c} x f_S(x; \alpha)\, dx$ is

$$\text{Stoy}(c, \alpha) = \frac{\alpha}{\alpha - 1}\frac{|c|}{\pi}\int_0^{\pi/2} g(\theta)\exp(-|c|^{\alpha/(\alpha-1)}v(\theta))d\theta,$$

and

$$g(\theta) = \frac{\sin(\alpha - 2)\theta}{\sin \alpha\theta} - \frac{\alpha\cos^2\theta}{\sin^2\alpha\theta}, \qquad v(\theta) = \left(\frac{\cos\theta}{\sin\alpha\theta}\right)^{\alpha/(\alpha-1)}\frac{\cos(\alpha - 1)\theta}{\cos\theta}.$$

To numerically confirm the exact integral expressions (A.122) and (A.309), we can perform the brute-force integration $I(\xi, \alpha) = \int_{-\infty}^{q_{S,\xi}} x f_S(x; \alpha)\, dx$, but we know that calculation of f_S for $|x|$ very large can be problematic. To address this, we use the asymptotic tail behavior of $S \sim S_{\alpha,0}(0, 1)$; as $x \to -\infty$, $f_S(x; \alpha) \approx K_\alpha(-x)^{-\alpha-1}$, where $K_\alpha := \alpha\pi^{-1}\sin(\pi\alpha/2)\Gamma(\alpha)$ and $a \approx b$ means that a/b converges to one in the limit.

Then, for some cutoff value ℓ, the integral can be approximated by

$$I(\xi, \alpha) \approx K_\alpha \int_{-\infty}^{\ell} x(-x)^{-\alpha-1}dx + \int_{\ell}^{q_{S,\xi}} x f_S(x; \alpha)dx$$

$$= K_\alpha \frac{(-\ell)^{1-\alpha}}{1 - \alpha} + \int_{\ell}^{q_{S,\xi}} x f_S(x; \alpha)dx. \qquad (A.310)$$

Some trial and error shows that a value of $\ell = -120$ results in very good performance.

For the asymmetric stable case, for $\alpha > 1$, $S \sim S_{\alpha,\beta}(0, 1)$, $0 < \xi < 1$, and $q_{S,\xi} = F_S^{-1}(\xi)$,

$$\text{ES}(S; \xi) = \frac{1}{\xi}\text{Stoy}(q_{S,\xi}, \alpha, \beta), \qquad (A.311)$$

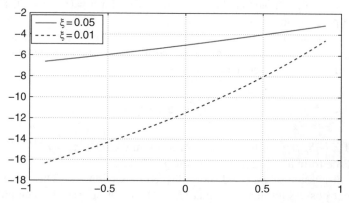

Figure A.5 *Expected shortfall for $\alpha = 1.7$ as a function of β, using $\xi = 0.05$ (solid) and $\xi = 0.01$ (dashed).*

where the tail component $\int_{-\infty}^{c} x f_S(x; \alpha, \beta)\,dx$ is

$$\text{Stoy}(c, \alpha, \beta) = \frac{\alpha}{\alpha - 1} \frac{|c|}{\pi} \int_{-\bar{\theta}_0}^{\pi/2} g(\theta) \exp(-|c|^{\alpha/(\alpha-1)} v(\theta))\,d\theta, \qquad (A.312)$$

with

$$g(\theta) = \frac{\sin(\alpha(\bar{\theta}_0 + \theta) - 2\theta)}{\sin(\alpha(\bar{\theta}_0 + \theta))} - \frac{\alpha \cos^2(\theta)}{\sin^2(\alpha(\bar{\theta}_0 + \theta))},$$

$$v(\theta) = (\cos(\alpha\bar{\theta}_0))^{1/(\alpha-1)} \left(\frac{\cos(\theta)}{\sin(\alpha(\bar{\theta}_0 + \theta))} \right)^{\alpha/(\alpha-1)} \frac{\cos(\alpha(\bar{\theta}_0 + \theta) - \theta)}{\cos(\theta)},$$

```
1  function [ES, VaR] = asymstableES(xi, a, b, mu, scale, method)
2  if nargin < 3, b=0; end,        if nargin < 4, mu=0; end
3  if nargin < 5, scale = 1; end, if nargin < 6, method = 1; end
4
5  % Get q, the quantile from the S(0,1) distribution
6  opt=optimset('Display', 'off', 'TolX', 1e-6);
7  q=fzero(@stabcdfroot, -6, opt, xi, a, b);   VaR=mu+scale*q;
8
9  if (q == 0)
10    t0 = (1 / a) * atan(b * tan(pi * a/2));
11    ES = ((2 * gamma((a-1) / a)) / (pi - 2*t0)) * (cos(t0) / cos(a*t0)^(1 / a));
12    return;
13 end
14 if (method==1), ES=(scale*Stoy(q,a,b)/xi)+mu;
15 else ES=(scale*stabletailcomp(q,a,b)/xi)+ mu;
16 end
17
18 function diff = stabcdfroot(x, xi, a, b)
19 if exist('stableqkcdf.m','file'), F = stableqkcdf(x, [a, b],1); % Nolan routine
20 else [~, F] = asymstab(x, a, b);
21 end
22 diff = F - xi;
23
24 function tailcomp = stabletailcomp(q, a, b)
25 % direct integration of x*f(x) and use of asymptotic tail behavior
26 K = (a / pi) * sin(pi * a / 2) * gamma(a) * (1 - b); % formula for K_
27 ell = -120; M = ell; display = 0;   term1 = K * (-M)^(1 - a) / (1 - a);
28 term3 = quadl(@stableCVARint, ell, q, 1e-5, display, a, b);
29 tailcomp = term1 + term3;
30
31 function [g] = stableCVARint(x, a, b)
32 if exist('stableqkpdf.m','file'), den = stableqkpdf(x, [a, b],1);
33 else den = asymstab(x, a, b)';
34 end
35 g = x.* den;
```

Program Listing A.7: Computes the VaR (left tail quantile) and associated expected short-fall (ES) corresponding to a probability level ξ, for the asymmetric stable Paretian distribution with tail index a, asymmetry parameter b, location mu, and scale `scale`. Set `method` to 1 (default) to use (A.311), otherwise attempts direct numeric integration and tail approximation. Function `Stoy` is given in Listing A.8.

```
1   function S = Stoy(cut, alpha, beta)
2   if nargin<3, beta=0; end
3   cut = -cut;  % we use a different sign convention
4   bbar = -sign(cut) * beta;
5   t0bar = (1 / alpha) * atan(bbar * tan(pi * alpha / 2));
6   % 'beta==0' => 'bbar==0' => 't0bar==0'
7   small = 1e-8; tol = 1e-8; abscut = abs(cut); display = 0;
8   integ = quadl(@stoyint, -t0bar+small, pi/2-small, tol, ...
9                 display, abscut, alpha, t0bar);
10  S = alpha / (alpha-1) / pi * abscut * integ;
11
12  function I = stoyint(t, cut, a, t0bar)
13  s = t0bar + t;
14  g = sin(a*s - 2*t) ./ sin(a*s) - a * cos(t).^2 ./ sin(a*s).^2;
15  v = (cos(a*t0bar)).^(1 / (a - 1)) .* (cos(t) ./ sin(a*s)).^(a / (a - 1)) ...
16      .* cos(a*s - t) ./ cos(t);
17  term = -(abs(cut)^(a / (a - 1)));
18  I = g .* exp(term .* v);
```

Program Listing A.8: Computes the integral (A.312).

and[3]

$$\bar{\theta}_0 = \frac{1}{\alpha} \arctan\left(\bar{\beta}\tan\left(\frac{\pi\alpha}{2}\right)\right) \quad \text{and} \quad \bar{\beta} = \text{sign}(c)\beta.$$

Figure A.5 plots the ES for $\xi = 0.01$ and $\xi = 0.05$ for $\alpha = 1.7$, as a function of β.

Observe that, for a given ξ, computation of the ES also requires determining the appropriate quantile (the VaR). Both are computed in the program in Listing A.7, which also implements the approach in (A.310).

Broda et al. (2017) derive a saddlepoint approximation (s.p.a.; see Section A.11) to the ES applicable to the (asymmetric) stable Paretian, and other cases (such as the noncentral t and generalized exponential distribution) for which the m.g.f. does not exist. Figure A.6 shows the ES, computed using one of the exact methods discussed above, and via the s.p.a. Its accuracy decreases as α decreases, though is quite accurate for $1.7 < \alpha < 2$, which is the range of interest for most applications involving financial asset returns data; see, for example, Paolella (2016b). The code to compute the exact and s.p.a. ES is given in Listing A.9.

A.17 PROBLEMS

A.1   We wish to prove that $\Gamma(1/2) = \sqrt{\pi}$ in two ways.
  (a) Use (A.7) and

$$2\int_0^\infty \exp\left(-\frac{1}{2}v^2\right) dv = \sqrt{2\pi}, \tag{A.313}$$

the latter as stated in Example I.A.79.

---

[3] The formulas for $\bar{\beta}$ and Stoy(c, α, β) have a minus sign in front in Stoyanov et al. (2006) because they use the positive sign convention for the VaR and ES.

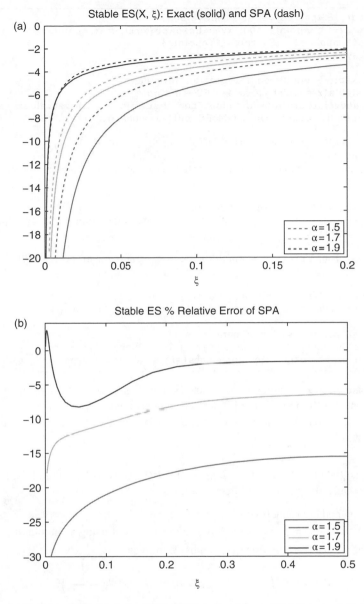

Figure A.6 *(a) The exact expected shortfall (solid lines) and its saddlepoint approximation (dashed lines) as a function of ξ for an $S_{\alpha,0}(0,1)$ random variable (truncated at $\xi = 0.2$ for visibility reasons), and three values of α. (b) The relative percentage error of the s.p.a., shown up to $\xi = 0.5$. The relative percentage error is symmetric about $\xi = 0.5$.*

```
1   function [cdf, ESspa, ES]=SPAstableES(alpha)
2   xmin=-10; xmax=10; npoints=100; xvec=linspace(xmin,xmax,npoints);
3   ESspa=zeros(1,npoints); ES= zeros(1,npoints);
4   [~,cdf] = asymstab(xvec,alpha); cdf=cdf';
5
6   if 1==1 % exact ES via Broda integration formula
7     pm = gp_stable(xvec,alpha,0,0,1); ES=-pm./cdf;
8   else % Stoyanov integration formula. They result in identical values.
9     for i=1:npoints, ES(i)=-asymstableES(cdf(i),alpha,0,0,1,1); end
10  end
11  optim=optimset('largescale','off','disp','none','TolFun',1e-7);
12  for loop=1:npoints
13    [~,ESspa(loop)]=fminunc(@(y1) y1+stablespa(y1,alpha)/cdf(loop), 1,optim);
14  end
15
16  function G=gp_stable(xvec,alpha,beta,mu,sig)
17  G=zeros(size(xvec)); [~,cfp0]=cf(0,alpha,beta,mu,sig); cfp0=1i*imag(cfp0);
18  for loop=1:numel(xvec)
19    x=xvec(loop); G(loop) = cfp0/2/1i + 1/(pi) * ...
20      quadgk(@(s) gpsint4(s,x,alpha,beta,mu,sig),0,inf,'RelTol',1e-10);
21  end
22
23  function I=gpsint4(s,x,alpha,beta,mu,sig)
24  [~,psip]=cf(s,alpha,beta,mu,sig); I=real((exp(-1i*s*x)).*(psip)./s);
25
26  function [psi,psip]=cf(s,alpha,beta,mu,sig)
27  K=-abs(sig*s).^alpha.*(1-1i*beta*sign(s)*tan(pi*alpha/2));
28  psi=exp(K+1i*mu*s); psip=(sig*alpha./abs(sig*s).*K+1i*mu).*psi;
```

Program Listing A.9: Computes the exact ES and its saddlepoint approximation for the stable Paretian distribution with tail index α. The exact calculation uses either (A.122) or (A.311). The functions `asymstableES` and `stablespa` are given in Listings A.7 and A.10, respectively.

    (b) Use the identity $B(a,b) = 2\int_0^{\pi/2}(\sin\theta)^{2a-1}(\cos\theta)^{2b-1}\,d\theta$ and the relationship between the gamma and beta functions.

A.2 (Gamma and beta functions)

    (a) Show that $I = \int_0^2 y^4(2-y)^3\,dy = 32/35$ without using a calculator.

    (b) Prove that the beta function can be written as

$$B(a,b) = \int_0^1 t^{a-1}(1-t)^{b-1}\,dt = \int_0^\infty s^{a-1}\left(\frac{1}{s+1}\right)^{a+b}ds. \qquad (A.314)$$

(Hint: Find the correct integral substitution; note that you want s to be a function of t such that $s(0) = 0$ and $\lim_{t\uparrow 1}s(t) = \infty$.)

    (c) First confirm that

$$\int_0^\infty e^{-mt}t^{z-1}\,dt = \frac{\Gamma(z)}{m^z}. \qquad (A.315)$$

Then, after substituting $m = s+1$ and $z = x+y$ in (A.315), use (A.314) to show that

$$B(x,y) = \frac{\Gamma(x)\Gamma(y)}{\Gamma(x+y)}.$$

```
1    function [Int,cdf,pm]=stablespa(y1vec,alpha,t1vec)
2      n=numel(y1vec);  cdf=zeros(size(y1vec));  Int=cdf;
3      for loop=1:n
4        if nargin<3
5          y1=y1vec(loop);  t1=fzero(@spe,[0+1e-7 sqrt(alpha/2)-1e-7]);
6          t1=abs(t1)*sign(y1);
7        else
8          t1=t1vec(loop);  y1=(2*t1)*exp(psi(1-t1^2)/2-psi(1-2*t1^2/alpha)/alpha);
9          y1=abs(y1)*sign(t1);
10       end
11       y2=y1/(2*t1);  t2=t1^2;  g=[y1/y2;2*log(y2)];  d1g=[1/y2;  0];
12       d2g=[-y1/y2^2;  2/y2];  d2g2=[2*y1/y2^3;-2/y2^2];
13       J=[d1g d2g];  Kpp=[2  0;  0 4*psi(1,1-2*t2/alpha)/alpha^2-psi(1,1-t2)];
14       k=det(Kpp)/det(J)^2*(d2g'*inv(Kpp)*d2g+[t1 t2]*d2g2);  %#ok<MINV>
15       c=[t1 t2]*d1g;  u=c*sqrt(k);  alpha0=0;
16       K=t1^2+gammaln(1-2*t2/alpha)-gammaln(1-t2);
17       w=sign(y1-alpha0)*sqrt(2*([t1 t2]*g-K));
18       cdf(loop)=normcdf(w)+normpdf(w)*(1/w-1/u);
19       Int(loop) = (-y1).*(1-normcdf(w))-normpdf(w).*((-y1) ...
20         .*(1./w)-(-y1).*(1./w.^3)-1./c./u);
21     end
22     pm=Int+y1vec.*(1-cdf);
23
24     function z=spe(t1)
25     z=-2*psi(1-2*t1^2/alpha)/alpha+psi(1-t1^2)-2*log(abs(y1/(2*t1)));
26     end
27   end
```

Program Listing A.10: Called by the function in Listing A.9 to compute the saddlepoint approximation to the ES for the stable Paretian distribution.

A.3 (Sampling)

(a) State (do not prove!) the binomial theorem and Vandermonde's theorem.

(b) Argue why the p.m.f. of a hypergeometric r.v. (A.39), including the indicator function, makes sense. State and prove precisely under what limiting conditions the hypergeometric p.m.f. approaches the binomial.

(c) Derive the Poisson p.m.f. as the limiting case of the binomial and thus state a Poisson approximation to the hypergeometric.

A.4 (Geometric generalization) Simultaneously throw d fair six-sided dice. Those showing a six are set aside and the remaining ones are again simultaneously thrown. This is continued until all d dice show a six. Let N be the required number of throws and denote its expected value by $\mathbb{E}_d[N]$. Observe that N is not a sum of d geometrically distributed r.v.s.

(a) Show that

$$\mathbb{E}_d[N] = \frac{6^d + \sum_{k=1}^{d-1} 5^k \binom{d}{k} \mathbb{E}_k[N]}{6^d - 5^d}$$

for $d = 2, 3, \ldots,$ and $\mathbb{E}_1[N] = 6$.   (Epstein, 1977, p. 141)

(b) Write a computer program to compute $\mathbb{E}_d[N]$ and plot it for a grid of values.

(c) For large d, the recursive program can take a long time to evaluate. Using the previously computed values of $\mathbb{E}_{150}[N]$ through $\mathbb{E}_{200}[N]$, approximate the expectation as a (nonlinear) function of d (using, say, least squares).

(d) While a computable formula for the expectation was obtainable without having to compute the mass function of N, we may not be so lucky if interest centers on higher moments of N. Instead of trying to algebraically determine the mass function, we can easily simulate it. Construct a program to do so and plot a histogram of the resulting values.

(e) It turns out that the mass function of N is not too difficult to determine. Define event A to be {after x throws, all d dice show sixes}, and event B to be {after $x - 1$ throws, at least one dice does not show a six}. Then

$$f_N(x; d) = \Pr(N = x ; d) = \Pr(A \cap B) = \Pr(A) + \Pr(B) - \Pr(A \cup B).$$

(i) Compute $\Pr(A^c \cap B^c)$ and $\Pr(A \cup B)$, which are easy.

(ii) Specify the relation between A^c and B, and the relation between B^c and A.

(iii) Draw a Venn diagram with the events.

(iv) Compute $\Pr(A)$ and $\Pr(B)$, from which $\Pr(N = x ; d)$ follows. Hint: First consider only one die $(d = 1)$ and work with the complement.

(v) Graphically compare your result to simulated values of the mass function.

(f) Using the mass function, a closed-form nonrecursive expression for the expected value can be obtained. In particular, simplify the definition $\mathbb{E}_d[N] = \sum_{x=1}^{\infty} x f_N(x; d)$ to obtain

$$\mathbb{E}_d[N] = \sum_{j=1}^{d} \binom{d}{j} \frac{(-1)^{j+1}}{1 - (5/6)^j}.$$

A.5 A miner is lost in a tunnel with three doors. Behind the first door is a tunnel that, after 3 hours, leads out of the mine. Behind the second and third doors are tunnels which, after 5 and 7 hours, respectively, lead back to the same position. Assume the doors all look the same and cannot be marked, and that the miner chooses randomly among them. How many hours do we expect the miner to require to get out of the mine? (Parzen, 1962, p. 50) Also derive the mass function of the number of hours spent in the mine.

A.6 Let Y be a continuous random variable with density f_Y symmetric about zero and c.d.f. F_Y. Clearly, $F_Y(Y) \sim \text{Unif}(0, 1)$ and $\mathbb{E}[F_Y(Y)] = 0.5$.

(a) Show that

$$\mathbb{E}\left[F_Y\left(\frac{Y}{\sigma}\right)\right] = 0.5, \quad \forall \sigma > 0. \tag{A.316}$$

(b) Show that, with $X = F_Y(Y/\sigma)$,

$$f_X(t; \sigma) = \sigma f_Y(\sigma F_Y^{-1}(t)) \frac{\partial F_Y^{-1}(t)}{\partial t} \mathbb{I}_{(0,1)}(t), \quad \forall \sigma > 0. \tag{A.317}$$

(c) Compare (A.317) graphically with a kernel density of simulated values for Y a Student's t with v degrees of freedom and a positive value of σ, and also plot the best-fitting beta distribution, showing that X is not beta distributed in general.

(d) Setting $\sigma = 1$ in (A.317) and the fact that, trivially, $X = F_Y(Y) \sim \text{Unif}(0,1)$ implies, for Y a continuous random variable with density f_Y symmetric about zero,

$$f_Y(F_Y^{-1}(t))\frac{\partial F_Y^{-1}(t)}{\partial t}\mathbb{1}_{(0,1)}(t) \sim \text{Unif}(0,1). \tag{A.318}$$

Show (A.318) directly.

A.7 Let $X \sim t(n)$, that is, Student's t with n degrees of freedom, the p.d.f. of which is given in (A.70). Consider the case with $n = 2$.
  i. Confirm that, for $X \sim t(2)$,

$$F_X(x) = \frac{1}{2}\left\{1 + \frac{x}{\sqrt{2+x^2}}\right\}, \tag{A.319}$$

  ii. Derive a closed-form expression for the quantile ξ_p, and (iii) compute the second raw moment.

A.8 Let $C \sim \chi_k^2$, with density

$$f(x;k) = \frac{1}{2^{k/2}\Gamma(k/2)}x^{k/2-1}e^{-x/2}\mathbb{1}_{(0,\infty)}(x).$$

  i. Calculate the density of $X \sim \sqrt{C/k}$. Hint: Let $c = kx^2$.
  ii. Simplify for $k = 1$. When $k = 1$, X is referred to as the *folded normal*. Compute the mean of X in this case.

A.9 This exercise develops a closed-form expression for $S = \sum_{j=0}^u ja^j$ in the following two ways..
  (a) Compare S and aS, and simplify.
  (b) Use the fact that

$$\frac{d}{da}\sum_{j=0}^u a^{j+1} = \sum_{j=0}^u (j+1)a^j.$$

A.10 Let $X_1, X_2, \ldots, X_n$ be i.i.d. continuous random variables with support $-\infty \le a < b \le \infty$, and let $Y = \min(X_i)$.
  (a) Argue in words why (A.174), that is,

$$F_Y(y) = 1 - [1 - F_X(y)]^n, \quad \text{for } y \in (a,b),$$

makes sense.
  (b) Now let $X_i \overset{\text{i.i.d.}}{\sim} \text{Unif}(0,1)$, $i = 1,2,\ldots,n$. Use the above formula to derive the c.d.f. of $Z = n \cdot \min(X_i)$.
  (c) Derive the p.d.f. of Z in the limit as $n \to \infty$.

A.11 Let $X \sim \exp(\lambda)$ and $Y = \exp(X)$. Derive the p.d.f. of Y and its mth moment.

A.12 The p.d.f. of the Student's t density with n degrees of freedom is given in (A.70). Show that, as $n \to \infty$, the kernel approaches that of the normal, and also that the constant of integration approaches that of the normal. Hint: Recall Stirling's approximation, $\Gamma(n) \approx \sqrt{2\pi} n^{n-1/2} \exp(-n)$.

A.13 (Expected shortfall) Let Z follow a so-called *double Weibull* distribution, with shape $\beta > 0$ and scale $\sigma > 0$. We write $Z \sim \mathrm{DWeib}(\beta, \sigma)$, where the p.d.f. is given by

$$f_{\mathrm{DWeib}}(z; \beta, \sigma) = f_Z(z; \beta, \sigma) = \frac{\beta}{2\sigma} \left| \frac{z}{\sigma} \right|^{\beta-1} \exp\left(-\left| \frac{z}{\sigma} \right|^{\beta} \right).$$

When $\beta = 1$, Z reduces to a Laplace random variable. From the symmetry of the density, the mean of Z is clearly zero, if it exists, and a location parameter μ could also be introduced in the usual way. One simple method of introducing asymmetry into the double Weibull is to allow for two different shape parameters, depending on the sign of z. We define the *asymmetric double Weibull* density as

$$f_{\mathrm{ADWeib}}(z; \beta^-, \beta^+, \sigma) = \begin{cases} f_{\mathrm{DWeib}}(z; \beta^-, \sigma), & \text{if } z < 0, \\ f_{\mathrm{DWeib}}(z; \beta^+, \sigma), & \text{if } z \geq 0, \end{cases}$$

where β^- and β^+ denote the shape parameters on the left ($z < 0$) and right support ($z \geq 0$), respectively.

Let $Z \sim \mathrm{ADWeib}(\beta^-, \beta^+, 1)$. Show that the c.d.f. is given by

$$F_Z(z; \beta^-, \beta^+, 1) = \begin{cases} \frac{1}{2} \exp(-(-z)^{\beta^-}), & \text{if } z < 0, \\ \frac{1}{2}, & \text{if } z = 0, \\ 1 - \frac{1}{2} \exp(-z^{\beta^+}), & \text{if } z > 0. \end{cases} \tag{A.320}$$

Then show that, for $\mathrm{ES}(Z, \xi)$ with $x = q_{Z,\xi} < 0$, the ES integral is

$$\int_{-\infty}^{x} z f_Z(z) \, \mathrm{d}z = -\frac{1}{2} \left(\Gamma\left(1 + \frac{1}{b}\right) - \Gamma_{(-x)^b}\left(1 + \frac{1}{b}\right) \right), \tag{A.321}$$

where $\Gamma_v(a) = \int_0^v x^{a-1} e^{-x} \, \mathrm{d}x$ is the incomplete gamma function (A.9).

A.14 For any random variable X, prove that the finiteness of $\mathbb{E}[X]$ is equivalent to finiteness of $\mathbb{E}[|X|]$. Note that finiteness of $\mathbb{E}[X]$ means that $|\mathbb{E}[X]| < \infty$. (The condition $\mathbb{E}[X] < \infty$ is not enough!) We thus wish to show that

$$|\mathbb{E}[X]| < \infty \Leftrightarrow \mathbb{E}[|X|] < \infty.$$

Hint: Let $X^+ = X\mathbb{I}(X \geq 0)$ and $X^- = -X\mathbb{I}(X < 0)$, so that both X^+ and X^- are nonnegative, $X = X^+ - X^-$, and $|X| = X^+ + X^-$. Recall the triangle inequality: for all a and b, $|a + b| \leq |a| + |b|$, which is equivalent to $|a - b| \leq |a| + |b|$.

A.15 For $m > 2$, let $B \sim \mathrm{Beta}((m-1)/2, (m-1)/2)$ independent of $X \sim \chi_m^2$ with p.d.f.

$$f_X(x; m) = \frac{1}{2^{m/2} \Gamma(m/2)} x^{m/2-1} e^{-x/2} \mathbb{I}_{(0,\infty)}(x).$$

Let $S = 2B - 1$, $Y = \sqrt{X}$, and $P = SY$. First, derive an integral expression for $f_P(p; m)$ and, for $m = 3$, simplify it and show that, for $m = 3$, $P \sim \mathrm{N}(0, 1)$. Next, verify using numeric integration that P is $N(0, 1)$ for *any* value of $m > 2$. This is proven in Ellison (1964, p. 90).

A.16 Let $X_i \overset{\text{i.i.d.}}{\sim} \text{Unif}(0, 1)$, $i = 1, \ldots n$, let s and t be values such that $0 < s < t < 1$, and define $N_n(s) = \sum_{i=1}^{n} \mathbb{I}_{[0,s]}(X_i)$, that is, $N_n(s)$ is the number of X_i that are less than or equal to s. Let $X = N_n(s)$, $Y = N_n(t)$, and $D = Y - X$. Prove that

$$(D \mid X = x) \sim \text{Bin}(n - x, (t - s)/1 - s).$$

Hint: What are $f_{X,Y}$ and f_X? From those, perform a bivariate transformation to get $f_{D,M}$ with $M = X$, and then compute $f_{D\mid M}$.

A.17 From Section A.7, recall the definitions of the m.g.f., c.g.f., c.f., and the inversion formula (A.99) for X a continuous r.v. with p.d.f. f_X. Consider an r.v. X such that its m.g.f. exists, and also $\varphi_X(t) = \mathbb{M}_X(it)$.

(a) (Problem II.1.12)

  (i) Derive the expression

$$f_X(x) = \frac{1}{2\pi i} \int_{-i\infty}^{i\infty} \exp\{\mathbb{K}_X(s) - sx\}\, ds. \qquad \text{(A.322)}$$

  (ii) The integral in (A.322) can also be expressed as

$$f_X(x) = \frac{1}{2\pi i} \int_{c-i\infty}^{c+i\infty} \exp\{\mathbb{K}_X(s) - sx\}\, ds, \qquad \text{(A.323)}$$

where c is an arbitrary real constant in the convergence strip of $\mathbb{K}_X$. This is called the Fourier–Mellin integral and is a standard result in the theory of Laplace transforms; see, for example, Schiff (1999, Ch. 4). Write out the double integral for the survivor function

$$\bar{F}_X(x) = 1 - F_X(x) = \int_x^{\infty} f_X(y)\, dy,$$

interchange the order of the integrals, and confirm that c needs to be positive for this to produce the valid integral expression

$$\bar{F}_X(x) = \frac{1}{2\pi i} \int_{c-i\infty}^{c+i\infty} \exp\{\mathbb{K}_X(s) - sx\}\, \frac{ds}{s}, \qquad c > 0. \qquad \text{(A.324)}$$

Similarly, derive the expression for the c.d.f. given by

$$F_X(x) = -\frac{1}{2\pi i} \int_{c-i\infty}^{c+i\infty} \exp\{\mathbb{K}_X(s) - sx\}\, \frac{ds}{s}, \qquad c < 0. \qquad \text{(A.325)}$$

(b) Let X be a continuous r.v. with support S_X, c.f. φ_X, and m.g.f. $\mathbb{M}_X$ related by $\varphi_X(s) = \mathbb{M}_X(is)$. We wish to derive an expression for $I_X(q; p) = \int_{-\infty}^{q} x^p f_X(x)\, dx$, $p = 0, 1, 2$. Substitute expression (A.323) into I_X, reverse the order of the integrals, and, for the inner integral for $p = 1$, apply integration by parts. The result for $p = 1$ is (A.121).

A.18 Let $X \sim N(\mu, 1)$ independent of $Y \sim \chi^2(m)$. Recall that $T = X/\sqrt{Y/m}$ is singly non-central t with degrees of freedom m and noncentrality parameter μ, written $T \sim t'(m, \mu)$.

(a) As in Ellison (1964, p. 92), let $X \sim N(\mu, \sigma^2)$ independent of $Y \sim \sqrt{\chi_m^2/m}$. Show that

$$\forall c \in \mathbb{R}, \quad \mathbb{E}[\Phi(X + cY)] = F_T\left(\frac{c}{\sqrt{1 + \sigma^2}}; m, \delta\right), \quad \delta = -\frac{\mu}{\sqrt{1 + \sigma^2}}, \tag{A.326}$$

where Φ is the standard normal c.d.f., and $T \sim t'(m, \delta)$.

Hint: Let $Z \sim N(0, 1)$ independent of X and Y and recall the law of total probability, which states that, for event A,

$$\Pr(A) = \int \Pr(A \mid B = b) f_B(b) \, db = \mathbb{E}_B[\Pr(A \mid B = b)].$$

Use this to justify that $\mathbb{E}[\Phi(X + cY)] = \Pr(Z \leq X + cY)$.

(b) There are two special cases of (A.326) of interest.

(i) Let X be degenerate at zero, so taking $\mu = \sigma = 0$ in (A.326) gives

$$\mathbb{E}[\Phi(cY)] = F_T(c; m, 0) = \Phi_m(c), \tag{A.327}$$

where $\Phi_m(c)$ denotes the c.d.f. of the Student's t distribution with m degrees of freedom, evaluated at c.

(ii) Show that taking $c = 0$ in (A.326) yields the result:

$$\text{If } X \sim N(\mu, \sigma^2), \text{ then } \mathbb{E}[\Phi(X)] = \Phi\left(\frac{\mu}{\sqrt{1 + \sigma^2}}\right). \tag{A.328}$$

(c) Recall the skew normal distribution, with density given in (A.115): $f_{\text{SN}}(z; \lambda) = 2\phi(z)\Phi(\lambda z)$. Section II.7.1.2 showed that $\int_{\mathbb{R}} f_{\text{SN}}(z; \lambda) \, dz = 1$. Show (A.117), that is, that the m.g.f. of a standard skew normal random variable X is

$$\mathbb{M}_X(t) = 2\exp\{t^2/2\}\Phi(t\delta), \quad \delta = \frac{\lambda}{\sqrt{1 + \lambda^2}}, \tag{A.329}$$

and use this to show

$$\mathbb{E}[X] = \delta\sqrt{2/\pi} \quad \text{and} \quad \mathbb{V}(X) = 1 - 2\delta^2/\pi. \tag{A.330}$$

A.19 We again consider the skew normal distribution, with density given in (A.115).

(a) The characteristic function (c.f.) $\varphi_X(t)$ of $X \sim \text{SN}(\lambda)$ was not given by Azzalini (1985). Formally evaluating $\mathbb{M}_X(it)$ using (A.117) is not valid, as the normal c.d.f. for complex arguments is not defined. However, the reader should verify that the normal c.d.f. can be expressed as

$$\Phi(x) = \frac{1}{2} + \frac{1}{2}\text{erf}\left(\frac{x}{\sqrt{2}}\right) = \frac{1}{2}\text{erfc}\left(-\frac{x}{\sqrt{2}}\right), \tag{A.331}$$

where erf is the so-called *error function*, and erfc is the *complementary error function*, given by

$$\text{erf}(x) = \frac{1}{\sqrt{\pi}} \int_{-x}^{x} \exp\{-t^2\}\, dt = \frac{2}{\sqrt{\pi}} \int_{0}^{x} \exp\{-t^2\}\, dt, \qquad (\text{A.332})$$

and

$$\text{erfc}(x) = 1 - \text{erf}(x) = \frac{2}{\sqrt{\pi}} \int_{x}^{\infty} \exp\{-t^2\}\, dt,$$

respectively. Based on (A.332), one can take the *imaginary error function* to be

$$\text{erfi}(z) = -i\,\text{erf}(iz), \quad z \in \mathbb{C} \backslash \mathbb{R}, \qquad (\text{A.333})$$

where $\mathbb{C}\backslash\mathbb{R}$ is the set of purely complex numbers. However, from (A.332) and (A.333), we have, and take as the definition,

$$\text{erfi}(z) := \frac{2}{\sqrt{\pi}} \int_{0}^{z} \exp\{t^2\}\, dt = \frac{2}{\sqrt{\pi}} \sum_{n=0}^{\infty} \frac{z^{2n+1}}{n!(2n+1)}, \quad z \in \mathbb{C}, \qquad (\text{A.334})$$

noting that the integral and infinite sum representations are valid for any complex number, and thus also strictly real numbers. (The erfi function is thus an *entire function* with *infinite growth*; see, for example, Bak and Newman, 2010, as well as the Wikipedia entry on erfi.)

Thus, using (A.117), (A.331), (A.333), and that $1/i = -i$ from (A.123), the c.f. of X can be evaluated as

$$\varphi_X(t) = \mathbb{M}_X(it) = 2 \exp\left\{-\frac{t^2}{2}\right\} \left[\frac{1}{2} + \frac{1}{2}\text{erf}\left(\frac{it\delta}{\sqrt{2}}\right)\right]$$

$$= \exp\left\{-\frac{t^2}{2}\right\} \left[1 + i\,\text{erfi}\left(\frac{t\delta}{\sqrt{2}}\right)\right], \qquad (\text{A.335})$$

with a similar expression also given in Pewsey (2000).

Another function associated with the error function is the *scaled complementary error function*,

$$\text{erfcx}(z) := \exp\{z^2\}\text{erfc}(z) = w(iz), \qquad (\text{A.336})$$

where w is the *Faddeeva* or *Kramp function*,

$$w(z) := \exp\{-z^2\}\text{erfc}(-iz) = \text{erfcx}(-iz)$$

$$= \exp\{-z^2\}\left(1 + i\frac{2}{\sqrt{\pi}} \int_{0}^{z} \exp\{t^2\}\, dt\right). \qquad (\text{A.337})$$

Observe how (A.335) cannot be used directly in conjunction with inversion formulas (A.104) and (A.108) to evaluate the p.d.f. and c.d.f. of a skew normal random variable, because the argument of the function erfi in (A.335), $t\delta/\sqrt{2}$, is real, and as $t \to \infty$, as required by (A.104) and (A.108), $\text{erfi}(t\delta/\sqrt{2})$ will (quickly) explode towards infinity. However, an idea is to truncate the upper integral limits in (A.104) and (A.108), which is valid because (i) the leading term

$\exp\{-t^2/2\}$ in (A.335) counteracts the growth of the erfi term; (ii) the integrals (A.104) and (A.108) exist and are thus convergent (see, for example, (I.A.67)); and (iii) we assume suitable conditions on the integrand such that the bounded convergence theorem applies (see, for example, (I.A.118) and Example I.A.58).

(i) The reader is to confirm this idea by programming the inversion formulas for the p.d.f. and c.d.f., for a standard skew normal random variable, using (A.335) and a suitable cutoff for the upper integral limit of the inversion formulas. The next issue concerns the evaluation of, say, the integral expression in (A.334), as required in (A.335). This implies nested integration for evaluation of the p.d.f. and c.d.f. via the inversion formula, and thus would be extraordinarily slow. However, fast numeric methods for erfi are available: In Matlab, as of release R2013A, there is the function `erfi`. Use it.

(ii) Use of (A.336) is beneficial because it is a scaled version of erfc, and does not explode as does erfi. Formulate the c.f. such that it uses erfcx. With its use, the upper limit in the integrals of the inversion formulas can be taken to be infinity, thus delivering the p.d.f. and c.d.f. to near machine precision (and useful for sums of independent skew normal r.v.s, for which there is no simple exact density or easily evaluated c.d.f. expression). Unfortunately, the function erfcx is not built into Matlab, but there is a package available for it using open-source C++ code (which needs to be linked with Matlab); see http://ab-initio.mit.edu/wiki/index.php/Faddeeva_Package. Based on that code, the function erfcx is built into the (free) Julia programming language – whose syntax is greatly inspired by that of Matlab and thus easy to get started with for Matlab users. It includes, for example, the numerical integration function `quadgk`, as we use in Matlab for evaluating the inversion formulas. The relevant Julia code is given in the solutions (and runs much faster than Matlab).[4]

(iii) Write a program to compute the p.d.f. and c.d.f. of sums of independent skew normal r.v.s via its c.f., for passed vectors for the location, scale, and asymmetry parameter.

(iv) Use (A.122), and the fact that

$$\frac{d}{dz}\mathrm{erfi}(z) = \frac{2}{\sqrt{\pi}}\exp\{z^2\}, \qquad (A.338)$$

to construct a program that evaluates the ES of a sum of independent SN r.v.s.

(b) From the c.g.f. based on (A.117), we can derive the saddlepoint approximation (s.p.a.) to the p.d.f. and c.d.f. of X. The saddlepoint equation is $x = \mathbb{K}'_X(t)$ and needs to be obtained numerically. The computation of $R(z) = \phi(z)/\Phi(z)$ becomes numerically problematic for $z < -36$, thus causing the saddlepoint root search to fail.[5] However, as $z \to -\infty$, $R(z) \approx -z$, or even more accurately (Feller, 1968, pp. 175, 193), $1/R(z) \approx z^{-3} - z^{-1}$. We use this approximation for $z < -36$. Note that if $\lambda = 0$, then $\delta = 0$, and the quantities reduce to $\mathbb{K}_X(t) = t^2/2$, $\mathbb{K}'_X(t) = t$, and

---

[4] The author is grateful to Simon Broda for pointing this out and writing the Julia code.
[5] In Matlab, $R(-38) = $ Inf and for more extreme values is NaN.

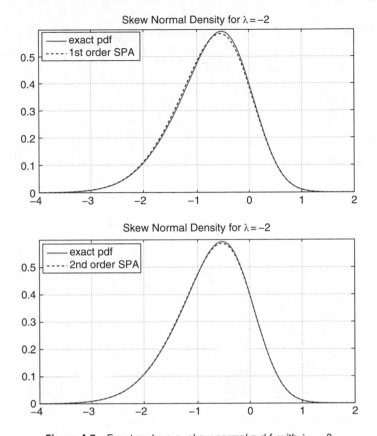

Figure A.7 *Exact and s.p.a. skew normal p.d.f. with $\lambda = -2$.*

$\mathbb{K}_X''(t) = 1$, as they should. In this case the saddlepoint approximation is exact, so we expect its accuracy to increase as $|\lambda| \to 0$.

The top panel of Figure A.7 shows the exact p.d.f. and first-order s.p.a. for $\lambda = -2$.

The second-order saddlepoint approximation requires $\mathbb{K}_X^{(3)}(t)$ and $\mathbb{K}_X^{(4)}(t)$. With $R = R(t\delta)$, these are

$$\mathbb{K}_X^{(3)}(t) = -\delta^3 R - \delta^3 tR' - 2\delta^2 RR',$$

$$\mathbb{K}_X^{(4)}(t) = -\delta^3 (tR'' + 2R') - 2\delta^2 (RR'' + (R')^2),$$

where $R''(t\delta) = -\delta^2 (R + tR') - 2\delta RR'$. The improvement in the p.d.f. approximation is noticeable, as depicted in the bottom panel of Figure A.7. Derive and/or confirm the expressions for $\mathbb{K}_X^{(i)}(t)$, program the s.p.a., and replicate the results in Figure A.7.

A.20 We wish to develop a saddlepoint approximation to the sum of independent skew normal random variables. Let $X = \sum_{i=1}^n X_i$ and $X_i \overset{\text{indep}}{\sim} \text{SN}(\lambda_i, \mu_i, \sigma_i), i = 1, \ldots, n$, where μ_i and σ_i are location and scale parameters, respectively, and let $\delta_i = \lambda_i / \sqrt{1 + \lambda_i^2}$.

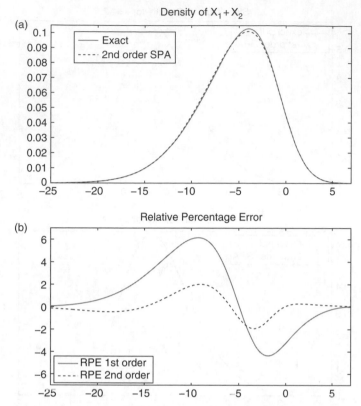

Figure A.8   *(a) The true (solid) and second-order s.p.a. (dashed) of the convolution of two indepen-dent skew normal r.v.s with $\lambda_1 = 1$, $\mu_1 = -4$, $\sigma_1 = 2$, and $\lambda_2 = -4$, $\mu_2 = 2$, and $\sigma_2 = 6$. (b) The relative percentage error of the first (solid) and second-order (dashed) s.p.a.*

Use the usual convolution formula (A.146) or inversion of the c.f. via (A.335) to get the exact density for two components. For illustration, we use $\lambda_1 = 1$, $\mu_1 = -4$, $\sigma_1 = 2$, and $\lambda_2 = -4$, $\mu_2 = 2$, $\sigma_2 = 6$. The top panel of Figure A.8 shows the exact and s.p.a. p.d.f., while the bottom panel shows the relative percentage error from the first and second order s.p.a.s. Make a program to compute the exact convolution if $n = 2$, and the first- and second-order s.p.a. for any n. Reproduce Figure A.8.

To get a sense of the relative computing times, for the grid of 320 values shown in the figure, the s.p.a. took 1.05 seconds to compute, while the convolution required 228 seconds, using a moderate tolerance level for the integration. Critically, observe that the time required for more than $n = 2$ components barely changes with the s.p.a., while nested integration for calculating the exact density via convolution becomes infeasible, and inversion of the c.f. is required, as discussed in Problem A.19.

A.21 Recall the definition of the (singly) noncentral t from Section A.14 and Problem A.18. Let $T \sim t'(k, \mu)$ be singly noncentral t. From Chapter II.10.4 we know that the p.d.f. can be expressed as

$$f_T(t; k, \mu) = K \int_0^\infty z^k \exp\left\{ -\frac{1}{2}[(tz - \mu)^2 + kz^2] \right\} \, dz, \quad K = \frac{2^{-k/2+1}k^{k/2}}{\Gamma(k/2)\sqrt{2\pi}},$$

$$(A.339)$$

and the c.d.f. as (A.219). There exists the following interesting relationship between the p.d.f. and c.d.f. of T:

$$f_T(t; k; \mu) = \frac{k}{t} \left[F_T \left(t\sqrt{\frac{k+2}{k}}; k+2, \mu \right) - F_T(t; k, \mu) \right], \quad t \neq 0, \quad \text{(A.340)}$$

or

$$f_T(t; k; \mu) = -\frac{k}{t} \left[F_T \left(-t\sqrt{\frac{k+2}{k}}; k+2, -\mu \right) - F_T(-t; k, -\mu) \right], \quad t \neq 0.$$

$$\text{(A.341)}$$

First confirm these numerically. (In Matlab, use `nctpdf` and `nctcdf`.) Then prove them.

A.22 In light of how a Student's t random variate is defined, it seems natural to define a skew Student's t random variable X as $X = S/Y$, where $S \sim \text{SN}(\lambda)$ and $Y = \sqrt{\chi_m^2/m}$. This is indeed the univariate version of the multivariate structure proposed and studied in Azzalini and Capitanio (2003). Derive the p.d.f.

$$f_X(x; \lambda, m) = 2\phi_m(x)\Phi_{m+1} \left(\lambda x \sqrt{\frac{m+1}{x^2 + m}} \right), \quad \text{(A.342)}$$

where $\phi_m(x)$ and $\Phi_m(x)$ denote the p.d.f. and c.d.f., respectively, of the Student's t distribution at x with m degrees of freedom, following Azzalini and Capitanio (2003), who state that "application of [(A.327)] ... and some simple algebra lead to the density." Readers interested in yet another asymmetric Student's t distribution can see Rosco et al. (2011).

A.23 There is a substantial literature on the use of skewness in asset pricing and portfolio optimization.[6] The concept of *co-skewness* is central to the discussion. Recall that, if $\mathbf{R} = (R_1, \ldots, R_n)'$ denotes the vector of n individual asset returns, then $\mathbb{E}[\mathbf{R}]$ is the n-length vector of means with ith element $\mu_i := \mathbb{E}[R_i]$, and $\text{Var}(\mathbf{R})$ is the $n \times n$ matrix of covariances with (i,j)th element

$$\sigma_{ij} = \text{Cov}\,(R_i, R_j) = \mathbb{E}[(R_i - \mu_i)(R_j - \mu_j)]$$

and $\sigma_i^2 = \sigma_{ii}$. For $n = 3$,

$$\boldsymbol{\Sigma} = \text{Var}(\mathbf{R}) = \mathbb{E}[(\mathbf{R} - \boldsymbol{\mu})(\mathbf{R} - \boldsymbol{\mu})'] = \begin{bmatrix} \sigma_1^2 & \sigma_{12} & \sigma_{13} \\ \sigma_{12} & \sigma_2^2 & \sigma_{23} \\ \sigma_{13} & \sigma_{23} & \sigma_3^2 \end{bmatrix}.$$

Let $\mathbf{a}$ be a vector of portfolio weights. The variance of portfolio $P = \mathbf{a}'\mathbf{R}$ is $\text{Var}(P) = \mathbf{a}'\boldsymbol{\Sigma}\mathbf{a}$.

The extension of the variance to third-order cross products, that is, the co-skewness, would involve an $n \times n \times n$ cube, with (i,j,k)th element $s_{ijk} = \mathbb{E}[(R_i - \mu_i)(R_j - \mu_j)(R_k - \mu_k)]$. It is more far more useful to work with a matrix version of this expression obtained by placing the n matrices of size $n \times n$ side by side in a particular, agreed-upon order. This is denoted by $\mathbf{M}_3$ (M is for moment,

---

[6] See Barone-Adesi et al. (2004), Vanden (2006), Smith (2007), Jondeau et al. (2007, Sec. 2.5), Martellini and Ziemann (2010), and the references therein.

and 3 indicates the order; $\mathbf{M}_4$ is used for kurtosis, etc.) and given by

$$\mathbf{M}_3 = \mathbb{E}[(\mathbf{R} - \boldsymbol{\mu})(\mathbf{R} - \boldsymbol{\mu})' \otimes (\mathbf{R} - \boldsymbol{\mu})'].$$

Recall: Let $\mathbf{A} = (a_{ij})$ be an $m \times n$ matrix and $\mathbf{B}$ a $p \times q$ matrix. Then the (right) Kronecker product of $\mathbf{A}$ and $\mathbf{B}$ is the $mp \times nq$ block matrix given by

$$\mathbf{A} \otimes \mathbf{B} = \begin{bmatrix} a_{11}\mathbf{B} & \cdots & a_{1n}\mathbf{B} \\ \vdots & & \vdots \\ a_{m1}\mathbf{B} & \cdots & a_{mn}\mathbf{B} \end{bmatrix}.$$

This is known as tensor notation and is used throughout the literature. Note $(\mathbf{R} - \boldsymbol{\mu})(\mathbf{R} - \boldsymbol{\mu})'$ is $n \times n$ and $(\mathbf{R} - \boldsymbol{\mu})(\mathbf{R} - \boldsymbol{\mu})' \otimes (\mathbf{R} - \boldsymbol{\mu})'$ is $n \times n^2$.

To illustrate, for $n = 4$, with $s_{ijk} = \mathbb{E}[(R_i - \mu_i)(R_j - \mu_j)(R_k - \mu_k)]$, $\mathbf{M}_3$ is given by

$$\mathbf{M}_3 = \mathbb{E}[(\mathbf{R} - \boldsymbol{\mu})(\mathbf{R} - \boldsymbol{\mu})' \otimes (\mathbf{R} - \boldsymbol{\mu})']$$

$$= \begin{bmatrix} S_{1jk} \mid S_{2jk} \mid S_{3jk} \mid S_{4jk} \end{bmatrix}, \quad S_{ijk} = \begin{bmatrix} s_{i11} & s_{i12} & s_{i13} & s_{i14} \\ s_{i21} & s_{i22} & s_{i23} & s_{i24} \\ s_{i31} & s_{i32} & s_{i33} & s_{i34} \\ s_{i41} & s_{i42} & s_{i43} & s_{i44} \end{bmatrix}.$$

Writing out the entire $\mathbf{M}_3$ matrix but just including the indices (leaving off the s) and marking the i index in bold, we have

$$\begin{bmatrix} \mathbf{1}11 & \mathbf{1}12 & \mathbf{1}13 & \mathbf{1}14 & \mathbf{2}11 & \mathbf{2}12 & \mathbf{2}13 & \mathbf{2}14 & \mathbf{3}11 & \mathbf{3}12 & \mathbf{3}13 & \mathbf{3}14 & \mathbf{4}11 & \mathbf{4}12 & \mathbf{4}13 & \mathbf{4}14 \\ \mathbf{1}21 & \mathbf{1}22 & \mathbf{1}23 & \mathbf{1}24 & \mathbf{2}21 & \mathbf{2}22 & \mathbf{2}23 & \mathbf{2}24 & \mathbf{3}21 & \mathbf{3}22 & \mathbf{3}23 & \mathbf{3}24 & \mathbf{4}21 & \mathbf{4}22 & \mathbf{4}23 & \mathbf{4}24 \\ \mathbf{1}31 & \mathbf{1}32 & \mathbf{1}33 & \mathbf{1}34 & \mathbf{2}31 & \mathbf{2}32 & \mathbf{2}33 & \mathbf{2}34 & \mathbf{3}31 & \mathbf{3}32 & \mathbf{3}33 & \mathbf{3}34 & \mathbf{4}31 & \mathbf{4}32 & \mathbf{4}33 & \mathbf{4}34 \\ \mathbf{1}41 & \mathbf{1}42 & \mathbf{1}43 & \mathbf{1}44 & \mathbf{2}41 & \mathbf{2}42 & \mathbf{2}43 & \mathbf{2}44 & \mathbf{3}41 & \mathbf{3}42 & \mathbf{3}43 & \mathbf{3}44 & \mathbf{4}41 & \mathbf{4}42 & \mathbf{4}43 & \mathbf{4}44 \end{bmatrix}.$$

The skewness of the portfolio $\mathbf{a}'\mathbf{R}$ is $s_P^3 = \mathbf{a}'\mathbf{M}_3(\mathbf{a} \otimes \mathbf{a})$. To confirm this for $n = 2$, we have $P = \mathbf{a}'\mathbf{R} = a_1 R_1 + a_2 R_2$, and, assuming $E[\mathbf{R}] = \mathbf{0}$ for simplicity, we have

$$\text{Skew}(P) = \text{Skew}(a_1 R_1 + a_2 R_2) = \mathbb{E}[(a_1 R_1 + a_2 R_2)^3]$$

$$= \mathbb{E}[R_1^3 a_1^3 + R_2^3 a_2^3 + 3 R_1 R_2^2 a_1 a_2^2 + 3 R_1^2 R_2 a_1^2 a_2]$$

$$= a_1^3 s_{111} + a_2^3 s_{222} + 3 a_1 a_2^2 s_{122} + 3 a_1^2 a_2 s_{112}.$$

We see this is indeed equal to

$$s_P^3 = \mathbf{a}'\mathbf{M}_3(\mathbf{a} \otimes \mathbf{a}) = \begin{pmatrix} a_1 & a_2 \end{pmatrix} \begin{pmatrix} s_{111} & s_{112} & s_{211} & s_{212} \\ s_{121} & s_{122} & s_{221} & s_{222} \end{pmatrix} \begin{pmatrix} a_1 a_1 \\ a_1 a_2 \\ a_2 a_1 \\ a_2 a_2 \end{pmatrix}$$

$$= a_1^3 s_{111} + a_2^3 s_{222} + a_1^2 a_2 s_{112} + a_1^2 a_2 s_{121} + a_1^2 a_2 s_{211} + a_1 a_2^2 s_{122} + a_1 a_2^2 s_{212}$$

$$+ a_1 a_2^2 s_{221}$$

$$= a_1^3 s_{111} + a_2^3 s_{222} + 3 a_1^2 a_2 s_{112} + 3 a_1 a_2^2 s_{122}.$$

Recall that Var($\mathbf{R}$) is symmetric, and so there are not n^2 unique elements. Without the diagonal, there are $n^2 - n$ elements, of which half are unique. Adding the diagonal back on gives

$$\frac{n^2 - n}{2} + n = \frac{n^2 + n}{2} = \frac{(n+1)n}{2} = \binom{n+1}{2}$$

unique elements. Another way to see this is to observe that the unique elements are on the diagonal and upper diagonal part of the matrix, of which there are $n + (n-1) + (n-2) + \cdots + 1 = (n+1)n/2$ elements.

We wish to determine the number of unique elements in $\mathbf{M}_3$ and $\mathbf{M}_4$.

(a) Working with the $n = 4$ case for illustration, observe that S_{1jk} has, like the covariance matrix, $\binom{n+1}{2} = \binom{5}{2} = 10$ unique elements. Now consider S_{2jk}. We would expect its first row (for which $j = 1$) and first column (for which $k = 1$) to have redundancies with elements in S_{1jk} (for which $i = 1$). (Of course, S_{ijk} is a symmetric matrix, so its jth row is just the transpose of the jth column.) Inspection shows that the first row of S_{2jk} is indeed equivalent to the second row of S_{1jk}. We see this clearly in the following expression for $\left[S_{1jk} \mid S_{2jk} \right]$:

111	112	113	114	211	212	213	214
121	122	123	124	221	222	223	224
131	132	133	134	231	232	233	234
141	142	143	144	241	242	243	244

Thus, from S_{2jk}, we only need to consider the principle submatrix obtained by "striking out" the first row and column. There are six unique elements here, or $\binom{n}{2}$. Continuing with S_{3jk}, we would expect its first row (and column) (for which $j = 1$ or $k = 1$) to have redundancies with elements in S_{1jk} (for which $i = 1$). In particular, the first row (and column) of S_{3jk} should be equivalent to the third row (and column) of S_{1jk}. Likewise, we expect the second row (and column) of S_{3jk} to be equivalent to the third row (and column) of S_{2jk}. We mark these in the following expression for $\left[S_{1jk} \quad S_{2jk} \quad S_{3jk} \right]$:

111	112	113	114	211	212	213	214	311	312	313	314
121	122	123	124	221	222	223	224	321	322	323	324
131	132	133	134	231	232	233	234	331	332	333	334
141	142	143	144	241	242	243	244	341	342	343	344

Thus, from S_{3jk}, we only need to consider the principle submatrix obtained by striking out the first two rows and columns, leaving three unique elements, or $\binom{n-1}{2}$. Continuing this argument, we see that $\mathbf{M}_3$ for $n = 4$ contains

$$\binom{n+1}{2} + \binom{n}{2} + \binom{n-1}{2} + \binom{n-2}{2} = \binom{5}{2} + \binom{4}{2} + \binom{3}{2} + \binom{2}{2}$$

$$= 10 + 6 + 3 + 1 = 20 = \binom{n+2}{3}$$

unique elements.

Apply the same argument to the general n case.

(b) Another way to determine the number of unique elements in $\mathbf{M}_3$ is as follows: First, consider again the variance–covariance matrix $\mathbb{V}(\mathbf{R})$. Above, we saw that the total number of unique elements in $\mathbb{V}(\mathbf{R})$ can be determined by noting that the unique elements are on the diagonal and upper (or lower) diagonal part of the matrix, of which there are $n + (n-1) + (n-2) + \cdots + 1 = (n+1)n/2$ elements.

What we have essentially done is to say that, for each of $i = 1, 2, \ldots, n$ rows, we want to count those elements in the jth column for which $j \leq i$. Thus, we want

$$\sum_{i=1}^{n} \sum_{j=1}^{i} 1 = \sum_{i=1}^{n} i = \frac{n(n+1)}{2}.$$

Apply this method to $\mathbf{M}_3$.

(c) The co-kurtosis matrix, $\mathbf{M}_4$, extends $\mathbf{M}_3$ in a natural way, and is defined as

$$\mathbf{M}_4 = \mathbb{E}[(R - \mu)(R - \mu)' \otimes (R - \mu)' \otimes (R - \mu)'],$$

with elements

$$\kappa_{ijk\ell} = \mathbb{E}[(R_i - \mu_i)(R_j - \mu_j)(R_k - \mu_k)(R_\ell - \mu_\ell)].$$

Determine the number of unique elements.

(d) Write a computer program that calculates the $\mathbf{M}_3$ matrix taking the redundancies into account, and for now, just fill the (i, j, k)th entry with $100i + 10j + k$. It should also confirm that the number of non-redundant terms to compute is $\binom{n+2}{3}$.

A.24 Let X and Y denote possibly multivariate random variables. Prove

$$\mathbb{E}[g(X)] = \mathbb{E}[g(Y)] \quad \Leftrightarrow \quad X \overset{d}{=} Y, \tag{A.343}$$

for all bounded, continuous functions g. Hint: Clearly, "half" of the proof (the $\Leftarrow$ part) is easy. To prove $\Rightarrow$, start with the univariate case, express the c.d.f. of X as $F_x(z) = \Pr(X \leq z) = \mathbb{E}[\mathbb{I}\{X \leq z\}]$, and, for fixed real number z and $\epsilon > 0$, define

$$g_\epsilon(t) = g_\epsilon(t; z) = \begin{cases} 1, & \text{if } t \leq z, \\ 1 - (t-z)/\epsilon, & \text{if } z < t < z + \epsilon, \\ 0, & \text{if } t \geq z + \epsilon. \end{cases}$$

Show that $\mathbb{E}[g_\epsilon(X)] = \frac{1}{\epsilon} \int_z^{z+\epsilon} F_X(x)\, dx$,

$$F_Y(z) - F_X(z) = \frac{1}{\epsilon} \int_z^{z+\epsilon} [F_X(x) - F_X(z)]\, dx - \frac{1}{\epsilon} \int_z^{z+\epsilon} [F_Y(y) - F_Y(z)]\, dy,$$

and apply the triangle inequality.

A.18 SOLUTIONS

A.1 (Proving that $\Gamma(1/2) = \sqrt{\pi}$)
  (a) From (I.1.39),

$$\Gamma(a) = 2 \int_0^\infty u^{2a-1} e^{-u^2}\, du$$

and with $v = \sqrt{2}u$, $u = 2^{-1/2}v$, $du = 2^{-1/2}\,dv$, we have

$$\Gamma(a) = 2 \int_0^\infty (2^{-1/2}v)^{2a-1} \exp(-2^{-1}v^2) 2^{-1/2}\, dv$$

$$= 2^{1-a} \int_0^\infty v^{2a-1} \exp\left(-\frac{1}{2}v^2\right) dv,$$

Using the result in Example I.A.79 that

$$2 \int_0^\infty \exp\left(-\frac{1}{2}v^2\right) dv = \sqrt{2\pi},$$

we have, for $a = 1/2$,

$$\Gamma\left(\frac{1}{2}\right) = \sqrt{2} \int_0^\infty \exp\left(-\frac{1}{2}v^2\right) dv = \frac{\sqrt{2}}{2} 2 \int_0^\infty \exp\left(-\frac{1}{2}v^2\right) dv$$

$$= \frac{\sqrt{2}}{2} \sqrt{2\pi} = \sqrt{\pi}.$$

  (b) We have

$$2 \int_0^{\pi/2} (\sin \theta)^{2a-1} (\cos \theta)^{2b-1}\, d\theta = \frac{\Gamma(a)\Gamma(b)}{\Gamma(a+b)},$$

and with $a = b = 1/2$,

$$2 \int_0^{\pi/2} d\theta = \frac{\Gamma(1/2)\Gamma(1/2)}{\Gamma(1)},$$

or $\Gamma(1/2) = \sqrt{\pi}$, as was to be shown.

A.2 (Gamma and beta functions)
  (a) Let $x = y/2$, so that $y = 2x$ and $dy = 2\,dx$. Then

$$I = 2 \int_0^1 (2x)^4 (2 - 2x)^3\, dx = 2^8 \int_0^1 x^4 (1 - x)^3\, dx$$

$$= 2^8 \frac{\Gamma(5)\Gamma(4)}{\Gamma(9)} = 2^8 \frac{4 \cdot 3 \cdot 2 \cdot 3 \cdot 2}{8 \cdot 7 \cdot 6 \cdot 5 \cdot 4 \cdot 3 \cdot 2} = \frac{2^5}{7 \cdot 5} = \frac{32}{35}.$$

(b) Using $s = t/(1 - t)$, so that $t = s/(s + 1)$ and $\mathrm{d}t / \mathrm{d}s = (s + 1)^{-2}$, the integral easily follows.

(c) (Lebedev, 1972, p. 13) The left-hand side of (A.315) is the gamma distribution, whose integration constant is the inverse of the right-hand side, and is derived in Example I.1.22. Setting $m = s + 1$ and $z = x + y$, (A.315) reads

$$\int_0^\infty e^{-(s+1)t} t^{x+y-1} \, \mathrm{d}t = \frac{\Gamma(x + y)}{(s + 1)^{x+y}} \quad \text{or} \quad \frac{1}{(s + 1)^{x+y}}$$

$$= \frac{1}{\Gamma(x + y)} \int_0^\infty e^{-(s+1)t} t^{x+y-1} \, \mathrm{d}t,$$

and substituting $(s + 1)^{x+y}$ into (A.314),

$$B(x, y) = \int_0^\infty s^{x-1} \left(\frac{1}{s + 1} \right)^{x+y} \mathrm{d}s = \int_0^\infty s^{x-1} \frac{1}{\Gamma(x + y)} \int_0^\infty e^{-(s+1)t} t^{x+y-1} \, \mathrm{d}t \, \mathrm{d}s$$

$$= \frac{1}{\Gamma(x + y)} \int_0^\infty \int_0^\infty e^{-st} e^{-t} s^{x-1} t^{x+y-1} \, \mathrm{d}t \, \mathrm{d}s$$

$$= \frac{1}{\Gamma(x + y)} \int_0^\infty \left(e^{-t} t^{x+y-1} \int_0^\infty e^{-st} s^{x-1} \, \mathrm{d}s \right) \mathrm{d}t,$$

and, as $\int_0^\infty e^{-st} s^{x-1} \, \mathrm{d}s = \Gamma(x)/t^x$ from (A.315), we have

$$B(x, y) = \frac{\Gamma(x)}{\Gamma(x + y)} \int_0^\infty e^{-t} t^{y-1} \, \mathrm{d}t = \frac{\Gamma(x)\Gamma(y)}{\Gamma(x + y)}.$$

A.3   (Sampling)

(a) The binomial theorem states that $(x + y)^n = \sum_{i=0}^n \binom{n}{i} x^i y^{n-i}$, $n \in \mathbb{N}$, while Vandermonde's theorem is, for $x, y, n \in \mathbb{N}$, $\binom{x+y}{n} = \sum_{i=0}^n \binom{x}{i}\binom{y}{n-i}$.

(b) Let $X \sim \mathrm{HGeo}(N, M, n)$ with p.m.f. (A.39). From the nature of the sampling (without replacement), $f_X(x)$ would approach that of a binomial distribution if the number of balls goes to infinity, so that removing a few of them does not change their relative frequency in the urn. That is, we require that $N \to \infty$ and $M \to \infty$ such that $N/(N + M) \to p$, for some constant p with $0 < p < 1$. To formally verify this, note that

$$\binom{N}{x} = \frac{N(N - 1) \cdots (N - x + 1)}{x!} \approx \frac{N^x}{x!},$$

where $a \approx b$ means that $a/b \to 1$ in the limit as $N \to \infty$. Similarly for the other two binomial coefficients, so that

$$\frac{\binom{N}{x}\binom{M}{n-x}}{\binom{N+M}{n}} \approx \frac{\dfrac{N^x}{x!} \dfrac{M^{n-x}}{(n - x)!}}{\dfrac{(N + M)^n}{n!}} = \frac{n!}{x!(n - x)!} \left(\frac{N}{N + M} \right)^x \left(\frac{M}{N + M} \right)^{n-x}$$

$$= \binom{n}{x} p^x (1 - p)^{n-x}.$$

The indicator function is $\mathbb{I}_{\{0,1,\dots,n\}}(x)$, which follows because $\lim_{M \to \infty} \max(0, n - M) = 0$ and $\lim_{N \to \infty} \min(n, N) = n$.

(c) If $X \sim \text{Bin}(n, p)$ and $np = \lambda$, then

$$\Pr(X = x) = \binom{n}{x} p^x (1-p)^{n-x} = \frac{n!}{(n-x)! x!} \left(\frac{\lambda}{n}\right)^x \left(1 - \frac{\lambda}{n}\right)^{n-x}$$

$$= \underbrace{\frac{n(n-1) \cdots (n-x+1)}{n^x}}_{\to 1} \frac{\lambda^x}{x!} \underbrace{\left(1 - \frac{\lambda}{n}\right)^n}_{\to e^{-\lambda}} \underbrace{\left(1 - \frac{\lambda}{n}\right)^{-x}}_{\to 1}$$

as $n \to \infty$ and $p \to 0$, so that, for large n and small p, $\Pr(X = x) \approx e^{-\lambda} \lambda^x / x!$.

Thus, if $X \sim \text{HGeo}(N, M, n)$, letting $N \to \infty$, $M \to \infty$, and $n \to \infty$ such that $N/(N+M) \to 0$ and $nN/(N+M) \to \lambda$,

$$f_{\text{HGeo}}(x; N, M, n) \approx f_{\text{Poi}}(x; \lambda).$$

A.4 (Geometric generalization)

(a) On the first throw, the probability of no sixes is $(5/6)^d$, so that $p = 1 - (5/6)^d$ is the probability of at least one six. The number of tosses (of all d dice) until at least one six occurs is clearly geometrically distributed, so the expected number of rolls until at least one six occurs is $1/p$ or

$$\frac{1}{p} = \frac{1}{1 - (5/6)^d} = \frac{6^d}{6^d - 5^d}.$$

On the throw that produced at least one six, the probability of getting exactly one six is

$$\Pr(\text{one six} \mid \text{at least one six}) = \frac{\Pr(\text{one six})}{p} = \frac{1}{p} \binom{d}{1} \frac{1}{6} \left(\frac{5}{6}\right)^{d-1} = d \frac{5^{d-1}}{6^d - 5^d}.$$

Similarly, on the throw that produced at least one six, the probability of getting exactly k sixes is

$$\frac{1}{p} \binom{d}{k} \left(\frac{1}{6}\right)^k \left(\frac{5}{6}\right)^{d-k} = \binom{d}{k} \frac{5^{d-k}}{6^d - 5^d}.$$

The critical step is now: If, say, T initial throws were required to produce at least one six, and k sixes occurred on that trial, then the expected total number of throws is $T + \mathbb{E}_{d-k}[N]$. That is, if k is fixed, we expect

$$\frac{1}{p} + \mathbb{E}_{d-k}[N]$$

throws. Taking the fact that k is random into account and that events $\{k = 1\}, \ldots, \{k = d\}$ are disjoint gives (with $j = d - k$ and reversing the order of summation),

$$\mathbb{E}_d[N] = \frac{1}{p} + \sum_{k=1}^{d} \binom{d}{k} \frac{5^{d-k}}{6^d - 5^d} \mathbb{E}_{d-k}[N]$$

$$= \frac{6^d}{6^d - 5^d} + \sum_{j=0}^{d-1} \binom{d}{j} \frac{5^j}{6^d - 5^d} \mathbb{E}_j[N]$$

$$= \frac{1}{6^d - 5^d} \left(6^d + \sum_{j=0}^{d-1} \binom{d}{j} 5^j \mathbb{E}_j[N]\right).$$

The result follows because $\mathbb{E}_0[N] = 0$.

```
1   function e = dsixes(d)
2   e=zeros(d,1); e(1)=6;
3   for m=2:d
4     s=0; for k=1:m, s=s + 5^k * c(m,k) * e(k); end
5     e(m)=(6^m + s) / (6^m - 5^m);
6   end
```

Program Listing A.11: Evaluates $\mathbb{E}_d[N]$.

```
1   function vec =dsixessim(d,S)
2   vec=zeros(S,1); m=d;
3   for i=1:S
4     z=0; while d>0, w=unidrnd(6,d,1); x =sum(w==6); d=d-x; z=z+1; end
5     d=m; vec(i)=z;
6   end
```

Program Listing A.12: Simulates the r.v. N S times, based on d dice.

(b) The function in Listing A.11 can be used to evaluate the mean of N.

(c) Some trial and error reveals that the regressors $1/d$ and $\sqrt{d}$ (and a constant) work very well, yielding (for the expectation values for $d = 150$ through $d = 200$) a parsimonious model with an R^2 measure of fit of 0.9999998 and coefficients

$$\mathbb{E}_d[N] \approx 26.4604689 + 0.554971041\sqrt{d} - 313.709051d^{-1}.$$

(d) See the function in Listing A.12.

(e) It is easy to see that $A^c \cap B^c = \emptyset$ so that, from De Morgan's law, $\Pr(A \cup B) = 1$. Also, A^c implies B, that is, $A^c \subset B$, and B^c implies A, that is, $B^c \subset A$. An attempt to illustrate the corresponding Venn diagram is made in Figure A.9.

For $\Pr(A)$, using the hint, the probability that a single die does *not* show a six after x throws is clearly $(5/6)^x$. The complement (i.e., after x throws, it shows a six) thus has probability $1 - (5/6)^x$. Now, the independence of the dice implies

$$\Pr(A) = (1 - (5/6)^x)^d.$$

(It might help to imagine that, at each toss, *all* the dice are thrown, but you keep track of those that have already displayed a six at least once.) Similarly, $\Pr(B) = 1 - (1 - (5/6)^{x-1})^d$. Thus, with $p = 5/6$,

$$\Pr(N = x \,; d) = (1 - p^x)^d - (1 - p^{x-1})^d \mathbb{I}_{\{1,2,\dots\}}(x), \quad d \in \mathbb{N}. \tag{A.344}$$

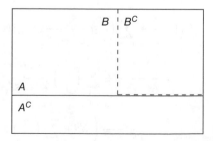

Figure A.9 *Venn diagram for events such that $A^c \subset B$, $B^c \subset A$, and $A^c \cap B^c = \emptyset$.*

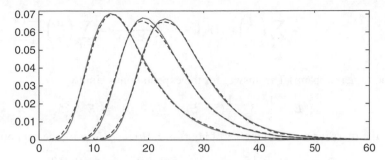

Figure A.10 *Mass functions (A.344) (solid lines) and kernel density estimates of the simulated density (dashed lines) based on 10,000 replications, for (from left to right) d = 10, d = 30, and d = 60.*

For $x = 1$, this reduces to $(1/6)^d$, which is clearly correct, while for $d = 1$,

$$\Pr(N = x \,;\, 1) = \left(\frac{5}{6}\right)^{x-1} - \left(\frac{5}{6}\right)^{x} = \frac{1}{6}\left(\frac{5}{6}\right)^{x-1} \mathbb{1}_{\{1,2,\dots\}}(x),$$

which is just the geometric distribution and also correct. Figure A.10 compares the derived mass function to kernel density estimates of simulated values, confirming its correctness.

(f) Using the binomial theorem,

$$\mathbb{E}_d[N] = \sum_{x=1}^{\infty} x((1 - p^x)^d - (1 - p^{x-1})^d)$$

$$= \sum_{x=1}^{\infty} x\left(\sum_{j=0}^{d} \binom{d}{j}((-p^x y^j) - \sum_{j=0}^{d} \binom{d}{j}(-p^{x-1})^j\right)$$

$$= \sum_{x=1}^{\infty} x\left(\sum_{j=0}^{d} \binom{d}{j}(-1)^j(p^{jx}) - \sum_{j=0}^{d} \binom{d}{j}(-1)^j p^{j(x-1)}\right)$$

$$= \sum_{x=1}^{\infty} x\left(\sum_{j=0}^{d} \binom{d}{j}(-1)^j p^{jx}(1 - p^{-j})\right).$$

Noting that the inner product is zero for $j = 0$, switching the sums, and using the fact that, for $q = p^j$,

$$\sum_{x=1}^{\infty} xq^x = \frac{q}{(1 - q)^2},$$

we have

$$\mathbb{E}_d[N] = \sum_{x=1}^{\infty} x\left(0 + \sum_{j=1}^{d} \binom{d}{j}(-1)^j p^{jx}(1 - p^{-j})\right)$$

$$= \sum_{j=1}^{d} \binom{d}{j}(-1)^j(1 - p^{-j}) \sum_{x=1}^{\infty} xp^{jx}$$

$$= \sum_{j=1}^{d} \binom{d}{j} (-1)^j (1 - p^{-j}) \frac{p^j}{(1 - p^j)^2} = \sum_{j=1}^{d} \binom{d}{j} \frac{(-1)^{j+1}}{1 - p^j}.$$

A.5 (The miner problem) The answer for the expectation is simply

$$E = \frac{1}{3} \cdot 3 + \frac{1}{3}(5 + E) + \frac{1}{3}(7 + E) \;\Rightarrow\; E = 15.$$

For the mass function of number of hours n, $f_N(n) = \Pr(N = n)$, the support is

$$S = \{3, 8, 10, 15 \ldots\} = 3 + \{5i + 7j\}, \quad i, j \in \{0, 1, 2, \ldots\}.$$

If we knew i and j, then

$$\Pr(N = 3 + \{5i + 7j\}) = \binom{i+j}{i} 3^{-(1+i+j)} \mathbb{I}\{i, j \in \{0, 1, 2, \ldots\}\},$$

but we do not, so we need to count them. For example, for $N = 73$, this could mean the miner took door #3 ten times, or door #2 fourteen times, or door #2 seven times and door #3 five times. Note the latter event can happen in $\binom{12}{7}$ ways. So, we need all the ways that 70 can be split up into multiples of 5 and 7. This can be done for general n (perhaps inelegantly, but it works) with a double sum over valid i and j, checking if $n = 3 + 7j + 5i$. That is,

$$f_N(n) = \frac{1}{3} \sum_{i=0}^{\lfloor (n-3)/5 \rfloor} \sum_{j=0}^{\lfloor (n-3)/7 \rfloor} \binom{i+j}{i} \left(\frac{1}{3}\right)^{(i+j)} \mathbb{I}\{n = 3 + 7j + 5i\}.$$

Then, as usual, $\mathbb{E}[N] = \sum_{n=3}^{\infty} n \Pr(N = n)$. The program in Listing A.13 gives the code for computing this.

A.6 Dropping the subscript Y from f_Y and f_Y for convenience, we have

$$\mathbb{E}[F(Y/\sigma)] = \int_{-\infty}^{\infty} F(y/\sigma) f(y) \, dy = \int_{-\infty}^{\infty} \int_{-\infty}^{y/\sigma} f(x) \, dx \, f(y) \, dy$$

```
1   function E=minerexpect(nup)
2   % expectation (and pmf below) of the miner with three doors problem.
3   nvec=(1:nup)'; f=minerpmf(nvec); E=sum(nvec.*f); figure, plot(nvec,f,'r-o')
4
5   function f=minerpmf(nvec) % nvec is vector of integer evaluation points
6   nlen=length(nvec); f=zeros(nlen,1);
7   for t=1:nlen
8     s=0; n=nvec(t); if n<3, continue, end
9     for ii=0:floor((n-3)/5), for jj=0:floor((n-3)/7) %#ok<ALIGN>
10      if n==(3+7*jj+5*ii), r=c(ii+jj,ii); s=s+r*(1/3)^(ii+jj); end
11    end, end
12    f(t)=s/3;
13  end
14  function c=c(n,k), c=round( gamma(n+1)./ (gamma(k+1).* gamma(n-k+1)));
```

Program Listing A.13: Computes the expectation for the miner problem, using the definition and the p.m.f.

$$= \int_{-\infty}^{\infty} \left[\int_{-\infty}^{0} f(x) \, dx + \int_{0}^{y/\sigma} f(x) \, dx \right] f(y) \, dy$$

$$= \frac{1}{2} + \int_{-\infty}^{\infty} \int_{0}^{y/\sigma} f(x) f(y) \, dx \, dy$$

$$= \frac{1}{2} + \int_{-\infty}^{0} \int_{y/\sigma}^{0} f(x) f(y) \, dx \, dy + \int_{0}^{\infty} \int_{0}^{y/\sigma} f(x) f(y) \, dx \, dy.$$

Substituting $z = y/\sigma$, $y = \sigma z$, $dy = \sigma \, dz$ in the first integral, and $z = -y/\sigma$, $y = -\sigma z$, $dy = -\sigma \, dz$ in the second, using the symmetry of f, and then $a = -z$, gives

$$\mathbb{E}[F(Y/\sigma)] = \frac{1}{2} + \int_{-\infty}^{0} \int_{z}^{0} f(x) f(\sigma z) \, dx \, dz + \int_{-\infty}^{0} \int_{0}^{-z} f(x) f(\sigma z) \, dx \, dz$$

$$= \frac{1}{2} + \int_{-\infty}^{0} f(\sigma z) \left[\int_{z}^{0} f(x) \, dx + \int_{0}^{-z} f(x) \, dx \right] dz$$

$$= \frac{1}{2} - \int_{0}^{\infty} f(\sigma a) \left[\int_{-a}^{0} f(x) \, dx + \int_{0}^{a} f(x) \, dx \right] da$$

$$= \frac{1}{2} - \int_{0}^{\infty} f(\sigma a)[1 - 2F(a)] \, da.$$

Finally, with $u = 1 - 2F(a)$, $dv = f(\sigma a)$, $du = -2f(a) \, da$, $v = \int_{0}^{\infty} f(\sigma a) \, da = 1/(2\sigma)$,

$$\mathbb{E}[F(Y/\sigma)] = \frac{1}{2} + \frac{1 - 2F(a)}{2\sigma} \bigg|_{0}^{\infty} + 2 \int_{0}^{\infty} \frac{f(a)}{2\sigma} \, da = \frac{1}{2} - \frac{1}{2\sigma} - 0 + \frac{1}{2\sigma} = \frac{1}{2},$$

For $X = F_Y(Y/\sigma)$, with density and c.d.f. f_X and F_X, respectively, the support of X is clearly $(0, 1)$. Then, for $0 < t < 1$,

$$F_X(t; \sigma) = \Pr(X \le t) = \Pr(F_Y(Y/\sigma) \le t) = \Pr(Y/\sigma \le F_Y^{-1}(t))$$

$$= \Pr(Y \le \sigma F_Y^{-1}(t)) = F_Y(\sigma F_Y^{-1}(t)) = \int_{-\infty}^{\sigma F_Y^{-1}(t)} f_Y(z) \, dz,$$

and, applying Leibniz's rule gives (A.317). Figure A.11 shows the desired graphics, for $Y \sim t_4$, based on the code given in Listing A.14.

Next, recall from calculus (see, for example, (I.A.32)) that, if f is a strictly increasing continuous function on a closed interval I, then the image of f is also a closed interval, and the inverse function g is defined as the function such that $g \circ f(x) = x$ and $f \circ g(y) = y$. It is also continuous and strictly increasing. If f is also differentiable in the interior of I with $f'(x) > 0$, then $g'(y) = 1/f'(g(y))$.

Thus, as F is strictly increasing,

$$\frac{\partial F_Y^{-1}(t)}{\partial t} = \frac{1}{f_Y(F_Y^{-1}(t))}, \tag{A.345}$$

and (A.318) is $\mathbb{I}_{(0,1)}(t) \sim \text{Unif}(0, 1)$. Relation (A.345) is also given in (2.28) in the context of studying p-values.

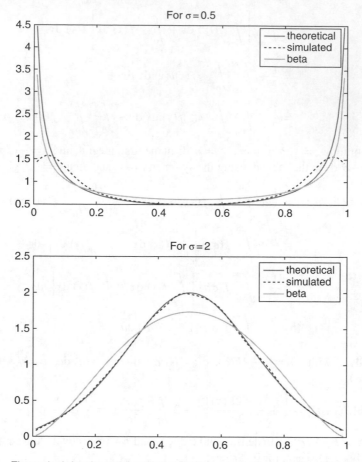

Figure A.11 *Theoretical density and kernel density from simulation, as well as fitted beta, for $Y \sim t_4$ and two values of σ.*

A.7   The p.d.f. for $n = 2$ is easily seen to be $(2 + x^2)^{-3/2}$, so we need to confirm that $dF_X(x)/dx = f_X(x)$, or

$$\frac{dF_X(x)}{dx} = \frac{d}{dx}\left(\frac{x}{2\sqrt{2 + x^2}}\right) = \frac{1}{2}\frac{(2 + x^2)^{1/2} - x\frac{1}{2}(2 + x^2)^{-1/2}2x}{2 + x^2} = \frac{1}{(2 + x^2)^{3/2}}.$$

Recalling the definition of the quantile, solving $p = F_X(\xi_p)$ for $\xi_p = F^{-1}(p)$ gives

$$(2p - 1)^2(2 + \xi_p^2) = \xi_p^2 \quad \text{or} \quad \xi_p^2 = \frac{(2p - 1)^2}{[2p(1 - p)]},$$

so that

$$\xi_p = \frac{2p - 1}{\sqrt{2p(1 - p)}}, \quad p \in (0, 1).$$

```
1   v=4; sigma=2; t=0.01:0.01:0.99;
2   s=1e-8; d=(tinv(t+s,v)-tinv(t-s,v))/(2*s); % numerical derivative
3   Z=tinv(t,v) * sigma; f=tpdf(Z,v) * sigma .* d;
4   figure, plot(t,f,'r-','linewidth',2)
5
6   %% Now simulate the true density.
7
8   % rng(2) % for Matlab version 2014 and higher
9   x=trnd(v,1e5,1); U=cdf('t', x/sigma, v);
10  hold on, [yy,xx]=ksdensity(U,t); plot(xx,yy,'b--','linewidth',2)
11
12  %% Now fit a beta distribution to the simulated data and plot it
13
14  phat = betafit(U); betaapprox=pdf('Beta',t,phat(1),phat(2));
15  plot(t,betaapprox,'g:','linewidth',2)
16  hold off, legend('theoretical','simulated','beta'), set(gca,'fontsize',16)
```

Program Listing A.14: Computes (A.317) and generates the graphics in Figure A.11.

The second moment is finite if and only if the integral $\int_0^\infty x^2(2 + x^2)^{-3/2}\, dx$ exists, but, from (I.A.54) and the second part of Example I.A.22,

$$\int_0^\infty \frac{x^2}{(2 + x^2)^{3/2}}\, dx > \int_0^\infty \frac{1}{x}\, dx > \lim_{c \to 0^+} \int_c^1 \frac{1}{x}\, dx = \infty.$$

A.8 Using the substitution $c = kx^2$, then $dc/dx = 2kx$ and

$$f_X(x) = \left| \frac{dc}{dx} \right| f_C(c) = 2kx \cdot \frac{2^{-k/2}}{\Gamma(k/2)} (kx^2)^{k/2-1} e^{-(kx^2)/2} \mathbb{1}_{(0,\infty)}(kx^2)$$

$$- \frac{2^{-k/2+1} k^{k/2}}{\Gamma(k/2)} x^{k-1} e^{-(kx^2)/2} \mathbb{1}_{(0,\infty)}(x).$$

For $k = 1$,

$$f_X(x) = \frac{\sqrt{2}}{\sqrt{\pi}} \exp\left\{ -\frac{1}{2} x^2 \right\} \mathbb{1}_{(0,\infty)}(x),$$

with expected value

$$\mathbb{E}[X] = \frac{\sqrt{2}}{\sqrt{\pi}} \int_0^\infty x \exp\left\{ -\frac{1}{2} x^2 \right\} dx = \frac{\sqrt{2}}{\sqrt{\pi}} \left(-e^{-\frac{1}{2} x^2} \right)\Big|_0^\infty = \frac{\sqrt{2}}{\sqrt{\pi}}.$$

A.9 (Closed-form expression for $S = \sum_{j=0}^u j a^j$.)

(a) Writing out S and aS and comparing,

$$S = a + 2a^2 + 3a^3 + \cdots + ua^u$$

$$aS = \qquad a^2 + 2a^3 + \cdots \qquad + ua^{u+1}$$

so that

$$S(1 - a) = S - aS = a + a^2 + a^3 + \cdots + a^u - ua^{u+1} = \frac{a - a^{u+1}}{1 - a} - ua^{u+1},$$

or

$$S = \frac{1}{1-a}\left(\frac{a - a^{u+1}}{1-a} - ua^{u+1}\right).$$

(b) Write

$$S + \sum_{j=0}^{u} a^j = \sum_{j=0}^{u}(j+1)a^j = \frac{d}{da}\sum_{j=0}^{u} a^{j+1} = \frac{d}{da}\left(a\sum_{j=0}^{u} a^j\right) = \frac{d}{da}\left(\frac{a - a^{u+2}}{1-a}\right)$$

$$= \frac{(1-a)(1-(u+2)a^{u+1}) - (a - a^{u+2})(-1)}{(1-a)^2},$$

so that

$$S = \frac{(1-a)(1-(u+2)a^{u+1}) + (a - a^{u+2})}{(1-a)^2} - \frac{1 - a^{u+1}}{1-a}\frac{(1-a)}{(1-a)}$$

$$= \frac{(1-a)(1-(u+2)a^{u+1}) + (a - a^{u+2}) - (1 - a^{u+1})(1-a)}{(1-a)^2}$$

$$= \frac{(1-a)(a^{u+1} - (u+2)a^{u+1}) + (a - a^{u+2})}{(1-a)^2}$$

$$= \frac{(1-a)(-(u+1)a^{u+1}) + (a - a^{u+2})}{(1-a)^2}$$

$$= \frac{1}{1-u}\left(-ua^{u+1} + \frac{a - a^{u+2}}{1-u} - a^{u+1}\frac{1-a}{1-u}\right)$$

$$= \frac{1}{1-a}\left(-ua^{u+1} + \frac{a - a^{u+1}}{1-a}\right).$$

A.10 With $F_X(x) = x$ for $X \sim \text{Unif}(0,1)$, we want, with $Y = \min(X_i)$,

$$F_Z(z) = \Pr(Z \le z) = \Pr(nY \le z) = \Pr(Y \le z/n) = 1 - (1 - z/n)^n, \quad 0 < z < n,$$

where the range follows because $Z \in (0, n)$. For $z < 0$, $F_Z(z) = 0$ and $F_Z(z) = 1$ for $z > n$. For the limit,

$$\lim_{n\to\infty} F_Z(z) = 1 - \exp(z),$$

so that in the limit, Z is exponential with rate 1, in which case $f_Z(z) = \exp(-z)$.

A.11 With $x = \log(y)$ and $dx = y^{-1}\,dy$, and noting that the support of Y is $(1, \infty)$,

$$f_Y(y) = \lambda \exp(-\lambda \log(y))\frac{1}{y} = \frac{\lambda}{y}\{\exp(\log(y))\}^{-\lambda} = \frac{\lambda}{y^{\lambda+1}}\mathbb{I}_{(0,\infty)}(y).$$

The mth moments are given, for $m - \lambda - 1 < -1$ or $m < \lambda$, by

$$\mathbb{E}[Y^m] = \lambda \int_1^{\infty} y^{m-\lambda-1}\,dy = \frac{\lambda}{m-\lambda}y^{m-\lambda}\Big|_1^{\infty} = \frac{\lambda}{\lambda - m}, \quad m < \lambda.$$

A.12 Recall that $e^{-k} = \lim_{n\to\infty}(1 + k/n)^{-n}$ for $k \in \mathbb{R}$, which, applied to the kernel of the density, gives

$$\lim_{n\to\infty}(1 + x^2/n)^{-\frac{n+1}{2}} = \lim_{n\to\infty}(1 + x^2/n)^{-n/2} = (\lim_{n\to\infty}(1 + x^2/n)^{-n})^{1/2}$$

$$= \exp\left\{-\frac{1}{2}x^2\right\}.$$

Applying Stirling's approximation to the integration constant in the second expression yields

$$K_n \approx n^{-\frac{1}{2}}\frac{\sqrt{2\pi}\left(\frac{n+1}{2}\right)^{\frac{n+1}{2}-1/2}\exp\left(-\frac{n+1}{2}\right)}{\sqrt{\pi}\sqrt{2\pi}\left(\frac{n}{2}\right)^{\frac{n}{2}-1/2}\exp\left(-\frac{n}{2}\right)} = \exp\left(-\frac{1}{2}\right)\frac{1}{\sqrt{2\pi}}\left(\frac{n+1}{n}\right)^{n/2}.$$

The result follows because $\lim_{n\to\infty}(1 + n^{-1})^{n/2} = e^{1/2}$.

A.13 (Expected shortfall for asymmetric double Weibull) For the c.d.f. $F_Z(x)$, for $x < 0$, substitute $r = (-z)^{\beta^-}$ to get

$$z = -(r^{1/\beta^-}), \quad dz = -\frac{1}{\beta^-}r^{1/\beta^--1}\,dr,$$

and

$$F_Z(x) = \int_{-\infty}^{x} f_Z(z; \beta^-, \beta^+, 1)\,dz = \frac{1}{2}\int_{-\infty}^{x}\beta^-(-z)^{\beta^--1}\exp(-(-z)^{\beta^-})\,dz$$

$$= \frac{1}{2}\int_{\infty}^{(-x)^{\beta^-}}\beta^- r^{(\beta^--1)/\beta^-}\exp(-r)\left(-\frac{1}{\beta^-}r^{\frac{1}{\beta^-}-1}\right)\,dr = \frac{1}{2}\int_{(-x)^{\beta^-}}^{\infty}\exp(-r)\,dr$$

$$= \frac{1}{2}\exp(-(-x)^{\beta^-}),$$

and $F_Z(0) = 1/2$. For $x \geq 0$, similarly to the previous calculation with $r = z^{\beta^+}$,

$$F_Z(x) = \frac{1}{2} + \int_{0}^{x} f_Z(z; \beta^-, \beta^+, 1)\,dz = \frac{1}{2} + \frac{1}{2}\int_{0}^{x}\beta^+ z^{\beta^+-1}\exp(-z^{\beta^+})\,dz$$

$$= \frac{1}{2} + \frac{1}{2}\int_{0}^{x^{\beta^+}}\exp(-r)\,dr = \frac{1}{2} + \frac{1}{2}(1 - \exp(-x^{\beta^+})) = 1 - \frac{1}{2}\exp(-x^{\beta^+}).$$

For $ES(Z, \xi)$ with $x = q_{Z,\xi} < 0$, the integral is, with $b = \beta^-$ and $r = (-z)^b$,

$$\int_{-\infty}^{x} z f_Z(z)\,dz = \int_{-\infty}^{x} z\frac{b}{2}(-z)^{b-1}\exp(-(-z)^b)\,dz$$

$$= -\frac{1}{2}\int_{(-x)^b}^{\infty} r^{1/b}\exp(-r)\,dr$$

$$= -\frac{1}{2}\left(\Gamma\left(1 + \frac{1}{b}\right) - \Gamma_{(-x)^b}\left(1 + \frac{1}{b}\right)\right).$$

A.14 (Finiteness of $\mathbb{E}[X]$ is equivalent to finiteness of $\mathbb{E}[|X|]$) Using the hint:
($\Leftarrow$) We have

$$|\mathbb{E}[X]| = |\mathbb{E}[X^+ - X^-]| = |\mathbb{E}[X^+] - \mathbb{E}[X^-]|$$

$$\leq |\mathbb{E}[X^+]| + |\mathbb{E}[X^-]| = \mathbb{E}[X^+] + \mathbb{E}[X^-]. \tag{A.346}$$

As $\mathbb{E}[|X|] < \infty$, we have $\mathbb{E}[|X|] = \mathbb{E}[X^+ + X^-] = \mathbb{E}[X^+] + \mathbb{E}[X^-] < \infty$, which
implies $\mathbb{E}[X^+] < \infty$ and $\mathbb{E}[X^-] < \infty$, and with (A.346) shows the result.

($\Rightarrow$) $|\mathbb{E}[X]| < \infty \Rightarrow -\infty < \mathbb{E}[X^+] - \mathbb{E}[X^-] < \infty$. This rules out that the difference is
of the form $\infty + \infty$ or $-\infty - \infty$; or that one term is (plus or minus) infinity and the
other is finite, so either both are finite, or the term is of the form $\infty - \infty$. But for the
latter case, recall from analysis that the difference of two infinities is not defined, so
it must be the case that $\mathbb{E}[X^+]$ and $\mathbb{E}[X^-]$ are finite; and as these are both nonnegative,
we have $\mathbb{E}[X^+] < \infty$ and $\mathbb{E}[X^-] < \infty$. Then $\mathbb{E}[|X|] = \mathbb{E}[X^+] + \mathbb{E}[X^-] < \infty$, as was to
be shown.

A.15 The p.d.f. of S is

$$f_S(s; m) = \left| \frac{db}{ds} \right| f_B(b)$$

$$= \frac{1}{2} \frac{1}{B\left(\frac{m-1}{2}, \frac{m-1}{2}\right)} \left(\frac{s+1}{2}\right)^{\frac{m-1}{2}-1} \left(1 - \frac{s+1}{2}\right)^{\frac{m-1}{2}-1} \mathbb{I}_{(0,1)}\left(\frac{s+1}{2}\right)$$

$$= \frac{2^{2-m}}{B\left(\frac{m-1}{2}, \frac{m-1}{2}\right)} (1 - s^2)^{\frac{m-3}{2}} \mathbb{I}_{(-1,1)}(s).$$

For $m = 3$, B is uniform, and $f_S(s; 3)$ easily reduces to what we expect, $(1/2)\mathbb{I}_{(-1,1)}(s)$.
   For the density of Y,

$$f_Y(y) = \left| \frac{dx}{dy} \right| f_X(x) = 2y \frac{1}{2^{m/2}\Gamma(m/2)} (y^2)^{m/2-1} e^{-y^2/2} \mathbb{I}_{(0,\infty)}(y^2)$$

$$= \frac{2^{1-m/2}}{\Gamma(m/2)} y^{m-1} \exp\{-y^2/2\} \mathbb{I}_{(0,\infty)}(y).$$

Then the density of $P = SY$ is

$$f_P(p; m) = \int_{-1}^{1} \frac{1}{s} f_S(s) f_Y(p/s) ds$$

$$= \frac{\Gamma(m-1) 2^{3(1-m/2)}}{\Gamma\left(\frac{m-1}{2}\right) \Gamma\left(\frac{m-1}{2}\right) \Gamma(m/2)}$$

$$\times \int_{0}^{1} \frac{1}{s} (1 - s^2)^{\frac{m-3}{2}} \left(\frac{p}{s}\right)^{m-1} \exp\{-(p/s)^2/2\} ds,$$

where the integral starts at zero because $f_Y(p/s)$ only has support on $(0, \infty)$.

For $m = 3$, recall $\Gamma(3/2) = \sqrt{\pi}/2$, then substitute $u = p/s$ and then (for $p > 0$) $v = -u^2/2$ to get

$$f_P(p; 3) = \frac{1}{\sqrt{2\pi}} \int_0^1 \frac{1}{s}\left(\frac{p}{s}\right)^2 \exp\left\{-\frac{(p/s)^2}{2}\right\} ds$$

$$= \frac{1}{\sqrt{2\pi}} \int_p^\infty u \exp\left\{-\frac{u^2}{2}\right\} du = \frac{1}{\sqrt{2\pi}} \int_{-\infty}^{-p^2/2} e^v dv$$

$$= \frac{1}{\sqrt{2\pi}} \exp\{-p^2/2\},$$

so that, for $m = 3$, P is indeed standard normal.

The general expression does not seem to simplify, though numerically integrating it for any $m > 2$ shows that $P \sim N(0, 1)$.

A.16 We first want the joint p.m.f. of X and Y, $f_{X,Y}(x, y)$. From the definition of X and Y, x of the X_i are in $[0, s]$, and y values in $[0, t]$, or $y - x$ values in $(s, t]$, and the remaining $n - y$ are greater than t.

Because the X_i are i.i.d., the joint distribution of X and Y is just trinomial, that is,

$$f_{X,Y}(x, y; n, s, t) = \binom{n}{x, y - x, n - y} s^x (t - s)^{y-x} (1 - t)^{n-y} \mathbb{1}(0 \le x \le y \le n),$$

and the marginal of X is binomial,

$$f_X(x; n, s) = \binom{n}{x} s^x (1 - s)^{n-x} \mathbb{1}(0 \le x \le n).$$

Let $D = Y - X$ and $M = X$, so that $X = M$ and $Y = D + M$. Then

$$f_{D,M}(d, m) = |\det \mathbf{J}| f_{X,Y}(x, y),$$

where

$$\mathbf{J} = \begin{bmatrix} \partial x/\partial d & \partial x/\partial m \\ \partial y/\partial d & \partial y/\partial m \end{bmatrix} = \begin{bmatrix} 0 & 1 \\ 1 & 1 \end{bmatrix}, \quad |\det \mathbf{J}| = 1,$$

so that $f_{D,M}(d, m) = f_{X,Y}(m, d + m)$ or

$$\binom{n}{m, d, n - (d + m)} s^m (t - s)^d (1 - t)^{n-(d+m)} \mathbb{1}(0 \le m \le d + m \le n).$$

Thus, the conditional distribution of D given M is

$$f_{D|M}(d \mid m) = \frac{f_{D,M}(d, m)}{f_M(m)}$$

or

$$\frac{\binom{n}{m}\binom{n-m}{d} s^m (t - s)^d (1 - t)^{n-(d+m)}}{\binom{n}{m} s^m (1 - s)^{n-m}} \mathbb{1}(0 \le m \le d + m \le n)\mathbb{1}(0 \le m \le n)$$

$$= \binom{n-m}{d} \frac{(t-s)^d (1-t)^{n-(d+m)}}{(1-s)^{n-m}} \frac{1}{(1-s)^d (1-s)^{-d}} \mathbb{1}(0 \le d \le n - m)$$

$$- \binom{n-m}{d} \left(\frac{t-s}{1-s}\right)^d \left(1 - \frac{t-s}{1-s}\right)^{(n-m)-d} \mathbb{1}(0 \le d \le n - m).$$

That is,

$$(D \mid M = m) \sim \text{Bin}(n - m, (t - s)/(1 - s))$$

or

$$(D \mid X = x) \sim \text{Bin}\left(n - x, \frac{t-s}{1-s}\right),$$

as was to be shown.

A.17 (Inversion formula)

    (a) (Problem II.1.12)

        (i) Substitute $s = it$ in the inversion theorem and, under the assumption that the m.g.f. exists and $\varphi_X(t) = \mathbb{M}_X(it)$, we have that $\varphi_X(s/i) = \mathbb{M}_X(s)$.

       (ii) By definition,

$$\bar{F}_X(x) = 1 - F_X(x) = \int_x^\infty f_X(y)\, dy$$

$$= \frac{1}{2\pi i} \int_x^\infty \int_{c-i\infty}^{c+i\infty} \exp\{\mathbb{K}_X(s) - sy\}\, ds\, dy.$$

Interchanging the order of integration,

$$\bar{F}_X(x) = \frac{1}{2\pi i} \int_{c-i\infty}^{c+i\infty} \int_x^\infty \exp\{\mathbb{K}_X(s) - sy\}\, dy\, ds$$

$$= \frac{1}{2\pi i} \int_{c-i\infty}^{c+i\infty} \left[\frac{\exp\{\mathbb{K}_X(s) - sy\}}{-s} \right]_x^\infty ds$$

$$= \frac{1}{2\pi i} \int_{c-i\infty}^{c+i\infty} \exp\{\mathbb{K}_X(s) - sx\} \frac{ds}{s}, \qquad (\text{A.347})$$

where the last equality holds as long as $\text{Re}(s) = c > 0$. If instead we choose $c < 0$, we can integrate over $(-\infty, x]$ to obtain

$$F_X(x) = \frac{1}{2\pi i} \int_{-\infty}^x \int_{c-i\infty}^{c+i\infty} \exp\{\mathbb{K}_X(s) - sy\}\, ds\, dy$$

$$= \frac{1}{2\pi i} \int_{c-i\infty}^{c+i\infty} \int_{-\infty}^x \exp\{\mathbb{K}_X(s) - sy\}\, dy\, ds$$

$$= \frac{1}{2\pi i} \int_{c-i\infty}^{c+i\infty} \left[\frac{\exp\{\mathbb{K}_X(s) - sy\}}{-s} \right]_{-\infty}^x ds$$

$$= -\frac{1}{2\pi i} \int_{c-i\infty}^{c+i\infty} \exp\{\mathbb{K}_X(s) - sx\} \frac{ds}{s}. \qquad (\text{A.348})$$

(b) For the $p = 1$ case, substituting (A.323) into the integral for the expected short-fall, $I_X(q, 1) = \int_{-\infty}^{q} x f_X(x)\, dx$, and reversing the order of the integrals gives

$$2\pi i I_X(q, 1) = \int_{c-i\infty}^{c+i\infty} \int_{-\infty}^{q} x \exp\{\mathbb{K}_X(s) - sx\}\, dx\, ds.$$

For the inner integral, with $u = x$ and $dv = \exp\{\mathbb{K}_X(s) - sx\}\, dx$, integration by parts and restricting $c < 0$ such that c is in the convergence strip of $\mathbb{M}_X$ (so that the real part of s is negative), gives

$$\int_{-\infty}^{q} x \exp\{\mathbb{K}_X(s) - sx\}\, dx$$

$$= x \frac{\exp\{\mathbb{K}_X(s) - sx\}}{-s}\bigg|_{-\infty}^{q} - \int_{-\infty}^{q} \frac{\exp\{\mathbb{K}_X(s) - sx\}}{-s}\, dx$$

$$= q \frac{\exp\{\mathbb{K}_X(s) - sq\}}{-s} + \frac{1}{s}\int_{-\infty}^{q} \exp\{\mathbb{K}_X(s) - sx\}\, dx$$

$$= -\frac{q}{s}\exp\{\mathbb{K}_X(s) - sq\} - \frac{1}{s^2}\exp\{\mathbb{K}_X(s) - sq\},$$

so that

$$2\pi i I_X(q, 1) = -\int_{c-i\infty}^{c+i\infty} \left(\frac{q}{s} + \frac{1}{s^2}\right)\exp\{\mathbb{K}_X(s) - sq\}\, ds.$$

Again using integration by parts with $u = \exp\{\mathbb{K}_X(s)\}$ and $dv = e^{-sq}(q/s + 1/s^2)\, ds$, so that $du = \exp\{\mathbb{K}_X(s)\}\mathbb{K}_X'(s)\, ds$ and $v = -\exp\{-qs\}/s$, gives

$$2\pi i I_X(q, 1) = -\int_{c-i\infty}^{c+i\infty} \exp\{\mathbb{K}_X(s) - qs\}\mathbb{K}_X'(s)\frac{ds}{s},$$

which is (A.121). Note that

$$\lim_{s \to c+i\infty} \left|\frac{\exp\{\mathbb{K}_X(s) - qs\}}{s}\right| = e^{-qc}\lim_{k \to \infty} \left|\frac{\mathbb{M}_X(c + ik)\exp\{-iqk\}}{c + ik}\right|$$

$$= e^{-qc}\lim_{k \to \infty} \frac{|\mathbb{M}_X(c + ik)|}{\sqrt{c^2 + k^2}} = 0.$$

For the $p = 2$ case, similarly, for $I_X(q, 2) = \int_{-\infty}^{q} x^2 f_X(x)\, dx$, we need to evaluate

$$2\pi i I_X(q, 2) = \int_{c-i\infty}^{c+i\infty} \int_{-\infty}^{q} x^2 \exp\{\mathbb{K}_X(s) - sx\}\, dx\, ds$$

and, with $u = x^2$ and $dv = \exp\{\mathbb{K}_X(s) - sx\}\, dx$,

$$\int_{-\infty}^{q} x^2 \exp\{\mathbb{K}_X(s) - sx\}\, dx$$

$$= x^2 \frac{\exp\{\mathbb{K}_X(s) - sx\}}{-s}\bigg|_{-\infty}^{q} - 2\int_{-\infty}^{q} x\frac{\exp\{\mathbb{K}_X(s) - sx\}}{-s}\, dx$$

$$= -\frac{q^2}{s} \exp\{\mathbb{K}_X(s) - sq\} + \frac{2}{s} \int_{-\infty}^{q} x \exp\{\mathbb{K}_X(s) - sx\} \, dx$$

$$= -\exp\{\mathbb{K}_X(s) - sq\} \left(\frac{q^2}{s} + \frac{2q}{s^2} + \frac{2}{s^3} \right).$$

With $u = \exp\{\mathbb{K}_X(s)\}$ and $dv = e^{-sq}(q^2/s + 2q/s^2 + 2/s^3) \, ds$, we have $du = \exp\{\mathbb{K}_X(s)\}\mathbb{K}'_X(s) \, ds$ and

$$v = \int e^{-sq}(q^2/s + 2q/s^2 + 2/s^3) \, ds = -\frac{e^{-qs}}{s^2}(1 + qs),$$

so that

$$2\pi i I_X(q, 2) = -\int_{c-i\infty}^{c+i\infty} \exp\{\mathbb{K}_X(s) - sq\} \left(\frac{q^2}{s} + \frac{2q}{s^2} + \frac{2}{s^3} \right) ds$$

$$= -\int_{c-i\infty}^{c+i\infty} (1 + sq) \exp\{\mathbb{K}_X(s) - qs\}\mathbb{K}'_X(s) \frac{ds}{s^2}.$$

A.18 (Noncentral Student's t and skew normal)

(a) From the hint, let $Z \sim \mathrm{N}(0, 1)$ independent of X and Y. Using the fact that

$$\Pr(A) = \int \Pr(A \mid B = b) f_B(b) \, db = \mathbb{E}_B[\Pr(A \mid B = b)],$$

we have

$$\mathbb{E}[\Phi(X + cY)] = \int_{-\infty}^{\infty} \int_{0}^{\infty} \Phi(x + cy) f_{X,Y}(x, y) \, dx \, dy$$

$$= \mathbb{E}_{X,Y}[\Pr(Z \leq X + cY \mid X = x, Y = y)]$$

$$= \Pr(Z \leq X + cY) = \Pr\left(\frac{Z - X}{Y} \leq c \right).$$

As $Z - X =: D \sim \mathrm{N}(-\mu, 1 + \sigma^2)$ and $D/\sqrt{1 + \sigma^2} \sim \mathrm{N}(\delta, 1)$, it follows that

$$\Pr\left(\frac{Z - X}{Y} \leq c \right) = \Pr\left(T \leq \frac{c}{\sqrt{1 + \sigma^2}} \right).$$

(b) For $c = 0$,

$$\mathbb{E}[\Phi(X)] = \Pr\left(\frac{Z - X}{Y} \leq c \right) = \Pr(D \leq 0)$$

$$= \Pr\left(\frac{D + \mu}{\sqrt{1 + \sigma^2}} \leq \frac{0 + \mu}{\sqrt{1 + \sigma^2}} \right) = \Phi\left(\frac{\mu}{\sqrt{1 + \sigma^2}} \right).$$

(c) For the m.g.f. of the skew normal, by completing the square in the exponent (adding and subtracting t^2), the m.g.f. is

$$\mathbb{M}_X(t) = \int_{-\infty}^{\infty} e^{tx} 2\phi(x)\Phi(\lambda x) \, dx = 2 \int_{-\infty}^{\infty} \frac{1}{\sqrt{2\pi}} \exp\left\{ -\frac{1}{2}x^2 + tx \right\} \Phi(\lambda x) \, dx$$

$$= 2 \int_{-\infty}^{\infty} \frac{1}{\sqrt{2\pi}} \exp\left\{ -\frac{1}{2}(x^2 - 2tx + t^2 - t^2) \right\} \Phi(\lambda x)\, dx$$

$$= 2\exp\{t^2/2\} \int_{-\infty}^{\infty} \frac{1}{\sqrt{2\pi}} \exp\left\{ -\frac{1}{2}(x - t)^2 \right\} \Phi(\lambda x)\, dx.$$

Let $X \sim \mathrm{N}(t, 1)$ and $Y = \lambda X \sim \mathrm{N}(\lambda t, \lambda^2)$. Then substituting $y = \lambda x$ and using (A.328), the latter integral can be expressed as

$$\int_{-\infty}^{\infty} \frac{1}{\sqrt{2\pi}\lambda} \exp\left\{ -\frac{1}{2}\left(\frac{y - t\lambda}{\lambda} \right)^2 \right\} \Phi(y)\, dy = \mathbb{E}[\Phi(Y)] = \Phi\left(\frac{\lambda t}{\sqrt{1 + \lambda^2}} \right),$$

yielding (A.117). Thus, $\mathbb{K}_X(t) = \ln \mathbb{M}_X(t) = \ln 2 + t^2/2 + \ln \Phi(t\delta)$ and

$$\mathbb{K}'_X(t) = t + \delta R(t\delta), \quad R(z) = \frac{\phi(z)}{\Phi(z)},$$

so that (recalling the basic results on cumulants from Chapter II.1) $\mathbb{E}[X] = \mathbb{K}'_X(0) = \delta\sqrt{2/\pi}$. As $\phi'(t) = -t\phi(t)$, we have $\phi'(t\delta) = -\delta^2 t\phi(t\delta)$. Thus,

$$R'(t\delta) = \frac{\Phi(t\delta)(-\delta^2 t\phi(t\delta)) - \delta(\phi(t\delta))^2}{(\Phi(t\delta))^2} = -\delta(\delta t R(t\delta) + R^2(t\delta)),$$

so that $\mathbb{K}''_X(t) = 1 + \delta R'(t\delta)$ and $\mathbb{V}(X) = \mathbb{K}''_X(0) = 1 - 2\delta^2/\pi$.

A.19 (Skew normal characteristic function and saddlepoint approximation)

(a) The program in Listing A.15 computes (an approximation to) the p.d.f. of a standard skew normal random variable via inversion of the c.f., in Matlab. The extension to sums of independent SN r.v.s is straightforward, as is the calculation of the ES, using (A.338). The program in Listing A.16 is similar, but is for the Julia programming language, and uses the scaled complementary error function.

(b) See the next solution.

```
1  function pdf=SNpdfcf(zvec,lambda)
2  % Calculates the pdf of the standard SN, via inversion of the c.f.
3  % Exact (to machine precision) pdf is:
4  % Exactpdf = 2*normpdf(zvec).*normcdf(lambda*zvec);
5
6  if verLessThan('matlab','8.10'), error('Need Release R2013A or higher'), end
7  up=11; % found by trial and error
8  delta=lambda/sqrt(1+lambda^2); zlen=length(zvec); pdf=zeros(zlen,1);
9  for j=1:zlen, z=zvec(j); pdf(j)=quadgk(@(t) intpdf(t,z,delta),0,up)/pi; end
10
11 function I=intpdf(t,z,delta)
12 arg=t*delta/sqrt(2); term=(1 + 1i*erfi(arg)); term=term(isfinite(term));
13 % Any Inf ==> zero
14 psi=exp(-t.^2/2) .* term; I = real( exp(-1i*t*z).*psi );
```

Program Listing A.15: Delivers the p.d.f. of a standard skew normal r.v., computed via inversion of the c.f., truncating the upper integral limit. Observe how infinite values of function erfi are set to zero, and also how we can check the version of Matlab.

```
1   function SNpdfcf(z,lambda)
2     delta=lambda/sqrt(1+lambda^2)
3     intpdf(t)=real(exp(-im*t*z)*(2*exp(-t^2/2)-exp(-t^2*(1-delta^2)/2)*erfcx
4       (im*t*delta/sqrt(2))))
5     I,_=quadgk(intpdf,0,Inf)
6     return I/pi
7   end
```

Program Listing A.16: Same as the function in Listing A.15, but in Julia, and using the scaled complementary error function erfcx, so that the upper limit of the c.f. inversion integral is infinity.

A.20 We have

$$\mathbb{K}_{X_i}(t) = \ln \mathbb{M}_{X_i}(t) = \ln 2 + t\mu_i + t^2\sigma_i^2/2 + \ln \Phi(t\sigma_i\delta_i),$$

$\mathbb{K}_X(t) = \sum_{i=1}^{n} \mathbb{K}_{X_i}(t)$, and a saddlepoint approximation to the p.d.f. and c.d.f. of X can be computed. In particular, with $R(z) = \phi(z)/\Phi(z)$, $\mathbb{K}'_{X_i}(t) = \mu_i + t\sigma_i^2 + \delta_i\sigma_i R(t\delta_i\sigma_i)$ and, as $\phi'(t) = -t\phi(t)$, we have $\phi'(t\delta_i\sigma_i) = -\delta_i^2\sigma_i^2 t\phi(t\delta_i\sigma_i)$, $R'(t\delta_i\sigma_i) = -\delta_i^2\sigma_i^2 tR(t\delta_i\sigma_i) - \delta_i\sigma_i R^2(t\delta_i\sigma_i)$, and $\mathbb{K}''_{X_i}(t) = \sigma_i^2 + \delta_i\sigma_i R'(t\delta_i\sigma_i)$.

Similarly, the second-order saddlepoint approximation requires $\mathbb{K}^{(3)}_{X_i}(t)$ and $\mathbb{K}^{(4)}_{X_i}(t)$; with $R = R(t\delta_i\sigma_i)$ and $R''(t\delta_i\sigma_i) = -\delta_i^2\sigma_i^2[tR' + R] - 2\delta_i\sigma_i RR'$, these are easily seen to be $\mathbb{K}^{(3)}_{X_i}(t) = \delta_i\sigma_i R''$ and $\mathbb{K}^{(4)}_{X_i}(t) = -\delta_i^3\sigma_i^3(tR'' + 2R') - 2\delta_i^2\sigma_i^2(RR'' + R'^2)$.

See the program in Listing A.17.

A.21 From (A.219), the c.d.f. is

$$F_T(t; k, \mu) = \frac{2^{-(k/2)+1}k^{k/2}}{\Gamma(k/2)} \int_0^\infty \Phi(tz; \mu, 1)z^{k-1} \exp\left\{-\frac{1}{2}kz^2\right\} dz$$

where

$$\Phi(t; \mu, 1) = \frac{1}{\sqrt{2\pi}} \int_{-\infty}^t \exp\left\{-\frac{1}{2}(x-\mu)^2\right\} dx,$$

while from (A.339), the p.d.f. is

$$f_T(t; k, \mu) = C(k) \int_0^\infty z^k \exp\left\{-\frac{1}{2}[(tz-\mu)^2 + kz^2]\right\} dz, \quad C(k) := \frac{2^{-(k/2)+1}k^{k/2}}{\Gamma(k/2)\sqrt{2\pi}}.$$

It is easy to verify that

$$C(k+2) = \left(\frac{k+2}{k}\right)^{(k/2)+1} C(k),$$

so that

$$F_T(t; k, \mu) = \sqrt{2\pi}C(k) \int_0^\infty \Phi(tz; \mu, 1)z^{k-1} \exp\left\{-\frac{1}{2}kz^2\right\} dz.$$

```
1   function [f,F]=skewnormsumspa(xvec,lambdavec,muvec,sigvec,acclevel)
2   % [f,F]=skewnormsumspa(xvec,lambdavec,muvec,sigvec,acclevel)
3   % SPA to sums of n independent SN r.v.s X_1,...,X_n, where X_i SN
4   %   with location mu_i, scale sig_i, skewness param lambda_i,
5   %   and muvec=(mu_1,...,mu_n), etc.
6   % Set acclevel=2 (default) to use the 2nd order SPA.
7   % If n=2, set acclevel=3 to use the exact convolution formula
8
9   if nargin<5, acclevel=2; end
10  xlen=length(xvec); F=zeros(xlen,1); f=F;
11  n=length(lambdavec); lam=reshape(lambdavec,n,1);
12  mu=reshape(muvec,n,1); sig=reshape(sigvec,n,1);
13  d = lam./sqrt(1+lam.^2); % delta
14  opt=optimset('Display','none','TolX',1e-6); guess=0; tol=1e-14;
15  if (n==2) çç (acclevel==3) % exact calc
16    lo1=mu(1); while dsn(lo1,mu(1),sig(1),lam(1))>tol, lo1=lo1-1; end
17    lo2=mu(2); while dsn(lo2,mu(2),sig(2),lam(2))>tol, lo2=lo2-1; end
18    hi1=mu(1); while dsn(hi1,mu(1),sig(1),lam(1))>tol, hi1=hi1+1; end
19    hi2=mu(2); while dsn(hi2,mu(2),sig(2),lam(2))>tol, hi2=hi2+1; end
20    lo=min(lo1,lo2); hi=max(hi1,hi2);
21    for xloop=1:xlen
22      x=xvec(xloop);
23      f(xloop) = quadl(@conv1,lo,hi,tol,[],x,mu,sig,lam);
24      if nargout>1, F(xloop) = quadl(@conv2,lo,hi,tol,[],x,mu,sig,lam); end
25    end, return
26  end
27  for xloop=1:xlen
28    x=xvec(xloop); t=fzero(@speeq,guess,opt,x,d,mu,sig); guess=t;
29    K = sum( log(2) + t*mu + 0.5*t^2*sig.^2 + log(normcdf(t*sig.*d)));
30    R=normrat(t*d.*sig); Rp = -d.^2.* sig.^2.* R * t - d.*sig.*R.^2;
31    kpp = sum( sig.^2 + d.*sig.*Rp);
32    f(xloop)= 1/sqrt(2*pi*kpp) * exp(K-x*t);
33    if nargout>1
34      w=sign(t)*sqrt(2*t*x-2*K); u=t*sqrt(kpp);
35      F(xloop)=normcdf(w)+normpdf(w)*(1/w - 1/u);
36    end
```

Program Listing A.17: Convolution for $n = 2$ and saddlepoint approximation (any n) for the sum of independent SN r.v.s. Continued in Listing A.18.

Then, with $R = \sqrt{(k+2)/k}$,

$$F_T(tR; k+2, \mu) = R^{k+2}\sqrt{2\pi}C(k)$$

$$\times \int_0^\infty \Phi(tRz; \mu, 1)z^{k+1} \exp\left\{-\frac{1}{2}(k+2)z^2\right\} dz$$

or

$$F_T(tR; k+2, \mu) = R^{k+2}C(k)$$

$$\times \int_0^\infty \int_0^{tRz} \exp\left\{-\frac{1}{2}(x-\mu)^2\right\} dx z^{k+1} \exp\left\{-\frac{1}{2}(k+2)z^2\right\} dz$$

$$= R^{k+2}C(k)\int_0^\infty (-1)(k+2)z \exp\left\{-\frac{1}{2}(k+2)z^2\right\} \dots$$

$$\dots \frac{(-1)}{k+2} z^k \int_0^{tRz} \exp\left\{ -\frac{1}{2}(x-\mu)^2 \right\} \, \mathrm{d}x \, \mathrm{d}z.$$

Now apply integration by parts with

$$u = \frac{-1}{k+2} z^k \int_0^{tRz} \exp\left\{ -\frac{1}{2}(x-\mu)^2 \right\} \, \mathrm{d}x, \quad \mathrm{d}v$$

$$= (-1)(k+2)z \exp\left\{ -\frac{1}{2}(k+2)z^2 \right\} \, \mathrm{d}z$$

to get v as

$$v = \exp\left\{ -\frac{1}{2}(k+2)z^2 \right\},$$

```
 1    if acclevel==2
 2      Rpp= -d.^2.* sig.^2.* (t*Rp+R) - 2*d.*sig.*R.*Rp; K3=sum( d.*sig.*Rpp);
 3      K4=sum(-d.^3.* sig.^3.* (2*Rp+t*Rpp) - 2*d.^2.* sig.^2.* (R.*Rpp+Rp.^2));
 4      kap3= K3/kpp^(1.5); kap4=K4/kpp^2;
 5      f(xloop) = f(xloop) * (1 + kap4/8 - 5*(kap3^2)/24);
 6      if nargout>1
 7        term = (kap4/8-5*kap3^2/24)/u - 1/u^3 - kap3/2/u^2 + 1/w^3;
 8        F(xloop)=F(xloop)-normpdf(w)*term;
 9      end
10    end
11  end
12
13  function diff=speeq(tvec,x,delta,mu,sig)
14    diff=zeros(size(tvec));
15    for i=1:length(tvec) % in case matlab wants to pass a vector of t values.
16      t=tvec(i); R = normrat(t*delta.*sig);
17      kp = sum( mu + t*sig.^2 + delta.*sig.*R); diff(i) = x-kp;
18    end
19
20  function rat=normrat(u)
21    if 1==2 % simple way, but numerically problematic
22      rat = normpdf(u)./normcdf(u);
23    else
24      rat=zeros(size(u)); oor = find(u<-36); % oor is out of range
25      ok = find(u>=-36); uok=u(ok); uoor=-u(oor);
26      rat(ok) = normpdf(uok)./normcdf(uok);
27      rat(oor) = 1./(1./uoor - 1./(uoor.^3));
28    end
29
30  function p=conv1(y,x,mu,sig,lam)
31    p=dsn(x-y,mu(1),sig(1),lam(1)).*dsn(y,mu(2),sig(2),lam(2));
32  function p=conv2(y,x,mu,sig,lam)
33    p=psn(x-y,mu(1),sig(1),lam(1)).*dsn(y,mu(2),sig(2),lam(2));
34  function d=dsn(x,location,scale,shape)
35    if nargin<4, shape=0; end; if nargin<3, scale=1; end;
36    if nargin<2, location=0; end;
37    z = (x-location)./scale; d = 2.*normpdf(z).* normcdf(shape.* z)./ scale;
```

Program Listing A.18: Continued from Listing A.17.

and, from the derivative of a product, and Leibniz's rule, du/dz as

$$\frac{-k}{k+2}z^{k-1}\int_0^{tRz}\exp\left\{-\frac{1}{2}(x-\mu)^2\right\}\,dx-\frac{1}{k+2}z^k tR\exp\left\{-\frac{1}{2}(tRz-\mu)^2\right\}.$$

Then $F_T(tR; k+2, \mu)$ is $R^{k+2}C(k)$ times $uv - \int v\,du$, that is, times

$$\left[\exp\left\{-\frac{1}{2}(k+2)z^2\right\}\frac{(-1)}{k+2}z^k\int_0^{tRz}\exp\left\{-\frac{1}{2}(x-\mu)^2\right\}\,dx\right]_{z=0}^{\infty} \tag{A.349}$$

$$+\int_0^{\infty}\exp\left\{-\frac{1}{2}(k+2)z^2\right\}\frac{k}{k+2}z^{k-1}\int_0^{tRz}\exp\left\{-\frac{1}{2}(x-\mu)^2\right\}\,dx\,dz$$

$$+\int_0^{\infty}\frac{1}{k+2}z^k tR\exp\left\{-\frac{1}{2}(k+2)z^2\right\}\exp\left\{-\frac{1}{2}(tRz-\mu)^2\right\}\,dz.$$

Note that the term in (A.349) is zero because $\int_0^z \exp\left\{-\frac{1}{2}(x-\mu)^2\right\}\,dx$ converges for all z and

$$\lim_{z\to\infty}z^k\exp\left\{-\frac{1}{2}(k+2)z^2\right\}=0.$$

Thus, after simplifying the k and $(k+2)$ terms, $F_T(tR; k+2, \mu)$ is

$$R^k C(k)\int_0^{\infty}\exp\left\{-\frac{1}{2}(k+2)z^2\right\}z^{k-1}\int_0^{tRz}\exp\left\{-\frac{1}{2}(x-\mu)^2\right\}\,dx\,dz$$

$$+R^{k+2}\frac{1}{\sqrt{(k+2)k}}C(k)\int_0^{\infty}t\exp\left\{-\frac{1}{2}(k+2)z^2\right\}z^k\exp\left\{-\frac{1}{2}(tRz-\mu)^2\right\}\,dz.$$

Using the change of variables

$$u = R\,z, \quad \Rightarrow \quad z = \frac{u}{R} \quad \Rightarrow \quad z^2 = \frac{k}{k+2}u^2,$$

$$du = R\,dz \quad \Rightarrow \quad dz = \frac{du}{R}$$

we get

$$F_T(tR; k+2, \mu) = C(k)\int_0^{\infty}\exp\left\{-\frac{1}{2}ku^2\right\}u^{k-1}\int_{-\infty}^{tu}\exp\left\{-\frac{1}{2}(x-\mu)^2\right\}\,dx\,du$$

$$+\frac{t}{k}C(k)\int_0^{\infty}t\exp\left\{-\frac{1}{2}ku^2\right\}u^k\exp\left\{-\frac{1}{2}(tu-\mu)^2\right\}\,du$$

$$= F_T(t; k, \mu) + \frac{t}{k}f_T(t; k, \mu),$$

or

$$f_T(t; k, \mu) = \frac{k}{t}[F_T(tR; k+2, \mu) - F_T(t; k, \mu)],$$

which is (A.340).

A.22 The density $f_X(x; \lambda, m)$ is

$$
F'_X(x; \lambda, m) = \frac{d}{dx} \Pr(S \le xY) = \frac{d}{dx} \int_0^\infty \Pr(S \le xY \mid Y = y) f_Y(y) \, dy
$$

$$
= \frac{d}{dx} \int_0^\infty F_X(xy) f_Y(y) \, dy = 2 \int_0^\infty \frac{d}{dx} \int_{-\infty}^{xy} \phi(t) \Phi(\lambda t) \, dt \, f_Y(y) \, dy
$$

$$
= 2 \int_0^\infty y \phi(xy) \Phi(\lambda xy) f_Y(y) \, dy = \mathbb{E}[Y \phi(xY) \Phi(\lambda xY)].
$$

Then, with

$$
f_Y(y; m) = \frac{2^{-m/2+1} m^{m/2}}{\Gamma(m/2)} y^{m-1} \exp(-my^2/2) \mathbb{1}_{(0,\infty)}(y) \tag{A.350}
$$

(see page II.80) and, using the substitution $s = y\sqrt{(x^2 + m)/(m + 1)}$, obtained after some trial and error, we have

$$
f_X(x; \lambda, m) = \frac{2}{\sqrt{2\pi}} \frac{2^{-m/2+1} m^{m/2}}{\Gamma(m/2)} \int_0^\infty y^m \exp\left\{ -\frac{1}{2} y^2 [x^2 + m] \right\} \Phi(\lambda xy) \, dy
$$

$$
= \frac{2^{-m/2+2} m^{m/2}}{\sqrt{2\pi} \Gamma(m/2)} \left(\frac{m+1}{x^2 + m} \right)^{(m+1)/2}
$$

$$
\times \int_0^\infty s^m \exp\left\{ -\frac{1}{2}(s^2(m+1)) \right\} \Phi\left(s\lambda x \sqrt{\frac{m+1}{x^2 + m}} \right) ds.
$$

Writing the terms preceding the integral as

$$
\frac{\Gamma((m+1)/2) m^{m/2}}{\sqrt{\pi} \Gamma(m/2)} (x^2 + m)^{-(m+1)/2} \times 2 \times \frac{2^{-(m+1)/2+1}}{\Gamma((m+1)/2)} (m+1)^{(m+1)/2},
$$

the result follows from using (A.350) with $m + 1$ degrees of freedom and (A.327).

A.23 **(a)** $\mathbf{M}_3$ has

$$
\sum_{i=1}^n \binom{n+2-i}{2}
$$

unique elements. To prove that

$$
\binom{n+2}{3} = \sum_{i=1}^n \binom{n+2-i}{2}, \tag{A.351}
$$

use induction. It is easy to see that it holds for $n = 1$ and $n = 2$. Assume it holds for n. Then, recalling that

$$
\binom{a}{b} = \binom{a-1}{b} + \binom{a-1}{b-1} \quad \text{and} \quad \sum_{i=1}^n i = \frac{(n+1)n}{2},
$$

we have, for the $n + 1$ case,

$$\sum_{i=1}^{(n+1)}\binom{(n+1)+2-i}{2} = \sum_{i=1}^{n+1}\binom{n+3-i}{2} = \sum_{i=1}^{n+1}\binom{n+2-i}{2} + \sum_{i=1}^{n+1}\binom{n+2-i}{1}$$

$$= \sum_{i=1}^{n}\binom{n+2-i}{2} + \sum_{i=1}^{n+1}(n+2-i)$$

$$= \binom{n+2}{3} + (n+2)(n+1) - \frac{(n+2)(n+1)}{2}$$

$$= \binom{n+2}{3} + \frac{(n+2)(n+1)}{2} = \binom{n+2}{3} + \binom{n+2}{2}$$

$$= \binom{n+3}{3}.$$

(b) For $\mathbf{M}_3$, we need to "run through" all $i = 1, \ldots, n$ but then only through the $j \le i$ and $k \le j$. Note that if you take any $i, j, k \in \{1, \ldots, n\}$, then $S_{i,j,k} = S_{i^*,j^*,k^*}$, where $i^* = \max(i, j, k)$, j^* is the second largest, and k^* is the last one remaining. Thus, $1 \le i^* \le n, j^* \le i^*$, and $k^* \le j^*$. So, we want

$$\sum_{i=1}^{n}\sum_{j=1}^{i}\sum_{k=1}^{j} 1 = \sum_{i=1}^{n}\sum_{j=1}^{i} j = \sum_{i=1}^{n} \frac{1}{2}i(i+1) = \frac{1}{2}\sum_{i=1}^{n}(i^2 + i)$$

$$= \frac{1}{2}\frac{n(n+1)(2n+1)}{6} + \frac{1}{2}\frac{n(n+1)}{2}.$$

Simplifying this yields

$$\frac{n(n+1)(n+2)}{6} = \binom{n+2}{3}.$$

(c) The number of unique elements in the co-kurtosis matrix $\mathbf{M}_4$ is

$$\sum_{i=1}^{n}\sum_{j=1}^{i}\sum_{k=1}^{j}\sum_{\ell=1}^{k} 1 = \sum_{i=1}^{n}\sum_{j=1}^{i}\sum_{k=1}^{j} k = \sum_{i=1}^{n}\sum_{j=1}^{i} \frac{j(j+1)}{2}$$

$$= \sum_{i=1}^{n} \frac{i(i+1)(i+2)}{6} = \frac{n(n+1)(n+2)(n+3)}{24} = \binom{n+3}{4},$$

where we used the fact that (see page I.21 for derivation)

$$\sum_{k=1}^{n} k^3 = \frac{n^2(n+1)^2}{4}.$$

(d) A Matlab program is given in Listing A.19.

A.24 (As in Severini, 2005, p. 31)
($\Leftarrow$) If X and Y have the same distribution, then clearly $\mathbb{E}[g(X)] = \mathbb{E}[g(Y)]$ for all functions g for which the expectations exist. As these expectations exist for bounded g, the first part of the result follows.

```
1    function M3 = coskewness(n)
2    % Constructs the co-skewness matrix M3, but with bogus entries.
3    % These should be replaced with data-based estimates.
4    % For portfolio weight vector a, the skewness is a'*M3*kron(a,a)
5
6    sk=zeros(n,n,n); count=0;
7    % the i=1 case
8    for j=1:n
9      for k=j:n
10       sk(1,j,k)=100*1+10*j+k; % replace later with real values.
11       count=count+1;
12       sk(1,k,j)=sk(1,j,k);
13     end
14   end
15   % now the other layers
16   for i=2:n
17     % Do the redundant ones.
18     % The first i-1 rows and columns are already determined
19     for j=1:(i-1)
20       sk(i,j,:) = sk(j,i,:); % the jth row. works without squeeze
21       % just copy the transpose into the jth column, but need to 'squeeze'
22       sk(i,:,j) = squeeze(sk(j,i,:))';
23     end
24     % the non-redundant entries
25     for j=i:n
26       for k=j:n
27         sk(i,j,k)=100*i+10*j+k; % replace later with real values.
28         count=count+1;
29         sk(i,k,j)=sk(i,j,k);
30       end
31     end
32   end
33   count, shouldbe = c(n+2,3)
34   M3=[]; for i=1:n, M3 = [M3 , squeeze(sk(i,:,:))]; end
```

Program Listing A.19: Constructs the co-skewness matrix $\mathbf{M}_3$.

($\Rightarrow$) Now suppose that $\mathbb{E}[g(X)] = \mathbb{E}[g(Y)]$ for all bounded continuous g. This implies that X and Y have the same support (neglecting sets with measure zero) because, if they do not, it is easy to construct a function g for which the expected values differ. The key is to express the c.d.f. of X,

$$F_X(z) = \Pr(X \leq z) = \mathbb{E}[\mathbb{I}\{X \leq z\}].$$

The indicator function is bounded but not continuous; however, it can be approximated by a bounded, continuous function to arbitrary accuracy. First suppose that X and Y are univariate. Fix a real number z and, for $\epsilon > 0$, define

$$g_\epsilon(t) = g_\epsilon(t;z) = \begin{cases} 1, & \text{if } t \leq z, \\ 1 - (t-z)/\epsilon, & \text{if } z < t < z + \epsilon, \\ 0, & \text{if } t \geq z + \epsilon. \end{cases}$$

Clearly, g_ϵ is bounded and continuous and

$$\mathbb{E}[g_\epsilon(X)] = F_X(z) + \int_z^{z+\epsilon} \left(1 - \frac{x-z}{\epsilon}\right) dF_X(x)$$

$$= F_X(z) + F_X(z+\epsilon) - F_X(z) - \frac{1}{\epsilon} \int_z^{z+\epsilon} (x-z) \, dF_X(x). \quad \text{(A.352)}$$

Using integration by parts with $u = x$, $dv = dF_X(x)$, $du = dx$, $v = F_X(x)$,

$$\int_z^{z+\epsilon} (x-z) \, dF_X(x) = \int_z^{z+\epsilon} x \, dF_X(x) - z \int_z^{z+\epsilon} dF_X(x)$$

$$= xF_X(x)\Big|_z^{z+\epsilon} - \int_z^{z+\epsilon} F_X(x) \, dx - z \int_z^{z+\epsilon} dF_X(x)$$

$$= (z+\epsilon)F_X(z+\epsilon) - zF_X(z)$$

$$- \int_z^{z+\epsilon} F_X(x) \, dx - z(F_X(z+\epsilon) - F_X(z))$$

$$= \epsilon F_X(z+\epsilon) - \int_z^{z+\epsilon} F_X(x) \, dx,$$

so that, from (A.352),

$$\mathbb{E}[g_\epsilon(X)] = F_X(z+\epsilon) - \frac{1}{\epsilon} \left[\epsilon F_X(z+\epsilon) - \int_z^{z+\epsilon} F_X(x) \, dx \right] = \frac{1}{\epsilon} \int_z^{z+\epsilon} F_X(x) \, dx.$$

This makes sense because, for small ϵ, $g_\epsilon(X) \approx \mathbb{I}(X \leq z)$ so that $\mathbb{E}[g_\epsilon(X)] \approx F_X(z)$, while, from the mean value theorem for integrals (see, for example, (I.A.58)), for some $c \in [z, z+\epsilon]$,

$$\mathbb{E}[g_\epsilon(X)] = \frac{1}{\epsilon} \int_z^{z+\epsilon} F_X(x) \, dx = \frac{1}{\epsilon}(z + c - z)F_X(c) \approx F_X(z).$$

Hence, for all $\epsilon > 0$, as $\mathbb{E}[g_\epsilon(X)] = \mathbb{E}[g_\epsilon(Y)]$ by assumption,

$$\frac{1}{\epsilon} \int_z^{z+\epsilon} F_X(x) \, dx = \frac{1}{\epsilon} \int_z^{z+\epsilon} F_Y(y) \, dy$$

or, equivalently,

$$F_Y(z) - F_X(z) = \frac{1}{\epsilon} \int_z^{z+\epsilon} [F_X(x) - F_X(z)] \, dx - \frac{1}{\epsilon} \int_z^{z+\epsilon} [F_Y(y) - F_Y(z)] \, dy.$$

As F_X and F_Y are nondecreasing, the previous two integrands are nonnegative, and so from the triangle inequality with this nonnegativity, $|x - y| \leq |x| + |y| = x + y$, that is,

$$|F_Y(z) - F_X(z)| \leq \frac{1}{\epsilon} \int_z^{z+\epsilon} [F_X(x) - F_X(z)] \, dx + \frac{1}{\epsilon} \int_z^{z+\epsilon} [F_Y(y) - F_Y(z)] \, dy.$$

As F_X is nondecreasing,

$$\frac{1}{\epsilon} \int_z^{z+\epsilon} F_X(x) \, dx \leq \frac{1}{\epsilon} \int_z^{z+\epsilon} F_X(z+\epsilon) \, dx = \frac{\epsilon}{\epsilon} F_X(z+\epsilon)$$

and

$$|F_Y(z) - F_X(z)| \leq [F_X(z + \epsilon) - F_X(z)] + [F_Y(z + \epsilon) - F_Y(z)].$$

Taking limits as $\epsilon \to 0$, and recalling that F_X and F_Y are right continuous, $|F_Y(z) - F_X(z)| \leq 0$. As the c.d.f.s of X and Y characterize $\Pr(X \in A)$ and $\Pr(Y \in A)$ for all measurable events A, this implies $X \overset{d}{=} Y$, as was to be shown.

For the multivariate case, it will be useful to let $X_0 = \rho_z(X) = (X - z)^+$, where $(x)^+ = x$ for $x > 0$ and zero otherwise. Then

$$g_\epsilon(t; z) = \begin{cases} 1, & \text{if } \rho_z(t) = 0, \\ 1 - \rho_z(t)/\epsilon, & \text{if } 0 < \rho_z(t) < \epsilon, \\ 0, & \text{if } \rho_z(t) \geq \epsilon, \end{cases}$$

and

$$\mathbb{E}[g_\epsilon(X)] = \Pr(X_0 = 0) + \int_0^\epsilon \left(1 - \frac{x_0}{\epsilon}\right) dF_{X_0}(x_0)$$

$$= \Pr(X \leq z) + F_{X_0}(\epsilon) - F_{X_0}(0) - \frac{1}{\epsilon} \int_0^\epsilon x_0 \, dF_{X_0}(x_0)$$

$$= F_X(z) + F_X(z + \epsilon) - F_X(z) - \frac{1}{\epsilon} \int_z^{z+\epsilon} (x - z) \, dF_X(x),$$

which follows because $x_0 > 0 \Leftrightarrow x > z$, and this is the same as (A.352). For r.v.s $X, Y \in \mathbb{R}^d$ and a given value of $z = (z_1, \ldots, z_d) \in \mathbb{R}^d$, let $A = (-\infty, z_1] \times \cdots \times (-\infty, z_d]$ and $\rho(t)$ be the Euclidean distance from $t = (t_1, \ldots, t_d)$ to A, defining $\rho(t)$ as

$$\rho^2(t) = \sum_{i=1}^d ((t_i - z_i)^+)^2.$$

For $\epsilon > 0$, let

$$g_\epsilon(t) \equiv g_\epsilon(t; z) = \begin{cases} 1, & \text{if } \rho(t) = 0, \\ 1 - \rho(t)/\epsilon, & \text{if } 0 < \rho(t) < \epsilon, \\ 0, & \text{if } \rho(t) \geq \epsilon. \end{cases}$$

As ρ is a continuous function of $\mathbb{R}^d$, g is also continuous, and clearly bounded.

Let $X_0 = \rho(X)$ and $Y_0 = \rho(Y)$, so that X_0 and Y_0 are scalar r.v.s with c.d.f.s F_{X_0} and F_{Y_0}, respectively. Note that

$$\mathbb{E}[g_\epsilon(X)] = F_X(z) + \int_0^\epsilon \left(1 - \frac{x}{\epsilon}\right) dF_{X_0}(x),$$

that is,

$$F_X(z) = \mathbb{E}[g_\epsilon(X)] - \int_0^\epsilon \left(1 - \frac{x}{\epsilon}\right) dF_{X_0}(x),$$

and $\mathbb{E}[g_\epsilon(\mathbf{X})] = \mathbb{E}[g_\epsilon(\mathbf{Y})]$ by assumption, so that

$$F_{\mathbf{Y}}(\mathbf{z}) - F_{\mathbf{X}}(\mathbf{z}) = \int_0^\epsilon \left(1 - \frac{x}{\epsilon}\right) dF_{X_0}(x) - \int_0^\epsilon \left(1 - \frac{y}{\epsilon}\right) dF_{Y_0}(y). \qquad (A.353)$$

The first integral is

$$\int_0^\epsilon \left(1 - \frac{x}{\epsilon}\right) dF_{X_0}(x) = F_{X_0}(\epsilon) - F_{X_0}(0) - \frac{1}{\epsilon} \int_0^\epsilon x \, dF_{X_0}(x)$$

$$= F_{X_0}(\epsilon) - F_{X_0}(0) - \frac{1}{\epsilon} \left[\epsilon F_{X_0}(\epsilon) - \int_0^\epsilon F_{X_0}(x) \, dx\right]$$

$$= -F_{X_0}(0) + \frac{1}{\epsilon} \int_0^\epsilon F_{X_0}(x) \, dx,$$

where we used integration by parts with $u = x$, $dv = dF_{X_0}(x)$, $du = dx$, $v = F_{X_0}(x)$ to get

$$\int_0^\epsilon x \, dF_{X_0}(x) = x F_{X_0}(x) \Big|_0^\epsilon - \int_0^\epsilon F_{X_0}(x) \, dx = \epsilon F_{X_0}(\epsilon) - \int_0^\epsilon F_{X_0}(x) \, dx.$$

Thus, (A.353) is

$$F_{\mathbf{Y}}(\mathbf{z}) - F_{\mathbf{X}}(\mathbf{z}) = \int_0^\epsilon \left(1 - \frac{x}{\epsilon}\right) dF_{X_0}(x) - \int_0^\epsilon \left(1 - \frac{y}{\epsilon}\right) dF_{Y_0}(y)$$

$$= -F_{X_0}(0) + \frac{1}{\epsilon} \int_0^\epsilon F_{X_0}(x) \, dx + F_{Y_0}(0) - \frac{1}{\epsilon} \int_0^\epsilon F_{Y_0}(y) \, dy$$

$$= \frac{1}{\epsilon} \int_0^\epsilon [F_{X_0}(x) - F_{X_0}(0)] \, dx - \frac{1}{\epsilon} \int_0^\epsilon [F_{Y_0}(y) - F_{Y_0}(0)] \, dy.$$

The result now follows as in the scalar random variable case.

References

Abad, P., S. Benito, and C. López (2014). A Comprehensive Review of Value at Risk Methodologies, *Spanish Review of Financial Economics* **12**(1), 15–32.

Abadir, K. M. and J. R. Magnus (2005). *Matrix Algebra*, Cambridge: Cambridge University Press.

Aban, I. B. and M. M. Meerschaert (2001). Shifted Hill's Estimator For Heavy Tails, *Communications in Statistics – Simulation and Computation* **30**(4), 949–962.

Abramowitz, M. and I. A. Stegun (1972), *Handbook of Mathematical Functions*, New York: Dover.

Acerbi, C. (2002). Spectral Measures of Risk: A Coherent Representation of Subjective Risk Aversion, *Journal of Banking & Finance* **26**(7), 1505–1518.

Acerbi, C. (2004). Coherent Representations of Subjective Risk–Aversion, in G. Szegö (edn.), *Risk Measures for the 21st Century*, chap. 10, Chichester: John Wiley & Sons.

Acerbi, C. and D. Tasche (2002). Expected Shortfall: A Natural Coherent Alternative to Value-at-Risk, *Economic Notes* **31**(2), 379–388.

Adcock, C. J. (2010). Asset Pricing and Portfolio Selection Based on the Multivariate Extended Skew-Student-*t* Distribution, *Annals of Operations Research* **176**(1), 221–234.

Adcock, C. J. (2014). Mean–Variance–Skewness Efficient Surfaces, Stein's Lemma and the Multivariate Extended Skew-Student Distribution, *European Journal of Operational Research* **234**(2), 392–401.

Adcock, C. J., M. Eling, and N. Loperfido (2015). Skewed Distributions in Finance and Actuarial Science: A Preview, *European Journal of Finance* **21**(13–14), 1253–1281.

Adcock, C. J. and K. Shutes (2005). An Analysis of Skewness and Skewness Persistence in Three Emerging Markets, *Emerging Markets Review* **6**, 396–418.

Adler, R. J. (1997). Discussion: Heavy Tail Modeling and Teletraffic Data, *Annals of Statistics* **25**(5), 1849–1852.

Adler, R. J., R. E. Feldman, and C. Gallagher (1998). Analysing Stable Time Series, in R. J. Adler, R. E. Feldman, and M. S. Taqqu (eds). *A Practical Guide to Heavy Tails*, 133–158, Boston: Birkhäuser.

Fundamental Statistical Inference: A Computational Approach, First Edition. Marc S. Paolella.
© 2018 John Wiley & Sons Ltd. Published 2018 by John Wiley & Sons Ltd.

Aggarwal, C. C. (2013). *Outlier Analysis*, New York: Springer-Verlag.

Agresti, A. (1992). A Survey of Exact Inference for Contingency Tables (with discussion), *Statistical Science* **7**, 131–177.

Agresti, A. and A. Gottard (2005). Comment: Randomized Confidence Intervals and the Mid–*P* Approach, *Statistical Science* **20**(4), 367–371.

Ahn, D.-H., R. F. Dittmar, and A. R. Gallant (2002). Quadratic Term Structure Models: Theory and Evidence, *Review of Financial Studies* **15**(1), 243–288.

Aitkin, M. and M. Stasinopoulos (1989). Likelihood Analysis of a Binominal Sample Size Problem, in L. J. Gleser, M. D. Perlman, S. J. Press, and A. R. Sampson (eds). *Contributions to Probability and Statistics: Essays in Honor of Ingram Olkin*, New York: Springer-Verlag.

Alexander, C. and E. Lazar (2006). Normal Mixture GARCH(1,1): Applications to Exchange Rate Modelling, *Journal of Applied Econometrics* **21**, 307–336.

Andersen, T. G., T. Bollerslev, P. Frederiksen, and M. Ø. Nielsen (2010). Continuous-Time Models, Realized Volatilities, and Testable Distributional Implications for Daily Stock Returns, *Journal of Applied Econometrics* **25**, 233–261.

Andersen, T. G., H.-J. Chung, and B. E. Sørensen (1999). Efficient Method of Moments Estimation of a Stochastic Volatility model: A Monte Carlo Study, *Journal of Econometrics* **91**, 61–87.

Andersen, T. G. and J. Lund (2003). Estimating Continuous-Time Stochastic Volatility Models of the Short-Term Interest Rate, *Journal of Econometrics* **77**(2), 343–377.

Anderson, D. R., K. P. Burnham, and W. L. Thompson (2000). Null Hypothesis Testing: Problems, Prevalence, and an Alternative, *Journal of Wildlife Management* **64**(4), 912–923.

Anderson, T. W. and D. A. Darling (1952). Asymptotic Theory of Certain "Goodness of Fit" Criteria Based on Stochastic Processes, *Annals of Mathematical Statistics* **23**, 193–212.

Anderson, T. W. and D. A. Darling (1954). A Test of Goodness of Fit, *Journal of the American Statistical Association* **49**, 765–769.

Andrews, D. W. K. (2001). Testing when a Parameter is on the Boundary of the Maintained Hypothesis, *Econometrica* **69**(3), 683–734.

Antonov, A., S. Mechkov, and T. Misirpashaev (2005). *Analytical Techniques for Synthetic CDOs and Credit Default Risk Measures*. Technical report, Numerix LLC, New York.

Asgharzadeh, A. and M. Abdi (2011). Confidence Intervals and Joint Confidence Regions for the Two-Parameter Exponential Distribution based on Records, *Communications of the Korean Statistical Society* **18**(1), 103–110.

Ash, R. B. and C. A. Doléans-Dade (2000). *Probability & Measure Theory*, 2nd edn, San Diego: Harcourt Academic Press.

Asimit, A. V. and J. Li (2016). Extremes for Coherent Risk Measures, *Insurance: Mathematics and Economics* **71**, 332–341.

Atienza, N., J. Garcia-Heras, and J. M. M. noz Pichardo (2006). A New Condition for Identifiability of Finite Mixture Distributions, *Metrika* **63**(2), 215–221.

Augustyniak, M. and M. Boudreault (2012). An Out-of-Sample Analysis of Investment Guarantees for Equity-linked Products: Lessons from the Financial Crisis of the Late-2000s, *North American Actuarial Journal* **16**, 183–206.

Azzalini, A. (1985). A Class of Distributions which Includes the Normal Ones, *Scandinavian Journal of Statistics* **12**, 171–178.

Azzalini, A. (1986). Further Results on a Class of Distributions which Includes the Normal Ones, *Statistica* **46**(2), 199–208. Errata: http://azzalini.stat.unipd.it/SN/errata86 .pdf.

Azzalini, A. and A. Capitanio (1999). Statistical Applications of the Multivariate Skew Normal Distribution, *Journal of the Royal Statistical Society, Series B* **61**(3), 579–602.

Azzalini, A. and A. Capitanio (2003). Distributions Generated by Perturbation of Symmetry with Emphasis on a Multivariate Skew *t*-Distribution, *Journal of the Royal Statistical Society, Series B* **65**, 367–389.

Azzalini, A. and A. Dalla Valle (1996). The Multivariate Skew-normal Distribution, *Biometrika* **83**(4), 715–726.

Bak, J. and D. J. Newman (2010). *Complex Analysis*, 3rd edn, New York: Springer-Verlag.

Bao, Y. (2007). The Approximate Moments of the Least Squares Estimator for the Stationary Autoregressive Model Under a General Error Distribution, *Econometric Theory* **23**, 1013–1021.

Bao, Y. and A. Ullah (2007). The Second-Order Bias and Mean Squared Error of Estimators in Time-Series Models, *Journal of Econometrics* **140**, 650–669.

Barndorff-Nielsen, O. (1991). Likelihood Theory, in D. V. Hinkley, N. Reid, and E. J. Snell (eds). *Statistical Theory and Modelling*, London: Chapman & Hall.

Barone-Adesi, G., P. Gagliardini, and G. Urga (2004). Testing Asset Pricing Models with Coskewness, *Journal of Business & Economic Statistics* **22**, 474–485.

Barry, A. E., L. E. Szucs, J. V. Reyes, Q. Ji, K. L. Wilson, and B. Thompson (2016). Failure to Report Effect Sizes: The Handling of Quantitative Results in Published Health Education and Behavior Research, *Health Education & Behavior* **43**(5), 518–527.

Bauwens, L., C. M. Hafner, and J. V. K. Rombouts (2007). Multivariate Mixed Normal Conditional Heteroskedasticity, *Computational Statistics & Data Analysis* **51**(7), 3551–3566.

Beirlant, J., Y. Goegebeur, J. Teugels, and J. Segers (2004). *Statistics of Extremes: Theory and Applications*, Chichester: John Wiley & Sons.

Bellini, F. and V. Bignozzi (2015). On Elicitable Risk Measures, *Quantitative Finance* **15**(5), 725–733.

Bera, A. K. (2003). The ET Interview: Professor C.R. Rao, *Econometric Theory* **19**(2), 331–400.

Bera, A. K. and C. M. Jarque (1981). Efficient Tests for Normality, Homoscedasticity and Serial Independence of Regression Residuals: Monte Carlo Evidence, *Economics Letters* **7**, 313–318.

Beran, J. and D. Schell (2012). On Robust Tail Index Estimation, *Computational Statistics & Data Analysis* **56**(11), 3430–3443.

Berger, J. O. and T. Sellke (1987). Testing a Point Null Hypothesis: The Irreconcilability of *P* Values and Evidence (with comments), *Journal of the American Statistical Association* **82**(397), 112–139.

Berkson, J. (1980). Minimum Chi-Square, Not Maximum Likelihood!, *Annals of Statistics* **8**(3), 457–487.

Bernardi, M. (2013). Risk Measures for Skew Normal Mixtures, *Statistics & Probability Letters* **83**(8), 1819–1824.

Berry, D. (2017). A *p*-value to die for, *Journal of the American Statistical Association* **112**(519), 895–897.

Bhattacharya, R., L. Lin, and V. Patrangenaru (2016). *A Course in Mathematical Statistics and Large Sample Theory*, New York: Springer-Verlag.

Bickel, P. J. and E. Levina (2008). Regularized Estimation of Large Covariance Matrices, *Annals of Statistics* **36**(1), 199–227.

Billio, M. and A. Monfort (2003). Kernel-Based Indirect Inference, *Journal of Financial Econometrics* **1**(3), 297–326.

Bishop, C. (2006). *Pattern Recognition and Machine Learning*, New York: Springer-Verlag.

Blair, G., K. Imai, and Y.-Y. Zhou (2015). Design and Analysis of the Randomized Response Technique, *Journal of the American Statistical Association* **110**(511), 1304–1319.

Blaker, H. (2000). Confidence Curves and Improved Exact Confidence Intervals for Discrete Distributions, *Canadian Journal of Statistics* **28**(4), 783–798. Corrigenda: (2001) **29**(4), 681.

Blom, G. (1958). *Statistical Estimates and Transformed Beta Variables*, New York: John Wiley & Sons.

Bolstad, W. M. and J. M. Curran (2017). *Introduction to Bayesian Statistics*, 3rd edn, Hoboken, NJ: John Wiley & Sons.

Borak, S., W. Härdle, and R. Weron (2005). Stable Distributions, in P. Čížek, W. Härdle, and R. Weron (eds). *Statistical Tools for Finance and Insurance*, 21–44, Springer Verlag.

Bowman, K. O. and L. R. Shenton (1975). Omnibus Test Contours for Departures from Normality Based on $\sqrt{b_1}$ and b_2, *Biometrika* **62**(2), 243–250.

Bowman, K. O. and L. R. Shenton (1988). *Properties of Estimators for the Gamma Distribution*, New York: Marcel Dekker.

Bratley, P., B. L. Fox, and L. E. Schrage (1987). *A Guide to Simulation*, 2nd edn, New York: Springer-Verlag.

Briggs, W. (2016). *Uncertainty: The Soul of Modeling, Probability & Statistics*, Switzerland: Springer-Verlag.

Brilhante, M. F., M. Ivette Gomes, and D. Pestana (2013). A Simple Generalisation of the Hill Estimator, *Computational Statistics & Data Analysis* **57**, 518–535.

Brockwell, P. J. and R. A. Davis (1991). *Time Series: Theory and Methods*, 2nd ed., New York: Springer-Verlag.

Broda, S. and M. S. Paolella (2007). Saddlepoint Approximations for the Doubly Noncentral t Distribution, *Computational Statistics & Data Analysis* **51**, 2907–2918.

Broda, S. A. (2011). Tail Probabilities and Partial Moments for Quadratic Forms in Multivariate Generalized Hyperbolic Random Vectors. Working paper.

Broda, S. A., K. Carstensen, and M. S. Paolella (2007). Bias-Adjusted Estimation in the ARX(1) Model, *Computational Statistics & Data Analysis* **51**(7), 3355–3367.

Broda, S. A., M. Haas, J. Krause, M. S. Paolella, and S. C. Steude (2013). Stable Mixture GARCH Models, *Journal of Econometrics* **172**(2), 292–306.

Broda, S. A., J. Krause, and M. S. Paolella (2017). Approximating Expected Shortfall for Heavy Tailed Distributions, *Econometrics and Statistics* (available online).

Broda, S. A. and M. S. Paolella (2009). CHICAGO: A Fast and Accurate Method for Portfolio Risk Calculation, *Journal of Financial Econometrics* **7**(4), 412–436.

Broda, S. A. and M. S. Paolella (2010). Saddlepoint Approximation of Expected Shortfall for Transformed Means. UvA Econometrics Discussion Paper 2010/08.

Broda, S. A. and M. S. Paolella (2011). Expected Shortfall for Distributions in Finance, in P. Čížek, W. Härdle, and Rafał Weron (eds). *Statistical Tools for Finance and Insurance*, 57–99, Berlin: Springer Verlag.

Brooks, C., A. D. Clare, J. W. Dalle-Molle, and G. Persand (2005). A Comparison of Extreme Value Theory Approaches for Determining Value at Risk, *Journal of Empirical Finance* **12**, 339–352.

Brown, L. D. (1986). *Fundamentals of Statistical Exponential Families with Applications in Statistical Decision Theory*, Lecture Notes–Monograph Series, Volume 9, Hayward, CA: Institute of Mathematical Statistics.

Brown, L. D., T. T. Cai, and A. DasGupta (2001). Interval Estimation for a Binomial Proportion (with discussion), *Statistical Science* **16**(2), 101–133.

Buckley, I., D. Saunders, and L. Seco (2008). Portfolio Optimization When Asset Returns Have the Gaussian Mixture Distribution, *European Journal of Operational Research* **185**, 1434–1461.

Bühlmann, P. and S. van de Geer (2011). *Statistics for High-Dimensional Data: Methods, Theory and Applications*, Heidelberg: Springer-Verlag.

Burnham, K. P. and D. Anderson (2002). *Model Selection and Multimodel Inference*, 2nd edn, New York: Springer-Verlag.

Butler, R. W. (2007). *An Introduction to Saddlepoint Methods*, Cambridge: Cambridge University Press.

Butler, R. W., P. L. Davies, and M. Jhun (1993). Asymptotics for the Minimum Covariance Determinant Estimator, *Annals of Statistics* **21**(3), 1385–1400.

Butler, R. W. and M. S. Paolella (2002). Saddlepoint Approximation and Bootstrap Inference for the Satterthwaite Class of Ratios, *Journal of the American Statistical Association* **97**, 836–846.

Butler, R. W. and A. T. A. Wood (2002). Laplace Approximations for Hypergeometric Functions with Matrix Arguments, *Annals of Statistics* **30**, 1155–1177.

Calzolari, G., G. Fiorentini, and E. Sentana (2004). Constrained Indirect Estimation, *Review of Economic Studies* **71**(249), 945–973.

Carr, P. and D. B. Madan (2009). Saddlepoint Methods for Option Pricing, *Journal of Computational Finance* **13**, 49–61.

Carrasco, M. and J.-P. Florens (2002). Simulation-Based Method of Moments and Efficiency, *Journal of Business and Economic Statistics* **20**(4), 482–492.

Casella, G. and R. L. Berger (1987). Rejoinder, *Journal of the American Statistical Association* **82**(397), 133–135.

Casella, G. and R. L. Berger (1990). *Statistical Inference*, Pacific Grove, CA: Wadsworth & Brooks/Cole.

Casella, G. and R. L. Berger (2002). *Statistical Inference*, 2nd edn, Pacific Grove, CA: Duxbury, Wadsworth.

Cavaliere, G., H. B. Nielsen, and A. Rahbek (2017). On the Consistency of Bootstrap Testing for a Parameter on the Boundary of the Parameter Space, *Journal of Time Series Analysis* **38**(4), 513–534.

Cerioli, A. (2010). Multivariate Outlier Detection with High-Breakdown Estimators, *Journal of the American Statistical Association* **105**(489), 147–156.

Cerioli, A., M. Riani, and A. C. Atkinson (2009). Controlling the Size of Multivariate Outlier Tests with the MCD Estimator of Scatter, *Statistics and Computing* **19**(3), 341–353.

Čencov, N. N. (1982). *Statistical Decision Rules and Optimal Inference*, Providence, RI: American Mathematical Society.

Chambers, J. M., C. L. Mallows, and B. W. Stuck (1976). A Method for Simulating Stable Random Variables, *Journal of the American Statistical Association* **71**, 340–344.

Chaudhuri, A. (2010). *Randomized Response and Indirect Questioning Techniques in Surveys*, Boca Raton, FL: CRC Press.

Chen, J. M. (2016). *Postmodern Portfolio Theory: Navigating Abnormal Markets and Investor Behavior*, New York: Palgrave Macmillan.

Chen, Q. and D. E. A. Giles (2008). General Saddlepoint Approximations: Application to the Anderson–Darling Test Statistic, *Communications in Statistics – Simulation and Computation* **37**, 789–804.

Choi, P. and K. Nam (2008). Asymmetric and Leptokurtic Distribution for Heteroscedastic Asset Returns: The S_U-Normal Distribution, *Journal of Empirical Finance* **15**, 41–63.

Chow, G. C. (1984). Maximum-Likelihood Estimation of Misspecified Models, *Economic Modelling* **1**(2), 134–138.

Christensen, R. (2005). Testing Fisher, Neyman, Pearson, and Bayes, *American Statistician* **59**(2), 121–126.

Christodoulakis, G. and D. Peel (2009). The Central Bank Inflation Bias in the Presence of Asymmetric Preferences and Non-Normal Shocks, *Economics Bulletin* **29**, 1608–1620.

Christoffersen, P. F. (2011). *Elements of Financial Risk Management*, 2nd edn, Amsterdam: Academic Press.

Christoffersen, P. F. and S. Gonçalves (2005). Estimation Risk in Financial Risk Management, *Journal of Risk* **7**(3), 1–28.

Ciuperca, G., A. Ridolfi, and J. Idier (2003). Penalized Maximum Likelihood Estimator for Normal Mixtures, *Scandinavian Journal of Statistics* **30**(1), 45–59.

Claeskens, G. and N. L. Hjort (2008). *Model Selection and Model Averaging*, Cambridge: Cambridge University Press.

Clarke, B. R. and C. R. Heathcote (1994). Robust Estimation of k-Component Univariate Normal Mixtures, *Annals of the Institute of Statistical Mathematics* **46**(1), 83–93.

Clementi, F., T. Di Matteo, and M. Gallegati (2006). The Power-Law Tail Exponent of Income Distributions, *Physica A* **370**, 49–53.

Cochrane, J. H. (2001). *Asset Pricing*, Princeton, NJ: Princeton University Press.

Coe, R. (2002). It's the Effect Size, Stupid: What Effect Size Is and Why it is Important. Paper presented at the Annual Conference of the British Educational Research Association, University of Exeter, England, 12–14 September.

Cogneau, P. and G. Hübner (2009a): The (more than) 100 Ways to Measure Portfolio Performance – Part 1: Standardized Risk-Adjusted Measures, *Journal of Performance Measurement* **13**(4), 56–71.

Cogneau, P. and G. Hübner (2009b). The (more than) 100 Ways to Measure Portfolio Performance – Part 2: Special Measures and Comparison, *Journal of Performance Measurement* **14**(1), 56–69.

Cohen, A. C. (1967). Estimation In Mixtures of Two Normal Distributions, *Technometrics* **9**(1), 15–28.

Coles, S. (2001). *An Introduction to Statistical Modeling of Extreme Values*, London: Springer-Verlag.

Covitz, D., N. Liang, and G. A. Suarez (2013). The Evolution of a Financial Crisis: Collapse of the Asset-Backed Commercial Paper Market, *Journal of Finance* **68**(3), 815–848.

Cox, D. R. (1977). The Role of Significance Tests, *Scandinavian Journal of Statistics* **4**, 49–63.

Cox, D. R. (2002). Karl Pearson and the Chi-Squared Test, in C. Huber-Carol, N. Balakrishnan, M. Nikulin, and M. Mesbah (eds). *Goodness-of-Fit Tests and Model Validity*, 3–8, New York: Springer-Verlag.

Csörgő, S. and L. Viharos (1998). Estimating the Tail Index, in B. Szyszkowicz (edn.), *Asymptotic Methods in Probability and Statistics: A Volume in Honour of Miklós Csörgő*, 833–882, Amsterdam: Elsevier.

Csörgőő, S. and D. M. Mason (1985). Central Limit Theorems for Sums of Extreme Values, *Mathematical Proceedings of the Cambridge Philosophical Society* **98**(3), 547–558.

D'Agostino, R. and E. S. Pearson (1973). Testing for Departures from Normality. Empirical Results for Distribution of b_2 and $\sqrt{b_1}$, *Biometrika* **60**(3), 613–622.

Danielsson, J., L. de Haan, L. Peng, and C. G. de Vries (2001). Using a Bootstrap Method to Choose the Sample Fraction in Tail Index Estimation, *Journal of Multivariate Analysis* **76**, 226–248.

Darling, D. A. (1957). The Kolmogorov-Smirnov, Cramér-von Mises Tests, *Annals of Mathematical Statistics* **28**(4), 823–838.

David, H. A. (1995). First (?) Occurrence of Common Terms in Mathematical Statistics, *American Statistician* **49**(2), 121–133. Corrigenda (1998): A second list, with corrections, **52**, 36–40.

Davis, M. H. A. (2016). Verification of Internal Risk Measure Estimates, *Statistics and Risk Modeling* **33**, 67–93.

Davis, R. and S. Resnick (1984). Tail Estimates Motivated by Extreme Value Theory, *Annals of Statistics* **12**(4), 1467–1487.

Davison, A. C. (2003). *Statistical Models*, Cambridge: Cambridge University Press.

Davison, A. C. and D. V. Hinkley (1997). *Bootstrap Methods and Their Application*, Cambridge Series on Statistical and Probabilistic Mathematics, Cambridge: Cambridge University Press.

Davison, A. C., D. V. Hinkley, and G. V. Young (2003). Recent Developments in Bootstrap Methodology, *Statistical Science* **18**, 141–157.

Day, N. E. (1969). Estimating the Components of a Mixture of Normal Distributions, *Biometrika* **56**, 463–474.

de Haan, L. and A. Ferreira (2006). *Extreme Value Theory: An Introduction*, New York: Springer-Verlag.

DeCarlo, L. T. (1997). On the Meaning and Use of Kurtosis, *Psychological Methods* **2**(3), 292–307.

DeMiguel, V., L. Garlappi, F. J. Nogales, and R. Uppal (2009a). A Generalized Approach to Portfolio Optimization: Improving Performance by Constraining Portfolio Norms, *Management Science* **55**(5), 798–812.

DeMiguel, V., L. Garlappi, and R. Uppal (2009b). Optimal Versus Naive Diversification: How Inefficient is the $1/N$ Portfolio Strategy?, *Review of Financial Studies* **22**(5), 1915–1953.

DeMiguel, V., A. Martin-Utrera, and F. J. Nogales (2013). Size Matters: Optimal Calibration of Shrinkage Estimators for Portfolio Selection, *Journal of Banking & Finance* **37**(8), 3018–3034.

Dempster, M. A. H. (edn.) (2002). *Risk Management: Value at Risk and Beyond*, Cambridge: Cambridge University Press.

Dhrymes, P. J. (1982). *Econometrics: Statistical Foundations and Applications*, New York: Springer-Verlag.

Diaconis, P. and S. Holmes (1994). Gray Codes for Randomization Procedures, *Statistics and Computing* **4**, 287–302.

Diebold, F. X. and C. Li (2006). Forecasting the Term Structure of Government Bond Yields, *Journal of Econometrics* **130**(2), 337–364.

Dobrev, D., T. D. Nesmith, and D. H. Oh (2017). Accurate Evaluation of Expected Shortfall for Linear Portfolios with Elliptically Distributed Risk Factors, *Journal of Risk and Financial Management* **10**(1). Article 5.

Dominicy, Y., P. Ilmonen, and D. Veredas (2017). Multivariate Hill Estimators, *International Statistical Review* **85**(1), 108–142.

Dowd, K. (2005). *Measuring Market Risk*, 2nd edn, Chichester: John Wiley & Sons.

Dridi, R., A. Guay, and E. Renault (2007). Indirect Inference and Calibration of Dynamic Stochastic General Equilibrium Models, *Journal of Econometrics* **136**(2), 397–430.

Drton, M. and H. Xiao (2016). Wald Tests of Singular Hypotheses, *Bernoulli* **22**(1), 38–59.

Du, Z. and J. C. Escanciano (2017). Backtesting Expected Shortfall: Accounting for Tail Risk, *Management Science* **63**(4), 940–958.

Dudewicz, E. J. and S. N. Mishra (1988). *Modern Mathematical Statistics*, New York: John Wiley & Sons.

DuMouchel, W. H. (1973). On the Asymptotic Normality of the Maximum-Likelihood Estimate when Sampling from a Stable Distribution, *Annals of Statistics* **1**(5), 948–957.

DuMouchel, W. H. (1975). Stable Distributions in Statistical Inference: 2. Information from Stably Distributed Samples, *Journal of the American Statistical Association* **70**(350), 386–393.

Durbin, J. (1973). *Distribution Theory for Tests Based on Sample Distribution Function*, CBMS-NSF Regional Conference Series in Applied Mathematics, Philadelphia: Society for Industrial and Applied Mathematics.

Dvoretzky, A., J. Kiefer, and J. Wolfowitz (1956). Asymptotic Minimax Character of the Sample Distribution Function and of the Classical Multinomial Estimator, *Annals of Mathematical Statistics* **27**(3), 642–669.

Ecochard, R. and D. G. Clayton (2000). Multivariate Parametric Random Effect Regression Models for Fecundability Studies, *Biometrics* **56**(4), 1023–1029.

Eeckhoudt, L., A. M. Fiori, and E. R. Gianin (2016). Loss-Averse Preferences and Portfolio Choices: An Extension, *European Journal of Operational Research* **249**(1), 224–230.

Efron, B. (1979). Bootstrap Methods: Another Look at the Jackknife, *Annals of Statistics* **7**, 1–26.

Efron, B. (1997). The Length Heuristic for Simultaneous Hypothesis Tests, *Biometrika* **84**, 143–157.

Efron, B. (2003). Second Thoughts on the Bootstrap, *Statistical Science* **18**, 135–140.

Efron, B. (2013). *Large–Scale Inference: Empirical Bayes Methods for Estimation, Testing, and Prediction*, Cambridge: Cambridge University Press.

Efron, B. (2014). Estimation and Accuracy After Model Selection, *Journal of the American Statistical Association* **109**(507), 991–1007.

Efron, B. and T. Hastie (2016). *Computer Age Statistical Inference: Algorithms, Evidence, and Data Science*, Cambridge: Cambridge University Press.

Efron, B. and D. V. Hinkley (1978). Assessing the Accuracy of the Maximum Likelihood Estimator: Observed versus Expected Fisher Information (with discussion), *Biometrika* **65**, 457–487.

Efron, B. and C. Morris (1977). Stein's Paradox in Statistics, *Scientific American* **236**(5), 119–127.

Efron, B. and R. J. Tibshirani (1993). *An Introduction to the Bootstrap*, New York: Chapman & Hall.

Ellison, B. E. (1964). Two Theorems for Inferences about the Normal Distribution with Applications in Acceptance Sampling, *Journal of the American Statistical Association* **59**(305), 89–95.

Embrechts, P. and M. Hofert (2014). Statistics and Quantitative Risk Management for Banking and Insurance, *Annual Review of Statistics and Its Application* **1**, 493–514.

Embrechts, P., C. Klüppelberg, and T. Mikosch (1997). *Modelling Extremal Events for Insurance and Finance*, Berlin: Springer-Verlag.

Embrechts, P. and R. Wang (2015). Seven Proofs for the Subadditivity of Expected Shortfall, *Dependence Modeling* **3**(1), 1–15.

Engle, R. F. (2002). Dynamic Conditional Correlation: A Simple Class of Multivariate Generalized Autoregressive Conditional Heteroskedasticity Models, *Journal of Business and Economic Statistics* **20**(3), 339–350.

Engle, R. F. (2009). *Anticipating Correlations: A New Paradigm for Risk Management*, Princeton, NJ: Princeton University Press.

Engle, R. F. and G. G. J. Lee (1996). Estimating Diffusion Models of Stochastic Volatility, in P. E. Rossi (edn.), *Modelling Stock Market Volatility: Bridging the Gap to Continuous Time*, 333–355, San Diego: Academic Press.

Epstein, B. and M. Sobel (1953). Life Testing, *Journal of the American Statistical Association* **48**, 486–502.

Epstein, R. A. (1977). *The Theory of Gambling and Statistical Logic*, revised edn., San Diego: Academic Press.

Evans, M., I. Guttman, and I. Olkin (1992). Numerical Aspects in Estimating the Parameters of a Mixture of Normal Distributions, *Journal of Computational and Graphical Statistics* **1**(4), 351–365.

Everitt, B. S. (1984). Maximum Likelihood Estimation of the Parameters in a Mixture of Two Univariate Normal Distributions: A Comparison of Different Algorithms, *The Statistician* **33**, 205–215.

Fama, E. (1963). Mandelbrot and the Stable Paretian Hypothesis, *Journal of Business* **36**, 420–429.

Fama, E. (1965a). The Behavior of Stock Market Prices, *Journal of Business* **38**, 34–105.

Fama, E. (1965b). Portfolio Analysis in a Stable Paretian Market, *Management Science* **11**, 404–419.

Fama, E. and R. Roll (1971). Parameter Estimates for Symmetric Stable Distributions, *Journal of the American Statistical Association* **66**(334), 331–338.

Fan, J., Y. Fan, and J. Lv (2008). High Dimensional Covariance Matrix Estimation Using a Factor Model, *Journal of Econometrics* **147**, 186–197.

Fauconnier, C. and G. Haesbroeck (2009). Outliers Detection with the Minimum Covariance Determinant Estimator in Practice, *Statistical Methodology* **6**(4), 363–379.

Feller, W. (1968). *An Introduction to Probability Theory and Its Applications*, vol. I, 3rd edn, New York: John Wiley & Sons.

Feller, W. (1971). *An Introduction to Probability Theory and Its Applications*, vol. II, 2nd edn, New York: John Wiley & Sons.

Ferguson, T. S. (1961). On the Rejection of Outliers, in J. Neyman (edn.), *Proceedings of the Fourth Berkeley Symposium on Mathematical Statistics and Probability*, vol. **1**, 253–287, Berkeley: University of California Press.

Ferguson, T. S. (1967). *Mathematical Statistics: A Decision Theoretic Approach*, New York: Academic Press.

Ferguson, T. S. (1996). *A Course in Large Sample Theory*, London: Chapman & Hall.

Fernández, C. and M. F. J. Steel (1998). On Bayesian Modelling of Fat Tails and Skewness, *Journal of the American Statistical Association* **93**, 359–371.

Fieller, E. C. (1954). Some Problems in Interval Estimation, *Journal of the Royal Statistical Society, Series B* **16**(2), 175–185.

Finkenstädt, B. and H. Rootzén (eds). (2004). *Extreme Values in Finance, Telecommunications, and the Environment*, Boca Raton, FL: Chapman & Hall/CRC.

Fiori, A. M. and D. Beltrami (2014). Right and Left Kurtosis Measures: Large Sample Estimation and an Application to Financial Returns, *Stat* **3**(1), 95–108.

Fiori, A. M. and M. Zenga (2009). Karl Pearson and the Origin of Kurtosis, *International Statistical Review* **77**(1), 40–50.

Fisher, R. A. (1922). On the Mathematical Foundations of Theoretical Statistics, *Philosophical Transactions of the Royal Society A* **222**, 309–368.

Fisher, R. A. (1926). The Arrangement of Field Experiments, *Journal of the Ministry of Agriculture* **33**, 503–513.

Fisher, R. A. (1929). The Statistical Method in Psychical Research, *Proceedings of the Society for Psychical Research* **39**, 189–192.

Flury, B. (1997). *A First Course in Multivariate Statistics*, New York: Springer-Verlag.

Fofack, H. and J. P. Nolan (1999). Tail Behavior, Modes and Other Characteristics of Stable Distribution, *Extremes* **2**(1), 39–58.

Fowlkes, E. B. (1979). Some Methods for Studying the Mixture of Two Normal (Lognormal) Distributions, *Journal of the American Statistical Association* **74**, 561–575.

Francioni, I. and F. Herzog (2012). Probability-Unbiased Value-at-Risk Estimators, *Quantitative Finance* **12**(5), 755–768.

Freedman, D. A. (2006). On the So-Called "Huber Sandwich Estimator" and "Robust Standard Errors", *American Statistician* **60**(4), 299–302.

Freedman, L. S. (1981). Watson's U_N^2 Statistic for a Discrete Distribution, *Biometrika* **68**(3), 708–711.

Fryer, J. G. and C. A. Robertson (1972). A Comparison of Some Methods for Estimating Mixed Normal Distributions, *Biometrika* **59**(3), 639–648.

Galambos, J. and S. Kotz (1978). *Characterizations of Probability Distributions: A Unified Approach with an Emphasis on Exponential and Related Models*, Berlin: Springer-Verlag.

Gallant, A. R., D. Hsieh, and G. Tauchen (1997). Estimation of Stochastic Volatility Models with Diagnostics, *Journal of Econometrics* **81**(1), 159–192.

Gallant, A. R. and R. E. McCulloch (2009). On the Determination of General Scientific Models With Application to Asset Pricing, *Journal of the American Statistical Association* **104**(485), 117–131.

Gallant, A. R. and G. Tauchen (1996). Which Moments to Match?, *Econometric Theory* **12**(4), 657–681.

Gambacciani, M. and M. S. Paolella (2017). Robust Normal Mixtures for Financial Portfolio Allocation, *Econometrics and Statistics* **3**, 91–111.

Gan, G. and L. J. Bain (1998). Some Results for Type I Censored Sampling from Geometric Distributions, *Journal of Statistical Planning and Inference* **67**(1), 85–97.

Gao, F. and F. Song (2008). Estimation Risk in GARCH VaR and ES Estimates, *Econometric Theory* **24**, 1404–1424.

Garcia, R., E. Renault, and D. Veredas (2009). Estimation of Stable Distributions by Indirect Inference, CORE Discussion Paper 2006/112, Université Catholique de Louvain.

Gel, Y. R. and J. L. Gastwirth (2008). A Robust Modification of the Jarque-Bera Test of Normality, *Economics Letters* **99**, 30–32.

Gel, Y. R., W. Miao, and J. L. Gastwirth (2007). Robust Directed Tests of Normality Against Heavy-Tailed Alternatives, *Computational Statistics & Data Analysis* **51**, 2734–2746.

Gelman, A., J. B. Carlin, H. S. Stern, D. B. Dunson, A. Vehtari, and D. B. Rubin (2013). *Bayesian Data Analysis*, 3rd edn, Boca Raton, FL: Chapman & Hall/CRC.

Gelman, A. and H. Stern (2006). The Difference between "Significant" and "Not Significant" is Not Itself Statistically Significant, *American Statistician* **60**(4), 328–331.

Genton, M. G. (edn.) (2004). *Skew-Elliptical Distributions and Their Applications: A Journey Beyond Normality*, Boca Raton, FL: Chapman & Hall/CRC.

Ghosh, M. and P. K. Sen (1989). Median Unbiasedness and Pitman Closeness, *Journal of the American Statistical Association* **84**(408), 1089–1091.

Ghosh, S. (1996). A New Graphical Tool To Detect Non-normality, *Journal of the Royal Statistical Society, Series B* **58**(4), 691–702.

Giannikis, D., I. D. Vrontos, and P. Dellaportas (2008). Modelling Nonlinearities and Heavy Tails via Threshold Normal Mixture GARCH Models, *Computational Statistics & Data Analysis* **52**(3), 1549–1571.

Gil-Peleaz, J. (1951). Note on the Inversion Theorem, *Biometrika* **38**, 481–482.

Giles, D. E. A. (2001). A Saddlepoint Approximation to the Distribution Function of the Anderson–Darling Test Statistic, *Communications in Statistics – Simulation and Computation* **30**, 899–905.

Giles, J. A. and D. E. A. Giles (1993). Pre-Test Estimation and Testing in Econometrics: Recent Developments, *Journal of Economic Surveys* **7**(2), 145–197.

Gilli, M. and E. Këllezi (2006). An Application of Extreme Value Theory for Measuring Financial Risk, *Computational Economics* **27**(2), 207–228.

Givens, G. H. and J. A. Hoeting (2013). *Computational Statistics*, 2nd edn, Hoboken, NJ: John Wiley & Sons.

Glasserman, P. and K.-K. Kim (2009). Saddlepoint Approximations for Affine Jump–Diffusion Models, *Journal of Economic Dynamics and Control* **33**, 15–36.

Gneiting, T. (2011). Making and Evaluating Point Forecasts, *Journal of the American Statistical Association* **106**(494), 746–762.

Gneiting, T., F. Balabdaoui, and A. E. Raftery (2007). Probabilistic Forecasts, Calibration and Sharpness, *Journal of the Royal Statistical Society, Series B* **69**(2), 243–268.

Golden, R. M., S. S. Henley, H. White, and T. M. Kashner (2016). Generalized Information Matrix Tests for Detecting Model Misspecification, *Econometrics* **4**(4). Article 46.

Goldie, C. M. and R. L. Smith (1987). Slow Variation with Remainder: Theory and Applications, *Quarterly Journal of Mathematics* **38**(1), 45–71.

Goodman, S. (2008). A Dirty Dozen: Twelve *P*-Value Misconceptions, *Seminars in Hematology* **45**(3), 135–140.

Goodman, S. N. and R. Royall (1988). Evidence and Scientific Research, *American Journal of Public Health* **78**, 1568–1574.

Gourieroux, C. and A. Monfort (1995). *Statistics and Econometric Models Volume 1: General Concepts, Estimation, Prediction and Algorithms*, Cambridge: Cambridge University Press.

Gourieroux, C., A. Monfort, and E. Renault (1993). Indirect Inference, *Journal of Applied Econometrics* **8**(1), 85–118.

Grace, A. W. and I. A. Wood (2012). Approximating the Tail of the Anderson–Darling Distribution, *Computational Statistics & Data Analysis* **56**, 4301–4311.

Gradshteyn, L. S. and I. M. Ryzhik (2007). *Table of Integrals, Series and Products*, 7th edn, Amsterdam: Academic Press.

Graybill, F. A. and H. K. Iyer (1994). *Regression Analysis: Concepts and Applications*, Belmont, CA: Duxbury, Wadsworth.

Greene, R. (2006). *The 33 Strategies of War*, London: Profile Books.

Grünwald, P. D. (2007). *The Minimum Description Length Principle*, Cambridge, MA: MIT Press.

Gupta, A. K., T. T. Nguyen, and J. A. T. Sanqui (2004). Characterization of the Skew-Normal Distribution, *Annals of the Institute of Statistical Mathematics* **56**(2), 351–360.

Gut, A. (2005). *Probability: A Graduate Course*, New York: Springer-Verlag.

Haas, M. (2005). Improved Duration-Based Backtesting of Value-at-Risk, *Journal of Risk* **8**(2), 17–38.

Haas, M. (2009). Value-at-Risk via Mixture Distributions Reconsidered, *Applied Mathematics and Computation* **215**(6), 2103–2119.

Haas, M. (2010). Skew-Normal Mixture and Markov-Switching GARCH Processes, *Studies in Nonlinear Dynamics & Econometrics* **14**(4). Article 1.

Haas, M. (2012). A Note on the Moments of the Skew-Normal Distribution, *Economics Bulletin* **32**(4), 3306–3312.

Haas, M., J. Krause, M. S. Paolella, and S. C. Steude (2013). Time-Varying Mixture GARCH Models and Asymmetric Volatility, *North American Journal of Economics and Finance* **26**, 602–623.

Haas, M., S. Mittnik, and M. S. Paolella (2004a). Mixed Normal Conditional Heteroskedasticity, *Journal of Financial Econometrics* **2**(2), 211–250.

Haas, M., S. Mittnik, and M. S. Paolella (2004b). A New Approach to Markov-Switching GARCH Models, *Journal of Financial Econometrics* **2**(4), 493–530.

Haas, M., S. Mittnik, and M. S. Paolella (2006). Modeling and Predicting Market Risk With Laplace-Gaussian Mixture Distributions, *Applied Financial Economics* **16**, 1145–1162.

Haas, M., S. Mittnik, and M. S. Paolella (2009). Asymmetric Multivariate Normal Mixture GARCH, *Computational Statistics & Data Analysis* **53**(6), 2129–2154.

Haas, M. and M. S. Paolella (2012). Mixture and Regime-Switching GARCH Models, in L. Bauwens, C. M. Hafner, and S. Laurent (eds). *Handbook of Volatility Models and their Applications*, chap. 3, Hoboken, NJ: John Wiley & Sons.

Hacking, I. (1965). *Logic of Statistical Inference*, Cambridge: Cambridge University Press.

Hacking, I. (1980). The Theory of Probable Inference: Neyman, Peirce and Braithwaite, in D. H. Mellor (edn.), *Science, Belief and Behavior: Essays in Honour of R. B. Braithwaite*, 141–160, Cambridge: Cambridge University Press.

Haeusler, E. and J. L. Teugels (1985). On Asymptotic Normality of Hill's Estimator for the Exponent of Regular Variation, *Annals of Statistics* **13**(2), 743–756.

Hall, P. (1982). On Some Simple Estimates of an Exponent of Regular Variation, *Journal of the Royal Statistical Society, Series B* **44**(1), 37–42.

Hallin, M., Y. Swand, T. Verdebout, and D. Veredas (2013). One-Step R-Estimation in Linear Models with Stable Errors, *Journal of Econometrics* **172**(2), 195–204.

Halvarsson, D. (2013). On the Estimation of Skewed Geometric Stable Distributions. Working paper No. 216, Royal Institute of Technology, Division of Economics, Stockholm, Sweden.

Hamilton, J. D. (1991). A Quasi-Bayesian Approach to Estimating Parameters for Mixtures of Normal Distributions, *Journal of Business and Economic Statistics* **9**(1), 21–39.

Hamilton, J. D. (1994). *Time Series Analysis*, Princeton, NJ: Princeton University Press.

Hamouda, O. and R. Rowley (1996). *Probability in Economics*, London: Routledge.

Hansen, N. and A. Ostermeier (1996). Adapting arbitrary normal mutation distributions in evolution strategies: The covariance matrix adaptation, in *Proceedings of the 1996 IEEE International Conference on Evolutionary Computation*, 312–317.

Hardin, J. and D. M. Rocke (2005). The Distribution of Robust Distances, *Journal of Computational and Graphical Statistics* **14**(4), 928–946.

Hardin, J. W. (2003). The Sandwich Estimator of Variance, in T. B. Fomby and R. C. Hill (eds). *Maximum Likelihood Estimation of Misspecified Models: Twenty Years Later*, New York: Elsevier.

Hartley, H. O. (1978). Contributions to the Discussion of Paper by R.E. Quandt and J. B. Ramsey, *Journal of the American Statistical Association* **73**, 738–741.

Hartz, C., S. Mittnik, and M. S. Paolella (2006). Accurate Value-at-Risk Forecasting Based on the Normal-GARCH Model, *Computational Statistics & Data Analysis* **51**(4), 2295–2312.

Harvey, C. R., M. Liechty, J. Liechty, and P. Muller (2010). Portfolio Selection with Higher Moments, *Quantitative Finance* **10**, 469–485.

Harvey, C. R. and A. Siddique (1999). Autoregressive Conditional Skewness, *Journal of Financial and Quantitative Analysis* **34**(4), 465–487.

Harvey, C. R. and A. Siddique (2000). Conditional Skewness in Asset Pricing Tests, *Journal of Finance* **55**(3), 1263–1295.

Harvill, J. L. (2008). Review of: Fundamental Probability: A Computational Approach, *American Statistician* **62**(2), 179–180.

Harvill, J. L. (2009). Review of: Intermediate Probability: A Computational Approach, *Journal of the American Statistical Association* **104**(487), 1285–1286.

Hastie, T., R. Tibshirani, and J. Friedman (2009). *The Elements of Statistical Learning: Data Mining, Inference, and Prediction*, 2nd edn, New York: Springer-Verlag.

Hastie, T., R. Tibshirani, and M. Wainwright (2015). *Statistical Learning with Sparsity: The Lasso and Generalizations*, Boca Raton, FL: CRC Press.

Hathaway, R. J. (1986). A Constrained EM Algorithm for Univariate Normal Mixtures, *Journal of Statistical Computation and Simulation* **23**, 211–230.

Heggland, K. and A. Frigessi (2004). Estimating Functions in Indirect Inference, *Journal of the Royal Statistical Society, Series B* **66**(2), 447–462.

Henze, N. (1986). A Probabilistic Representation of the "Skew-normal" Distribution, *Scandinavian Journal of Statistics* **13**, 271–275.

Heyde, C. C. and S. G. Kou (2004). On the Controversy over Tailweight of Distributions, *Operations Research Letters* **32**, 399–408.

Hill, B. M. (1975). A Simple General Approach to Inference About the Tail of a Distribution, *Annals of Statistics* **3**(5), 1163–1174.

Hinkley, D. V. (1987). Comment, *Journal of the American Statistical Association* **82**(397), 128–129.

Hogg, R. V., J. McKean, and A. T. Craig (2014). *Introduction to Mathematical Statistics*, 7th edn, Harlow, England: Pearson Education.

Holm, E. (2016). Warren Buffett's Epic Rant against Wall Street, *Wall Street Journal* May 2.

Hotelling, H. (1953). New Light on the Correlation Coefficient and Its Transforms, *Journal of the Royal Statistical Society, Series B* **15**, 193–232.

Hoyt, J. P. (1969). Two Instructive Examples of Maximum Likelihood Estimates, *American Statistician* **23**(2), 14.

Huang, X. and C. W. Oosterlee (2011). Saddlepoint Approximations for Expectations and an Application to CDO Pricing, *SIAM Journal on Financial Mathematics* **2**, 692–714.

Hubbard, R. and M. J. Bayarri (2003). Confusion over Measures of Evidence (p's) versus Errors (α's) in Classical Statistical Testing, *American Statistician* **57**(3), 171–182. Comments by K. N. Berk and M. A. Carlton, and Rejoinder.

Hubbard, R. and R. M. Lindsay (2008). Why P-Values Are Not a Useful Measure of Evidence in Statistical Significance Testing, *Theory & Psychology* **18**(1), 69–88.

Huber, P. J. (1967). The Behavior of Maximum Likelihood Estimates under Nonstandard Conditions, in L. L. Cam and J. Neyman (eds). *Proceedings of the Fifth Berkeley Symposium on Mathematical Statistics and Probability*, vol. **1**, 221–233, Berkeley: University of California Press.

Huber, P. J. and E. M. Ronchetti (2009). *Robust Statistics*, 2nd edn, Hoboken, NJ: John Wiley & Sons.

Hubert, M., P. J. Rousseeuw, and S. Van Aelst (2008). High-Breakdown Robust Multivariate Methods, *Statistical Science* 92–119.

Hubert, M., P. J. Rousseeuw, and T. Verdonck (2012). A Deterministic Algorithm for Robust Location and Scatter, *Journal of Computational and Graphical Statistics* **21**(3), 618–637.

Hwang, J. T. G. and M.-C. Yang (2001). An Optimality Theory for Mid p-Values in 2×2 Contingency Tables, *Statistica Sinica* **11**, 807–826.

Ioannidis, J. P. A. (2005). Why Most Published Research Findings Are False, *PLoS Med* **2**(8).

Jagannathan, R. and T. Ma (2003). Risk Reduction in Large Portfolios: Why Imposing the Wrong Constraints Helps, *Journal of Finance* **58**, 1651–1684.

Jarque, C. M. and A. K. Bera (1980). Efficient Tests for Normality, Homoskedasticity and Serial Independence of Regression Residuals, *Economics Letters* **6**, 255–259.

Jarque, C. M. and A. K. Bera (1987). A Test for Normality of Observations and Regression Residuals, *International Statistical Review* **55**(2), 163–172.

Jensen, J. L. (1995). *Saddlepoint Approximations*, Oxford: Oxford University Press.

Johnson, D. H. (1999). The Insignificance of Statistical Significance Testing, *Journal of Wildlife Management* **63**(3), 763–772.

Johnson, N. L. (1949). Systems of Frequency Curves Generated by Method of Translation, *Biometrika* **36**, 149–176.

Johnson, N. L. (1978). Comment, *Journal of the American Statistical Association* **73**(364), 750.

Johnson, N. L., S. Kotz, and N. Balakrishnan (1995). *Continuous Univariate Distributions, Volumes 1 and 2*, 2nd edn, New York: John Wiley & Sons.

Jondeau, E. (2016). Asymmetry in Tail Dependence of Equity Portfolios, *Computational Statistics & Data Analysis* **100**, 351–368.

Jondeau, E., S.-H. Poon, and M. Rockinger (2007). *Financial Modeling under Non-Gaussian Distributions*, New York: Springer-Verlag.

Jones, M. C. (2007). Connecting Distributions with Power Tails on the Real Line, the Half Line and the Interval, *International Statistical Review* **75**(1), 58–69.

Jorion, P. (1986). Bayes-Stein Estimation for Portfolio Analysis, *Journal of Financial and Quantitative Analysis* **21**, 279–292.

Kallenberg, O. (2002). *Foundations of Modern Probability*, 2nd edn, New York: Springer-Verlag.

Kamdem, J. S. (2005). Value-at-Risk and Expected Shortfall for Linear Portfolios with Elliptically Distributed Risk Factors, *International Journal of Theoretical and Applied Finance* **8**, 537–551.

Kan, R. and G. Zhou (2007). Optimal Portfolio Choice with Parameter Uncertainty, *Journal of Financial and Quantitative Analysis* **42**(3), 621–656.

Kiefer, N. M. (1978a). Comment, *Journal of the American Statistical Association* **73**(364), 744–745.

Kiefer, N. M. (1978b). Discrete Parameter Estimation of a Switching Regression Model, *Econometrica* **46**, 427–434.

Kim, K.-i. (2016). Higher Order Bias Correcting Moment Equation for M-Estimation and Its Higher Order Efficiency, *Econometrics* **4**, 1–19.

Kim, M. and S. Lee (2016). On the Tail Index Inference for Heavy-Tailed GARCH-Type Innovations, *Annals of the Institute of Statistical Mathematics* **68**(2), 237–267.

Kim, S. and K.-K. Kim (2017). Saddlepoint Methods for Conditional Expectations with Applications to Risk Management, *Bernoulli* **23**(3), 1481–1517.

Kim, S.-H. and A. S. Cohen (1998). On the Behrens-Fisher Problem: A Review, *Journal of Educational and Behavioral Statistics* **23**(4), 356–377.

Kim, Y. S., S. T. Rachev, M. L. Bianchi, and F. J. Fabozzi (2010). Computing VaR and AVaR in Infinitely Divisible Distributions, *Probability and Mathematical Statistics* **30**(2), 223–245.

Kim, Y. S., S. T. Rachev, M. L. Bianchi, I. Mitov, and F. J. Fabozzi (2011). Time Series Analysis for Financial Market Meltdowns, *Journal of Banking & Finance* **35**, 1879–1891.

Kirk, R. E. (1996). Practical Significance: A Concept Whose Time Has Come, *Educational and Psychological Measurement* **56**(5), 746–759.

Koch-Medina, P. and C. Munari (2016). Unexpected Shortfalls of Expected Shortfall: Extreme Default Profiles and Regulatory Arbitrage, *Journal of Banking & Finance* **62**, 141–151.

Kogon, S. M. and D. B. Williams (1998). Characteristic Function Based Estimation of Stable Distribution Parameters, in R. J. Adler, R. E. Feldman, and M. S. Taqqu (eds). *A Practical Guide to Heavy Tails: Statistical Techniques for Analyzing Heavy Tailed Distributions*, Boston: Birkhäuser.

Konishi, S. and G. Kitagawa (2008). *Information Criteria and Statistical Modeling*, New York: Springer-Verlag.

Kotz, S. and N. L. Johnson (eds). (1992). *Breakthroughs in Statistics: Volume 1: Foundations and Basic Theory*, New York: Springer-Verlag.

Kotz, S., T. Kozubowski, and K. Podgorski (2001). *The Laplace Distribution and Generalizations: A Revisit with Application to Communication, Economics, Engineering and Finance*, Boston: Birkhäuser.

Kou, S. and X. Peng (2016). On the Measurement of Economic Tail Risk, *Operations Research* **64**(5), 1056–1072.

Koutrouvelis, I. A. (1980). Regression-Type Estimation of the Parameters of Stable Laws, *Journal of the American Statistical Association* **75**(372), 918–928.

Koutrouvelis, I. A. and S. G. Meintanis (1999). Testing for Stability Based on the Empirical Characteristic Function with Applications to Financial Data, *Journal of Statistical Computation and Simulation* **64**(4), 275–300.

Kozubowski, T. J. (2000). Exponential Mixture Representation of Geometric Stable Distributions, *Annals of the Institute of Statistical Mathematics* **52**(2), 231–238.

Kozubowski, T. J. and S. T. Rachev (1999). Univariate Geometric Stable Laws, *Journal of Computational Analysis and Applications* **1**(2), 177–217.

Krämer, W. and G. Gigerenzer (2005). How to Confuse with Statistics or: The Use and Misuse of Conditional Probabilities, *Statistical Science* **20**, 223–230.

Krantz, D. H. (1999). The Null Hypothesis Testing Controversy in Psychology, *Journal of the American Statistical Association* **94**(448), 1372–1381.

Krause, J. and M. S. Paolella (2014). A Fast, Accurate Method for Value at Risk and Expected Shortfall, *Econometrics* **2**, 98–122.

Kshirsagar, A. M. (1961). Some Extensions of the Multivariate Generalization *t* distribution and the Multivariate Generalization of the Distribution of the Regression Coefficient, *Proceedings of the Cambridge Philosophical Society* **57**, 80–85.

Küchler, U. and S. Tappe (2013). Tempered Stable Distributions and Processes, *Stochastic Processes and their Applications* **123**(12), 4256–4293.

Kuester, K., S. Mittnik, and M. S. Paolella (2006). Value-at-Risk Prediction: A Comparison of Alternative Strategies, *Journal of Financial Econometrics* **4**, 53–89. Reproduced in: *The Foundations of Credit Risk Analysis*, W. Semmler and L. Bernard (eds). Chapter 14, Cheltenham: Edward Elgar Publishing, 2007.

Kumar, K. D., E. H. Nicklin, and A. S. Paulson (1979). Comment on "Estimating Mixtures of Normal Distributions and Switching Regressions", *Journal of the American Statistical Association* **74**(365), 52–55.

Kusuoka, S. (2001). On Law Invariant Coherent Risk Measures, *Advances in Mathematical Economics* **3**, 83–95.

Landsman, Z. M. and E. A. Valdez (2003). Tail Conditional Expectations for Elliptical Distributions, *North American Actuarial Journal* **7**(4), 55–71.

Lau, H.-S. and A. H.-L. Lau (1993). The Reliability of the Stability-Under-Addition Test for the Stable-Paretian Hypothesis, *Journal of Statistical Computation and Simulation* **48**, 67–80.

Le Cam, L. (1990). Maximum Likelihood: An Introduction, *International Statistical Review* **58**(2), 153–171.

Lebedev, N. N. (1972). *Special Functions and Their Applications*, Mineola, NY: Dover.

Ledoit, O. and M. Wolf (2003). Improved Estimation of the Covariance Matrix of Stock Returns with an Application to Portfolio Selection, *Journal of Empirical Finance* **10**, 603–621.

Ledoit, O. and M. Wolf (2004). Honey, I Shrunk the Sample Covariance Matrix, *Journal of Portfolio Management* **30**(4), 110–119.

Lehmann, E. L. (1959). *Testing Statistical Hypotheses*, New York: John Wiley & Sons.

Lehmann, E. L. and G. Casella (1998). *Theory of Point Estimation*, 2nd edn, New York: Springer Verlag.

Lin, J.-W. and A. I. McLeod (2008). Portmanteau Tests for ARMA Models with Infinite Variance, *Journal of Time Series Analysis* **29**(3), 600–617.

Lin, T. I. (2009). Maximum Likelihood Estimation for Multivariate Skew Normal Mixture Models, *Journal of Multivariate Analysis* **100**, 257–265.

Lin, T. I., J. C. Lee, and S. Y. Yen (2007). Finite Mixture Modelling using the Skew Normal Distribution, *Statistica Sinica* **17**, 909–927.

Lindley, D. V. (1968). Discussion of Nelder (1968), *Journal of the Royal Statistical Society, Series A* **131**(3), 320–321.

Lindley, D. V. (1993). The Analysis of Experimental Data: The Appreciation of Tea and Wine, *Teaching Statistics* **15**, 22–25.

Lindley, D. V. (1999). Comment on Bayarri and Berger, in J. M. Bernardo, J. O. Berger, A. P. Dawid, and A. F. M. Smith (eds). *Bayesian Statistics*, vol. **6**, 75, Oxford: Oxford University Press.

Lindsay, B. G. and P. Basek (1993). Multivariate Normal Mixtures: A Fast Consistent Method of Moments, *Journal of the American Statistical Association* **88**(422), 468–476.

Lindsey, J. K. (1999). Some Statistical Heresies, *The Statistician* **48**, 1–40.

Lombardi, M. J. and G. Calzolari (2008). Indirect Estimation of α-Stable Distributions and Processes, *Econometrics Journal* **11**(1), 193–208.

Lombardi, M. J. and D. Veredas (2009). Indirect Estimation of Elliptical Stable Distributions, *Computational Statistics & Data Analysis* **53**(6), 2309–2324.

Longin, F. (edn.) (2017). *Extreme Events in Finance: A Handbook of Extreme Value Theory and its Applications*, Hoboken, NJ: John Wiley & Sons.

Lopuhaa, H. P. and P. J. Rousseeuw (1991). Breakdown Points of Affine Equivariant Estimators of Multivariate Location and Covariance Matrices, *Annals of Statistics* 229–248.

Loretan, M. and P. C. B. Phillips (1994). Testing the Covariance Stationarity of Heavy–Tailed Time Series, *Journal of Empirical Finance* **1**, 211–248.

Lukacs, E. (1970). *Characteristic Functions*, 2nd edn, London: Griffin.

Lütkepohl, H. (1993). *Introduction to Multiple Time Series Analyis*, 2nd edn, Berlin: Springer-Verlag.

Lyness, J. N. (1969). Notes on the Adaptive Simpson Quadrature Routine, *Journal of the Association for Computing Machinery* **16**(3), 483–495.

MacKinnon, J. G. and A. A. Smith (1998). Approximate Bias Correction in Econometrics, *Journal of Econometrics* **85**, 205–30.

Magnello, M. E. (2009). Karl Pearson and the Establishment of Mathematical Statistics, *International Statistical Review* **77**(1), 3–29.

Malmquist, S. (1950). On a Property of Order Statistics from a Rectangular Distribution, *Skandinavisk Aktuarietidskrift* **33**, 214–222.

Mammen, E. and S. Nandi (2004). Bootstrap and Resampling, in J. E. Gentle, W. Härdle, and Y. Mori (eds). *Handbook of Computational Statistics*, 467–496, Heidelberg: Springer-Verlag.

Mandelbrot, B. (1963). The Variation of Certain Speculative Prices, *Journal of Business* **36**(4), 394–419.

Manganelli, S. and R. F. Engle (2004). A Comparison of Value-at-Risk Models in Finance, in G. Szegö (edn.), *Risk Measures for the 21st Century*, chap. 9, Chichester: John Wiley & Sons.

Mardia, K. V. (1971). Measures of Multivariate Skewness and Kurtosis with Applications, *Biometrika* **57**(3), 519–530.

Mardia, K. V. (1974). Applications of Some Measures of Multivariate Skewness and Kurtosis in Testing Normality and Robustness Studies, *Sankhyā: The Indian Journal of Statistics, Series B* **36**(2), 115–128.

Markowitz, H. (1952). Portfolio Selection, *Journal of Finance* **7**(1), 77–91.

Maronna, R., D. Martin, and V. Yohai (2006). *Robust Statistics: Theory and Methods*, Chichester: John Wiley & Sons.

Marsaglia, G. and J. C. W. Marsaglia (2004). Evaluating the Anderson–Darling distribution, *Journal of Statistical Software* **9**(2), 1–5.

Martellini, L. and V. Ziemann (2010). Improved Estimates of Higher-Order Co-Moments and Implications for Portfolio Selection, *Review of Financial Studies* **23**, 1467–1502.

Martin, R. (2006). The Saddlepoint Method and Portfolio Optionalities, *Risk Magazine* **19**(12), 93–95.

Martínez, E. H., H. Varela, H. W. Gómez, and H. Bolfarine (2008). A Note on the Likelihood and Moments of the Skew-Normal Distribution, *SORT* **32**, 57–66.

Mason, D. M. and T. S. Turova (1994). Weak Convergence of the Hill Estimator Process, in J. Galambos, J. Lechner, and E. S. and (eds). *Extreme Value Theory and Applications*, 419–431, Dordrecht: Kluwer Academic Publishers.

Massart, P. (1990). The Tight Constant in the Dvoretzky–Kiefer–Wolfowitz Inequality, *Annals of Probability* **18**, 1269–1283.

Matsui, M. and A. Takemura (2008). Goodness-of-Fit Tests for Symmetric Stable Distributions – Empirical Characteristic Function Approach, *TEST* **17**(3), 546–566.

McCabe, B. P. M. and S. J. Leybourne (2000). A General Method of Testing for Random Parameter Variation in Statistical Models, in R. D. H. Heijmans, D. S. G. Pollock, and A. Satorra (eds). *Innovations in Multivariate Statistical Analysis: A Festschrift for Heinz Neudecker*, 75–85, Amsterdam: Kluwer.

McCulloch, J. H. (1986). Simple Consistent Estimators of Stable Distribution Parameters, *Communications in Statistics – Simulation and Computation* **15**(4), 1109–1136.

McCulloch, J. H. (1997a). Financial Applications of Stable Distributions, in G. Maddala and C. Rao (eds). *Handbook of Statistics*, vol. 14, Amsterdam: Elsevier Science.

McCulloch, J. H. (1997b). Measuring Tail Thickness in Order to Estimate the Stable Index α: A Critique, *Journal of Business and Economic Statistics* **15**(1), 74–81.

McCulloch, J. H. (1998). Linear Regression with Stable Disturbances, in R. J. Adler, R. E. Feldman, and M. S. Taqqu (eds). *A Practical Guide to Heavy Tails*, 359–376, Boston: Birkhäuser.

McLachlan, G. J. and T. Krishnan (2008). *The EM Algorithm and Extensions*, 2nd edn, Hoboken, NJ: John Wiley & Sons.

McLachlan, G. J. and D. Peel (2000). *Finite Mixture Models*, New York: John Wiley & Sons.

McNeil, A. J., R. Frey, and P. Embrechts (2005). *Quantitative Risk Management: Concepts, Techniques, and Tools*, Princeton, NJ: Princeton University Press.

McNeil, A. J., R. Frey, and P. Embrechts (2015). *Quantitative Risk Management: Concepts, Techniques, and Tools*, revised edn., Princeton, NJ: Princeton University Press.

McQuarrie, A. D. R. and C.-L. Tsai (1998). *Regression and Time Series Model Selection*, River Edge, NJ: World Scientific.

McShane, B. B. and D. Gal (2016). Blinding Us to the Obvious? The Effect of Statistical Training on the Evaluation of Evidence, *Management Science* **62**(6), 1707–1718.

McShane, B. B. and D. Gal (2017). Statistical Significance and the Dichotomization of Evidence (with Comments), *Journal of the American Statistical Association* **112**(519), 885–895.

Meeden, G. (1987). Estimation when Using a Statistic that is not Sufficient, *American Statistician* **41**(2), 135–136.

Meehl, P. E. (1978). Theoretical Risks and Tabular Asterisks: Sir Karl, Sir Ronald, and the Slow Progress of Soft Psychology, *Journal of Consulting and Clinical Psychology* **46**(4), 806–384.

Meintanis, S. G. (2005). Consistent Tests for Symmetric Stability with Finite Mean Based on the Empirical Characteristic Function, *Journal of Statistical Planning and Inference* **128**, 373–380.

Mendell, N. R., S. J. Finch, and H. C. Thode (1993). Where is the Likelihood Ratio Test Powerful for Detecting Two Component Normal Mixtures, *Biometrics* **49**, 907–915.

Meng, X.-L. (1997). The EM Algorithm, in S. Kotz (edn.), *Encyclopedia of Statistical Sciences, Update Volume 1*, New York: John Wiley & Sons.

Michael, J. R. (1983). The Stabilized Probability Plot, *Biometrika* **70**(1), 11–17.

Mikosch, T., T. Gadrich, C. Klüppelberg, and R. J. Adler (1995). Parameter Estimation for ARMA Models with Infinite Variance Innovations, *Annals of Statistics* **23**(1), 305–326.

Miller, R. G. (1997). *Beyond ANOVA: Basics of Applied Statistics*, Boca Raton, FL: Chapman & Hall.

Misiorek, A. and R. Weron (2004). Heavy-Tailed Distributions in VaR Calculations, in J. E. Gentle, W. Härdle, and Y. Mori (eds). *Handbook of Computational Statistics*, 1025–1059, Heidelberg: Springer-Verlag.

Mittelhammer, R. C. (1996). *Mathematical Statistics for Economics and Business*, New York: Springer-Verlag.

Mittnik, S. and M. S. Paolella (1999). A Simple Estimator for the Characteristic Exponent of the Stable Paretian Distribution, *Mathematical and Computer Modelling* **29**, 161–176.

Mittnik, S. and M. S. Paolella (2003). Prediction of Financial Downside Risk with Heavy Tailed Conditional Distributions, in S. T. Rachev (edn.), *Handbook of Heavy Tailed Distributions in Finance*, Amsterdam: Elsevier Science.

Mittnik, S., M. S. Paolella, and S. T. Rachev (1998). A Tail Estimator for the Index of the Stable Paretian Distribution, *Communications in Statistics – Theory and Methods* **27**(5), 1239–1262.

Mittnik, S., M. S. Paolella, and S. T. Rachev (2000). Diagnosing and Treating the Fat Tails in Financial Returns Data, *Journal of Empirical Finance* **7**, 389–416.

Mittnik, S., M. S. Paolella, and S. T. Rachev (2002). Stationarity of Stable Power-GARCH Processes, *Journal of Econometrics* **106**, 97–107.

Molenberghs, G. and G. Verbeke (2007). Likelihood Ratio, Score, and Wald Tests in a Constrained Parameter Space, *American Statistician* **61**(1), 22–27.

Monfardini, C. (1998). Estimating Stochastic Volatility through Indirect Inference, *Econometrics Journal* **1**(1), 113–128.

Mood, A. M., F. A. Graybill, and D. C. Boes (1974). *Introduction to the Theory of Statistics*, 3rd edn, New York: McGraw-Hill.

Moore, D. S. (1971). Maximum Likelihood and Sufficient Statistics, *American Mathematical Monthly* **78**, 50–52.

Moors, J. J. A. (1986). The Meaning of Kurtosis: Darlington Reexamined, *American Statistician* **40**(4), 283–284.

Mulkay, M. and G. N. Gilbert (1981). Putting Philosophy to Work: Karl Popper's Influence on Scientific Practice, *Philosophy of the Social Sciences* **11**(3), 389–407.

Murakami, H. (2009). Saddlepoint Approximations to the Limiting Distribution of the Modified Anderson–Darling Test Statistic, *Communications in Statistics – Simulation and Computation* **38**(10), 2214–2219.

Murphy, K. P. (2012). *Machine Learning: A Probabilistic Perspective*, Cambridge, MA: MIT Press.

Nadarajah, S., B. Zhang, and S. Chan (2013). Estimation Methods for Expected Shortfall, *Quantitative Finance* **14**(2), 271–291.

Neyman, J. and E. S. Pearson (1928). On the Use and Interpretation of Certain Test Criteria for Purposes of Statistical Inference: Part I, *Biometrika* **20A**, 175–240.

Neyman, J. and E. L. Scott (1948). Consistent Estimates Based on Partially Consistent Observations, *Econometrica* **16**(1), 1–32.

Ng, S. K., T. Krishnan, and G. J. McLachlan (2004). The EM Algorithm, in J. E. Gentle, W. Härdle, and Y. Mori (eds). *Handbook of Computational Statistics*, 137–168, Heidelberg: Springer-Verlag.

Nguyen, H. T. (2016). On Evidential Measures of Support for Reasoning with Integrate Uncertainty: A Lesson from the Ban of P-values in Statistical Inference, in V.-N. Huynh, M. Inuiguchi, B. Le, B. N. Le, and T. Denoeux (eds). *Integrated Uncertainty in Knowledge Modeling and Decision Making: 5th International Symposium, IUKM 2016*, 3–15, Cham, Switzerland: Springer.

Nguyen, T. and G. Samorodnitsky (2012). Tail inference: Where does the tail begin?, *Extremes* **15**(4), 437–461.

Nolan, J. P. (1997). Numerical Calculation of Stable Densities and Distribution Functions, *Communications in Statistics: Stochastic Models* **13**(4), 759–774.

Nolan, J. P. (1999). Fitting Data and Assessing Goodness-of-fit with Stable Distributions, in *Proceedings of the Conference on Applications of Heavy Tailed Distributions in Economics, Engineering and Statistics*, American University, Washington, DC.

Nolan, J. P. (2018). *Stable Distributions – Models for Heavy Tailed Data*, Boston:Birkhäuser. Forthcoming; Chapter 1 online at http://fs2.american.edu/jpnolan/www/stable/stable.html.

Nolan, J. P. and D. O. Revah (2013). Linear and Nonlinear Regression with Stable Errors, *Journal of Econometrics* **172**(2), 186–194.

Noorbaloochi, S. and G. Meeden (1983). Unbiasedness as the Dual of Being Bayes, *Journal of the American Statistical Association* **78**, 619–623.

Noughabi, H. A. and N. R. Arghami (2013). General Treatment of Goodness-of-Fit Tests Based on Kullback–Leibler Information, *Journal of Statistical Computation and Simulation* **83**(8), 1556–1569.

Olkin, I. and J. W. Pratt (1958). Unbiased Estimation of Certain Correlation Coefficients, *Annals of Mathematical Statistics* **29**, 201–211.

Osborne, M. F. M. (1959). Brownian Motion in the Stock Market, *Operations Research* **7**(2), 145–173.

Pal, N. and J. C. Berry (1992). On Invariance and Maximum Likelihood Estimation, *American Statistician* **46**, 209–212.

Panier, H. H. (edn.) (1998). *Financial Economics: With Applications to Investments, Insurance, and Pensions*, Schaumburg, IL: The Actuarial Foundation. Authors: P. P. Boyle, S. H. Cox, D. Dufresne, H. U. Gerber, H. H. Müller, H. W. Pedersen, S. R. Pliska, M. Sherris, E. S. Shiu, K. S. Tan.

Paolella, M. S. (2001). Testing the Stable Paretian Assumption, *Mathematical and Computer Modelling* **34**, 1095–1112.

Paolella, M. S. (2006). *Fundamental Probability: A Computational Approach*, Chichester: John Wiley & Sons.

Paolella, M. S. (2007). *Intermediate Probability: A Computational Approach*, Chichester: John Wiley & Sons.

Paolella, M. S. (2015a). Multivariate Asset Return Prediction with Mixture Models, *European Journal of Finance* **21**(13–14), 1214–1252.

Paolella, M. S. (2015b). New Graphical Methods and Test Statistics for Testing Composite Normality, *Econometrics* **3**, 532–560.

Paolella, M. S. (2016a). Asymmetric Stable Paretian Distribution Testing, *Econometrics and Statistics* **1**, 19–39.

Paolella, M. S. (2016b). Stable-GARCH Models for Financial Returns: Fast Estimation and Tests for Stability, *Econometrics* **4**(2). Article 25.

Paolella, M. S. and P. Polak (2015a). ALRIGHT: Asymmetric LaRge-Scale (I)GARCH with Hetero-Tails, *International Review of Economics and Finance* **40**, 282–297.

Paolella, M. S. and P. Polak (2015b). COMFORT: A Common Market Factor Non-Gaussian Returns Model, *Journal of Econometrics* **187**(2), 593–605.

Paolella, M. S. and P. Polak (2015c). Portfolio Selection with Active Risk Monitoring, Research paper, Swiss Finance Institute.

Paolella, M. S. and P. Polak (2018a). COBra: Copula-Based Portfolio Optimization, in V. Kreinovich, S. Sriboonchitta, and N. Chakpitak (eds) *Studies in Computational Intelligence: Predictive Econometrics and Big Data*, Springer.

Paolella, M. S. and P. Polak (2018b). Density and Risk Prediction with Non-Gaussian COMFORT Models. Submitted.

Pardo, L. (2006). *Statistical Inference Based on Divergence Measures*, Boca Raton, FL: Chapman & Hall/CRC.

Parzen, E. (1962). *Stochastic Processes*, San Francisco: Holden-Day.

Pastorello, S., E. Renault, and N. Touzi (2000). Statistical Inference for Random-Variance Option Pricing, *Journal of Business and Economic Statistics* **18**(3), 358–367.

Paulson, A. S., E. W. Holcomb, and R. A. Leitch (1975). The Estimation of the Parameters of the Stable Laws, *Biometrika* **62**(1), 163–170.

Pawitan, Y. (2001). *In All Likelihood: Statistical Modelling and Inference Using Likelihood*, Oxford: Oxford University Press.

Pearson, E. S. (1974). Memories of the Impact of Fisher's Work in the 1920s, *International Statistical Review* **42**(1), 5–8.

Pelletier, D. and W. Wei (2016). The Geometric-VaR Backtesting Method, *Journal of Financial Econometrics* **14**(4), 725–745.

Perman, M. and J. Wellner (2014). An Excursion Approach to Maxima of the Brownian Bridge, *Stochastic Processes and their Applications* **124**, 3106–3120.

Pewsey, A. (2000). The Wrapped Skew-Normal Distribution on the Circle, *Communications in Statistics: Theory and Methods* **29**, 2459–2472.

Pflug, G. C. (2000). Some Remarks on the Value-at-Risk and on the Conditional Value-at-Risk, in S. P. Uryasev (edn.), *Probabilistic Constrained Optimization: Methodology and Applications*, 272–281, Amsterdam: Kluwer.

Philippou, A. N. and R. C. Dahiya (1970). Some Instructive Examples Where the Maximum Likelihood Estimator of the Population Mean is Not the Sample Mean, *American Statistician* **24**(3), 26–27.

Pictet, O. V., M. M. Dacorogna, and U. A. Müller (1998). Hill, Bootstrap and Jacknife Estimators for Heavy Tails, in R. J. Adler, R. E. Feldman, and M. S. Taqqu (eds). *A Practical Guide to Heavy Tails*, Boston: Birkhäuser.

Pillai, N. S. and X.-L. Meng (2016). An Unexpected Encounter with Cauchy and Lévy, *Annals of Statistics* **44**(5), 2089–2097.

Plackett, R. L. (1983). Karl Pearson and the Chi-squared Test, *International Statistical Review* **51**, 59–72.

Platen, E. and D. Heath (2006). *A Benchmark Approach to Quantitative Finance*, Berlin: Springer.

Poirier, D. J. (1995). *Intermediate Statistics and Econometrics, A Comparative Approach*, Cambridge, MA: The MIT Press.

Politis, D. N., J. P. Romano, and M. Wolf (1999). *Subsampling*, New York: Springer.

Press, S. J. (1972). Estimation in Univariate and Multivariate Stable Distributions, *Journal of the American Statistical Association* **67**(340), 842–846.

Pritsker, M. (1997). Evaluating Value at Risk Methodologies: Accuracy versus Computational Time, *Journal of Financial Services Research* **12**(2), 201–242.

Puig, P. and M. A. Stephens (2000). Tests of Fit for the Laplace Distribution With Applications, *Technometrics* **42**(4), 417–424.

Quandt, R. E. and J. B. Ramsey (1978). Estimating Mixtures of Normal Distributions and Switching Regressions, With Comments, *Journal of the American Statistical Association* **73**(364), 730–752.

Quenouille, M. H. (1956). Notes on Bias in Estimation, *Biometrika* **43**(3–4), 353–360.

Rachev, S. T. and S. Mittnik (2000). *Stable Paretian Models in Finance*, New York: John Wiley & Sons.

Rahman, M. and S. Chakrobartty (2004). Tests for Uniformity: A Comparative Study, *Journal of Korean Data & Information Science Society* **15**(1), 211–218.

Rao, C. R. (2002). Karl Pearson Chi-Squared Test – The Dawn of Statistical Inference, in C. Huber-Carol, N. Balakrishnan, M. Nikulin, and M. Mesbah (eds). *Goodness-of-Fit Tests and Model Validity*, 9–24, New York: Springer-Verlag.

Ravishanker, N. and D. K. Dey (2002). *A First Course in Linear Model Theory*, Boca Raton, FL: Chapman & Hall/CRC.

Rayner, G. D. (2002). Partitioning the Pearson-Fisher Chi-Squared Goodness-of-Fit Statistic, in C. Huber-Carol, N. Balakrishnan, M. Nikulin, and M. Mesbah (eds). *Goodness-of-Fit Tests and Model Validity*, 45–56, New York: Springer-Verlag.

Rayner, J. C. W. and D. J. Best (1990). Smooth Tests of Goodness of Fit: An Overview, *International Statistical Review* **58**(1), 9–17.

Rayner, J. C. W. and G. D. Rayner (1998). S-Sample Smooth Goodness of Fit Testing: Rederivation and Monte Carlo Assessment, *Biometrical Journal* **40**(6), 651–663.

Rayner, J. C. W., O. Thas, and D. J. Best (2009). *Smooth Tests of Goodness of Fit*, 2nd edn, Singapore: John Wiley & Sons.

Read, C. B. (1982). Median Unbiased Estimator, in S. Kotz and N. L. Johnson (eds). *Encyclopedia of Statistical Sciences*, vol. 1, 424–426, New York: John Wiley & Sons.

Redner, R. A. and H. F. Walker (1984). Mixture Densities, Maximum Likelihood and the EM Algorithm, *SIAM Review* **26**(2), 195–239.

Reid, N. (1988). Saddlepoint Methods and Statistical Inference (with discussion), *Statistical Science* **3**, 213–238.

Reinhart, A. (2015). *Statistics Done Wrong: The Woefully Complete Guide*, San Francisco: No Starch Press.

Reiss, R.-D. (1989). *Approximate Distributions of Order Statistics: With Applications to Nonparametric Statistics*, New York: Springer-Verlag.

Reiss, R.-D. and M. Thomas (2007). *Statistical Analysis of Extreme Values: With Applications to Insurance, Finance, Hydrology and Other Fields*, 3rd edn, Basel: Birkhäuser.

Rényi, A. (1953). On the Theory of Order Statistics, *Acta Mathematica Academiae Scientiarum Hungaricae* **4**, 191–231.

Resnick, S. (1999). *A Probability Path*, Boston: Birkhäuser.

Resnick, S. and H. Rootzén (2000). Self-Similar Communication Models and Very Heavy Tails, *Annals of Applied Probability* **10**(3), 753–778.

Richardson, S. and P. J. Green (1997). On Bayesian Analysis of Mixtures with an Unknown Number of Components, *Journal of the Royal Statistical Society, Series B* **59**(4), 731–792.

Robert, C. P. (1996). Mixtures of Distributions: Inference and Estimation, in W. Gilks, S. Richardson, and D. Spiegelhalter (eds). *Markov Chain Monte Carlo in Practice*, 441–464, London: Chapman & Hall.

Robert, C. P. (2007). *The Bayesian Choice*, 2nd edn, New York: Springer-Verlag.

Roberts, C. and S. Geisser (1966). A Necessary and Sufficient Condition for the Square of a Random Variable to be Gamma, *Biometrika* **53**, 275–277.

Robinson, D. H. and H. Wainer (2002). On the Past and Future of Null Hypothesis Significance Testing, *Journal of Wildlife Management* **66**(2), 262–271.

Roccioletti, S. (2016). *Backtesting Value at Risk and Expected Shortfall*, Wiesbaden: Springer Fachmedien Wiesbaden.

Rochon, J., G. Matthias, and M. Kieser (2012). To Test or Not to Test: Preliminary Assessment of Normality When Comparing Two Independent Samples, *BMC Medical Research Methodology* **12**, 81.

Rockafellar, R. T. and S. P. Uryasev (2000). Optimization of Conditional Value at Risk, *Journal of Risk* **2**, 21–41.

Rockafellar, R. T. and S. P. Uryasev (2002). Conditional Value-at-Risk for General Loss Distributions, *Journal of Banking & Finance* **26**(7), 1443–1471.

Rockafellar, R. T., S. P. Uryasev, and M. Zabarankin (2006a). Master Funds in Portfolio Analysis with General Deviation Measures, *Journal of Banking & Finance* **30**(2), 743–778.

Rockafellar, R. T., S. P. Uryasev, and M. Zabarankin (2006b). Optimality Conditions in Portfolio Analysis with General Deviation Measures, *Mathematical Programming* **108**(2), 515–540.

Rockafellar, R. T., S. P. Uryasev, and M. Zabarankin (2007). Equilibrium with Investors using a Diversity of Deviation Measures, *Journal of Banking & Finance* **31**(11), 3251–3268.

Roeder, K. (1994). A Graphical Technique For Determining the Number of Components in a Mixture of Normals, *Journal of the American Statistical Association* **89**(426), 487–495.

Roeder, K. and L. Wasserman (1997). Practical Bayesian Density Estimation Using Mixtures of Normals, *Journal of the American Statistical Association* **92**(439), 894–902.

Rogers, L. C. G. and O. Zane (1999). Saddlepoint Approximations to Option Prices, *Annals of Applied Probability* **9**, 493–503.

Rohatgi, V. K. (1976). *An Introduction to Probability Theory and Mathematical Statistics*, New York: John Wiley & Sons.

Rohatgi, V. K. and A. K. M. E. Saleh (2015). *An Introduction to Probability Theory and Mathematical Statistics*, 3rd edn, Hoboken, NJ: John Wiley & Sons.

Romano, J. P. and A. F. Siegel (1986). *Counterexamples in Probability and Statistics*, Belmont, CA: Wadsworth & Brooks/Cole.

Rosco, J. F., M. C. Jones, and A. Pewsey (2011). Skew *t* Distributions via the Sinh-Arcsinh Transformation, *TEST* **20**(3), 630–652.

Rosenblatt, M. (1952). Remarks on a Multivariate Transformation, *Annals of Mathematical Statistics* **23**, 470–472.

Rosenkrantz, W. A. (2000). Confidence Bands for Quantile Functions: A Parametric and Graphic Alternative for Testing Goodness of Fit, *American Statistician* **54**(3), 185–190.

Rousseeuw, P. J. (1984). Least Median of Squares Regression, *Journal of the American Statistical Association* **79**(388), 871–880.

Rousseeuw, P. J. and M. Hubert (2011). Robust Statistics for Outlier Detection, *Wiley Interdisciplinary Reviews: Data Mining and Knowledge Discovery* **1**(1), 73–79.

Rousseeuw, P. J. and A. M. Leroy (1987). *Robust Regression and Outlier Detection*, New York: John Wiley & Sons.

Rousseeuw, P. J. and K. Van Driessen (1999). A Fast Algorithm for the Minimum Covariance Determinant Estimator, *Technometrics* **41**(3), 212–223.

Royall, R. (1997). *Statistical Evidence: A Likelihood Paradigm*, London: Chapman & Hall.

Royall, R. (2000). On the Probability of Observing Misleading Statistical Evidence, *Journal of the American Statistical Association* **95**(451), 760–768.

Rubinstein, R. Y. and D. P. Kroese (2017). *Simulation and the Monte Carlo Method*, 3rd edn, Hoboken, NJ: John Wiley & Sons.

Samorodnitsky, G. and M. S. Taqqu (1994). *Stable Non-Gaussian Random Processes: Stochastic Models with Infinite Variance*, London: Chapman & Hall.

Sampson, A. and B. Spencer (1976). Sufficiency, Minimal Sufficiency, and the Lack Thereof, *American Statistician* **30**(1), 34–35. Corrigenda: **31**(1), 54.

Santos, A. A. P., F. J. Nogales, and E. Ruiz (2013). Comparing Univariate and Multivariate Models to Forecast Portfolio Value-at-Risk, *Journal of Financial Econometrics* **11**(2), 400–441.

Satterthwaite, F. E. (1946). An Approximate Distribution of Estimates of Variance Components, *Biometrics Bulletin* **2**, 110–114.

Savage, L. J. (1961). The Foundations of Statistics Reconsidered, in J. Neyman (edn.), *Proceedings of the Fourth Berkeley Symposium on Mathematical Statistics and Probability*, vol. 1, 575–586, Berkeley: University of California Press.

Schäfer, J. and K. Strimmer (2005). A Shrinkage Approach to Large-Scale Covariance Matrix Estimation and Implications for Functional Genomics, *Statistical Applications in Genetics and Molecular Biology* **4**(1).

Schervish, M. J. (1995). *Theory of Statistics*, New York: Springer-Verlag.

Schiff, J. L. (1999). *The Laplace Transform – Theory and Applications*, New York: Springer-Verlag.

Schlattmann, P. (2009). *Medical Applications of Finite Mixture Models*, Heidelberg: Springer-Verlag.

Schmidt, P. (1982). An Improved Version of the Quandt–Ramsey MGF Estimator for Mixtures of Normal Distributions and Switching Regressions, *Econometrica* **50**(2), 501–516.

Schott, J. R. (2005). *Matrix Analysis for Statistics*, 2nd edn, New York: John Wiley & Sons.

Sen, P. K. and J. M. Singer (1993). *Large Sample Methods in Statistics: An Introduction with Applications*, Boca Raton, FL: Chapman & Hall/CRC.

Sentana, E., G. Calzolari, and G. Fiorentini (2008). Indirect Estimation of Large Conditionally Heteroskedastic Factor Models, with an Application to the Dow 30 Stocks, *Journal of Econometrics* **146**, 10–25.

Severini, T. A. (2000). *Likelihood Methods in Statistics*, New York: Oxford University Press.

Severini, T. A. (2005). *Elements of Distribution Theory*, Cambridge: Cambridge University Press.

Shao, J. (2003). *Mathematical Statistics*, New York: Springer-Verlag.

Shao, J. and D. Tu (1995). *The Jackknife and Bootstrap*, New York: Springer-Verlag.

Singh, R. K. (1988). Estimation of Error Variance in Linear Regression Models with Errors Having a Multivariate Student-t Distribution with Unknown Degrees of Freedom, *Economic Letters* **27**, 47–53.

Singleton, K. (2001). Estimation of Affine Pricing Models Using the Empirical Characteristic Function, *Journal of Econometrics* **102**, 111–141.

Slim, S., Y. Koubaa, and A. BenSaïda (2016). Value-at-Risk Under Lévy GARCH models: Evidence from Global Stock Markets, *Journal of International Financial Markets, Institutions & Money* **46**, 30–53.

Smith, A. A. (1993). Estimating Nonlinear Time Series Models Using Simulated Vector Autoregressions, *Journal of Applied Econometrics* **8**, S63–S84.

Smith, D. (2007). Conditional Coskewness and Asset Pricing, *Journal of Empirical Finance* **14**, 91–119.

Snedecor, G. W. and W. G. Cochran (1967). *Statistical Methods*, Ames: Iowa State University Press.

Sorkin, A. R. (2017). Buffett Asks Big Money: Why Pay High Fees?, *New York Times* February 27.

Spokoiny, V. and T. Dickhaus (2015). *Basics of Modern Mathematical Statistics*, New York: Springer-Verlag.

Stein, C. (1981). Estimation of the Mean of a Multivariate Normal Distribution, *Annals of Statistics* **9**, 1135–1151.

Stengos, T. and X. Wu (2010). Information-Theoretic Distribution Test With Application To Normality, *Econometric Reviews* **29**(3), 307–329.

Stigler, S. M. (1972). Completeness and Unbiased Estimation, *American Statistician* **26**(2), 28–29.

Stigler, S. M. (1999). *Statistics on the Table: The History of Statistical Concepts and Methods*, Cambridge, MA: Harvard University Press.

Storn, R. and K. Price (1995). Differential Evolution – a Simple and Efficient Adaptive Scheme for Global Optimization over Continuous Spaces, Tech. Rep. TR-95-012, International Computer Science Institute.

Stoyanov, S., G. Samorodnitsky, S. Rachev, and S. Ortobelli (2006). Computing the portfolio conditional value-at-risk in the alpha-stable case, *Probability and Mathematical Statistics* **26**, 1–22.

Stuart, A. and J. K. Ord (1994). *Kendall's Advanced Theory of Statistics, Vol. 1: Distribution Theory*, 6th edn, London: Edward Arnold.

Stuart, A., J. K. Ord, and S. F. Arnold (1999). *Kendall's Advanced Theory of Statistics, Vol. 2A: Classical Inference and the Linear Model*, 6th edn, London: Edward Arnold.

Suissa, S. and J. J. Shuster (1984). Are Uniformly Most Powerful Unbiased Tests Really Best?, *American Statistician* **38**(3), 204–206.

Sukhatme, P. V. (1937). Tests of Significance for Samples of the χ^2 Population with Two Degrees of Freedom, *Annals of Eugenics* **8**, 52–56.

Sullivan, G. M. and R. Feinn (2012). Using Effect Size – or Why the P Value Is Not Enough, *Journal of Graduate Medical Education* **4**(3), 279–282.

Sun, Y., D. Wierstra, T. Schaul, and J. Schmidhuber (2009a). Efficient Natural Evolution Strategies., in F. Rothlauf (edn.), *GECCO '09: Proceedings of the 11th Annual Conference on Genetic and Evolutionary Computation*, 539–546, New York: ACM.

Sun, Y., D. Wierstra, T. Schaul, and J. Schmidhuber (2009b). Stochastic Search using the Natural Gradient., in A. P. Danyluk, L. Bottou, and M. L. Littman (eds). *Proceedings of the 26th Annual International Conference on Machine Learning, ACM International Conference Proceeding Series*, vol. 382, 146, New York: ACM.

Szegö, G. (edn.) (2004). *Risk Measures for the 21st Century*, Chichester: John Wiley & Sons.

Takemura, A., M. Akimichi, and S. Kuriki (2006). Skewness and Kurtosis as Locally Best Invariant Tests of Normality. http://arxiv.org/abs/math.ST/0608499.

Tardiff, R. M. (1981). L'Hospital's Rule and the Central Limit Theorem, *American Statistician* **35**, 43.

Teicher, H. (1963). Identifiability of Finite Mixtures, *Annals of Mathematical Statistics* **34**, 1265–1269.

The Economist (2013). Trouble at the Lab, *The Economist* October 19.

Thode, H. C. (2002). *Testing for Normality*, New York: Marcel Dekker.

Tian, G.-L., K. W. Ng, and M. Tan (2008). EM-Type Algorithms for Computing Restricted MLEs in Multivariate Normal Distributions and Multivariate *t*-Distributions, *Computational Statistics & Data Analysis* **52**(10), 4768–4778.

Tibshirani, R. (1996). Regression Shrinkage and Selection via the LASSO, *Journal of the Royal Statistical Society, Series B* **58**, 267–288.

Titterington, D. M., A. F. M. Smith, and U. E. Makov (1985). *Statistical Analysis of Finite Mixture Distributions*, New York: Wiley.

Torabi, H., N. H. Montazeri, and A. Grané (2016). A Test for Normality Based on the Empirical Distribution Function, *Statistics and Operations Research Transactions* **40**(1), 55–88.

Trafimow, D. and M. Marks (2015). Editorial, *Basic and Applied Social Psychology* **37**(1), 1–2.

Tsonias, E. G. (2000). Efficient Posterior Integration in Stable Paretian Models, *Statistical Papers* **41**(3), 305–325.

Tsukahara, H. (2009). One-Parameter Families of Distortion Risk Measures, *Mathematical Finance* **19**, 691–705.

Tsukahara, H. (2014). Estimation of Distortion Risk Measures, *Journal of Financial Econometrics* **12**(1), 213–235.

Tukey, J. (1962). The Future of Data Analysis, *Annals of Mathematical Statistics* **33**(1), 1–67.

Tukey, J. W. (1969). Analyzing Data: Sanctification or Detective Work?, *American Psychologist* **24**, 83–91.

Tukey, J. W. (1978). Discussion of Granger on Seasonality, in A. Zellner (edn.), *Seasonal Analysis of Economic Time Series*, 50–53, Washington: U.S. Dept. of Commerce: National Bureau of Economic Research, Bureau of the Census.

Vanden, J. (2006). Option Coskewness and Capital Asset Pricing, *Review of Financial Studies* **19**, 1279–1320.

Vandewalle, B., J. Beirlant, A. Christmann, and M. Hubert (2007). A Robust Estimator for the Tail Index of Pareto-Type Distributions, *Computational Statistics & Data Analysis* **51**(12), 6252–6268.

Venkataraman, S. (1997). Value at Risk for a Mixture of Normal Distributions: The Use of Quasi-Bayesian Estimation Techniques, *Economic Perspective: Federal Reserve Bank of Chicago* **21**(2), 2–13.

Venkatesh, S. S. (2013). *The Theory of Probability: Explorations and Applications*, Cambridge: Cambridge University Press.

Verboven, S. and M. Hubert (2005). LIBRA: A MATLAB Library for Robust Analysis, *Chemometrics and Intelligent Laboratory Systems* **75**(2), 127–136.

Vuong, Q. H. (1989). Likelihood Ratio Tests for Model Selection and Non-Nested Hypotheses, *Econometrica* **57**(2), 307–333.

Wackerly, D. D. (1976). On Deriving a Complete Sufficient Statistic, *American Statistician* **30**(1), 37–38.

Warner, S. L. (1965). Randomized Response: A Survey Technique for Eliminating Evasive Answer Bias, *Journal of the American Statistical Association* **60**, 63–69.

Wasserstein, R. L. and N. A. Lazar (2016). The ASA's Statement on *p*-Values: Context, Process, and Purpose, *American Statistician* **70**(2), 129–133.

Watson, G. S. (1964). A Note on Maximum Likelihood, *Sankhyā* **26**(2), 303–304.

Webber, W. F. (2001). Comment on Rosenkrantz (2000), *American Statistician* **55**(2), 171–172.

Weerahandi, S. (1993). Generalized Confidence Intervals, *Journal of the American Statistical Association* **88**(423), 899–905.

Weerahandi, S. (1995). *Exact Statistical Methods for Data Analysis*, New York: Springer-Verlag.

Welch, B. L. (1947). The Generalization of "Student's" Problem When Several Different Population Variances Are Involved, *Biometrika* **34**(1–2), 28–35.

Weron, R. (2001). Levy-Stable Distributions Revisited: Tail Index > 2 Does Not Exclude the Levy-Stable Regime, *International Journal of Modern Physics C* **28**(2), 165–171.

White, H. (1982). Maximum Likelihood Estimation of Misspecified Models, *Econometrica* **50**(1), 1–25.

Wiersema, U. F. (2008). *Brownian Motion Calculus*, Chichester: John Wiley & Sons.

Wierstra, D., T. Schaul, J. Peters, and J. Schmidhuber (2008). Natural Evolution Strategies., in *IEEE Congress on Evolutionary Computation*, 3381–3387, IEEE.

Wilks, S. S. (1938). The Large-Sample Distribution of the Likelihood Ratio for Testing Composite Hypotheses, *Annals of Mathematical Statistics* **9**(1), 60–62.

Wirjanto, T. S. and D. Xu (2013). A Mixture-of-Normal Distribution Modeling Approach in Financial Econometrics: A Selected Review, Mimeo.

Wolf, M. and D. Wunderli (2015). Bootstrap Joint Prediction Regions, *Journal of Time Series Analysis* **36**(3), 352–376.

Wu, C. and J. C. Lee (2007). Estimation of a Utility-Based Asset Pricing Model Using Normal Mixture GARCH(1,1), *Economic Modelling* **24**, 329–349.

Wu, C. F. J. (1986). Jackknife, Bootstrap, and Other Resampling Methods in Regression Analysis, *Annals of Statistics* **14**, 1261–1295.

Yakowitz, S. J. and J. D. Spragins (1968). On the Identifiability of Finite Mixtures, *Annals of Mathematical Statistics* **39**(1), 209–214.

Yang, J., T. R. Hurd, and X. Zhang (2006). Saddlepoint Approximation Method for Pricing CDOs, *Journal of Computational Finance* **10**, 1–20.

Yang, Y. (2005). Can the Strengths of AIC and BIC be Shared? A Conflict between Model Indentification and Regression Estimation, *Biometrika* **92**(4), 937–950.

Yen, V. C. and A. H. Moore (1988). Modified Goodness-of-Fit Test for the Laplace Distribution, *Communications in Statistics – Simulation and Computation* **17**, 275–281.

Young, G. A. and R. L. Smith (2005). *Essentials of Statistical Inference*, Cambridge: Cambridge University Press.

Yu, J. (2004). Empirical Characteristic Function Estimation and Its Applications, *Econometric Reviews* **23**(2), 93–123.

Zacks, S. (1971). *The Theory of Statistical Inference*, New York: John Wiley & Sons.

Zaman, A. (1996). *Statistical Foundations for Econometric Techniques*, Bingley, W. Yorks: Emerald Group Publishing.

Zehna, P. W. (1966). Invariance of Maximum Likelihood Estimations, *Annals of Mathematical Statistics* **37**(3), 744.

Zheng, W. and Y. K. Kwok (2014). Saddlepoint Approximation Methods for Pricing Derivatives on Discrete Realized Variance, *Applied Mathematical Finance* **21**, 1–31.

Ziliak, S. T. and D. N. McCloskey (2008). *The Cult of Statistical Significance: How the Standard Error Costs Us Jobs, Justice, and Lives*, Ann Arbor, MI: University of Michigan Press.

Zolotarev, V. M. (1986). *One Dimensional Stable Distributions*, Translations of Mathematical Monographs, vol. **65**, Providence, RI: American Mathematical Society. Translated from the original Russian version (1983).

Index

Fundamental Statistical Inference: A Computational Approach, First Edition. Marc S. Paolella.
© 2018 John Wiley & Sons Ltd. Published 2018 by John Wiley & Sons Ltd.